CAMBRIDGE STUDIES IN
ADVANCED MATHEMATICS 105

ADDITIVE COMBINATORICS

Additive combinatorics is the theory of counting additive structures in sets. While this theory has been developing for many decades, the field has seen exciting advances and dramatic changes in direction in recent years thanks to its connections with other areas of mathematics, such as number theory, ergodic theory and graph theory.

Now in paperback, this graduate-level textbook will quickly allow students and researchers easy entry into this fascinating field. Here, for the first time, the authors bring together in a self-contained and systematic manner the many different tools and ideas that are used in the modern theory, presenting them in an accessible, coherent and intuitively clear way, and providing immediate applications to problems in additive combinatorics. The power of these tools is well demonstrated in the presentation of recent advances such as the Green–Tao theorem on arithmetic progressions: Erdős distance problems; and the newly developing field of sum-product estimates. The text is supplemented by a large number of exercises and other material which has not previously appeared elsewhere.

TERENCE TAO is a Professor in the Department of Mathematics at the University of California, Los Angeles. He was awarded the Fields Medal in 2006 for his contributions to partial differential equations, combinatorics, harmonic analysis and additive number theory.

VAN H. VU is a Professor in the Department of Mathematics at Rutgers University, New Jersey.

T0297332

CAMBRIDGE STUDIES IN ADVANCED MATHEMATICS

All the titles listed below can be obtained from good booksellers or from Cambridge University Press.
For a complete series listing visit: http://www.cambridge.org/series/sSeries.asp?code=CSAM

Already published

73 B. Bollobás *Random graphs (2nd Edition)*
74 R. M. Dudley *Real analysis and probability (2nd Edition)*
75 T. Sheil-Small *Complex polynomials*
76 C. Voisin *Hodge theory and complex algebraic geometry, I*
77 C. Voisin *Hodge theory and complex algebraic geometry, II*
78 V. Paulsen *Completely bounded maps and operator algebras*
79 F. Gesztesy & H. Holden *Soliton equations and their algebro-geometric solutions, I*
81 S. Mukai *An introduction to invariants and moduli*
82 G. Tourlakis *Lectures in logic and set theory, I*
83 G. Tourlakis *Lectures in logic and set theory, II*
84 R. A. Bailey *Association schemes*
85 J. Carlson, S. Müller-Stach & C. Peters *Period mappings and period domains*
86 J. J. Duistermaat & J. A. C. Kolk *Multidimensional real analysis, I*
87 J. J. Duistermaat & J. A. C. Kolk *Multidimensional real analysis, II*
89 M. C. Golumbic & A. N. Trenk *Tolerance graphs*
90 L. H. Harper *Global methods for combinatorial isoperimetric problems*
91 I. Moerdijk & J. Mrčun *Introduction to foliations and Lie groupoids*
92 J. Kollár, K. E. Smith & A. Corti *Rational and nearly rational varieties*
93 D. Applebaum *Lévy processes and stochastic calculus (1st Edition)*
94 B. Conrad *Modular forms and the Ramanujan conjecture*
95 M. Schechter *An introduction to nonlinear analysis*
96 R. Carter *Lie algebras of finite and affine type*
97 H. L. Montgomery & R. C. Vaughan *Multiplicative number theory, I*
98 I. Chavel *Riemannian geometry (2nd Edition)*
99 D. Goldfeld *Automorphic forms and L-functions for the group GL(n,R)*
100 M. B. Marcus & J. Rosen *Markov processes, Gaussian processes, and local times*
101 P. Gille & T. Szamuely *Central simple algebras and Galois cohomology*
102 J. Bertoin *Random fragmentation and coagulation processes*
103 E. Frenkel *Langlands correspondence for loop groups*
104 A. Ambrosetti & A. Malchiodi *Nonlinear analysis and semilinear elliptic problems*
105 T. Tao & V. H. Vu *Additive combinatorics*
106 E. B. Davies *Linear operators and their spectra*
107 K. Kodaira *Complex analysis*
108 T. Ceccherini-Silberstein, F. Scarabotti & F. Tolli *Harmonic analysis on finite groups*
109 H. Geiges *An introduction to contact topology*
110 J. Faraut *Analysis on Lie groups: An Introduction*
111 E. Park *Complex topological K-theory*
112 D. W. Stroock *Partial differential equations for probabilists*
113 A. Kirillov, Jr *An introduction to Lie groups and Lie algebras*
114 F. Gesztesy *et al. Soliton equations and their algebro-geometric solutions, II*
115 E. de Faria & W. de Melo *Mathematical tools for one-dimensional dynamics*
116 D. Applebaum *Lévy processes and stochastic calculus (2nd Edition)*
117 T. Szamuely *Galois groups and fundamental groups*
118 G. W. Anderson, A. Guionnet & O. Zeitouni *Introduction to random matrices*
119 C. Perez-Garcia & W. H. Schikhof *Locally convex spaces over non-Archimedean valued fields*
120 P. K. Friz & N. B. Victoir *Multidimensional Stochastic Processes as Rough Paths*
121 T. Ceccherini-Silberstein, F. Scarabotti & F. Tolli *Representation Theory of the Symmetric Groups*
122 S. Kalikow & R. McCutcheon *An Outline of Ergodic Theory*

Additive Combinatorics

TERENCE TAO AND VAN H. VU

CAMBRIDGE
UNIVERSITY PRESS

CAMBRIDGE UNIVERSITY PRESS
Cambridge, New York, Melbourne, Madrid, Cape Town, Singapore,
São Paulo, Delhi, Dubai, Tokyo, Mexico City

Cambridge University Press
The Edinburgh Building, Cambridge CB2 8RU, UK

Published in the United States of America by Cambridge University Press, New York

www.cambridge.org
Information on this title: www.cambridge.org/9780521136563

First published 2006
Reprinted with corrections 2007
Paperback edition 2010

A catalogue record for this publication is available from the British Library

Library of Congress Cataloging in Publication data
Tao, Terence, 1975–
Additive combinatorics / Terence Tao, Van H. Vu. – Pbk. ed.
p. cm. – (Cambridge studies in advanced mathematics ; 105)
Includes bibliographical references and index.
ISBN 978-0-521-13656-3
1. Additive combinatorics. I. Vu, Van, 1970– II. Title.
QA164.T35 2010
511.6 – dc22 2009031809

ISBN 978-0-521-85386-6 Hardback
ISBN 978-0-521-13656-3 Paperback

To our families

Contents

Prologue

This book arose out of lecture notes developed by us while teaching courses on additive combinatorics at the University of California, Los Angeles and the University of California, San Diego. Additive combinatorics is currently a highly active area of research for several reasons, for example its many applications to additive number theory. One remarkable feature of the field is the use of tools from many diverse fields of mathematics, including elementary combinatorics, harmonic analysis, convex geometry, incidence geometry, graph theory, probability, algebraic geometry, and ergodic theory; this wealth of perspectives makes additive combinatorics a rich, fascinating, and multi-faceted subject. There are still many major problems left in the field, and it seems likely that many of these will require a combination of tools from several of the areas mentioned above in order to solve them.

The main purpose of this book is to gather all these diverse tools in one location, present them in a self-contained and introductory manner, and illustrate their application to problems in additive combinatorics. Many aspects of this material have already been covered in other papers and texts (and in particular several earlier books [168], [257], [113] have focused on some of the aspects of additive combinatorics), but this book attempts to present as many perspectives and techniques as possible in a unified setting.

Additive combinatorics is largely concerned with the additive structure[1] of sets. To clarify what we mean by "additive structure", let us introduce the following definitions.

Definition 0.1 An *additive group* is any abelian group Z with group operation $+$. Note that we can define a multiplication operation $nx \in Z$ whenever $n \in \mathbf{Z}$ and

[1] We will also occasionally consider the multiplicative structure of sets as well; we will refer to the combined study of such structures as *arithmetic combinatorics*.

$x \in Z$ in the usual manner: thus $3x = x + x + x, -2x = -x - x$, etc. An *additive set* is a pair (A, Z), where Z is an additive group, and A is a finite non-empty subset of Z. We often abbreviate an additive set (A, Z) simply as A, and refer to Z as the *ambient group* of the additive set. If A, B are additive sets in Z, we define the *sum set*

$$A + B := \{a + b : a \in A, \ b \in B\}$$

and *difference set*

$$A - B := \{a - b : a \in A, \ b \in B\}.$$

Also, we define the *iterated sumset* kA for $k \in \mathbf{Z}^+$ by

$$kA := \{a_1 + \cdots + a_k : a_1, \ldots, a_k \in A\}.$$

We caution that the sumset kA is usually distinct from the dilation $k \cdot A$ of A, defined by

$$k \cdot A := \{ka : a \in A\}.$$

For us, typical examples of additive groups Z will be the integers $\mathbf{Z}$, a cyclic group $\mathbf{Z}_N$, a Euclidean space $\mathbf{R}^n$, or a finite field geometry F_p^n. As the notation suggests, we will eventually be viewing additive sets as "intrinsic" objects, which can be embedded inside any number of different ambient groups; this is somewhat similar to how a manifold can be thought of intrinsically, or alternatively can be embedded into an ambient space. To make these ideas rigorous we will need to develop the theory of *Freiman homomorphisms*, but we will defer this to Section 5.3.

Additive sets may have a large or small amount of additive structure. A good example of a set with little additive structure would be a randomly chosen subset A of a finite additive group Z with some fixed cardinality. At the other extreme, examples of sets with very strong additive structure would include arithmetic progressions

$$a + [0, N) \cdot r := \{a, a + r, \ldots, a + (N - 1)r\}$$

where $a, r \in Z$ and $N \in \mathbf{Z}^+$; or d-dimensional generalized arithmetic progressions

$$a + [0, N) \cdot v := \{a + n_1 v_1 + \cdots + n_d v_d : 0 \leq n_j < N_j \text{ for all } 1 \leq j \leq d\}$$

where $a \in Z$, $v = (v_1, \ldots, v_d) \in Z^d$, and $N = (N_1, \ldots, N_d) \in (\mathbf{Z}^+)^d$; or d-dimensional cubes

$$a + \{0, 1\}^d \cdot v = \{a + \epsilon_1 v_1 + \cdots + \epsilon_d v_d : \epsilon_1, \ldots, \epsilon_d \in \{0, 1\}\};$$

or the subset sums $FS(A) := \{\sum_{a \in B} a : B \subseteq A\}$ of a finite set A.

A fundamental task in this subject is to give some quantitative measures of additive structure in a set, and then investigate to what extent these measures are equivalent to each other. For example, one could try to quantify each of the following informal statements as being some version of the assertion "A has additive structure":

- $A + A$ is small;
- $A - A$ is small;
- $A - A$ can be covered by a small number of translates of A;
- kA is small for any fixed k;
- there are many quadruples $(a_1, a_2, a_3, a_4) \in A \times A \times A \times A$ such that $a_1 + a_2 = a_3 + a_4$;
- there are many quadruples $(a_1, a_2, a_3, a_4) \in A \times A \times A \times A$ such that $a_1 - a_2 = a_3 - a_4$;
- the convolution $1_A * 1_A$ is highly concentrated;
- the subset sums $FS(A) := \{\sum_{a \in B} a : B \subseteq A\}$ have high multiplicity;
- the Fourier transform $\widehat{1_A}$ is highly concentrated;
- the Fourier transform $\widehat{1_A}$ is highly concentrated in a cube;
- A has a large intersection with a generalized arithmetic progression, of size comparable to A;
- A is contained in a generalized arithmetic progression, of size comparable to A;
- A (or perhaps $A - A$, or $2A - 2A$) contains a large generalized arithmetic progression.

The reader is invited to investigate to what extent these informal statements are true for sets such as progressions and cubes, and false for sets such as random sets. As it turns out, once one makes the above assertions more quantitative, there are a number of deep and important equivalences between them; indeed, to oversimplify tremendously, all of the above criteria for additive structure are "essentially" equivalent. There is also a similar heuristic to quantify what it would mean for two additive sets A, B of comparable size to have a large amount of "shared additive structure" (e.g. A and B are progressions with the same step size v); we invite the reader to devise analogs of the above criteria to capture this concept.

Making the above heuristics precise and rigorous will require some work, and in fact will occupy large parts of Chapters 2, 3, 4, 5, 6. In deriving these basic tools of the field, we shall need to develop and combine techniques from elementary combinatorics, additive geometry, harmonic analysis, and graph theory; many of these methods are of independent interest in their own right, and so we have devoted some space to treating them in detail.

Of course, a "typical" additive set will most likely behave like a random additive set, which one expects to have very little additive structure. Nevertheless, it is a deep and surprising fact that as long as an additive set is dense enough in its ambient group, it will always have *some* level of additive structure. The most famous example of this principle is *Szemerédi's theorem*, which asserts that every subset of the integers of positive upper density will contain arbitrarily long arithmetic progressions; we shall devote all of Chapter 11 to this beautiful and important theorem. A variant of this fact is the very recent *Green–Tao theorem*, which asserts that every subset of the prime numbers of positive upper *relative* density also contains arbitrarily long arithmetic progressions; in particular, the primes themselves have this property. If one starts with an even sparser set A than the primes, then it is not yet known whether A will necessarily contain long progressions; however, if one forms sum sets such as $A + A$, $A + A + A$, $2A - 2A$, $FS(A)$ then these sets contain extraordinarily long arithmetic progressions (see in particular Section 4.7 and Chapter 12). This basic principle – that sumsets have much more additive structure than general sets – is closely connected to the equivalences between the various types of additive structure mentioned previously; indeed results of the former type can be used to deduce results of the latter type, and conversely.

We now describe some other topics covered in this text. In Chapter 1 we recall the simple yet powerful *probabilistic method*, which is very useful in additive combinatorics for constructing sets with certain desirable properties (e.g. thin additive bases of the integers), and provides an important conceptual framework that complements more classical deterministic approaches to such constructions. In Chapter 6 we present some ways in which graph theory interacts with additive combinatorics, for instance in the theory of sum-free sets, or via Ramsey theory. Graph theory is also decisive in establishing two important results in the theory of sum sets, the Balog–Szemerédi–Gowers theorem and the Plünnecke inequalities. Two other important tools from graph theory, namely the crossing number inequality and the Szemerédi regularity lemma, will also be covered in Chapter 8 and Sections 10.6, 11.6 respectively. In Chapter 7 we view sum sets from the perspective of random walks, and give some classical and recent results concerning the distribution of these sum sets, and in particular recent applications to random matrices. Last, but not least, in Chapter 9 we describe some algebraic methods, notably the combinatorial Nullstellensatz and Chevalley–Warning type methods, which have led to several deep arithmetical results (often with very sharp bounds) not obtainable by other means.

Acknowledgements

The authors would like to thank Shimon Brooks, Robin Chapman, Michael Cowling, Andrew Granville, Ben Green, Timothy Gowers, Harald Helfgott, Martin Klazar, Mariah Hamel, Vsevolod Lev, Roy Meshulam, Melvyn Nathanson, Imre Ruzsa, Roman Sasyk, and Benny Sudakov for helpful comments and corrections, and the Australian National University and the University of Edinburgh for their hospitality while portions of this book were being written. Parts of this work were inspired by the lecture notes of Ben Green [144], the expository article of Imre Ruzsa [297], and the book by Melvyn Nathanson [257]. TT is also particularly indebted to Roman Sasyk and Hillel Furstenberg for explaining the ergodic theory proof of Szemerédi's theorem. VV would like to thank Endre Szemerédi for many useful discussions on mathematics and other aspects of life. Last, and most importantly, the authors thank their wives, Laura and Huong, without whom this book would not be finished.

General notation

The following general notational conventions will be used throughout the book.

Sets and functions

For any set A, we use

$$A^d := A \times \cdots \times A = \{(a_1, \ldots, a_d) : a_1, \ldots, a_d \in A\}$$

to denote the Cartesian product of d copies of A: thus for instance $\mathbf{Z}^d$ is the d-dimensional integer lattice. We shall occasionally denote A^d by $A^{\oplus d}$, in order to distinguish this Cartesian product from the d-fold product set $A^{\cdot d} = A \cdot \ldots \cdot A$ of A, or the d-fold powers $A^{\wedge}d := \{a^d : a \in A\}$ of A.

If A, B are sets, we use $A \backslash B := \{a \in A : a \notin B\}$ to denote the set-theoretic difference of A and B; and B^A to denote the space of functions $f : A \to B$ from A to B. We also use $2^A := \{B : B \subset A\}$ to denote the power set of A. We use $|A|$ to denote the cardinality of A. (We shall also use $|x|$ to denote the magnitude of a real or complex number x, and $|v| = \sqrt{v_1^2 + \cdots + v_d^2}$ to denote the magnitude of a vector $v = (v_1, \ldots, v_d)$ in a Euclidean space $\mathbf{R}^d$. The meaning of the absolute value signs should be clear from context in all cases.)

If $A \subset Z$, we use $1_A : Z \to \{0, 1\}$ to denote the indicator function of A: thus $1_A(x) = 1$ when $x \in A$ and $1_A(x) = 0$ otherwise. Similarly if P is a property, we let $\mathbf{I}(P)$ denote the quantity 1 if P holds and 0 otherwise; thus for instance $1_A(x) = \mathbf{I}(x \in A)$.

We use $\binom{n}{k} = \frac{n!}{k!(n-k)!}$ to denote the number of k-element subsets of an n-element set. In particular we have the natural convention that $\binom{n}{k} = 0$ if $k > n$ or $k < 0$.

Number systems

We shall rely frequently on the integers $\mathbf{Z}$, the positive integers $\mathbf{Z}^+ := \{1, 2, \ldots\}$, the natural numbers $\mathbf{N} := \mathbf{Z}_{\geq 0} = \{0, 1, \ldots\}$, the reals $\mathbf{R}$, the positive reals

$\mathbf{R}^+ := \{x \in \mathbf{R} : x > 0\}$, the non-negative reals $\mathbf{R}_{\geq 0} := \{x \in \mathbf{R} : x \geq 0\}$, and the complex numbers $\mathbf{C}$, as well as the circle group $\mathbf{R}/\mathbf{Z} := \{x + \mathbf{Z} : x \in \mathbf{R}\}$.

For any natural number $N \in \mathbf{N}$, we use $\mathbf{Z}_N := \mathbf{Z}/N\mathbf{Z}$ to denote the cyclic group of order N, and use $n \mapsto n \bmod N$ to denote the canonical projection from $\mathbf{Z}$ to $\mathbf{Z}_N$. If q is a prime power, we use F_q to denote the finite field of order q (see Section 9.4). In particular if p is a prime then F_p is identifiable with $\mathbf{Z}_p$.

If x is a real number, we use $\lfloor x \rfloor$ to denote the greatest integer less than or equal to x.

Landau asymptotic notation

Let n be a positive variable (usually taking values on $\mathbf{N}$, $\mathbf{Z}^+$, $\mathbf{R}_{\geq 0}$, or $\mathbf{R}^+$, and often assumed to be large) and let $f(n)$ and $g(n)$ be real-valued functions of n.

- $g(n) = O(f(n))$ means that f is non-negative, and there is a positive constant C such that $|g(n)| \leq Cf(n)$ for all n.
- $g(n) = \Omega(f(n))$ means that f, g are non-negative, and there is a positive constant c such that $g(n) \geq cf(n)$ for all sufficiently large n.
- $g(n) = \Theta(f(n))$ means that f, g are non-negative and both $g(n) = O(f(n))$ and $g(n) = \Omega(f(n))$ hold; that is, there are positive constants c and C such that $cf(n) \geq g(n) \geq Cf(n)$ for all n.
- $g(n) = o_{n\to\infty}(f(n))$ means that f is non-negative and $g(n) = O(a(n)f(n))$ for some $a(n)$ which tends to zero as $n \to \infty$; if f is strictly positive, this is equivalent to $\lim_{n\to\infty} g(n)/f(n) = 0$.
- $g(n) = \omega_{n\to\infty}(f(n))$ means that f, g are non-negative and $f(n) = o_{n\to\infty}(g(n))$.

In most cases the asymptotic variable n will be clear from context, and we shall simply write $o_{n\to\infty}(f(n))$ as $o(f(n))$, and similarly write $\omega_{n\to\infty}(f(n))$ as $\omega(f(n))$. In some cases the constants c, C and the decaying function $a(n)$ will depend on some other parameters, in which case we indicate this by subscripts. Thus for instance $g(n) = O_k(f(n))$ would mean that $g(n) \leq C_k f(n)$ for all n, where C_k depends on the parameter k; similarly, $g(n) = o_{n\to\infty;k}(f(n))$ would mean that $g(n) = O(a_k(n)f(n))$ for some $a_k(n)$ which tends to zero as $n \to \infty$ for each fixed k.

The notation $g(n) = \tilde{O}(f(n))$ has been used widely in the combinatorics and theoretical computer science community in recent years; $g(n) = \tilde{O}(f(n))$ means that there is a constant c such that $g(n) \leq f(n) \log^c n$ for all sufficiently large n. We can define, in a similar manner, $\tilde{\Omega}$ and $\tilde{\Theta}$, though this notation will only be used occasionally here. Here and throughout the rest of the book, log shall denote the natural logarithm unless specified by subscripts, thus $\log_x y = \frac{\log y}{\log x}$.

Progressions

We have already encountered the concept of a generalized arithmetic progression. We now make this concept more precise.

Definition 0.2 (Progressions) For any integers $a \leq b$, we let $[a, b]$ denote the discrete closed interval $[a, b] := \{n \in \mathbf{Z} : a \leq n \leq b\}$; similarly define the half-open discrete interval $[a, b)$, etc. More generally, if $a = (a_1, \ldots, a_d)$ and $b = (b_1, \ldots, b_d)$ are elements of $\mathbf{Z}^d$ such that $a_j \leq b_j$, we define the *discrete box*

$$[a, b] := \{(n_1, \ldots, n_d) \in \mathbf{Z}^d : a_j \leq n_j \leq b_j \text{ for all } 1 \leq j \leq d\},$$

and similarly

$$[a, b) := \{(n_1, \ldots, n_d) \in \mathbf{Z}^d : a_j \leq n_j < b_j \text{ for all } 1 \leq j \leq d\},$$

etc. If Z is an additive group, we define a *generalized arithmetic progression* (or just *progression* for short) in Z to be any set[1] of the form $P = a + [0, N] \cdot v$, where $a \in Z$, $N = (N_1, \ldots, N_d)$ is a tuple, $[0, N] \subset \mathbf{Z}^d$ is a discrete box, $v = (v_1, \ldots, v_d) \in Z^d$, the map $\cdot : \mathbf{Z}^d \times Z^d \to Z$ is the dot product

$$(n_1, \ldots, n_d) \cdot (v_1, \ldots, v_d) := n_1 v_1 + \cdots + n_d v_d,$$

and $[0, N] \cdot v := \{n \cdot v : n \in [0, N]\}$. In other words,

$$P = \{a + n_1 v_1 + \cdots + n_d v_d : 0 \leq n_j \leq N_j \text{ for all } 1 \leq j \leq d\}.$$

We call a the *base point* of P, $v = (v_1, \ldots, v_d)$ the *basis vectors* of P, N the *dimension* of P, d the *dimension* or *rank* of P, and $\mathrm{vol}(P) := |[0, N]| = \prod_{j=1}^{d}(N_j + 1)$ the *volume* of P. We say that the progression P is *proper* if the map $n \mapsto n \cdot v$ is injective on $[0, N]$, or equivalently if the cardinality of P is equal to its volume (as opposed to being strictly smaller than the volume, which can occur if the basis vectors are linearly dependent over $\mathbf{Z}$). We say that P is *symmetric* if $-P = P$; for instance $[-N, N] \cdot v = -N \cdot v + [0, 2N] \cdot v$ is a symmetric progression.

Other notation

There are a number of other definitions that we shall introduce at appropriate junctures and which will be used in more than one chapter of the book. These include the probabilistic notation (such as **E**(), **P**(), **I**(), **Var**(), **Cov**()) that we introduce

[1] Strictly speaking, this is an abuse of notation; the arithmetic progression should really be the sextuple (P, d, N, a, v, Z), because the set P alone does not always uniquely determine the base point, step, ambient space or even length (if the progression is improper) of the progression P. However, as it would be cumbersome continually to use this sextuple, we shall usually just P to denote the progression.

at the start of Chapter 1, and measures of additive structure such as the doubling constant $\sigma[A]$ (Definition 2.4), the Ruzsa distance $d(A, B)$ (Definition 2.5), and the additive energy $E(A, B)$ (Definition 2.8). We also introduce the concept of a partial sum set $A \overset{G}{+} B$ in Definition 2.28. The Fourier transform and the averaging notation $\mathbf{E}_{x\in Z} f(x)$, $\mathbf{P}_Z A$ is defined in Section 4.1, Fourier bias $\|A\|_u$ is defined in Definition 4.12, Bohr sets $\mathrm{Bohr}(S, \rho)$ are defined in Definition 4.17, and $\Lambda(p)$ constants are defined in Definition 4.26. The important notion of a Freiman homomorphism is defined in Definition 5.21. The notation for group theory (e.g. $\mathrm{ord}(x)$ and $\langle x\rangle$) is summarized in Section 3.1, while the notation for finite fields is summarized in Section 9.4.

1

The probabilistic method

In additive number theory, one frequently faces the problem of showing that a set A contains a subset B with a certain property $\mathcal{P}$. A very powerful tool for such a problem is Erdős' probabilistic method. In order to show that such a subset B exists, it suffices to prove that a properly defined random subset of A satisfies $\mathcal{P}$ with positive probability. The power of the probabilistic method has been justified by the fact that in most problems solved using this approach, it seems impossible to come up with a *deterministically* constructive proof of comparable simplicity.

In this chapter we are going to present several basic probabilistic tools together with some representative applications of the probabilistic method, particularly with regard to additive bases and the primes. We shall require several standard facts about the distribution of primes $P = \{2, 3, 5, \ldots\}$; so as not to disrupt the flow of the chapter we have placed these facts in an appendix (Section 1.10).

Notation. We assume the existence of some sample space (usually this will be finite). If E is an event in this sample space, we use $\mathbf{P}(E)$ to denote the probability of E, and $\mathbf{I}(E)$ to denote the indicator function (thus $\mathbf{I}(E) = 1$ if E occurs and 0 otherwise). If E, F are events, we use $E \wedge F$ to denote the event that E, F both hold, $E \vee F$ to denote the event that at least one of E, F hold, and $\bar{E}$ to denote the event that E does not hold. In this chapter all random variables will be assumed to be real-valued (and usually denoted by X or Y) or set-valued (and usually denoted by B). If X is a real-valued random variable with discrete support, we use

$$\mathbf{E}(X) := \sum_x x\mathbf{P}(X = x)$$

to denote the expectation of X, and

$$\mathbf{Var}(X) := \mathbf{E}(|X - \mathbf{E}(X)|^2) = \mathbf{E}(X^2) - \mathbf{E}(X)^2$$

to denote the variance. Thus for instance

$$\mathbf{E}(\mathbf{I}(E)) = \mathbf{P}(E); \qquad \mathbf{Var}(\mathbf{I}(E)) = \mathbf{P}(E) - \mathbf{P}(E)^2. \tag{1.1}$$

If F is an event of non-zero probability, we define the conditional probability of another event E with respect to F by:

$$\mathbf{P}(E|F) := \frac{\mathbf{P}(E \wedge F)}{\mathbf{P}(F)}$$

and similarly the conditional expectation of a random variable X by

$$\mathbf{E}(X|F) := \frac{\mathbf{E}(X\mathbf{I}(F))}{\mathbf{E}(\mathbf{I}(F))} = \sum_x x\mathbf{P}(X = x|F).$$

A random variable is *boolean* if it takes values in $\{0, 1\}$, or equivalently if it is an indicator function $\mathbf{I}(E)$ for some event E.

1.1 The first moment method

The simplest instance of the probabilistic method is the *first moment method*, which seeks to control the distribution of a random variable X in terms of its expectation (or first moment) $\mathbf{E}(X)$. Firstly, we make the trivial observation (essentially the pigeonhole principle) that $X \le \mathbf{E}(X)$ with positive probability, and $X \ge \mathbf{E}(X)$ with positive probability. A more quantitative variant of this is

Theorem 1.1 (Markov's inequality) *Let X be a non-negative random variable. Then for any positive real $\lambda > 0$*

$$\mathbf{P}(X \ge \lambda) \le \frac{\mathbf{E}(X)}{\lambda}. \tag{1.2}$$

Proof Start with the trivial inequality $X \ge \lambda\mathbf{I}(X \ge \lambda)$ and take expectations of both sides. □

Informally, this inequality asserts that $X = O(\mathbf{E}(X))$ with high probability; for instance, $X \le 10\mathbf{E}(X)$ with probability at least 0.9. Note that this is only an *upper tail estimate*; it gives an upper bound for how likely X is to be much larger than $\mathbf{E}(X)$, but does not control how likely X is to be much smaller than $\mathbf{E}(X)$. Indeed, if all one knows is the expectation $\mathbf{E}(X)$, it is easy to see that X could be as small as zero with probability arbitrarily close to 1, so the first moment method cannot give any non-trivial lower tail estimate. Later on we shall introduce more refined methods, such as the second moment method, that give further upper and lower tail estimates.

To apply the first moment method, we of course need to compute the expectations of random variables. A fundamental tool in doing so is *linearity of expectation*, which asserts that

$$\mathbf{E}(c_1 X_1 + \cdots + c_n X_n) = c_1 \mathbf{E}(X_1) + \cdots + c_n \mathbf{E}(X_n) \tag{1.3}$$

whenever $X_1, \ldots, X_n$ are random variables and $c_1, \ldots, c_n$ are real numbers. The power of this principle comes from there being no restriction on the independence or dependence between the X_is. A very typical application of (1.3) is in estimating the size $|B|$ of a subset B of a given set A, where B is generated in some random manner. From the obvious identity

$$|B| = \sum_{a \in A} \mathbf{I}(a \in B)$$

and (1.3), (1.1) we see that

$$\mathbf{E}(|B|) = \sum_{a \in A} \mathbf{P}(a \in B). \tag{1.4}$$

Again, we emphasize that the events $a \in B$ do not need to be independent in order for (1.4) to apply.

A weaker version of the linearity of expectation principle is the *union bound*

$$\mathbf{P}(E_1 \vee \cdots \vee E_n) \leq \mathbf{P}(E_1) + \cdots + \mathbf{P}(E_n) \tag{1.5}$$

for arbitrary events $E_1, \ldots, E_n$ (compare this with (1.3) with $X_i := \mathbf{I}(E_i)$ and $c_i := 1$). This trivial bound is still useful, especially in the case when the events $E_1, \ldots, E_n$ are rare and not too strongly correlated (see Exercise 1.1.3). A related estimate is as follows.

Lemma 1.2 (Borel–Cantelli lemma) *Let $E_1, E_2, \ldots$ be a sequence of events (possibly infinite or dependent), such that $\sum_n \mathbf{P}(E_n) < \infty$. Then for any integer M, we have*

$$\mathbf{P}(\textit{Fewer than } M \textit{ of the events } E_1, E_2, \ldots \textit{ hold}) \geq 1 - \frac{\sum_n \mathbf{P}(E_n)}{M}.$$

In particular, with probability 1 *at most finitely many of the events $E_1, E_2, \ldots$ hold.*

Another useful way of phrasing the Borel–Cantelli lemma is that if $F_1, F_2, \ldots$ are events such that $\sum_n (1 - \mathbf{P}(F_n)) < \infty$, then, with probability 1, all but finitely many of the events F_n hold.

Proof By monotone convergence it suffices to prove the claim when there are only finitely many events. From (1.3) we have $\mathbf{E}(\sum_n \mathbf{I}(E_n)) = \sum_n \mathbf{P}(E_n)$. If one now applies Markov's inequality with $\lambda = M$, the claim follows. □

1.1.1 Sum-free sets

We now apply the first moment method to the theory of sum-free sets. An additive set A is called *sum-free* iff it does not contain three elements x, y, z such that $x + y = z$; equivalently, A is sum-free iff $A \cap 2A = \emptyset$.

Theorem 1.3 *Let A be an additive set of non-zero integers. Then A contains a sum-free subset B of size $|B| > |A|/3$.*

Proof Choose a prime number $p = 3k + 2$, where k is sufficiently large so that $A \subset [-p/3, p/3]\backslash\{0\}$. We can thus view A as a subset of the cyclic group $\mathbf{Z}_p$ rather than the integers $\mathbf{Z}$, and observe that a subset B of A will be sum-free in $\mathbf{Z}_p$ if and only if[1] it is sum-free in $\mathbf{Z}$.

Now choose a random number $x \in \mathbf{Z}_p\backslash\{0\}$ uniformly, and form the random set

$$B := A \cap (x \cdot [k+1, 2k+1]) = \{a \in A : x^{-1}a \in \{k+1, \ldots, 2k+1\}\}.$$

Since $[k+1, 2k+1]$ is sum-free in $\mathbf{Z}_p$, we see that $x \cdot [k+1, 2k+1]$ is too, and thus B is a sum-free subset of A. We would like to show that $|B| > |A|/3$ with positive probability; by the first moment method it suffices to show that $\mathbf{E}(|B|) > |A|/3$. From (1.4) we have

$$\mathbf{E}(|B|) = \sum_{a \in A} \mathbf{P}(a \in B) = \sum_{a \in A} \mathbf{P}(x^{-1}a \in [k+1, 2k+1]).$$

If $a \in A$, then a is an invertible element of $\mathbf{Z}_p$, and thus $x^{-1}a$ is uniformly distributed in $\mathbf{Z}_p\backslash\{0\}$. Since $|[k+1, 2k+1]| > \frac{p-1}{3}$, we conclude that $\mathbf{P}(x^{-1}a \in [k+1, 2k+1]) > \frac{1}{3}$ for all $a \in A$. Thus we have $\mathbf{E}(|B|) > \frac{|A|}{3}$ as desired. □

Theorem 1.3 was proved by Erdős in 1965 [86]. Several years later, Bourgain [37] used harmonic analysis arguments to improve the bound slightly. It is surprising that the following question is open.

Question 1.4 *Can one replace $n/3$ by $(n/3) + 10$?*

Alon and Kleitman [10] considered the case of more general additive sets (not necessarily in $\mathbf{Z}$). They showed that in this case A always contains a sum-free subset of $2|A|/7$ elements and the constant $2/7$ is best possible.

Another classical problem concerning sum-free sets is the Erdős–Moser problem. Consider a finite additive set A. A subset B of A is *sum-free* with respect to A if $2^*B \cap A = \emptyset$, where $2^*B := \{b_1 + b_2 | b_1, b_2 \in B, b_1 \neq b_2\}$. Erdős and Moser asked for an estimate of the size of the largest sum-free subset of any given set A of cardinality n. We will discuss this problem in Section 6.2.1.

[1] This trick can be placed in a more systematic context using the theory of *Freiman homomorphisms*: see Section 5.3.

Exercises

1.1.1 If X is a non-negative random variable, establish the identity

$$\mathbf{E}(X) = \int_0^\infty \mathbf{P}(X > \lambda)\, d\lambda \tag{1.6}$$

and more generally for any $0 < p < \infty$

$$\mathbf{E}(X^p) = p \int_0^\infty \lambda^{p-1} \mathbf{P}(X > \lambda)\, d\lambda. \tag{1.7}$$

Thus the probability distribution function $\mathbf{P}(X > \lambda)$ controls all the moments $\mathbf{E}(X^p)$ of X.

1.1.2 When does equality hold in Markov's inequality?

1.1.3 If $E_1, \ldots, E_n$ are arbitrary probabilistic events, establish the lower bound

$$\mathbf{P}(E_1 \vee \cdots \vee E_n) \geq \sum_{i=1}^n \mathbf{P}(E_i) - \sum_{1 \leq i < j \leq n} \mathbf{P}(E_i \wedge E_j);$$

this bound should be compared with (1.5), and can be thought of as a variant of the second moment method which we discuss in the next section. (Hint: consider the random variable $\sum_{i=1}^n \mathbf{I}(E_i) - \sum_{1 \leq i < j \leq n} \mathbf{I}(E_i)\mathbf{I}(E_j)$.) More generally, establish the *Bonferroni inequalities*

$$\mathbf{P}(E_1 \vee \cdots \vee E_n) \geq \sum_{A \subset [1,n]: 1 \leq |A| \leq k} (-1)^{|A|+1} \mathbf{P}\left(\bigwedge_{i \in A} E_i\right)$$

when k is even, and

$$\mathbf{P}(E_1 \vee \cdots \vee E_n) \leq \sum_{A \subset [1,n]: 1 \leq |A| \leq k} (-1)^{|A|+1} \mathbf{P}\left(\bigwedge_{i \in A} E_i\right)$$

when k is odd.

1.1.4 Let X be a non-negative random variable. Establish the *popularity principle* $\mathbf{E}(X\mathbf{I}(X > \frac{1}{2}\mathbf{E}(X))) \geq \frac{1}{2}\mathbf{E}(X)$. In particular, if X is bounded by some constant M, then $\mathbf{P}(X > \frac{1}{2}\mathbf{E}(X)) \geq \frac{1}{2M}\mathbf{E}(X)$. Thus while there is in general no lower tail estimate on the event $X \leq \frac{1}{2}\mathbf{E}(X)$, we can say that the majority of the expectation of X is generated outside of this tail event, which does lead to a lower tail estimate if X is bounded.

1.1.5 Let A, B be non-empty subsets of a finite additive group Z. Show that there exists an $x \in Z$ such that

$$1 - \frac{|A \cup (B + x)|}{|Z|} \leq \left(1 - \frac{|A|}{|Z|}\right)\left(1 - \frac{|B|}{|Z|}\right),$$

and a $y \in Z$ such that

$$1 - \frac{|A \cup (B+y)|}{|Z|} \geq \left(1 - \frac{|A|}{|Z|}\right)\left(1 - \frac{|B|}{|Z|}\right).$$

1.1.6 Consider a set A as above. Show that there exists a subset $\{v_1, \ldots, v_d\}$ of Z with $d = O(\log \frac{|Z|}{|A|})$ such that

$$|A + [0,1]^d \cdot (v_1, \ldots, v_d)| \geq |Z|/2.$$

1.1.7 Consider a set A as above. Show that there exists a subset $\{v_1, \ldots, v_d\}$ of Z with $d := O(\log \frac{|Z|}{|A|} + \log\log(10 + |Z|))$ such that

$$A + [0,1]^d \cdot (v_1, \ldots, v_d) = Z.$$

1.1.8 Let A be an additive set. Show that there exists a subset $X \subset 2A - A$ of cardinality $|X| = O(\frac{|2A-A|}{A}(1 + \log|A|))$ such that $2A \subset A + X$. (Hint: translate A randomly by independent elements of $2A - A$, and use the first moment method.)

1.2 The second moment method

The first moment method allows one to control the order of magnitude of a random variable X by its expectation $\mathbf{E}(X)$. In many cases, this control is insufficient, and one also needs to establish that X usually does not deviate too greatly from its expected value. These types of estimates are known as *large deviation inequalities*, and are a fundamental set of tools in the subject. They can be significantly more powerful than the first moment method, but often require some assumptions concerning independence or approximate independence.

The simplest such large deviation inequality is *Chebyshev's inequality*, which controls the deviation in terms of the variance $\mathbf{Var}(X)$:

Theorem 1.5 (Chebyshev's inequality) *Let X be a random variable. Then for any positive λ*

$$\mathbf{P}\big(|X - \mathbf{E}(X)| > \lambda \mathbf{Var}(X)^{1/2}\big) \leq \frac{1}{\lambda^2}. \tag{1.8}$$

Proof We may assume $\mathbf{Var}(X) > 0$ as the case $\mathbf{Var}(X) = 0$ is trivial. From Markov's inequality we have

$$\mathbf{P}(|X - \mathbf{E}(X)|^2 > \lambda^2 \mathbf{Var}(X)) \leq \frac{\mathbf{E}(|X - \mathbf{E}(X)|^2)}{\lambda^2 \mathbf{Var}(X)} = \frac{1}{\lambda^2}$$

and the claim follows. □

Thus Chebyshev's inequality asserts that $X = \mathbf{E}(X) + O(\mathbf{Var}(X)^{1/2})$ with high probability, while in the converse direction it is clear that $|X - \mathbf{E}(X)| \geq \mathbf{Var}(X)^{1/2}$ with positive probability. The application of these facts is referred to as the *second moment method*. Note that Chebyshev's inequality provides both upper tail and lower tail bounds on X, with the tail decaying like $1/\lambda^2$ rather than $1/\lambda$. Thus the second moment method tends to give better distributional control than the first moment method. The downside is that the second moment method requires computing the variance, which is often trickier than computing the expectation.

Assume that $X = X_1 + \cdots + X_n$, where X_is are random variables. In view of (1.3), one might wonder whether

$$\mathbf{Var}(X) = \mathbf{Var}(X_1) + \cdots + \mathbf{Var}(X_n). \tag{1.9}$$

This equality holds in the special case when the X_is are pairwise independent (and in particular when they are jointly independent), but does not hold in general. For arbitrary X_is, we instead have

$$\mathbf{Var}(X) = \sum_{i=1}^{n} \mathbf{Var}(X_i) + \sum_{i,j \in [1,n]: i \neq j} \mathbf{Cov}(X_i, X_j), \tag{1.10}$$

where the *covariance* $\mathbf{Cov}(X_i, X_j)$ is defined as

$$\mathbf{Cov}(X_i, X_j) := \mathbf{E}((X_i - \mathbf{E}(X_i))(X_j - \mathbf{E}(X_j)) = \mathbf{E}(X_i X_j) - \mathbf{E}(X_i)\mathbf{E}(X_j).$$

Applying (1.9) to the special case when $X = |B|$, where B is some randomly generated subset of a set A, we see from (1.1) that if the events $a \in B$ are pairwise independent for all $a \in A$, then

$$\mathbf{Var}(|B|) = \sum_{a \in A} \mathbf{P}(a \in B) - \mathbf{P}(a \in B)^2 \tag{1.11}$$

and in particular we see from (1.4) that

$$\mathbf{Var}(|B|) \leq \mathbf{E}(|B|). \tag{1.12}$$

In the case when the events $a \in B$ are not pairwise independent, we must replace (1.11) by the more complicated identity

$$\mathbf{Var}(|B|) = \sum_{a \in A} \mathbf{P}(a \in B) - \mathbf{P}(a \in B)^2 + \sum_{a,a' \in A: a \neq a'} \mathbf{Cov}(\mathbf{I}(a \in B), \mathbf{I}(a' \in B)). \tag{1.13}$$

1.2.1 The number of prime divisors

Now we present a nice application of the second moment method to classical number theory. To this end, let[1]

$$\nu(n) := \sum_{p \leq n} \mathbf{I}(p|n)$$

[1] We shall adopt the convention that whenever a summation is over the index p, then p is understood to be prime.

denote the number of prime divisors of n. This function is among the most studied objects in classical number theory. Hardy and Ramanujan in the 1920s showed that "almost" all n have about $\log\log n$ prime divisors. We give a very simple proof of this result, found by Turán in 1934 [369].

Theorem 1.6 *Let $\omega(n)$ tend to infinity arbitrarily slowly. Then*

$$|\{x \in [1,n] : |\nu(x) - \log\log x| > \omega(n)\sqrt{\log\log n}\}| = o(n). \qquad (1.14)$$

Informally speaking, this result asserts that for a "generic" integer x, we have $\nu(x) = \log\log x + O(\sqrt{\log\log x})$ with high probability.

Proof Let x be chosen uniformly at random from the interval $\{1, 2, \ldots, n\}$. Our task is now to show that

$$\mathbf{P}(|\nu(x) - \log\log x| > \omega(n)\sqrt{\log\log n}) = o(1).$$

Due to a technical reason, instead of $\nu(x)$ we shall consider the related quantity $|B|$, where

$$B := \{p \text{ prime} : p \leq n^{1/10}, p|x\}.$$

Since x cannot have 10 different prime divisors larger than $n^{1/10}$, it follows that $|B| \leq \nu(x) \leq |B| + 10$. Thus, to prove (1.14), it suffices to show

$$\mathbf{P}(||B| - \log\log n| \geq \omega(n)\sqrt{\log\log n}) = o(1).$$

Note that $\log\log x = \log\log n + O(1)$ with probability $1 - o(1)$. In light of Chebyshev's inequality, this will follow from the following expectation and variance estimates:

$$\mathbf{E}(|B|), \mathbf{Var}(|B|) = \log\log n + O(1).$$

It remains to verify the expectation and variance estimate. From linearity of expectation (1.4) we have

$$\mathbf{E}(|B|) = \sum_{p \leq n^{1/10}} \mathbf{P}(p|x)$$

while from the variance identity (1.13) we have

$$\mathbf{Var}(|B|) = \sum_{p \leq n^{1/10}} (\mathbf{P}(p|x) - \mathbf{P}(p|x)^2) + \sum_{p,q \leq n^{1/10}: p \neq q} \mathbf{Cov}(\mathbf{I}(p|x), \mathbf{I}(q|x)).$$

Observe that $\mathbf{I}(p|x)\mathbf{I}(q|x) = \mathbf{I}(pq|x)$. Since $\mathbf{P}(d|x) = \frac{1}{d} + O(\frac{1}{n})$ for any $d \geq 1$, we conclude that

$$\mathbf{P}(p|x) = \frac{1}{p} + O\left(\frac{1}{n}\right)$$

and

$$\mathbf{Cov}(\mathbf{I}(p|x), \mathbf{I}(q|x)) = \frac{1}{pq} + O\left(\frac{1}{n}\right) - \left(\frac{1}{p} + O\left(\frac{1}{n}\right)\right)\left(\frac{1}{q} + O\left(\frac{1}{n}\right)\right) = O\left(\frac{1}{n}\right).$$

We thus conclude that

$$\mathbf{E}(|B|) = \sum_{p \leq n^{1/10}} \frac{1}{p} + O\left(n^{-9/10}\right)$$

and

$$\mathbf{Var}(|B|) = \sum_{p \leq n^{1/10}} \left(\frac{1}{p} - \frac{1}{p^2}\right) + O\left(n^{-8/10}\right).$$

The expectation and variance estimates now follow from Mertens' theorem (see Proposition 1.51) and the convergence of the sum $\sum_k \frac{1}{k^2}$. □

Exercises

1.2.1 When does equality hold in Chebyshev's inequality?

1.2.2 If X and Y are two random variables, verify the *Cauchy–Schwarz inequality* $|\mathbf{Cov}(X, Y)| \leq \mathbf{Var}(X)^{1/2}\mathbf{Var}(Y)^{1/2}$ and the *triangle inequality* $\mathbf{Var}(X + Y)^{1/2} \leq \mathbf{Var}(X)^{1/2} + \mathbf{Var}(Y)^{1/2}$. When does equality occur?

1.2.3 Prove (1.10).

1.2.4 If $\phi : \mathbf{R} \to \mathbf{R}$ is a convex function and X is a random variable, verify *Jensen's inequality* $\mathbf{E}(\phi(X)) \geq \phi(\mathbf{E}(X))$. If ϕ is strictly convex, when does equality occur?

1.2.5 Generalize Chebyshev's inequality using higher moments $\mathbf{E}(|X - \mathbf{E}(X)|^p)$ instead of the variance.

1.2.6 By obtaining an upper bound on the fourth moment, improve Theorem 1.6 to

$$\frac{1}{N}|\{x \in [1, N] : |\nu(x) - \log\log N| > K\sqrt{\log\log N}\}| = O(K^{-4}).$$

Can you generalize this to obtain a bound of $O_m(K^{-m})$ for any even integer $m \geq 2$, where the constant in the $O()$ notation is allowed to depend on m?

1.3 The exponential moment method

Chebyshev's inequality shows that if one has control of the second moment $\mathbf{Var}(X) = \mathbf{E}(|X - \mathbf{E}(X)|^2)$, then a random variable X takes the value $\mathbf{E}(X) + O(\lambda \mathbf{Var}(X)^{1/2})$ with probability $1 - O(\lambda^{-2})$. If one uses higher moments, one can obtain better decay of the tail probability than $O(\lambda^{-2})$. In particular, if one can control *exponential* moments[1] such as $\mathbf{E}(e^{tX})$ for some real parameter t, then one can obtain exponential decay in upper and lower tail probabilities, since Markov's inequality yields

$$\mathbf{P}(X \geq \lambda) = \mathbf{P}(e^{tX} \geq e^{t\lambda}) \leq \frac{\mathbf{E}(e^{tX})}{e^{t\lambda}} \tag{1.15}$$

for $t > 0$ and $\lambda \in \mathbf{R}$, and similarly

$$\mathbf{P}(X \leq -\lambda) = \mathbf{P}(e^{-tX} \geq e^{t\lambda}) \leq \frac{\mathbf{E}(e^{-tX})}{e^{t\lambda}} \tag{1.16}$$

for the same range of t, λ. The quantity $\mathbf{E}(e^{tX})$ is known as an *exponential moment* of X, and the function $t \mapsto \mathbf{E}(e^{tX})$ is known as the *moment generating function*, thanks to the Taylor expansion

$$\mathbf{E}(e^{tX}) = 1 + t\mathbf{E}(X) + \frac{t^2}{2!}\mathbf{E}(X^2) + \frac{t^3}{3!}\mathbf{E}(X^3) + \cdots .$$

The application of (1.15) or (1.16) is known as the *exponential moment method*. Of course, to use it effectively one needs to be able to compute the exponential moments $\mathbf{E}(e^{tX})$. A preliminary tool for doing so is

Lemma 1.7 *Let X be a random variable with $|X| \leq 1$ and $\mathbf{E}(X) = 0$. Then for any $-1 \leq t \leq 1$ we have $\mathbf{E}(e^{tX}) \leq \exp(t^2\mathbf{Var}(X))$.*

Proof Since $|tX| \leq 1$, a simple comparison of Taylor series gives the inequality

$$e^{tX} \leq 1 + tX + t^2X^2.$$

Taking expectations of both sides and using linearity of expectation and the hypothesis $\mathbf{E}(X) = 0$ we obtain

$$\mathbf{E}(e^{tX}) \leq 1 + t^2\mathbf{Var}(X) \leq \exp(t^2\mathbf{Var}(X))$$

as desired. □

This lemma by itself is not terribly effective as it requires both X and t to be bounded. However the power of this lemma can be amplified considerably when applied to random variables X which are *sums* of bounded random variables, $X = X_1 + \cdots + X_n$, provided that we have the very strong assumption of *joint* independence between the $X_1, \ldots, X_n$. More precisely, we have

[1] To avoid questions of integrability or measurability, let us assume for sake of discussion that the random variable X here only takes finitely many values; this is the case of importance in combinatorial applications.

Theorem 1.8 (Chernoff's inequality) *Assume that $X_1, \dots, X_n$ are jointly independent random variables where $|X_i - \mathbf{E}(X_i)| \le 1$ for all i. Set $X := X_1 + \cdots + X_n$ and let $\sigma := \sqrt{\mathbf{Var}(X)}$ be the standard deviation of X. Then for any $\lambda > 0$*

$$\mathbf{P}(|X - \mathbf{E}(X)| \ge \lambda\sigma) \le 2\max\left(e^{-\lambda^2/4}, e^{-\lambda\sigma/2}\right). \tag{1.17}$$

Informally speaking, (1.17) asserts that $X = \mathbf{E}(X) + O(\mathbf{Var}(X)^{1/2})$ with high probability, and $X = \mathbf{E}(X) + O(\log^{1/2} n \mathbf{Var}(X)^{1/2})$ with extremely high probability ($1 - O(n^{-C})$ for some large C). The bound in Chernoff's theorem provides a huge improvement over Chebyshev's inequality when λ is large. However the joint independence of the X_i is essential (Exercise 1.3.8). Later on we shall develop several variants of Chernoff's inequality in which there is some limited interaction between the X_i.

Proof By subtracting a constant from each of the X_i we may normalize $\mathbf{E}(X_i) = 0$ for each i. Observe that $\mathbf{P}(|X| \ge \lambda\sigma) = \mathbf{P}(X \ge \lambda\sigma) + \mathbf{P}(X \le -\lambda\sigma)$. By symmetry, it thus suffices to prove that

$$\mathbf{P}(X \ge \lambda\sigma) \le e^{-t\lambda\sigma/2} \tag{1.18}$$

where $t := \min(\lambda/2\sigma, 1)$.

Applying (1.15) we have

$$\mathbf{P}(X \ge \lambda\sigma) \le e^{-t\lambda\sigma}\mathbf{E}\left(e^{tX_1} \cdots e^{tX_n}\right).$$

Since the X_i are jointly independent, so are the e^{tX_i}. Using this and Lemma 1.7 we obtain

$$\mathbf{E}\left(e^{tX_1} \cdots e^{tX_n}\right) = \mathbf{E}\left(e^{tX_1}\right) \cdots \mathbf{E}\left(e^{tX_n}\right) \le \exp(t^2\mathbf{Var}(X_1)) \cdots \exp(t^2\mathbf{Var}(X_n)).$$

On the other hand, from (1.9) we have

$$\mathbf{Var}(X_1) + \cdots + \mathbf{Var}(X_n) = \sigma^2.$$

Putting all this together, we obtain

$$\mathbf{P}(X \ge \lambda\sigma) \le e^{-t\lambda\sigma}e^{t^2\sigma^2}.$$

Since $t \le \lambda/2\sigma$, the claim follows. □

Now let us consider a special, but important case when X_is are independent *boolean* (or *Bernoulli*) variables.

Corollary 1.9 *Let $X = t_1 + \cdots + t_n$ where the t_i are independent boolean random variables. Then for any $\epsilon > 0$*

$$\mathbf{P}(|X - \mathbf{E}(X)| \ge \epsilon\mathbf{E}(X)) \le 2e^{-\min(\epsilon^2/4, \epsilon/2)\mathbf{E}(X)}. \tag{1.19}$$

Applying this with $\epsilon = 1/2$ *(for instance), we conclude in particular that*

$$\mathbf{P}(X = \Theta(\mathbf{E}(X))) \geq 1 - 2e^{-\mathbf{E}(X)/16}. \tag{1.20}$$

Proof From (1.1) we have that $|t_i - \mathbf{E}(t_i)| \leq 1$ and $\mathbf{Var}(t_i) \leq \mathbf{E}(t_i)$. Summing this using (1.3), (1.9), we conclude that $\mathbf{Var}(X) \leq \mathbf{E}(X)$ (cf. (1.12)). The claim now follows from Theorem 1.8 with $\lambda := \epsilon\mathbf{E}(X)/\sigma$. □

As an immediate consequence of Corollary 1.9 and (1.4) we obtain the following concentration of measure property for the distribution of certain types of random sets.

Corollary 1.10 *Let A be a set (possibly infinite), and let $B \subset A$ be a random subset of A with the property that the events $a \in B$ are independent for every $a \in A$. Then for any $\epsilon > 0$ and any finite $A' \subseteq A$ we have*

$$\mathbf{P}\left(\left||B \cap A'| - \sum_{a \in A'} p_a\right| \geq \epsilon \sum_{a \in A'} p_a\right) \leq 2e^{-\min(\epsilon^2/4, \epsilon/2)\sum_{a\in A'} p_a}$$

where $p_a := \mathbf{P}(a \in B)$. In particular

$$\mathbf{P}\left(\frac{1}{2}\sum_{a \in A'} p_a \leq |B \cap A'| \leq \frac{3}{2}\sum_{a \in A'} p_a\right) \geq 1 - 2e^{-\sum_{a\in A'} p_a/16}.$$

1.3.1 Sidon's problem on thin bases

We now apply Chernoff's inequality to the study of thin bases in additive combinatorics.

Definition 1.11 (Bases) Let $B \subset \mathbf{N}$ be an (infinite) set of natural numbers, and let $k \in \mathbf{Z}_+$. We define the *counting function* $r_{k,B}(n)$ for any $n \in \mathbf{N}$ as

$$r_{k,B}(n) := |\{(b_1, \ldots, b_k) \in B^k : b_1 + \cdots + b_k = n\}|.$$

We say that B is a *basis of order k* if every sufficiently large positive integer can be represented as sum of k (not necessarily distinct) elements of B, or equivalently if $r_{k,B}(n) \geq 1$ for all sufficiently large n. Alternatively, B is a basis of order k if and only if $\mathbf{N}\backslash kB$ is finite.

Examples 1.12 The squares $\mathbf{N}^\wedge 2 = \{0, 1, 4, 9, \ldots\}$ are known to be a basis of order 4 (Legendre's theorem), while the primes $P = \{2, 3, 5, 7, \ldots\}$ are conjectured to be a basis of order 3 (Goldbach's conjecture) and are known to be a basis of order 4 (Vinogradov's theorem). Furthermore, for any $k \geq 1$, the kth powers $\mathbf{N}^\wedge k = \{0^k, 1^k, 2^k, \ldots\}$ are known to be a basis of order $C(k)$ for some finite $C(k)$ (Waring's conjecture, first proven by Hilbert). Indeed in this case, the powerful Hardy–Littlewood circle method yields the stronger result that

$r_{m,\mathbf{N}^{\wedge}k}(n) = \Theta_{m,k}(n^{\frac{m}{k}-1})$ for all large n, if m is sufficiently large depending on k (see for instance [379] for a discussion). On the other hand, the powers of k $k^{\wedge}\mathbf{N} = \{k^0, k^1, k^2, \ldots\}$ and the infinite progression $k \cdot \mathbf{N} = \{0, k, 2k, \ldots\}$ are not bases of any order when $k > 1$.

The function $r_{k,B}$ is closely related to the density of the set B. Indeed, we have the easy inequalities

$$\sum_{n \leq N} r_{k,B}(n) \leq |B \cap [0, N]|^k \leq \sum_{n \leq kN} r_{k,B}(n) \tag{1.21}$$

for any $N \geq 1$; this reflects the obvious fact that if $n = b_1 + \cdots + b_k$ is a decomposition of a natural number n into k natural numbers $b_1, \ldots, b_k$, then $n \leq N$ implies that $b_1, \ldots, b_k \in [0, N]$, and conversely $b_1, \ldots, b_k \in [0, N]$ implies $n \leq kN$. In particular if B is a basis of order k then

$$|B \cap [0, N]| = \Omega(N^{1/k}). \tag{1.22}$$

Let us say that a basis B of order k is *thin* if $r_{k,B}(n) = O(\log n)$ for all large n. This would mean that $|B \cap [0, N]| = N^{1/k + o_k(1)}$, thus the basis B would be nearly as "thin" as possible given (1.22). In the 1930s, Sidon asked the question of whether thin bases actually exist (or more generally, any basis which is "high quality" in the sense that $r_{k,B}(n) = n^{o(1)}$ for all n). As Erdős recalled in one of his memoirs, he thought he could provide an answer within a few days. It took a little bit longer. In 1956, Erdős [92] positively answered Sidon's question.

Theorem 1.13 *There exists a basis $B \subset \mathbf{Z}^+$ of order 2 so that $r_{2,B}(n) = \Theta(\log n)$ for every sufficiently large n. In particular, there exists a thin basis of order 2.*

Remark 1.14 A very old, but still unsolved conjecture of Erdős and Turán [98] states that if $B \subset \mathbf{N}$ is a basis of order 2, then $\limsup_{n\to\infty} r_{2,B}(n) = \infty$. In fact, Erdős later conjectured that $\limsup_{n\to\infty} r_{2,B}(n)/\log n > 0$ (so that the thin basis constructed above is essentially as thin as possible). Nothing is known concerning these conjectures (though see Exercise 1.3.10 for a much weaker result).

Proof Define[1] a set $B \subset \mathbf{Z}^+$ randomly by requiring the events $n \in B$ (for $n \in \mathbf{Z}^+$) to be jointly independent with probability

$$\mathbf{P}(n \in B) = \min\left(C\sqrt{\frac{\log n}{n}}, 1\right)$$

[1] Strictly speaking, to make this argument rigorous one needs an infinite probability space such as Wiener space, which in turn requires a certain amount of measure theory to construct. One can avoid this by proving a "finitary" version of Theorem 1.13 to provide a thin basis for an interval $[1, N]$ for all sufficiently large N, and then gluing those bases together; we leave the details to the interested reader. A similar remark applies to other random subsets of $\mathbf{Z}^+$ which we shall construct later in this chapter.

where $C > 0$ is a large constant to be chosen later. We now show that $r_{2,B}(n) = \Theta(\log n)$ for all sufficiently large n with positive probability (indeed, it is true with probability 1). Writing

$$r_{2,B}(n) = \sum_{i+j=n} \mathbf{I}(i \in B)\mathbf{I}(j \in B) = \Theta\left(\sum_{1 \le i < n/2} \mathbf{I}(i \in B)\mathbf{I}(n-i \in B)\right) + O(1)$$

we see that it suffices to show that the probability

$$\mathbf{P}\left(\sum_{1 \le i < n/2} \mathbf{I}(i \in B)\mathbf{I}(n-i \in B) = \Theta(\log n) \text{ for all but finitely many } n\right)$$

is positive (if the constants in the $\Theta()$ notation are chosen appropriately). By the Borel–Cantelli lemma (Lemma 1.2) and the convergence of $\sum_{n=1}^{\infty} \frac{1}{n^2}$, it thus suffices to show that

$$\mathbf{P}\left(\sum_{1 \le i < n/2} \mathbf{I}(i \in B)\mathbf{I}(n-i \in B) = \Theta(\log n)\right) = 1 - O\left(\frac{1}{n^2}\right)$$

for all large n.

By linearity of expectation (1.3), we have for $n > 1$

$$\begin{aligned}\mathbf{E}\left(\sum_{1 \le i < n/2} \mathbf{I}(i \in B)\mathbf{I}(n-i \in B)\right) &= \sum_{1 \le i < n/2} C^2 \sqrt{\frac{\log i}{i}} \sqrt{\frac{\log(n-i)}{n-i}} + O_C(1) \\ &= \Theta\left(C^2 \frac{\log^{1/2} n}{n^{1/2}} \sum_{1 \le i < n/2} \frac{\log^{1/2} i}{i^{1/2}}\right) + O_C(1) \\ &= \Theta(C^2 \log n) + O_C(1).\end{aligned}$$

In particular, by choosing C large enough, we may take

$$32 \log n \le \mathbf{E}\left(\sum_{1 \le i < n/2} \mathbf{I}(i \in B)\mathbf{I}(n-i \in B)\right) \le \kappa \log n$$

for all $n > 1$ and some $\kappa > 32$.

Observe that the restriction $i < n/2$ ensures that the boolean random variables $\mathbf{I}(i \in B)\mathbf{I}(n-i \in B)$ are jointly independent. If we now apply Corollary 1.9 with $\epsilon := 1/2$, we conclude that

$$\mathbf{P}\left(\sum_{1 \le i < n/2} \mathbf{I}(i \in B)\mathbf{I}(n-i \in B) \le \frac{\kappa}{16} \log n\right) \le 2/n^2,$$

and

$$\mathbf{P}\left(\sum_{1\le i<n/2} \mathbf{I}(i \in B)\mathbf{I}(n-i \in B) \ge \frac{3\kappa}{2}\log n\right) \le 2/n^2.$$

The claim follows. □

It is quite natural to ask whether Theorem 1.13 can be generalized to arbitrary k. Using the above approach, in order to obtain a basis B such that $r_{k,B}(n) = \Theta(\log n)$, we should set $\mathbf{P}(n \in B) = cn^{1/k-1}\log^{1/k} n$ for all sufficiently large n. As before, we have

$$r_{k,B}(n) = \sum_{x_1+\cdots+x_k=n} \mathbf{I}(x_1 \in B)\cdots\mathbf{I}(x_k \in B). \tag{1.23}$$

Although $r_{k,B}(n)$ does have the right expectation $\Theta(\log n)$, we face a major problem: the variables $\mathbf{I}(x_1 \in B), \ldots, \mathbf{I}(x_k \in B)$ with $k > 2$ are no longer independent. In fact, a typical number x appears in quite many ($\Omega(n^{k-2})$) solutions of $x_1 + \cdots + x_k = n$. This dashes the hope that one can use Theorem 1.8 to conclude the argument.

It took a long time to overcome this problem of dependency. In 1990, Erdős and Tetali [97] successfully generalized Theorem 1.13 for arbitrary k:

Theorem 1.15 *For any fixed k, there is a subset $B \subset \mathbf{N}$ such that $r_{k,B}(n) = \Theta(\log n)$ for all sufficiently large n. In particular, there exists a thin basis of order k for any k.*

We shall discuss this theorem later in a later section. Let us now turn instead to another application.

1.3.2 Complementary bases

Given a set $A \subset \mathbf{N}$ and an integer $k \ge 1$, a set $B \subset \mathbf{N}$ is a *complementary basis of order k of A* if every sufficiently large natural number can be written as a sum of an element in A and k elements in B (not necessarily distinct), or equivalently if $\mathbf{N}\backslash(A + kB)$ is finite.

As in the theory of bases, it is convenient to introduce the counting function

$$r_{A+B+\cdots+B}(n) := |\{(a, b_1, \ldots, b_k) \in A \times B^k : n = a + b_1 + \cdots + b_k\}|$$

and observe (analogously to (1.21)) that

$$\sum_{n\le N} r_{A+B+\cdots+B}(n) \le |A \cap [0, N]||B \cap [0, N]|^k \le \sum_{n\le(k+1)N} r_{A+B+\cdots+B}(n).$$

Now consider the set $P = \{2, 3, 5, \ldots\}$ of primes, and let B be a complementary basis for P of order 1. Recall that $|P \cap [0, N]| = \Theta(n/\log n)$ (Exercise 1.10.4 from the Appendix (Section 1.10)). From the preceding inequality we thus have the lower bound

$$|B \cap [0, n]| = \Omega(\log n)$$

for all large n. It is not known whether this bound can actually be attained. However, Erdős showed that P has a complementary base of size $O(\log^2 n)$ [92, 170]:

Theorem 1.16 *P has a complementary base $B \subset \mathbf{Z}^+$ of order 1 such that $|B \cap [0, n]| = O(\log^2 n)$ for all sufficiently large n.*

Proof Again B is created in a random manner, setting the events $n \in B$ to be jointly independent with probability

$$\mathbf{P}(n \in B) = \min\left(C\frac{\log n}{n}, 1\right)$$

for some large constant C. From Corollary 1.10 we have

$$\mathbf{P}(|B \cap [0, n]| > 10C \log^2 n) = O\left(\frac{1}{n^2}\right)$$

(say) for each n, and hence by the Borel–Cantelli lemma (Lemma 1.2) we have with probability 1 that $|B \cap [0, n]| = O(\log^2 n)$ for all sufficiently large n. Thus it suffices to show that with probability 1, $r_{P+B}(n) > 0$ for all sufficiently large n. By the Borel–Cantelli lemma again, it will suffice to show that

$$\mathbf{P}(r_{P+B}(n) > 0) = 1 - O\left(\frac{1}{n^2}\right)$$

for all large n. To show this, we write $r_{P+B}(n) = |B \cap (n - P)|$. From linearity of expectation (1.4) we have

$$\mathbf{E}(|B \cap (n - P)|) = C \sum_{p \in P \cap [1,n)} \frac{\log(n - p)}{n - p} + O_C(1).$$

We now use the estimate

$$\sum_{p \in P \cap [1,n)} \frac{\log(n - p)}{n - p} = \Omega(\log n)$$

for all sufficiently large n (see Proposition 1.54 in the Appendix); if we choose C large enough, we thus conclude that

$$\mathbf{E}(|B \cap (n - P)|) > 8 \log n$$

for all sufficiently large n. From Corollary 1.10 (or Corollary 1.9), the desired claim follows. □

Exercises

1.3.1 Let ε be the uniform distribution on $\{-1, +1\}$, and let $\varepsilon_1, \ldots, \varepsilon_n$ for even n be independent trials of ε. For any odd $\lambda > 0$, prove the *reflection principle*

$$\mathbf{P}\left(\max_{1 \le j \le n} \sum_{i=1}^{j} \varepsilon_i \ge \lambda\right) = 2\mathbf{P}\left(\sum_{i=1}^{n} \varepsilon_i \ge \lambda\right).$$

Hint: Let $A \subset \{-1, 1\}^n$ be the set of n-tuples $(\varepsilon_1, \ldots, \varepsilon_n)$ such that $\sum_{i=1}^n \varepsilon_i \ge \lambda$, and let $B \subset \{-1, 1\}^n$ be the set of n-tuples $(\varepsilon_1, \ldots, \varepsilon_n)$ such that $\sum_{i=1}^n \varepsilon_i < \lambda$ but $\sum_{i=1}^j \varepsilon_i \ge \lambda$ for some $1 \le j < n$. Create a "reflection map" which exhibits a bijection between A and B.

1.3.2 With the same notation as the previous exercise, show that

$$\mathbf{P}\left(\max_{1 \le j \le n} \sum_{i=1}^{j} a_i\varepsilon_i \ge \lambda\right) \le 2\mathbf{P}\left(\sum_{i=1}^{n} a_i\varepsilon_i \ge \lambda\right)$$

for all non-negative real numbers $a_1, \ldots, a_n$.

1.3.3 By considering the case when $X_1, \ldots, X_n \in \{-1, 1\}$ are independent variables taking values $+1$ and -1 with equal probability 1/2, show that Theorem 1.8 cannot be improved except for the constant in the exponent.

1.3.4 Let the hypotheses be as in Theorem 1.8, but with the X_i *complex*-valued instead of real-valued. Show that

$$\mathbf{P}(|X - \mathbf{E}(X)| \ge \lambda\sigma) \le 4 \max\left(e^{-\lambda^2/8}, e^{\lambda\sigma/2\sqrt{2}}\right)$$

for all $\lambda > 0$. (Hint: if $|z| \ge \lambda\sigma$, then either $|\mathrm{Re}(z)| \ge \frac{1}{\sqrt{2}}\lambda\sigma$ or $|\mathrm{Im}(z)| \ge \frac{1}{\sqrt{2}}\lambda\sigma$.) The constants here can be improved slightly.

1.3.5 (Hoeffding's inequality) Let $X_1, \ldots, X_n$ be jointly independent random variables, taking finitely many values, with $a_i \le X_i \le b_i$ for all i and some real numbers $a_i < b_i$. Let $X := X_1 + \cdots + X_n$. Using the exponential moment method, show that for $\lambda > 0$

$$\mathbf{P}\left(|X - \mathbf{E}(X)| \ge \lambda \left(\sum_{i=1}^{n} |b_i - a_i|^2\right)^{1/2}\right) \le 2e^{-2\lambda^2}.$$

1.3.6 (Azuma's inequality) Let $X_1, \ldots, X_n$ be random variables taking finitely many values with $|X_i| \le 1$ for all i. We do not assume that the X_i are jointly independent, however we do require that the X_i form a *martingale difference sequence*, by which we mean that $\mathbf{E}(X_i|X_1 = x_1, \ldots, X_{i-1} =$

$x_{i-1}) = 0$ for all $1 \le i \le n$ and all $x_1, \dots, x_{i-1}$. Using the exponential moment method, for $\lambda > 0$ establish the large deviation inequality

$$\mathbf{P}(|X_1 + \cdots + X_n| \ge \lambda\sqrt{n}) \le 2e^{-\lambda^2/4}. \tag{1.24}$$

1.3.7 Let n be a sufficiently large integer, and color each of the elements in $[1, n]$ red or blue, uniformly and independently at random (so each element is red with probability $1/2$ and blue with probability $1/2$). Show that the following statements hold with probability at least 0.9:
(a) there is a red arithmetic progression of length at least $\frac{\log n}{10}$;
(b) there is no monochromatic arithmetic progression of length exceeding $10 \log n$;
(c) the number of red elements and the number of blue elements in $[1, n]$ differ by $O(n^{1/2})$;
(d) in every arithmetic progression in $[1, n]$, the numbers of red and blue elements differ by $O(n^{1/2} \log^{1/2} n)$.

1.3.8 Let us color the elements of $[1, n]$ red or blue as in the preceding exercise. For each $A \subset [1, n]$, let t_A denote the parity of the red elements in A; thus $t_A = 1$ if there are an odd number of red elements in A, and $t_A = 0$ otherwise. Let $X = \sum_{A \subseteq [1,n]} t_A$. Show that the t_A are pairwise (but not necessarily jointly) independent, that $\mathbf{E}(X) = 2^{n-1} - 1/2$, and that $\mathbf{Var}(X) = 2^{n-2} - 1/4$. Furthermore, show that $\mathbf{P}(X = 0) = 2^{-n}$. This shows that Chernoff's inequality can fail dramatically if one only assumes pairwise independence instead of joint independence (though Chebyshev's inequality is of course still valid in this case).

1.3.9 For any $k \ge 1$, find a basis $B \subset \mathbf{N}$ of order k such that $|B \cap [0, n]| = \Theta_k(n^{1/k})$ for all large n. (This can be done constructively, without recourse to the probabilistic method, for instance by taking advantage of the base k representation of the integers.)

1.3.10 Prove that there do not exist positive integers $k, m \ge 1$, and a set $B \subset \mathbf{N}$ such that $r_{k,B}(n) = m$ for all sufficiently large n; thus a base of order k cannot be perfectly regular. (Hint: consider the complex-analytic function $\sum_{n \in B} z^n$, defined for $|z| < 1$, and compute the kth power of this function. It is rather challenging to find an elementary proof of this fact that does not use complex analysis, or the closely related tools of Fourier analysis.)

1.3.11 With the hypotheses of Theorem 1.8, establish the moment estimates

$$\mathbf{E}(|X - \mathbf{E}(X)|^p)^{1/p} = O(\sqrt{p}\sigma + p)$$

for all $p \ge 1$.

1.3.12 With the hypotheses of Corollary 1.9, establish the inequality

$$\mathbf{E}\left(\binom{X}{n}\right) \leq \frac{1}{n!}\mathbf{E}(X)^n$$

for all $n \in \mathbf{N}$. (Hint: expand $\binom{X}{n}$ as $\sum_{i_1<\dots<i_n} t_{i_1}\cdots t_{i_n}$). Use this (and Stirling's formula (1.52)) to derive an inequality similar to that in Corollary 1.9 in the case $\epsilon > 1$. For a generalization of this inequality, see Lemma 1.40 below.

1.4 Correlation inequalities

Chernoff's inequality is useful for controlling quantities of the form $t_1 + \cdots + t_n$ where $t_1, \dots, t_n$ are independent variables. In many applications, however, one needs to instead control more complicated polynomial expressions of $t_1, \dots, t_n$, such as monotone quantities.

Definition 1.17 (Monotone increasing variables) Let $t_1, \dots, t_n$ be jointly independent boolean random variables. A random variable $X = X(t_1, \dots, t_n)$ is *monotone increasing* if we have

$$X(t_1, \dots, t_n) \geq X(t_1', \dots, t_n') \text{ whenever } t_i \geq t_i' \text{ for all } 1 \leq i \leq n$$

or equivalently if X is monotone increasing in each of the variables t_i separately. We call X *monotone decreasing* if $-X$ is monotone increasing. We say that an event A is *monotone increasing* (resp. decreasing) if the indicator $\mathbf{I}(A)$ is monotone increasing (resp. decreasing).

Example 1.18 If $P(t_1, \dots, t_n)$ is any polynomial of $t_1, \dots, t_n$ with non-negative coefficients, then P is monotone increasing and $-P$ is monotone decreasing, and the event $P(t_1, \dots, t_n) \geq k$ is monotone increasing for any fixed k.

It is reasonable to think that any two increasing (resp. decreasing) variables or events are, in some way, positively correlated; intuitively, if both X and Y are monotone increasing (resp. decreasing), then the event that X is large (resp. small) should boost up the chance that Y is also large (resp. small). This intuition was materialized by Fortuin, Kasteleyn and Ginibre [104], motivated by problems in statistical mechanics:

Theorem 1.19 (FKG inequality) *Let $n \geq 0$, and let X and Y be two monotone increasing variables. Then*

$$\mathbf{E}(XY) \geq \mathbf{E}(X)\mathbf{E}(Y)$$

or equivalently

$$\mathbf{Cov}(X, Y) \geq 0.$$

The same inequality holds for the case both X and Y are monotone decreasing.

Proof By replacing X, Y with $-X, -Y$ if necessary, we may assume that X and Y are both monotone increasing.

We use induction on n. The base case $n = 0$ is trivial since in this case X and Y are deterministic. Now assume inductively that $n \geq 1$ and the claim has already been proven for $n - 1$. We may assume that $\mathbf{P}(t_n = 0)$ and $\mathbf{P}(t_n = 1)$ are non-zero since otherwise the claim follows immediately from the induction hypothesis. Observe that the covariance $\mathbf{Cov}(X, Y)$ is unaffected if we shift X and Y by constants. Thus we may normalize

$$\mathbf{E}(X|t_n = 0) = \mathbf{E}(Y|t_n = 0) = 0 \tag{1.25}$$

where $\mathbf{E}(X|t_n = 0)$ denotes the conditional expectation of X relative to the event $t_n = 0$. By monotonicity of X, Y in the t_n variable and the joint independence of the t_i we then have

$$\mathbf{E}(X|t_n = 1), \mathbf{E}(Y|t_n = 1) \geq 0. \tag{1.26}$$

Observe that, conditioning on the event $t_n = 0$, the random variables X, Y are monotone increasing functions of $t_1, \ldots, t_{n-1}$. Thus by the induction hypothesis

$$\mathbf{E}(XY|t_n = 0) \geq \mathbf{E}(X|t_n = 0)\mathbf{E}(Y|t_n = 0) = 0$$

and similarly

$$\mathbf{E}(XY|t_n = 1) \geq \mathbf{E}(X|t_n = 1)\mathbf{E}(Y|t_n = 1).$$

By Bayes' formula we thus have

$$\begin{aligned}\mathbf{E}(XY) &= \mathbf{E}(XY|t_n = 0)\mathbf{P}(t_n = 0) + \mathbf{E}(XY|t_n = 1)\mathbf{P}(t_n = 1)\\ &\geq \mathbf{E}(X|t_n = 1)\mathbf{E}(Y|t_n = 1)\mathbf{P}(t_n = 1).\end{aligned}$$

On the other hand, from (1.25) and another application of the total probability formula we have

$$\mathbf{E}(X)\mathbf{E}(Y) = \mathbf{E}(X|t_n = 1)\mathbf{P}(t_n = 1)\mathbf{E}(Y|t_n = 1)\mathbf{P}(t_n = 1).$$

Since $\mathbf{P}(t_n = 1) \leq 1$, the claim now follows from (1.26). □

From (1.1) and an easy induction we have an immediate corollary to Theorem 1.19:

Corollary 1.20 *Let A and B be two increasing events, then*

$$\mathbf{P}(A \wedge B) \geq \mathbf{P}(A)\mathbf{P}(B).$$

More generally, if $A_1, \ldots, A_k$ are increasing events, then

$$\mathbf{P}(A_1 \wedge \cdots \wedge A_k) \geq \mathbf{P}(A_1) \cdots \mathbf{P}(A_k).$$

1.4.1 Asymptotic complementary bases

Now we are going to use the FKG inequality to prove a result of Ruzsa [293] concerning asymptotic complementary bases.

Definition 1.21 (Asymptotic complementary bases) Let $A \subset \mathbf{N}$ be a set of natural numbers and $k \geq 1$. We define the *lower density* $\underline{\sigma}(A)$ and *upper density* $\overline{\sigma}(A)$ of A to be the numbers

$$\underline{\sigma}(A) := \liminf_{n\to\infty} \frac{|A \cap [0, n)|}{n}; \overline{\sigma}(A) := \limsup_{n\to\infty} \frac{|A \cap [0, n)|}{n}.$$

If $\epsilon > 0$ and $X \subset \mathbf{N}$, we say that X is a $(1 - \varepsilon)$*-complementary base of order* k of A if $\underline{\sigma}(A + kX) \geq 1 - \varepsilon$, and that X is an *asymptotic complementary base of order* k of A if $\underline{\sigma}(A + kX) = 1$.

Theorem 1.22 *[293] Let $P = \{2, 3, 5, \ldots\}$ be the primes. For any $0 < \epsilon < 1$, there is an $(1 - \epsilon)$-complementary base $X \subset \mathbf{Z}^+$ of order 1 of P with $|X \cap [1, n]| = O_\epsilon(\log n)$ for all large n.*

It follows that (the proof is left as an exercise)

Corollary 1.23 *For any function $\omega(n)$ tending to infinity with n, there is an asymptotic complementary base $X \in \mathbf{Z}^+$ of order 1 of P with $|X \cap [1, n]| \leq \omega(n) \log n$ for all large n.*

Corollary 1.23 improves an earlier result of Kolountzakis [214], and should also be compared with Theorem 1.16 (note that every complementary basis is automatically an asymptotic complementary basis). Since P has density $\Theta(n/\log n)$, it is clear that an asymptotic complementary base of P should have density $\Omega(\log n)$. Thus, Corollary 1.23 is nearly best possible.

Proof of Theorem 1.22 The theorem follows from the following finite statement.

Lemma 1.24 *For every $\varepsilon > 0$, and all natural numbers n which are sufficiently large depending on ε, there exists a set $B \subset [n^{2/3}, 2n^{2/3}]$ with $|B| = O_\varepsilon(\log n)$ such that*

$$|[1, x] \backslash (P + B)| \leq \varepsilon x, \tag{1.27}$$

for all $n^{3/4} \leq x \leq n$.

The deduction of Theorem 1.22 from Lemma 1.24 is straightforward and is left as an exercise. To prove Lemma 1.24, we use the probabilistic method. We choose $B \subset [n^{2/3}, 2n^{2/3}]$ randomly, by letting the events $l \in B$ with $l \in [n^{2/3}, 2n^{2/3}]$ be jointly independent with probability

$$\mathbf{P}(l \in B) = \frac{K \log n}{n^{2/3}}$$

where $K = K_\varepsilon$ is a large constant to be chosen later. From Corollary 1.10 we have

$$\mathbf{P}(|B| \geq 100K \log n) < \frac{1}{n} \tag{1.28}$$

(say).

Now let $J := \lfloor \frac{1}{4} \log_2 n \rfloor$. If $j \in [0, J]$, we say that j is *good* if

$$\big|[2n^{2/3}, n/2^j] \backslash (P + B)\big| \leq \frac{\varepsilon}{2} \frac{n}{2^j}.$$

It is easy to verify that if all the elements of $[0, J]$ are good, then (1.27) holds (recall that we assume n large depending on ε). In view of (1.28), it thus suffices to show that

$$\mathbf{P}(j \text{ is good for all } j \in [0, J]) \geq \frac{1}{n}. \tag{1.29}$$

Let us first estimate the probability that a single $j \in [0, J]$ is good. Fixing $j \in [0, J]$, we observe for each $m \in [2n^{2/3}, n/2^j]$ that

$$\begin{aligned}
\mathbf{P}(m \notin P + B) &= \mathbf{P}(m - p \notin B \text{ for all } p \in m - [n^{2/3}, 2n^{2/3}]) \\
&= \prod_{p \in m - [n^{2/3}, 2n^{2/3}]} \mathbf{P}(m - p \notin B) \\
&= \prod_{p \in m - [n^{2/3}, 2n^{2/3}]} \left(1 - \frac{K \log n}{n^{2/3}}\right) \\
&\leq \exp\left(-|P \cap m - [n^{2/3}, 2n^{2/3}]| \frac{K \log n}{n^{2/3}}\right)
\end{aligned}$$

where we have used the independence of the events $l \in B$. By Theorem 1.53, we conclude

$$\mathbf{P}(m \notin P + B) \leq \exp(-\Omega(K)).$$

Summing this over $m \in [2n^{2/3}, n/2^j]$ and using linearity of expectation (1.4), we conclude

$$\mathbf{E}\big(\big|[2n^{2/3}, n/2^j] \backslash (P + B)\big|\big) \leq \exp(-\Omega(K)) \frac{n}{2^j}.$$

If we choose K sufficiently large depending on ε, we thus see from Markov's inequality that

$$\mathbf{P}(j \text{ is good}) \geq \frac{1}{2}.$$

Now we come to the final and most important observation: For any fixed j, the event that j is good is a monotone increasing random variable, with respect to indicator variables $t_l := \mathbf{I}(l \in B)$. Thus, by Corollary 1.20,

$$\mathbf{P}(j \text{ is good for all } j \in [0, J]) \geq \prod_{j \in [0,J]} \mathbf{P}(j \text{ is good}) \geq 2^{-J-1}.$$

Since $J = \frac{1}{4}\log_2 n + O(1)$ and n is assumed to be large, the claim (1.29) follows. □

Exercises

1.4.1 Deduce Theorem 1.22, from Lemma 1.24. (Hint: the convergence of the geometric series $1 + g + g^2 + \cdots$ for $|g| < 1$ may be useful at one point.)

1.4.2 Deduce Corollary 1.23, from Theorem 1.22.

1.4.3 Let the notation and assumptions be as in Theorem 1.19. Suppose that each of the independent variables $t_1, \ldots, t_n$ attain the values 0 and 1 with positive probability. Show that equality holds in Theorem 1.19 if and only if X and Y depend on disjoint subsets of the random variables $t_1, \ldots, t_n$.

1.5 The Lovász local lemma

Let $(A_i)_{i \in V}$ be a finite collection of events in a probabilistic space; we will later view the index set V as the vertex set of a graph. In many situations, it is desirable to show that there is a chance that the complementary events $(\bar{A}_i)_{i \in V}$ hold simultaneously, i.e. that $\mathbf{P}(\bigwedge_{i \in V} \bar{A}_i) > 0$. This is particularly useful when the A_i are bad events that we would like to avoid.

If the A_i are mutually independent, then the problem is trivial, as we have

$$\mathbf{P}\left(\bigwedge_{i \in V} \bar{A}_i\right) = \prod_{i \in V} \mathbf{P}(\bar{A}_i) = \prod_{v \in V}(1 - \mathbf{P}(A_i)), \tag{1.30}$$

which is positive if $\mathbf{P}(A_i)$ are all strictly less than one. On the other hand, mutual independence is a very strong assumption which rarely holds.

One may expect that something similar to (1.30) is still true if we allow a sufficiently "local" dependence among the A_is, so that we still have good control on $\mathbf{P}(A_i)$ even after conditioning on most of the events $\bar{A}_j$. This is indeed possible, as shown by Lovász in 1975 in a joint paper with Erdős [93]. We present a modern version of this lemma as follows.

Lemma 1.25 (Lovász local lemma) *Let V be a finite set, and for each $i \in V$ let A_i be a probabilistic event. Assume that there is a directed graph $G(V, E)$ (without loops) on the vertex set V (which is known as the* dependency graph *of the A_i); and a sequence of numbers $0 \leq x_i < 1$ for each $i \in V$ such that the estimate*

$$\mathbf{P}\left(A_i | \bigwedge_{j\in S} \bar{A}_j\right) \leq x_i \prod_{(i,j)\in E} (1 - x_j) \tag{1.31}$$

holds whenever $i \in V$; and $S \subseteq V\backslash\{i\}$ is such that $\bigwedge_{j\in S} \bar{A}_j$ has non-zero probability and $(i, j) \notin E$ for all $j \in S$. Then for any disjoint $S, S' \subseteq V$ we have

$$\mathbf{P}\left(\bigwedge_{i\in S} \bar{A}_i | \bigwedge_{i\in S'} \bar{A}_i\right) \geq \prod_{i\in S}(1 - x_i) > 0. \tag{1.32}$$

In particular we have

$$\mathbf{P}\left(\bigwedge_{i\in V} \bar{A}_i\right) \geq \prod_{i\in V}(1 - x_i) > 0.$$

The graph G is usually referred to as the *dependency* graph of the A_i. Note that (1.31) will hold if we have

$$\mathbf{P}(A_i) \leq x_i \prod_{(i,j)\in E} (1 - x_j)$$

and each A_i is mutually independent to all of the A_j with $(i, j) \notin E$ and $j \neq i$. This was in fact the hypothesis stated in the original formulation of the lemma. However, there are situations where these rather strong mutual independence hypotheses are not available and one needs the full strength of Lemma 1.25. Alon and Spencer's book [12] Chapter 5 contains many interesting applications.

Proof of Lemma 1.25 We shall induct on the total cardinality $|S| + |S'|$. If $|S| + |S'| = 0$ then S, S' are empty, and the claim (1.32) is trivial. Now assume inductively that $|S| + |S'| \geq 1$, and the claim has already been proven for smaller values of $|S| + |S'|$. Note that the case $|S| = 0$ is trivial. To establish the claim for $|S| \geq 1$, it suffices to do so for the case $|S| = 1$. Indeed, if $|S| \geq 1$, then we can split $S = \{j\} \cup (S\backslash\{j\})$ for some $j \in S$. From the definition of conditional probability we have

$$\mathbf{P}\left(\bigwedge_{i\in S} \bar{A}_i | \bigwedge_{i\in S'} \bar{A}_i\right) = \mathbf{P}\left(\bar{A}_j | \bigwedge_{i\in S'\cup S\backslash\{j\}} \bar{A}_i\right) \mathbf{P}\left(\bigwedge_{i\in S\backslash\{j\}} \bar{A}_i | \bigwedge_{i\in S'} \bar{A}_i\right)$$

and the claim (1.32) then follows by applying the induction hypothesis to estimate the second factor.

Thus it remains to verify the $|S| = 1$ case of (1.32). Writing $S = \{i\}$, we reduce to showing that

$$\mathbf{P}\left(A_i | \bigwedge_{j \in S'} \bar{A}_j\right) \leq x_i.$$

We split $S' = S_1 \cup S_2$ where $S_1 := \{j \in S | (i, j) \in E\}$ are those indices j which are adjacent to i in the dependency graph, and $S_2 := S' \backslash S_1$. From the definition of conditional probability again we have

$$\mathbf{P}\left(A_i | \bigwedge_{j \in S'} \bar{A}_j\right) = \frac{\mathbf{P}\left(A_i, \bigwedge_{j \in S_1} \bar{A}_j | \bigwedge_{j \in S_2} \bar{A}_j\right)}{\mathbf{P}\left(\bigwedge_{j \in S_1} \bar{A}_j | \bigwedge_{j \in S_2} \bar{A}_j\right)}.$$

Note that by induction hypothesis, $\bigwedge_{j \in S_2} \bar{A}_j$ occurs with positive probability. From (1.31) we have

$$\mathbf{P}\left(A_i, \bigwedge_{j \in S_1} \bar{A}_j | \bigwedge_{j \in S_2} \bar{A}_j\right) \leq \mathbf{P}\left(A_i | \bigwedge_{j \in S_2} \bar{A}_j\right) \leq x_i \prod_{j \in V : (i,j) \in E} (1 - x_j).$$

On the other hand, from the induction hypothesis (since $|S_1| + |S_2| < 1 + |S'|$) we have

$$\mathbf{P}\left(\bigwedge_{j \in S_1} \bar{A}_j | \bigwedge_{j \in S_2} \bar{A}_j\right) \geq \prod_{j \in S_1} (1 - x_j) \geq \prod_{j \in V : (i,j) \in E} (1 - x_j).$$

Combining the two, we obtain the claim. □

In practice, the following corollary of Lemma 1.25 is sometimes easier to apply.

Corollary 1.26 *Let $d \geq 1$ and $0 < p < 1$ be numbers such that*

$$p \leq \frac{1}{e(d+1)},$$

where $e = 2.718\ldots$ is the base of the natural logarithm. Let V be a finite set, and for each $i \in V$ let A_i be a probabilistic event with $\mathbf{P}(A_i) \leq p$. Assume also that each A_i is mutually independent of all but at most d of the other events A_j. Then

$$\mathbf{P}\left(\bigwedge_{i \in V} \bar{A}_i\right) \geq \left(1 - \frac{1}{d+1}\right)^{|V|} > 0.$$

If $d = 0$, then Corollary 1.26 follows from (1.30). For $d \geq 1$, the corollary follows from Lemma 1.25 by setting $x_i = \frac{1}{d+1}$ and using the fact that $(1 - \frac{1}{d+1})^d > \frac{1}{e}$. The constant e is best possible as shown by Shearer.

1.5.1 Colorings of the real line

We now give an application of Corollary 1.26. This is the original result from the paper [93] of Erdős and Lovász, which motivated the development of the local lemma.

Let us use k colors $[1, k]$ to color the real numbers. (Thus, a coloring is a map from $\mathbf{R}$ to $[1, k]$.) A subset T of $\mathbf{R}$ is called *colorful* if it contains all k colors.

Theorem 1.27 *Let m and k be two positive integers satisfying*

$$e(m(m-1)+1)k\left(1-\frac{1}{k}\right)^m \leq 1. \tag{1.33}$$

Then for any set S of real numbers with $|S| = m$, and any set $X \subset \mathbf{R}$ (possibly infinite), there is a k-coloring of $\mathbf{R}$ such that the translates $x + S$ of S are colorful for every $x \in X$.

Proof We first prove this theorem in the special case when X is finite, and then use a compactness argument to handle the general case (of course, the theorem is strongest when $X = \mathbf{R}$). The point is that the bound (1.33) does not depend on the cardinality of X.

Fix X to be finite; thus $X + S$ is also finite. Note that we only need to color the real numbers in $X + S$, since the real numbers outside of $X + S$ are irrelevant. For each element y in $X + S$, we color it randomly and independently: y receives each of the colors in $[1, k]$ with the same probability $1/k$. Let A_x be the event that the translate $x + S$ is not colorful. We need to show that

$$\mathbf{P}\left(\bigwedge_{x\in X} \bar{A}_x\right) > 0.$$

In order to apply Corollary 1.26, we first estimate $\mathbf{P}(A_x)$. If x_S is not colorful, then at least one color is missing. The probability that a particular color (say 1) is missing is $(1-\frac{1}{k})^{|x+S|} = (1-\frac{1}{k})^m$. As there are k colors, we conclude

$$\mathbf{P}(A_x) \leq k\left(1-\frac{1}{k}\right)^m.$$

(In fact we have a strict inequality as there is a positive chance that more than one color is missing.) Next, observe that if two translates $x + S$ and $x' + S$ are disjoint, then the events A_x and $A_{x'}$ are independent. On the other hand, $x + S$ and $x' \in S$ intersect if and only if there are two elements $s_1, s_2 \in S$ such that $x + s_1 = x' + s_2$. It follows that $x' = x + (s_1 - s_2)$. Since that number of (ordered) pairs (s_1, s_2) with $s_1 \neq s_2$ and $s_1, s_2 \in S$ is $m(m-1)$, we conclude that each A_x is independent from all but at most $m(m-1)$ events $A_{x'}$. Set $p = k(1-\frac{1}{k})^m$ and $d = m(m-1)$. The condition (1.33) guarantees that the condition of Corollary 1.26 is met and this corollary implies that $\mathbf{P}(\bigwedge_{x\in X} \bar{A}_x) > 0$, as desired.

A routine way of passing from a finite statement to an infinite one is to use a compactness argument and that is what we do next. The space of colorings of $\mathbf{R}$ can be identified with the product space $[1, k]^{\mathbf{R}}$, which is compact in the product topology by Tychonoff's theorem. In this product space, for each $x \in \mathbf{R}$ we set

K_x to be the set of all k-colorings such that $x + S$ is colorful. It is easy to see that each K_x is closed. The finite statement proved above asserts that any finite collection of the K_x has a non-empty intersection. It follows, by compactness, that all K_x, $x \in \mathbf{R}$, have a non-empty intersection. Any element in this intersection is a coloring desired by the theorem. □

Exercise

1.5.1 Show that there exists a positive constant c such that the following holds. For every sufficiently large n, there is a graph on n points which does not contain the following two objects: a triangle and an independent set of size $c\sqrt{n}\log n$. (An independent set is a set of vertices, no two of which are connected by an edge.)

1.6 Janson's inequality

Let $t_1, \ldots, t_n$ be jointly independent boolean random variables. In Corollary 1.9 we established a large deviation inequality for the polynomial $t_1 + \cdots + t_n$. In many applications, it is also of interest to obtain large deviation inequalities for more general polynomials $P(t_1, \ldots, t_n)$ of the boolean variables $t_1, \ldots, t_n$. One particularly important case is that of a *boolean polynomial*

$$X := \sum_{A \in \mathcal{A}} \prod_{j \in A} t_j,$$

where $\mathcal{A}$ is some collection of non-empty subsets of $[1, n]$. Observe that boolean polynomials are automatically positive and monotone increasing, and hence any two boolean polynomials are positively correlated via the FKG inequality (Theorem 1.19). More generally, if X and Y are boolean polynomials, then $f(X)$ and $f(Y)$ will be positively correlated whenever f is a monotone increasing or decreasing function. In particular, we see that

$$\mathbf{E}\big(e^{-s(X+Y)}\big) \geq \mathbf{E}\big(e^{-sX}\big)\mathbf{E}\big(e^{-sY}\big) \tag{1.34}$$

for any real number s. Using this fact, the exponential moment method, and some additional convexity arguments, Janson [190] derived a powerful bound for the lower tail probability $\mathbf{P}(X \leq \mathbf{E}(X) - T)$:

Theorem 1.28 (Janson's inequality) *Let $t_1, \ldots, t_n$, $\mathcal{A}$, X be as above. Then for any $0 \leq T \leq \mathbf{E}(X)$ we have the lower tail estimate*

$$\mathbf{P}(X \leq E(X) - T) \leq \exp\left(-\frac{T^2}{2\Delta}\right)$$

where

$$\Delta = \sum_{A,B\in\mathcal{A}:A\cap B\neq\emptyset} \mathbf{E}\left(\prod_{j\in A\cup B} t_j\right).$$

In particular, we have

$$\mathbf{P}(X=0) \le \exp\left(-\frac{\mathbf{E}(X)^2}{2\Delta}\right).$$

Remark 1.29 Informally, Janson's inequality asserts that if $\Delta = O(\mathbf{E}(X)^2)$, then $X = \Omega(\mathbf{E}(X))$ with large probability. In the case where $\mathcal{A}$ is just the collection of singletons $\{1\}, \ldots, \{n\}$, then $X = t_1 + \cdots + t_n$, $\Delta = \mathbf{E}(X)$, and the above claim is then essentially (the lower half of) Corollary 1.9.

The quantity Δ is somewhat inconvenient to work with directly. Using the independence of the t_j, one can rewrite it as

$$\Delta = \sum_{A\in\mathcal{A}} \mathbf{E}\left(\prod_{j\in A} t_j\right) \sum_{B\in\mathcal{A}:A\cap B\neq\emptyset} \mathbf{E}\left(\prod_{j\in B\setminus A} t_j\right).$$

Since $\mathbf{E}(X) = \sum_{A\in\mathcal{A}} \mathbf{E}(\prod_{j\in A} t_j)$, we thus have

$$\Delta \le \mathbf{E}(X) \sup_{A\in\mathcal{A}} \sum_{B\in\mathcal{A}:A\cap B\neq\emptyset} \mathbf{E}\left(\prod_{j\in B\setminus A} t_j\right). \tag{1.35}$$

We record a particular consequence of this estimate concerning quadratic boolean polynomials that we shall use shortly.

Corollary 1.30 *Let $t_1, \ldots, t_n$ be as above, and let $X = \sum_{1\le i\le j\le n: i\sim j} t_i t_j$, where $i \sim j$ is some symmetric relation on $[1, n]$. Then we have*

$$\mathbf{P}(X=0) \le \exp\left(-\frac{\mathbf{E}(X)}{2 + 4\sup_i \sum_{j:i\sim j} \mathbf{E}(t_j)}\right).$$

Proof We take $\mathcal{A} := \{\{i, j\} : i \sim j\}$. For any $A \in \mathcal{A}$, it is easy to verify that

$$\sum_{B\in\mathcal{A}:A\cap B\neq\emptyset} \mathbf{E}\left(\prod_{j\in B\setminus A} t_j\right) \le 1 + 2\sup_i \sum_{j:i\sim j} \mathbf{E}(t_j)$$

and so the claim follows from (1.35) and Theorem 1.28. □

Before presenting the proof of Theorem 1.28, let us give an application. This application again concerns complementary bases of primes, but this time of order 2 rather than 1. The following result (which should be compared with Theorems 1.16 and 1.22) in the case $k = 2$ was recently proved by Vu [376].

Theorem 1.31 *For any $k \geq 2$, P has a complementary base $B \in \mathbf{Z}^+$ of order k with $|B \cap [1, n]| = O(\log n)$ for all large n.*

Proof It suffices to establish the claim when $k = 2$. To construct B we shall again use the probabilistic method. More precisely, we let $B \subset \mathbf{Z}^+$ be a random set with the events $n \in B$ being independent with probability

$$\mathbf{P}(n \in B) = \min\left(\frac{c}{n}, 1\right)$$

for all $n \in \mathbf{Z}^+$, where c is a positive constant to be determined. As before, we will not discuss the measure-theoretic issues associated with requiring infinitely-many independent random variables, as they can be dealt with by a suitable finitization of this argument. Let t_n be the boolean random variable $t_n := \mathbf{I}(n \in B)$. By Corollary 1.10 we have

$$\mathbf{P}(|B \cap [1, m]| \leq 10c \log m) = 1 - O\left(\frac{1}{m^2}\right)$$

for all large m, and hence by the Borel–Cantelli lemma (Lemma 1.2) we have with probability 1 that

$$|B \cap [1, m]| = O_c(\log m) \text{ for all sufficiently large } m > 1. \qquad (1.36)$$

Now for each $n \in \mathbf{Z}^+$, consider the counting function

$$\begin{aligned} r_{P+B+B}(n) &= |\{(p, i, j) \in P \times B \times B : n = p + i + j\}| \\ &= \sum_{p<n} \sum_{i+j=n-p} t_i t_j. \end{aligned}$$

This is of course a random variable for each n. In view of (1.36), it will suffice to show that with probability 1, we have $r_{P+B+B}(n) \neq 0$ for all but finitely many n. From the Borel–Cantelli lemma, it thus suffices to show that

$$\mathbf{P}(r_{P+B+B}(n) = 0) = O\left(\frac{1}{n^2}\right)$$

for all large n, if c is chosen large enough.

Fix n to be large. It will be convenient to work with a reduced version of $r_{P+B+B}(n)$, namely the boolean polynomial

$$Y_n := \sum_{i>j\geq n^{2/3}: i+j \in n-P} t_i t_j.$$

Clearly we have $Y_n \leq r_{P+B+B}(n)$, and so it suffices to show that

$$\mathbf{P}(Y_n = 0) = O\left(\frac{1}{n^2}\right).$$

We now apply Corollary 1.30 (using the relation $i \sim j$ if $i \neq j$ and $i + j \in n - P$) to give

$$\mathbf{P}(Y_n = 0) \leq \exp\left(-\frac{\mathbf{E}(Y_n)}{2 + 4\sup_{i \geq n^{2/3}} \sum_{j \geq n^{2/3}: i+j \in n-P} \mathbf{E}(t_j)}\right).$$

By construction of the t_j, and Proposition 1.54 from the Appendix, we have for any $i \geq n^{2/3}$

$$\sum_{j \geq n^{2/3}: i+j \in n-P} \mathbf{E}(t_j) = \sum_{p \leq n-i-n^{2/3}} \min\left(\frac{c}{n-i-p}, 1\right)$$
$$= O(c).$$

On the other hand, from linearity of expectation (1.3) and independence, we have

$$\begin{aligned}
\mathbf{E}(Y_n) &= \sum_{i>j \geq n^{2/3}: i+j \in n-P} \mathbf{E}(t_i t_j) \\
&= \sum_{i>j \geq n^{2/3}: i+j \in n-P} \frac{c^2}{ij} \\
&= c^2 \sum_{p \leq n-2n^{2/3}} \sum_{i>j \geq n^{2/3}: i+j=n-p} \frac{1}{ij} \\
&= c^2 \sum_{p \leq n-2n^{2/3}} \Omega\left(\frac{\log(n-p)}{n-p}\right) \\
&= \Omega(c^2 \log n),
\end{aligned}$$

where in the last line we again used Proposition 1.54 from the Appendix. Putting all of these estimates together we obtain

$$\mathbf{P}(Y_n = 0) \leq \exp(-\Omega(c \log n))$$

and the claim follows by choosing c to be suitably large. □

Now we are going to prove Theorem 1.28.

Proof of Theorem 1.28 We shall use the exponential moment method. By a limiting argument we may assume that $\mathbf{P}(t_j = 0), \mathbf{P}(t_j = 1) > 0$ for all j. We introduce the moment generating function $F(t) := \mathbf{E}(e^{-tX})$ for any $t > 0$. By (1.16) we have

$$\mathbf{P}(X \leq \mathbf{E}(X) - T) \leq \frac{F(t)}{e^{-t(\mathbf{E}(X)-T)}}.$$

Taking logarithms, we see that we only need to establish the inequality

$$\log F(t) + t(\mathbf{E}(X) - T) \leq -\frac{T^2}{2\Delta}$$

for some $t > 0$. Unlike the situation in Theorem 1.8, the summands in X are not necessarily independent, so we cannot factorize $F(t) = \mathbf{E}(e^{-tX})$ easily. Janson found a beautiful argument to get around this difficulty. Since $F(0) = 1$, we see from the fundamental theorem of calculus that

$$\log F(t) = \int_0^t \frac{F'(s)}{F(s)}\, ds.$$

Direct calculation shows that

$$\begin{aligned} F'(s) &= -\mathbf{E}(Xe^{-sX}) \\ &= -\sum_{A \in \mathcal{A}} \mathbf{E}\left(e^{-sX} \prod_{j \in A} t_j\right) \\ &= -\sum_{A \in \mathcal{A}} \mathbf{E}(e^{-sX} | E_A) \mathbf{P}(E_A), \end{aligned}$$

where E_A is the event that $t_j = 1$ for all $j \in A$. Thus it suffices to show that

$$\sum_{A \in \mathcal{A}} \mathbf{P}(E_A) \int_0^t \frac{\mathbf{E}(e^{-sX} | E_A)}{F(s)}\, ds - t(\mathbf{E}(X) - T) \geq \frac{T^2}{2\Delta}$$

for some $t > 0$.

We now exploit the fact that some of the factors of e^{-sX} are independent of E_A. For each $A \in \mathcal{A}$, we split X as $Y_A + Z_A$, which are the boolean polynomials

$$Y_A := \sum_{B \in \mathcal{A}: A \cap B \neq \emptyset} \prod_{j \in B} t_j; \quad Z_A = \sum_{B \in \mathcal{A}: A \cap B = \emptyset} \prod_{j \in B} t_j.$$

By (1.34) (conditioning on the variables in E_A), we conclude

$$\mathbf{E}(e^{-sX} | E_A) \geq \mathbf{E}(e^{-sY_A} | E_A) \mathbf{E}(e^{-sZ_A} | E_A).$$

On the other hand, Z_A is independent from E_A and is bounded from above by X; thus

$$\mathbf{E}(e^{-sZ_A} | E_A) = \mathbf{E}(e^{-sZ_A}) \geq \mathbf{E}(e^{-sX}) = F(s).$$

Combining all these estimates, we have reduced to showing that

$$\sum_{A \in \mathcal{A}} \mathbf{P}(E_A) \int_0^t \mathbf{E}(e^{-sY_A} | E_A)\, ds - t(\mathbf{E}(X) - T) \geq \frac{T^2}{2\Delta}$$

for some $t > 0$.

Next, we exploit the convexity of the function $x \mapsto e^{-sx}$ via Jensen's inequality (Exercise 1.2.4), concluding that

$$\mathbf{E}(e^{-sY_A}|E_A) \geq e^{-s\mathbf{E}(Y_A|E_A)}.$$

From linearity of expectation we have $\sum_{A\in\mathcal{A}} \mathbf{P}(E_A) = \mathbf{E}(X)$, and so another application of Jensen's inequality gives

$$\sum_{A\in\mathcal{A}} \mathbf{P}(E_A)e^{-s\mathbf{E}(Y_A|E_A)} \geq \mathbf{E}(X)e^{-s\sum_{A\in\mathcal{A}} \frac{\mathbf{P}(E_A)}{\mathbf{E}(X)}\mathbf{E}(Y_A|E_A)}.$$

On the other hand, from the definition of conditional probability we have

$$\sum_{A\in\mathcal{A}} \mathbf{P}(E_A)\mathbf{E}(Y_A|E_A) = \sum_{A\in\mathcal{A}} \sum_{B\in\mathcal{A}:A\cap B\neq\emptyset} \mathbf{E}\left(\mathbf{I}(E_A)\prod_{j\in B} t_j\right) = \Delta.$$

We thus have

$$\sum_{A\in\mathcal{A}} \mathbf{P}(E_A)\int_0^t \mathbf{E}(e^{-sY_A}|E_A)\,ds - t(\mathbf{E}(X) - T) \tag{1.37}$$

$$\geq \mathbf{E}(X)\int_0^t e^{-s\Delta/\mathbf{E}(X)}\,ds - t(\mathbf{E}(X) - T)$$

$$= \frac{\mathbf{E}(X)^2}{\Delta}(1 - e^{-t\Delta/\mathbf{E}(X)}) - t(\mathbf{E}(X) - T). \tag{1.38}$$

If we set $t := T/\Delta$, then $t\Delta/\mathbf{E}(X) = T/\mathbf{E}(X) \leq 1$, and we have

$$\begin{aligned} 1 - e^{-t\Delta/\mathbf{E}(X)} &= 1 - e^{-T/\mathbf{E}(X)} \\ &\geq T/\mathbf{E}(X) - T^2/2\mathbf{E}(X)^2 \end{aligned}$$

and hence

$$\begin{aligned} \sum_{A\in\mathcal{A}} \mathbf{P}(E_A)\int_0^t \mathbf{E}(e^{-sY_A}|E_A)\,ds &- t(\mathbf{E}(X) - T) \\ &\geq \frac{T\mathbf{E}(X)}{\Delta} - \frac{T^2}{2\Delta} - \frac{T}{\Delta}(\mathbf{E}(X) - T) \\ &= \frac{T^2}{2\Delta} \end{aligned}$$

as desired. □

Remark 1.32 Choosing $t = T/\Delta$ might be convenient, but may not be optimal. One can have a slightly better bound by optimizing the right hand side of (1.38) over t.

Remark 1.33 The proof of Janson's inequality is not symmetric. In other words, it cannot be extended to give a bound for the upper tail probability $\mathbf{P}(X \geq \mu + T)$. This probability will be addressed in the next section.

Exercises

1.6.1 By refining the argument, show that the complementary base B constructed in the proof of Theorem 1.31 has (with high probability) the property that $r_{P+B+B}(n) = \Omega(\log n)$ for all sufficiently large n.

1.6.2 Define a random graph $G(n, p)$ on the vertex set $[1, n]$ as follows. For each pair i, j ($1 \leq i < j \leq n$) draw an edge between i and j with probability p, independently.
   (a) Prove that if $p = o(n^{-1})$, then with probability $1 - o(1)$, $G(n, p)$ does not contain a triangle.
   (b) Assume that $p = n^{-1+\epsilon}$ for some small positive constant ϵ. Bound the probability that G does not contain a triangle.

1.6.3 Prove that for any $k \geq 2$ there is a basis B of order k with with $|B \cap [1, n]| = O(n^{1/2} \log^{1/k} n)$ for all large n.

1.7 Concentration of polynomials

In previous sections, we often considered a polynomial $Y = Y(t_1, \ldots, t_n)$ of n independent random variables $t_1, \ldots, t_n$, and wished to control the tail distribution of Y. For instance Chernoff's inequality shows that the polynomial $t_1 + \cdots + t_n$ is concentrated around its mean, while Janson's inequality shows that the values of certain polynomials (especially those of low degree) could very rarely be significantly less than the mean.

In this section, we present some further results of this type, that assert that certain polynomials with small degrees are strongly concentrated. These results can be seen as generalizing Chernoff's bound, and also provide (in certain cases) the missing half (upper tail bound) of Janson's inequality.

To motivate the results, let us first give a classical result which works for any function Y (not just a polynomial) provided that the Lipschitz constant of Y is small.

Lemma 1.34 (Lipschitz concentration inequality) *Let $Y : \{0, 1\}^n \to \mathbf{R}$ be a function such that $|Y(t) - Y(t')| \leq K$ whenever $t, t' \in \{0, 1\}^n$ differ in only one coordinate. Then if $t_1, \ldots, t_n$ are independent boolean variables, we have*

$$\mathbf{P}(|Y(t_1, \ldots, t_n) - \mathbf{E}(Y(t_1, \ldots, t_n))| \geq \lambda K \sqrt{n}) \leq 2e^{-\lambda^2/2}$$

for all $\lambda > 0$.

Remark 1.35 This inequality asserts that if each t_i can only influence the random variable $Y(t_1, \ldots, t_n)$ by at most $O(K)$, then $Y(t_1, \ldots, t_n)$ itself is concentrated in an interval of length $O(K\sqrt{n})$ around its mean. It should be compared with Hoeffding's inequality, which deals with the case $Y(t_1, \ldots, t_n) := t_1 + \cdots + t_n$, and also with Corollary 1.30.

Proof By dividing Y by K we may renormalize $K = 1$. Introduce the partially-conditioned random variables $Y_0, Y_1(t_1), \ldots, Y_n(t_1, \ldots, t_n) = Y(t_1, \ldots, t_n)$ by $Y_j(t_1, \ldots, t_j) := \mathbf{E}(Y|t_1, \ldots, t_j)$; thus Y_j is the conditional expectation of Y with the first j boolean variables t_j fixed. In particular $Y_0 = \mathbf{E}(Y)$ and $Y_n = Y(t_1, \ldots, t_n)$. We can thus write

$$Y(t_1, \ldots, t_n) - \mathbf{E}(Y(t_1, \ldots, t_n)) = X_1 + \cdots + X_n$$

where $X_j := Y_j - Y_{j-1}$. One then easily verifies (using the Lipschitz property) that $|X_j| \leq 1$ and $X_1, \ldots, X_n$ form a martingale difference sequence in the sense of Exercise 1.3.6. The claim then follows from Azuma's inequality (1.24). □

The above lemma is very useful when one has uniform Lipschitz control on Y, for instance if $Y = Y(t_1, \ldots, t_n)$ is a polynomial for which the partial derivatives $\frac{\partial Y}{\partial t_i}$ are small for *all* $t_1, \ldots, t_n$ in the unit cube. However in many applications (especially to thin bases), these partial derivatives will only be small on the *average*. Fortunately there are analogs of the above lemma which apply in this case, though they also require some average control on higher derivatives of Y. To state the results we need some notation. Let $Y = Y(t_1, \ldots, t_n)$ be a polynomial of n real variables. We say that Y is *totally positive* if all of its coefficients are non-negative, and furthermore that Y is *regular* if all the coefficients are between zero and one. We also say that Y is *simplified* if all of its monomials are square-free (i.e. do not contain any factor of t_i^2), and *homogeneous* if all the monomials have the same degree. Thus for instance a boolean polynomial is automatically regular and simplified, though not necessarily homogeneous. Given any multi-index $\alpha = (\alpha_1, \ldots, \alpha_n) \in \mathbf{Z}_+^n$, we define the partial derivative $\partial^\alpha Y$ as

$$\partial^\alpha Y := \left(\frac{\partial}{\partial t_1}\right)^{\alpha_1} \cdots \left(\frac{\partial}{\partial t_n}\right)^{\alpha_n} Y(t_1, \ldots, t_n),$$

and denote the order of α as $|\alpha| := \alpha_1 + \cdots + \alpha_n$. For any order $d \geq 0$, we denote $\mathbf{E}_d(Y) := \max_{\alpha:|\alpha|=d} \mathbf{E}(\partial^\alpha Y)$; thus for instance $\mathbf{E}_0(Y) = \mathbf{E}(Y)$, and $\mathbf{E}_d(Y) = 0$ if d exceeds the degree of Y. These quantities are vaguely reminiscent of Sobolev norms for the random variable Y. We also define $\mathbf{E}_{\geq d}(Y) := \max_{d' \geq d} \mathbf{E}_{d'}(Y)$.

The following result is due to Kim and Vu [203].

Theorem 1.36 *Let $k \geq 1$, and let $Y = Y(t_1, \ldots, t_n)$ be a totally positive polynomial of n independent boolean variables $t_1, \ldots, t_n$. Then there exists a constant $C_k > 0$ depending only on k such that*

$$\mathbf{P}\left(|Y - \mathbf{E}(Y)| \geq C_k \lambda^{k-1/2} \sqrt{\mathbf{E}_{\geq 0}(Y)\mathbf{E}_{\geq 1}(Y)}\right) = O_k\left(e^{-\lambda/4 + (k-1)\log n}\right)$$

for all $\lambda > 0$.

Informally Theorem 1.36 asserts that when the derivatives of Y are smaller on average than Y itself, and the degree of Y is small, then Y is concentrated around its mean, and in fact we have $Y = (1 + O_k(\sqrt{\frac{\mathbf{E}_{\geq 1}(Y)}{\mathbf{E}_{\geq 0}(Y)}}\log^{k-1/2} n)\mathbf{E}(Y)$ with high probability.

In applications in additive number theory, we frequently deal with the case when Y is roughly of size $\log n$. In this case, the error term $e^{(k-1)\log n}$ renders Theorem 1.36 ineffective. We, however, have a variant which is designed to handle this case:

Theorem 1.37 *[378] Let $k, n \geq 1$ and $\beta, \gamma, \epsilon > 0$. If $Y = Y(t_1, \ldots, t_n)$ is a regular polynomial (not necessarily simplified) of n independent boolean variables $t_1, \ldots, t_n$, which is homogeneous of degree k and obeys the expectation bounds*

$$Q \log n \leq \mathbf{E}(Y) \leq n/Q; \qquad \mathbf{E}_1(Y), \ldots, \mathbf{E}_{k-1}(Y) \leq n^{-\gamma}$$

for some sufficiently large $Q = Q(k, \epsilon, \beta, \gamma)$ (independent of n), then

$$\mathbf{P}(|Y - \mathbf{E}(Y)| \geq \epsilon \mathbf{E}(Y)) \leq n^{-\beta}.$$

In the next section, we will use this theorem to prove Theorem 1.15.

The next theorem deals with the case when the expectation of Y is less than one. In this case it is convenient to remove the constant term from any derivative of Y which appears. More precisely, introduce the renormalized derivative $\partial_*^\alpha Y(t) := \partial^\alpha Y(t) - \partial^\alpha Y(0)$.

Theorem 1.38 *Let $Y = Y(t_1, \ldots, t_n)$ be a simplified regular polynomial of n independent boolean variables (not necessarily homogeneous) such that $\mathbf{E}(\partial_*^\alpha Y) \leq n^{-\gamma}$ for some $\gamma > 0$ and all α. Then, for any $\beta > 0$, we have the bound $\mathbf{P}(Y \geq K_{\beta,\gamma}) < n^{-\beta}$ for some $K_{\beta,\gamma}$ which is independent of n and Y.*

Notice that the assumption implies that Y has small expectation. Taking α to be all zero, we have $\mathbf{E}(Y) \leq n^{-\gamma}$.

The proof of Theorem 1.36 relies on the so-called "divide and conquer martingale" technique, together with the exponential moment method. It is not too technical but requires lots of introduction. We thus skip it and refer the reader to [203]. The proof of Theorem 1.37 is more complicated. Besides the above-mentioned martingale technique, it also requires some non-trivial combinatorial considerations. Theorem 1.38 is a by-product of this proof (for details see [378]). These theorems have a wide range of applications in several areas and we refer the reader to [377] for a survey.

1.7.1 $B_h[g]$ sets

Let us conclude this section by an application of Theorem 1.38. A set $A \subset \mathbf{N}$ is called a $B_h[g]$ *set* or a $B_h[g]$ sequence if for any positive integer m, the equation $m = x_1 + \cdots + x_h$, $x_1 \leq x_2 \leq \cdots \leq x_h$, $x_i \in A$, has at most g solutions; up to a factor of $h!$, this is equivalent to requiring that $r_{h,A}(m)$ be bounded by g for all m. $B_h[g]$ sets were studied by Erdős and Turán in [98]. From (1.21) we see that if A is a $B_h[g]$ set, then $|A \cap [0,n]| = O_{h,g}(n^{1/h})$ for all n. In the converse direction, Erdős and Turán proved

Theorem 1.39 *For any $h \geq 1$ and $\epsilon > 0$, there exists a set $A \subset \mathbf{Z}^+$ with $|A \cap [0,n]| = \Omega_h(n^{1/h-\epsilon})$ for all large n, which is a $B_h[g]$ set for some $g = g_{h,\epsilon}$ (or in other words, $r_{h,A}(n)$ is uniformly bounded in n).*

Proof By using Theorem 1.38 we can give a short proof of this theorem. Without loss of generality we may assume $\epsilon < 1/h$. As before, we construct A randomly, letting the events $n \in A$ be independent with probability $\mathbf{P}(n \in A) = n^{1/h-1-\epsilon}$. A simple application of Corollary 1.9 and the Borel–Cantelli lemma also gives $|A \cap [0,n]| = \Omega_{h,\epsilon}(n^{1/h-\epsilon})$ for all but finitely many n with probability 1. Thus it will suffice to show that A is a $B_h[g]$ set with probability 1 (perhaps after removing finitely many elements), for some suitably large $g = g_h$ depending only on h.

Let t_n denote the indicator variables $t_n := \mathbf{I}(n \in A)$. For each m, we observe that the random variable

$$Y_m = Y_m(t_1, \ldots, t_m) = \sum_{n_1 \leq \cdots \leq n_h : n_1 + \cdots + n_h = m} t_{n_1} \cdots t_{n_h}$$

will become a regular polynomial of degree h in the $t_1, \ldots, t_m$ once we use the identity $t_i^a = t_i$ for $a = 2, 3, \ldots$ to make the monomials square-free. To show that A is a $B_h[g]$ set after removing finitely many elements, it will suffice to show that $Y_m \leq g$ for all but finitely many m; by the Borel–Cantelli lemma, it is enough to establish the upper tail estimate

$$\mathbf{P}(Y_m > g) \leq m^{-2}$$

for all large m. From linearity of expectation and independence we have

$$\begin{aligned}
\mathbf{E}(Y_m) &= \sum_{n_1 \leq \cdots \leq n_h : n_1 + \cdots + n_h = m} n_1^{1/h-1-\epsilon} \cdots n_h^{1/h-1-\epsilon} \\
&\leq O_h\left(m^{1/h-1-\epsilon} \sum_{n_1, \ldots, n_{h-1} \leq m} n_1^{1/h-1-\epsilon} \cdots n_{h-1}^{1/h-1-\epsilon} \right) \\
&\leq m^{1/h-1-\epsilon} O_h\left(\left(\sum_{n \leq m} n^{1/h-1-\epsilon} \right)^{h-1} \right) \\
&\leq O_h(m^{-h\epsilon}).
\end{aligned}$$

This already gives some non-trivial bound on $\mathbf{P}(Y_m > g)$ from Markov's inequality, but does not give the required decay in m. However, a similar computation to the above (which we leave as an exercise) establishes that $\mathbf{E}(\partial_*^\alpha Y_m) = O_h(m^{-1/h})$ for all non-zero α. The claim now follows from Theorem 1.38. □

The study of $B_h[g]$ sets is a popular topic in additive combinatorics. A detailed discussion of this topic is beyond the scope of our book. Let us, however, mention one new result of Cilleruelo, Ruzsa and Trujillo from [62]. Many other recent results can be found in [62, 191, 213, 61, 145, 272].

Let $A \subset [1, N]$ be a $B_h[g]$ set. A simple counting argument (related to (1.21)) gives $\binom{|A|+h-1}{h} \leq ghN$, which in turn yields the trivial bound $|A| \leq (ghh!N)^{1/h}$. Cilleruelo, Ruzsa and Trujillo gave the first non-trivial bounds for the case $g \geq 2$. They prove that $|A| \leq 1.864(gN)^{1/2} + 1$ when $h = 2$, and that

$$|A| \leq (1 + \cos^h(\pi/h))^{-1/h}(hh!gN)^{1/h}$$

when $h > 2$. The proofs made use of harmonic analysis methods via the consideration of the trigonometric polynomials $f(t) = \sum_{a \in X} e^{iat}$. The authors also constructed sets to establish for any g, the existence of a $B_2[g]$ set $A \subset [1, N]$ with

$$|A| \geq \left(\frac{g + [g/2]}{\sqrt{g + 2[g/2]}} + o_g(1) \right) N^{1/2}.$$

Exercises

1.7.1 Consider the random graph $G(n, p)$ defined in Exercise 1.6.2, and set $p := n^{-1+\epsilon}$. Let Y be the number of triangles in $G(n, p)$. Give an upper bound and a lower bound for

$$\mathbf{P}\left(Y \geq \frac{3}{2}\mathbf{E}(Y) \right).$$

1.7.2 Verify the bound $\mathbf{E}(\partial_*^\alpha Y_m) = O_h(n^{-1/h})$ claimed in the Proof of Theorem 1.39.

1.8 Thin bases of higher order

We now return to the study of thin bases B and their associated counting functions $r_{k,B}(n)$, initiated in Section 1.3. However, in this section we can use Theorem 1.37 to present a proof of Theorem 1.15, which asserted for each $k \geq 1$ the existence of a base B of order k with $r_{k,B}(n) = O_k(\log n)$ for all large n. This was proven in the $k = 2$ case (see Theorem 1.13) using Chernoff's inequality, but that method does not directly apply for higher k because $r_{k,B}(n)$ cannot be easily expressed as the sum of independent random variables.

We begin with a simple lemma on boolean polynomials that shows that if $\mathbf{E}(X)$ is not too large, then at most points $(t_1, \ldots, t_n)$ of the sample space, the polynomial X does not contain too many independent terms (cf. Exercise 1.3.12).

Lemma 1.40 *Let $X = \sum_{A \in \mathcal{A}} \prod_{j \in A} t_j$ be a boolean polynomial of n independent boolean variables $t_1, \ldots, t_n$, let $B \subseteq [1, n]$ be the random set $B := \{j \in [1, n] : t_j = 1\}$, and let $D \in \mathbf{N}$ be the random variable, defined as the largest number of disjoint sets in $\mathcal{A}$ which are contained in B. Then for any integer $K \geq 1$ we have*

$$\mathbf{P}(D \geq K) \leq \frac{\mathbf{E}(X)^K}{K!}.$$

Proof Observe that for $A_1, \ldots, A_k$ disjoint,

$$\mathbf{I}(D \geq K) \leq \frac{1}{K!} \sum_{A_1, \ldots, A_K \in \mathcal{A}, \text{disjoint}} \prod_{j \in A_1} t_j \ldots \prod_{j \in A_k} t_j.$$

Taking expectations of both sides and using linearity of expectation (1.3) followed by independence, we conclude

$$\mathbf{P}(D \geq K) \leq \frac{1}{K!} \sum_{A_1, \ldots, A_K \in \mathcal{A}} \mathbf{E}\left(\prod_{j \in A_1} t_j\right) \ldots \mathbf{E}\left(\prod_{j \in A_k} t_j\right).$$

But by linearity of expectation again, the left-hand side is just $\mathbf{E}(X)^K / K!$, and the claim follows. □

This lemma is particularly useful when combined with the *sunflower lemma* of Erdős and Rado [95]. A collection of sets $A_1, \ldots, A_l$ forms a *sunflower* if the pairwise intersections $A_i \cap A_j$ for $i \neq j$ are all the same (the A_i are called the *petals* of the flower). We allow this common pairwise intersection to be empty.

Lemma 1.41 (Sunflower lemma) *If $\mathcal{A}$ is a collection of sets, each of size at most k, and $|\mathcal{A}| > (l - 1)^k k!$, then $\mathcal{A}$ contains l sets forming a sunflower.*

This lemma can be proven by elementary combinatorics and is left as an exercise. It has the following consequence for the counting function $r_{k,B}(n)$.

Corollary 1.42 *Let $B \subset \mathbf{Z}^+$ and $k \geq 2$, and for each $n \in \mathbf{Z}^+$ let $D_{k,n}$ be the largest number of disjoint multisets[2] $\{x_1, \ldots, x_k\}$ of elements of B which sum to n. Then*

$$r_{k,B}(n) \leq k! k^k \max\left((D_{k,n} - 1)^k, \left(\sup_{m<n} r_{k-1,B}(m) - 1\right)^k\right).$$

Proof Fix n, and consider the collection $\mathcal{A}$ of sets which arise from taking the multisets $\{x_1, \ldots, x_k\}$ of elements of B which sum to n and then removing repeated

[2] A multiset is a set which is allowed to have repeated elements

elements. Clearly $r_{k,B}(n) \leq k^k|\mathcal{A}|$. Also observe that any sunflower in $\mathcal{A}$ has cardinality at most $D_{k,n}$ (if the petals are disjoint) or $\sup_{m<n} r_{k-1,B}(m)$ (if the petals are not disjoint); the latter follows by taking one of the elements in the common intersection of the sunflower and removing it once from each of the associated multisets. The claim then follows from the sunflower lemma. □

Using the above methods, we can now give a preliminary result towards proving Theorem 1.15.

Proposition 1.43 *Let $k \geq 2$, and let $B \subset \mathbf{Z}^+$ be a random subset of $\mathbf{Z}^+$, defined by letting $x \in B$ be independent with probability*

$$\mathbf{P}(x \in B) = \min\left(Cx^{1/k-1}\log^{1/k} x, 1\right)$$

for some positive constant $C > 1$. Then with probability 1, we have $\sup_n r_{k',B}(n) = O_{C,k,k',B}(1)$ for all $1 \leq k' < k$.

Proof We induct on k. The case $k = 1$ is obvious. Now suppose that $1 < k' < k$ and the claim has already been proven for $k' - 1$. Applying Corollary 1.42, we conclude that, with probability 1,

$$r_{k',B}(n) = O_{C,k,k',B}\left(\left(D_{k',n} + 1\right)^{k'}\right). \tag{1.39}$$

On the other hand, if we apply Lemma 1.40 with $t_x := \mathbf{I}(x \in B)$ for $1 \leq x \leq n$, and $\mathcal{A} = \mathcal{A}_n$ equal to all the sets which arise from the multisets $\{x_1, \ldots, x_{k'}\}$ that sum to n, then we observe that

$$\mathbf{P}(D_{k',n} \geq K) \leq \frac{\mathbf{E}\left(\sum_{A\in\mathcal{A}_n}\prod_{j\in A} t_j\right)^K}{K!}$$

for any $K \in \mathbf{Z}^+$. However, from linearity of expectation (1.3) and independence we have

$$\begin{aligned}
\mathbf{E}\left(\sum_{A\in\mathcal{A}_n}\prod_{j\in A} t_j\right) &= \sum_{A\in\mathcal{A}_n}\prod_{j\in A}\min\left(Cj^{1/k-1}\log^{1/k} j, 1\right)\\
&\leq O_{C,k,k'}\left(\sum_{j_1\leq\cdots\leq j_{k'}:j_1+\ldots+j_{k'}=n} j_1^{1/k-1}\cdots j_{k'}^{1/k-1}\right)\log n\\
&\leq O_{C,k,k'}\left(\sum_{j_1,\ldots,j_{k'-1}\in[1,n]} j_1^{1/k-1}\cdots j_{k'-1}^{1/k-1}\right)n^{1/k-1}\log n\\
&= O_{C,k,k'}\left(\left(\sum_{j\in[1,n]} j^{1/k-1}\right)^{k'-1}\right)n^{1/k-1}\log n\\
&= O_{C,k,k'}\left(n^{k'/k-1}\log n\right).
\end{aligned}$$

Since $k' < k$, we thus see that, by choosing K depending on k sufficiently large (e.g. $K = 2k + 1$), we have

$$\mathbf{P}(D_{k',n} \geq K) = O_{C,k,k',K}\left(\frac{1}{n^2}\right).$$

Applying the Borel–Cantelli lemma (Lemma 1.2) we see that with probability 1, we have $D_{k',n} < K$ for all but finitely many n. Combining this with (1.39) we obtain the claim. □

Now we prove Theorem 1.15. It will suffice to show that

Proposition 1.44 *Let $k \geq 2$, and let $B \subset \mathbf{Z}^+$ be a random subset of $\mathbf{Z}^+$, defined by letting $x \in B$ be independent with probability*

$$\mathbf{P}(x \in B) = \min(Cx^{1/k-1}\log^{1/k} x, 1)$$

for some positive constant $C > 1$. If C is sufficiently large depending on k, then with probability 1, we have $r_{k,B}(n) = \Theta_{C,k}(\log n)$ for all but finitely many n. In particular, B is a thin basis of order k with probability 1.

Proof We shall estimate $r_{k,B}(n)$ in terms of two related expressions:

$$R(n) := \{(x_1, \ldots, x_k) \in B : x_1 + \cdots + x_k = n; n^{0.1} < x_1 < x_2 < \cdots < x_k\} \tag{1.40}$$

$$E(n) := \{(x_1, \ldots, x_k) \in B : x_1 + \cdots + x_k = n; x_1 = x_2 \text{ or } x_1 \leq n^{0.1}\}. \tag{1.41}$$

It is clear (using the symmetry of $x_1 + \cdots + x_k$ under permutations) that

$$k!R(n) \leq r_{k,B}(n) \leq k!R(n) + k^2 E(n).$$

We view $R(n)$ as the main term and $E(n)$ as the error term; this reflects the intuitive fact that for most representations $n = x_1 + \cdots + x_k$, the x_i will be distinct and comparable in magnitude to n. It will suffice to show that with probability 1 we have

$$E(n) = O_{C,k,B}(1); \qquad R(n) = \Theta_{C,k,B}(\log n)$$

for all but finitely many n.

Let us deal first with the error term $E(n)$. We argue as in the proof of Proposition 1.43. Let $\mathcal{A}_n$ denote those sets which arise from the multisets $\{x_1, \cdots, x_k\}$ with $x_1 + \cdots + x_k = n$ and either $x_1 = x_2$ or $x_1 \leq n^{0.1}$. By arguing as in Corollary 1.42, we have

$$E(n) \leq k!k^k \max\left((D_n - 1)^k, \left(\sup_{m<n} r_{k-1,B}(m) - 1\right)^k\right)$$

where D_n is the largest number of disjoint sets that one can find in $\mathcal{A}_n$. Applying Proposition 1.43, we conclude that

$$E(n) = O_{C,k,B}(D_n + 1)^k$$

with probability 1. On the other hand, from Lemma 1.40, we have for any K that

$$\mathbf{P}(D_n \geq K) \leq \frac{\mathbf{E}(\sum_{A \in \mathcal{A}_n} \prod_{j \in A} t_j)^K}{K!}.$$

By arguing as in Proposition 1.43, one can establish

$$\mathbf{E}\left(\sum_{A \in \mathcal{A}_n} \prod_{j \in A} t_j\right) \leq O_k(n^{-1/k} n^{-0.9/k} \log n)$$

and thus, for a suitably large constant K depending only on k,

$$\mathbf{P}(D_n \geq K) = O_k(1/n^2).$$

From the Borel–Cantelli lemma we conclude that, with probability 1,

$$E(n) = O_{C,k,B}(1)$$

for all but finitely many n, and so the contribution of $E(n)$ is negligible.

Now we estimate the main term $R(n)$. Observe that we can write $R(n)$ as a homogeneous boolean polynomial $Y = Y(t_1, \ldots, t_n)$ of degree k; more explicitly, we have

$$Y(t_1, \ldots, t_n) = \sum_{A \in \mathcal{A}'_n} \prod_{j \in A} t_j$$

where $\mathcal{A}'_n$ is the collection of all sets $\{x_1, \ldots, x_k\}$ where $x_1 + \cdots + x_k = n$ and $n^{0.1} < x_1 < x_2 < \cdots < x_k$. Repeating the computations in Proposition 1.43 we see that

$$\mathbf{E}(Y) = \Theta_k(C \log n)$$

when n is sufficiently large depending on C, k. To conclude the proof it would thus suffice by the Borel–Cantelli lemma to establish the large deviation inequality

$$\mathbf{P}\left(|Y - \mathbf{E}(Y)| > \frac{1}{2}\mathbf{E}(Y)\right) = O_{C,k}\left(\frac{1}{n^2}\right)$$

for all large n. Applying Theorem 1.37 (and choosing C sufficiently large), we see that it suffices to show the derivative estimates

$$\mathbf{E}_1(Y), \ldots, \mathbf{E}_{k-1}(Y) \leq n^{-\gamma}$$

for all large n and some $\gamma > 0$. In other words, we need to establish

$$\mathbf{E}\left(\left(\frac{\partial}{\partial t_1}\right)^{\alpha_1} \cdots \left(\frac{\partial}{\partial t_n}\right)^{\alpha_n} Y(t_1, \ldots, t_n)\right) \leq n^{-\gamma}$$

whenever n is large and $1 \leq \alpha_1 + \cdots + \alpha_n \leq k - 1$. From the definition of $\mathcal{A}'_n$ we see that we may take $\alpha_j = 0$ for all $j \leq n^{0.1}$, and all the other α_j equal to 0 or 1, since the above partial derivative vanishes otherwise. One can then compute the partial derivative and reduce our problem to showing that

$$\mathbf{E}\left(\sum_{A \in \mathcal{A}'_n : A \supset A_0} \prod_{j \in A \backslash A_0} t_j \right) \leq n^{-\gamma}$$

whenever A_0 is any subset of $[n^{0.1}, n]$ of cardinality $1 \leq |A_0| \leq k - 1$ (this is the set of indices where $\alpha_j = 1$). Applying linearity of expectation and independence, and noting that $j \in [n^{0.1}, n]$ for all $j \in A \backslash A_0$, we conclude that

$$\begin{aligned}
\mathbf{E}\left(\sum_{A \in \mathcal{A}'_n : A \supset A_0} \prod_{j \in A \backslash A_0} t_j \right) &\leq \sum_{A \in \mathcal{A}'_n : A \supset A_0} O_{C,k}\left(n^{1/k-1} \log^{1/k} n\right)^{k-|A_0|} \\
&\leq O_k\left(n^{k-|A_0|-1}\right) O_{C,k}\left(n^{1/k-1} \log^{1/k} n\right)^{k-|A_0|} \\
&\leq O_{C,k}\left(n^{-1/k} \log n\right)
\end{aligned}$$

and the claim follows for large n. □

Remark 1.45 The proof above is from [378] and is based on the proof of Theorem 1.48 in [379]. The original proof in [98] was different and did not use Theorem 1.37.

Exercises

1.8.1 Let $A \in \mathbf{Z}^+$ be a set of n different integers. Prove that A contains a subset B of cardinality $\Omega(\log n)$ with the following property. No two elements of B add up to an element of A (thus $r_{2,B}(m)$ vanishes for all $m \in A$, or equivalently $A \cap 2B = \emptyset$).

1.8.2 Prove Lemma 1.41. (Hint: first use the pigeonhole principle to show that if $|\mathcal{A}| > (l-1)k$, then either $\mathcal{A}$ contains l disjoint sets, or that there exist at least $|\mathcal{A}|/(l-1)k$ sets in $\mathcal{A}$ which all have a common element x_0. Then use induction on k.)

1.9 Thin Waring bases

Recall that a thin basis of order k is a set $B \subset \mathbf{N}$ such that $r_{k,B}(n) = O(\log n)$ for all large n. Theorem 1.15, proved above, asserts that $\mathbf{N}$ contains a thin basis of any order. Given the abundance of classical bases such as the squares and primes, it is then natural to pose the following question:

Question 1.46 *Let A be any fixed basis of order k. Does A contain a thin subbasis B?*

Note that Sidon's original question can be viewed as the $k = 2$, $A = \mathbf{N}$ case of this question. From (1.21) we know that a thin basis B enjoys the bounds

$$|B \cap [0, N]| = \Omega_k(N^{1/k}); \qquad |B \cap [0, N]| = O_k(N^{1/k} \log^{1/k} N)$$

for all large N. Thus we can consider the following weaker version of Question 1.46:

Question 1.47 *Let A be any fixed basis of order k. Does A contain a subbasis B with* $|B \cap [0, N]| = O_k(N^{1/k} \log^{1/k} N)$ *for all large N?*

Question 1.47 has been investigated intensively for the Waring bases $\mathbf{N}^\wedge r = \{0^r, 1^r, 2^r, \ldots\}$, especially when $r = 2$ [90, 56, 387, 388, 384, 331]. For these bases it is known that if k is sufficiently large depending on r, then $\mathbf{N}^\wedge r$ is a basis of order k, and furthermore that

$$r_{k,\mathbf{N}^\wedge r}(n) = \Theta_{k,r}(n^{\frac{k}{r}-1}); \tag{1.42}$$

note that this is consistent with (1.21).

Choi, Erdős and Nathanson proved in [56] that $\mathbf{N}^\wedge 2$, the set of squares, contains a subbasis B of order 4, with $|B \cap [0, N]| = O_\varepsilon(N^{1/3} + \varepsilon)$ for all $N > 1$ and all $\varepsilon > 0$. This was generalized by Zöllner [387, 388], who showed that for any $k \geq 4$ there was a subbasis $B \subset \mathbf{N}^\wedge 2$ of order k with $|B \cap [0, N]| = O_{k,\varepsilon}(N^{1/k+\varepsilon})$ for any $\varepsilon > 0$ and $N > 1$. This bound was then sharpened further to $|B \cap [0, N]| = O_k(N^{1/k} \log^{1/k} N)$; from (1.21) we know that this is sharp except for the logarithmic factor. A short proof of Wirsing's result for the case $k = 4$ was given by Spencer in [331]. For $r \geq 3$, much less was known. In 1980, Nathanson [259] proved that $\mathbf{N}^\wedge r$ contains a subbasis of some order with density $o(N^{1/r})$. In the same paper, he posed a special case of Question 1.47, when $A = \mathbf{N}^\wedge r$.

In [379], Vu positively answered Question 1.46 (and hence Question 1.47) for the case $A = \mathbf{N}^\wedge r$ for any $r \geq 1$:

Theorem 1.48 *For any fixed r there is an integer* k_0 *such that the following holds. For any* $k \geq k_0$, *the set* $\mathbf{N}^\wedge r$ *of all rth powers contains a thin basis B of order k. In particular, from (1.21) we have* $|B \cap [0, n)| = O_k(N^{1/k} \log^{1/k} N)$ *for all large N.*

Remark 1.49 The sharp concentration result in Theorem 1.37 was first developed in order to prove Theorem 1.48.

Just as Theorem 1.15 followed from Proposition 1.44, Theorem 1.48 is an immediate consequence of

Proposition 1.50 *Let $k, r \geq 2$, and let B be a random subset of $(\mathbf{Z}^+)^\wedge r$, defined by letting $x^r \in B$ be independent with probability*

$$\mathbf{P}(x^r \in B) = \min\left(Cx^{\frac{r}{k}-1} \log^{1/k} x, 1\right)$$

for some positive constant $C > 1$. If k is sufficiently large depending on r, and C is sufficiently large depending on k, r, then with probability 1 we have $r_{k,B}(n) = \Theta_{C,k,r,B}(\log n)$ for all but finitely many n. In particular, B is a thin basis of order k with probability 1.

Proof (Sketch) As in the proof of Proposition 1.44, it suffices to show that with probability 1 we have

$$E(n) = O_{C,k,r,B}(1); \qquad R(n) = \Theta_{C,k,r,B}(\log n)$$

for all but finitely many n, where $R(n)$ and $E(n)$ were defined in (1.40), (1.41). The contribution of $E(n)$ can be dealt with by similar arguments to the previous section and is left as an exercise, so we focus on $R(n)$. As before we can write $R(n)$ as a boolean polynomial $Y_n = Y_n(t_1, \ldots, t_m)$, where $m = \lfloor n^{1/r} \rfloor$, $t_x = \mathbf{I}(x^r \in B)$, and

$$Y_n = \sum_{A \in \mathcal{A}_n} \prod_{x \in A} t_x$$

where $\mathcal{A}_n$ is the collection of sets $\{x_1, \ldots, x_k\}$ of positive integers with $x_1^r + \cdots + x_k^r = n$ and $n^{0.1} < x_1^r < \cdots < x_k^r$. Given the framework presented in the last section, the substantial difficulty remaining is to estimate the expectations of Y_n and its partial derivatives. In the following, we shall focus on the expectation of Y_n, establishing in particular that

$$\mathbf{E}(Y_n) = \Theta_{k,r}(C^k \log n).$$

This is the main estimate, and the remainder of the argument proceeds as in Proposition 1.44. Notice that

$$\mathbf{E}(Y_n) = C^k \sum_{x_1 < \cdots < x_k : \{x_1, \ldots, x_k\} \in \mathcal{A}_n} \prod_{j=1}^{k} x_j^{\frac{r}{k}-1} \log^{1/k} x_j;$$

since all the x_j range between $n^{1/10r}$ and $n^{1/r}$, it thus suffices to show that

$$\sum_{x_1 < \cdots < x_k : \{x_1, \ldots, x_k\} \in \mathcal{A}_n} x_1^{\frac{r}{k}-1} \ldots x_k^{\frac{r}{k}-1} = \Theta_{r,k}(1). \tag{1.43}$$

This bound implies, but is a little bit stronger than, the standard bound (1.42), as the estimate also asserts some improved bound on the counting function $r_{k,\mathbf{N}^\wedge r}(n)$ when one or more of the summands are restricted to be small (so that the corresponding weight $x^{\frac{r}{k}-1}$ is large).

The proof of (1.43) is a standard but lengthy application of the Hardy–Littlewood circle method, and is beyond the scope of this book. The reader may consult [379] for the full proof. □

Wooley [382] shown that one can set $k_0 = O(r \log r)$. This is (up to a constant factor) also the best current bound for k in (1.42). His proof also relies on Theorem 1.37, but the number-theoretic part is different.

Exercise

1.9.1 In the proof of Proposition 1.50, verify that with probability 1 one has $E(n) = O_{C,k,r,B}(1)$ for all but finitely many n.

1.10 Appendix: the distribution of the primes

Several results in this chapter relied on facts concerning the distribution of the primes

$$P = \{2, 3, 5, \ldots\}.$$

The distribution of this set is of course a very well-studied subject in analytic number theory, with one of the fundamental results being the *prime number theorem*

$$|P \cap [1, n]| = (1 + o(1))\frac{n}{\log n}. \quad (1.44)$$

An equivalent formulation is that if p_k denotes the kth prime, then $p_k = (1 + o(1))k \log k$. The famous *Riemann hypothesis*, which is still unsolved, is equivalent to the stronger statement that

$$|P \cap [1, n]| = \int_2^n \frac{dx}{\log x} + O_\varepsilon\left(n^{1/2+\varepsilon}\right) \quad (1.45)$$

for any $\varepsilon > 0$, or equivalently that $p_k = k \log k + O_\varepsilon(k^{1/2+\varepsilon})$ for any $\varepsilon > 0$.

The prime number theorem is rather deep and will not be proven here. In this Appendix we present some related results, most of which have surprisingly elementary and beautiful proofs. As they are number-theoretical rather than probabilistic in nature we have chosen to place these results in an appendix to this chapter.

We begin with some classical estimates of Chebyshev and Mertens. As is customary, when summing over a variable p, p is understood to denote a prime.

Proposition 1.51 (Elementary prime number estimates) *Let $n \geq 1$ be an integer. Then we have the estimates*

$$\sum_{p \leq n} \log p = O(n) \tag{1.46}$$

$$\sum_{p \leq n} \frac{\log p}{p} = \log n + O(1) \tag{1.47}$$

$$\sum_{p \leq n} \frac{1}{p} = \log \log n + O(1). \tag{1.48}$$

Remark 1.52 With the prime number theorem, we can improve (1.46) to $\sum_{p \leq n} \log p = (1 + o(1))n$, but it is not necessary to do so for our applications here.

Proof We first prove (1.46). Without loss of generality we may take n to be a power of two. Consider the binomial $\binom{2n}{n}$. From Pascal's formula we know that $\binom{2n}{n} \leq 4^n$. On the other hand, it is clear that every prime between n and $2n$ will divide $\binom{2n}{n}$. Thus

$$\prod_{n < p \leq 2n} p \leq 4^n.$$

Taking logarithms we conclude

$$\sum_{n < p \leq 2n} \log p = O(n).$$

Applying this bound to $n/2$, $n/4$, and so forth, and then summing the geometric series, the claim (1.46) follows.

Now we prove (1.47). This is a similar argument but based around the factorial $n!$ instead of $\binom{2n}{n}$. Observe that the only primes dividing $n!$ are those less than or equal to n. For each prime $p \leq n$, there are $\lfloor n/p \rfloor$ numbers (between 1 and n) divisible by p, $\lfloor n/p^2 \rfloor$ numbers (between 1 and n) divisible by p^2 and so on. Thus

$$n! = \prod_{p \leq n} p^{\lfloor n/p \rfloor + \lfloor n/p^2 \rfloor + \cdots}. \tag{1.49}$$

Taking the logarithm of both sides and applying Stirling's formula (Exercise 1.10.1) we obtain

$$n \log n + O(n) = \sum_{p \leq n} (\lfloor n/p \rfloor + \lfloor n/p^2 \rfloor + \cdots) \log p.$$

Since

$$\lfloor n/p \rfloor + \lfloor n/p^2 \rfloor + \cdots = \frac{n}{p} + O(1) + O\left(\frac{n}{p^2}\right),$$

we conclude, after some rearranging, that

$$\sum_{p\leq n} n\frac{\log p}{p} = n\log n + O(n) + \sum_{p\leq n} O(\log p) + \sum_{p\leq n} O\left(\frac{n\log p}{p^2}\right).$$

Since $\sum_k \frac{\log k}{k^2}$ is convergent, the last term is $O(n)$. The claim now follows from (1.46).

We shall deduce (1.48) from (1.47) using Abel's summation technique, rewriting one partial sum over primes as an average of others. Observe from the fundamental theorem of calculus that

$$\begin{aligned}\frac{1}{p} &= \frac{\log p}{p}\frac{1}{\log p}\\ &= \frac{\log p}{p}\int_1^\infty \mathbf{I}(t > p)\frac{dt}{t\log^2 t}\end{aligned}$$

and hence

$$\sum_{p\leq n}\frac{1}{p} = \sum_{p\leq n}\frac{\log p}{p}\int_1^\infty \mathbf{I}(t > p)\frac{dt}{t\log^2 t}.$$

Swapping the sum and integral, we obtain

$$\sum_{p\leq\min(n,t)}\frac{1}{p} = \int_1^\infty\left(\sum_{p\leq t}\frac{\log p}{p}\right)\frac{dt}{t\log^2 t}.$$

Applying (1.47), we obtain

$$\sum_{p\leq n}\frac{1}{p} = \int_2^\infty (\log\min(n,t) + O(1))\frac{dt}{t\log^2 t}.$$

Direct computation then shows that $\int_2^\infty \log\min(n,t)\frac{dt}{t\log^2 t} = \log\log n + O(1)$ and $\int_2^\infty \frac{dt}{t\log^2 t} = O(1)$, and the claim follows. □

We now turn to a deeper fact concerning the distribution of primes in intervals.

Theorem 1.53 *For all sufficiently large n, we have $|P\cap[n-x,n)| = \Theta(\frac{x}{\log n})$ for all $n^{2/3} < x < n$.*

Results of this type first appeared by Hoheisel [183]; the result as claimed is due to Ingham [188]. Note that this theorem follows immediately from the Riemann hypothesis (1.45). However, this theorem can be proven without using the Riemann hypothesis, rather some weaker (but still very non-trivial) facts on the distribution of zeroes of the Riemann zeta function: see [170]. We remark that if one only seeks the upper bound on $|P\cap[n-x,n)|$ then one can use relatively elementary sieve theory methods to establish the claim. The constant $2/3$ has been lowered (the current record is $7/12$, see [187], [178]). However, for the applications here, any exponent less than 1 will suffice.

We now combine this theorem with the Abel summation method to establish some further estimates on sums involving primes.

Proposition 1.54 *Let n be a large integer. Then we have the estimates*

$$\sum_{p \in P \cap [1, n-n^{2/3})} \frac{1}{n-p} = \Theta(1) \tag{1.50}$$

$$\sum_{p \in P \cap [1, n-n^{2/3})} \frac{\log(n-p)}{n-p} = \Theta(\log n). \tag{1.51}$$

Proof We begin by proving (1.50). From the fundamental theorem of calculus we have

$$\frac{1}{n-p} = \int_1^\infty 1_{p \in [n-x, n-n^{2/3})} \frac{1}{x^2}\, dx$$

for all $p \in P \cap [1, n-n^{2/3})$, and hence

$$\sum_{p \in P \cap [1, n-n^{2/3})} \frac{1}{n-p} = \int_1^\infty \left|P \cap \left[n-x, n-n^{2/3}\right)\right| \frac{dx}{x^2}.$$

The integrand vanishes when $x \leq n^{2/3}$. When $n^{2/3} < x \leq 2n^{2/3}$, Theorem 1.53 shows that the integrand is $O(\frac{1}{n^{2/3}\log n})$, while for $x \geq n^{2/3}$ another application of Theorem 1.53 shows that the integrand is $\Theta(\frac{1}{x \log n})$ when $x \leq n$ and $\Theta(\frac{n}{x^2 \log n})$ when $x > n$. Putting all these estimates together we obtain (1.50). The estimate (1.51) then follows immediately from (1.50) since $\log(n-p) = \Theta(\log n)$ when $p \in [1, n-n^{2/3}]$. □

Exercises

1.10.1 By approximating the sum $\sum_{m=1}^n \log m$ by the integral $\int_1^n \log x\, dx$, prove *Stirling's formula*

$$\log n! = n \log n - n + O(\log n) \tag{1.52}$$

for all $n > 1$.

1.10.2 Using Proposition 1.51, show that there is a constant c so that there is always a prime between n and cn for every positive integer n.

1.10.3 By being more careful in the proof of (1.46), show that

$$\sum_{p<n} \log p \leq 2n \log 2 + O\left(n^{1/2}\right)$$

and

$$\sum_{n\leq p<2n} \log p + \sum_{p\leq 2n/3} \log p \geq 2n\log 2 - O\left(n^{1/2}\right),$$

and conclude *Bertrand's postulate*, namely that for every sufficiently large integer n there exists a prime between n and $2n$. (This argument is due to Ramanujan. Bertrand's postulate in fact holds for all integers n, as the case of small n can be verified directly.)

1.10.4 Without using the prime number theorem, prove that $|P \cap [1,n]| = \Theta(\frac{n}{\log n})$; this is known as *Chebyshev's theorem*. This theorem is of course superseded by the prime number theorem $\pi(n) = (1+o(1))\frac{n}{\log n}$, but has the advantage of having a short elementary proof.

1.10.5 Prove that $p_k = \Theta(k \log k)$, where p_k denotes the kth prime. Again, this is superseded by the prime number theorem $p_k = (1+o(1))k\log k$.

1.10.6 Define the *von Mangoldt function* $\Lambda : \mathbf{Z}^+ \to \mathbf{R}$ by setting $\Lambda(n) := \log p$ if $n > 1$ is a power of a prime p, and $\Lambda(n) = 0$ otherwise. Show that

$$\sum_{d|n} \Lambda(d) = \log n \tag{1.53}$$

for all integers $n \geq 1$. Use this to prove that

$$\left(\sum_{n=1}^{\infty} \frac{\Lambda(n)}{n^s}\right)\left(\sum_{n=1}^{\infty} \frac{1}{n^s}\right) = \sum_{n=1}^{\infty} \frac{\log n}{n^s}$$

for all real numbers $s > 1$. Also, use (1.53) to give an alternative proof of (1.49).

1.10.7 Using the preceding exercise, show that

$$\sum_{n=1}^{\infty} \frac{\log p}{p^s} = \frac{1}{s-1} + O(1)$$

for all $s > 1$; integrate this to conclude

$$\sum_{n=1}^{\infty} \frac{1}{p^s} = \log \frac{1}{s-1} + O(1) \tag{1.54}$$

for all $s > 1$. Show that these estimates can also be deduced from Proposition 1.51 via Abel's method. Conversely, use (1.54) and (1.46) to give an alternative proof of (1.48).

1.10.8 Using Abel's summation method, show that the prime number theorem $\pi(x) = (1+o(1))\frac{x}{\log x}$ is equivalent to the estimate $\sum_{n\leq x} \Lambda(n) = (1+o(1))x$.

1.10.9 By being more careful in the proof of (1.48), show that

$$\sum_{p<n} \frac{1}{p} = \log\log n + C + O\left(\frac{1}{\log n}\right)$$

for some absolute constant C. Use this to deduce *Mertens' theorem*

$$\prod_{p<n}\left(1-\frac{1}{p}\right) = (1+o(1))\frac{C'}{\log n} \tag{1.55}$$

for some other absolute constant C' and all $n > 1$. (In fact one has $C' = e^{-\gamma}$, where $\gamma = 0.577\ldots$ is Euler's constant.)

2

Sum set estimates

Many classical problems in additive number theory revolve around the study of sum sets for *specific* sets A, B (though one typically works with infinite sets rather than finite ones). For instance, if $\mathbf{N}^\wedge 2 := \{0, 1, 4, 9, 16, \ldots\}$ is the set of square numbers, then it is a famous theorem of Lagrange that $4\mathbf{N}^\wedge 2 = \mathbf{N}$, i.e. every natural number is the sum of four squares; if $P := \{2, 3, 5, 7, 11, \ldots\}$ is the set of prime numbers, then it is a famous theorem of Vinogradov that $(2 \cdot \mathbf{N} + 1)\backslash 3P$ is finite (i.e. every sufficiently large odd number is the sum of three primes); in fact it is conjectured that this exceptional set consists only of 1, 3, and 5. The corresponding result for $(2 \cdot \mathbf{N})\backslash 2P$ remains open; the infamous *Goldbach conjecture* asserts that $2\mathbf{P}$ contains every even integer greater than 2, but this conjecture remains far from resolution.

In this text, we shall not focus on these types of problems, which rely heavily on the specific number-theoretic structure of the sets involved. Instead, we shall focus instead on the analysis of sum sets $A + B$ and related objects for more general sets A, B. To simplify the discussion we shall focus primarily on *additive sets* A, B, which are finite and non-empty subsets of an additive group such as $\mathbf{Z}$; thus our theory will not cover infinite sets such as the squares $\mathbf{N}^\wedge 2$ or the primes P directly, although one can certainly use this theory to analyze those sets simply by considering finite truncations, say to an interval $[0, N]$.

A fundamental problem in this field is the *inverse sum set problem*: if $A + B$ or $A - B$ is small, what can one say about A and B? A more specific question is as follows: if A is a finite non-empty subset of integers such that $|A + A| = K|A|$ for some small number K, what can one say about A? Here and in the rest of the text we use $|A|$ to denote the cardinality of a finite set A. The number $K := |A + A|/|A|$ is referred to as the *doubling constant* of A and will be denoted in this text by $\sigma[A]$. It is easy to see that this constant is at least 1, but it can be much larger; for instance, if A is a geometric progression such as $A = 2^\wedge[0, N) = \{1, 2, 2^2, \ldots, 2^{N-1}\}$

then one can easily verify that $\sigma[A] = (N+1)/2$, so the doubling constant can be arbitrarily large; indeed for "generic" sparse sets A we will have $\sigma[A] = (|A|+1)/2$.

At the other extreme, if A is an arithmetic progression $A := a + [0, N) \cdot r = \{a, a+r, a+2r, \ldots, a+(N-1)r\}$ of length N then one can check that A has doubling constant $\sigma[A] = 2 - \frac{1}{N}$. Thus arithmetic progressions are examples of sets with small doubling constant. One can perturb this example to produce a number of other examples of sets with small doubling constant; for instance if A is the above arithmetic progression, and we let A' be a subset of A of cardinality $N/2$ (say), then one can easily check that A' has doubling constant at most 4. Another example comes from adding an arbitrary integer n to A; then the set $A \cup \{n\}$ also has doubling constant at most 4.

One can generalize the concept of an arithmetic progression, to create more sets with small doubling constant. Consider the set

$$A := a + [0, (N_1, N_2)) \cdot (v_1, v_2) = \{a + n_1 v_1 + n_2 v_2 : 0 \leq n_1 < N_1; 0 \leq n_2 < N_2\},$$

where a, v_1, v_2 are integers, N_1, N_2 are positive integers and n_1, n_2 are understood to lie in the integers; this is an example of a *generalized arithmetic progression of rank* 2. One can verify that such sets have a doubling constant of at most 4. Note that such sets can look quite different from an ordinary arithmetic progression if N_1, N_2 are large and v_1, v_2 are very widely separated.

We have just remarked that generalized arithmetic progressions have small doubling constant. One of the fundamental theorems in this subject is *Freiman's theorem*, which asserts a partial converse to this claim. Freiman's theorem shows that any finite subset of the integers with small doubling constant can be efficiently contained in a generalized arithmetic progression (of bounded rank). This theorem is very useful, but is rather deep, and we will defer its proof to Section 5.4. It also has the drawback that some of the constants in this theorem depend exponentially on the doubling constant $\sigma[A]$. As such, it tends to only be useful in contexts where the doubling constant $\sigma[A]$ is of the order of $\log |A|$ or smaller.

Roughly speaking, one can classify results in inverse sum set theory by the range of $\sigma[A]$ for which the results are non-trivial. The case $\sigma[A] = 1$ is group theory (see Proposition 2.7). When $\sigma[A]$ is very small, e.g. $\sigma[A] < 2$ or $\sigma[A] < 3$, we have a complete characterization of the inverse problem, characterizing A in terms of groups and arithmetic progressions (see Corollary 5.6, Theorem 5.11). When $\sigma[A] = O(\log |A|)$, the best result is Freiman's theorem, which characterizes A in terms of generalized arithmetic progressions. When $\sigma[A] = O(|A|^{\varepsilon})$ for some small ε, we have Proposition 2.26 (as well as many of the other results in this chapter), which characterizes A in terms of approximate groups. In the remaining cases $|A|^\varepsilon \ll \sigma[A] \leq |A|$, some of the estimates here are still useful, but our understanding is still quite poor.

We will not prove Freiman's theorem in this chapter. However, we will develop the more elementary theory of *sum set estimates*, which can be used as substitutes for Freiman's theorem in some cases and are also of interest in their own right; this theory will also be needed in the proof of Freiman's theorem later on. These estimates are obtained by very simple combinatorial considerations, and rely on simple arithmetic facts such as $a - c = (a - b) + (b - c)$ and $a + b = a' + b' \iff a - b' = a' - b$. Because of the simplicity of the techniques used here, the results in this section are quite general, being applicable to any additive group and even to a large extent to non-abelian groups (see Section 2.7); we will wait until Chapter 5 until developing sum set estimates which exploit the specific structure of the ambient group (though see also Section 3.4). Also, the bounds obtained here are fairly reasonable, for instance the dependence of constants on the doubling constant $\sigma[A]$ is only polynomial in all the results in this section (in contrast to the exponential dependence on $\sigma[A]$ in Freiman's theorem). In some cases, though, the results in this section will be superseded by more precise results proven using advanced techniques, which we will address in later sections; for instance, in Section 6.5 we shall develop the theory of *Plünnecke inequalities*, which give more precise control on iterated sum sets and also handle the case when A and B have very different sizes, a case which is not treated efficiently by the tools in this section.

There are a large number of results in this chapter, but we point out a couple of specific results proven here which have a very large number of applications. The first is *Ruzsa's triangle inequality*, Lemma 2.6, which allows us to define a "metric" on the space of additive sets and which measures how small their sum sets are. Then there is Corollary 2.12, which links the size of $|A + B|$ and $|A - B|$ for arbitrary additive sets A, B. This generalizes to the iterated sum set estimates in Corollary 2.23 and Corollary 2.24. Another very useful class of tools are the *covering lemmas* – Ruzsa's covering lemma (Lemma 2.14), Green–Ruzsa's covering lemma (Lemma 2.17), and Chang's covering lemma (Lemma 5.31), which gives conditions under which one set A can be efficiently covered by translates of another set B. These results are collected together in Proposition 2.26 and Proposition 2.27, which characterize sets with small sum set in terms of *approximate groups*. Last, but certainly not least, there is the *Balog–Szemerédi–Gowers theorem*, which generalizes the previous results to the setting when one has only partial information on a sum set (or equivalently, one only controls the "additive energy" between two sets); see Theorem 2.29 and Theorem 2.31. We also develop an asymmetric version of this theorem in Section 2.6.

2.1 Sum sets

We now systematically study the sum sets $A + B$ and difference sets $A - B$ of two additive sets A, B in an ambient group Z as defined in Definition 0.1, as well as the iterated sum sets nA. We should caution the reader that the iterated sum set nA is in general not the same as the dilate $n \cdot A := \{n \cdot a : a \in A\}$ though we do have the inclusion $n \cdot A \subseteq nA$. Similarly the difference set $A - B$ should not be confused with the set-theoretic difference $A \backslash B := \{x \in A : x \notin B\}$. We also write $A + x = A + \{x\}$ for the translate of A by an element $x \in Z$.

Since addition of group elements is associative and commutative, one can easily verify the same is true for addition of sets. We should caution however that the sum set operation is not invertible: for instance, $A + B - B$ contains A but is generally not equal to A. Similarly, when $n > m$, then $nA - mA$ will contain $(n - m)A$ but will generally be larger.

A very fundamental question in this topic is the following: under what conditions is $A + B$ "small", and under what conditions is it "large"? More precisely, we will be interested in the cardinality $|A + B|$ of the sum set $A + B$. We have the following trivial estimates:

Lemma 2.1 (Trivial sum set estimates) *Let A, B be additive sets with common ambient group Z, and let $x \in Z$. Then we have the identities $|A + x| = |-A| = |A|$, the inequalities*

$$\max(|A|, |B|) \leq |A + B|, |A - B| \leq |A||B| \tag{2.1}$$

and the inequalities

$$|A| \leq |A + A| \leq \frac{|A|(|A| + 1)}{2}. \tag{2.2}$$

More generally, for any integer $n \geq 1$, we have $|(n + 1)A| \geq |nA|$ and

$$|nA| \leq \binom{|A| + n - 1}{n} = \frac{|A|(|A| + 1) \cdots (|A| + n - 1)}{n!}. \tag{2.3}$$

We remark that the lower bound in (2.1) can be improved for specific groups Z, or when A and B have large "dimension"; see Theorem 3.16, Lemma 5.3, Theorem 5.17, Corollary 5.13, Theorem 5.4.

Proof We shall just prove (2.3), as all the other inequalities either follow from this inequality or are trivial. We argue by induction on $|A|$. If $|A| = 1$ then both sides of (2.3) are equal to 1. If $|A| > 1$, then we can write $A = B \cup \{x\}$ where B is a non-empty set with $|B| = |A| - 1$. Then

$$nA = \bigcup_{j=0}^{n} (jB + (n - j) \cdot x)$$

and hence by the induction hypothesis and Pascal's triangle identity

$$|nA| \leq \sum_{j=0}^{n} |jB| \leq \sum_{j=0}^{n} \binom{|A| - 1 + j - 1}{j} = \binom{|A| + n - 1}{n}$$

as claimed. (We adopt the convention that $0B = \{0\}$.) □

Observe from the above facts that the magnitude of sum sets such as $A + B$, $A - B$, kA are unaffected if one translates A or B by an arbitrary amount. This gives much of the theory of sum sets a "translation-invariant" or "affine" flavor. We will sometimes take advantage of this translation invariance to normalize one of the sets, for instance to contain the origin 0.

For "generic" additive sets A and B, the cardinalities of the sum sets considered in Lemma 2.1 are much more likely to be closer to the upper bounds listed above than the lower bounds; see for instance Exercise 2.1.1. This suggests that the lower bounds are only attainable, or close to being attainable, when the sets A and B have a considerable amount of structure; we shall develop this theme in the remainder of this chapter, by introducing tools such as doubling and difference constants, Ruzsa distance, additive energy, and K-approximate groups to quantify some of these notions of "structure". For now, we at least settle the question of when the lower bound in (2.1) is attained.

Proposition 2.2 (Exact inverse sum set theorem) *Suppose that A, B are additive sets with common ambient group Z. Then the following are equivalent:*

- $|A + B| = |A|$;
- $|A - B| = |A|$;
- $|A + nB - mB| = |A|$ *for at least one pair of integers* $(n, m) \neq (0, 0)$;
- $|A + nB - mB| = |A|$ *for all integers* n, m;
- *there exists a finite subgroup G of Z such that B is contained in a coset of G, and A is a union of cosets of G.*

Proof We shall just show that the first claim implies the fifth; the remaining claims are either similar or easy and are left to the exercises. By translating B if necessary we may assume that B contains 0. Then $A + B \supset \{0\} + A = A$, but since $|A + B| = |A|$ we have $A + B = A$. In particular $A + b = A$ for all $b \in B$. Thus if we define the *symmetry group* $\mathrm{Sym}_1(A)$ (also known as the *period* of A) to be the set $\mathrm{Sym}_1(A) := \{h \in Z : A + h = A\}$, then we have $B \subseteq \mathrm{Sym}_1(A)$. We leave as an exercise for the reader the verification that $\mathrm{Sym}_1(A)$ is a finite group, and A is the union of cosets of $\mathrm{Sym}_1(A)$; the claim then follows by setting $G := \mathrm{Sym}_1(A)$. □

We shall study the symmetry group $\mathrm{Sym}_1(A)$, as well as the more general symmetry sets $\mathrm{Sym}_\alpha(A)$, more systematically in Section 2.6.

As to when the upper bound is attained, we do not have as explicit a description, but we can give a number of equivalent formulations of the condition.

Proposition 2.3 *Suppose that A, B are additive sets with common ambient group Z. Then the following are equivalent:*

- $|A + B| = |A||B|$;
- $|A - B| = |A||B|$;
- $|\{(a, a', b, b') \in A \times A \times B \times B : a + b = a' + b'\}| = |A||B|$;
- $|\{(a, a', b, b') \in A \times A \times B \times B : a - b = a' - b'\}| = |A||B|$;
- $|A \cap (x - B)| = 1$ *for all* $x \in A + B$;
- $|A \cap (B + y)| = 1$ *for all* $y \in A - B$;
- $(A - A) \cap (B - B) = \{0\}$.

We leave the easy proof of this proposition to the exercises. For a partial generalization of it, see Corollary 2.10 below.

In Proposition 2.2 and Proposition 2.3, the sets $A + B$ and $A - B$ have the same size (see also Exercise 2.1.6). However, this is not true in general. A basic example is the set $A = \{0, 1, 3\} \subset \mathbf{Z}$; then $A + A = \{0, 1, 2, 3, 4, 6\}$ has six elements and $A - A = \{-3, -2, -1, 0, 1, 2, 3\}$ has seven elements. More generally, if $A = \{0, 1, 3\}^d \subset \mathbf{Z}^d$, then $A + A$ has 6^d elements and $A - A$ has 7^d. Thus $A - A$ can be larger than $A + A$ by an arbitrarily large amount. In the converse direction, the set $A := \{(0, 0), (1, 0), (2, 0), (3, 1), (4, 0), (5, 1), (6, 1), (7, 0), (8, 1), (9, 1)\} \in \mathbf{Z}_{10} \times \mathbf{Z}_2$ is such that $A + A = \mathbf{Z}_{10} \times \mathbf{Z}_2$ has 20 elements, but $A - A = \mathbf{Z}_{10} \times \mathbf{Z}_2 \backslash \{(0, 1)\}$ has only 19 elements; one can amplify this example as before by raising to the power d. Despite these examples, however, there are still several relationships between the size of $|A + A|$ and $|A - A|$; see in particular (2.11) below.

Exercises

2.1.1 Let $N, M \geq 1$ be integers, and let A and B be sets of cardinality N and M respectively chosen uniformly at random from the real interval $\{x \in \mathbf{R} : 0 \leq x \leq 1\}$. Show that with probability 1 we have $|A + B| = |A||B|$ and $|nA| = \binom{|A|+n-1}{n}$ for all $n \geq 1$.

2.1.2 Prove the remaining claims in Proposition 2.2.

2.1.3 Let A be an additive set. Show that A is a group if and only if $2A = A$.

2.1.4 Prove Proposition 2.3.

2.1.5 [289] Find an additive set A of integers such that $|A - A| < |A + A|$. (Hint: there are several ways to proceed. One way is to tile the lattice $\mathbf{Z}^2$ with the $\mathbf{Z}_{10} \times \mathbf{Z}_2$ example given above, and somehow truncate and then project this back to $\mathbf{Z}$.)

2.1.6 Let A, B be additive sets in a *finite* additive group Z, such that $|A| + |B| > |Z|$. Prove that $A + B = A - B = Z$. Give an example to show that the condition $|A| + |B| > |Z|$ cannot be improved.

2.1.7 Show that for any additive set A, the symmetry group $\mathrm{Sym}_1(A)$ of A as defined in the proof of Proposition 2.2 is a finite group contained in $A - A$, obeys the identity $A = A + \mathrm{Sym}_1(A)$, and that A is a union of cosets of $\mathrm{Sym}_1(A)$. (We shall define a more general notion of *symmetry sets* $\mathrm{Sym}_\alpha(A)$ of an additive set in Section 2.6.)

2.1.8 Let $d \geq 1$. Give an example of an additive set A of *integers* such that $|A + A| = 6^d$ and $|A - A| = 7^d$. (see also Lemma 5.25.)

2.2 Doubling constants

The traditional way to measure the additive structure inside an additive set A is via *doubling constants* $\sigma[A]$, which we now define. We will shortly develop two other measures of additive structure, namely the *additive energy* $E(A, A)$, and the concept of a *K-approximate group*, which are also useful, and are closely related to the doubling constant.

Definition 2.4 (Doubling constant) For an additive set A, the *doubling constant* $\sigma[A]$ is defined to be the quantity

$$\sigma[A] := \frac{|2A|}{|A|} = \frac{|A + A|}{|A|}.$$

Similarly we define the *difference constant* $\delta[A]$ as

$$\delta[A] := \frac{|A - A|}{|A|}.$$

From (2.2) we thus have the bounds

$$1 \leq \sigma[A] \leq \frac{|A| + 1}{2} \text{ and } 1 \leq \delta[A] \leq |A| - 1 + \frac{1}{|A|}.$$

The upper bound here is quite easy to attain; for instance if $A = 2^\wedge[0, N) = \{1, 2, 2^2, \ldots, 2^{N-1}\} \subset \mathbf{Z}$, then $|A| = N$, $|A + A| = \frac{N(N+1)}{2}$, and $|A - A| = N(N - 1) + 1$, hence $\sigma[A] = \frac{N+1}{2}$ and $\delta[A] = N - 1 + \frac{1}{N}$. In the converse direction, Proposition 2.2 shows that $\sigma[A] = 1$ (or $\delta[A] = 1$) if and only if A is a coset of a group; we shall elaborate upon this in Proposition 2.7 below.

An additive set A with the maximal value of doubling constant $\sigma[A] = (|A| + 1)/2$ (or equivalently, with maximal difference constant $\delta[A] = |A| - 1 + \frac{1}{|A|}$) is known as a *Sidon set* or a *B_2 set*. Informally, this means that all the pairwise sums of A are distinct, excluding the trivial equalities coming from the identity $a + b = b + a$; see Exercise 2.2.1. We will revisit Sidon sets in Section 4.5.

There are various senses in which this behavior is "generic"; for instance, if A is a set of N real numbers chosen uniformly at random from the unit interval $\{x \in \mathbf{R} : 0 \leq x \leq 1\}$, then we see from Exercise 2.1.1 that A is a Sidon set with probability 1, and so $|A + A| = \frac{N(N+1)}{2}$; the point is that if $\{a, b\} \neq \{c, d\}$ then $a + b$ and $c + d$ will "generically" be distinct. A more interesting question is to understand the conditions under which the doubling constant $\sigma[A]$ (or difference constant $\delta[A]$) can be *small*.

As mentioned earlier, $\sigma[A] = 1$ if and only if A is the coset of a finite subgroup G of Z. We thus expect that if A has a doubling constant which is small, but not actually equal to 1, then it should behave "approximately" like a group (up to translations); we shall see several manifestations of this heuristic throughout this book, when we develop more tools with which to analyze the doubling constant. Indeed, the study of sets of small doubling constant can be thought of as a kind of "approximate group theory", with the inverse sum set theorems of Chapter 5 then being analogous to a classification theorem for groups.

The study of sets with close to maximal doubling appears to be hopeless at present. A probabilistic construction of Ruzsa [291] shows that there exist large additive sets A with $|A - A|$ very close to the maximal value of $|A|^2$, but $|A + A| < |A|^{2-c}$ for some explicit absolute constant $c > 0$; and similarly with the roles of $A - A$ and $A + A$ reversed.

Exercises

2.2.1 Let A be an additive set. Show that A is a Sidon set if and only if, for any $a, b, c, d \in A$, we have $a + b \neq c + d$ unless $\{a, b\} = \{c, d\}$.

2.2.2 Let Z be an additive group, let $a, r \in Z$, and let $N \geq 1$ be an integer. Let $P = \{a, a + r, \ldots, a + (N - 1)r\}$ be an arithmetic progression in Z. Show that $\sigma[P] \leq 2 - \frac{1}{N}$, with equality if and only if $\mathrm{ord}(r) \geq 2N - 1$, where $\mathrm{ord}(r)$ is the order of the group element r in Z.

2.2.3 If $\phi : Z' \to Z$ is a surjective group homomorphism whose kernel $\ker(\phi) := \phi^{-1}(\{0\})$ is finite, and A is an additive set in Z, show that $\sigma[\phi^{-1}(A)] = \sigma[A]$.

2.2.4 If A, A' are additive sets in Z, Z' respectively, show that $\sigma[A \times A'] = \sigma[A]\sigma[A']$. In particular $\sigma[A^{\oplus d}] = \sigma[A]^d$ for all $d \geq 1$.

2.2.5 Let A be any additive set. Show that a non-empty subset of A can have doubling constant at most $\sqrt{\sigma[A]|A|}$. Give examples that show that this bound cannot be improved except by an absolute constant. What is the analogous statement for the difference constant?

2.2.6 [100] Let A be any additive set. Show that a Sidon set contained in A can have cardinality at most $2\sqrt{\sigma[A]|A|}$. (Thus sets with small doubling constant cannot contain very large Sidon sets.) What is the analogous statement for the difference constant?

2.2.7 [294] Let p be a prime, let $\theta \in \mathbf{Z}_p \backslash 0$ be a multiplicative generator of $\mathbf{Z}_p$, and let $Z := \mathbf{Z}_{p-1} \times \mathbf{Z}_p$. Let $A \subset Z$ be the set $A := \{(t, \theta^t) : t = 1, \ldots, p-1\}$. Show that A is a Sidon set, and compare this to Exercise 2.2.6. Modify this construction to give an example of a Sidon set $A \subset [0, N]$ for a large integer N such that $|A|$ is comparable to $N^{1/2}$. A similar example can be given by using the discrete parabola $\{(t, t^2) : t \in \mathbf{Z}_p\}$ in $\mathbf{Z}_p \times \mathbf{Z}_p$. For a survey of other constructions of Sidon sets, see [264].

2.2.8 Let N be a large integer. Give examples of finite non-empty sets A, B of integers such that $|A| = |B| = N$ and $\sigma[A], \sigma[B] \leq 2$, but $\sigma[A \cup B] \geq \frac{N}{2}$. This example shows that doubling constants can behave very badly under set union (see however Exercise 2.3.17). On the other hand, establish the inequality $\sigma[A \cup B] \leq \sigma[A] + |B|$; thus adding a *small* set to A will not significantly affect the doubling constant.

2.2.9 Let N be a large integer. Give examples of finite non-empty sets A, B of integers such that $|A| = |B| = N$ and $\sigma[A], \sigma[B] \leq 10$, but $\sigma[A \cap B] \geq \frac{1}{10} N^{1/2}$. (Hint: concatenate a Sidon set with an arithmetic progression.) Compare this result against Exercise 2.2.6. This example shows that doubling constants can behave badly under set intersection (but see Exercise 2.4.7).

2.2.10 Let A be an additive set in Z, and let $\pi : Z \to Z'$ be a group homomorphism. Show by example that $\sigma[\pi(A)]$ is not necessarily less than or equal to $\sigma[A]$. (Hint: this is surprisingly delicate. One way is to start with an additive set C in some additive group Z_0 with $\sigma[C] > \delta[C]$, and consider the additive set $A := ((-C)^n \times \{0\} \times G) \cup (C^n \times X \times \{0\})$ in $Z_0^n \times Z \times G$, where $n \geq 1$ is large, G is a very large finite group, and X is a Sidon set of medium size in a group Z.) See however Exercise 2.3.8 and Exercise 6.5.17.

2.2.11 Let A be an additive set in Z, and let G be a finite subgroup of Z. Show by example that $\sigma[A + G]$ is not necessarily less than or equal to $\sigma[A]$. (Hint: use the previous exercise.)

2.3 Ruzsa distance and additive energy

The doubling constant measures the amount of *internal* additive structure of a single additive set A. We now introduce two useful quantities measuring the amount of common additive structure *between* two additive sets A, B – the Ruzsa distance and the additive energy.

Definition 2.5 (Ruzsa distance) Let A and B be two additive sets with a common ambient group Z. We define the *Ruzsa distance* $d(A, B)$ between these two sets to be the quantity

$$d(A, B) := \log \frac{|A - B|}{|A|^{1/2}|B|^{1/2}}.$$

Thus for instance $d(A, A) = \log \delta[A]$.

We now justify the terminology "Ruzsa distance".

Lemma 2.6 (Ruzsa triangle inequality) *[297] The Ruzsa distance $d(A, B)$ is non-negative, symmetric, and obeys the triangle inequality*

$$d(A, C) \leq d(A, B) + d(B, C)$$

for all additive sets A, B, C with common ambient group Z.

Proof The non-negativity follows from (2.1). The symmetry follows since $B - A = -(A - B)$. Now we prove the triangle inequality, which we can rewrite as

$$|A - C| \leq \frac{|A - B||B - C|}{|B|}.$$

From the identity

$$a - c = (a - b) + (b - c)$$

we see that every element $a - c$ in $A - C$ has at least $|B|$ distinct representations of the form $x + y$ with $(x, y) \in (A - B) \times (B - C)$. The claim then follows. □

For an approximate version of this inequality in which one replaces complete difference sets with nearly complete difference sets (using at least 75% of the differences), see Exercise 2.5.4.

The Ruzsa distance thus satisfies all the axioms of a metric except one; we do not have that $d(A, A) = 0$ for all sets A (also, we have $d(G + x, G + y) = 0$ whenever $G + x, G + y$ are cosets of a group G). Indeed we have a precise characterization on when this Ruzsa distance vanishes:

Proposition 2.7 *Suppose that (A, Z) is an additive set. Then the following are equivalent:*

- $\sigma[A] = 1$ *(i.e. $|A + A| = |A|$);*
- $\delta[A] = 1$ *(i.e. $|A - A| = |A|$, or $d(A, A) = 0$);*
- $d(A, B) = 0$ *for at least one additive set B;*

- $|nA - mA| = |A|$ *for at least one pair of non-negative integers* n, m *with* $n + m \geq 2$;
- $|nA - mA| = |A|$ *for all non-negative integers* n, m;
- A *is a coset of a finite subgroup* G *of* Z.

Proof Apply Proposition 2.2 and the Ruzsa triangle inequality. □

Later on in this chapter we shall generalize this proposition to the case when the Ruzsa distance, difference constant, or doubling constant are a little larger than 0, 0, or 1 respectively, but still fairly small; see Proposition 2.26.

Despite the non-vanishing of the distance $d(A, A)$ in general, it is still a useful heuristic to view the Ruzsa distance as behaving like a metric[1]. Now we relate the difference constant to the doubling constant. From the definition of Ruzsa distance and doubling constant we have the identity

$$d(A, -A) = \log \sigma[A]. \tag{2.4}$$

In particular, from Lemma 2.6 we have

$$\log \delta[A] = d(A, A) \leq 2 \log \sigma[A]$$

and hence we obtain the estimate

$$\delta[A] \leq \sigma[A]^2 \tag{2.5}$$

or in other words that $|A - A| \leq \frac{|A+A|^2}{|A|}$. A similar argument gives the more general estimate

$$|B - B| \leq \frac{|A + B|^2}{|A|} \tag{2.6}$$

for any two additive sets A, B with common ambient group Z.

It turns out that we can conversely bound the doubling constant of a set by its difference constant; see (2.11) below.

Having introduced the Ruzsa distance, we now turn to the closely related notion of *additive energy* $E(A, B)$ between two additive sets.

Definition 2.8 (Additive energy) If A and B are two additive sets with ambient group Z, we define the *additive energy* $E(A, B)$ between A and B to be the quantity

$$E(A, B) := |\{(a, a', b, b') \in A \times A \times B \times B : a + b = a' + b'\}|.$$

[1] One could artificially convert the Ruzsa distance into a genuine metric by identifying A with $A + x$ for all x, and redefining $d(A, A)$ to be zero, or alternatively by introducing the metric space $X := \{A \times \{j\} : A \subseteq Z; 0 < |A| < \infty; j \in \{1, 2\}\}$ – consisting of two copies of each finite non-empty subset of Z (again identifying A with its translations) – with the metric $d_X(A \times \{j\}, B \times \{k\})$ defined to equal $d(A, B)$ if $A \times \{j\} \neq B \times \{k\}$ and equal to 0 otherwise. However there appears to be no significant advantage in working in such an artificial setting.

We observe the trivial bounds

$$|A||B| \leq E(A, B) \leq |A||B| \min(|A||B|). \tag{2.7}$$

The lower bound follows since $a + b = a' + b'$ whenever $(a, b) = (a', b')$. To see the upper bound, observe that if one fixes a, a', b, then $b' = a + a' - b$ is completely determined, and hence $E(A, B) \leq |A|^2|B|$. A similar argument gives $E(A, B) \leq |A||B|^2$. Note that Proposition 2.3 addresses the case when $E(A, B) = |A||B|$.

We will analyze the additive energy more comprehensively in Section 4.2, when we have developed the machinery of Fourier transforms, and in Section 2.5, when we have developed the Balog–Szemerédi–Gowers theorem. For now we concentrate on the elementary properties of this energy. We first observe the symmetry property $E(A, B) = E(B, A)$ and the translation invariance property $E(A + x, B + y) = E(A, B)$ for all $x, y \in Z$. From the trivial observation

$$a + b = a' + b' \iff a - b' = a' - b$$

we also see that $E(A, B) = E(A, -B)$, and similarly if we reflect A to $-A$.

The additive energy reflects the extent to which A intersects with translates of B or $-B$, as the following simple identities show:

Lemma 2.9 *Let A, B be additive sets with ambient group Z. Then we have the identities*

$$|A||B| = \sum_{x \in A+B} |A \cap (x - B)| = \sum_{y \in A-B} |A \cap (B + y)|$$

and

$$\begin{aligned} E(A, B) &= \sum_{x \in A+B} |A \cap (x - B)|^2 \\ &= \sum_{y \in A-B} |A \cap (B + y)|^2 \\ &= \sum_{z \in (A-A) \cap (B-B)} |A \cap (z + A)||B \cap (z + B)|. \end{aligned}$$

In particular, if we let $r_{A+B}(n)$ denote the number of representations of n as $a + b$ for some $a \in A$ and $b \in B$, and define $r_{A-B}(n)$ similarly, then we have

$$|A||B| = \sum_n r_{A+B}(n) = \sum_n r_{A-B}(n); \; E(A, B) = \sum_n r_{A+B}(n)^2 = \sum_n r_{A-B}(n)^2.$$

Proof A simple counting argument yields

$$|A||B| = \sum_{x \in A+B} |\{(a, b) \in A \times B : a + b = x\}| = \sum_{x \in A+B} |A \cap (x - B)|;$$

By replacing B with $-B$ we similarly obtain $|A||B| = \sum_{y\in A-B} |A \cap (B+y)|$. This gives the first set of identities. For the second set we compute

$$
\begin{aligned}
&\sum_{x\in A+B} |A \cap (x-B)|^2 \\
&\quad = \sum_{x\in A+B} |\{(a,b) \in A \times B : a+b = x\}|^2 \\
&\quad = \sum_{x\in A+B} |\{(a,a',b,b') \in A \times A \times B \times B : a+b = a'+b' = x\}| \\
&\quad = |\{(a,a',b,b') \in A \times A \times B \times B : a+b = a'+b'\}| \\
&\quad = |\{(a,a',b,b') \in A \times A \times B \times B : a-b' = a'-b\}| \\
&\quad = \sum_{y\in A-B} |\{(a,b') \in A \times B : a-b' = a'-b\}|^2 \\
&\quad = \sum_{y\in A-B} |A \cap (B+y)|^2
\end{aligned}
$$

and

$$
\begin{aligned}
&\sum_{z\in (A-A)\cap(B-B)} |A \cap (z+A)||B \cap (z+B)| \\
&\quad = \sum_{z\in (A-A)\cap(B-B)} |\{(a,a',b,b') \in A \times A \times B \times B : z = a-a' = b'-b\}| \\
&\quad = |\{(a,a',b,b') \in A \times A \times B \times B : a-a' = b'-b\}| \\
&\quad = |\{(a,a',b,b') \in A \times A \times B \times B : a+b = a'+b'\}|
\end{aligned}
$$

and the claims follow from the definition of $E(A,B)$. The last identity follows since $r_{A+B}(n) = |A \cap (n-B)|$ and $r_{A-B}(n) = |A \cap (B+n)|$. □

As a consequence of this Lemma we have the following inequalities, which assert that pairs of sets with small Ruzsa distance have large additive energy, and pairs with large additive energy have large intersection (after translating and possibly reflecting one of the sets).

Corollary 2.10 *Let A, B be additive sets. Then there exists $x \in A+B$ and $y \in A-B$ such that*

$$
|A \cap (x-B)|, |A \cap (B+y)| \geq \frac{E(A,B)}{|A||B|} \geq \frac{|A||B|}{|A \mp B|} \tag{2.8}
$$

for either choice of sign $\pm$. In particular all of the above quantities are bounded by $|(A-A) \cap (B-B)|$. Finally we have the Cauchy–Schwarz inequality

$$
E(A,B) \leq E(A,A)^{1/2} E(B,B)^{1/2}. \tag{2.9}
$$

Proof From Lemma 2.9 and Cauchy–Schwarz we have

$$\frac{E(A, B)}{|A||B|} \geq \frac{|A||B|}{|A \pm B|}.$$

Also, from the last part of Lemma 2.9 we have

$$E(A, B) \leq |A||B| \max_{x \in A+B} r_{A+B}(x), |A||B| \max_{y \in A-B} r_{A-B}(y)$$

which establishes (2.8). To bound $|A \cap (x - B)|$ and $|A \cap (B + y)|$, observe that if $z \in A \cap (x - B)$, then $A \cap (x - B) \subset z + ((A - A) \cap (B - B))$, hence $|A \cap (x - B)| \leq |(A - A) \cap (B - B)|$, and similarly $|A \cap (B + y)| \leq |(A - A) \cap (B - B)|$. Finally, (2.9) follows from the formula $E(A, B) = \sum_{z \in (A-A) \cap (B-B)} |A \cap (z + A)||B \cap (z + B)|$ from Lemma 2.9 and the Cauchy–Schwarz inequality. □

Another connection in a similar spirit is

Lemma 2.11 *Let A, B be additive sets. Then for any $x \in A + B$ we have $|A \cap (x - B)| \leq \frac{|A-B|^2}{|A+B|}$.*

Proof (Lev Vsevolod, private communication) We can rewrite the inequality as

$$|\{(a, b, c) \in A \times B \times (A + B) : a + b = x\}| \leq |(A - B) \times (A - B)|.$$

Now for each (a, b, c) in the set on the left-hand side, we can write $c = a_c + b_c$ for some $a_c \in A, b_c \in B$, and then form the pair $(a - b_c, a_c - b) \in (A - B) \times (A - B)$. Using the identity $c = x - (a - b_c) + (a_c - b)$ we can verify that this map is injective. The claim follows. □

Corollary 2.12 *Let A, B be additive sets with ambient group Z. Then there exists $x \in A + B$ such that*

$$\frac{|A - B|^2}{|A \cap (x - B)|} \leq \frac{|A - B|^2|A||B|}{E(A, B)} \leq \frac{|A - B|^3}{|A||B|}. \tag{2.10}$$

Furthermore we have

$$d(A, -B) \leq 3d(A, B).$$

Proof The inequalities in (2.10) follow from (2.8), and the final inequality $d(A, -B) \leq 3d(A, B)$ then follows from Lemma 2.11 and the definition of Ruzsa distance. □

From (2.10) and and (2.5) we obtain the inequalities

$$\delta[A]^{1/2} \leq \sigma[A] \leq \delta[A]^3 \tag{2.11}$$

which were first observed in [289]. Thus an additive set has small doubling constant if and only if its difference constant is small. It is not known whether the lower bound is best possible. However, the upper bound can be improved to $\sigma[A] \leq \delta[A]^2$ using Plünnecke inequalities; see Exercise 6.5.15.

We now show how the Ruzsa distance can be used to control iterated sum sets. We begin with a lemma which controls iterated sum sets of "most" of $A + B$.

Lemma 2.13 *Let A and B be additive sets in a common ambient group. Then there exists $S \subset A + B$ such that*

$$|\{(a, b) \in A \times B : a + b \in S\}| \geq |A||B|/2 \tag{2.12}$$

and such that

$$|A + B + nS| \leq \frac{2^n |A + B|^{2n+1}}{|A|^n |B|^n} \tag{2.13}$$

for all integers $n \geq 0$.

Note that (2.12) gives a lower bound on $|S|$, namely

$$|S| \geq \max(|A|, |B|)/2. \tag{2.14}$$

Proof If we define S to be the set of all $x \in A + B$ such that

$$|\{(a, b) \in A \times B : a + b = x\}| \geq \frac{|A||B|}{2|A + B|}$$

then we have

$$|\{(a, b) \in A \times B : a + b \in (A + B)\backslash S\}| < |A + B| \frac{|A||B|}{2|A + B|}$$

which gives (2.12).

Now we prove (2.13). A typical element of $A + B + nS$ can be written as

$$a_0 + s_1 + s_2 + \cdots + s_n + b_{n+1}$$

where $a_0 \in A$, $b_{n+1} \in B$, and $s_1, \ldots, s_n \in S$. By definition of S, we can expand this in at least $(\frac{|A||B|}{2|A+B|})^n$ different ways as

$$a_0 + (b_1 + a_1) + (b_2 + a_2) + \cdots + (b_n + a_n) + b_{n+1}$$

where $b_i \in B$, $a_i \in A$, and $b_i + a_i = s_i$ for all $1 \leq i \leq n$. We regroup this as the sum of $n + 1$ elements from $A + B$,

$$(a_0 + b_1) + (a_1 + b_2) + \cdots + (a_n + b_{n+1})$$

and observe that for fixed $a_0, s_1, \ldots, s_n, b_{n+1}$, the quantities $a_0 + b_1, a_1 + b_2, \ldots, a_n + b_{n+1}$ completely determine all the variables $a_0, \ldots, a_n, b_1, \ldots, b_{n+1}$. Thus we have shown that every element of $A + B + nS$ has at least $(\frac{|A||B|}{2|A+B|})^n$ representations of the form $t_0 + \cdots + t_n$ where each $t_i \in A + B$. The claim then follows. □

This result can then be used, together with the Ruzsa triangle inequality, to deduce control on iterated sum sets of A and B; see Exercise 2.3.10. However we will pursue an approach that gives slightly better bounds in the next section (and an even better result will be developed in Section 6.5).

Exercises

2.3.1 If $\phi : Z' \to Z$ is a surjective group homomorphism whose kernel $\ker(\phi) := \phi^{-1}(\{0\})$ is finite, and A, B are additive sets in Z, show that $d(\phi^{-1}(A), \phi^{-1}(B)) = d(A, B)$. Also show that $d(A + x, B + y) = d(A, B)$ for any $x, y \in Z$.

2.3.2 If A, B, C, D are additive sets in Z, show that

$$d(A, B) - \frac{1}{2}\log |C||D| \leq d(A + C, B + D) \leq d(A, B) + \log |C - D|$$

and

$$d(A, B \cup C) \leq \max(d(A, B), d(A, C)) + \frac{1}{2}\log 2.$$

If A', B' are additive sets in Z', show that

$$d(A \times A', B \times B') = d(A, B) + d(A', B').$$

2.3.3 Let A, B be additive sets with common ambient group. Show that $d(A, B) \leq \frac{1}{2}\log |A| + \frac{1}{2}\log |B|$, and that $d(A, B) = \frac{1}{2}\log |A| + \frac{1}{2}\log |B|$ if and only if $d(A, -B) = \frac{1}{2}\log |A| + \frac{1}{2}\log |B|$.

2.3.4 Let A, B, C be additive sets in Z. Show that

$$d(A, C) \leq d(A, B) + \frac{1}{2}\log\frac{|B|}{|C|} \tag{2.15}$$

whenever $C \subseteq B$; this shows that the Ruzsa distance $d(A, B)$ is stable under refinement of one or both of the sets A, B. By combining this inequality with the triangle inequality $d(A, -B) \leq d(A, (x - A) \cap B) + d((x - A) \cap B, -B)$, give another proof of Lemma 2.11.

2.3.5 Show that for any $n \geq 1$, there exists an additive set A such that $|A| = 4^n$, $|A + A| = 10^n$, and $|2A - A| = 28^n$. Thus it is not possible to obtain an estimate of the form $|2A - A| = O(\sigma^2[A]|A|)$.

2.3.6 Let A, B be additive sets with common ambient group. Show that $e^{-2d(A,B)}|A| \leq |B| \leq e^{2d(A,B)}|A|$. Thus sets which are close in the Ruzsa distance are necessarily close in cardinality also. Of course the converse is far from true.

2.3.7 Let A, B be additive sets with common ambient group Z. Show that $d(A, B) = 0$ if and only if A, B are cosets of the same finite subgroup G of Z. (We shall generalize this result later; see Proposition 2.27.)

2.3.8 Let A be an additive set in an additive group Z, and let G be a finite subgroup of Z. Show that $\sigma[A+G] \leq \frac{|3A|}{|A|}$. (Hint: apply the Ruzsa triangle inequality to $2A$, $-A$, and G.) Conclude that if $\pi : Z \to Z'$ is a group homomorphism then $\sigma[\pi(A)] \leq \frac{|3A|}{|A|}$. One cannot replace the tripling constant $\frac{|3A|}{|A|}$ with the doubling constant; see Exercise 2.2.10. See however Exercise 6.5.17.

2.3.9 Let K be a large integer, and let $A = B = \{e_1, \ldots, e_K\}$ be the standard basis of $\mathbf{Z}^K$. Show that if S is any subset of $A+B$ obeying (2.12) then

$$|A+B+nS| = \Omega_n\left(\frac{|A+B|^{2n+1}}{|A|^n|B|^n}\right)$$

where we are using the Landau notation $\Omega()$. This shows that Lemma 2.13 cannot be significantly improved (except possibly by improving the bound (2.14)).

2.3.10 Let A, B be additive sets with common ambient group such that $|A+B| \leq K|A|^{1/2}|B|^{1/2}$ for some $K \geq 1$. Using Lemma 2.13 and many applications of the Ruzsa triangle inequality, establish the estimate

$$|n_1A - n_2A + n_3B - n_4B| = O_{n_1,n_2,n_3,n_4}\left(K^{O_{n_1,n_2,n_3,n_4}(1)}|A|^{1/2}|B|^{1/2}\right)$$

for all integers n_1, n_2, n_3, n_4. In particular, establish the bounds

$$\begin{aligned} &d(n_1A - n_2A + n_3B - n_4B, n_5A - n_6A + n_7B - n_8B) \\ &\quad \leq O_{n_1,\ldots,n_8}(1 + d(A, B)) \end{aligned}$$

for all integers $n_1, \ldots, n_8$. We shall improve this bound slightly in Corollary 2.23 and Corollary 2.24; see also Corollary 2.19 for the "tensor power trick" that can eliminate lower order terms such as the implicit constant preceding the $K^{O_{n_1,n_2,n_3,n_4}(1)}$ factor.

2.3.11 Let G and H be subgroups of Z. Show that

$$d(G, H) = \log \frac{|G|^{1/2}|H|^{1/2}}{|G \cap H|}$$

Conclude that $d(G, H) = d(G, G+H) + d(G+H, H) = d(G, G \cap H) + d(G \cap H, H)$. Also, if K is another subgroup of Z, prove the contractivity properties $d(G+K, H+K) \leq d(G, H)$ and $d(G \cap K, H \cap K) \leq d(G, H)$. Note that the Ruzsa distance, when restricted to subgroups of Z, is indeed a genuine metric, thanks to Proposition 2.7. See also Exercises 2.4.7 and 2.4.8 below.

2.3.12 Let A be an additive set. Show that

$$\sigma[A \cup (-A)] \leq 2\sigma[A] + \sigma[A]^2.$$

Thus a set with small doubling can be embedded in a symmetric set (i.e. a set B such that $-B = B$) with small doubling which has at most twice the cardinality.

2.3.13 [289] Let A be an additive set. Prove the inequalities $|A - A| \leq |A + A|^{3/2}$ and $|A + A| \leq |A - A|^{3/2}$. (Hint: use (2.11), Corollary 2.12 and (2.1).)

2.3.14 [26] Let A be an additive set. Show that there exists an element $x \in A + A$ such that the set $F := A \cap (x - A)$ has size $|F| \geq |A|/\sigma[A]$ and doubling constant $\sigma[F] \leq \sigma[A]^2$. Thus every additive set A of small doubling contains a large symmetric subset F of small doubling, where by symmetry here we mean that $F = x - F$.

2.3.15 Let A, B be additive sets with common ambient group Z. Show that $\delta[A] \leq e^{2d(A,B)}$ and $\sigma[A] \leq e^{6d(A,B)}$. Thus only sets with small doubling constant can be close to other sets in the Ruzsa metric. (The 6 can be lowered to a 4, see Exercise 6.5.15.)

2.3.16 Let A, B be additive sets with common ambient group Z. Show that $\sigma[A \cup B] \leq e^{d(A,B)} + 2e^{4d(A,B)}$. Thus a pair of sets which are close in the Ruzsa metric can be embedded in a slightly larger set with small doubling. In the converse direction, establish the estimate

$$d(A, B) \leq \log \sigma[A \cup B] + \frac{1}{2} \log \frac{|A \cup B|}{|A|} + \frac{1}{2} \log \frac{|A \cup B|}{|B|}.$$

2.3.17 Let A, B be additive sets with common ambient group Z, such that $\sigma[A], \sigma[B] \leq K$ for some $K \geq 1$, and such that $A \cap B$ is non-empty. Show that

$$\sigma[A \cup B] \leq 2K + K^3 \frac{\min(|A|, |B|)}{|A \cap B|}.$$

Thus the union of sets with small doubling remains small doubling provided that those two sets had substantial intersection.

2.3.18 [40], [41] Let $K \geq 1$, and let A_1, A_2, A_3 be additive sets with common ambient group Z, such that

$$|A_j \cap A_3| \geq \frac{1}{K}|A_j| \text{ and } |A_j + A_j| \leq K|A_j|$$

for all $j = 1, 2, 3$. Prove that $|A_1 + A_2| \leq K^6|A_3|$. Hint: use the triangle inequality

$$d(A_1, -A_2) \leq d(A_1, -(A_1 \cap A_3)) + d(-(A_1 \cap A_3), A_2 \cap A_3) + d(A_2 \cap A_3, -A_2)$$

2.3.19 Suppose that A and B are subgroups of Z, and let $x = y = 0$. Show that all the inequalities in (2.8) are in fact equalities.

2.3.20 Let A, B, C be additive sets in an ambient group Z. Show that

$$\max(E(A, B), E(A, C))^{1/2} \leq E(A, B \cup C)^{1/2} \leq E(A, B)^{1/2} + E(A, C)^{1/2}.$$

(Hint: use Lemma 2.9 and the triangle inequality for the l^2 norm.)

2.3.21 Let A, B, C be additive sets in an ambient group Z with $|A| = |B| = |C| = N$. Give examples of such sets where $E(A, B)$ and $E(A, C)$ are comparable to N^2 and $E(B, C)$ is comparable to N^3, or where $E(A, B)$ and $E(A, C)$ are comparable to N^3 and $E(B, C)$ are comparable to N^2. These examples show that there is no hope of any useful "triangle inequality" connecting $E(A, B)$, $E(B, C)$, and $E(A, C)$.

2.3.22 Suppose A, B are additive sets in an ambient group Z. Show that $E(A, B) = |A|^2|B|$ holds if and only if $|A + B| = |B|$. One can thus use Proposition 2.2 to determine when the upper bound in (2.7) is obtained. Conclude in particular that $E(A, B) = |A|^{3/2}|B|^{3/2}$ if and only if $d(A, B) = 0$, which in turn occurs if and only if A and B are cosets of the same finite group G.

2.3.23 Give an example of an additive set $A \subset \mathbf{Z}$ of cardinality $|A| = N$ such that $E(A, A) \geq \frac{1}{100} N^3$ but $d(A, A) \geq \frac{1}{100} \log N$. Compare this with (2.8) (and with Corollary 2.31 below).

2.3.24 Let A be an additive set. Show that there exists a subset A' of A of cardinality $|A'| \geq \frac{1}{2\sigma[A]}|A|$ and an element $a_0 \in A'$ such that $|(a + A) \cap (a_0 + A)| \geq \frac{1}{2\sigma[A]}|A|$ for all $a \in A'$. (Hint: first obtain a lower bound for $E(A, A)$.)

2.4 Covering lemmas

We now describe some covering lemmas, which roughly speaking have the following flavor: if A and B have similar additive structure (for instance, if their Ruzsa distance is small) then one can cover A by a small number translates of B (or some modification of B).

Lemma 2.14 (Ruzsa's covering lemma) *[300] For any additive sets A, B with common ambient group Z, there exists an additive set $X_+ \subseteq B$ with*

$$B \subseteq A - A + X_+; \qquad |X_+| \leq \frac{|A + B|}{|A|}; \qquad |A + X_+| = |A||X_+|$$

and similarly there exists an additive set $X_- \subseteq B$ *with*

$$B \subseteq A - A + X_-; \qquad |X_-| \le \frac{|A - B|}{|A|}; \qquad |A - X_-| = |A||X_-|.$$

In particular, B *can be covered by* $\min(\frac{|A+B|}{|A|}, \frac{|A-B|}{|A|})$ *translates of* $A - A$.

Remark 2.15 One useful side benefit of this covering lemma is that there exist at least $\frac{|B|}{|A-A|}$ disjoint translates $A + b$ of A with $b \in B$, as can be seen by restricting b to X_+.

Proof It suffices to prove the claim concerning $A + B$, since the claim concerning $A - B$ follows by replacing B with $-B$ and X_+ with $-X_-$ (note that $A - A$ is symmetric around the origin). Consider the family $\{A + b : b \in B\}$ of translates of A by elements of B. All of these translates have volume $|A|$ and are contained inside $A + B$. Thus if we take a maximal disjoint sub-family of these translates, i.e. $\{A + x : x \in X_+\}$ for some $X_+ \subseteq B$, then X_+ can have cardinality at most $\frac{|A+B|}{|A|}$. Also we have $|A + X_+| = |A||X_+|$ by construction. Now for any element $b \in B$, we see that $A + b$ cannot be disjoint from every member of $\{A + x : x \in X_+\}$ as this would contradict the maximality of X_+. Thus $A + b$ must intersect $A + X_+$, which implies that b is in $A - A + X_+$. Since $b \in B$ was arbitrary, we thus have $B \subseteq A - A + X_+$ and the claim follows. □

Covering lemmas such as the one above are convenient for a number of reasons. Firstly, they allow for easy computation of iterated sum sets. For instance, if one knows that

$$A + B \subseteq A + X$$

then one can immediately deduce that

$$A + nB \subseteq A + nX \text{ for all } n \ge 0.$$

This is advantageous if X is substantially smaller than B. Also, a covering property such as $A + B \subseteq A + X$ is preserved under Freiman homomorphisms, whereas bounds such as $|A + A| \le K|A|$ are only preserved by Freiman isomorphisms (see Chapter 5, in particular Exercise 5.3.13).

Remark 2.16 Observe that we are covering B by $A - A$ rather than by A. This reflects the fact that $A - A$ is a "smoother" set than A, and tends to contain fewer "holes" that would render it unsuitable for covering other sets. Later on we shall see that higher-order sum-difference sets such as $2A - 2A$ are even smoother, in that they tend to contain very large arithmetic progressions; see Section 4.7 and Chapter 12 for further discussion.

One can modify Ruzsa's covering lemma in a number of ways. For instance, one can ensure the covering of B by translates of $A - A$ has very high multiplicity (at the cost of increasing the number of covers by a factor of 2).

Lemma 2.17 (Green–Ruzsa covering lemma) *[154] Let A and B be additive sets with common ambient group. Then there exists an additive set $X \subseteq B$ with $|X| \leq 2\frac{|A+B|}{|A|} - 1$ such that for every $y \in B$ there are at least $|A|/2$ triplets $(x, a, a') \in X \times A \times A$ with $x + a - a' = y$. More informally, $A - A + X$ covers B with multiplicity at least $|A|/2$. Furthermore, we have*

$$B - B \subseteq A - A + X - X.$$

Similar claims hold if $\frac{|A+B|}{|A|}$ is replaced by $\frac{|A-B|}{|A|}$.

Proof Again it suffices to prove the claim for $\frac{|A+B|}{|A|}$. We perform the following algorithm. Initialize X to be the empty set, so that $X + A - A$ is also the empty set. We now run the following loop. If we cannot find any element y in B which is "sufficiently disjoint from $X + A - A$" in the sense that $|(y + A) \cap (X + A)| \leq |A|/2$, we terminate the algorithm. Otherwise, if there is such an element y, we add it to X, and then repeat the algorithm.

Every time we add an element to X, the size of $|X + A|$ increases by at least $|A|/2$, by construction, and at the first stage it increases by $|A|$. However, $X + A$ must always lie within the set $B + A$. Thus this algorithm terminates after at most $\frac{2|A+B|}{|A|} - 1$ steps.

Now let y be any element of B. By construction, we have $|(y + A) \cap (X + A)| > |A|/2$, and hence y has at least $|A|/2$ representations of the form $x + a - a'$ for some $(x, a, a') \in X \times A \times A'$, as desired.

Finally, if y and y' are two elements of B, then we have

$$|\{a \in A : y + a \in X + A\}| = |(y + A) \cap (X + A)| > |A|/2$$

and similarly we have $|\{a \in A : y' + a \in X + A\}| > |A|/2$. Thus by the pigeonhole principle there exists $a \in A$ such that $y + a \in X + A$ and $y' + a \in X + A$, thus $y - y' = (y + a) - (y' + a) \in X + A - (X + A) = A - A + X - X$. Since $y, y' \in B$ is arbitrary, we have $B - B \subseteq A - A + X - X$ as claimed. □

In Section 5.4 we develop yet another covering lemma (Lemma 5.31), in which the covering set X is not arbitrary, but is in fact a cube.

We now give an application of the Green–Ruzsa covering lemma, namely a variant of (2.6) which controls quadruple sums rather than double sums.

Proposition 2.18 *Let A, B be additive sets in an ambient group Z. Then*

$$|2B-2B| \le 16\frac{|A+B|^4|A-A|}{|A|^4}.$$

Proof Applying the Green–Ruzsa covering lemma, we may find a set X of cardinality $|X| \le 2\frac{|A+B|}{|A|}$ such that $A-A+X$ covers B with multiplicity at least $|A|/2$.

Now let z be any element of $B-B$. By definition, we have $z = b_1 - b_2$ for some $b_1, b_2 \in B$. By construction of X, we can find at least $|A|/2$ triplets $(x, a_1, a_2) \in X \times A \times A$ such that $b_2 = x + a_1 - a_2$, and thus

$$|\{(x, a_1, a_2) \in X \times A \times A : z = b_1 - a_1 + a_2 - x\}| \ge |A|/2.$$

Making the change of variables $c := b_1 + a_2 \in A + B$, we conclude that

$$|\{(x, c, a_1) \in X \times (A+B) \times A : z = c - a_1 - x\}| \ge |A|/2.$$

Similarly, if z' is another element of $B-B$, we have

$$|\{(x', c', a_1') \in X \times (A+B) \times A : z' = c' - a_1' - x'\}| \ge |A|/2,$$

and hence

$$|\{(x, x', c, c', a_1, a_1') \in X \times X \times (A+B) \times (A+B) \times A \times A :$$
$$z = c - a_1 - x, \quad z' = c' - a_1' - x'\}| \ge |A|^2/4.$$

Now write $d := a_1 - a_1' \in A - A$, and observe that if $z = c - a_1 - x$ and $z' = c' - a_1' - x'$ then

$$z - z' = c - c' - d - x + x'.$$

Also, if one fixes z, z', c, c', d, x, x', then a_1 and a_1' are determined by the equations $a_1 = c - x - z$, $a_1' = c' - x' - z'$. Thus we have

$$|\{(x, x', c, c', d) \in X \times X \times (A+B) \times (A+B) \times (A-A) :$$
$$z - z' = c - c' - d - x + x'\}| \ge |A|^2/4.$$

Note that $z - z'$ is an arbitrary element of $(B-B)-(B-B) = 2B-2B$. Thus we have shown that an arbitrary element of $2B-2B$ has at least $|A|^2/4$ representations of the form $c - c' - d - x + x'$ where $(x, x', c, c', d) \in X \times X \times (A+B) \times (A+B) \times (A-A)$. The claim then follows since $|X| \le 2\frac{|A+B|}{|A|}$. □

We can eliminate the factor of 16 by the following elegant "tensor power trick" of Ruzsa [297]:

Corollary 2.19 *Let A, B be additive sets in an ambient group Z. Then*

$$|2B-2B| \le \frac{|A+B|^4|A-A|}{|A|^4}.$$

Proof Fix A, B, and let M be a large integer parameter. We consider the M-fold Cartesian product $A^{\oplus M} := A \times \cdots \times A$, which is a subset of the additive group $Z^{\oplus M} := Z \oplus \cdots \oplus Z$; similarly consider $B^{\oplus M}$. Then one easily verifies

$$\begin{aligned} 2B^{\oplus M} - 2B^{\oplus M} &= (2B - 2B)^{\oplus M}; \\ A^{\oplus M} + B^{\oplus M} &= (A + B)^{\oplus M}; \\ A^{\oplus M} - A^{\oplus M} &= (A - A)^{\oplus M}. \end{aligned}$$

Thus by applying Lemma 2.18 with A, B replaced by $A^{\oplus M}, B^{\oplus M}$ we obtain

$$|2B - 2B|^M \leq 16 \frac{|A + B|^{4M} |A - A|^M}{|A|^{4M}}.$$

Taking Mth roots of both sides and letting $M \to \infty$, we obtain the result. □

Specializing Corollary 2.19 to the case $B := -A$, we obtain

Corollary 2.20 *Let A be an additive set. Then*

$$|2A - 2A| \leq \frac{|A - A|^5}{|A|^4}$$

or, in other words,

$$d(A - A, A - A) \leq 4d(A, A).$$

Remark 2.21 One can improve these estimates slightly by using the machinery of Plünnecke inequalities; see Corollary 6.28.

Combining Corollary 2.20 with the Ruzsa covering lemma (Lemma 2.14 with $B = 2A - A$) we obtain

Corollary 2.22 *For any additive set A, $2A - A$ can be covered by $\delta[A]^5$ translates of $A - A$.*

This then shows that $3A - A$ is covered by $\delta[A]^5$ translates of $2A - A$, and hence by $\delta[A]^{10}$ translates of $A - A$. Continuing in this fashion, an easy induction then shows

$$mA - nA \text{ can be covered by } \delta[A]^{5(m+n-2)} \text{ translates of } A - A \tag{2.16}$$

for all $m, n \geq 1$. In particular we have

$$|mA - nA| \leq \delta[A]^{5(m+n-1)} |A| \text{ for all } m, n \geq 1. \tag{2.17}$$

From this (and the trivial estimates $|kA| \geq |A|$ for any $k \geq 1$) we obtain

Corollary 2.23 (Symmetric sum set estimates, preliminary version) *Let A be an additive set. Then we have the estimates*

$$d(n_1 A - n_2 A, n_3 A - n_4 A) \leq 5(n_1 + n_2 + n_3 + n_4) d(A, A)$$

for any non-negative integers n_1, n_2, n_3, n_4. (The constant 5 is not best possible; we will improve it later.)

Thus if A has small difference constant, then in fact all iterated sum sets of A are close to each other in the Ruzsa metric. Another consequence of the corollary is that

$$\sigma[n_1 A - n_2 A] \leq \sigma[A]^{10(n_1+n_2)}$$

for all non-negative integers n_1, n_2. The factor of 10 is not best possible; we shall obtain improvements to this constant later when we develop the machinery of Plünnecke inequalities in Section 6.5. However, the linear growth in n_1 and n_2 is necessary; see Exercise 2.4.9.

By combining the above corollary with the Ruzsa triangle inequality one can obtain similar estimates for pairs of sets:

Corollary 2.24 (Asymmetric sum set estimates, preliminary version) *Let A, B be additive sets with common ambient group Z. Then we have the estimates*

$$d(n_1 A - n_2 A + n_3 B - n_4 B, n_5 A - n_6 A + n_7 B - n_8 B) = O((n_1 + \cdots + n_8) d(A, B))$$

for any $n_1, \ldots, n_8 \in \mathbf{N}$.

The proof is left as an exercise.

We can use the above machinery to place additive sets with small difference or doubling constant inside a more structured set, namely an "approximate group".

Definition 2.25 (Approximate groups) Let $K \geq 1$. An additive set H is said to be a K*-approximate group* if it is symmetric (so $H = -H$), contains the origin, and $H + H$ can be covered by at most K translates of H.

Observe that a 1-approximate group is necessarily a finite group, and conversely every finite group is a 1-approximate group.

We can summarize many of the preceding results by giving the following partial generalization of Proposition 2.7.

Proposition 2.26 *Let A be an additive set and let $K \geq 1$. Then the following statements are equivalent up to constants, in the sense that if the jth property holds for some absolute constant C_j, then the kth property will also hold for some absolute constant C_k depending on C_j:*

(i) $\sigma[A] \leq K^{C_1}$ (i.e. $|A + A| \leq K^{C_1}|A|$);
(ii) $\delta[A] \leq K^{C_2}$ (equivalently, $d(A, A) \leq C_2 \log K$ or $|A - A| \leq K^{C_2}|A|$);
(iii) $d(A, B) \leq C_3 \log K$ for at least one additive set B;
(iv) $|nA - mA| \leq K^{C_4(n+m)}|A|$ for all non-negative integers n, m;

(v) *there exists a* K^{C_5}*-approximate group* H *such that* $A \subseteq x + H$ *for all* $x \in A$*, and furthermore* $|A| \geq K^{-C_5}|H|$.

Proof The equivalence of the first three properties follows from the Ruzsa triangle inequality and (2.11). The equivalence of the fourth property with (say) the second follows from Corollary 2.24. To see that the fifth property implies (say) the first, observe that if the former holds, then

$$|A + A| \leq |H + H| \leq K^{C_5}|H| \leq K^{2C_5}|A|.$$

To deduce the fifth from the fourth, take $H = A - A$ and apply the Ruzsa covering lemma. □

Thus, in a qualitative sense, we have reduced the study of additive sets with small difference or doubling constant to the study of approximate groups, or precisely to the study of dense subsets of translates of approximate groups. This is a fairly satisfactory state of affairs, except for the fact that we do not have a good characterization of which sets are approximate groups. The well known structure theorem for finite groups (see Corollary 3.8 below) asserts that every finite group is the product of finite cyclic groups; we shall eventually be able to obtain a somewhat similar characterization of approximate groups, showing that they are efficiently contained in a generalized arithmetic progression. For some other properties of approximate groups, see the exercises below.

There is an asymmetric counterpart to Proposition 2.26, whose proof we leave as an exercise.

Proposition 2.27 *Let* A, B *be additive sets in an ambient group* Z*, and let* $K \geq 1$*. Then the following statements are equivalent up to constants, in the sense that if the* jth *property holds for some absolute constant* C_j*, then the* kth *property will also hold for some absolute constant* C_k *depending on* C_j*:*

(i) $d(A, B) \leq C_1 \log K$;
(ii) $d(A, -B) \leq C_2 \log K$;
(iii) $|A + B| \leq K^{C_3} \min(|A|, |B|)$;
(iv) $|A - B| \leq K^{C_4} \min(|A|, |B|)$;
(v) $|n_1 A - n_2 A + n_3 B - n_4 B| \leq K^{C_5(n_1+n_2+n_3+n_4)}|A|$ *for all non-negative integers* n_1, n_2, n_3, n_4;
(vi) $\sigma[A], \sigma[B] \leq K^{C_6}$*, and there exists* $x \in Z$ *such that* $|A \cap (B + x)| \geq K^{-C_6}|A|^{1/2}|B|^{1/2}$;
(vii) $\sigma[A], \sigma[B] \leq K^{C_7}$*, and* $E(A, B) \geq K^{-C_7}|A|^{3/2}|B|^{3/2}$;
(viii) *there exists a* K^{C_8}*-approximate group* H *such that* $A \subseteq H + a$ *and* $B \subseteq H + b$ *for all* $a \in A$*,* $b \in B$*, and furthermore* $|A|, |B| \geq K^{-C_8}|H|$.

Observe that Exercise 2.3.7 is essentially the $K = 1$ case of this Proposition.

Proposition 2.27 gives a satisfactory characterization of pairs of sets with small Ruzsa distance, in terms of approximate groups, provided that one is ready to lose some absolute constants in the exponents. Note however that it is restricted to treating those sets A, B which are comparable in magnitude up to powers of K (cf. Exercise 2.3.6). A partial analogue of this proposition exists in the case when A and B are very different in magnitude, but the theory here is not as satisfactory; see Section 2.6.

Exercises

2.4.1 Let Z be a finite additive group, and let A be a random subset of Z such that the events $a \in A$ are independent with probability $3/4$ for all $a \in Z$. Show that with probability $1 - o_{|Z|\to\infty}(1)$, $|A| > |Z|/2$ (so in particular $A + A = A - A = Z$, by Exercise 2.1.6), but that it is not possible to cover Z using fewer than $\frac{1}{10}\log|Z|$ translates of A. (Hint: if X is an additive set with $|X| \leq \frac{1}{10}\log|Z|$, use Lemma 2.14 to find an additive set Y with $|Y| = \Theta(|Z|/\log^2|Z|)$ such that the translates $y - X$ are disjoint for all $y \in Y$. Compute the probability that A is disjoint from at least one of the sets $y - X$, and conclude an upper bound for the probability that $A + X = Z$. Now take the union bound over all choices of X.) This shows that we cannot replace $A - A$ by A in Lemma 2.14 without admitting some sort of logarithmic loss.

2.4.2 Let A be an additive set in a group Z, and let $\phi : Z \to Z'$ be a group homomorphism. Establish the inequalities

$$|A| \leq |\phi(A)| \sup_{x\in Z'} |A \cap \phi^{-1}(x)| \leq |2A|.$$

(Hint: use the Ruzsa covering lemma to cover A by translates of a subset of $\phi^{-1}(0)$.) In particular equality is attained in both inequalities when A is the coset of a group.

2.4.3 Prove Corollary 2.24. What value of the implicit constant in the $O()$ notation do you get?

2.4.4 Let A be an additive set such that $|2A - 2A| < 2|A|$. Conclude that $A - A$ is a group. (Hint: use Lemma 2.14.) From this and Corollary 2.20 we see that if $|A - A| < 2^{1/5}|A|$, then $A - A$ is a group. The constant $2^{1/5}$ can be improved to $\frac{3}{2}$; see Exercise 2.6.5 below.

2.4.5 Let G be a K-approximate group for some integer $K \geq 1$. Show that $|nG| \leq \binom{K+n-1}{n}|G|$ for all integers $n \geq 1$. Conclude in particular the bounds

$$|nG| \leq \min(K^n, n^{K-1})|G| \text{ for all } n \geq 1;$$

thus the numbers $|nG|$ grow exponentially in n for $n \leq K$ but settle down to become polynomial growth for $n > K$. In fact for any additive set, $|nA|$ *is* a polynomial in n for sufficiently large n; see [261] for a proof of this fact and some further discussion.

2.4.6 Let A be an additive set with doubling constant $\sigma[A] = K$ for some $K \geq 1$. Show that

$$|nA| \leq \min(K^{Cn}, n^{K^C - 1})|A|$$

for all $n \geq 1$ and some absolute constant $C > 0$. (Note that if K is very close to 1, then one can use Exercise 2.4.4 to obtain a much stronger bound.)

2.4.7 Let G be a K-approximate group in an ambient group Z, and let H be a K'-approximate group in Z. Show that $G + H$ is a KK'-approximate group. Show that $2G \cap 2H$ is a $(KK')^3$-approximate group. (Hint: first show that $(2G \cap 2H) - (2G \cap 2H) \subset (G + X) \cap (H + Y)$ for some X, Y of cardinality at most K^3 and $(K')^3$ respectively, and then show that each set of the form $(G + x) \cap (H + y)$ is contained in a translate of $2G \cap 2H$.) Modify Exercise 2.2.9 to show that this type of statement fails quite badly if the set $2G \cap 2H$ is replaced by $G \cap H$. Also, establish the cardinality bounds

$$\frac{|G||H|}{|G + H|} \leq |2G \cap 2H| \leq (KK')^3 \frac{|G||H|}{|G + H|}.$$

(Hint: use (2.8) for the lower bound, and the Ruzsa triangle inequality for the upper bound.) Conclude the estimates

$$d(G, H) \leq d(G, G + H) + d(G + H, H) \leq d(G, H) + \log KK'$$

and

$$d(G, H) \leq d(G, 2G \cap 2H) + d(2G \cap 2H, H) \leq d(G, H) + 3 \log KK',$$

and compare this with Exercise 2.3.11.

2.4.8 For each $j = 1, 2, 3$, let G_j be a K_j-approximate group in an ambient group Z. Using the Ruzsa triangle inequality, show that

$$|G_1 + G_2 + G_3| \leq K_2 \frac{|G_1 + G_2||G_2 + G_3|}{|G_2|}.$$

Conclude that

$$d(G_1 + G_2, G_1 + G_2 + G_3) \leq d(G_2, G_2 + G_3) + \log K_1 K_2.$$

Similarly for permutations. Conclude from this and the preceding exercise that

$$d(G_1, G_2) \leq d(G_1 + G_3, G_2 + G_3) + 2 \log K_1 K_2 K_3$$

and compare this with Exercise 2.3.11. (A corresponding statement exists for intersections but is somewhat tricky to establish.)

2.4.9 For any integers $K, n_1, n_2 \geq 1$, give an example of an additive set A with $\sigma[A] = K$ and $\sigma[n_1 A - n_2 A] = \Omega_{n_1,n_2}(K^{n_1+n_2})$.

2.4.10 Let A, B be additive sets in a common ambient group Z. Show that $\sigma[A + B] \leq (\sigma[A]\sigma[B])^C$ where $C \geq 1$ is an absolute constant. (Hint: use Proposition 2.26 to place A and B inside translates of approximate groups. To obtain lower bounds on $|A + B|$, use the inequality

$$|A + B| \geq \frac{|A||B|}{|(A - A) \cap (B - B)|}$$

from (2.8).)

2.4.11 Prove Proposition 2.27. (Hint: to construct the approximate group H, one possible choice is $H = A - A + B - B$.)

2.4.12 Try to improve upon the constant 5 in (2.17), by using the Ruzsa triangle inequality instead of the Ruzsa covering lemma. This exercise demonstrates that the triangle inequality is slightly sharper than the covering lemma when one wants cardinality bounds, but the covering lemmas of course give much more information than just cardinality.

2.4.13 [209] Let A, B be additive sets in an ambient group Z, and let G be the group generated by A. Show that there exists an additive set $B' \in B$ such that B' is contained in a coset of G, and such that $|A + B'| \leq \frac{|B'|}{|B|}|A + B|$.

2.4.14 Let A, B, A', B' be additive sets with common ambient group Z. Establish the inequality $d(A + A', B + B') = O(d(A, B) + d(A', B'))$. (Hint: argue as in Exercise 2.4.10.) Conclude that if $\phi : Z \to Z'$ is a group homomorphism, then $d(\phi(A), \phi(B)) = O(d(A, B))$. Thus group homomorphisms are "Lipschitz" with respect to the Ruzsa distance.

2.5 The Balog–Szemerédi–Gowers theorem

In the previous sections we have only considered complete sum sets $A + B$ and complete difference sets $A - B$. In many applications one only controls a partial collection of sums and differences. Fortunately, there is a very useful tool, the *Balog–Szemerédi–Gowers theorem*, which allows one to pass from control of partial sum and difference sets to control of complete sum and difference sets (after refining the sets slightly). We begin with some notation.

Definition 2.28 (Partial sum sets) If A, B are additive sets with common ambient group Z, and G is a subset of $A \times B$, we define the *partial sum set*

$$A \overset{G}{+} B := \{a + b : (a, b) \in G\}$$

and the *partial difference set*

$$A \overset{G}{-} B := \{a - b : (a, b) \in G\}.$$

One may like to think of G as a bipartite graph connecting A and B. Note that when $G = A \times B$ is complete, then the notion of partial sum set and partial difference set collapse to just the complete sum set and difference set.

Partial sum sets and partial difference sets are not as nice to work with algebraically as complete sum sets. In particular, the above machinery of sum set estimates do not directly yield any conclusion if one only assumes that the cardinality $|A \overset{G}{+} B|$ of a partial sum set is small. Note that even when G is very large, it is possible for $|A \overset{G}{+} B|$ to be small while $|A + B|$ is large; see exercises. Fortunately, the Balog–Szemerédi–Gowers theorem, which we will present shortly, does allow us to conclude information on complete sum sets from information on partial sum sets, if we are willing to refine A and B by a small factor (i.e. replace A and B by subsets A' and B' which are only slightly smaller than A and B).

The first result in this direction was by Balog and Szemerédi [16], using the regularity lemma. A different, more effective proof, was found by Gowers [137] (with a slight refinement by Bourgain [38]), in particular with dependence of constants that are only polynomial in nature. Here we present a modern formulation of the theorem, following [340].

Theorem 2.29 (Balog–Szemerédi–Gowers theorem) *Let A, B be additive sets in an ambient group Z, and let $G \subseteq A \times B$ be such that*

$$|G| \geq |A||B|/K \text{ and } |A \overset{G}{+} B| \leq K'|A|^{1/2}|B|^{1/2}$$

for some $K \geq 1$ and $K' > 0$. Then there exists subsets $A' \subseteq A$, $B' \subseteq B$ such that

$$|A'| \geq \frac{|A|}{4\sqrt{2}K} \tag{2.18}$$

$$|B'| \geq \frac{|B|}{4K} \tag{2.19}$$

$$|A' + B'| \leq 2^{12}K^5(K')^3|A|^{1/2}|B|^{1/2}. \tag{2.20}$$

In particular we have

$$d(A', -B') \leq 5 \log K + 3 \log K' + O(1).$$

The proof of this theorem is graph-theoretical. It is elementary, but a little lengthy and so we postpone it to Section 6.4. One can of course combine this theorem with Corollary 2.24 and Proposition 2.26 to gain more information on the iterated sum and difference sets of A' and B'. It is likely that the factor of $2^{12}K^5(K')^3$ in (2.20) can be improved. However, the bounds (2.18), (2.19) cannot be significantly improved; see exercises.

To apply the Balog–Szemerédi–Gowers theorem, it is convenient to introduce the following lemma connecting large additive energy to small partial sum sets or small partial difference sets.

Lemma 2.30 *Let A, B be additive sets in an ambient group Z, and let G be a non-empty subset of $A \times B$. Then*

$$E(A, B) \geq \frac{|G|^2}{|A \overset{G}{+} B|}, \frac{|G|^2}{|A \overset{G}{-} B|}.$$

Conversely, if $E(A, B) \geq |A|^{3/2}|B|^{3/2}/K$ for some $K \geq 1$, then there exists $G \subseteq A \times B$ such that

$$|G| \geq |A||B|/2K; \qquad |A \overset{G}{+} B| \leq 2K|A|^{1/2}|B|^{1/2}.$$

and similarly there exists $H \subseteq A \times B$ such that

$$|H| \geq |A||B|/2K; \qquad |A \overset{H}{-} B| \leq 2K|A|^{1/2}|B|^{1/2}.$$

Proof Observe that

$$\sum_{x \in A \overset{G}{+} B} |\{(a, b) \in G : a + b = x\}| = |G|$$

and hence by Cauchy–Schwarz

$$\sum_{x \in A \overset{G}{+} B} |\{(a, b) \in G : a + b = x\}|^2 \geq \frac{|G|^2}{|A \overset{G}{+} B|}.$$

But the left-hand side is equal to

$$|\{(a, a', b, b') \in A \times A \times B \times B : a + b = a' + b'; (a, b), (a', b') \in G\}|$$

which was less than $E(A, B)$. This proves that $E(A, B) \geq |G|^2/|A \overset{G}{+} B|$; using the symmetry $E(A, B) = E(A, -B)$ we thus also obtain $E(A, B) \geq |G|^2/|A \overset{G}{-} B|$.

Now assume $E(A, B) \geq |A|^{3/2}|B|^{3/2}/K$. Then by Lemma 2.9 we have

$$\sum_{x \in A+B} |A \cap (x - B)|^2 \geq \frac{|A|^{3/2}|B|^{3/2}}{K}.$$

If we set $S := \{x \in A + B : |A \cap (x - B)| \geq |A|^{1/2}|B|^{1/2}/2K\}$, we then have (by Lemma 2.9 again)

$$\sum_{x \in S} |A \cap (x - B)|^2 \geq \frac{|A|^{3/2}|B|^{3/2}}{K} - \frac{|A||B||A|^{1/2}|B|^{1/2}}{2K} = \frac{|A|^{3/2}|B|^{3/2}}{2K}.$$

Now observe from Lemma 2.9 again that

$$\frac{|S||A|^{1/2}|B|^{1/2}}{2K} \leq \sum_{x \in S} |A \cap (x - B)| \leq |A||B|$$

and hence

$$|S| \leq 2K|A|^{1/2}|B|^{1/2}.$$

Now let $G := \{(a, b) \in A \times B : a + b \in S\}$, then clearly $A \overset{G}{+} B \subseteq S$ and hence

$$|A \overset{G}{+} B| \leq 2K|A|^{1/2}|B|^{1/2}.$$

Furthermore we have

$$\begin{aligned}
|G| &= \sum_{x \in S} |\{(a, b) \in A \times B : a + b = x\}| \\
&= \sum_{x \in S} |A \cap (x - B)| \\
&\geq \sum_{x \in S} \frac{|A \cap (x - B)|^2}{|A|^{1/2}|x - B|^{1/2}} \\
&\geq \frac{|A|^{3/2}|B|^{3/2}/2K}{|A|^{1/2}|B|^{1/2}} \\
&= |A||B|/2K.
\end{aligned}$$

This gives the desired set G. The construction of H follows by using the symmetry $E(A, B) = E(A, -B)$. □

Combining this Lemma with the Balog–Szemerédi–Gowers theorem, we can obtain a characterization of pairs of sets with large additive energy.

Theorem 2.31 (Balog–Szemerédi–Gowers theorem, alternative version) *Let A, B be additive sets in an ambient group Z, and let $K \geq 1$. Then the following statements are equivalent up to constants, in the sense that if the jth property holds for some absolute constant C_j, then the kth property will also hold for some absolute constant C_k depending on C_j:*

(i) $E(A, B) \geq K^{-C_1}|A|^{3/2}|B|^{3/2}$;
(ii) there exists $G \subset A \times B$ *such that* $|G| \geq K^{-C_2}|A||B|$ *and* $|A \overset{G}{+} B| \leq K^{C_2}|A|^{1/2}|B|^{1/2}$;

(iii) there exists $G \subset A \times B$ *such that* $|G| \geq K^{-C_3}|A||B|$ *and* $|A \stackrel{G}{-} B| \leq K^{C_3}|A|^{1/2}|B|^{1/2}$*;*

(iv) there exists subsets $A' \subseteq A$, $B' \subseteq B$ *with* $|A'| \geq K^{-C_4}|A|$, $|B'| \geq K^{-C_4}|B|$, *and* $d(A', B') \leq C_4 \log K$*;*

(v) there exists a K^{C_5}*-approximate group* H *and* $x, y \in Z$ *such that* $|A \cap (H + x)|, |B \cap (H + y)| \geq K^{-C_5}|H|$ *and* $|A|, |B| \leq K^{C_5}|H|$.

We leave the proof of this theorem to the exercises. Theorem 2.31 should be compared with Exercise 2.3.22, which is the $K = 1$ case of this Theorem. As with Proposition 2.27, this Theorem is restricted to sets A, B which are close in cardinality (see exercises). We shall address the question of sets A, B of widely differing cardinalities in the next section.

Exercises

2.5.1 Let A, B be additive sets with common ambient group Z such that $E(A, B) \geq K^{-1}|A|^{3/2}|B|^{3/2}$. Show that $K^{-2}|A| \leq |B| \leq K^2|A|$, and show by means of an example that these bounds cannot be improved.

2.5.2 Give an example of an additive set $A \subset \mathbf{Z}$ of cardinality N, and a set $G \subset A \times A$ of cardinality $N^2/4$, such that $|A \stackrel{G}{+} A| \leq N$ but $|A + A| \geq N^2/8$. (Hint: concatenate a Sidon set with an arithmetic progression.)

2.5.3 Let $N \gg K \gg 1$ be large integers, with N a multiple of K. Give an example of sets $A, B \subset \mathbf{Z}$ of cardinality $|A| = |B| = N$ and a subset $G \subset A \times B$ of cardinality $|G| = |A||B|/K$ with the property that $|A \stackrel{G}{+} B| \leq 2N$, but such that $|A'' + B''| \geq N^2/K^2$ whenever $A'' \subset A$ and $B'' \subset B$ is such that $|A''| \geq 2|A|/K$ and $|B''| \geq 2|B|/K$. (Hint: take B to be a long progression, and take A to be a short progression concatenated with some generic integers.) This shows that the conditions (2.18), (2.19) in Theorem 2.29 cannot be significantly improved.

2.5.4 Let A, B, C be additive sets in an ambient group Z, let $0 < \varepsilon < 1/4$, and let $G \subset A \times B$, $H \subset B \times C$ be such that $|G| \geq (1 - \varepsilon)|A||B|$ and $|H| \geq (1 - \varepsilon)|B||C|$. Show that there exists subsets $A' \subseteq A$ and $C' \subseteq C$ with $|A'| \geq (1 - \varepsilon^{1/2})|A|$ and $|C'| \geq (1 - \varepsilon^{1/2})|C|$ such that $|A' - C'| \leq |A \stackrel{G}{-} B||B \stackrel{H}{-} C|/(1 - 2\varepsilon^{1/2})|B|$. (Hint: show that at most $\varepsilon^{1/2}|B|$ elements of B have a G-degree of less than $(1 - \varepsilon^{1/2})|A|$, and similarly at most $\varepsilon^{1/2}|B|$ elements have a H-degree of less than $(1 - \varepsilon^{1/2})|C|$.) This result can be used as a substitute for the Balog–Szemerédi–Gowers theorem in the case when the graph G is extremely dense; it has the advantage that it does not require A, B, C to be comparable in size and it does not lose any constants in the limit $\varepsilon \to 0$; indeed it collapses to Ruzsa's triangle inequality in that limit.

2.5.5 Prove Theorem 2.31. (Hint: for K large, e.g. $K \geq 1.1$, one can use the Balog–Szemerédi–Gowers theorem and Proposition 2.27. For K small, e.g. $1 \leq K < 1.1$, one can use Exercise 2.5.4 as a substitute for the Balog–Szemerédi–Gowers theorem.)

2.5.6 [80] Let A, B be additive sets with common ambient group such that $|A| = |B| = N$ and $|A + A| \leq KN$. Suppose also that $|A \overset{G}{+} B| \leq KN$, where $G \subset A \times B$ is a bipartite graph such that every element of B is connected to at least $K^{-1}N$ elements of A. Show that $|A + B| \leq K^{O(1)}N$ and $|B + B| \leq K^{O(1)}N$. (Hint: write the elements of $A + B$ in the form $x - y + z$ where $x \in A + A$, $y \in A + A$, and $z \in A \overset{G}{+} B$.)

2.5.7 [80] Let A be an additive set such that $|A \overset{G}{+} A| \leq K|A|$, where $G \subset A \times A$ is such that every element of A is connected via G to at least $K^{-1}|A|$ elements of A. Show that one can partition A into $O(K^{O(1)})$ subsets $A_1, \ldots, A_m$ such that $|A_i + A_i| = O(K^{O(1)}|A|)$ for each $1 \leq i \leq m$. (Hint: use the Balog–Szemerédi–Gowers theorem and an iteration argument to obtain most of the subsets, and then Exercise 2.5.6 to deal with the remainder.)

2.6 Symmetry sets and imbalanced partial sum sets

The Balog–Szemerédi–Gowers theorem is a very powerful tool when studying two additive sets A, B with additive energy $E(A, B)$ close to $|A|^{3/2}|B|^{3/2}$; however from (2.7) we see that this situation only occurs when $|A|$ and $|B|$ are comparable in size. This leaves open the question of what happens in the case $|A| \gg |B|$ (say) and $E(A, B)$ is close to the upper bound of $|A||B|^2$ given by (2.7). A special sub-case of this (thanks to (2.8)) is the case when $|A + B|$ or $|A - B|$ is comparable to $|A|$. Note that Proposition 2.2 already gives an answer to this question in the extreme case when $|A + B| = |A|$ or $|A - B| = |A|$ (or equivalently if $E(A, B) = |A||B|^2$; see Exercise 2.3.22). However, an example of Ruzsa [297] shows that things become bad when $|A|$ and $|B|$ are very widely separated; see the exercises.

If however we are prepared to endure logarithmic-type losses in the ratio $|A|/|B|$ (or more precisely losses of the form $(|A|/|B|)^{\varepsilon}$ where ε can be chosen to be small), then one can recover a reasonable theory. In analogy with Proposition 2.2, one expects that if $|A + B|$ is comparable to $|A|$, or if $E(A, B)$ is close to $|A||B|^2$, then there should be an approximate group H such that A is approximately the union of translates of H, and B is approximately contained in a single translate of H. To achieve this will be the main objective of this section.

In the extreme case when $|A+B| = |A|$ or $E(A, B) = |A||B|^2$, the approximate group H was in fact an exact group and in the proof of Proposition 2.2 it was constructed as the symmetry group $\mathrm{Sym}_1(A)$ of the larger additive set A. In the general case this symmetry group is likely to be trivial. However, a more general notion is still useful.

Definition 2.32 (Symmetry sets) Let (A, Z) be an additive set. For any non-negative real number $\alpha \geq 0$, define the *symmetry set* $\mathrm{Sym}_\alpha(A) \subseteq Z$ *at threshold* α to be the set

$$\mathrm{Sym}_\alpha(A) := \{h \in Z : |A \cap (A+h)| \geq \alpha|A|\}.$$

Note that $\mathrm{Sym}_1(A) = \{h \in Z : A + h = A\}$ is the same symmetry group applied in the proof of Proposition 2.2. The other symmetry sets are not groups in general, but nevertheless they are still symmetric (so $-\mathrm{Sym}_\alpha(A) = \mathrm{Sym}_\alpha(A)$) and contain the origin, and they obey the nesting property $\mathrm{Sym}_\alpha(A) \subseteq \mathrm{Sym}_\beta(A)$ for $\alpha \geq \beta$. It is also clear that $\mathrm{Sym}_\alpha(A) \subseteq A - A$ for all $0 < \alpha \leq 1$. Note that as $\mathrm{Sym}_\alpha(A)$ is empty for $\alpha > 1$ and equal to all of Z for $\alpha \leq 0$, we shall mostly restrict ourselves to the non-trivial region where $0 < \alpha \leq 1$.

We now relate the size of these symmetry sets to the additive energy. From Lemma 2.9 we have

$$E(A, A) = \sum_{h \in A-A} |A \cap (A+h)|^2$$

and hence for any $0 < \alpha \leq 1$ and the crude bounds $|A \cap (A+h)| \leq |A|$ when $h \in \mathrm{Sym}_\alpha(A)$ and $|A \cap (A+h)| \leq \alpha|A|$ when $h \notin \mathrm{Sym}_\alpha(A)$, we have

$$\alpha^2|A|^2|\mathrm{Sym}_\alpha(A)| \leq E(A, A) \leq \alpha^2|A|^2|A-A| + |A|^2|\mathrm{Sym}_\alpha(A)|,$$

which indicates that $\mathrm{Sym}_\alpha(A)$ should be large whenever the energy is large. In particular, from (2.7) we have

$$|\mathrm{Sym}_\alpha(A)| \leq |A|/\alpha^2. \tag{2.21}$$

Now let A, B be additive sets in an additive group Z. From Lemma 2.9 again, we have

$$E(A, B) = \sum_{b, b' \in B} |A \cap (A + b - b')|$$

and hence for any $0 < \alpha \leq 1$ we have

$$E(A, B) \leq |B|^2\alpha|A| + |A||\{(b, b') \in B : b - b' \in \mathrm{Sym}_\alpha(A)\}|.$$

In particular, if $E(A, B) \geq 2\alpha|A||B|^2$, then we conclude that there is a set $G \subset B \times B$ of cardinality $|G| \geq \alpha|B|^2$ such that

$$B \stackrel{G}{-} B \subseteq \mathrm{Sym}_\alpha(A). \tag{2.22}$$

At first glance it seems that one may now be able to apply the symmetric Balog–Szemerédi–Gowers theorem. However, the fact that A is much larger than B means that $B \stackrel{G}{-} B$ may be much larger than B (compare (2.22) to (2.21)). To get around this difficulty we need to iterate this construction, and exploit the fact that $\mathrm{Sym}_\alpha(A)$ behaves like a group. This is already clear when $\alpha = 1$, when $\mathrm{Sym}_1(A)$ is indeed a genuine group; the following lemma shows that this behavior persists in an approximate sense for α less than 1.

Lemma 2.33 *Let A be an additive set. Then we have*

$$\mathrm{Sym}_{1-\varepsilon}(A) + \mathrm{Sym}_{1-\varepsilon'}(A) \subseteq \mathrm{Sym}_{1-\varepsilon-\varepsilon'}(A) \tag{2.23}$$

whenever $\varepsilon, \varepsilon' > 0$. Furthermore, if $0 < \alpha \leq 1$ and $S \subseteq \mathrm{Sym}_\alpha(A)$ is a non-empty set, then there exists a set $G \subseteq S^2$ with

$$|G| \geq \alpha^2|S|^2/2 \tag{2.24}$$

such that

$$S \stackrel{G}{-} S \subseteq \mathrm{Sym}_{\alpha^2/2}(A). \tag{2.25}$$

Proof To verify the first claim, observe that if $x \in \mathrm{Sym}_{1-\varepsilon}(A)$ and $y \in \mathrm{Sym}_{1-\varepsilon'}(A)$ then

$$|(A + x)\backslash A| = |A| - |A \cap (A + x)| \leq \varepsilon|A|$$

and

$$|(A + x)\backslash(A + x + y)| = |A| - |A \cap (A + y)| \leq \varepsilon'|A|,$$

and hence

$$|A \cap (A + x + y)| \geq |(A + x) \cap A \cap (A + x + y)| \geq (1 - \varepsilon - \varepsilon')|A|$$

which proves (2.23).

Now we prove the second claim. By definition of S, we see that for each $x \in S$ there exist at least $\alpha|A|$ elements $a \in A$ such that $a + x \in A$. Summing this over all x we see that

$$\sum_{a\in A} |\{x \in S : a + x \in A\}| \geq \alpha|A||S|.$$

Applying Cauchy–Schwarz we conclude that

$$\sum_{x,y\in S\times S} |\{a \in A : a+x, a+y \in A\}| = \sum_{a\in A} |\{x \in S : a+x \in A\}|^2 \geq \alpha^2|A||S|^2.$$

If we set $G \subseteq S \times S$ to be all the pairs (x, y) such that

$$|\{a \in A : a+x, a+y \in A\}| \geq \alpha^2|A|/2$$

then we have

$$|A||G| \geq \sum_{(x,y)\in G} |\{a \in A : a+x, a+y \in A\}| \geq \alpha^2|A||S|^2 - \frac{\alpha^2|A|}{2}|S|^2$$

which gives (2.24). Also, if $(x, y) \in G$ then $|A \cap (A + x - y)| \geq \alpha^2|A|/2$ by definition of G, which gives (2.25). □

Before we proceed with the main theorem, we need a technical lemma that uniformizes the size of the fibers $\{(a, a') \in G : a - a' = x\}$ of $A \overset{G}{-} A$.

Lemma 2.34 (Dyadic pigeonhole principle) *Let A be an additive set, and let $G \subset A \times A$ be such that $|G| \geq \alpha|A|^2$ and $|A \overset{G}{-} A| \leq L|A|$ for some $0 < \alpha < 1$ and $L \geq 1$. Then there exists a subset G' of G with*

$$|G'| = \Omega\left(\frac{\alpha}{1 + \log\frac{1}{\alpha} + \log L}|A|^2\right)$$

and

$$|\{(a, a') \in G' : a - a' = x\}| \geq \frac{|G'|}{2|A \overset{G'}{-} A|}$$

for all $x \in A \overset{G'}{-} A$.

It is important to note that the dependence on L only enters in a logarithmic manner.

Proof Let D be the set of all x such that

$$|\{(a, a') \in G : a - a' = x\}| \geq \frac{\alpha|A|^2}{2L|A|} = \frac{\alpha}{2L}|A|$$

(thus D is the set of "popular differences") and set $\tilde{G}$ to be the pairs (a, a') in G such that $a - a' \in D$. Then we have $|G\backslash\tilde{G}| \leq \frac{\alpha}{2L}|A||A \overset{G}{-} A| \leq \alpha|A|^2/2$, and hence $|\tilde{G}| \geq \alpha|A|^2/2$. On the other hand, we have the crude upper bound

$$|\{(a, a') \in \tilde{G} : a - a' = x\}| \leq \sum_{a'\in A} |\{a \in A : a = x + a'\}| \leq |A|.$$

Thus if we let M be the least integer such that $2^{-M} < \frac{\alpha}{2L}$, we can partition $\tilde{G} = G_1 \cup \cdots \cup G_M$ where $G_m := \{(a, a') \in \tilde{G} : a - a' \in D_m\}$ and

$$D_m := \{x \in A \overset{\tilde{G}}{-} A : 2^{-m}|A| < |\{(a, a') \in \tilde{G} : a - a' = x\}| \leq 2^{-m+1}|A|\}.$$

By the pigeonhole principle, there exists $1 \leq m \leq M$ such that

$$|G_m| \geq \frac{1}{M}|G| \geq \frac{\alpha}{C\left(1 + \log\frac{1}{\alpha} + \log L\right)}|A|^2.$$

By the definition of D_m, we have

$$\frac{|G_m|}{2^{-m+1}|A|} \leq |D_m| \leq \frac{|G_m|}{2^{-m}|A|};$$

since $D_m = A \overset{G_m}{-} A$, we thus see that

$$|\{(a, a') \in G_m : a - a' = x\}| \geq 2^{-m}|A| \geq \frac{|G'|}{2|A \overset{G'}{-} A|}$$

for all $x \in A \overset{G_m}{-} A$. The claim then follows by setting $G' := G_m$. □

Now we give the main theorem of this section.

Theorem 2.35 (Asymmetric Balog–Szemerédi–Gowers theorem) *Let A, B be additive sets in an additive group Z such that $E(A, B) \geq 2\alpha|A||B|^2$ and $|A| \leq L|B|$ for some $L \geq 1$ and $0 < \alpha \leq 1$. Let $\varepsilon > 0$. Then there exists a $O_\varepsilon(\alpha^{-O_\varepsilon(1)}L^\varepsilon)$-approximate group H in Z, an additive set X in Z of cardinality $|X| = O_\varepsilon(\alpha^{-O_\varepsilon(1)}L^\varepsilon|A|/|H|)$ such that $|A \cap (X + H)| = \Omega_\varepsilon(\alpha^{O_\varepsilon(1)}L^{-\varepsilon}|A|)$, and an $x \in Z$ such that $|B \cap (x + H)| = \Omega_\varepsilon(\alpha^{O_\varepsilon(1)}L^{-\varepsilon}|B|)$.*

Observe in the converse direction that if the conclusions of this theorem are true, then $E(A, B) = \Omega_\varepsilon(\alpha^{O_\varepsilon(1)}L^{-O(\varepsilon)}|A||B|^2)$ (Exercise 2.6.3 at the end of this section). Thus this theorem is sharp up to polynomial losses in α and L^ε, where ε can be made arbitrary small; the example in Exercise 2.6.1 can be adapted to show that this loss is necessary (Exercise 2.6.2).

Proof A direct application of Theorem 2.31 will lose far too many powers of L. The trick is to embed B in a long increasing sequence of sets $B_0, B_1, B_2, \ldots$, with each B_j being (roughly speaking) a partial difference set of the previous one, and use the pigeonhole principle to show that at some stage the ratio $|B_{j+1}|/|B_j|$ is bounded by a small power of L. One can then apply Theorem 2.31 with acceptable losses and conclude the theorem. (This method of proof is inspired by a similar argument in [40].)

We turn to the details. It will be convenient to use a variant of the Landau $O()$ and $\Omega()$ notation which can absorb factors of α and $\log L$ (which we think of as being relatively close to 1). If X, Y are non-negative quantities and j is a parameter, let us say that $X = \tilde{O}_j(Y)$ or $Y = \tilde{\Omega}_j(X)$ if one has an estimate of the form

$$X \leq C(j)\alpha^{-C(j)} Y \log^{C(j)} L$$

for some $C(j) > 0$ depending only on j.

Let $J = J(\varepsilon) \gg 1$ be a large integer to be chosen later. Let $1 > \alpha_1 > \cdots > \alpha_{J+1} > 0$ be the sequence defined recursively by $\alpha_1 := \alpha$ and $\alpha_{j+1} := \alpha_j^2/2$ for all $1 \leq j \leq J$. From induction we see that $\alpha_j = \tilde{\Omega}_j(1)$. We claim that we can find a sequence $B_0, B_1, \ldots, B_J, B_{J+1}$ of additive sets in Z with the following properties.

- $B_0 = B$, and for all $1 \leq j \leq J+1$ we have

$$B_j \subseteq \mathrm{Sym}_{\alpha_j}(A). \tag{2.26}$$

- For all $0 \leq j \leq J+1$, we have

$$\alpha_j^{-2} L|B| \geq |B_j| = \tilde{\Omega}_j(|B|). \tag{2.27}$$

- For all $0 \leq j \leq J$, there exists $G_j \subseteq B_j \times B_j$ such that

$$|G_j| = \tilde{\Omega}_j(|B_j|^2) \tag{2.28}$$

and

$$B_{j+1} = B_j \overset{G_j}{-} B_j. \tag{2.29}$$

Furthermore, for all $x \in B_{j+1}$ we have

$$|\{(b, b') \in G_j : b - b' = x\}| = \tilde{\Omega}_j\left(\frac{|B_j|^2}{|B_{j+1}|}\right). \tag{2.30}$$

We construct the B_j as follows. We set $B_0 := B$. From (2.22) followed by Lemma 2.34 we can construct $G_0 \subseteq B_0 \times B_0$ and $B_1 := B_0 \overset{G_0}{-} B_0$ obeying (2.26), (2.28), (2.29), (2.30). Since each element in $B_0 \overset{G_0}{-} B_0$ can be represented as a difference of a pair in G in at most $|B_0|$ ways, we have

$$|B_1| = |B_0 \overset{G_0}{-} B_0| \geq |G_0|/|B_0| = \tilde{\Omega}_j(|B|),$$

which is the lower bound in (2.27); the upper bound follows from (2.26) and (2.21).

Next, suppose inductively that $B_j \subseteq \mathrm{Sym}_{\alpha_j}(A)$ has already been chosen for some $1 \leq j \leq J$. Applying Lemma 2.33 (with $S := B_j$) followed by Lemma 2.34, and using the cardinality bounds already obtained in (2.27) and the construction $\alpha_{j+1}^2 := \alpha_j^2/2$ of the α_j, we can thus find $G_j \subseteq B_j \times B_j$ and $B_{j+1} := B_j \overset{G_j}{-} B_j$

obeying (2.26), (2.28), (2.29), (2.30). This closes the induction and so we can construct the B_j for all $0 \le j \le J+1$, and similarly obtain the G_j for all $1 \le j \le J$.

Now for the crucial step (which explains why we iterated the above procedure so many times). From (2.27) and the pigeonhole principle, there exists $1 \le j \le J$ such that

$$|B_{j+1}| = \tilde{O}_J\big(L^{O(1/J)}|B_j|\big);$$

the point is that we have managed to replace L by the substantially smaller quantity $L^{O(1/J)}$. If we now apply (2.29), (2.28), and Theorem 2.31, we can thus find a $\tilde{O}_J(L^{O(1/J)})$-approximate group H of cardinality

$$|H| = \tilde{O}_J\big(L^{O(1/J)}|B_j|\big) \tag{2.31}$$

and an $x_j \in Z$ such that

$$|B_j \cap (H + x_j)| = \tilde{\Omega}_J\big(L^{-C_0/J}|B_j|\big) \tag{2.32}$$

for some absolute constant C_0. It remains to relate H to B and to A. We begin with B. From (2.32) and (2.30) (with j replaced by $j-1$) we have

$$|\{(b, b') \in G_{j-1} : b - b' \in B_j \cap (H + x_j)\}| = \tilde{\Omega}_J\big(L^{-C_0/J}|B_{j-1}|^2\big),$$

so in particular

$$|\{(b, b') \in B_{j-1} \times B_{j-1} : b \in H + x_j + b'\}| = \tilde{\Omega}_J\big(L^{-C_0/J}|B_{j-1}|^2\big).$$

Thus by the pigeonhole principle, there exists a b' such that

$$|\{b \in B_{j-1} : b - b' \in H + x_j + b'\}| = \tilde{\Omega}_J\big(L^{-C_0/J}|B_{j-1}|\big).$$

Thus if we set $x_{j-1} := x_j + b'$ then we have

$$|B_{j-1} \cap (H + x_{j-1})| = \tilde{\Omega}_J\big(L^{-C_0/J}|B_{j-1}|\big). \tag{2.33}$$

We now repeat this argument with j replaced by $j-1$ and (2.32) replaced by (2.33). Iterating this at most J times, we eventually locate an $x = x_0 \in Z$ such that

$$|B \cap (H + x)| = \tilde{\Omega}_J\big(L^{-C_0/J}|B|\big),$$

which gives the desired control on B if J is sufficiently large depending on ε.

It remains to control A. From (2.32), (2.31) and (2.26) we have

$$|\{y \in H + x_j : y \in \mathrm{Sym}_{\alpha_j}(A)\}| = \tilde{\Omega}_J\big(L^{-O(1/J)}|H|\big)$$

and thus by definition of $\mathrm{Sym}_{\alpha_j}(A)$ and α_j

$$|\{(a, y) \in A \times (H + x_j) : a + y \in A\}| = \tilde{\Omega}_J\big(L^{-O(1/J)}|H||A|\big).$$

We rewrite this as

$$\sum_{x\in x_j+A} |A\cap(H+x)| = \tilde{\Omega}_J\big(L^{-O(1/J)}|H||A|\big).$$

We can therefore find a subset X_0 of $x_j + A$ with

$$|X_0| = \tilde{\Omega}_J\big(L^{-O(1/J)}|A|\big) \tag{2.34}$$

such that

$$|A\cap(H+x)| = \tilde{\Omega}_J\big(L^{-O(1/J)}|H|\big) \text{ for all } x\in X_0.$$

Now we use an argument similar to that used to prove Ruzsa's covering lemma (Lemma 2.14). Let X be a subset of X_0 such that the sets $\{H + x : x \in X\}$ are all disjoint, and which is maximal with respect to set inclusion. Then we have

$$|A\cap(H+X)| = \sum_{x\in X}|A\cap(H+x)| = \tilde{\Omega}_J\big(L^{-O(1/J)}|H||X|\big). \tag{2.35}$$

On the other hand, if $y \in X_0$, then by maximality of X there exists $x \in X$ such that $x + H$ intersects $y + H$. In other words, X_0 is covered by $X + H - H$, and hence (since H is a $\tilde{O}(L^{O(1/J)})$-approximate group)

$$|X_0| \le |X||H-H| = \tilde{O}\big(|X|L^{O(1/J)}|H|\big). \tag{2.36}$$

Combining (2.34), (2.35), (2.36) we see that X obeys all the desired properties, if J is chosen sufficiently small depending on ε. □

The above theorem can also be put in a form resembling Theorem 2.29:

Corollary 2.36 *Let A, B be additive sets with common ambient group such that $E(A, B) \ge 2\alpha|A||B|^2$ and $|A| \le L|B|$ for some $L \ge 1$ and $0 < \alpha \le 1$. Let $\varepsilon > 0$. Then there exists subsets $A' \subseteq A$ and $B' \subseteq B$ such that*

$$|A'| = \Omega_\varepsilon\big(\alpha^{O_\varepsilon(1)}L^{-\varepsilon}|A|\big)$$
$$|B'| = \Omega_\varepsilon\big(\alpha^{O_\varepsilon(1)}L^{-\varepsilon}|B|\big)$$
$$|A'+nB'-mB'| = O_\varepsilon\big(\alpha^{-O_\varepsilon(1)}L^{\varepsilon}\big)^{n+m}|A|$$

for all integers $n, m \ge 0$.

Proof Apply Theorem 2.35 and set $A' := A\cap(X+H)$ and $B' := B\cap(x+H)$. □

Because of (2.8), the above results give some partial results concerning the situation when $|A + B| \le K|A|$ and $|A|$ is much larger than $|B|$, but these results will be rather weak. We will give a better result concerning this problem in Section 6.5, once we develop the Plünnecke inequalities.

Exercises

2.6.1 [297] Let n be a large integer, and let $Z := \mathbf{Z}^{2n}$. Let A be the additive set

$$A := \{(x_1, x_2, \ldots, x_{2n}) \in \mathbf{Z}^{2n} : x_1 + \cdots + x_{2n} = n; x_1, \ldots, x_{2n} \geq 0\}$$

and let $B := \{e_1, \ldots, e_{2n}\}$. Show that $|B| = 2n$, that $|A| = (27/4)^{n+o(1)}$, that $|A + B| = O(|A|)$, but that $|A - B| \geq n|A|$. (You may find Stirling's formula (1.52) to be useful.)

2.6.2 Modify Exercise 2.6.1 to show that one cannot take $\varepsilon = 0$ in Theorem 2.35.

2.6.3 Let A, B be additive sets and let $\varepsilon > 0, 0 < \alpha < 1$, and $L \geq 1$ be such that the conclusions of Theorem 2.35 are satisfied. Conclude that $E(A, B) = \Omega_\varepsilon(\alpha^{O_\varepsilon(1)} L^{-O(\varepsilon)} |A||B|^2)$.

2.6.4 Let A be an additive set. By modifying the proof of Lemma 2.13, establish the inequality

$$|A - A + n\mathrm{Sym}_\alpha(A)| \leq \frac{\delta[A]^{n+1}}{\alpha^n} |A|$$

for all integers $n \geq 0$ and all $0 < \alpha < 1$.

2.6.5 [220] Let A be an additive set such that $A - A$ is not a group. Show that there exists $h \in A - A$ such that $1 \leq |A \cap (A + h)| \leq |A|/2$. (Hint: argue by contradiction, and analyze $\mathrm{Sym}_\alpha(A)$ for some α slightly greater than 1/2.) Conclude in particular that if $|A - A| < \frac{3}{2}|A|$, then $A - A$ is a group. Note that the example $A = \{0, 1\} \subset \mathbf{Z}$ shows that the constant $\frac{3}{2}$ cannot be improved; one can also make this example larger, for instance by taking the Cartesian product of $\{0, 1\}$ with a finite group. For a more refined estimate on $A - A$, see Theorem 5.5 and Corollary 5.6.

2.6.6 Let A, B be additive sets with common ambient group such that $|A + B| \leq K|A|$ and $|A| \leq L|B|$ for some $K, L \geq 1$. Let $\varepsilon > 0$. Show that there exists a $O_\varepsilon(K^{O_\varepsilon(1)} L^\varepsilon)$-approximate group H such that B is contained in a translate of H, and that A is contained in at most $O_\varepsilon(K^{O_\varepsilon(1)} L^\varepsilon |A|/|H|)$ translates of H; compare this with Proposition 2.2. (Hint: Apply Theorem 2.35 and the Ruzsa covering lemma (Lemma 2.14).)

2.6.7 Let A be an additive set, and let B be a subset of A such that $|B| \geq (1 - \varepsilon)|A|$ for some $0 < \varepsilon < 1$. Prove that

$$\mathrm{Sym}_{\alpha/(1-\varepsilon)}(B) \subseteq \mathrm{Sym}_\alpha(A) \subseteq \mathrm{Sym}_{(\alpha-2\varepsilon)/(1-\varepsilon)}(B)$$

for every $\alpha \in \mathbf{R}$.

2.6.8 Let A be an additive set. Refine (2.21) slightly to

$$|\mathrm{Sym}_\alpha(A)| \leq 1 + \frac{|A|(|A|-1)}{\alpha} \text{ for all } \alpha > 0.$$

2.6.9 [350] Let A, B be additive sets in $\mathbf{Z}$, such that B consists entirely of positive numbers. Show that there exists $b \in B$ such that

$$|A \cap (A+b)| < \frac{|A|-1}{|B|}\frac{|A|}{2}.$$

(Hint: use Exercise 2.6.8, and exploit the fact that only half of the elements of $\mathrm{Sym}_\alpha(A)\backslash\{0\}$ are positive.)

2.6.10 [44] Let A be an additive set such that $|A+A| \leq K|A|$ for some $K \geq 1$. Let G be the group generated by $\mathrm{Sym}_{\frac{2}{9K}}(A)$. Show that there exists a coset $x+G$ of G such that $|A \cap (x+G)| \geq |A|/3$. (Hint: suppose for contradiction that $|A \cap (x+G)| < |A|/3$ for all x. Use the greedy algorithm to partition $A = A' \cup A''$ where $|A|/3 \leq |A'|, |A''| \leq 2|A|/3$ and such that $A' - A''$ is disjoint from G (and thus disjoint from $\mathrm{Sym}_{\frac{2}{9K}}(A)$). Use this to obtain an upper bound on $E(A', A'')$ and use (2.8) to obtain a contradiction.)

2.7 Non-commutative analogues

Many of the above arguments carry over to the non-commutative setting, though one of course now needs to take care with the ordering of multiplication. We sketch some of the main points here and leave the details as exercises. For further details see [362].

Definition 2.37 A *multiplicative group* is any group G (not necessarily abelian) with group operation $\cdot$, with inversion operation $x \mapsto x^{-1}$, and identity element 1. An *multiplicative set* is a pair (A, G), where G is a multiplicative group, and A is a finite non-empty subset of G. We often abbreviate a multiplicative set (A, G) simply as A, and refer to G as the *ambient group.*

If A and B are multiplicative sets with common ambient group G, we define their product set

$$A \cdot B := \{ab : a \in A, b \in B\}$$

and the inverse set

$$A^{-1} := \{a^{-1} : a \in A\}.$$

We also define right translates $A \cdot x$ and left translates $x \cdot A$ for $x \in G$ in the usual manner. Note that $x \cdot A \neq A \cdot x$ and $A \cdot B \neq B \cdot A$ in general, although we do have $|A| = |x \cdot A| = |A \cdot x| = |A^{-1}|$. We also define iterated product sets $A^{\cdot n} := A \cdot \ldots \cdot A$ for $n \geq 1$, with the conventions that $A^{\cdot 0} := \{1\}$ and $A^{\cdot -n} := (A^{\cdot n})^{-1} = (A^{-1})^{\cdot n}$.

We remark that $A \cdot B$ and $B \cdot A$ may have widely different cardinalities; for instance if H is a finite subgroup of G and x is an element of G that does not lie in the normalizer $N(H) := \{x \in G : xH = Hx\}$ of H, then $H \cdot (x \cdot H)$ and $(x \cdot H) \cdot H$ can have very different cardinalities. However, we still have the analogue of (2.1):

$$\max(|A|, |B|) \leq |A \cdot B|, |B \cdot A| \leq |A||B|;$$

see exercises.

We define the *(left-invariant) Ruzsa distance* $d(A, B)$ between two multiplicative sets:

$$d(A, B) := \log \frac{|A \cdot B^{-1}|}{|A|^{1/2}|B|^{1/2}}.$$

This distance still obeys the Ruzsa triangle inequality, mainly thanks to the identity $(ab^{-1})(bc^{-1}) = ac^{-1}$. It is left-invariant in each variable, thus $d(x \cdot A, B) = d(A, x \cdot B) = d(A, B)$, and is jointly right-invariant, $d(A \cdot x, B \cdot x) = d(A, B)$, but is not separately right-invariant in each variable. Also it is not reflection invariant; the metric $d^*(A, B) := d(A^{-1}, B^{-1})$ is the *right-invariant Ruzsa distance*, which we will not use here.

Define a *multiplicative K-approximate group* to be any multiplicative set H which is symmetric (so $H = H^{-1}$), contains the identity, and is such that there exists a set X of cardinality $|X| \leq K$ such that we have the inclusions

$$H \cdot H \subseteq X \cdot H \subseteq H \cdot X \cdot X; \qquad H \cdot H \subseteq H \cdot X \subseteq X \cdot X \cdot H.$$

We can characterize when $d(A, B)$ is zero:

Proposition 2.38 *Let A, B be multiplicative sets in an ambient group G. Then $d(A, B) = 0$ if and only if A and B are both left cosets of the same finite subgroup H, thus $A = x \cdot H$ and $B = y \cdot H$ for some $x, y \in G$.*

We leave the proof as an exercise. Observe that $d(A, B) = 0$ does not necessarily imply that A or B has small doubling; if x or y lie outside the normalizer of H then A^2 or B^2 can be significantly larger than A or B. Similarly we see that $d(A, B) = 0$ does not imply that $d(A, B^{-1}) = 0$. So there does not appear to be an analogue of Corollary 2.12. However, with some care and a few new arguments, we can still obtain the analogues of the results from Sections 2.4 and 2.5. Let us start by the analogue of Ruzsa's covering lemma, which can be proved by the same argument.

Lemma 2.39 *Let A, B be multiplicative sets in an ambient group G such that $|A \cdot B| \leq K|A|$. Then there exists a finite set X in B of cardinality at most K such that $B \subset A^{-1} \cdot A \cdot X$.*

From Section 2.4, we know that if A is a subset of a commutative group G and $|A + A| \leq K|A|$, then $|nA - mA| \leq O(K^{O(m+n)}|A|)$ for any n, m. This no longer holds in a non-commutative setting. Consider for instance $A := H \cup \{x\}$ where H is a subgroup of G and x lies outside the normalizer $N(H)$ of H. Then $A \cdot A = H \cup (x \cdot H) \cup (H \cdot x) \cup \{x^2\}$, so $|A \cdot A| \leq 3|A| - 2$; but $A \cdot A \cdot A$ contains $H \cdot x \cdot H$ which can be as large as $|H|^2 = (|A| - 1)^2$. Interestingly, it turns out that if we assume that $|A \cdot A \cdot A|$ is small, then the problem disappears and we can obtain the following analogue of Proposition 2.26.

Proposition 2.40 *Let A be a multiplicative set in a group G, and let $K \geq 1$. Then the following statements are equivalent up to constants, in the sense that if the* jth *property holds for some positive absolute constant C_j, then the* kth *property will also hold for some absolute constant C_k depending on C_j:*

(i) $|A \cdot A \cdot A| \leq K^{C_1}|A|$;
(ii) We have $|A^{\epsilon_1} \cdots A^{\epsilon_n}| \leq K^{C_2 n}|A|$ *for all* $n \geq 1$ *and all signs* $\epsilon_1, \ldots, \epsilon_n \in \{-1, 1\}$;
(iii) there exists a K^{C_3}-approximate group H containing A where $|H| \leq K^{C_3}|A|$.

Proof First we show that (i) implies (ii). Assuming (i), we have $|A \cdot A| \leq |A \cdot A \cdot A| \leq K^{C_1}|A|$. It follows that $d(A, A^{-1})$ (which equals $d(A^{-1}, A)$) and $d(A \cdot A, A^{-1})$ are $O(\log K)$. By the triangle inequality $d(A \cdot A, A) = O(\log K)$, which implies $|A \cdot A \cdot A^{-1}| \leq K^{O(1)}|A|$ and $d(A, A \cdot A^{-1}) = O(\log K)$. Again by the triangle inequality, we have $d(A \cdot A^{-1}, A^{-1}) = O(\log K)$, which implies $|A \cdot A^{-1} \cdot A| \leq K^{O(1)}|A|$. By a similar argument, we can show that $|A^{-1} \cdot A \cdot A| \leq K^{O(1)}|A|$. With these bounds (and taking inverse) we obtain the statment of (ii) for $n = 3$. From here, it is easy to finish the proof by induction on n, with $n = 3$ being the base case. (For $n = 2$, the statement in (ii) is trivial.)

Next, we prove that (ii) implies (iii). Set $H' = A \cup \{1\} \cup A^{-1}$ and $H = H' \cdot H' \cdot H'$. Clearly H is symmetric and contains A. By (ii), $|H| \leq K^{O(1)}|A|$. It thus remains to show that H is a $K^{O(1)}$- approximate group. Notice that $|H' \cdot H \cdot H| \leq K^{O(1)}|A|$. By the covering lemma, we have a set Y of cardinality $K^{O(1)}$ in $H \cdot H$ such that

$$H \cdot H \subset H'^{-1} \cdot H' \cdot Y.$$

Notice that the right-hand side is a subset of $H \cdot Y$. Now set $X = Y \cup Y^{-1}$. Since both H and X are symmetric $H \cdot H$ is contained in both $H \cdot X$ and $X \cdot H$.

Moreover, as $X \subset H \cdot H$,

$$H \cdot X \subset H \cdot H \cdot H \subset X \cdot H \cdot H \subset X \cdot X \cdot H$$

completing the proof.

The remaining implications are straightforward and left as an exercise. □

Now we are going to prove we can still obtain (iii) under the assumption that $d(A, B) = O(\log K)$ for some set B. We will need the following variant of Lemma 2.13, whose proof we leave as an exercise.

Lemma 2.41 *Let A be a multiplicative set. Then there exists a symmetric set $S \subset A^{-1} \cdot A$ such that $|S| \geq |A|/2$ and*

$$|A \cdot S^{\cdot n} \cdot A^{-1}| \leq \frac{2^n |A \cdot A^{-1}|^{n+1} |A^{-1} \cdot A|^n}{|A|^{2n}}$$

for all integers $n \geq 0$.

A similar argument (see [Proposition 4.5, 362]) gives

Proposition 2.42 *Let A be a multiplicative set such that $d(A, A) \leq \log K$ for some $K \geq 1$. Then there exists a symmetric set S such that $|S| \geq |A|/2K$ and*

$$|A \cdot S^{\cdot n} \cdot A^{-1}| \leq 2^n K^{(2n+1)} |A|$$

for all integers $n \geq 0$.

Proposition 2.43 *Let A, B be multiplicative sets in a group G, and let $K \geq 1$. Then the following statements are equivalent up to constants, in the sense that if the jth property holds for some absolute constant C_j, then the kth property will also hold for some absolute constant C_k depending on C_j:*

(i) $d(A, B) \leq C_1(1 + \log K)$;
(ii) there exists a $C_2 K^{C_2}$-approximate group H such that $|H| \leq C_2 K^{C_2} |A|$, $A \subset X \cdot H$ and $B \subset Y \cdot H$ for some multiplicative sets X, Y of cardinality at most $C_2 K^{C_2}$.

Proof We only need to prove that (i) implies (ii), as the reverse implication is trivial. Notice that (i) implies $d(A, A) = O(\log K)$. Thus, we have a symmetric set S of cardinality $K^{O(1)}|A|$ such that

$$|A \cdot S^{\cdot 3} \cdot A^{-1}| \leq K^{O(1)} |A|.$$

This implies that $|A \cdot S| \leq K^{O(1)}|A|$ and thus $d(A, S) = O(\log K)$. Furthermore, $|S^3| \leq K^{O(1)}|S|$ so we can find a $O(K^{O(1)})$-approximate group H of size $K^{O(1)}|A|$ containing S. This, in particular, implies that $d(S, H^{-1}) = O(\log K)$. By the triangle inequality, $d(A, H^{-1}) = O(\log K)$, which yields $|A \cdot H| \leq K^{O(1)}|A|$. By the covering lemma, there is a set Y of cardinality $K^{O(1)}$ such that $A \subset Y \cdot H \cdot H^{-1}$. But as H is an approximate group, $H^{-1} = H$ and $H \cdot H \subset Z \cdot H$ for some set Z of size $K^{O(1)}$. Thus, $A \subset (Y \cdot Z) \cdot H$, where $|Y \cdot Z| \leq |Y||Z| = K^{O(1)}$. The conclusion for B can be proved similarly. □

Let us now consider the non-commutative version of Balog–Szemerédi–Gowers theorem. Theorem 2.29 still holds when the ambient group Z is non-commutative. The proof of this theorem is purely graph-theoretical (see Section 6.4) and has little to do with the commutativity of the group.

Theorem 2.44 (Balog–Szemerédi–Gowers theorem, non-commutative version) *Let A, B be multiplicative sets in an ambient group Z, and let $G \subseteq A \times B$ be such that*

$$|G| \geq |A||B|/K \text{ and } |A \overset{G}{\cdot} B| \leq K'|A|^{1/2}|B|^{1/2}$$

for some $K \geq 1$ and $K' > 0$. Then there exists subsets $A' \subseteq A$, $B' \subseteq B$ such that

$$|A'| \geq \frac{|A|}{4\sqrt{2}K} \tag{2.37}$$

$$|B'| \geq \frac{|B|}{4K} \tag{2.38}$$

$$|A' \cdot B'| \leq 2^{12}K^4(K')^3|A|^{1/2}|B|^{1/2}. \tag{2.39}$$

In particular we have

$$d(A', B'^{-1}) \leq 5 \log K + 3 \log K' + O(1).$$

Define the *multiplicative energy* $E(A, B)$ between two multiplicative sets A, B with common ambient group to be

$$E(A, B) := |\{(a, a', b, b') \in A \times A \times B \times B : ab = a'b'\}|. \tag{2.40}$$

A significant difficulty here is that $E(A, B)$ obeys far fewer symmetries in the non-commutative case than in the commutative case; indeed, the only symmetry available is that $E(A, B) = E(B^{-1}, A^{-1})$. However in the case when $B = A^{-1}$ we have a crucial additional identity $E(A, A^{-1}) = E(A^{-1}, A)$ (see exercises), which can be thought of as a very weak, restricted form of commutativity.

The following variant of Lemma 2.30 holds, with basically the same proof.

Lemma 2.45 *Let A, B be multiplicative sets in an ambient group Z, and let G be a non-empty subset of $A \times B$. Then*

$$E(A, B) \geq \frac{|G|^2}{|A \overset{G}{\cdot} B|}.$$

Conversely, if $E(A, B) \geq |A|^{3/2}|B|^{3/2}/K$ for some $K \geq 1$, then there exists $G \subseteq A \times B$ such that

$$|G| \geq |A||B|/2K; \quad |A \overset{G}{\cdot} B| \leq 2K|A|^{1/2}|B|^{1/2}.$$

Finally, notice that by the triangle inequality

$$d(A', A') \leq d(A', B'^{-1}) + d(B'^{-1}, A') = 2d(A', B'^{-1}),$$

which means that if $d(A', B'^{-1})$ is small, then $d(A', A')$ is also small. From here, we can use the same arguments for the commutative case to deduce

Corollary 2.46 *Let A, B be multiplicative sets in an ambient group Z such that $E(A, B) \geq |A|^{3/2}|B|^{3/2}/K$ for some $K > 1$. Then there exists a subset $A' \subset A$ such that $|A'| = \Omega(K^{-O(1)}|A|)$ and $|A' \cdot (A')^{-1}| = O(K^{O(1)}|A|)$ for some absolute constant C.*

Combining this with the identity $E(A, A^{-1}) = E(A^{-1}, A)$ we obtain the following weak commutativity property between A and A^{-1}:

Corollary 2.47 *Let A be a multiplicative set such that $|A \cdot A| \leq K|A|$ for some $K \geq 1$. Then there exists a subset $A' \subset A$ such that $|A'| = \Omega(K^{-O(1)}|A|)$ and $|A' \cdot (A')^{-1}| = O(K^{O(1)}|A|)$.*

It is now not too hard to obtain the following theorem.

Theorem 2.48 *Let A, B be multiplicative sets in a group G, and let $K \geq 1$. Then the following statements are equivalent up to constants, in the sense that if the* jth *property holds for some absolute constant C_j, then the* kth *property will also hold for some absolute constant C_k depending on C_j:*

(i) $E(A, B) \geq C_1^{-1}K^{-C_1}|A|^{3/2}|B|^{3/2}$;
(ii) there exists a subset $G \subset A \times B$ with $|G| \geq C_2^{-1}K^{-C_2}|A||B|$ such that $|A \overset{G}{\cdot} B| \leq C_2K^{C_2}|A|^{1/2}|B|^{1/2}$;
(iii) there exists a $C_3K^{C_3}$-approximate group H and $x, y \in G$ such that $|H| \leq C_3K^{C_3}|A|^{1/2}|B|^{1/2}$ *and*

$$|A \cap (x \cdot H)|, |B \cap (H \cdot y)| \geq C_3^{-1}K^{-C_3}|H|.$$

We leave the proofs of these statements to the exercises. Despite these characterizations, there is much left to be done in the study of product sets in non-commutative groups. For instance we do not currently have a satisfactory version of Freiman's theorem in general. However there has been some progress in the case of very small doubling [172] and also in certain special groups such as $SL_2(\mathbf{Z})$ or free groups; see for instance [78], [182].

Exercises

2.7.1 Prove a multiplicative version of Lemma 2.1.

2.7.2 Prove a multiplicative version of Lemma 2.6.

2.7.3 Prove Proposition 2.38.

2.7.4 Let (A, G) be a multiplicative set. Prove that $|A \cdot A| = |A|$ if and only if A is a normal coset of H, i.e. $A = x \cdot H = H \cdot x$ for some $x \in N(H)$.

2.7.5 Let A be a symmetric multiplicative set, so $A = A^{-1}$, and let $\sigma_n[A]$ denote the n-fold doubling numbers $|A^n|/|A|$. Using the Ruzsa triangle inequality, show that $\sigma_{m+n-2}[A] \leq \sigma_m[A]\sigma_n[A]$ for all $m, n \geq 2$.

2.7.6 Let A and B be multiplicative sets. Establish the identities $E(A, B) = E(B^{-1}, A^{-1})$ and $E(A, A^{-1}) = E(A^{-1}, A)$, and the inequality $E(A, B) \geq \frac{|A|^2|B|^2}{|A \cdot B|}$.

2.7.7 Let A, B, C be additive sets in an ambient group Z, let $0 < \varepsilon < 1/4$, and let $G \subset A \times B^{-1}$, $H \subset B \times C^{-1}$ be such that $|G| \geq (1 - \varepsilon)|A||B|$ and $|H| \geq (1 - \varepsilon)|B||C|$. By modifying the solution of Exercise 2.5.4, show that there exists subsets $A' \subseteq A$ and $C' \subseteq C$ with $|A'| \geq (1 - \varepsilon^{1/2})|A|$ and $|C'| \geq (1 - \varepsilon^{1/2})|C|$ such that $|A' \cdot (C')^{-1}| \leq \frac{|A \stackrel{G}{\cdot} B^{-1}||B \stackrel{H}{\cdot} C^{-1}|}{(1-2\varepsilon^{1/2})|B|}$.

2.7.8 Let A be a multiplicative set such that $|A \cdot A^{-1}| \leq K|A|$ and $|A^{-1} \cdot A| \leq K|A|$. Show that there exists a subset $\tilde{A}$ of A such that $|\tilde{A}| \geq |A|/2K$ and

$$|\tilde{A} \cdot \tilde{A}^{-1} \cdot \ldots \cdot \tilde{A}^{(-1)^{n+1}}| \leq 2^{n-2}K^{2n-3}|A|$$

for all $n \geq 2$, where the product consists of n factors alternating between $\tilde{A}$ and $\tilde{A}^{-1}$.

2.7.9 If A and B are multiplicative sets in a group G, show that there exist sets $X_1, X_2 \subseteq A$ such that $|X_1| \leq \frac{|A \cdot B|}{|B|}$, $|X_2| \leq \frac{|B \cdot A|}{|B|}$, and $A \subseteq X_1 \cdot B \cdot B^{-1}$ and $A \subseteq B^{-1} \cdot B \cdot X_2$, by modifying the proof of Lemma 2.14.

2.7.10 Prove Lemma 2.41.

2.7.11 Show that the direct analogue of Proposition 2.18 fails in the non-commutative case, even when $A = B = A^{-1}$.

2.7.12 Let A, B be multiplicative sets in an ambient group G, and let $\tilde{A}$ be the set

$$\tilde{A} := \left\{ a \in A : |\{(a', b, b') \in A \times B \times B : a = a'b'b^{-1}\}| \geq \frac{|A||B|^2}{2|A \cdot B|} \right\}.$$

Establish the bounds

$$|\tilde{A}| \geq \frac{|A|^2}{2|A \cdot B|}$$

and

$$|A \cdot \tilde{A}^{-1} \cdot \tilde{A} \cdot A^{-1}| \le 4\frac{|A \cdot B|^4 |A^{-1} \cdot A|}{|A|^4}.$$

Compare this against Exercise 2.7.11. Hint: if $x := a_1 a_2^{-1} a_3 a_4^{-1}$ be a typical element of $A \cdot \tilde{A}^{-1} \cdot \tilde{A} \cdot A^{-1}$, obtain at least $(\frac{|A||B|^2}{2|A \cdot B|})^2$ representations of the form

$$x = [a_1 b_2](b_2')^{-1}[(a_2')^{-1} a_3'] b_3' [a_4 b_2]^{-1}$$

where $a_1 b_2, a_4 b_2 \in A \cdot B$, $b_2', b_3' \in B$, and $(a_2')^{-1} a_3' \in A^{-1} \cdot A$.

2.7.13 Prove Theorem 2.48.

2.8 Elementary sum-product estimates

We now discuss some results concerning the sum set and product set of a subset A of a commutative ring Z, thus combining both the additive and multiplicative theory of the preceding sections (but keeping the multiplication commutative, for simplicity). The question here is to analyze the extent to which a set A can be approximately closed under addition and multiplication simultaneously. Of course, one way that this can happen is if A is a subring of Z; it appears that up to trivial changes (such as removing some elements, adding a small number of new elements, or dilating the set), this is essentially the only such example, although we currently only have a satisfactory and complete formalization of this principle when Z is a field (Theorem 2.55). In some ways the theory here is in fact easier than the sum set theory, because one can exploit two rather different structures arising from the smallness of $A + A$ and the smallness of $A \cdot A$ to obtain a conclusion. As in the rest of this chapter, our discussion is for general fields, with a particular emphasis on the finite field $\mathbf{Z}_p$. We remark that for the field $\mathbf{R}$ much better results are known, see Sections 8.3, 8.5.

In this section Z will always denote a commutative ring, and Z^* will denote the elements of Z which are not zero-divisors; these form a multiplicative cancellative commutative monoid in Z. The situation is significantly better understood in the case that Z is a field (see in particular Theorem 2.55 below); in such cases we shall emphasize this by writing the field as F instead of Z, and $F^\times$ instead of $F^* = F \backslash \{0\}$ to emphasize that $F^\times$ is now a multiplicative group. A fundamental concept in the field setting is that of a *quotient set*, which is the arithmetic equivalent of the concept of a quotient field of a division ring.

Definition 2.49 (Quotient set) Let A be a finite subset of a field F such that $|A| \geq 2$. Then the *quotient set* $Q[A]$ of A is defined to be

$$Q[A] := \frac{A - A}{(A - A)\backslash 0} := \left\{ \frac{a-b}{c-d} : a, b, c, d \in A; c \neq d \right\}.$$

We also set $Q[A]^\times := Q[A]\backslash 0$ to be the invertible elements in $Q[A]$.

Observe that $Q[A]$ contains both 0 and 1, and is symmetric under both additive and multiplicative inversion, thus $Q[A] = -Q[A]$ and $Q[A]^\times = (Q[A]^\times)^{-1}$. It is also invariant under translations and dilations of A, thus $Q[A] = Q[A + x] = Q[\lambda \cdot A]$ for all $x \in F$ and $\lambda \in F^\times$. Geometrically, $Q[A]$ can be viewed as the set of slopes of lines connecting points in $A \times A$.

The relevance of the quotient set to sum-product estimates lies in the trivial but fundamental observation:

Lemma 2.50 *Let A be a finite subset of a field F such that $|A| \geq 2$, and let $x \in F$. Then $|A + x \cdot A| = |A|^2$ if and only if $x \notin Q[A]$.*

Proof We have $|A + x \cdot A| = |A|^2$ if and only if the map $(a, b) \mapsto a + xb$ is injective on $A \times A$, which is true if and only if $a + xb \neq c + xd$ for all distinct $(a, b), (c, d) \in A \times A$, which after some algebra is equivalent to asserting that $x \notin Q[A]$. □

This has an immediate corollary:

Corollary 2.51 *If A is a subset of a finite field F such that $|A| > |F|^{1/2}$, then $Q[A] = F$.*

Note that the condition $|A| > |F|^{1/2}$ is absolutely sharp, as can be seen by considering the case when A is a subfield of F of index 2.

Lemma 2.50 has another important consequence: it gives a criterion under which $Q[A]$ is a subfield of F.

Corollary 2.52 *Let A be a finite subfield of a field F such that $|A| \geq 2$ and*

$$|A + Q[A] \cdot Q[A] \cdot A|, |A + (Q[A] + Q[A]) \cdot A| < |A|^2.$$

Then $Q[A]$ is a subfield of F.

This corollary may be compared with Exercise 2.6.5.

Proof From Lemma 2.50 and the hypotheses we see that $Q[A] \cdot Q[A] \subseteq Q[A]$ and $Q[A] + Q[A] \subseteq Q[A]$. In particular $Q[A]^\times \cdot Q[A]^\times = Q[A]^\times$. Since $Q[A]$ is finite and contains 0, 1, we see from Proposition 2.7 that $Q[A]$ is an additive group, and similarly from the multiplicative version of this Proposition we see that $Q[A]^\times$ is a multiplicative group. The claim follows. □

In order to use this corollary, one needs to control rational expressions of A such as $A + Q[A] \cdot Q[A] \cdot A$. In analogy with sum set estimates such as Corollary 2.23, one might first expect that once $|A + A| \leq K|A|$ and $|A \cdot A| \leq K|A|$, then all polynomial or rational expressions of A are controlled in cardinality by $CK^C|A|$. This however is not the case, even if one normalizes A to contain 0 and 1. To see this, consider $A = G \cup \{x\}$ where G is a subfield of F and $x \notin G$. Then one easily verifies $|A + A|, |A \cdot A| < 2|A|$ but $|A \cdot A + A \cdot A| \geq (|A| - 1)^2$, since $A \cdot A + A \cdot A$ contains $G + x \cdot G$, which has size $|G|^2$ by Lemma 2.50. This example is similar to one appearing in the preceding section, and it is resolved in a similar way, namely by passing from A to a subset of A.

Lemma 2.53 (Katz–Tao lemma) *[199], [41] Let Z be a commutative ring, and let $A \subseteq Z^*$ be a finite non-empty subset such that $|A + A| \leq K|A|$ and $|A \cdot A| \leq K|A|$ for some $K \geq 1$. Then there exists a subset A' of A such that $|A'| \geq |A|/2K - 1$ and $|A' \cdot A' - A' \cdot A'| = O(K^{O(1)}|A'|)$.*

Note that this lemma works in arbitrary commutative rings, not just in fields. The requirement that none of the elements of A be zero-divisors is not serious in the case of a field, since one can simply remove the origin 0 from A if necessary, but is a non-trivial requirement in other commutative rings.

Proof We use an argument from [41]. We may assume that $A > 10K$ (for instance) since the claim is trivial otherwise. Consider the dilates $\{a \cdot A : a \in A\}$ of A. Since $a \in Z^*$, $a \cdot A$ has the same cardinality as A. In particular we have

$$\sum_{x \in A \cdot A} \sum_{a \in A} 1_{a \cdot A}(x) = |A|^2.$$

Since $|A \cdot A| \leq K|A|$, we may apply Cauchy–Schwarz and conclude

$$\sum_{x \in A \cdot A} \left(\sum_{a \in A} 1_{a \cdot A}(x) \right)^2 \geq |A|^3/K.$$

We rearrange this as

$$\sum_{a,b \in A} |(a \cdot A) \cap (b \cdot A)| \geq |A|^3/K.$$

By the pigeonhole principle we can thus find a $b \in A$ such that

$$\sum_{a \in A} |(a \cdot A) \cap (b \cdot A)| \geq |A|^2/K.$$

Fix this b. Setting A' to be the set of all $a \in A$ such that

$$|(a \cdot A) \cap (b \cdot A)| \geq |A|/2K$$

we conclude that

$$\sum_{a\in A'} |(a\cdot A)\cap(b\cdot A)| \geq |A|^2/2K$$

and hence $|A'| \geq |A|/2K$. By shrinking A' by one if necessary we may assume $b \notin A'$. Now recall the Ruzsa distance $d(A, B) := \log \frac{|A-B|}{|A|^{1/2}|B|^{1/2}}$, and observe that $d(a\cdot A, a\cdot B) = d(A, B)$ whenever a is not a zero-divisor. Then $d(A, A) \leq 2d(A, -A) = 2\log K$, and hence

$$d(a\cdot A, a\cdot A) = d(b\cdot A, b\cdot A) = d(A, A) \leq 2\log K \text{ for all } a \in A'.$$

Since $(a\cdot A)\cap(b\cdot A)$ is a large subset of $a\cdot A$ and $b\cdot A$, one can compute

$$d(a\cdot A, a\cdot A\cap b\cdot A), d(b\cdot A, a\cdot A\cap b\cdot A) = O(1+\log K)$$

and hence by the Ruzsa triangle inequality

$$d(a\cdot A, b\cdot A) = O(1+\log K) \text{ for all } a \in A'. \tag{2.41}$$

Dilating this, we obtain

$$d(a_1a_2\cdot A, ba_2\cdot A), d(ba_2\cdot A, b^2\cdot A) = O(1+\log K) \text{ for all } a_1, a_2 \in A'$$

and hence by the Ruzsa triangle inequality

$$d(a_1a_2\cdot A, b^2\cdot A) = O(1+\log K) \text{ for all } a_1, a_2 \in A'. \tag{2.42}$$

To proceed further we need to "invert" elements in A. For any $a \in A$ let $\hat{a} := \prod_{a'\in A\setminus\{a\}} a' \in Z^*$. By dilating (2.41) (with a replaced by a_3) by $a_1a_2\prod_{a'\in A\setminus\{a_3,b\}} a'$ for $a_1, a_2, a_3 \in A'$, we obtain

$$d(a_1a_2\hat{b}\cdot A, a_1a_2\hat{a}_3\cdot A) = O(1+\log K) \text{ for all } a_1, a_2, a_3 \in A'.$$

Meanwhile, from dilating (2.42) we have

$$d(a_1a_2\hat{b}\cdot A, b^2\hat{b}\cdot A) = O(1+\log K) \text{ for all } a_1, a_2, a_3 \in A'.$$

Applying the Ruzsa triangle inequality, we thus have

$$d(a_1a_2\hat{a}_3\cdot A, a_1'a_2'\hat{a}_3'\cdot A) = O(1+\log K) \text{ for all } a_1, a_2, a_3, a_1', a_2', a_3' \in A'$$

and hence

$$|a_1a_2\hat{a}_3\cdot A - a_1'a_2'\hat{a}_3'\cdot A| = O(K^{O(1)})|A|.$$

Therefore we have

$$\sum_{x,y\in A'\cdot A'\cdot \hat{A}'} |x\cdot A - y\cdot A| = O(K^{O(1)})|A||A'\cdot A'\cdot \hat{A}'|^2,$$

where $\hat{A}' := \{\hat{a} : a \in A'\}$. But since $|A \cdot A| \leq K|A|$ and $|A'| \geq |A|/2K - 1$, we see from the multiplicative version of sum set estimates (working in the formal multiplicative group generated by the cancellative commutative monoid Z^*) that $|A' \cdot A' \cdot \hat{A}'| = O(K^{O(1)}|A|)$. We thus have

$$\sum_{x,y \in A' \cdot A' \cdot \hat{A}'} |x \cdot A - y \cdot A| \leq O(K^{O(1)}|A'|^3).$$

We rewrite the left-hand side as

$$\sum_{z \in Z} |\{(x, y) : \exists\, a, b \in A' \text{ such that } z = xa - yb\}|.$$

Write $\omega := \prod_{a \in A} a$, and observe that whenever $a_1, a_2, a_3, a_4 \in A'$, the number $\omega(a_1a_2 - a_3a_4)$ has at least $|A'|^2$ representations of the form $xa - yb$ with $x, y \in A' \cdot A' \cdot \hat{A}'$ and $a, b \in A'$, with (x, y) distinct, thanks to the identity

$$\omega(a_1a_2 - a_3a_4) = (a_1a_2\hat{a})a - (a_3a_4\hat{b})b.$$

Thus

$$|\omega \cdot (A' \cdot A' - A' \cdot A')| = O(K^{O(1)}|A'|)$$

and the claim follows since $\omega \in Z^*$. □

A modification of the above argument also gives the following statement, which can be viewed as a variant of Corollary 2.23 for the sum-product setting; we leave the proof to Exercise 2.8.1.

Lemma 2.54 *[43] Let Z be a commutative ring, and let $A \subseteq Z^*$ be a finite non-empty set such that $|A \cdot A - A \cdot A| \leq K|A|$. Then we have $|A^k - A^k| \leq K^{O(k)}|A|$ for all $k \geq 1$, where $A^k = A \cdot \ldots \cdot A$ is the k-fold product set of A.*

We can now classify those finite subsets of fields with small additive doubling and multiplicative doubling constant, up to polynomial losses:

Theorem 2.55 (Freiman theorem for sum-products) *Let A be a finite non-empty subset of a field F, and let $K \geq 1$. Then the following statements are equivalent up to constants, in the sense that if the* jth *property holds for some absolute constant C_j, then the* kth *property will also hold for some absolute constant C_k depending on C_j:*

(i) $|A + A| \leq C_1K^{C_1}|A|$ and $|A \cdot A| \leq C_1K^{C_1}|A|$;
(ii) either $|A| \leq C_2K^{C_2}$, or else there exists a subfield G of F, a non-zero element $x \in F$, and a set X in F such that $|G| \leq C_2K^{C_2}|A|$, $|X| \leq C_2K^{C_2}$, and $A \subseteq x \cdot G \cup X$.

This is a slight strengthening of a result in [43], [44].

Proof We shall only show the forward implication, leaving the easy backward implication to Exercise 2.8.2. By relabeling $C_1 K^{C_1}$ as K, we may thus assume that $|A+A| \leq K|A|$ and $|A \cdot A| \leq K|A|$. We may assume that $|A| \geq C_0 K^{C_0}$ for some large absolute constant C_0, since the claim is trivial otherwise. We may also remove 0 from A without any difficulty, thus we may assume $A \subseteq F^*$. Applying Lemma 2.53 and Lemma 2.54, we may find a subset A' of A with $|A'| = \Omega(K^{-O(1)}|A|)$ and $|(A')^k - (A')^k| = O(K)^{O(k)}|A'|$ for all $k \geq 1$. By Corollary 2.23 this implies that

$$|n(A')^k - m(A')^k| \leq O(K)^{O_{k,n,m}(1)}|A'| \text{ for all } n, k, m \geq 1. \tag{2.43}$$

Dilating A with a non-zero factor if necessary, we may assume $1 \in A'$ (noting that the hypothesis and conclusion of the theorem are invariant under such dilations). We may now add 0 back to A' and A without affecting (2.43).

Now we apply Corollary 2.52. Let $D := (A' - A')\backslash\{0\}$ and $G := Q[A'] = (A' - A')/D$. Using lowest common denominators, we observe that

$$A' + G \cdot G \cdot A' \subseteq \frac{(A' \cdot D \cdot D - (A' - A') \cdot (A' - A') \cdot A')}{D^2} \subseteq \frac{(4(A')^3 - 4(A')^3)}{D^2}.$$

on the other hand, from (2.43) we have

$$|(4(A')^3 - 4(A')^3) \cdot D^2| = O(K^{O(1)}|A'|),$$

so by the multiplicative version of Corollary 2.12 we see that

$$|A' + G \cdot G \cdot A'| = O(K^{O(1)}|A'|) < |A'|^2$$

if C_0 is sufficiently large. A similar argument gives $|A' + (G + G) \cdot A'| = O(K^{O(1)}|A'|) < |A'|^2$. Applying Corollary 2.52 we see that G is in fact a field.

Now let x be a non-zero element of A', and let y be an element of A'. Then $(a - y)/x \in Q[A'] = G$ for all $a \in A'$, thus

$$A' \subseteq x \cdot G + y.$$

Thus

$$x \cdot G + y = A' + x \cdot G \subseteq A' + A' \cdot Q[A']$$

and hence

$$(x \cdot G + y)^2 \subseteq (A' + A' \cdot Q[A'])^2.$$

But an argument using (2.43) and Corollary 2.12 as before gives $|(A' + A' \cdot Q[A'])^2| = O(K^{O(1)}|A'|) \leq O(K^{O(1)}|G|)$. Direct computation shows that $|(x \cdot G + y)^2| \geq |G|^2$ unless $y \in x \cdot G$. Thus (if C_0 is sufficiently large) we can take $y \in x \cdot G$. Because A' contains 1, we thus have $A' \subseteq G$.

Since $|A + A'| \leq K|A| = O(K^{O(1)}|A'|)$, we may apply Ruzsa's covering lemma (Lemma 2.14) and cover A by $O(K^{O(1)})$ translates of $A' - A'$, and hence by $O(K^{O(1)})$ translates of G. A similar argument using the multiplicative version of this lemma (and temporarily removing the non-invertible 0 element from A if necessary) covers A by $O(K^C)$ dilates of G. On the other hand, we have $|(G \cdot x) \cap (G + y)| \leq 1$ whenever $x \neq 1$. Thus we have $|A \backslash G| = O(K^{O(1)})$, and the claim follows. □

This theorem implies that at least one of $A + A$ or $A \cdot A$ is large if A does not intersect with a subfield of F:

Corollary 2.56 (Sum-product estimate) *[43],[44] Let A be a finite non-empty subset of a field F, and suppose that $K \geq 1$ is such that there is no finite subfield G of F of cardinality $|G| \leq K|A|$ and no $x \in F$ such that $|A \backslash (x \cdot G)| \leq K$. Then we have either $|A| = O(K^{O(1)})$ or $|A + A| + |A \cdot A| = \Omega(K^c|A|)$ for some absolute constant $c > 0$.*

Remark 2.57 In the particular case when F has no finite subfields we thus obtain $|A + A| + |A \cdot A| = \Omega(|A|^{1+\varepsilon})$ for some absolute constant $\varepsilon > 0$; this result was first obtained (when $F = \mathbf{R}$) by Erdős and Szemerédi [91]. In the setting of the real line it is was in fact conjectured in [91] that one can take ε arbitrarily close to 1 in the above estimate. For the most recent value of ε, see Theorem 8.15.

In the particular case of the field $F = F_p$ of prime order, which has no subfields other than $\{1\}$ and F_p, one obtains

Corollary 2.58 (Sum-product estimate for F_p) *[43],[44] Let A be a non-empty subset of F_p. Then*

$$|A + A| + |A \cdot A| = \Omega(\min(|A|, |F_p|/|A|)^c|A|)$$

for some absolute constant $c > 0$.

If H is any non-empty subset of F_p, then we have $kH^k + kH^k,\ kH^k \cdot kH^k \subset k^2H^{k^2}$ for all $k \geq 2$. Thus we have

$$|k^2H^{k^2}| = \Omega(\min(|kH^k|, p/|kH^k|)^c|kH^k|)$$

for some absolute constant $c > 0$. We can iterate this estimate (starting with $k = 2$ and squaring repeatedly) to establish

Corollary 2.59 *Let H be any non-empty subset of F_p, and let $A, \delta > 0$. Then there exists an integer $k = k(A, \delta) \geq 1$ such that*

$$|kH^k| = \Omega_{A,\delta}(\min(|H|^A, p^{1-\delta})).$$

We leave the proof of this corollary as an exercise. By using Lemma 4.10 from Chapter 4 one can in fact set $\delta = 0$ here, though we will not need this fact here.

In the special case when H is a multiplicative subgroup of F_p, we have $H^k = H$, and hence Corollary 2.59 gives

$$|kH| = \Omega_{A,\delta}(\min(|H|^A, p^{1-\delta})).$$

Thus multiplicative subgroups have rather rapid additive expansion. It turns out that one can do something similar for approximate groups:

Theorem 2.60 *[40] Let H be a non-empty subset of F_p such that $|H^2| \leq K|H|$, and let $A, \delta > 0$. Then there exists an integer $k = k(A, \delta) \geq 1$ such that*

$$|kH| = \Omega_{A,\delta}(K^{-O_{A,\delta}(1)} \min(|H|^A, p^{1-\delta})).$$

This result can be deduced from Corollary 2.59 and the following proposition; we leave the precise deduction as an exercise.

Proposition 2.61 *Let F be an arbitrary field, and let $H \subset F^\times$ be a finite non-empty subset of invertible field elements such that $|H^2| \leq K|H|$ for some $K \geq 1$. Let $k \geq 1$ and $L \geq 1$ be such that kH obeys the following "additive non-expansion" property: we have $|2kH| \leq L|kH''|$ for any subset H'' of H of cardinality $|H''| \geq \frac{1}{2K}|H|$. Then there exists a subset H' of H of cardinality $|H'| \geq \frac{1}{2K}|H|$ such that*

$$|j(H')^j| = O_j((1 + \log|H|)^{j^2} K^{O(j^2)} L^{O(j^2)} |kH|)$$

for all $j \geq 1$.

Proof From the multiplicative version of Exercise 2.3.24 we can find $H' \subset H$ with $|H'| \geq \frac{1}{2K}|H|$ and $h_0 \in H'$ such that $|(h \cdot H) \cap (h_0 \cdot H)| \geq \frac{1}{2K}|H|$ for all $h \in H'$. By dilation we may normalize $h_0 = 1$. From the additive non-expansion property we conclude that

$$|2kH| \leq L|k((h \cdot H) \cap H)| \leq L|A_h| \text{ for all } h \in H',$$

where $A_h := k(h \cdot H) \cap kH$. Since

$$|kH + A_h| \leq |2kH|; \quad |k(h \cdot H) + A_h| \leq |2k(h \cdot H)| = |2kH|$$

we thus obtain the Ruzsa distance estimates

$$d(kH, -A_h), d(k(h \cdot H), -A_h) \leq \log L$$

and hence by the triangle inequality

$$d(kH, k(h \cdot H)) \leq 2\log L. \tag{2.44}$$

Now we turn to controlling $j(H')^j$ for some j. We first observe that

$$|(H')^2| \leq |H^2| \leq K|H| \leq 2K^2|H'|$$

and thus by the multiplicative analog of Exercise 2.3.10 we have

$$|(H')^2 \cdot (H')^{-1}| = O\big(K^{O(1)}|H'|\big).$$

We can then apply the multiplicative version of Exercise 1.1.8 to obtain a set $X \subset (H')^2 \cdot (H')^{-1}$ of cardinality $|X| = O(K^{O(1)}(1 + \log|H|))$ such that $(H')^2 \subset X \cdot H'$, and thus $(H')^j \subset X^{j-1} \cdot H'$. Thus by the pigeonhole principle we can bound

$$|j(H')^j| \leq |j(X^{j-1}H')| \leq |X|^{j(j-1)}|x_1 \cdot H' + \cdots + x_j \cdot H'|$$

for some $x_1, \ldots, x_j \in X^{j-1}$; it thus suffices to show that

$$|x_1 \cdot H' + \cdots + x_j \cdot H'| = O_j\big(L^{O(j^2)}|kH|\big).$$

Since xH' is contained in a translate of $k(xH')$, we have the somewhat crude estimate

$$|x_1 \cdot H' + \cdots + x_j \cdot H'| \leq |jB|$$

where $B := k(x_1 \cdot H) \cup \cdots \cup k(x_j \cdot H)$. But the x_i are all products of $O(j)$ elements from H' and $(H')^{-1}$. From repeated application of (2.44) and the triangle inequality we conclude that

$$d(k(x_i \cdot H), k(x_{i'} \cdot H)) \leq O(j \log L) \text{ for all } 1 \leq i, i' \leq j$$

and hence

$$d(B, B) \leq O(j \log L) + O(\log j).$$

From Exercise 2.3.10 we conclude that $|jB| = O_j(L^{O(j^2)}|B|)$, and the claim follows. □

By combining Theorem 2.60 with the asymmetric Balog–Szemerédi–Gowers theorem, we can show that multiplicative subgroups of F_p cannot have high additive energy:

Corollary 2.62 *Let H be a multiplicative subgroup of F_p such that $|H| \geq p^\delta$ for some $0 < \delta \leq 1$. Then there exists an $\varepsilon = \varepsilon(\delta) > 0$, depending only on δ, such that $E(A, H) \leq p^{-\varepsilon}|A||H|^2$ for all $A \subseteq F_p$ with $1 \leq |A| \leq p^{1-\delta}$, if p is sufficiently large and depending on δ.*

Proof Let $\varepsilon' = \varepsilon'(\delta) > 0$ be a small number to be chosen later, and let $\varepsilon = \varepsilon(\varepsilon', \delta') > 0$ be an even smaller number to be chosen later. Suppose for contradiction that there existed a set A such that $E(A, H) \geq p^{-\varepsilon}|A||H|^2$. Applying Corollary 2.36 (with $L := p$ and ε replaced by ε') we can find (if ε is sufficiently small and depending on ε') a subset H' of H with cardinality

$$|H'| = \Omega_{\varepsilon'}\big(p^{-\varepsilon'/2}|H|\big)$$

such that

$$|kH'| \leq |A + kH'| = O_{\varepsilon',k}\big(p^{k\varepsilon'/2}|A|\big)$$

for all k. Since H is a multiplicative subgroup, we see that

$$|H' \cdot H'| \leq |H^2| = |H| = O_{\varepsilon'}\big(p^{\varepsilon'/2}|H'|\big).$$

Since $|H| \geq p^{\delta}$, we also see (if ε' is sufficiently small depending on δ) that $|H|^A \geq p^{1-\delta/2}$ for some A depending only on δ. We can thus apply Corollary 2.60 (with δ replaced by $\delta/2$) and conclude that for a sufficiently large k depending on δ we have

$$|kH'| = \Omega_{\varepsilon',\delta}\big(p^{1-\delta/2-O_\delta(\varepsilon')}\big).$$

This gives a contradiction if ε' is sufficiently small and depending on δ, and p is sufficiently large. □

We shall apply this to exponential sums over multiplicative subgroups; see Theorem 4.41. For a variant of this estimate, see Lemma 9.44.

It seems of interest to obtain estimates of this type for more general commutative rings, and possibly even to non-commutative rings by combining these arguments with those in the preceding section. In this direction, Bourgain has established

Theorem 2.63 *[41] Let p be a large prime, and let A be a subset of the commutative ring $F_p \times F_p$ (endowed with the product structure $(a, b) \cdot (c, d) = (ac, bd)$) be such that $|A| \geq p^{\delta}$ and $|A + A|$, $|A \cdot A| \leq p^{\varepsilon}|A|$ for some $\delta, \varepsilon > 0$. Then there exists a set G of $F_p \times F_p$ such that $|G| \leq p^{O_\delta(\varepsilon)}|A|$ and $|A \cap G| \geq p^{-O_\delta(\varepsilon)}|A|$, where G is one of the following objects:*

- *the whole space $G = F_p \times F_p$;*
- *a horizontal line $G = F_p \times \{a\}$ for some $a \in F_p$;*
- *a vertical line $G = \{a\} \times F_p$ for some $a \in F_p$;*
- *a line $G = \{(x, ax) : x \in F_p\}$ for some $a \in F_p^{\times}$.*

We sketch a proof of this proposition in the exercises. This is not as complete a characterization of sets with small sum-product as Theorem 2.55 – in particular, it does not address the case of very small A – but is already sufficient to control a number of exponential sums of importance in number theory and cryptography. See [41], [40].

The problem of obtaining good sum-product estimates when the ambient commutative ring is the integers $\mathbf{Z} = Z$ has attracted a lot of interest. In this case it has been conjectured by Erdős and Szemerédi [91] that

$$|kA| + |A^k| = \Omega_{k,\varepsilon}(|A|^{k-\varepsilon}) \tag{2.45}$$

for all $\varepsilon > 0$, all $k \geq 2$ and all additive sets $A \subset \mathbf{Z}$. Even the $k = 2$ case is open (and considered very difficult); this $k = 2$ case has currently been verified for all $\varepsilon > \frac{8}{11}$, see Theorem 8.15. In another direction towards (2.45), a recent result of Bourgain and Chang [42] has shown that for every $m > 1$ there exists an integer $k = k(m) \geq 1$ such that

$$|kA| + |A^k| = \Omega_m(|A|^m) \tag{2.46}$$

for all additive sets $A \subset \mathbf{Z}$. This last result is rather deep, in particular using an intricate "induction on scales" argument, coupled with some quantitative Freiman-type theorems.

Exercises

2.8.1 [41] Modify the proof of Lemma 2.53 to prove Lemma 2.54. (Hint: first use multiple applications of the triangle inequality to obtain control on $|x \cdot A - y \cdot A|$ for all $x, y \in A^k \cdot \hat{A}$.)

2.8.2 Prove the remaining implication in Theorem 2.55.

2.8.3 Deduce Corollary 2.56 and Corollary 2.58 from Theorem 2.55.

2.8.4 [44], [43] Let A, A', B be non-empty subsets of a field F such that $0 \notin B$. Using the first moment method, show that there exists $\xi \in B$ such that

$$E(A, \xi \cdot A') \leq \frac{|A|^2|A'|^2}{|B|} + |A||A'|$$

and conclude from (2.8) that

$$|A + \xi \cdot A'| \geq \frac{|A||A'||B|}{|A||A'| + |B|}.$$

2.8.5 [44] Let A be a subset of a finite field F such that $|A| > |F|^{1/2}$. Show that $|(A - A) \cdot A + (A - A) \cdot A| \geq \sup_{x \in F} |A + x \cdot A| \geq \frac{|F|}{2}$ and then conclude that

$$F = (A - A) \cdot A + (A - A) \cdot A + (A - A) \cdot A + (A - A) \cdot A.$$

(Hints: the first inequality follows easily from Corollary 2.51. For the second inequality, use Exercise 2.8.4.)

2.8.6 (Croot, personal communication) Let A be a subset of a finite field F such that $|A| > |F|^{1/k}$ for some integer $k \geq 2$. Show that $|Q[A]| \geq |F|^{1/(k-1)}$; this clearly generalizes Corollary 2.51. (Hint: exploit the fact that the maps $(a_1, \ldots, a_k) \mapsto x_1a_1 + \cdots + x_ka_k$ fail to be injective for *arbitrary* $x_1, \ldots, x_k \in F$.)

2.8.7 [43] Let A be a subset of a field F such that $|A| \geq |F|^{\varepsilon}$ for some $\varepsilon > 0$. Show that there exists an integer $k = k(\varepsilon) > 1$ depending only on ε such that $k(A^k) - k(A^k) = G$ for some subfield G of F. (Use Exercise 2.8.5 or Lemma 4.10.)

2.8.8 [41] Let F_p be a field of prime order p and $Z = F_p \times F_p$. Let $A \subseteq Z$ be such that $|A \cap (\{a\} \times F_p)| \geq p^{\delta}$ and $|A \cap (\{b\} \times F_p)| \geq p^{\delta}$ for some $0 < \delta < 1$ and $a, b \in F_p$. Show that for some $k = k(\delta) > 0$ we have $k(A^k) - k(A^k) = Z$. (Hint: use Exercise 2.8.7.)

2.8.9 [41] Let F_p, Z, be as in Exercise 2.8.8, and let $\pi_1 : Z \to F_p$, $\pi_2 : Z \to F_p$ be the coordinate projections. Suppose that $A \subseteq Z$ is such that $|\pi_1(A)|, |\pi_2(A)| \geq p^{\delta}$ for some $0 < \delta < 1$ and such that at least one of π_1, π_2 is not injective. Show that for some $k = k(\delta) > 0$ we have $k(A^k) - k(A^k) = Z$. (Hint: by Exercise 2.8.8 it suffices to find some k' such that $k'(A^{k'}) - k'(A^{k'})$ contains a large intersection with either a horizontal line or a vertical line.)

2.8.10 [41] Let F_p, Z, π_1, π_2 be as in Exercises 2.8.8, 2.8.9. Suppose that $A \subseteq Z$ is such that $|\pi_1(A)|, |\pi_2(A)| \geq p^{\delta}$ for $0 < \delta < 1$. Show that either A is contained in a line $\{(x, ax) : x \in F_p\}$ for some $a \in F_p^{\times}$, or else $k(A^k) - k(A^k) = Z$ for some $k = k(\delta) > 0$. (Hint: by Exercise 2.8.7 one can reduce to the case where $\pi_1(A) = \pi_2(A) = F_p$. Now divide into two cases depending on whether π_1 or π_2 is injective on $2A - 2A$ or not.)

2.8.11 [41] Use Exercise 2.8.10 and Lemmas 2.53, 2.54 to deduce Theorem 2.63. (You will have to take a small amount of care concerning the zero-divisors $\{0\} \times F_p \cup F_p \times \{0\}$.)

2.8.12 Let Z be a commutative ring, and A_1, A_2, A_3, A_4 be subsets of $Z^{\times}$ such that $|A_1| = |A_2| = |A_3| = |A_4| = N$ and $|A_1 \cdot A_2 - A_3 \cdot A_4| \leq KN$. Show that $|A_j \cdot A_j - A_j \cdot A_j| \leq K^{O(1)}N$ for all $j = 1, 2, 3, 4$. This lemma allows one to extend several of the above results to the setting where the single set A is replaced by a number of sets of comparable cardinality.

2.8.13 Prove Corollary 2.59.

2.8.14 Use Corollary 2.59 and Proposition 2.61 to prove Theorem 2.60. (Hint: start with k equal to a large power of 2, and set L equal to a small power of $|H|$. If the hypotheses of Proposition 2.61 are satisfied, then one can lower bound $|kH|$ by $|j(H')^j|$, which can be controlled using

Corollary 2.59. If not, we can lower bound $|2kH|$ by $L|kH'|$ for some large subset H' of H; now replace k by $k/2$ and H by H' and argue as before. Continuing this process, one eventually obtains a good lower bound on $|kH|$ or $|2kH|$, either by combining Proposition 2.61 with Corollary 2.59, or by accumulating enough powers of L.)

2.8.15 [40] Prove the following variant of Corollary 2.62: for any $\delta > 0$ there exists $\varepsilon > 0$ such that whenever H, A are subsets of F_p with $|H| \geq p^{\delta}$, $|H \cdot H| \leq p^{\varepsilon}|H|$, and $1 < |A| < p^{1-\delta}$, then $E(A, H) = O_{\delta}(p^{-\varepsilon}|A||H|^2)$. In particular we have $|A + H| = \Omega_{\delta}(p^{\varepsilon}|H|)$.

2.8.16 [18] Let A be an additive set in F_p such that $|A| < p^{1-\delta}$ for some $\delta > 0$. Show that there exists an $\varepsilon > 0$ depending on δ such that $|\{(a, b, c, d, e, f) \in A^6 : ab + c = de + f\}| = O_{\varepsilon,\delta}(|A|^{5-\varepsilon})$. (Hint: use the Balog–Szemerédi–Gowers theorem in both the additive and multiplicative forms, together with Corollary 2.58.) This estimate is used in [18] to show that iterations of the map $X \mapsto X_1 \cdot X_2 + X_3$ on random variables in F_p (where X_1, X_2, X_3 are independent trials of X) converge in a certain sense to the uniform distribution, which has applications to random number generation.

3

Additive geometry

In Chapter 2 we studied the elementary theory of sum sets $A + B$ for general subsets A, B of an arbitrary additive group Z. In order to progress further with this theory, it is important first to understand an important subclass of such sets, namely those with a strong geometric and additive structure. Examples include (generalized) arithmetic progressions, convex sets, lattices, and finite subgroups. We will term the study of such sets (for want of a better name) *additive geometry*; this includes in particular the classical convex geometry of Minkowski (also known as *geometry of numbers*). Our aim here is to classify these sets and to understand the relationship between their geometrical structure, their dimension (or rank), their size (or volume, or measure), and their behavior under addition or subtraction. Despite looking rather different at first glance, it will transpire that progressions, lattices, groups, and convex bodies are all related to each other, both in a rigorous sense and also on the level of heuristic analogy. For instance, progressions and lattices play a similar role in arithmetic combinatorics that balls and subspaces play in the theory of normed vector spaces. In later sections, by combining methods of additive geometry, sum set estimates, Fourier analysis, and Freiman homomorphisms, we will be able to prove *Freiman's theorem*, which shows that all sets with small doubling constant can be efficiently approximated by progressions and similarly structured sets.

Closely related to all of these additive geometric sets are *Bohr sets*, which are in many ways the dual object to progressions, but we shall postpone the discussion of these sets (and their relationship with progressions) to Section 4.4, once we have introduced the Fourier transform.

3.1 Additive groups

We first review the theory of additive groups, which we introduced in Definition 0.1, obtaining in particular the classification theorem for finitely generated additive groups (Corollary 3.9). This is a fundamental result in additive group theory, but it will also motivate similar results concerning other additively structured sets such as progressions, Bohr sets, and the intersection of convex sets and lattices.

Typical examples of additive groups include the integers $\mathbf{Z}$, the reals $\mathbf{R}$, the lattices $\mathbf{Z}^d$, the Euclidean spaces $\mathbf{R}^d$, the torus groups $\mathbf{R}^d/\mathbf{Z}^d$, and the cyclic groups $\mathbf{Z}_N := \mathbf{Z}/N \cdot \mathbf{Z}$. Note that the direct sum $Z \oplus Z'$ of two additive groups is again an additive group. We now make an important distinction between torsion groups and torsion-free groups.

Definition 3.1 (Torsion) If Z is an additive group and $x \in Z$, we let $\mathrm{ord}(x)$ be the least integer $n \geq 1$ such that $n \cdot x = 0$, or $\mathrm{ord}(x) = +\infty$ if no such integer exists. We say that Z is a *torsion group* if $\mathrm{ord}(x)$ is finite for all $x \in Z$, and we say that it is an *r-torsion group* for some $r \geq 1$ if $\mathrm{ord}(x)$ divides r for all $x \in Z$. We say that Z is *torsion-free* if $\mathrm{ord}(x) = +\infty$ for all $x \in Z$.

Examples 3.2 The groups $\mathbf{Z}$, $\mathbf{R}$, $\mathbf{Z}^d$, $\mathbf{R}^d$ are torsion-free, whereas any finite group such as $\mathbf{Z}_N$ is a torsion group.

A *group homomorphism* $\phi : Z \to Z'$ between two additive groups Z, Z' is any map which preserves addition, negation, and zero (thus $\phi(x + y) = \phi(x) + \phi(y)$, $\phi(-x) = -\phi(x)$, and $\phi(0) = 0$ for all $x, y \in Z$). If ϕ is also invertible, then the inverse ϕ^{-1} is automatically a group homomorphism, and we say that ϕ is an *group isomorphism*, and Z and Z' are *group isomorphic*. Since all of our notions here shall be defined in terms of the addition, negation, and zero operations, they will all be preserved by group isomorphism, and so we will treat group isomorphic groups to be essentially equivalent. Later on we shall develop a weaker notion of *Freiman homomorphism* and *Freiman isomorphism* which is more suitable for the study of "approximate groups" (sets that are "almost" closed under addition); see Section 5.3.

If G is a subgroup of an additive group Z, then we can form the *quotient group*

$$Z/G := \{x + G : x \in Z\}$$

formed by taking all the cosets of G; this is easily verified to be a group (though it is no longer a subgroup of Z). For instance, the cyclic group $\mathbf{Z}_N = \mathbf{Z}/(N \cdot \mathbf{Z})$ is the quotient of the integers $\mathbf{Z}$ by the subgroup $N \cdot \mathbf{Z}$. Observe that the map $\pi : Z \to Z/G$ defined by $\pi(x) := x + G$ is a surjective homomorphism.

The sumset $G + H$ and intersection $G \cap H$ of two subgroups are still subgroups. Indeed, the arbitrary intersection of a family of subgroups is still a subgroup. Hence, given any subset X of Z, we can define the *span* $\langle X \rangle$ of X to be the smallest subgroup of Z which contains X; equivalently, $\langle X \rangle$ is the space of all finite $\mathbf{Z}$-linear combinations of elements of X. Thus for instance if $x \in Z$, then $\langle x \rangle$ is a group with cardinality $\mathrm{ord}(x)$. We say that an additive group Z is *finitely generated* if it can be written as the span $Z = \langle X \rangle$ of some finite set X. Clearly, every additive set X is contained in at least one finitely generated group, namely $\langle X \rangle$. Thus in the theory of additive sets one can usually reduce to the case when the ambient group Z is finitely generated (though it is sometimes convenient to work in some selected non-finitely generated additive groups, such as $\mathbf{Q}$, $\mathbf{R}$, or $\mathbf{R}^d$). In Corollary 3.9, we shall completely classify all finitely generated additive groups up to isomorphism.

Let $v = (v_1, \ldots, v_d) \in Z^d$ denote a d-tuple of elements in Z. We can rewrite the span $\langle v \rangle := \langle \{v_1, \ldots, v_d\} \rangle$ of this d-tuple in the following manner. For any element $n = (n_1, \ldots, n_d) \in \mathbf{Z}^d$ we define the *dot product* $n \cdot v$ in the usual manner as

$$n \cdot v := n_1 v_1 + \cdots + n_d v_d.$$

The map $n \mapsto n \cdot v$ is then a homomorphism from $\mathbf{Z}^d$ to Z, and its image $\mathbf{Z}^d \cdot v$ is precisely the span of v:

$$\langle v \rangle = \mathbf{Z}^d \cdot v.$$

The notion of a *progression*, introduced in Definition 0.2, is a truncated version of the concept of a span, in which the infinite lattice $\mathbf{Z}^d$ is replaced instead by a box. Alternatively, one can think of lattices as infinite progressions.

3.1.1 Lattices

We now study a special type of additive group, namely the lattices in Euclidean space.

Definition 3.3 (Lattices) A *lattice* Γ in $\mathbf{R}^d$ is any additive subgroup of the Euclidean space $\mathbf{R}^d$ which is discrete (i.e. every point in Γ is isolated). We define the *rank* k of Γ to be the dimension of the linear space spanned by the elements of Γ, thus $0 \leq k \leq d$. If $k = d$, we say that Γ has *full rank*. If Γ' is another lattice in $\mathbf{R}^d$ which is contained in Γ, we say that Γ' is a *sub-lattice* of Γ.

Thus for instance $\mathbf{Z}^d$ is a lattice of full rank in $\mathbf{R}^d$. More generally, a typical example of a lattice of rank k is the set $\mathbf{Z}^k \cdot v$, where $v = (v_1, \ldots, v_k)$ is a collection of linearly independent vectors in $\mathbf{R}^d$ for some $0 \leq k \leq d$. In fact, this is the only possible type of lattice, as we shall see in Lemma 3.4. We observe that if

$T : \mathbf{R}^d \to \mathbf{R}^d$ is an invertible linear transformation on $\mathbf{R}^d$, and Γ is a lattice, then $T(\Gamma)$ is also a lattice with the same rank as Γ.

If Γ is a lattice, then the quotient space $\mathbf{R}^d/\Gamma$ is a smooth manifold with a natural Lebesgue (or Haar) measure induced from $\mathbf{R}^d$. If Γ has full rank, it is easy to see that $\mathbf{R}^d/\Gamma$ is also compact, and thus has a volume $\mathrm{mes}(\mathbf{R}^d/\Gamma)$, which we refer to as the *covolume* of Γ.

Next, we classify all lattices in $\mathbf{R}^d$. Call a vector v in Γ *irreducible* if $v/n \notin \Gamma$ for any integer $n \geq 2$.

Lemma 3.4 (Fundamental theorem of lattices) *If Γ is a lattice in $\mathbf{R}^d$ of rank k, then there exist linearly independent vectors $v_1, \ldots, v_k$ in $\mathbf{R}^d$ such that $\Gamma = \mathbf{Z}^k \cdot v$. In particular every lattice of rank k is finitely generated and is isomorphic (via an invertible linear transformation from the linear span of Γ to $\mathbf{R}^k$) to the standard lattice $\mathbf{Z}^k$. Furthermore, if w is an irreducible vector in Γ, we may choose the above representation $\Gamma = \mathbf{Z}^k \cdot v$ so that $v_1 = w$.*

Proof We first observe that we may assume that the vectors in Γ span $\mathbf{R}^d$, else we could pass from $\mathbf{R}^d$ to a smaller vector space and continue the argument. In other words, we may assume that the rank k of Γ is equal to d. We may also clearly assume that $d \geq 1$, since the $d = 0$ case is vacuously true.

Observe that Γ contains at least one irreducible vector w, since one can start with any non-zero vector v in Γ and take w to be the smallest vector of the form v/n (such a vector must exist since Γ is discrete). Now let w be an irreducible vector. By the full rank assumption, we can find d linearly independent vectors $v_1, \ldots, v_d$ in Γ with $v_1 = w$, so in particular the volume $|v_1 \wedge \cdots \wedge v_d|$ of the parallelepiped spanned by $v_1, \ldots, v_d$ is strictly positive. Since Γ contains $\mathbf{Z}^d \cdot (v_1, \ldots, v_d)$, we obtain an upper bound for the covolume:

$$|v_1 \wedge \cdots \wedge v_d| \geq \mathrm{mes}(\mathbf{R}^d/\Gamma).$$

We now use the method of descent. If $\mathbf{Z}^d \cdot (v_1, \ldots, v_d)$ is equal to Γ then we are done. Otherwise, the half-open parallelepiped $\{\sum_{i=1}^d t_i v_i : 0 \leq t_i < 1\}$ generated by the vectors $v_1, \ldots, v_d$, being a fundamental domain of $\mathbf{Z}^d \cdot (v_1, \ldots, v_d)$, must contain a non-zero lattice point x in Γ. Write $x = \sum_{i=1}^d t_i v_i$; note that at least one of $t_2, \ldots, t_d$ must be non-zero otherwise we would have $tw \in \Gamma$ for some $0 < t < 1$, which (by the Euclidean algorithm) contradicts the irreducibility of w. By permuting the indices $1, \ldots, d$ if necessary we may assume that $t_d > 0$. We may also assume that $t_d \leq 1/2$ since we could replace w by $v_1 + \cdots + v_d - x$ otherwise. Then the volume $|v_1 \wedge \cdots \wedge v_{d-1} \wedge x|$ is at most half that of $|v_1 \wedge \cdots \wedge v_d|$, but is still non-zero. We thus replace v_d by x and repeat the above argument. Because of our absolute lower bound on the volume of parallelepipeds, this argument must eventually terminate, at which point we have found the desired presentation for Γ. Note that this procedure will never alter v_1 and hence v_1 is equal to w as desired. □

Corollary 3.5 (Splitting lemma) *Let Γ be a lattice of rank k, and let v be an irreducible vector in Γ. Then there exists a sub-lattice Γ' of Γ of rank $k-1$ such that Γ is the direct sum of $\mathbf{Z} \cdot v$ and Γ', i.e. $\Gamma = \mathbf{Z} \cdot v + \Gamma'$ and $\mathbf{Z} \cdot v \cap \Gamma' = \{0\}$.*

Proof Apply Lemma 3.4 with $v_1 := v$, and set $\Gamma' := \mathbf{Z}^{k-1} \cdot (v_2, \ldots, v_k)$; the claim then follows from the linear independence of $v_1, \ldots, v_k$. □

Corollary 3.6 (Fundamental theorem of finitely-generated torsion-free additive groups) *Let Z be a finitely generated torsion-free additive group. Then Z is isomorphic to $\mathbf{Z}^d$ for some $d \geq 0$.*

Proof We shall use the homomorphism theorems (Exercise 3.1.1). Since Z is finitely generated, we may find elements $v_1, \ldots, v_n$ in Z such that $\mathbf{Z}^n \cdot (v_1, \ldots, v_n) = Z$. Now let Γ be the set $\{n \in \mathbf{Z}^n : n \cdot (v_1, \ldots, v_n) = 0\}$; then Γ is a sub-lattice of $\mathbf{Z}^n$ and Z is isomorphic to $\mathbf{Z}^n / \Gamma$. In particular, $\mathbf{Z}^n / \Gamma$ is torsion-free. We shall show that this implies that $\mathbf{Z}^n / \Gamma$ is isomorphic to some $\mathbf{Z}^d$, as desired. We induce on n, the case $n = 0$ being trivial. If $\Gamma = \{0\}$ we are done, so suppose Γ contains a non-zero vector $v \in \Gamma$, which we may assume without loss of generality to be irreducible in Γ. It is also irreducible in $\mathbf{Z}^n$, for if $v = m \cdot w$ for some $w \in \mathbf{Z}^d$ and $m > 1$, then $w + \Gamma$ would be a non-zero element of $\mathbf{Z}^n / \Gamma$ such that $m \cdot (w + \Gamma) = 0 + \Gamma$, contradicting the torsion-free assumption. By Lemma 3.4 or Corollary 3.5, this implies that $\mathbf{Z}^n / (\mathbf{Z} \cdot v)$ is isomorphic to $\mathbf{Z}^{n-1}$. Since $\mathbf{Z}^n / \Gamma$ is isomorphic to $(\mathbf{Z}^n / (\mathbf{Z} \cdot v)) / (\Gamma / (\mathbf{Z} \cdot v))$, the claim then follows from the induction hypothesis. □

3.1.2 Quotients of lattices

Let G be a finitely generated additive group generated by d elements $v_1, \ldots, v_d \in G$. If we write $v := (v_1, \ldots, v_d)$, and let $\Gamma \subset \mathbf{Z}^d$ be the lattice $\Gamma := \{n \in \mathbf{Z}^d : n \cdot v = 0\}$, it is easy to see that G is isomorphic to the quotient $\mathbf{Z}^d / \Gamma$. Thus it is of interest to understand the quotient of two lattices. A basic tool for doing so is

Theorem 3.7 (Smith normal form) *Let Γ and Γ' be two lattices of full rank in $\mathbf{R}^d$ such that Γ' is a sub-lattice of Γ. Then there exist linearly independent vectors $v_1, \ldots, v_d$ in Γ such that*

$$\Gamma = \mathbf{Z}^d \cdot (v_1, \ldots, v_d)$$

and

$$\Gamma' = \mathbf{Z}^d \cdot (N_1 v_1, \ldots, N_d v_d),$$

where $1 \leq N_1 \leq \cdots \leq N_d$ *are positive integers such that* N_j *divides* N_{j+1} *for all* $j = 1, \ldots, d-1$.

Note that by applying an invertible linear transformation one can set $(v_1, \ldots, v_d)$ equal to the standard basis $(e_1, \ldots, e_d)$, so that Γ becomes just the standard lattice $\mathbf{Z}^d$, while Γ' is the sub-lattice of $\mathbf{Z}^d$ of vectors whose jth coordinate is a multiple of N_j for $j = 1, \ldots, d$.

Proof We induce on d. For $d = 0$ the statement is vacuously true, so suppose $d \geq 1$ and the claim has already been proven for $d - 1$. Given any non-zero vector $v \in \Gamma$, define the *index* of v to be the largest positive integer n such that $v/n \in \Gamma$; note that the index is finite since Γ is discrete. Note that the index of v is n if and only if $v = nw$ for some irreducible vector w in Γ.

Since Γ' has full rank, it contains non-zero vectors, each of which has an index. Let N_1 denote the minimum index of all such vectors. By the well-ordering principle, this index is attained, and thus there exists an irreducible vector $v_1 \in \Gamma$ such that $N_1 v_1 \in \Gamma'$.

Using Lemma 3.4, we may apply an invertible linear transformation to map Γ to $\mathbf{Z}^d$, in such a way that v_1 is now equal to the standard basis vector e_1. Now let $(n_1, \ldots, n_d)$ be any vector in Γ'. Observe that $n_1, \ldots, n_d$ are integers; furthermore, n_1 must be a multiple of N_1, otherwise by subtracting a multiple of $N_1 e_1$ we could ensure that $|n_1| < N_1$, which contradicts the definition of N_1 as the minimal index of Γ'. Thus we may factorize $\Gamma' = N_1 \mathbf{Z} \cdot e_1 + \Gamma''$, where Γ'' is some sub-lattice of $\mathbf{Z}^{d-1}$ (which we think of as the span of $e_2, \ldots, e_d$). Note that if $x \in \Gamma''$, then $(N_1, x) \in \Gamma'$, and hence (since (N_1, x) must have index at least N_1), x must be a multiple of N_1. Thus Γ'' actually lies in $N_1 \cdot \mathbf{Z}^{d-1}$, and we may therefore write $\Gamma' = N_1(\mathbf{Z} \cdot e_1 + \Gamma''')$ for some sub-lattice Γ''' of $\mathbf{Z}^{d-1}$. Note that Γ''' must have rank $d - 1$ since Γ' has rank d.

We now invoke the inductive hypothesis, and, by applying an invertible linear transformation to $\mathbf{Z}^{d-1}$ if necessary, we may assume that

$$\Gamma''' = \{(n_2 M_2, \ldots, n_d M_d) : n_2, \ldots, n_d \in \mathbf{Z}\}$$

for some $1 \leq M_2 \leq \cdots \leq M_d$ such that M_j divides M_{j+1} for all $j = 2, \ldots, d-1$. The claim follows by setting $N_j := N_1 M_j$ for $j = 2, \ldots, d$. □

We can now obtain the well-known classification of finite and finitely generated additive groups:

Corollary 3.8 (Fundamental theorem of finite additive groups) *Every finite additive group G is isomorphic to the direct sum of a finite number of cyclic groups* $\mathbf{Z}_N = \mathbf{Z}/(N \cdot \mathbf{Z})$.

Proof Let $g_1, \ldots, g_d$ be a finite set of generators for G. Then the map $\phi : \mathbf{Z}^d \mapsto G$ defined by $\phi(n) := n \cdot (g_1, \ldots, g_d)$ is a surjection, and thus G is isomorphic to $\mathbf{Z}^d/\phi^{-1}(0)$, which is a subgroup of $\mathbf{R}^d/\phi^{-1}(0)$. The kernel $\phi^{-1}(0)$ is clearly a lattice of some rank $0 \leq k \leq d$, and hence by Lemma 3.4 is generated by k linearly independent vectors $v_1, \ldots, v_k$ in $\mathbf{Z}^d$. Observe that we must have full rank $k = d$, otherwise $\mathbf{Z}^d/\phi^{-1}(0)$ (and hence G) will be infinite. Using the Smith normal form, we can after applying an isomorphism write $\phi^{-1}(0)$ as the lattice generated by $N_1 e_1, \ldots, N_d e_d$ for some integers $N_1, \ldots, N_d \geq 1$; this makes G isomorphic to $G \equiv \mathbf{Z}/N_1\mathbf{Z} \oplus \cdots \oplus \mathbf{Z}/N_d\mathbf{Z}$, as desired (indeed we even obtain a normal form in which N_j divides N_{j+1} for $j = 1, \ldots, d-1$). □

Corollary 3.9 (Fundamental theorem of finitely generated additive groups) *Every finitely generated additive group G is isomorphic to the direct sum of a finite number of cyclic groups $\mathbf{Z}/(N \cdot \mathbf{Z})$, and a lattice $\mathbf{Z}^d$ for some $d \geq 0$.*

Proof Let $\tilde{G} := \{x \in G : nx = 0 \text{ for some } n > 0\}$ be the torsion group of G; then by Corollary 3.8, $\tilde{G}$ is the direct sum of cyclic groups. The quotient group $G/\tilde{G}$ is torsion-free and is thus isomorphic to $\mathbf{Z}^d$ for some $d \geq 0$ by Corollary 3.6. If we let $\tilde{e}_1, \ldots, \tilde{e}_d$ be arbitrary representatives in G of the standard basis $e_1, \ldots, e_d$ of $\mathbf{Z}^d$, we thus see that G is the direct sum of $\tilde{G}$ and $\mathbf{Z} \cdot \tilde{e}_1, \ldots, \mathbf{Z} \cdot \tilde{e}_d$, and the claim follows. □

Exercises

3.1.1 (Homomorphism theorems) If $\phi : Z \to Z'$ is a homomorphism between groups, show that the range $\phi(Z)$ is a group which is isomorphic to the quotient group $Z/\phi^{-1}(0)$. If G, H are subgroups of Z, show that $(G + H)/G$ is isomorphic to $H/(G \cap H)$. If furthermore $G \subset H$, show that H/G is a subgroup of Z/G and that $(Z/G)/(H/G)$ is isomorphic to Z/H. If G' is a subgroup of Z', show that $(Z \oplus Z')/(G \oplus G')$ is isomorphic to $(Z/G) \oplus (Z'/G')$.

3.1.2 (Lagrange's theorem) Show that if G is a subgroup of a finite additive group Z, then $|Z/G| = |Z|/|G|$ (and in particular $|G|$ must divide $|Z|$). By considering the groups $\langle x \rangle$ for various $x \in Z$, conclude that every finite additive group Z is an $|Z|$-torsion group; in particular, $\mathrm{ord}(x)$ divides $|Z|$ for all $x \in Z$.

3.1.3 Show that if x is any element of a additive group Z, then the group $\langle x \rangle = \mathbf{Z} \cdot x$ has cardinality $\mathrm{ord}(x)$. More generally, if $v = (v_1, \ldots, v_d) \in Z^d$, show that the group $\mathbf{Z}^d \cdot v$ has cardinality at most $\mathrm{ord}(v_1) \cdots \mathrm{ord}(v_d)$, but at least as large as the least common multiple of $\mathrm{ord}(v_j)$.

3.1.4 Let Z be an additive group. Show that Z is an N-torsion group if and only if for every $x \in Z$, the torsion of x is a divisor of N. Show that Z is torsion-free if and only if Z contains no finite subgroups other than the trivial subgroup $\{0\}$.

3.1.5 Let $Z = Z_1 \oplus Z_2$ be a direct sum of additive groups and $r \geq 1$. Show that Z is torsion-free (resp. r-torsion) if and only if Z_1 and Z_2 are torsion-free (resp. r-torsion).

3.1.6 Prove that $\mathbf{Q}$ and $\mathbf{R}$ are not finitely generated.

3.1.7 If x, y are elements of an additive group Z with finite order, show that $x + y$ also has finite order, and that $\mathrm{ord}(x + y)$ divides the least common multiple of $\mathrm{ord}(x)$ and $\mathrm{ord}(y)$. Conclude that the set $\mathrm{tor}(Z) := \{x \in Z : \mathrm{ord}(x) < \infty\}$ is a torsion group; we refer to it as the *torsion subgroup* of Z. It is clearly the largest subgroup of Z which is a torsion group. Show that the quotient group $Z/\mathrm{tor}(Z)$ is torsion-free, and is in fact the largest quotient which is torsion-free (in the sense that all other torsion-free quotients are quotients of $Z/\mathrm{tor}(Z)$).

3.1.8 Show that Corollary 3.5 fails whenever v is not irreducible.

3.2 Progressions

We now study a basic example of an additive set, namely that of a *generalized arithmetic progression* (or *progression* for short), as defined in Definition 0.2. These will be model examples of additive sets with large amounts of additive structure; they can be viewed as a hybrid between a lattice and a convex set. (For a more quantitative realization of this heuristic, see Lemma 3.36 below.)

Note that progressions with the same set of basis vectors add very easily,

$$(a + [0, N] \cdot v) + (a' + [0, N'] \cdot v) = (a + a') + [0, N + N'] \cdot v \qquad (3.1)$$

(so in particular the rank and basis vectors do not change), whereas progressions with different basis vectors add via the formula

$$(a + [0, N] \cdot v) + (a' + [0, N'] \cdot v') = (a + a') + [0, N \oplus N'] \cdot (v \oplus v'). \qquad (3.2)$$

Note the progression on the right-hand side of (3.2) is likely to be highly improper if v and v' share some basis vectors in common. Also one can replace the box $[0, N]$ by another one and also obtain a progression:

$$a + [N, M] \cdot v = (a + N \cdot v) + [0, M - N] \cdot v.$$

Similarly if one uses boxes such as $[N, M)$, etc. In particular, the negation of a progression is also a progression:

$$-(a + [0, N] \cdot v) = (-a) + [0, N] \cdot (-v) = (-a - N \cdot v) + [0, N] \cdot v. \qquad (3.3)$$

From this and (3.2) we see that the sum or difference of two progressions is again a progression. Finally, we make the easy observation that the Cartesian product of two progressions is again a progression.

We now show that, up to errors of $O(1)^d$, that progressions of rank d are essentially closed under addition.

Lemma 3.10 *Let $P = a + [0, N] \cdot v$ be a progression of rank d in an additive group Z; we do not require that P be proper (see Definition 0.2). Then for any integers $n < m$ and any $b \in Z$, we can cover $b + [nN, mN] \cdot v$ by $(m - n)^d$ translates of P. In particular for any $n, m \geq 0$ with $(n, m) \neq (0, 0)$, we can cover $nP - mP$ by $(n + m)^d$ translates of P, and in particular*

$$|nP - mP| \leq (n + m)^d |P|.$$

Furthermore, $nP - mP$ is also a progression of rank d and volume at most $\mathrm{vol}(nP - mP) \leq (n + m)^d \mathrm{vol}(P)$.

Proof The first claim is clear since

$$\begin{aligned}&[n \cdot N, m \cdot N] \cdot v \\ &\quad = [0, N] \cdot v + [(n, \dots, n), (m - 1, \dots, m - 1)] \cdot (N_1 v_1, \dots, N_d v_d).\end{aligned}$$

From (3.1) we have

$$nP - mP = (na - ma - mN \cdot v) + [0, (n + m)N] \cdot v$$

from which the remaining claims follow. □

From this lemma we see in particular that if P is a symmetric progression of rank d and contains the origin (e.g. if $P = [-N, N] \cdot v$), then P is a 2^d-approximate group in the sense of Definition 2.25. Indeed one can think of (symmetric) progressions of small rank as substitutes for subgroups in torsion-free settings (since torsion-free groups cannot contain finite subgroups). They also are the arithmetic analogue of boxes (or more generally, parallelepipeds) in Euclidean space, and in fact many of the results from real-variable harmonic analysis regarding covering by boxes (in physical space, Fourier space, or both) will have analogues for progressions.

In the special case when the rank d is equal to 1, a generalized arithmetic progression is the same as an *ordinary arithmetic progression* (or *arithmetic progression* for short)

$$P = a + [0, N] \cdot v = \{a + nv : 0 \leq n \leq N\}$$

with base point $a \in Z$, basis vector or *step* $v \in Z$, and length $N + 1$. Note again that the cardinality of P may be less than $N + 1$ if P is not proper, though in a torsion-free group this is only possible if the step v is zero.

We record a trivial lemma that asserts that the sum set of a progression and a small set can be contained (somewhat inefficiently) in another progression.

Lemma 3.11 *If P is a progression of rank d, and $P + w_1, \ldots, P + w_K$ are translates of P, then all the translates $P + w_1, \ldots, P + w_K$ can be contained inside a single progression of rank $d + K - 1$ and volume $2^{K-1}\mathrm{vol}(P)$.*

Proof Write $P = a + [0, N] \cdot v$. By translation invariance we may set $w_K = 0$. Then the claim follows by using the progression $a + [0, N] \cdot v + [0, 1]^{K-1} \cdot (w_1, \ldots, w_{K-1})$. □

Thus if one adds a small number of elements to a progression, one can still place the combined set inside a progression of slightly larger rank and volume, although the volume can grow exponentially in $|A|$. This is unavoidable: see Exercise 3.2.2. Because of this exponential loss, it is sometimes better not to invoke this lemma, and deal with multiple shifts of a single progression rather than trying to contain everything inside a single progression. Note that we have not guaranteed that the progressions in Lemma 3.11 are proper; we will return to this point in Section 3.6.

Exercises

3.2.1 Let $N = (N_1, \ldots, N_d)$ be a collection of non-negative integers. Show that every proper ordinary arithmetic progression of length $(N_1 + 1) \cdots (N_d + 1)$ is equal (as a set) to a proper generalized arithmetic progression of dimension N. (This example shows that the rank of a progression cannot be uniquely determined from the set of its elements, even if we restrict the progression to be proper.)

3.2.2 Let $K \geq 1$ and $d \geq 0$ be integers, and $P = a + [0, N] \cdot v$ be a rank d progression in an additive group Z for some basis vectors $v = (v_1, \ldots, v_d)$, and let $X = \{e_1, \ldots, e_K\}$ be a set of K elements in Z. Suppose that the elements $v_1, \ldots, v_d, e_1, \ldots, e_K$ are linearly independent over $\mathbf{Z}$. Show that any progression which contains $P + X$ must necessarily have rank at least $d + K - 1$ and volume at least $2^{K-1}\mathrm{vol}(P)$, which shows that Lemma 3.11 is sharp.

3.2.3 Show that in a torsion-free additive group, the intersection of two ordinary arithmetic progressions is again an ordinary arithmetic progression. What happens if the torsion-free hypothesis is removed? What happens if one or both of the progressions is allowed to have rank greater than one?

3.2.4 Show that every finite additive group is also a proper progression.

3.2.5 Let P be a progression of rank d. Show that P contains an arithmetic progression Q with $|Q| \geq |P|^{1/d}$, and furthermore that Q is proper if P is, and Q can be chosen to be symmetric around the origin if P is.

3.2.6 Let P be a proper progression of rank d, and let A be a subset of P such that $|A| \leq \varepsilon|P|$ for some $0 < \varepsilon < 1$. Show that $P\backslash A$ contains a proper progression Q of rank d with $|Q| \geq C^{-d}/\varepsilon$ for some absolute constant C.

3.2.7 Let A be an additive set in an ambient group Z, and let $v \in Z$. Show that $|(A+v)\backslash A| \leq 1$ if and only if A is equal to a proper arithmetic progression of step v, union a finite (possibly zero) number of translates of the group $\langle v \rangle$. In particular, if $|A| < \mathrm{ord}(v)$, then $|(A+v)\backslash A| > 0$, and $|(A+v)\backslash A| = 1$ if and only if A is a proper arithmetic progression of step v.

3.3 Convex bodies

We now review some of the theory of convex bodies in $\mathbf{R}^d$, which are in some sense the continuous analogue of generalized arithmetic progressions. This is of course a vast field, and we shall restrict ourselves with just a small sample of results, relating to the additive theory of such sets, to covering lemmas, and the relationship between addition and volume.

We shall use $\mathrm{mes}(A)$ to denote the volume of a set A in $\mathbf{R}^d$; to avoid issues with measurability we shall mostly concern ourselves with bounded open sets A. If $A \in \mathbf{R}^d$ and $\lambda \in \mathbf{R}$, we use $\lambda \cdot A$ to denote the dilation $\lambda \cdot A := \{\lambda x : x \in A\}$. Observe that $\mathrm{mes}(\lambda A) = |\lambda|^d \mathrm{mes}(A)$.

Recall that a set A in $\mathbf{R}^d$ is *convex* if we have $(1-\theta)x + \theta y \in A$ whenever $x, y \in A$ and $0 \leq \theta \leq 1$; equivalently, a set is convex if and only if

$$a \cdot A + b \cdot A = (a+b) \cdot A$$

for all non-negative reals $a, b \geq 0$ (Exercise 3.3.3). In particular we have $nA = n \cdot A$ for any positive integer n. We call A a *convex body* if it is convex, open, non-empty, and bounded. In particular we see that if A is a convex body, then

$$\mathrm{mes}(A+A) = \mathrm{mes}(2 \cdot A) = 2^d \mathrm{mes}(A), \tag{3.4}$$

so convex bodies have small doubling constant. As for $A - A$, we can use

Lemma 3.12 *[297] For any bounded open subsets A, B, C of $\mathbf{R}^d$ (not necessarily convex), we have*

$$\mathrm{mes}(A-C)\mathrm{mes}(B) \leq \mathrm{mes}(A-B)\mathrm{mes}(B-C).$$

This is proven by modifying the proof of Lemma 2.6 appropriately and is left as an exercise. From this Lemma (with $A = C$ and $B = -A$) and (3.4) we obtain

$$\mathrm{mes}(A - A) \leq 4^d \mathrm{mes}(A); \tag{3.5}$$

compare these bounds with Lemma 3.10. For a slight refinement of (3.5), see Exercise 3.4.6. In the converse direction, the Brunn–Minkowski inequality (Theorem 3.16 below) will give $\mathrm{mes}(A - A) \geq 2^d \mathrm{mes}(A)$.

Call a convex body A *symmetric* if $A = -A$; thus for us symmetry will always be with respect to the origin. The following theorem of John essentially classifies all convex bodies (symmetric and non-symmetric) up to a (dimension-dependent) constant factor.

Theorem 3.13 (John's theorem) *[194] Let A be a convex body in $\mathbf{R}^d$. Then there exists an invertible linear transformation $T : \mathbf{R}^d \to \mathbf{R}^d$ on $\mathbf{R}^d$ and a point $x_0 \in A$ such that*

$$B_d \subseteq T(A - x_0) \subseteq d \cdot B_d,$$

where B_d is the unit ball $\{(x_1, \ldots, x_d) \in \mathbf{R}^d : x_1^2 + \cdots + x_d^2 < 1\}$. If A is symmetric, then we can improve these inclusions to

$$B_d \subseteq T(A) \subseteq \sqrt{d} \cdot B_d.$$

The constants d and $\sqrt{d}$ are sharp; see the exercises.

Proof We will use a variational argument. Define an *ellipsoid* to be any set E of the form $E = L(B_d) + x_0$, where B_d is the unit ball, $x_0 \in \mathbf{R}^d$, and L is a (possibly degenerate) linear transformation in $\mathbf{R}^d$; we allow the ellipsoid to be degenerate for compactness reasons. Since A is open and bounded, it is easy to see that the set of all ellipsoids E contained in A is a compact set (with respect to the usual topology on L and x_0). Also the volume of the ellipsoid E is $\mathrm{mes}(E) = |\det(L)|\mathrm{mes}(B_d)$, which is clearly a continuous function of E. Thus there exists an ellipsoid $E = L(B_d) + x_0$ in A which maximizes the volume $\mathrm{mes}(E)$; since A is open, this volume is non-zero, and hence L is invertible. By applying L^{-1} if necessary (observing that the conclusion of the lemma is invariant under invertible linear transformations) we may thus assume that E is a translate $E = B_d + y_0$ of the unit ball, where $y_0 = L^{-1}(x_0)$.

Let us now restrict to the case where A is symmetric. Observe that if A contains $B_d + y_0$ then it also contains $B_d - y_0$ by symmetry, and hence contains B_d, which is in the convex hull of $B_d + y_0$ and $B_d - y_0$. To conclude the proof of the lemma in this case we need to show that A is contained in $\sqrt{d} \cdot B_d$. Suppose for contradiction that A was not contained in $\sqrt{d} \cdot B_d$; without loss of generality (and using the hypothesis that A is open) we may then suppose that $re_1 \in A$ for some $r > \sqrt{d}$, where e_1 is the first basis vector. Observe now from elementary geometry that if ω is any point on the boundary of B_d making an angle $\angle(\omega, e_1) < \arctan(\sqrt{r^2-1})$, then the line segment connecting ω to re_1 is disjoint from (and not tangent to) B_d, and, since B_d and re_1 both lie in the convex set A, we thus see that ω also lies in the open set A. By symmetry, the same is true if $\angle(\omega, -e_1) < \arctan(\sqrt{r^2-1})$.

We now perturb the ball B_d by an epsilon. Now let $\delta > 0$ be a small number, let $\varepsilon > 0$ be an even smaller one, and consider the ellipsoid $L_{\varepsilon,\delta}(B_d)$, where

$$L_{\varepsilon,\delta}(x_1, \ldots, x_d) := ((1 + (d-1+\delta)\varepsilon)x_1, (1-\varepsilon)x_2, \ldots, (1-\varepsilon)x_d)).$$

When $\varepsilon = 0$, $L_{\varepsilon,\delta}(B_d)$ is just B_d. Now consider how $L_{\varepsilon,\delta}(B_d)$ evolves in ε. The determinant of this transformation is $(1 + (d-1+\delta)\varepsilon)(1-\varepsilon)^{d-1}$, which has a positive ε-derivative at $\varepsilon = 0$. Thus $L_{\varepsilon,\delta}(B_d)$ has larger volume than B for sufficiently small ε (depending on δ). Now we check which points on the surface of $L_{\varepsilon,\delta}(B_d)$ expand away from the origin, and which ones contract. A simple computation shows that for any $\omega = (\omega_1, \ldots, \omega_d)$ on the boundary of B_d, the derivative

$$\frac{d}{d\varepsilon}\|L_{\varepsilon,\delta}(\omega)\|^2\Big|_{\varepsilon=0},$$

where $\|(y_1, \ldots, y_d)\|^2 := y_1^2 + \cdots + y_d^2$, is negative unless

$$(d-1+\delta)\omega_1^2 - \omega_2^2 - \cdots - \omega_d^2 \geq 0,$$

or in other words that

$$\angle(\omega, \pm e_1) \leq \arctan(\sqrt{d-1+\delta}).$$

But if δ is small enough depending on r, this region is contained entirely within the interior of A by the previous discussion. Thus for ε small enough $L_{\varepsilon,\delta}(B_d)$ is completely contained inside A but has larger volume, contradicting the maximality of B_d, and we are done.

Now suppose that A is not symmetric. In this case we may translate so that $y_0 = 0$. Thus again we have $B_d \subseteq A$, and the task is to show that $A \subseteq d \cdot B_d$. Suppose again for contradiction that $re_1 \in A$ for some $r > d$; again this means that every point ω in the boundary of B_d with $\angle(\omega, e_1) < \arctan(\sqrt{r^2-1})$ will lie in the interior of A. Now let $\delta, \varepsilon > 0$ and consider the ellipsoid

$$\{L_{\varepsilon,\delta}(x_1, \ldots, x_d) + (d-1+\delta)\varepsilon e_1 : (x_1, \ldots, x_d) \in B_d\};$$

again, this ellipsoid has larger volume than B_d if ε is sufficiently small. Also, we see that

$$\frac{d}{d\varepsilon}\|L_{\varepsilon,\delta}(\omega) + (d-1+\delta)\varepsilon e_1\|^2\Big|_{\varepsilon=0}$$

is negative unless

$$(d-1+\delta)\omega_1^2 + (d-1+\delta)\omega_1 - \omega_2^2 - \cdots - \omega_d^2 \geq 0,$$

which can be rewritten (using $\|\omega\| = 1$) as

$$((d+\delta)\omega_1 - 1)(\omega_1 + 1) \geq 0,$$

or equivalently

$$\angle(\omega, e_1) \leq \arctan(\sqrt{(d+\delta)^2 - 1}).$$

We now argue as in the symmetric case to obtain again the desired contradiction, if δ is chosen so that $d + \delta < r$. □

As a corollary of Theorem 3.13 we see that if A is a convex body, we can cover $A + A$ or $A - A$ by a relatively small number of copies of A:

$$A \pm A \text{ can be covered by } O(d)^d \text{ translates of } A. \tag{3.6}$$

This follows immediately from the geometric observation that $d \cdot B_d + d \cdot B_d = 2d \cdot B_d$ can be covered by $O(d)^d$ translates of B_d. If A is symmetric, we can improve this somewhat. In the special case when A is a cube or a box, it is clear that $A \pm A$ can be covered by 2^d translates of A (cf. Lemma 3.10), but one cannot hope for this in general; for instance if A is a disk in $\mathbf{R}^2$ then one needs six copies of A to cover $A \pm A$. In the general case, we will need the following continuous version of

Lemma 3.14 (Ruzsa's covering lemma) *[300], [250] For any bounded subsets A, B of $\mathbf{R}^d$ with positive measure (not necessarily convex), we can cover B by at most* $\min(\frac{\mathrm{mes}(A+B)}{\mathrm{mes}(A)}, \frac{\mathrm{mes}(A-B)}{\mathrm{mes}(A)})$ *translates of $A - A$.*

The proof of this lemma is nearly identical to that of Lemma 2.14 and is left as an exercise. As a consequence we can improve (3.6) for symmetric convex bodies:

Corollary 3.15 *Let $A \subset \mathbf{R}^d$ be a convex body, and let $\lambda, \mu > 0$ be real. Then $\lambda \cdot A$ can be covered by at most $(\lambda + 1)^d$ translates of $A - A$, and $\lambda \cdot A - \mu \cdot A$ can be covered by $(2\max(\lambda, \mu) + 1)^d$ translates of $A - A$. If A is symmetric, then $\lambda \cdot A$ can be also covered by $(2\lambda + 1)^d$ translates of A.*

Proof The first claim follows from Lemma 3.14 since $\mathrm{mes}(\lambda \cdot A + A) = (\lambda + 1)^d \mathrm{mes}(A)$. To prove the second claim, we may take $\lambda \geq \mu$. The first claim implies that $2\lambda \cdot A$ can be covered by $(2\lambda + 1)^d$ translates of $A - A = 2 \cdot A$, and the third claim follows by rescaling by $1/2$. Finally, the second claim follows by applying the third claim to $A - A$. □

Observe that all the bounds obtained here tend to be exponential in d or worse. Thus when using the theory of convex bodies to obtain explicit estimates, it is often important to keep the dimension d as low as possible, even at the cost of making some other parameters larger than would otherwise be necessary. See [250] for further discussion of sum set and covering estimates for convex bodies.

We have not yet seen what happens to the sum or difference of two unrelated convex bodies A and B. The relationship here is given by the *Brunn–Minkowski inequality*, which we turn to next.

Exercises

3.3.1 Prove Lemma 3.12.

3.3.2 Prove Lemma 3.14.

3.3.3 Verify that the two definitions of convexity given are indeed equivalent.

3.3.4 Let A be an open bounded subset of $\mathbf{R}^d$. Show that A is convex if and only if $2A = 2 \cdot A$, and that A is convex and symmetric if and only if $2A = -2 \cdot A$.

3.3.5 For any $s > 0$ let $\Gamma(s) := \int_0^\infty e^{-x} x^{s-1}\, dx$ denote the Gamma function. Show that $\Gamma(s+1) = s\Gamma(s)$ for all $s > 0$, that $\Gamma(d) = (d-1)!$ for all $d \geq 1$, that $\Gamma(1/2) = \sqrt{\pi}$, and we have the Stirling formula

$$\log \Gamma(s) = s \log s - s + O(\log s) \tag{3.7}$$

for all large s. (Hint: use (1.52) and the monotonicity of the Γ function.)

3.3.6 Let B_d be the unit ball in $\mathbf{R}^d$. By evaluating the integral $\int_{\mathbf{R}^d} e^{-\pi |x|^2}\, dx$ in both Cartesian and polar coordinates, and using the preceding exercise, establish the volume formula

$$\operatorname{mes}(B_d) = \frac{\Gamma(3/2)^d 2^d}{\Gamma(d/2+1)} = (2\pi e + o(1))^{d/2} d^{-d/2}. \tag{3.8}$$

3.3.7 Let O_d be the octahedron given by the convex hull of $\pm e_1, \ldots, \pm e_d$ in $\mathbf{R}^d$. Show that $\operatorname{mes}(O_d) = 2^d/d! = (2e + o(1))^d d^{-d}$. Thus in large dimension the octahedron becomes considerably smaller than the circumscribing ball B_d which contains it, which in turn is considerably smaller than the circumscribing cube.

3.3.8 Show that the constants d and $\sqrt{d}$ in Theorem 3.13 cannot be improved. (For the non-symmetric case, take A to be a d-simplex (the convex hull of d points in $\mathbf{R}^d$); for the symmetric case, take A to be a cube.)

3.3.9 If A and A' are two symmetric convex bodies in $\mathbf{R}^d$, show that there exists an invertible linear transformation $T : \mathbf{R}^d \to \mathbf{R}^d$ such that

$$A \subseteq T(A') \subset d \cdot A.$$

State and prove a similar result in the case when A and A' are not necessarily symmetric.

3.3.10 Let A, B be open bounded sets. Show that

$$\mathrm{mes}((A - A) \cap (B - B)) \geq \frac{\mathrm{mes}(A)\mathrm{mes}(B)}{\mathrm{mes}(A \pm B)}$$

for either choice of sign $\pm$, by developing a continuous analogue of the arguments used to prove (2.8). (Alternatively, one can try to discretize A and B to replace them with finite sets, and then use (2.8) directly.)

3.3.11 [26] Let A be a symmetric convex body in $\mathbf{R}^d$, which contains the ball $\rho \cdot B$ of radius $\rho > 0$ centered at the origin. Let V be any r-dimensional subspace of $\mathbf{R}^d$. Show that $\mathrm{mes}_r(A \cap V) \leq \frac{d!}{r!(2\rho)^{d-r}} \mathrm{mes}_d(A)$, where mes_r denotes r-dimensional measure. (Hint: first show that if $r < d$, then there exists an $r + 1$-dimensional space V_1 containing V such that $\mathrm{mes}_{r+1}(A \cap V_1) \geq \frac{2\rho}{r+1} \mathrm{mes}_r(A \cap V)$. Then continue inductively.)

3.4 The Brunn–Minkowski inequality

The purpose of this section is to prove the following lower bound for the volume $\mathrm{mes}(A + B)$ of a sum set.

Theorem 3.16 (Brunn–Minkowski inequality) *If A and B are non-empty bounded open subsets of $\mathbf{R}^d$, then*

$$\mathrm{mes}(A + B)^{1/d} \geq \mathrm{mes}(A)^{1/d} + \mathrm{mes}(B)^{1/d}.$$

This inequality is sharp (Exercise 3.4.2). The theorem also applies if A and B are merely measurable (as opposed to being bounded and open), though one must then also assume that $A + B$ is measurable; we will not prove this here. In general, there is no upper bound for $\mathrm{mes}(A + B)$; consider for instance the case when A is the x-axis and B is the y-axis in $\mathbf{R}^2$, then A, B both have measure zero but $A + B$ is all of $\mathbf{R}^2$. One can easily modify this example to show that there is no upper bound for $\mathrm{mes}(A + B)$ in terms of $\mathrm{mes}(A)$ and $\mathrm{mes}(B)$ when A, B are bounded open sets. See [128] for a thorough survey of the Brunn–Minkowski inequality and related topics.

To prove this theorem, it suffices to prove the following dimension-independent version:

Theorem 3.17 *If A and B are non-empty bounded open subsets of $\mathbf{R}^d$, and $0 < \theta < 1$, then*

$$\mathrm{mes}((1 - \theta) \cdot A + \theta \cdot B) \geq \mathrm{mes}(A)^{1-\theta} \mathrm{mes}(B)^{\theta}.$$

To see why Theorem 3.17 implies the Brunn–Minkowski inequality, apply Theorem 3.17 with A and B replaced by $\mathrm{mes}(A)^{-1/d} \cdot A$ and $\mathrm{mes}(B)^{-1/d} \cdot B$ to obtain

$$\mathrm{mes}\left(\frac{1-\theta}{\mathrm{mes}(A)^{1/d}} \cdot A + \frac{\theta}{\mathrm{mes}(B)^{1/d}} \cdot B\right) \geq 1$$

for any $0 < \theta < 1$. Setting

$$\theta := \frac{\mathrm{mes}(B)^{1/d}}{\mathrm{mes}(A)^{1/d} + \mathrm{mes}(B)^{1/d}}$$

we obtain the result. Conversely, one can easily deduce Theorem 3.17 from the Brunn–Minkowski inequality (Exercise 3.4.1).

It remains to prove Theorem 3.17. We begin by first proving

Lemma 3.18 (One-dimensional Brunn–Minkowski inequality) *If A and B are non-empty bounded open subsets of* $\mathbf{R}$*, then* $\mathrm{mes}(A + B) \geq \mathrm{mes}(A) + \mathrm{mes}(B)$.

Proof The hypotheses and conclusion of this lemma are invariant under independent translations of A and B, so we can assume that $\sup(A) = 0$ and $\inf(B) = 0$, hence in particular A and B are disjoint. But then we see that $A + B$ contains both A and B separately, and we are done. □

Using this Lemma, we deduce

Proposition 3.19 (One-dimensional Prékopa–Leindler inequality) *Let $0 < \theta < 1$, and let $f, g, h : \mathbf{R} \to [0, \infty)$ be lower semi-continuous, compactly supported non-negative functions on* $\mathbf{R}$ *such that*

$$h((1-\theta)x + \theta y) \geq f(x)^{1-\theta} g(y)^{\theta}$$

for all $x, y \in \mathbf{R}$. Then we have

$$\int_{\mathbf{R}} h \geq \left(\int_{\mathbf{R}} f\right)^{1-\theta} \left(\int_{\mathbf{R}} g\right)^{\theta}.$$

Proof By multiplying f, g, h by appropriate positive constants we may normalize $\int_{\mathbf{R}} f = \int_{\mathbf{R}} g = 1$.

Let $1 > \lambda > 0$ be arbitrary. Observe that if $f(x) > \lambda$ and $g(y) > \lambda$, then by hypothesis $h((1-\theta)x + \theta y) > \lambda$. Thus we have

$$\{z \in \mathbf{R} : h(z) > \lambda\} \subseteq (1-\theta) \cdot \{x \in \mathbf{R} : f(x) > \lambda\} + \theta \cdot \{y \in \mathbf{R} : g(y) > \lambda\}.$$

Since f, g, h are lower semi-continuous and compactly supported, all the sets above are open and bounded, hence by Lemma 3.18

$$\mathrm{mes}(\{z \in \mathbf{R} : h(z) > \lambda\}) \geq (1-\theta)\mathrm{mes}(\{x \in \mathbf{R} : f(x) > \lambda\}) + \theta\,\mathrm{mes}(\{y \in \mathbf{R} : g(y) > \lambda\}).$$

Integrating this for $\lambda \in [0, \infty)$ and using Fubini's theorem (cf. (1.6)), the claim follows from the arithmetic mean–geometric mean inequality. □

Now we iterate this to higher dimensions.

Proposition 3.20 (Higher-dimensional Prékopa–Leindler inequality) *Let $0 < \theta < 1$, $d \geq 1$, and let $f, g, h : \mathbf{R}^d \to [0, \infty)$ be lower semi-continuous, compactly supported non-negative functions on $\mathbf{R}^d$ such that*

$$h((1-\theta)x + \theta y) \geq f(x)^{1-\theta} g(y)^\theta$$

for all $x, y \in \mathbf{R}^d$. Then we have

$$\int_{\mathbf{R}} h \geq \left(\int_{\mathbf{R}} f\right)^{1-\theta} \left(\int_{\mathbf{R}} g\right)^\theta.$$

Proof We induce on d. When $d = 1$ this is just Proposition 3.19. Now assume inductively that $d > 1$ and the claim has already been proven for all smaller dimensions d. Define the one-dimensional function $h_d : \mathbf{R} \to [0, \infty)$ by

$$h_d(x_d) := \int_{\mathbf{R}^{d-1}} h(x_1, \ldots, x_d)\, dx_1 \cdots dx_{d-1},$$

and similarly define f_d, g_d. One can easily check (using Fatou's lemma) that these functions are lower semi-continuous and compactly supported. Also, applying the inductive hypothesis at dimension $d - 1$ we see that

$$h_d((1-\theta)x_d + \theta y_d) \geq f_d(x_d)^{1-\theta} g_d(y_d)^\theta$$

for all $x_d, y_d \in \mathbf{R}$. If we then apply the one-dimensional Prékopa–Leindler inequality, we obtain the desired result. □

If we apply Proposition 3.20 with $f := 1_A$, $g := 1_B$, and $h := 1_{(1-\theta)A+\theta B}$ we obtain Theorem 3.17, and the Brunn–Minkowski inequality follows.

Exercises

3.4.1 Show that Theorem 3.16 implies Theorem 3.17.

3.4.2 Show that equality in Theorem 3.17 can occur when A is convex, and $B = \lambda \cdot A + x_0$ for some $\lambda, x_0 \in \mathbf{R}^n$. Conversely, if A and B are non-empty bounded open subsets of $\mathbf{R}^d$, show that the preceding situation is in fact the only case in which equality can be attained. (The case when A and B are merely measurable is a bit trickier, and is of course only true up to sets of measure zero; see [128] for further discussion).

3.4.3 Let A be a convex body in $\mathbf{R}^d$. Using Theorem 3.17, show that the cross-sectional areas $f(x_d) := \mathrm{mes}(\{x' \in \mathbf{R}^{d-1} : (x', x_d) \in A\})$ are a log-concave function of x_d, i.e. $f((1-\lambda)x_d + \lambda y_d) \geq f(x_d)^{1-\lambda} f(y_d)^{\lambda}$ for all $0 \leq \lambda \leq 1$ and $x_d, y_d \in \mathbf{R}$; this is known as *Brunn's inequality*.

3.4.4 Let A be a bounded open set with smooth boundary ∂A, and let B be a ball with the same volume as A. Prove the *isoperimetric inequality* $\text{mes}(\partial A) \geq \text{mes}(\partial B)$. (Hint: Use the Brunn–Minkowski inequality to estimate $\frac{\text{mes}(A+\varepsilon\cdot B)-\text{mes}(A)}{\varepsilon}$ for $\varepsilon > 0$ small, and then let $\varepsilon \to 0$.)

3.4.5 Let A, B be symmetric convex bodies in $\mathbf{R}^d$. Show by examples that there is no upper bound for $\text{mes}(A+B)$ in terms of $\text{mes}(A)$, $\text{mes}(B)$, and d alone, except in the $d=1$ case. However, by using Lemma 3.12, show that $\text{mes}(A+B) \leq 4^d \frac{\text{mes}(A)\text{mes}(B)}{\text{mes}(A\cap B)}$.

3.4.6 [282] Let A be a convex body. Use the Brunn's inequality to show that $\text{mes}(A \cap (x+A)) \geq (1-r)^d \text{mes}(A)$ whenever $0 \leq r \leq 1$ and $x \in r\cdot(A-A)$. Conclude that

$$\begin{aligned} \text{mes}(A)^2 &= \int_{A-A} \text{mes}(A \cap (x+A))\, dx \\ &\geq \int_0^1 d(1-r)^{d-1}\text{mes}(A)\text{mes}(r\cdot(A-A))\, dr \\ &= \frac{1}{\binom{2d}{d}}\text{mes}(A)\text{mes}(A-A) \end{aligned}$$

whence one obtains the *Rogers–Shepard inequality* $\text{mes}(A-A) \leq \binom{2d}{d}\text{mes}(A)$. Show that this inequality is sharp when A is a simplex. Use Stirling's formula to compare this inequality with (3.5).

3.4.7 [162] Let A, B be additive sets in $\mathbf{Z}^d$. Use the Brunn–Minkowski inequality to show that $|A + B + \{0,1\}^d| \geq 2^d \min(|A|, |B|)$. (Hint: consider $A + [0,1]^d$ and $B + [0,1]^d$.)

3.4.8 [162] Let A, B be additive sets in $\mathbf{R}^d$. Show that $|A + B + \{0,1\}^d| \geq 2^d \min(|A|, |B|)$. (Hint: partition $\mathbf{R}^d$ into cosets of $\mathbf{Z}^d$, locate the coset with the largest intersection with A or B, and apply the preceding exercise.)

3.4.9 Let A be an open bounded set in $\mathbf{R}^d$. Show that $\text{mes}(A+A) \geq 2^d\text{mes}(A)$, with equality if and only if A is convex. (Hint: $A+A$ contains $2\cdot A$.)

3.5 Intersecting a convex set with a lattice

In previous sections we have studied lattices, which are discrete but unbounded, and convex sets, which are bounded but continuous. We now study the intersection $B \cap \Gamma$ of a convex set B and a lattice Γ in a Euclidean space $\mathbf{R}^d$, which is then necessarily a finite set. A model example of such set is the discrete box $[0, N)$ for some $N = (N_1, \ldots, N_d)$, which is the intersection of the convex body $\{(x_1, \ldots, x_d) : -1 < x_i < N_i \text{ for all } 1 \leq i \leq d\}$ with the Euclidean lattice $\mathbf{Z}^d$. One of the main objectives of this section shall to show a "discrete John's lemma" which shows that all intersections $B \cap \Gamma$ can be approximated in a certain sense by a discrete box.

We begin with some elementary estimates.

Lemma 3.21 *Let Γ be a lattice in $\mathbf{R}^d$. If $A \subset \mathbf{R}^d$ is an arbitrary bounded set and $P \subset \mathbf{R}^d$ is a finite non-empty set, then*

$$|A \cap (\Gamma + P)| \leq |(A - A) \cap (\Gamma + P - P)|. \tag{3.9}$$

If B is a symmetric convex body, then

$$(k \cdot B) \cap \Gamma \text{ can be covered by } (4k+1)^d \text{ translates of } B \cap \Gamma \tag{3.10}$$

for all $k \geq 1$. If furthermore Γ' is a sub-lattice of Γ of finite index $|\Gamma/\Gamma'|$, then we have

$$|B \cap \Gamma'| \leq |B \cap \Gamma| \leq 9^d |\Gamma/\Gamma'| |B \cap \Gamma'|. \tag{3.11}$$

Proof We first prove (3.9). We may of course assume that $A \cap (\Gamma + P)$ contains at least one element a. But then $A \cap (\Gamma + P) \subseteq ((A - A) \cap (\Gamma + P - P)) + a$, and the claim follows. Now we prove (3.10). By the preceding argument we can cover $|(\frac{1}{2} \cdot B + x) \cap \Gamma|$ by a translate of $B \cap \Gamma$ for any $x \in \mathbf{R}^d$. But by Corollary 3.15 we can cover $k \cdot B$ by $(4k+1)^d$ translates of $\frac{1}{2} \cdot B$, and the claim (3.10) follows.

Finally, we prove (3.11). The lower bound is trivial. For the upper bound, observe that Γ is the union of $|\Gamma/\Gamma'|$ translates of Γ, so it suffices to show that $|B \cap (\Gamma' + x)| \leq 9^d |B \cap \Gamma'|$ for all $x \in \mathbf{R}^d$. But by (3.9) and (3.10) we have

$$|B \cap (\Gamma' + x)| \leq |(2 \cdot B) \cap \Gamma'| \leq 9^d |B \cap \Gamma'|$$

as desired. □

Next, we recall a result of Gauss concerning the intersection of a large convex body with a lattice of full rank.

Lemma 3.22 *Let $\Gamma \subset \mathbf{R}^d$ be a lattice of full rank, let $v_1, \ldots, v_d \in \Gamma$ be a set of generators for Γ, and let B be a convex body in $\mathbf{R}^d$. Then for large $R > 0$, we have*

$$|(R \cdot B) \cap \Gamma| = (R^d + O_{\Gamma,B,d}(R^{d-1})) \frac{\operatorname{mes}(B)}{|v_1 \wedge \cdots \wedge v_d|}.$$

Here $|v_1 \wedge \cdots \wedge v_d|$ denotes the volume of the parallelepiped with edges $v_1, \ldots, v_d$.

Proof We use a "volume-packing argument". Since Γ has full rank, $v_1, \dots, v_d$ are linearly independent. By applying an invertible linear transformation we may assume that $v_1, \dots, v_d$ is just the standard basis $e_1, \dots, e_d$, so that $\Gamma = \mathbf{Z}^d$. Now let Q be the unit cube centered at the origin. Observe that the sets $\{x + Q : x \in (R \cdot B) \cap \mathbf{Z}^d\}$ are disjoint up to sets of measure zero, and their union differs from $R \cdot B$ only in the $\sqrt{d}$-neighborhood of the surface of $R \cdot B$, which has volume $O_{\Gamma,B,d}(R^{d-1})$. The claim follows. □

Remark 3.23 The task of improving the error term $O_{\Gamma,B,d}(R^{d-1})$ for various lattices and convex bodies (e.g. Gauss' circle problem) is a deep and important problem in number theory and harmonic analysis, but we will not discuss this issue in this book; our only concern here is that the error term is strictly lower order than the main term.

If Γ is a lattice, we define a *fundamental parallelepiped* for Γ to be any parallelepiped whose edges $v_1, \dots, v_d$ generate Γ. From the above lemma we conclude that all fundamental parallelepipeds have the same volume; indeed this volume is nothing more than the covolume $\mathrm{mes}(\mathbf{R}^d/\Gamma)$ of Γ. Thus for instance $\mathrm{mes}(\mathbf{R}^d/\mathbf{Z}^d) = 1$.

By another volume-packing argument we can establish

$$\mathrm{mes}(\mathbf{R}^d/\Gamma)|\Gamma/\Gamma'| = \mathrm{mes}(\mathbf{R}^d/\Gamma') \tag{3.12}$$

whenever $\Gamma' \subseteq \Gamma \subset \mathbf{R}^d$ are two lattices of full rank; see the exercises. In particular we see that the quotient group $|\Gamma/\Gamma'|$ is finite.

Yet another volume-packing argument gives the following continuous and periodic analogue of (2.8).

Lemma 3.24 (Volume-packing lemma) *Let $\Gamma \subset \mathbf{R}^d$ be a lattice of full rank, let V be a bounded open subset of $\mathbf{R}^d$, and let P be a finite non-empty set in $\mathbf{R}^d$. Then*

$$|(V - V) \cap (\Gamma + P - P)| \geq \frac{\mathrm{mes}(V)|P|}{\mathrm{mes}(\mathbf{R}^d/\Gamma)}.$$

In particular, we have

$$|(V - V) \cap \Gamma| \geq \frac{\mathrm{mes}(V)}{\mathrm{mes}(\mathbf{R}^d/\Gamma)}.$$

Proof Let B be the unit ball on $\mathbf{R}^d$, and let $R > 0$ be a large number. Consider the integral of the function

$$f(x) := \sum_{y \in \Gamma \cap (R \cdot B)} \sum_{p \in P} 1_{V+y+p}(x).$$

On the one hand we can compute this integral using Lemma 3.22 as

$$\begin{aligned}\int_{\mathbf{R}^d} f(x)\,dx &= \sum_{y\in\Gamma\cap(R\cdot B)}\sum_{p\in P}\mathrm{mes}(V+y)\\ &= |\Gamma\cap(R\cdot B)||P||\mathrm{mes}(V)|\\ &= (R^d + O_{\Gamma,B,d}(R^{d-1}))|P|\frac{\mathrm{mes}(B)\mathrm{mes}(V)}{\mathrm{mes}(\mathbf{R}^d/\Gamma)}\end{aligned}$$

On the other hand, from (3.9) we have

$$f(x) \leq |(x-V)\cap(\Gamma+P)| \leq |(V-V)\cap(\Gamma+P-P)|.$$

Furthermore, $f(x)$ is only non-zero when x lies in $R\cdot B+V+P\subset(R+O_{V,P}(1))\cdot B$, which has volume $(R^d+O_{V,P,d}(R^{d-1}))\mathrm{mes}(B)$. Thus

$$\int_{\mathbf{R}^d} f(x)\,dx \leq |(V-V)\cap(\Gamma+P-P)|R^d + O_{V,P,d}(R^{d-1}).$$

Combining these inequalities, dividing by R^d, and taking limits as $R\to\infty$, we obtain the result. □

To see the utility of this lemma, let us pause to establish the following classical result in number theory, which we will need later in this book. Let $\|x\|_{\mathbf{R}/\mathbf{Z}}$ denote the distance from x to the nearest integer.

Corollary 3.25 (Kronecker approximation theorem) *Let $\alpha_1,\ldots,\alpha_d$ be real numbers, and let $0<\theta_1,\ldots,\theta_d\leq 1/2$. Then for any $N>0$, we have*

$$|\{n\in(-N,N): \|n\alpha_j\|_{\mathbf{R}/\mathbf{Z}}<\theta_j \text{ for all } j=1,\ldots,d\}| \geq N\theta_1\cdots\theta_d.$$

In particular, if $N\theta_1\cdots\theta_d\geq 1$, then there exists an integer $0<n<N$ such that $\|n\alpha_j\|_{\mathbf{R}/\mathbf{Z}}\leq\theta_j$ for all $j=1,\ldots,d$.

Proof Apply Lemma 3.24 with $\Gamma:=\mathbf{Z}^d$,

$$V:=\{(t_1,\ldots,t_d): 0<t_j<\theta_j \text{ for all } 1\leq j\leq d\},$$

and P equal to the arithmetic progression $P=[0,N)\cdot(\alpha_1,\ldots,\alpha_d)$ in $\mathbf{R}^d$. □

Even when B is symmetric, it is possible for $|B\cap\Gamma|$ to be extremely large compared with $\frac{\mathrm{mes}(B)}{2^d\mathrm{mes}(\mathbf{R}^d/\Gamma)}$; consider for instance $\Gamma:=\mathbf{Z}^2$ and $B:=\{(x,y): -1/N^2<x<1/N^2; -N<y<N\}$. However, if $B\cap\Gamma$ has full rank, then we can complement the lower bound (3.14) with an upper bound:

Lemma 3.26 *Let Γ be a lattice of full rank in $\mathbf{R}^d$, and let B be a symmetric convex body in $\mathbf{R}^d$ such that the vectors in $B\cap\Gamma$ linearly span $\mathbf{R}^d$. Then*

$$|B\cap\Gamma| \leq \frac{3^d d!\,\mathrm{mes}(B)}{2^d\mathrm{mes}(\mathbf{R}^d/\Gamma)}. \tag{3.13}$$

This bound is with a factor of $3^d/(2d+1)$ of being sharp, as can be seen by the example where $\Gamma = \mathbf{Z}^d$ and B is (a slight enlargement of) the octahedron with vertices $\pm e_1, \ldots, \pm e_d$. Indeed this example motivates the volume-packing argument used in the proof.

Proof By hypothesis, $B \cap \Gamma$ contains a d-tuple $(v_1, \ldots, v_d)$ of linearly independent vectors. Since $B \cap \Gamma$ is finite, we can choose $v_1, \ldots, v_d$ in order to minimize the volume $\mathrm{mes}(O) = \frac{2^d}{d!}|v_1 \wedge \cdots \wedge v_d|$ of the octahedron with vertices $\pm v_1, \ldots, \pm v_d$. Since B is symmetric and convex, we see that $O \subseteq B$. Also O does not contain any elements of Γ other than $v_1, \ldots, v_d$, since otherwise one could replace one of $v_1, \ldots, v_d$ with this element and reduce the volume of O, a contradiction. Thus we see that the sets $\{x + \frac{1}{2} \cdot O : x \in B \cap \Gamma\}$ are all disjoint and are contained in $B + \frac{1}{2} \cdot O \subseteq \frac{3}{2} \cdot B$. Thus

$$|B \cap \Gamma| \leq \frac{\mathrm{mes}(\frac{3}{2} \cdot B)}{\mathrm{mes}(\frac{1}{2} \cdot O)} = \frac{3^d d!}{2^d |v_1 \wedge \cdots \wedge v_d|} \mathrm{mes}(B).$$

Since $|v_1 \wedge \cdots \wedge v_d| \geq \mathrm{mes}(\mathbf{R}^d / \Gamma)$, the claim follows. □

A special case of the volume-packing lemma gives

Lemma 3.27 (Blichfeldt's lemma) *Let $\Gamma \subset \mathbf{R}^d$ be a lattice of full rank, and let V be an open set in $\mathbf{R}^d$ such that* $\mathrm{mes}(V) > \mathrm{mes}(\mathbf{R}^d / \Gamma)$. *Then there exists distinct $x, y \in V$ such that $x - y \in \Gamma$.*

Now let us apply Lemma 3.24 to the case $V = \frac{1}{2} \cdot B$ and $P = \{0\}$, where B is a symmetric convex body; we obtain the lower bound

$$|B \cap \Gamma| \geq \frac{\mathrm{mes}(B)}{2^d \mathrm{mes}(\mathbf{R}^d / \Gamma)}, \tag{3.14}$$

which is the classical Minkowski's first theorem. The assumption of symmetry is essential. Consider for instance $\Gamma := \mathbf{Z}^2$ and a convex set of the form $B := \{(x, y) : 1/3 < x < 2/3; -N < y < N\}$ for arbitrarily large N.

Theorem 3.28 (Minkowski's first theorem) *Let Γ be a lattice of full rank, and let B be a symmetric convex body such that* $\mathrm{mes}(B) \geq 2^d \mathrm{mes}(\mathbf{R}^d / \Gamma)$. *Then the closure of B must contain at least one non-zero element of Γ (in fact it contains at least two, by symmetry). If we have strict inequality,* $\mathrm{mes}(B) > 2^d \mathrm{mes}(\mathbf{R}^d / \Gamma)$, *then we can replace the closure of B with the interior of B in the above statement.*

Proof Apply (3.14) to $(1 + \epsilon)B$ and let ϵ go to zero. □

The constant in Minkowski's first theorem is sharp. We may apply an invertible linear transformation to set $\Gamma := \mathbf{Z}^d$, and then the example of the cube $A :=$

$\{(t_1, \ldots, t_d) : -1 < t_j < 1 \text{ for all } j = 1, \ldots, d\}$ shows that the constant 2^d cannot be improved. Nevertheless, it is possible to improve Minkowski's first theorem by generalizing it to a "multiparameter" version as follows.

Definition 3.29 (Successive minima) Let Γ be a lattice in $\mathbf{R}^d$ of rank k, and let B be a convex body in $\mathbf{R}^d$. We define the *successive minima* $\lambda_j = \lambda_j(B, \Gamma)$ for $1 \leq j \leq k$ of B with respect to Γ as

$$\lambda_j := \inf\{\lambda > 0 : \lambda \cdot B \text{ contains } k \text{ linearly independent elements of } \Gamma\}.$$

Note that $0 < \lambda_1 \leq \cdots \leq \lambda_k < \infty$.

Thus, for instance, if $\Gamma = \mathbf{Z}^d$ and B is the box

$$B := \{(t_1, \ldots, t_d) : |t_j| < a_j \text{ for all } j = 1, \ldots, d\}$$

for some $a_1 \geq a_2 \geq \cdots \geq a_d > 0$, then $\lambda_j = 1/a_j$ for $j = 1, \ldots, d$. Note that the assumption that Γ has rank k ensures that the λ_j are both finite and non-zero.

Theorem 3.30 (Minkowski's second theorem) *Let Γ be a lattice of full rank in $\mathbf{R}^d$, and let B be an symmetric convex body in $\mathbf{R}^d$, with successive minima $0 < \lambda_1 \leq \cdots \leq \lambda_d$. Then there exists d linearly independent vectors $v_1, \ldots, v_d \in \Gamma$ with the following properties:*

- *for each $1 \leq j \leq d$, v_j lies in the boundary of $\lambda_j \cdot B$, but $\lambda_j \cdot B$ itself does not contain any vectors in Γ outside of the span of $v_1, \ldots, v_{j-1}$;*
- *the octahedron with vertices $\pm v_j$ contains no elements of Γ in its interior, other than the origin;*
- *we have*

$$\frac{2^d |\Gamma/(\mathbf{Z}^d \cdot (v_1, \ldots, v_d))|}{d!} \leq \frac{\lambda_1 \cdots \lambda_d \mathrm{mes}(B)}{\mathrm{mes}(\mathbf{R}^d/\Gamma)} \leq 2^d; \tag{3.15}$$

in particular, the sub-lattice $\mathbf{Z}^d \cdot (v_1, \ldots, v_d)$ of Γ has bounded index:

$$|\Gamma/(\mathbf{Z}^d \cdot (v_1, \ldots, v_d))| \leq d!. \tag{3.16}$$

One can state (3.15) rather crudely as

$$\lambda_1 \cdots \lambda_d \mathrm{mes}(B) = d^{O(d)} \mathrm{mes}(\mathbf{R}^d/\Gamma)$$

thus relating the successive minima to the volume of the body B and the covolume of the lattice Γ.

Note that if B contains no non-zero elements of Γ then $\lambda_j \geq 1$ for all j, so Minkowski's second theorem implies Minkowski's first theorem. Conversely, we shall see from the proof that Minkowski's second theorem can be obtained from Minkowski's first theorem by a non-isotropic dilation. The basis $v_1, \ldots, v_d$ is sometimes referred to as a *directional basis* for A with respect to Γ, although one should caution that this basis does not quite generate Γ (the index in (3.16) is bounded but not necessarily equal to 1).

Proof By definition of λ_1, we may find a vector $v_1 \in \Gamma$ such that v_1 lies in the closure of $\lambda_1 \cdot B$, but that $\lambda \cdot B$ contains no non-zero elements of Γ for any $\lambda \leq \lambda_1$. By definition of λ_2, we can then find a vector $v_2 \in \Gamma$, linearly independent from v_1, such that v_2 lies in the closure of $\lambda_2 B$, but that $\lambda \cdot B$ contains no elements of Γ outside of the span of v_1 for any $\lambda \leq \lambda_2$. Continuing inductively we can eventually find a linearly independent set $v_1, \ldots, v_d$ in Γ such that v_j lies in the boundary of $\lambda_j \cdot B$, but $\lambda_j \cdot A$ itself does not contain any vectors in Γ outside of the span of $v_1, \ldots, v_{j-1}$, for all $1 \leq j \leq n$.

The set $v_1, \ldots, v_d$ is a basis of $\mathbf{R}^d$; by applying an invertible linear transformation we may assume it is the standard basis $e_1, \ldots, e_d$ (this changes both B and Γ, but one may easily verify that the conclusion of the theorem remains unchanged). In particular this forces Γ to contain $\mathbf{Z}^d$, hence by (3.12)

$$\mathrm{mes}(\mathbf{R}^d / \Gamma) = \mathrm{mes}(\mathbf{R}^d/\mathbf{Z}^d)/|\Gamma/\mathbf{Z}^d| = 1/|\Gamma/\mathbf{Z}^d| \leq 1. \tag{3.17}$$

Let O^d be the open octahedron whose vertices are $\pm e_1, \ldots, \pm e_d$. We need to verify that O^d contains no lattice points from Γ other than the origin. Suppose for contradiction that $O^d \cap \Gamma$ contained $w = t_1 e_1 + \cdots + t_j e_j$ where $1 \leq j \leq d$ and $t_j \neq 0$. Then $(1+\varepsilon)w$ would be a linear combination of $\pm e_1, \ldots, \pm e_j$ for some $\varepsilon > 0$. All of these points lie in the closure of $\lambda_j \cdot B$, hence w lies in the interior of $\lambda_j \cdot B$, but does not lie in the span of $e_1, \ldots, e_{j-1}$. But this contradicts the construction of $v_j = e_j$. Hence $O^d \cap \Gamma = \{0\}$.

Next, observe that $\pm v_j = \pm e_j$ lies on the boundary of $\lambda_j \cdot B$ for each $1 \leq j \leq d$. Thus B contains the open octahedron whose vertices are $\pm e_1/\lambda_1, \ldots, \pm e_d/\lambda_d$. This octahedron is easily verified to have volume $\frac{2^d}{d!\lambda_1 \cdots \lambda_d}$; indeed one can rescale to the case when all the λ_j are equal to 1, and then one can decompose the octahedron into 2^d simplices, each of which has volume $1/d!$. This establishes the lower bound in (3.15).

Now we establish the upper bound in (3.15). We need the following lemma.

Lemma 3.31 (Squeezing lemma) *Let K be a symmetric convex body in $\mathbf{R}^d$, let A be an open subset of K, let V be a k-dimensional subspace of $\mathbf{R}^d$, and let $0 < \theta \leq 1$. Then there exists an open subset A' of K such that* $\mathrm{mes}(A') = \theta^k \mathrm{mes}(A)$ *and* $(A' - A') \cap V \subseteq \theta \cdot (A - A) \cap V$.

Note that we do not assume any convexity on A or A'. Indeed the squeezing operation we define in the proof below does not preserve the convexity of A.

Proof Without loss of generality we may take $V = \mathbf{R}^k$, and write $\mathbf{R}^d = \mathbf{R}^k \times \mathbf{R}^{d-k}$. Let $\pi : \mathbf{R}^d \to \mathbf{R}^{d-k}$ be the orthogonal projection map, which restricts to a map $\pi : K \to \pi(K)$. Let $f : \pi(K) \to K$ be any continuous right-inverse of π; thus for instance $f(y)$ could be the center of mass of $\pi^{-1}(y)$.

A point $w \in K$ can be written as $w = (x, y)$, using the decomposition $\mathbf{R}^d = \mathbf{R}^k \times \mathbf{R}^{d-k}$. Consider the map Φ which maps $w = (x, y)$ to $\theta w + (1-\theta) f(y)$ and set $A' = \Phi(A)$. Since both w and $f(y)$ belong to K and K is convex, it follows that A' is an open subset of K. Furthermore, the second coordinate of $\Phi(w)$ is y as is that of $f(y)$. By applying Cavalieri's principle (or Fubini's theorem) we see that $\mathrm{mes}(A') = \theta^k \mathrm{mes}(A)$ (the map contracts A by a factor θ with respect to $V = \mathbf{R}^k$).

Consider a point $v = \Phi(w) - \Phi(w')$, where $w = (x, y), w' = (x', y')$ are points from A. If $v \in V$, then the second coordinate of v is zero, which means $y = y'$. Then by the definition of Φ, $v = \theta(w - w')$. Thus $v \in \theta \cdot (A - A)$, concluding the proof of Lemma 3.31. □

We apply the squeezing lemma iteratively, starting with $A_0 := \frac{\lambda_d}{2} \cdot B$, to create open sets $A_1, \ldots, A_{d-1} \subseteq A_0$ such that

$$\mathrm{mes}(A_j) = \left(\frac{\lambda_j}{\lambda_{j+1}}\right)^j \mathrm{mes}(A_{j-1})$$

and

$$(A_j - A_j) \cap \mathbf{R}^j \subseteq \frac{\lambda_j}{\lambda_{j+1}} \cdot (A_{j-1} - A_{j-1}) \cap \mathbf{R}^j$$

for all $1 \leq j \leq d-1$, where $\mathbf{R}^j$ is the span of $e_1, \ldots, e_j$. In every application of the squeezing lemma, A_0 plays the role of the mother set K.

Using the definition of A_0, it is easy to check that

$$\mathrm{mes}(A_{d-1}) = \lambda_1 \cdots \lambda_d 2^{-d} \mathrm{mes}(B). \tag{3.18}$$

Furthermore, by induction one can show

$$(A_{d-1} - A_{d-1}) \cap \mathbf{R}^j \subseteq \frac{\lambda_j}{\lambda_d} \cdot (A_{j-1} - A_{j-1}) \cap \mathbf{R}^j.$$

On the other hand, $A_{j-1} \subset A_0 = (\lambda_d/2) \cdot B$. Since B is symmetric, $\frac{\lambda_d}{2} \cdot B - \frac{\lambda_d}{2} \cdot B = \lambda_d \cdot B$. It follows that

$$(A_{d-1} - A_{d-1}) \cap \mathbf{R}^j \subset \lambda_j \cdot B \cap \mathbf{R}^j$$

for all $1 \leq j \leq d$.

By the definition of the successive minima, $\lambda_j \cdot B \cap \mathbf{R}^j$ does not contain any lattice point in Γ, except for those in $\mathbf{R}^{j-1}$. This implies that $A_{d-1} - A_{d-1}$ does not contain any point in Γ other than the origin. Applying Blichtfeld's lemma, we conclude that

$$\mathrm{mes}(A_{d-1}) \leq \mathrm{mes}(\mathbf{R}^d / \Gamma),$$

which when combined with (3.18) gives the upper bound in (3.15). □

We now give several applications of this theorem. First we "factorize" a convex body B as the finitely overlapping sum of a subset of Γ and and a dilate of a small convex body B', up to some scaling factors of $O(d)^{O(1)}$:

Lemma 3.32 *Let B be a symmetric convex body in $\mathbf{R}^d$, and let Γ be a lattice in $\mathbf{R}^d$. Then there exists a symmetric convex body $B' \subseteq B$ such that B' contains no non-zero elements of Γ, and such that $B \subseteq O(d^{3/2}) \cdot B' + ((O(d^{3/2}) \cdot B) \cap \Gamma$. In particular, the projection of B in $\mathbf{R}^d / \Gamma$ is contained in the projection of $O(d^{3/2}) \cdot B'$. Furthermore, we have the bounds*

$$\frac{\mathrm{mes}(B)}{O(d)^{5d/2}|B \cap \Gamma|} \leq \mathrm{mes}(B') \leq O(1)^d \frac{\mathrm{mes}(B)}{|B \cap \Gamma|}. \tag{3.19}$$

Proof By using John's theorem and an invertible linear transformation we may assume that $B_d \subseteq B \subseteq \sqrt{d} \cdot B_d$, where B_d is the unit ball. We may assume that the vectors in $B \cap \Gamma$ generate Γ, since otherwise we could replace Γ by the lattice generated by $B \cap \Gamma$.

Let us temporarily assume that Γ has full rank, and thus that the linear span of $B \cap \Gamma$ is $\mathbf{R}^d$. Thus if we let $\lambda_1 \leq \cdots \leq \lambda_d$ be the successive minima of B, then we have $\lambda_j \leq 1$ for all j.

Now we take a directional basis $v_1, \ldots, v_d$ of Γ, and let B' be the open octahedron with vertices $\pm v_j$; this octahedron then contains no non-zero elements of Γ, and is also contained in B (since $\pm v_j/\lambda_j$ already lies on the boundary of B). Observe that $d \cdot B'$ contains a parallelepiped with edges $v_1, \ldots, v_d$, and hence $d \cdot B' + \Gamma = \mathbf{R}^d$. Thus

$$B \subseteq d \cdot B' + ((B - d \cdot B') \cap \Gamma) \subseteq d \cdot B' + (((d+1) \cdot B) \cap \Gamma)$$

as desired (with about $d^{1/2}$ room to spare). In particular we have

$$\mathrm{mes}(B) \leq \mathrm{mes}(d \cdot B')|(d+1) \cdot B \cap \Gamma| \leq (d(4d+5))^d \mathrm{mes}(B')|B \cap \Gamma|$$

thanks to (3.10); this proves the lower bound in (3.19) (with a factor of $d^{d/2}$ to spare). Conversely, the sets $\{x + \frac{1}{2} \cdot B' : x \in B \cap \Gamma\}$ are disjoint (since B' contains no non-zero elements of Γ) and contained in $2 \cdot B$, hence

$$|B \cap \Gamma| \mathrm{mes}\left(\frac{1}{2} \cdot B'\right) \leq \mathrm{mes}(2 \cdot B)$$

which gives the upper bound in (3.19). This concludes the proof when Γ has full rank.

Now suppose that Γ has rank $r < d$, then after a rotation we may assume that Γ is contained in $\mathbf{R}^r \times \{0\} \subset \mathbf{R}^r \times \mathbf{R}^{d-r}$. The point is that the behavior in the $d - r$ dimensions orthogonal to $\mathbf{R}^r$ is rather trivial and can be easily dealt with as follows. Let $\tilde{B} \subset \mathbf{R}^r$ be the intersection of B with $\mathbf{R}^r \times \{0\}$, identifying $\mathbf{R}^r \times \{0\}$ with $\mathbf{R}^r$ in the usual manner. Then by John's theorem we have the inclusions

$$\frac{1}{2} \cdot (\tilde{B} \times B_{d-r}) \subseteq B \subseteq \sqrt{d} \cdot (\tilde{B} \times B_{d-r}).$$

Applying the previous arguments to $\tilde{B}$ to obtain a set $\tilde{B}' \subseteq \tilde{B}$, and then defining $B' := \frac{1}{2} \cdot (\tilde{B}' \times B_{d-r})$, we can verify the claim in this case (losing some additional factors of $d^{1/2}$ and $d^{d/2}$); we omit the details. □

In this theorem, we did not use the full strength of Minkowski's second theorem (in particular we did not need the upper bound). The notion of a directional vector was, however, useful.

As another consequence of Minkowski's second theorem, we show how to find large proper progressions inside sets of the form $B \cap \Gamma$.

Lemma 3.33 *Let B be a convex symmetric body in $\mathbf{R}^d$, and let Γ be a lattice in $\mathbf{R}^d$. Then there exists a proper progression P in $B \cap \Gamma$ of rank at most d such that $|P| \geq O(d)^{-7d/2}|B \cap \Gamma|$.*

Proof Applying John's theorem (Theorem 3.13) and (3.10) followed by a linear transformation, we may reduce to the case where B is the unit ball $B = B_d$ in $\mathbf{R}^d$, provided that we also reduce the $7d/2$ exponent to $3d$. We may assume that $B \cap \Gamma$ spans $\mathbf{R}^d$, since otherwise we may restrict B to the linear span of $B \cap \Gamma$, which is then isomorphic to a Euclidean space of some lower dimension. In particular this means Γ has full rank, and that the successive minima $0 < \lambda_1 \leq \cdots \leq \lambda_d$ of B with respect to Γ cannot exceed 1. Let $v_1, \ldots, v_d \in \Gamma \cap B$ be the corresponding directional basis. Let Q denote the parallelepiped

$$Q := \{t_1 v_1 + \cdots + t_d v_d : 0 \leq t_j < 1/2 \text{ for all } j \in [1, d]\}.$$

By (3.16), Since each translate of $Q - Q$ is a fundamental domain for $\mathbf{Z}^d \cdot (v_1, \ldots, v_d)$, it contains at most $d!$ elements of Γ. By Lemma 2.14, we can cover B by at most $\frac{\mathrm{mes}(B+Q)}{\mathrm{mes}(Q)}$ translates of $Q - Q$, and thus

$$|B \cap \Gamma| \leq d! \frac{\mathrm{mes}(B + Q)}{\mathrm{mes}(Q)}.$$

Since the $v_1, \ldots, v_d$ lie in the unit ball B, we see that $Q \subseteq \frac{d}{2} \cdot B$ and hence $B + Q \subseteq (\frac{d}{2} + 1) \cdot B$. Crudely bounding $d! = O(d^d)$, we thus conclude that

$$|B \cap \Gamma| \leq O(d)^{2d}/\text{mes}(Q).$$

From (3.15) we have

$$\lambda_1 \cdots \lambda_d \leq O(1)^d \text{mes}(\mathbf{Z}^d/\Gamma) \leq O(1)^d \text{mes}(Q)$$

and thus

$$|B \cap \Gamma| \leq O(d)^{2d}/\lambda_1 \cdots \lambda_d.$$

The claim now follows by setting $P := [-N, N] \cdot (v_1, \ldots, v_d)$, where $N_j := 1/2d\lambda_j$ for $j \in [1, d]$; note that one can easily verify that P is contained in $B \cap \Gamma$. □

We now give an alternative approach that gives results similar to Lemma 3.33. We first need a lemma to modify the directional basis given by Minkowski's second theorem (which only spans a sub-lattice of Γ, see (3.16)) into a genuine basis.

Theorem 3.34 (Mahler's theorem) *Let Γ be a lattice of full rank in $\mathbf{R}^d$, and let B be an symmetric convex body in $\mathbf{R}^d$, with successive minima $0 < \lambda_1 \leq \cdots \leq \lambda_d$. Let $v_1, \ldots, v_d$ be a directional basis for Γ. Then there exists a basis $w_1, \ldots, w_d$ of Γ such that w_1 lies in the closure of $\lambda_1 \cdot B$, and w_i lies in the closure of $\frac{i\lambda_i}{2} \cdot B$ for all $2 \leq i \leq d$. Furthermore, if V_i is the linear span of $v_1, \ldots, v_i$, then $w_1, \ldots, w_i$ forms a basis for $\Gamma \cap V_i$.*

The basis $w_1, \ldots, w_d$ is sometimes known as a *Mahler basis* for Γ.

Proof We choose $w_1 := v_1$; clearly w_1 forms a basis for $\Gamma \cap V_1$. Now suppose inductively that $2 \leq i \leq d$ and $w_1, \ldots, w_{i-1}$ have already been chosen with the desired properties. The lattice $\Gamma \cap V_i$ has one higher rank than $\Gamma \cap V_{i-1}$ and hence there exists a vector w_i in $\Gamma \cap (V_i \backslash V_{i-1})$ which, together with $\Gamma \cap V_{i-1}$, generates $\Gamma \cap V_i$; in particular, $w_1, \ldots, w_i$ will generate $\Gamma \cap V_i$. Since $v_1, \ldots, v_i$ linearly span V_i, we may write $w_i = t_1 v_1 + \cdots + t_{i-1} v_{i-1} + t_i v_i$ for some real numbers $t_1, \ldots, t_i$ with $t_i \neq 0$. Since v_i lies in $\Gamma \cap V_i$, we must have $t_i = \pm 1/n$ for some integer n. If $|t_i| = 1$, then $\Gamma \cap V_i$ is generated by $\Gamma \cap V_{i-1}$ and v_i, and we can take $w_i := v_i$. Thus we may assume $|t_i| \leq 1/2$. Also, by subtracting integer multiples of $v_1, \ldots, v_{i-1}$ from w_i if necessary (which will not affect the fact that $\Gamma \cap V_i$ is generated by $\Gamma \cap V_{i-1}$ and w_i) we may assume that $|t_j| \leq 1/2$ for all $1 \leq j < i$. But since each v_j lies in the closure of $\lambda_j \cdot B$ and hence $\lambda_i \cdot B$, we conclude by convexity that w_i lies in the closure of $\frac{i\lambda_i}{2} \cdot B$, and so we can continue the iterative construction. Setting $i = d$ we obtain the remaining claims in the theorem. □

As an application we give

Corollary 3.35 *Let Γ be a lattice of full rank in $\mathbf{R}^d$. Then there exists linearly independent vectors $w_1, \ldots, w_d \in \Gamma$ which generate Γ, and such that*

$$\operatorname{mes}(\mathbf{R}^d/\Gamma) = |w_1 \wedge \cdots \wedge w_d| \geq \Omega(d^{-3d/2})|w_1|\cdots|w_d|. \tag{3.20}$$

Proof Let $w_1, \ldots, w_d$ be a Mahler basis for Γ with respect to the unit ball B, and let $\lambda_1, \ldots, \lambda_d$ be the successive minima. Then by Theorem 3.34 we have

$$|w_1|\cdots|w_d| \leq \lambda_1 \prod_{i=2}^{d} \frac{i\lambda_i}{2}.$$

Applying (3.15) we obtain

$$|w_1|\cdots|w_d| \leq \frac{2d!}{\operatorname{mes}(B)} \operatorname{mes}(\mathbf{R}^d/\Gamma).$$

On the other hand, from (3.8) we have

$$\operatorname{mes}(B) = \frac{\Gamma(3/2)^d 2^d}{\Gamma(d/2+1)} = (2\pi e + o(1))^{d/2} d^{-d/2}.$$

Crudely bounding $d! = O(d^d)$, the claim follows. □

As a consequence, we can give a "discrete John's theorem" to characterize the intersection of a convex symmetric body with a lattice.

Lemma 3.36 (Discrete John's theorem) *Let B be a convex symmetric body in $\mathbf{R}^d$, and let Γ be a lattice in $\mathbf{R}^d$ of rank r. Then there exists a r-tuple $w = (w_1, \ldots, w_r) \in \Gamma^r$ of linearly independent vectors in Γ and and a r-tuple $N = (N_1, \ldots, N_r)$ of positive integers such that*

$$(r^{-2r} \cdot B) \cap \Gamma \subseteq (-N, N) \cdot w \subseteq B \cap \Gamma \subseteq (-O(r)^{(3r/2)}N, O(r)^{(3r/2)}N) \cdot w.$$

Notice that the fact $(-N, N) \cdot w \subseteq B \cap \Gamma$ is similar to the conclusion of Lemma 3.33. However, the generalized arithmetic progression in Lemma 3.33 has higher density.

Proof We first observe, using John's theorem and an invertible linear transformation, that we may assume without loss of generality that $B_d \subseteq B \subseteq d \cdot B_d$, where B_d is the unit ball in $\mathbf{R}^d$. We may assume that Γ has full rank $r = d$, for if $r < d$ then we may simply restrict B to the linear span of Γ, which can then be identified with $\mathbf{R}^r$. We may assume $d \geq 2$ since the claim is easy otherwise.

Now let $w = (w_1, \ldots, w_d)$ be as in Lemma 3.35. For each j, let L_j be the least integer greater than $1/d|w_j|$. Then from the triangle inequality we see that $|l_1 w_1 + \cdots + l_d w_d| < 1$ whenever $|l_j| < L_j$, and so $(-L, L) \cdot w$ is contained in B_d and hence in B.

Now let $x \in B \cap \Gamma$. Since w generates Γ, we have $x = l_1 w_1 + \cdots + l_d w_d$ for some integers $l_1, \ldots, l_d$; since $B \subseteq d \cdot B_d$, we have $|x| \leq d$. Applying Cramer's rule to solve for $l_1, \ldots, l_d$ and (3.20), we have

$$\begin{aligned} |l_j| &= \frac{|x \wedge w_1 \cdots w_{j-1} \wedge w_{j+1} \wedge w_d|}{|w_1 \wedge \cdots \wedge w_d|} \leq \frac{|x||w_1| \cdots |w_d|}{|w_j||w_1 \wedge \cdots \wedge w_d|} \\ &= \frac{|x|\, O(d)^{(3d/2)}}{|w_j|} \leq \frac{O(d)^{3d/2}}{|w_j|}, \end{aligned}$$

which is certainly at most $O(d)^{3d/2} L_j$. So $x \in (-O(d)^{3d/2} L, O(d)^{3d/2} L) \cdot w$, which is what we wanted to prove. A more-or-less identical argument gives the inclusion $(d^{-2d} \cdot B) \cap \Gamma \subseteq (-L, L) \cdot w$. □

It would be of interest to see if the constant $O(r)^{3r/2}$ could be significantly improved here, for instance to $e^{O(r)}$ or even $r^{O(1)}$. Progress on this issue may well have applications to improvements for Freiman's theorem (see Chapter 5), which can be viewed as a variant of the above theorem in which the set $B \cap \Gamma$ is replaced by a more general set of small doubling.

Exercises

3.5.1 Prove (3.12).

3.5.2 Let α be an irrational number, and let I be any open interval in $\mathbf{R}$. Show that $\mathbf{Z} \cdot \alpha$ and $I + \mathbf{Z}$ have non-empty intersection. (In other words, the integer multiples of α are dense in $\mathbf{R}/\mathbf{Z}$.)

3.5.3 Let Γ be a lattice in $\mathbf{R}^d$, and let A be a convex body (possibly asymmetric). Show that $\sigma[A \cap \Gamma] \leq O(1)^d$.

3.5.4 Let $v_1, \ldots, v_d$ be any vectors in a lattice $\Gamma \subset \mathbf{R}^d$ of full rank. Show that $|v_1 \wedge \cdots \wedge v_d|$ is an integer multiple of the covolume $\mathrm{mes}(\mathbf{R}^d / \Gamma)$.

3.5.5 Let Γ be a lattice of full rank in $\mathbf{R}^d$, let B be a symmetric convex body, and let $v_1, \ldots, v_d$ be a directional basis with successive minima $\lambda_1 \leq \cdots \leq \lambda_d$. Let O be the open octahedron with vertices $\pm v_j/\lambda_j$. Show that $O \subseteq B \subseteq O(d)^d \cdot O$. Thus Minkowski's second theorem can be used to give a rather weak version of John's theorem.

3.5.6 Let Γ be a lattice of full rank in $\mathbf{R}^d$, let B be a symmetric convex body, and let $\lambda_1 \leq \cdots \leq \lambda_d$ be the successive minima of B. Establish the bounds

$$O(d)^{-O(d)} \prod_{1 \leq i \leq d} \max\left(1, \frac{1}{\lambda_i}\right) \leq |B \cap \Gamma| \leq O(d)^{O(d)} \prod_{1 \leq i \leq d} \max\left(1, \frac{1}{\lambda_i}\right). \tag{3.21}$$

3.5.7 Generalize Lemma 3.32 and Lemma 3.36 to the case when B is an asymmetric convex body.

3.5.8 Let A be a bounded open subset of $\mathbf{R}^d$, and let B, C be open subsets of A. Prove that

$$\mathrm{mes}((B-B)\cap(C-C)) \geq \frac{\mathrm{mes}(B)\mathrm{mes}(C)\mathrm{mes}(A)}{\mathrm{mes}(A-B)\mathrm{mes}(A-C)}.$$

(Hint: use the volume-packing argument to locate a large set of the form $(x+B)\cap(y+C)$ where $x \in A-B$ and $y \in A-C$.)

3.5.9 Let B the the unit ball in $\mathbf{R}^5$, and let Γ be the lattice generated by the five basis vectors $e_1, \ldots, e_5$ and by $\frac{1}{2}(e_1 + \cdots + e_5)$. Show that in this case the directional basis for Γ does not actually generate Γ.

3.6 Progressions and proper progressions

In this section we work in a fixed additive group Z, which may or may not be torsion-free.

Recall from Definition 0.2 that a progression $P = a + [0, N] \cdot v$ is *proper* if the map $n \mapsto n \cdot v$ is injective on $[0, N]$. Not all progressions are proper; however it turns out that, just as John's theorem (Theorem 3.13) shows that all convex sets are in some sense comparable to ellipsoids, all progressions are comparable to proper progressions. This is most obvious in the rank 1 case, in which every arithmetic progression is equal (as a set) to a proper arithmetic progression:

Lemma 3.37 *Let $a + [0, N] \cdot v$ be an arithmetic progression in an additive group Z. Then there exists an $n > 0$ such that $a + [0, n) \cdot v$ is a proper arithmetic progression and $a + [0, n) \cdot v = a + [0, N] \cdot v$.*

Proof If $a + [0, N] \cdot v$ is already proper, then we are done. Otherwise, there exist distinct $n_1, n_2 \in [0, N]$ such that $a + n_1 \cdot v = a + n_2 \cdot v$. In particular, there exists $n \in [1, N]$ such that $n \cdot v = 0$. Let n be the least integer in $[1, N]$ with this property. Then $a + [0, n) \cdot v$ is necessarily proper, and by the Euclidean algorithm it is clear that $a + [0, n) \cdot v = a + [0, N] \cdot v$. □

We now consider the higher rank case; as with John's theorem, the constants will deteriorate worse than exponentially in d. We first show the easier of the two containments, namely that every progression contains a large proper progression of equal or lesser rank.

Theorem 3.38 *Let P be a progression of rank d in an additive group Z. Then P contains a proper progression of rank at most d and volume at least $O(d)^{-5d}|P|$.*

Remark 3.39 For a result of similar flavor (but proven by completely different methods), see Theorem 4.42 below. Note that the $d = 1$ case already follows from Lemma 3.37 (with a constant of 1 instead of $O(d)^{-5d}$).

Proof The idea is to pass to a convex body, apply Lemma 3.32 to obtain a "proper" subset of this body, and then use Lemma 3.33 to pass back to a progression.

By translating and enlarging P slightly we may assume $P = [-N, N] \cdot v$. We may assume that none of the components N_j of N are equal to 0 or 1, since otherwise we could refine P by at worst a factor of 3^d to eliminate those dimensions. Now consider the set $\Gamma := \{n \in \mathbf{Z}^d : n \cdot v = 0\}$, which is clearly a sub-lattice of $\mathbf{Z}^d$, and let A be the symmetric convex box

$$A := \{(x_1, \ldots, x_d) \in \mathbf{R}^d : -N_j \leq x_j \leq N_j \text{ for all } 1 \leq j \leq d\}.$$

By Lemma 3.32, we may find a symmetric convex subset A' of A such that $A' - A'$ is disjoint from $\Gamma - \{0\}$, and such that $A \subset O(d)^{3/2} \cdot A' + \Gamma$ for some $x \in \mathbf{R}^d$. From Corollary 3.15, we thus see that A can be covered by $O(d)^{3d/2}$ translates of $\frac{1}{2} \cdot A' + \Gamma$. Since $[-N, N] = A \cap \mathbf{Z}^d$ and $\Gamma \subseteq \mathbf{Z}^d$, we conclude that $[-N, N]$ can be covered by $O(d)^{3d/2}$ sets of the form $[(\frac{1}{2} \cdot A' + x) \cap \mathbf{Z}^d] + \Gamma$. Taking inner products with v, we conclude that $P = [-N, N] \cdot v$ can be covered by $O(d)^{3d/2}$ sets of the form $[(\frac{1}{2} \cdot A' + x) \cap \mathbf{Z}^d] \cdot v$. By the pigeonhole principle, there must thus exist an x such that

$$\left|\left(\frac{1}{2} \cdot A' + x\right) \cap \mathbf{Z}^d\right| \geq \Omega\left(\frac{1}{d}\right)^{3d/2} |P|$$

and hence by (3.9)

$$|A' \cap \mathbf{Z}^d| \geq \Omega\left(\frac{1}{d}\right)^{3d/2} |P|.$$

We now apply Lemma 3.33 to find a proper progression $\tilde{P} \subseteq A' \cap \mathbf{Z}^d \subseteq [0, N]$ of rank at most d such that

$$|\tilde{P}| \geq O(d)^{-7d/2} |A' \cap \mathbf{Z}^d| \geq \Omega\left(\frac{1}{d}\right)^{5d} |P|.$$

The set $\tilde{P} \cdot v$ is then clearly a progression of rank at most d contained in P; it is proper since $A' - A'$ is disjoint from $\Gamma - \{0\}$, so in particular $|\tilde{P} \cdot v| = |\tilde{P}|$. The claim follows. □

Now we show the more difficult containment, that every progression can be contained inside a proper progression of equal or lesser rank, but somewhat larger volume.

Theorem 3.40 *Let P be a progression of rank d in an additive group Z. Then P is contained in a proper progression Q of rank at most d and volume at most $d^{C_0 d^3}|P|$ for some absolute constant $C_0 > 0$. Also, Q is contained in a translate of $d^{C_0 d^2} P$. If $d \geq 2$, P is not proper, and Z is torsion-free then Q can be chosen to have rank at most $d - 1$. Finally, if Z is torsion-free and P is symmetric, then one can ensure that Q is symmetric also.*

Remark 3.41 Theorems of this type first appeared in the literature in [26], and later in some unpublished work of Gowers–Walters and Ruzsa. The version we give here is taken from [365].

Comparison with Theorem 3.38 suggests that the factor $d^{C_0 d^3}$ is probably not best possible, but we do not know what the correct constant here should be. This theorem can be thought of as the analogue of Corollary 3.8 or Corollary 3.9, but for progressions rather than finitely generated additive groups.

Proof This claim is analogous to the basic linear algebra statement that every linear space spanned by d vectors is equal to a linear space with a *basis* of at most d vectors. Recall that the proof of that fact proceeds by a descent argument, showing that if the d spanning vectors were linearly dependent, then one could exploit that dependence to "drop rank" and span the same linear space with $d - 1$ vectors. Our proof of Theorem 3.40 shall be based on a similar strategy.

We shall work only in the case when Z is torsion-free; the general case is proven similarly but contains a few additional technicalities, and we leave it as an exercise (Exercise 3.6.3).

We induce on d. When $d = 1$ the claim follows from Lemma 3.37. Now suppose inductively that $d \geq 2$, and the claim has already been proven for $d - 1$ (for arbitrary groups Z and arbitrary progressions P). Let $P = a + [0, N] \cdot v$ be a progression in Z of rank d, where $N = (N_1, \ldots, N_d)$ and $v = (v_1, \ldots, v_d)$; we may translate P so that the base point a equals 0. If P is proper, then we are done. Similarly, if one of the N_j is equal to zero, then we are done by induction hypothesis. Suppose instead that P is not proper and all the N_j are at least 1; then there exist distinct $n, n' \in [0, N]$ such that $n \cdot v = n' \cdot v$. If we then let $\Gamma_0 \subseteq \mathbf{Z}^d$ denote the lattice $\{m \in \mathbf{Z}^d : m \cdot v = 0\}$, then we see that $\Gamma_0 \cap [-N, N]$ contains at least one non-zero element, namely $n' - n$.

Let $m = (m_1, \ldots, m_d)$ be a non-zero element of $\Gamma_0 \cap [-N, N]$, thus

$$m_1 \cdot v_1 + \cdots + m_d \cdot v_d = 0. \tag{3.22}$$

We may assume without loss of generality that m is irreducible in Γ_0. Since Z is torsion-free, this also implies that m is irreducible in $\mathbf{Z}^d$ (i.e. that the $m_1, \ldots, m_d$ have no common divisor). The strategy shall be to contain P inside a progression

Q of rank $d-1$ and size

$$|Q| \leq d^{O(d^2)}|P|, \tag{3.23}$$

such that Q is contained in a translate of $d^{O(d)}P$. If we can achieve this, then by the induction hypothesis we can contain Q inside a proper progression R of rank at most $d-1$ and cardinality

$$|R| \leq (d-1)^{C_0(d-1)^3}(O(d))^{O(d^2)}|P|$$

and which is contained in a translate of $d^{C_0(d-1)^2}d^{O(d)}P$. If C_0 is sufficiently large, we will have completed the induction.

It remains to cover P by a progression of rank at most $d-1$ with the bound (3.23) and contained in a translate of $d^{O(d)}P$. Observe that m lies in $[-N, N]$, so the rational numbers $m_1/N_1, \ldots, m_d/N_d$ lie between -1 and 1. Without loss of generality we may assume that m_d/N_d has the largest magnitude, thus

$$|m_d|/N_d \geq |m_j|/N_j \tag{3.24}$$

for all $1 \leq j \leq d$. By replacing v_d with $-v_d$ if necessary, we may also assume that m_d is positive.

To exploit the cancellation in (3.22) we introduce the rational vector $q \in \frac{1}{m_d} \cdot \mathbf{Z}^{d-1}$ by the formula

$$q := \left(-\frac{m_1}{m_d}, \ldots, -\frac{m_{d-1}}{m_d}\right).$$

Since m is irreducible in $\mathbf{Z}^d$, we see, for any integer n, that $n \cdot q$ lies in $\mathbf{Z}^{d-1}$ if and only if n is a multiple of m_d, because $(m_1, \ldots, m_d)$ is irreducible in $\mathbf{Z}^d$.

Next, let $\Gamma \subset \mathbf{R}^{d-1}$ denote the lattice $\Gamma := \mathbf{Z}^{d-1} + \mathbf{Z} \cdot q$. Since q is rational, this is indeed a lattice; since it contains $\mathbf{Z}^{d-1}$, it is certainly full rank. We define the homomorphism $f : \Gamma \to Z$ by the formula

$$f((n_1, \ldots, n_{d-1}) + n_d q) := (n_1, \ldots, n_d) \cdot v;$$

the condition (3.22) ensures that this homomorphism is indeed well defined, in the sense that different representations $v = (n_1, \ldots, n_{d-1}) + n_d q$ of the same vector $v \in \Gamma$ give the same value of $f(v)$. We also let $B \subseteq \mathbf{R}^{d-1}$ denote the convex symmetric body

$$B := \{(t_1, \ldots, t_{d-1}) \in \mathbf{R}^{d-1} : -3N_j < t_j < 3N_j \text{ for all } 1 \leq j \leq d-1\}.$$

We now claim the inclusions

$$P \subseteq f(B \cap \Gamma) \subseteq 5P - 5P.$$

To see the first inclusion, let $n \cdot v \in P$ for some $n \in [0, N]$, then we have $n \cdot v = f((n_1, \ldots, n_{d-1}) + n_d q)$; from (3.24) we see that the jth coefficient of $(n_1, \ldots, n_{d-1}) + n_d q$ has magnitude at most $3N_j$, and thus $n \cdot v$ lies in $f(B \cap \Gamma)$ as claimed. To see the second inclusion, let $(n_1, \ldots, n_{d-1}) + n_d q$ be an element of $B \cap \Gamma$. By subtracting if necessary an integer multiple of m_d from n_d (and thus adding integer multiples of $m_1, \ldots, m_{d-1}$ to $n_1, \ldots, n_{d-1}$) we may assume that $|n_d| \leq |m_d|/2$. By (3.24) and the definition of B, this forces $|n_j| \leq 5N_j$ for all $1 \leq j \leq d$, and hence

$$f((n_1, \ldots, n_{d-1}) + n_d q) = (n_1, \ldots, n_d) \cdot v \subseteq [-5N, 5N] \cdot v = 5P - 5P.$$

Next, we apply Theorem 3.36 to find vectors $w_1, \ldots, w_{d-1} \in \Gamma$ and $M := (M_1, \ldots, M_{d-1})$ such that

$$(-M, M) \cdot w \subseteq B \cap \Gamma \subseteq (-d^{O(d)} M, d^{O(d)} M) \cdot w.$$

Applying the homomorphism f, we obtain

$$(-M, M) \cdot f(w) \subseteq f(B \cap \Gamma) \subseteq (-d^{O(d)} M, d^{O(d)} M) \cdot f(w)$$

where $f(w) := (f(w_1), \ldots, f(w_{d-1}))$. Observe that $(-d^{O(d)} M, d^{O(d)} M) \cdot f(w)$ is a progression of rank $d-1$ which contains $f(B \cap \Gamma)$ and hence contains P. Furthermore, by two applications of Lemma 3.10 we have

$$\begin{aligned} |(-d^{O(d)} M, d^{O(d)} M) \cdot f(w)| &\leq (O(d))^{O(d^2)} |f(B \cap \Gamma)| \\ &\leq (O(d))^{O(d^2)} |5P - 5P| \\ &\leq (O(d))^{O(d^2)} O(1)^d |P| \end{aligned}$$

which proves (3.23). Also, since $(-M, M) \cdot f(w)$ is contained in $f(B \cap \Gamma)$, which is contained in $5P - 5P$, which is a translate of $10P$, we see that $(-d^{O(d)} M, d^{O(d)} M) \cdot f(w)$ is contained in a translate of $d^{O(d)} P$. This completes the induction and proves the theorem. When P is symmetric, one can easily modify the above argument to ensure that all progressions in the above construction are also symmetric; we leave this modification to the interested reader. □

Exercises

3.6.1 Let $P = a + [0, N] \cdot v$ be a progression of rank d in some additive group Z, and let $\Gamma := \{n \in \mathbf{Z}^d : n \cdot v = 0\}$ be the associated sub-lattice of $\mathbf{Z}^d$. Prove the inequalities

$$\frac{|[0, N]|}{|P|} \leq |[-N, N] \cap \Gamma| \leq 3^d \frac{|[0, N]|}{|P|}.$$

Thus the ratio between the volume and cardinality of a progression P is essentially controlled by the quantity $|[-N, N] \cap \Gamma|$. (Hints: for the lower bound, first use Cauchy–Schwarz to obtain a lower bound for $\{(n, n') \in [0, N] : n \cdot v = n' \cdot v\}$. For the upper bound, consider the multiplicity of the map $f : [-N, 2N] \to Z$ defined by $f(n) := n \cdot v$.)

3.6.2 Let $[0, N]$ be a box in $\mathbf{Z}^d$, and let Γ be a sub-lattice of $\mathbf{Z}^d$. Show that $|[-kN, kN] \cap \Gamma| \leq (2k)^d |[-N, N] \cap \Gamma|$ for all integers $k \geq 1$.

3.6.3 Prove Theorem 3.40 in the case when Z is not necessarily torsion-free. (The main new difficulty is that the vector m is not always irreducible in $\mathbf{Z}^d$; in such a case one will have to "quotient out" a finite cyclic group from P before proceeding with the rest of the argument. However, this will only introduce additional factors of C^d into the inductive bound (3.23), which is acceptable.) Note that the second part of the Theorem does not extend to the torsion case, as can already be seen by considering $P = Z = \mathbf{Z}_2$.

3.6.4 Prove an extension of Theorem 3.40 in the torsion-free case in which one requires that kQ is also proper for some fixed constant $k \geq 1$ (of course, the bounds on Q will depend on k). Note that the torsion-free hypothesis is now essential, as can be seen by considering the case when $P = [1, N] \cdot 1$ in $\mathbf{Z}_N$.

3.6.5 [349] Let N_1, N_2, a_1, a_2 be positive integers such that $0 < a_2 < N_1/5$ and $0 < a_1 < N_2/5$, and a_1, a_2 are coprime. Use the Chinese remainder theorem to show the inclusion

$$\left[\frac{1}{5}(a_1N_1 + a_2N_2), \frac{4}{5}(a_1N_1 + a_2N_2)\right] \subseteq [0, (N_1, N_2)] \cdot (a_1, a_2).$$

Conclude that if P is any progression of rank 2 in the integers of dimensions N_1, N_2 and steps v_1, v_2 with $0 < v_2 < N_1/5$ and $0 < v_1 < N_2/5$, then P contains a proper arithmetic progression of length $3(N_1v_1 + N_2v_2)/5\gcd(v_1, v_2)$ and spacing $\gcd(v_1, v_2)$.

3.6.6 [349] Let A be an additive set in an ambient group Z. Show that there exists $d = O(\log |A|)$ and distinct elements $v_1, \ldots, v_d \in A$ such that the cube $[0, 1]^d \cdot (v_1, \ldots, v_d)$ has cardinality at least $\frac{1}{4}|A|$. (Hint: Using (2.21), show that if S is any additive set in Z such that $|S| < \frac{1}{4}|A|$, then there exists $a \in A$ such that $|S \cup (S + a)| \geq \frac{3}{2}|S|$. Then use the greedy algorithm.)

4

Fourier-analytic methods

In Chapter 1 we have already seen the power of the probabilistic method in additive combinatorics, in which one understands the additive structure of a random object by means of computing various averages or moments of that object. In this chapter we develop an equally powerful tool, that of *Fourier analysis*. This is another way of computing averages and moments of additively structured objects; it is similar to the probabilistic method but with an important new ingredient, namely that the quantities being averaged are now "twisted" or "modulated" by some very special complex-valued phase functions known as *characters*. This gives rise to the concept of a *Fourier coefficient* of a set or function, which measures the *bias* that object has with respect to a certain character. These coefficients serve two major purposes in this theory. Firstly, one can exploit the *orthogonality* between different characters to obtain non-trivial bounds on these coefficients; this orthogonality plays a role somewhat similar to the role of independence in probability theory. Secondly, Fourier coefficients are very good at controlling the operation of *convolution*, which is the analog of the sum set operation, but for functions instead of sets. Because of this, the Fourier transform is ideal for studying certain arithmetic quantities, most notably the additive energy introduced in Definition 2.8.

Using Fourier analysis, one can essentially divide additive sets A into two classes. At one extreme are the *pseudo-random* sets, whose Fourier transform is very small (except at 0); we shall introduce the *linear bias* $\|A\|_u$ and the $\Lambda(p)$ *constants* to measure this pseudo-randomness. Such sets are very "mixing" with respect to set addition (or to locating progressions of length three), and as the terminology implies, they behave more or less like random sets. At the other extreme are the *almost periodic sets*, which include arithmetic progressions, Bohr sets, and other sets with small doubling constant or large additive energy. The behavior of these sets with respect to set addition or progressions of length three is almost completely described by a rather small *spectrum* $\mathrm{Spec}_\alpha(A)$, defined as the set of frequencies where the Fourier transform of 1_A is large. We shall rely on this dichotomy between randomness and structure in a number of ways, most strikingly in proving Roth's celebrated theorem (which we discuss in Chapter 10) that subsets of integers of positive upper density contain infinitely many progressions of length 3. (Progressions of higher length cannot be treated by linear Fourier techniques, requiring either higher order Fourier analysis or other approaches; see Chapter 11.)

Fourier analysis can be performed on any additive group Z (and even on non-abelian groups). However, we shall only need this transform on finite groups, where the theory is slightly simpler technically. Thus we shall restrict our attention exclusively to the finite case. The cases $Z = \mathbf{Z}$, $Z = \mathbf{R}/\mathbf{Z}$, and $Z = \mathbf{R}$ are also of importance to additive combinatorics (in particular leading to the *Hardy–Littlewood circle method* in analytic number theory), but it turns out that the finite Fourier theory forms an acceptable substitute for these infinite Fourier theories in our applications.

4.1 Basic theory

Let Z be a finite additive group (for instance, Z could be a cyclic group $Z = \mathbf{Z}_N$). In this section we recall the basic theory of the finite Fourier transform on such groups.

Fourier analysis relies on the duality between a group Z and its *Pontryagin dual* $\hat{Z}$, which can be defined as the space of homomorphisms from Z to the circle group $\mathbf{R}/\mathbf{Z}$. In the case of finite groups, it turns out that a group Z and its Pontryagin dual $\hat{Z}$ are always isomorphic, and so it shall be convenient to identify the two in order to simplify the theory slightly. This can be done by means of a non-degenerate bilinear form:

Definition 4.1 (Bilinear forms) A *bilinear form* on an additive group Z is a map $(\xi, x) \mapsto \xi \cdot x$ from $Z \times Z$ to $\mathbf{R}/\mathbf{Z}$, which is a homomorphism in each of the variables ξ, x separately. We say that the form is *non-degenerate* if for every non-zero ξ the map $x \mapsto \xi \cdot x$ is not identically zero, and similarly for every non-zero x the map $\xi \mapsto \xi \cdot x$ is not identically zero. We say the form is *symmetric* if $\xi \cdot x = x \cdot \xi$.

Examples 4.2 If Z is a cyclic group $\mathbf{Z}_N$ then the bilinear form $x \cdot \xi := x\xi/N$ is symmetric and non-degenerate. If Z is a standard vector space F^n over a finite field F, then the bilinear form $(x_1, \ldots, x_n) \cdot (\xi_1, \ldots, \xi_n) := \phi(x_1\xi_1 + \cdots + x_n\xi_n)$ is symmetric and non-degenerate whenever $\phi : F \to \mathbf{R}/\mathbf{Z}$ is any non-trivial homomorphism from F to $\mathbf{R}/\mathbf{Z}$ (e.g. if $F = \mathbf{Z}_p$ we can take $\phi(x) := x/p$). This particular choice has the useful additional property that $a\xi \cdot x = \xi \cdot ax$ for all $a \in F$ and $x, \xi \in Z$.

Lemma 4.3 (Existence of bilinear forms) *Every finite additive group Z has at least one non-degenerate symmetric bilinear form.*

Proof From Corollary 3.8 we know that every finite additive group is the direct sum of cyclic groups. We have already seen in Example 4.2 that each cyclic group has a symmetric non-degenerate bilinear form. Finally, observe that if Z_1 and Z_2 have symmetric non-degenerate bilinear forms, then the direct sum $Z_1 \oplus Z_2$ also has a symmetric non-degenerate bilinear form, defined by $(\xi_1, \xi_2) \cdot (x_1, x_2) := \xi_1 \cdot x_1 + \xi_2 \cdot x_2$. The claim follows. □

Remark 4.4 A given additive group Z generally has multiple bilinear forms (see Exercise 4.1.10), but from the point of view of Fourier analysis they are all equivalent[1]. The symmetry property has some minor aesthetic advantages but is not essential to the Fourier theory, as the physical space variable and the frequency space variable usually play completely different roles.

Henceforth we fix a finite additive group Z, equipped with a non-degenerate symmetric bilinear form $\xi \cdot x$; in practice we shall usually use one of the two examples from Example 4.2.

To perform Fourier analysis, it will be convenient to adopt the following "ergodic" notation. Let $\mathbf{C}^Z$ denote the space of all complex-valued functions $f : Z \to \mathbf{C}$. If $f \in \mathbf{C}^Z$, we define the *mean* or *expectation* of f to be the quantity

$$\mathbf{E}_Z(f) = \mathbf{E}_{x \in Z} f(x) := \frac{1}{|Z|} \sum_{x \in Z} f(x).$$

Similarly, if $A \subseteq Z$, we define the *density* or *probability* of A as

$$\mathbf{P}_Z(A) = \mathbf{P}_{x \in Z}(x \in A) := \mathbf{E}_Z(1_A) = \frac{|A|}{|Z|}.$$

We can generalize this notation to other finite non-empty domains than Z, thus for instance $\mathbf{E}_{x \in A, y \in B} f(x, y) := \frac{1}{|A||B|} \sum_{x \in A, y \in B} f(x, y)$. This notation not only suggests the connections between Fourier analysis, ergodic theory, and probability, but is also useful in concealing from view a number of normalizing powers of $|Z|$ which would otherwise clutter the estimates. Generally, we shall use this ergodic notation for the physical variable, but use the discrete notation $\sum_{\xi \in Z} f(\xi)$ and $|A|$ (without the normalizing $|Z|$ factor) for the frequency variable. We shall also rely

[1] One way of viewing this is that the identification between $\hat{Z}$ and Z is non-canonical, and one should really be placing the frequency variable in $\hat{Z}$ instead of Z. This is ultimately the more correct viewpoint; however since we shall usually be working in very concrete situations such as cyclic groups $\mathbf{Z}_N$, where one does have a standard identification, we have chosen to rely on the bilinear form approach here rather than the abstract approach.

heavily on the exponential map $e : \mathbf{R}/\mathbf{Z} \to \mathbf{C}$, defined by

$$e(\theta) := e^{2\pi i \theta}. \tag{4.1}$$

The following two orthogonality properties form the foundation for Fourier analysis.

Lemma 4.5 (Orthogonality properties) *For any $\xi, \xi' \in Z$ we have*

$$\mathbf{E}_{x \in Z} e(\xi \cdot x)\overline{e(\xi' \cdot x)} = \mathbf{I}(\xi = \xi')$$

and for any $x, x' \in Z$ we have

$$\sum_{\xi \in Z} e(\xi \cdot x)\overline{e(\xi \cdot x')} = |Z|\mathbf{I}(x = x').$$

Proof We prove the first identity only, as the second is similar. Since $e(\xi \cdot x)\overline{e(\xi' \cdot x)} = e((\xi - \xi') \cdot x)$, it will suffice to show the claim in the $\xi' = 0$ case, i.e. it suffices to show

$$\mathbf{E}_{x \in Z} e(\xi \cdot x) = \mathbf{I}(\xi = 0).$$

This is clear in the case $\xi = 0$. If $\xi \neq 0$, then by non-degeneracy there exists $h \in Z$ such that $e(\xi \cdot h) \neq 1$. Shifting x by h we then have

$$\mathbf{E}_{x \in Z} e(\xi \cdot x) = \mathbf{E}_{x \in Z} e(\xi \cdot (x + h)) = e(\xi \cdot h)\mathbf{E}_{x \in Z} e(\xi \cdot x)$$

and hence $\mathbf{E}_{x \in Z} e(\xi \cdot x) = 0 = \mathbf{I}(\xi = 0)$ as desired. □

For every $\xi \in Z$, we can define the associated *character* $e_\xi \in \mathbf{C}^Z$ by $e_\xi(x) := e(\xi \cdot x)$. The above lemma then shows that the e_ξ are an orthonormal system in $\mathbf{C}^Z$, with respect to the complex Hilbert space structure

$$\langle f, g \rangle_{\mathbf{C}^Z} := \mathbf{E}_Z(f\overline{g}) = \mathbf{E}_{x \in Z} f(x)\overline{g(x)}.$$

Since the number $|Z|$ of characters equals the dimension $|Z|$ of the space, we see that this system is in fact a *complete* orthonormal system. This motivates

Definition 4.6 (Fourier transform) If $f \in \mathbf{C}^Z$, we define the *Fourier transform* $\hat{f} \in \mathbf{C}^Z$ by the formula

$$\hat{f}(\xi) := \langle f, e_\xi \rangle_{\mathbf{C}^Z} = \mathbf{E}_{x \in Z} f(x)\overline{e(\xi \cdot x)}.$$

We refer to $\hat{f}(\xi)$ as the *Fourier coefficient* of f at the frequency (or *mode*) ξ.

Since the e_ξ are a complete orthonormal basis, we have the *Parseval identity*

$$(\mathbf{E}_Z|f|^2)^{1/2} = \left(\sum_{\xi \in Z} |\hat{f}(\xi)|^2\right)^{1/2} \tag{4.2}$$

the *Plancherel theorem*

$$\langle f, g\rangle_{\mathbf{C}^Z} = \sum_{\xi\in Z} \hat{f}(\xi)\overline{\hat{g}(\xi)} \tag{4.3}$$

and the *Fourier inversion formula*

$$f = \sum_{\xi\in Z} \hat{f}(\xi)e_\xi. \tag{4.4}$$

In particular we see that two functions are equal if and only if their Fourier coefficients match at every frequency. In other words, the Fourier transform is a bijection from $\mathbf{C}^Z$ to $\mathbf{C}^Z$ (in fact it is a unitary isometry, thanks to (4.2), (4.3)).

From Lemma 4.5 we see that the Fourier coefficients of a character e_ξ are just a Kronecker delta function:

$$\widehat{e_\xi}(\xi') = \mathbf{I}(\xi = \xi').$$

In particular $\hat{1}(\xi) = \mathbf{I}(\xi = 0)$.

A special role in the additive theory of the Fourier transform is played by the *zero frequency* $\xi = 0$. This is because the zero Fourier coefficient is same concept as expectation:

$$\hat{f}(0) = \langle f, 1\rangle_{\mathbf{C}^Z} = \mathbf{E}_Z(f). \tag{4.5}$$

If S is any subset of Z, define the *orthogonal complement* $S^\perp \subseteq Z$ of S to be the set

$$S^\perp := \{\xi \in Z : \xi \cdot x = 0 \text{ for all } x \in S\}.$$

One can easily verify that $S^\perp$ is a subgroup of Z. Also one has the pleasant identity

$$\widehat{1_G} = \mathbf{P}_Z(G)1_{G^\perp} \tag{4.6}$$

whenever G is a subgroup; see Exercise 4.1.6. Applying (4.2) we see in particular that

$$|G||G^\perp| = |Z|. \tag{4.7}$$

We now introduce the fundamental notion of *convolution*, which links the Fourier transform to the theory of sum sets.

Definition 4.7 (Convolution) If $f, g \in L^2(Z)$ are random variables, we define their *convolution* $f * g$ to be the random variable

$$f * g(x) = \mathbf{E}_{y\in Z} f(x-y)g(y) = \mathbf{E}_{y\in Z} f(y)g(x-y).$$

We also define the *support* $\operatorname{supp}(f)$ of f to be the set $\operatorname{supp}(f) = \{f \neq 0\} = \{x \in Z : f(x) \neq 0\}$.

The significance of convolution to sum sets lies in the obvious inclusion

$$\mathrm{supp}(f * g) \subseteq \mathrm{supp}(f) + \mathrm{supp}(g)$$

and particularly in the identity

$$A + B = \mathrm{supp}(1_A * 1_B).$$

Indeed we have the more precise statement

$$1_A * 1_B(x) := \mathbf{P}_Z(A \cap (x - B)). \tag{4.8}$$

The relevance of the Fourier transform to convolution lies in the easily verified identity

$$\widehat{f * g} = \hat{f} \cdot \hat{g} \tag{4.9}$$

Applying (4.9) at the zero frequency we have the basic formula

$$\mathbf{E}_Z(f * g) = (\mathbf{E}_Z f) \cdot (\mathbf{E}_Z g). \tag{4.10}$$

In particular, if f or g has mean zero, then so does $f * g$.

As one consequence of (4.9) we see that convolution is bilinear, symmetric, and associative. We also have a dual version of (4.9), namely the formula

$$\widehat{fg}(\xi) = \sum_{\eta \in Z} \hat{f}(\eta)\hat{g}(\xi - \eta) \tag{4.11}$$

which converts pointwise product back to convolution; we leave the verification of these identities as an exercise.

In the exercises below, Z is a fixed finite additive group, with a fixed symmetric non-degenerate bilinear form $\cdot$.

Exercises

4.1.1 Let $\hat{Z}$ be the additive group consisting of all the homomorphisms from Z to $\mathbf{R}/\mathbf{Z}$. Show that the identification of a frequency $\xi \in Z$ with the homomorphism $x \mapsto \xi \cdot x$ gives an isomorphism from Z to $\hat{Z}$.

4.1.2 Define a *character* to be any map $\chi : Z \to \mathbf{C}$ with $\chi(0) = 1$ and $\chi(x + y) = \chi(x)\chi(y)$ for all $x, y \in Z$. Show that the set of all characters is precisely $\{e_\xi : \xi \in Z\}$.

4.1.3 Show that for any $\xi \in Z$, e_ξ takes values in the $|Z|$th roots of unity.

4.1.4 Define a *linear phase function* to be any map $\phi : Z \to \mathbf{R}/\mathbf{Z}$ with the property that

$$\phi(x + h_1 + h_2) - \phi(x + h_1) - \phi(x + h_2) + \phi(x) = 0 \text{ for all } x, h_1, h_2 \in Z.$$

Show that $\phi : Z \to \mathbf{R}/\mathbf{Z}$ is a linear phase function if and only if there exists $\xi \in Z$ and $c \in \mathbf{R}/\mathbf{Z}$ such that $\phi(x) = \xi \cdot x + c$ for all c. (The map ϕ is also a *Freiman homomorphism of order* 2; see Definition 5.21.)

4.1.5 Let x be an element of Z chosen uniformly at random. Show that the random variables $\{e_\xi(x) : \xi \in Z\}$ are pairwise independent, and have variance 1 and mean zero for $\xi \neq 0$, and variance 0 and mean 1 for $\xi = 0$. Use this and (1.9), (4.4) to give an alternative proof of (4.2).

4.1.6 Prove (4.6).

4.1.7 Let $f : Z \to \mathbf{C}$. If H is a subgroup of Z, and $g := f 1_H$, show that

$$\hat{g}(\xi) = \mathbf{E}_{\eta \in H^\perp} \hat{f}(\xi + \eta) \text{ for all } \xi \in Z$$

and conclude in particular the *Poisson summation formula*

$$\mathbf{E}_{x \in H} f(x) = \sum_{\xi \in H^\perp} \hat{f}(\xi).$$

In the converse direction, if $h = f * \frac{1}{\mathbf{P}_Z(H)} 1_H$ is the average of f on cosets of H, i.e.

$$h(x) := \mathbf{E}_{y \in H} f(x + y),$$

show that $\hat{h} = \hat{f} \cdot 1_{H^\perp}$.

4.1.8 If $\phi : Z \to Z$ is a group isomorphism of Z, then there exists a unique group isomorphism $\phi^\dagger : Z \to Z$, called the *adjoint* of ϕ, such that $\xi \cdot \phi(x) = \phi^\dagger(\xi) \cdot x$ for all $x, \xi \in Z$. Furthermore if $g(x) = f(\phi(x))$ for all $x \in Z$ then $\hat{g}(x) = f((\phi^\dagger)^{-1}(x))$ for all $x \in Z$.

4.1.9 If $\phi : Z \to Z$ and $\psi : Z \to Z$ are group isomorphisms, show that $(\phi \circ \psi)^\dagger = \psi^\dagger \circ \phi^\dagger$.

4.1.10 Let $\bullet : Z \times Z \to \mathbf{R}/\mathbf{Z}$ and $\tilde{\bullet} : Z \times Z \to \mathbf{R}/\mathbf{Z}$ be two non-degenerate symmetric bilinear forms on a finite additive group Z. Show that there exists a self-adjoint group isomorphism $\phi : Z \to Z$ such that $\xi \tilde{\bullet} x = \xi \bullet \phi(x) = \phi^\dagger(\xi) \bullet x$ for all $x, \xi \in Z$. This shows that all Fourier transforms are equivalent up to isomorphisms of either the x or ξ variable.

4.1.11 Prove (4.9) and (4.11).

4.1.12 Let x be an element of Z chosen uniformly at random, and let $\xi_1, \ldots, \xi_n \in Z$. Show that the random variables $e_{\xi_1}(x), \ldots, e_{\xi_n}(x)$ are jointly independent if and only if the group $\langle \xi_1, \ldots, \xi_n \rangle$ generated by $\xi_1, \ldots, \xi_n$ has order $\mathrm{ord}(\xi_1) \ldots \mathrm{ord}(\xi_n)$.

4.1.13 Let G, H be two subgroups of Z. Show that $(G + H)^\perp = G^\perp \cap H^\perp$, $(G \cap H)^\perp = G^\perp + H^\perp$, and $d(G^\perp, H^\perp) = d(G, H)$, where d is the Ruzsa distance defined in Definition 2.5. This may help explain the symmetric nature of $G + H$ and $G \cap H$ in the estimates in Exercise 2.3.11.

4.1.14 Let G, H be two subgroups of Z and let x be an element of Z chosen randomly. Show that the indicators $\mathbf{I}(x \in G)$ and $\mathbf{I}(x \in H)$ have non-negative correlation, i.e. $\mathbf{Cov}(\mathbf{I}(x \in G), \mathbf{I}(x \in H)) \geq 0$; establish this both by Fourier-analytic means and by direct computation. Show that equality occurs if and only if $G + H = Z$.

4.1.15 Show that for any subgroup G of Z, we have $(G^\perp)^\perp = G$, and for any random variable f, we have $\widehat{\widehat{f}}(x) = |Z|^{-1} f(-x)$. More generally, for any $A \subset Z$, we have $\langle A \rangle = (A^\perp)^\perp$, where $\langle A \rangle$ is the group generated by A.

4.1.16 If Z and Z' are finite groups, formulate a rigorous version of the statement that the Fourier transform on $Z \times Z'$ is the composition of the Fourier transform on Z and the Fourier transform on Z'.

4.2 L^p theory

We now turn to the analytic theory of the Fourier transform and of convolutions, starting with the L^p theory, and then apply it to the problem of locating arithmetic progressions inside sum sets.

If $f \in \mathbf{C}^Z$ and $0 < p < \infty$, we define the $L^p(Z)$ *norm* of f to be the quantity

$$\|f\|_{L^p(Z)} := (\mathbf{E}_Z |f|^p)^{1/p} = (\mathbf{E}_{x \in Z} |f(x)|^p)^{1/p}.$$

Thus for instance $\|f\|_{L^2(Z)}$ is just the Hilbert space magnitude of f. We also define

$$\|f\|_{L^\infty(Z)} = \sup_{x \in Z} |f(x)|.$$

Similarly we define

$$\|f\|_{l^p(Z)} := \left(\sum_{\xi \in Z} |f(\xi)|^p \right)^{1/p}$$

for $0 < p < \infty$ and

$$\|f\|_{l^\infty(Z)} := \sup_{\xi \in Z} |f(\xi)|.$$

We have the following two basic L^p estimates on the Fourier transform and on convolution.

Theorem 4.8 *Let $f, g : Z \to \mathbf{C}$ be functions on an additive group Z. Then for any $1 \leq p \leq 2$ we have the* Hausdorff–Young inequality

$$\|\widehat{f}\|_{l^{p'}(Z)} \leq \|f\|_{L^p(Z)} \tag{4.12}$$

where the dual exponent p' to p is defined by $\frac{1}{p} + \frac{1}{p'} = 1$. Also, whenever $1 \leq p, q, r \leq \infty$ are such that $\frac{1}{p} + \frac{1}{q} = \frac{1}{r} + 1$, we have the Young inequality

$$\|f * g\|_{L^r(Z)} \leq \|f\|_{L^p(Z)} \|g\|_{L^q(Z)}. \tag{4.13}$$

Both inequalities follow easily from Riesz–Thorin complex interpolation theorem. With this theorem, one only needs to verify the extremal (and easy) cases. The Riesz–Thorin theorem, however, is beyond the scope of this book. On the other hand, one can also have an elementary proof, using combinatorial arguments (see Exercise 4.2.3).

Recall the additive energy $E(A, B)$ between two additive sets A, B in Z, defined in Definition 2.8. From that definition one can easily check that

$$E(A, B) = |Z|^3 \|1_A * 1_B\|^2_{L^2(Z)}.$$

By (4.2) and (4.9) we obtain the fundamental identity

$$E(A, B) = |Z|^3 \mathbf{E}_Z (1_A * 1_B)^2 = |Z|^3 \sum_{\xi \in Z} |\hat{1}_A(\xi)|^2 |\hat{1}_B(\xi)|^2. \tag{4.14}$$

This formula may illuminate some of the properties of the additive energy that were obtained in Section 2.3, such as the symmetries $E(A, B) = E(B, A) = E(A, -B)$ and the Cauchy–Schwarz inequality (2.9); see Exercise 4.2.7.

For the purposes of additive combinatorics, the Fourier transform is most useful when applied to characteristic functions $f = 1_A$, and in this case one can say quite a bit about the Fourier transform and its relation to the additive energy $E(A, A)$.

Lemma 4.9 *Let A be a subset of a finite additive group Z, and let $\widehat{1_A} : Z \to \mathbf{C}$ be the Fourier transform of the characteristic function of A. Then we have the identities:*

$$\|\widehat{1_A}\|_{l^\infty(Z)} = \sup_{\xi \in Z} |\widehat{1_A}(\xi)| = \widehat{1_A}(0) = \mathbf{P}_Z(A); \tag{4.15}$$

$$\|\widehat{1_A}\|^2_{l^2(Z)} = \sum_{\xi \in Z} |\widehat{1_A}(\xi)|^2 = \mathbf{P}_Z(A); \tag{4.16}$$

$$\widehat{1_A}(\xi) = \overline{\widehat{1_A}(-\xi)}; \tag{4.17}$$

$$\|\widehat{1_A}\|^4_{l^4(Z)} = \sum_{\xi \in Z} |\widehat{1_A}(\xi)|^4 = \frac{E(A, A)}{|Z|^3}; \tag{4.18}$$

$$\widehat{1_A}(\xi) = \sum_{\eta \in Z} \widehat{1_A}(\eta) \widehat{1_A}(\xi - \eta). \tag{4.19}$$

This lemma follows easily from the estimates that have already been established; see Exercise 4.2.4.

We now present a simple application of the Fourier transform in the setting of a finite field F.

Lemma 4.10 *[41] Let F be a finite field, and let A be a subset of $F\backslash\{0\}$ such that $|A| > |F|^{3/4}$. Then*

$$3(A \cdot A) = A \cdot A + A \cdot A + A \cdot A = F.$$

Proof We give F a symmetric non-degenerate bilinear form of the type in Example 4.2. Let $f : F \to \mathbf{R}$ denote the non-negative function

$$f := \mathbf{E}_{a \in A} 1_{a \cdot A}.$$

Observe that $\operatorname{supp}(f) = A \cdot A$ and $\hat{f}(0) = \mathbf{E}_F f = \mathbf{P}_F(A)$. Taking Fourier transforms we obtain

$$\hat{f}(\xi) = \mathbf{E}_{a \in A} \widehat{1_A}(\xi a)$$

for any $\xi \in F$. If $\xi \neq 0$, then we observe that the frequencies ξa are all distinct as a varies. Using Cauchy–Schwarz and then (4.16), we then obtain

$$|\hat{f}(\xi)| \leq \frac{1}{|A|}|A|^{1/2}\mathbf{P}_F(A)^{1/2} = 1/|F|^{1/2} \text{ for } \xi \neq 0.$$

Now let $x \in F$ be arbitrary. We use (4.4) and (4.9) to compute

$$\begin{aligned}
f * f * f(x) &= \operatorname{Re} f * f * f(x) \\
&= \operatorname{Re} \sum_{\xi \in F} \hat{f}(\xi)^3 e(\xi \cdot x) \\
&\geq \operatorname{Re} \hat{f}(0)^3 - \sum_{\xi \in F\backslash\{0\}} |\hat{f}(\xi)|^3 \\
&\geq \mathbf{P}_F(A)^3 - \sum_{\xi \in F} |F|^{-1/2} |\hat{f}(\xi)|^2 \\
&= \mathbf{P}_F(A)^3 - |F|^{(-1/2)} \|f\|_{L^2(F)^2} \\
&\geq \mathbf{P}_F(A)^3 - |F|^{(-1/2)} [\mathbf{E}_{a \in A} \|1_{a \cdot A}\|_{L^2(F)}]^2 \\
&= \mathbf{P}_F(A)^3 - |F|^{-1/2} \mathbf{P}_F(A) \\
&> 0
\end{aligned}$$

since $\mathbf{P}_F(A) > |F|^{-1/4}$ by hypothesis. Since $\operatorname{supp}(f * f * f) = 3(A \cdot A)$ and x was arbitrary, we are done. □

Remark 4.11 Lemma 4.10 is a simple example of a *sum-product estimate* – an assertion that a combination of a sum and product of a set A is necessarily much larger than A itself. It can be viewed as a quantitative reflection of the fact that a set A of cardinality greater than $|F|^{3/4}$ has difficulty behaving like a subfield of F. It should be compared with the results in Section 2.8.

Exercises

4.2.1 Let $1 \le p < \infty$. By exploiting the convexity of the function $x \mapsto |x|^p$, establish the convexity of the set $\{f \in \mathbf{C}^Z : \|f\|_{L^p(Z)} \le 1\}$, and conclude the *triangle inequality*

$$\|f+g\|_{L^p(Z)} \le \|f\|_{L^p(Z)} + \|g\|_{L^p(Z)}.$$

Argue similarly for the $p = \infty$ case and with L^p replaced by l^p.

4.2.2 Let $1 < p < \infty$, and let p' be the dual exponent, thus $1/p + 1/p' = 1$. By exploiting the convexity of the function $x \mapsto e^x$, establish the preliminary inequality

$$\mathbf{E}_{x\in Z}|f(x)||g(x)| \le 1 \text{ whenever } \|f\|_{L^p(Z)}, \|g\|_{L^{p'}(Z)} \le 1,$$

and then conclude *Hölder's inequality*

$$\|fg\|_{L^r(Z)} \le \|f\|_{L^p(Z)}\|g\|_{L^q(Z)}$$

whenever $0 < p, q, r \le \infty$ are such that $\frac{1}{p} + \frac{1}{q} = \frac{1}{r}$. Similarly with the L^p norms replaced by l^p norms.

4.2.3 The purpose of this exercise is to give a proof of Theorem 4.8 that does not require complex interpolation. First use (4.2), the trivial bound

$$\|\hat{f}\|_{l^\infty(\mathbf{Z})} \le \|f\|_{L^1(\mathbf{Z})}, \tag{4.20}$$

and Hölder's inequality to establish the weaker estimate

$$\|\hat{f}\|_{l^{p'}(Z)} = O_p(\|f\|_{L^p(Z)})$$

whenever $f \in \mathbf{C}^Z$ is supported on a set A and obeys an estimate of the form $|f(x)| = \Theta(\lambda)$ for all $x \in A$ and some threshold λ. Then, prove the even weaker estimate

$$\|\hat{f}\|_{l^{p'}(Z)} = O_p(\|f\|_{L^p(Z)} \log(1 + |Z|))$$

for arbitrary $f \in \mathbf{C}^Z$ by applying the previous inequality to a dyadic decomposition of f, followed by the triangle inequality. Finally, remove the $O_p(\log(1 + |Z|))$ factor to establish (4.12) by replacing Z with a large power Z^M of Z, and similarly replacing f with a large tensor power (as in Corollary 2.19) and letting $M \to \infty$. Argue similarly to establish (4.13).

4.2.4 Prove Lemma 4.9.

4.2.5 Let A be an additive set in a finite additive group Z. Show that $\hat{1}_A$ is real-valued if and only if A is symmetric.

4.2.6 (Law of large numbers for finite groups) Let $f : Z \to \mathbf{R}_{\ge 0}$ be such that $\mathbf{E}_Z f = 1$ and $f(0) \ne 0$, and let H be the subgroup of Z generated by $\mathrm{supp}(f)$. Show that $|\hat{f}(\xi)| \le 1$, with equality if and only if $\xi \in H^\perp$.

Next, define the iterated convolutions $f^{(n)}$ for $n = 1, 2, \ldots$ inductively by $f^{(1)} := f$ and $f^{(n+1)} := f * f^{(n)}$, and show that $\lim_{n\to\infty} f^{(n)} = \frac{1}{\mathbf{P}_Z(H)} 1_H$. What can happen when the hypothesis $f(0) \neq 0$ is dropped?

4.2.7 Use Fourier-analytic methods to give another proof of Corollary 2.10.

4.2.8 Use Fourier-analytic methods to give another proof of Proposition 2.7.

4.2.9 Let f be a random variable which is not identically zero. By using (4.2) and (4.20), establish the *uncertainty principle*

$$|\mathrm{supp}(f)||\mathrm{supp}(\hat{f})| \geq |Z|. \tag{4.21}$$

Prove that equality occurs if and only if $f(x) = ce(\xi \cdot x) 1_{H+x_0}(x)$ for some complex number $c \in \mathbf{C}$, some subgroup H of Z, and some $\xi, x_0 \in Z$. This inequality can be improved for certain groups Z: see Theorem 9.52.

4.2.10 Let $f \in \mathbf{C}^Z$ be normalized so that $\|f\|^2_{L^2(Z)} = \sum_{\xi \in Z} |\hat{f}(\xi)|^2 = 1$. By differentiating the Hausdorff–Young inequality in p, establish the *entropy uncertainty principle*

$$\mathbf{E}_{x\in Z}|f(x)|^2 \log \frac{1}{|f(x)|^2} + \sum_{\xi \in Z} |\hat{f}(\xi)|^2 \log \frac{1}{|\hat{f}(\xi)|^2} \geq 0,$$

where we adopt the convention that $0 \log \frac{1}{0} = 0$. (Hint: differentiate the Hausdorff–Young inequality in p at $p = 2$, using the fact that equality holds at that endpoint.) Using Jensen's inequality, show that this inequality implies (4.21).

4.3 Linear bias

One common way to apply the Fourier transform to the theory of sum sets or to arithmetic progressions is to introduce the notion of *Fourier bias* of that set (also known as *linear bias* or *pseudo-randomness*). Roughly speaking, this notion separates sets into two extremes, ones which are highly uniform (and behave like random sets, especially with regard to iterated sum sets), and ones which are highly non-uniform (and behave like arithmetic progressions).

Definition 4.12 (Fourier bias) Let Z be a finite additive group. If A is a subset of Z, we define the *Fourier bias* $\|A\|_u$ of the set A to be the quantity

$$\|A\|_u := \sup_{\xi \in Z \setminus \{0\}} |\hat{1}_A(\xi)|.$$

This quantity is always non-negative, with $\|A\|_u = 0$ if and only if A is equal to Z or the empty set (Exercise 4.3.1). It obeys the symmetries $\|A\|_u = \|-A\|_u = \|A + h\|_u = \|Z \backslash A\|_u$ for any $h \in Z$ (Exercise 4.3.2). We warn that this quantity is not monotone: $A \subseteq B$ does not imply $\|A\|_u \leq \|B\|_u$. However, the Fourier bias does obey a triangle inequality (Exercise 4.3.3). The Fourier bias $\|A\|_u$ can be as large as the density $\mathbf{P}_Z(A)$, but is usually smaller (Exercise 4.3.4). Sets A with Fourier bias less than α are sometimes called *α-uniform* or *α-pseudo-random*; sets with small Fourier bias are called *linearly uniform*, *Gowers uniform of order 1*, or *pseudo-random*.

The connection between Fourier bias and sum sets can be described by the following lemma.

Lemma 4.13 (Uniformity implies large sum sets) *Let $n \geq 3$, and let $A_1, \ldots, A_n$ be additive sets in a finite additive group Z. Then for any $x \in Z$ we have*

$$\left|\frac{1}{|Z|^{n-1}}|\{(a_1, \ldots, a_n) \in A_1 \times \cdots \times A_n : x = a_1 + \cdots + a_n\}| - \mathbf{P}_Z(A_1) \cdots \mathbf{P}_Z(A_n)\right|$$
$$\leq \|A_1\|_u \cdots \|A_{n-2}\|_u \mathbf{P}_Z(A_{n-1})^{1/2} \mathbf{P}_Z(A_n)^{1/2}.$$

In particular, if we have

$$\|A_1\|_u \cdots \|A_{n-2}\|_u < \mathbf{P}_Z(A_1) \cdots \mathbf{P}_Z(A_{n-2}) \mathbf{P}_Z(A_{n-1})^{1/2} \mathbf{P}_Z(A_n)^{1/2}$$

then $A_1 + \cdots + A_n = Z$.

Of course, a similar result is true if we permute the $A_1, \ldots, A_n$. Note that the quantity $\mathbf{P}_Z(A_1) \cdots \mathbf{P}_Z(A_n)$ is the quantity one would expect for $\frac{1}{|Z|^{n-1}}|\{(a_1, \ldots, a_n) \in A_1 \times \cdots \times A_n : x = a_1 + \cdots + a_n\}|$ if the events $a_1 \in A_1, \ldots, a_n \in A_n$ were jointly independent conditioning on $x = a_1 + \cdots + a_n$. This may help explain why uniformity is sometimes referred to as pseudo-randomness.

Proof By (4.9), the function $1_{A_1} * \cdots * 1_{A_n}$ has Fourier transform $\widehat{1_{A_1}} \cdots \widehat{1_{A_n}}$. Applying the Fourier inversion formula (4.4), (4.15), the Cauchy–Schwarz inequality and (4.16) we thus see that

$$\begin{aligned}
1_{A_1} * \cdots * 1_{A_n}(x) &= \mathrm{Re} 1_{A_1} * \cdots * 1_{A_n}(x) \\
&= \mathrm{Re} \sum_{\xi \in Z} \widehat{1_{A_1}}(\xi) \cdots \widehat{1_{A_n}}(\xi) e(x \cdot \xi) \\
&\geq \widehat{1_{A_1}}(0) \cdots \widehat{1_{A_n}}(0) - \sum_{\xi \in Z \setminus \{0\}} |\widehat{1_{A_1}}(\xi)| \cdots |\widehat{1_{A_n}}(\xi)| \\
&\geq \mathbf{P}_Z(A_1) \cdots \mathbf{P}_Z(A_n) - \|A_1\|_u \cdots \|A_{n-2}\|_u \sum_{\xi \in Z} |\widehat{1_{A_{n-1}}}(\xi)||\widehat{1_{A_n}}(\xi)| \\
&\geq \mathbf{P}_Z(A_1) \cdots \mathbf{P}_Z(A_n) - \|A_1\|_u \cdots \|A_{n-2}\|_u \|\widehat{1_{A_{n-1}}}\|_{l^2(Z)} \|\widehat{1_{A_n}}(\xi)\|_{l^2(Z)} \\
&= \mathbf{P}_Z(A_1) \cdots \mathbf{P}_Z(A_n) - \|A_1\|_u \cdots \|A_{n-2}\|_u \mathbf{P}_Z(A_{n-1})^{1/2} \mathbf{P}_Z(A_n)^{1/2}.
\end{aligned}$$

A similar argument gives

$$1_{A_1} * \cdots * 1_{A_n}(x) \leq \mathbf{P}_Z(A_1) \cdots \mathbf{P}_Z(A_n) + \|A_1\|_u \cdots \|A_{n-2}\|_u \mathbf{P}_Z(A_{n-1})^{1/2} \mathbf{P}_Z(A_n)^{1/2}.$$

Since by definition of convolution

$$1_{A_1} * \cdots * 1_{A_n}(x) = |Z|^{1-n} |\{(a_1, \ldots, a_n) \in A_1 \times \cdots \times A_n : x = a_1 + \cdots + a_n\}|,$$

and the lemma follows. □

We now give an application of the above machinery to the finite field Waring problem. We first need a standard lemma.

Lemma 4.14 (Gauss sum estimate) *Let F be a finite field of odd order, and let $A := F\wedge 2 = \{a^2 : a \in F\}$ be the set of squares in F. Then $\|A\|_u \leq \frac{1}{2|F|} + \frac{1}{2|F|^{1/2}}$.*

Proof Let $\xi \in F \backslash \{0\}$. Since every non-zero element in A has exactly two representations of the form a^2, we have

$$\hat{1}_A(\xi) = \frac{1}{|F|} \sum_{x \in A} e(-\xi \cdot x) = \frac{1}{2|F|} + \frac{1}{2|F|} \sum_{a \in F} e(-\xi \cdot a^2).$$

On the other hand, we may square

$$\begin{aligned} \left| \sum_{a \in F} e(-\xi \cdot a^2) \right|^2 = \left| \sum_{a \in F} e(\xi \cdot a^2) \right|^2 &= \sum_{a,b \in F} e(\xi \cdot (a^2 - b^2)) \\ &= \sum_{a,h \in F} e(\xi \cdot (a^2 - (a+h)^2)) \\ &= \sum_{h \in F} e(-\xi \cdot h^2) \sum_{a \in F} e(\xi \cdot 2ah). \end{aligned}$$

If $h \neq 0$, then $2h \neq 0$, and $\sum_{a \in F} e(\xi \cdot 2ah) = \sum_{c \in F} e(\xi \cdot c) = 0$ thanks to Lemma 4.5. On the other hand, if $h = 0$, then $\sum_{a \in F} e(\xi \cdot 2ah) = |F|$. We conclude that $|\sum_{a \in F} e(\xi \cdot a^2)|^2 = |F|$, and the claim follows. □

By combining this lemma with Lemma 4.13, one immediately obtains

Corollary 4.15 *Let F be a finite field of odd order, and let $A = F\wedge 2$ be the set of squares in F. Then $kA = F$ for all $k \geq 3$. Indeed, for any $x \in F$, the number of representations of x as a sum $x = a_1 + \cdots + a_k$ with $a_1, \ldots, a_k \in A$ is $(2^{(-k)}(1 + |F|^{(-1)}) + O_k(|F|^{-(k-2)/2}))|F|^{k-1}$.*

We leave the verification of this corollary as an exercise. It shows that the sum sets kA are more or less uniformly distributed for $k \geq 3$. Note that when $k = 2$, one can still prove that $2A = F$, but the sum sets can be quite irregular; for instance, if -1 is not a square in F, then 0 only has one representation as the sum of two elements in A.

We now present a lemma which asserts, roughly speaking, that if B is a randomly-chosen subset of A, then $\|B\|_u$ is approximately equal to $\frac{|B|}{|A|}\|A\|_u$; thus the Fourier bias decreases proportionally when passing to random subsets.

Lemma 4.16 *[149] Let A be an additive set in a finite additive group Z, and let $0 < \tau \leq 1$. Let B be a random subset of A defined by letting the events $a \in B$ be independent with probability τ. Then for any $\lambda > 0$ we have*

$$\mathbf{P}(|\|B\|_u - \tau\|A\|_u| \geq \lambda\sigma) \leq 4|Z| \max\left(e^{-\lambda^2/8}, e^{-\lambda\sigma/2\sqrt{2}}\right),$$

where $\sigma^2 := |A|\tau(1-\tau)/|Z|^2$.

The lemma is an easy consequence of Chernoff's inequality and is left as an exercise. Applying it with $\lambda = C \log^{1/2}|Z|$ for some large C, and assuming $|A|\tau(1-\tau) \gg \log|Z|$, we see in particular that

$$\mathbf{P}\big(\|B\|_u = \tau\|A\|_u + O\big(\sigma \log^{1/2}|Z|\big)\big) = 1 - O(|Z|^{-100})$$

(for instance). In particular if we set $A = Z$ then we have $\|B\|_u = \tau Z + O(\tau(1-\tau)\frac{\log^{1/2}|Z|}{|Z|})$ with high probability; thus random subsets of Z tend to be extremely uniform. Note that $\mathbf{P}_Z(B) \approx \tau$ with high probability, thanks to Corollary 1.10.

A major application of Fourier bias is in the study of arithmetic progressions of length 3. We will study this application in detail in Chapter 10.

Exercises

4.3.1 Let A be a subset of a finite additive group Z. Show that $\|A\|_u = 0$ if and only if $A = Z$ or $A = \emptyset$.

4.3.2 Let A be a subset of a finite additive group Z. Show that $\|A\|_u = \|-A\|_u = \|T^h A\|_u = \|Z \backslash A\|_u$ for any $h \in Z$. More generally, if $\phi : Z \to Z'$ is any isomorphism from one additive group to another, show that $\|\phi(A)\|_u = \|A\|_u$. In a similar spirit, show that the Fourier bias of a set A does not depend on the choice of symmetric non-degenerate bilinear form.

4.3.3 Let A, B be disjoint subsets of a finite additive group Z. Show that $|\|A\|_u - \|B\|_u| \leq \|A \cup B\|_u \leq \|A\|_u + \|B\|_u$.

4.3.4 Let A be an additive set in a finite additive group Z. Show that $\|A\|_u \leq \mathbf{P}_Z(A)$, with equality if and only if A is contained in a coset of a proper subgroup of Z.

4.3.5 Let A and A' be subsets of finite additive groups Z and Z' respectively. Show that $\|A \times A'\|_u = \|A\|_u \|A'\|_u$.

4.3.6 Let A be a subset of a finite additive group Z. Show that $\|A\|_u = \sup_\phi |\langle 1_A, e(\phi)\rangle_{\mathbf{C}^Z}|$, where $\phi : Z \to \mathbf{R}/\mathbf{Z}$ ranges over all non-constant linear phase functions (as defined in Exercise 4.1.4).

4.3.7 Let A, B be additive sets in a finite additive group Z. Show that

$$E(A, B) \leq \frac{|A|^2|B|^2}{|Z|} + |Z|^2\|A\|_u^2|B|.$$

Using (2.8), conclude that if $\|A\|_u \leq \alpha \mathbf{P}_Z(A)$, then

$$|A + B| \geq \frac{1}{2}\min\left(|Z|, \frac{1}{\alpha^2}|B|\right). \quad (4.22)$$

Thus α-uniform sets tend to expand sum sets by a factor of roughly α^{-2} (unless this is impossible due to the trivial bound $|A + B| \leq |Z|$).

4.3.8 Let A be an additive set in a finite additive group Z. Show that

$$\|A\|_u^4 \leq \frac{1}{|Z|^3}E(A, A) - \mathbf{P}_Z(A)^4 \leq \|A\|_u^2\mathbf{P}_Z(A). \quad (4.23)$$

Thus uniform sets have additive energy $E(A, A)$ close to the minimal value of $\mathbf{P}_Z(A)^4|Z|^3$, and vice versa.

4.3.9 Let A be an additive set in a finite additive group Z, and let $n \geq 3$ be an integer. Using Lemma 4.13, show that if $nA \neq Z$, then $\mathbf{P}_Z(A)^{1+\frac{1}{n-2}} \leq \|A\|_u \leq \mathbf{P}_Z(A)$. This estimate is especially useful when n is very large, as it shows that 1_A has a very large non-trivial Fourier coefficient.

4.3.10 Prove Corollary 4.15. Also show the identity $A \cdot 2A = 2A$ and conclude that $2A = F$ (using the fact that $3A = F$ to show that $2A \neq A$).

4.3.11 Use Chernoff's inequality (in the form of Exercise 1.3.4) to prove Lemma 4.16.

4.3.12 [149] Let A, B be additive sets in a finite additive group Z. Use Lemma 4.13 to establish the inequality

$$\|S\|_u \geq \mathbf{P}_Z(A)^{1/2}\mathbf{P}_Z(B)^{1/2}\mathbf{P}_Z(S)$$

whenever S is disjoint from $A + B$. In particular, this inequality holds when $S = Z\backslash(A + B)$. This shows that complements of sum sets are "hereditarily non-uniform".

4.3.13 Let A be a subset of a cyclic group $\mathbf{Z}_p$ of prime order. Show that for any arithmetic progression P in $\mathbf{Z}_p$, we have the uniform distribution estimate

$$\mathbf{P}_{\mathbf{Z}_p}(A \cap P) = \mathbf{P}_{\mathbf{Z}_p}(A)\mathbf{P}_{\mathbf{Z}_p}(P) + O(\varepsilon) + O\left(\log\frac{1}{\varepsilon}\|A\|_u\right)$$

for any $0 < \varepsilon \leq 1$. (Hint: apply a change of variables to make $P = [-N, N]$ for some N. Approximate the indicator 1_P by something a bit smoother (smoothed out at scale εp) and then compute the Fourier expansion. Apply Plancherel's theorem (4.3) with this smoothed out function and $1_A - \mathbf{P}(A)$.) This inequality is a crude form of the famous *Erdős–Turán inequality* in discrepancy theory, and is related to the Weyl criterion for uniform distribution modulo one.

4.3.14 Let $A = \mathbf{Z}_p^2$ be the set of squares in a cyclic group of prime order. Show that for any arithmetic progression P in $\mathbf{Z}_p$, we have

$$|A \cap P| = \frac{1}{2}|P| + O(\sqrt{p}\log p).$$

(Hint: use Lemma 4.14 and the preceding exercise.) This is a special case of the *Polya–Vinogradov inequality* from analytic number theory.

4.3.15 Let F be a finite field, let Z be a vector space over F, and let $M : Z \to Z$ be a linear transformation. Show that if $\dim_F(Z) \geq 3$, then there exists a non-zero $x \in Z$ such that $Mx \cdot x = 0$. (Hint: reduce to the case when M has full rank, and then modify Lemma 4.14. One can also solve this problem by purely algebraic methods.)

4.3.16 [160] Let W be a vector space over a finite field F of odd order, and let $M : W \to W$ be a linear transformation. Show that there exists a subspace U of W with dimension $\dim_F(U) \geq \frac{1}{2}\dim_F(W) - \frac{3}{2}$ such that M is null on U, i.e. $Mx \cdot y = 0$ for all $x, y \in U$. (Hint: take a maximal space U which is null with respect to M. If the orthogonal complement $U^\perp := \{y \in W : Mx \cdot y = 0 \text{ for all } x \in U\}$ is at least three dimensions larger than U, then use the previous lemma.) For a purely algebraic proof of this fact, see Exercise 9.4.11.

4.4 Bohr sets

In many applications of the Fourier-analytic method, one starts with some additive set A and concludes some information about the Fourier transform $\hat{1}_A$ of A (for instance, one may obtain some bound on the Fourier bias $\|A\|_u$). One would then like to pass from this back to some new combinatorial information on the original set A. For some special groups (e.g. finite field geometries F_p^n) one can do this quite directly (see for instance Lemma 10.15). However, to convert Fourier information on general groups to combinatorial information we need the notion of a *Bohr set* (also known as *Bohr neighborhoods* in the literature). We first define a "norm" $\|\theta\|_{\mathbf{R}/\mathbf{Z}}$ on the circle group by defining $\|\theta + \mathbf{Z}\|_{\mathbf{R}/\mathbf{Z}} = |\theta|$ whenever $-1/2 < \theta \leq 1/2$; in other words, $\|\theta\|_{\mathbf{R}/\mathbf{Z}}$ is the distance from θ (or more precisely, any representative of the coset θ) to the integers. We observe the elementary bounds

$$4\|\theta\|_{\mathbf{R}/\mathbf{Z}} \leq |e(\theta) - 1| \leq 2\pi\|\theta\|_{\mathbf{R}/\mathbf{Z}} \tag{4.24}$$

which follow from elementary trigonometry and the observation that the sinc function $\sin(x)/x$ varies between 1 and $2/\pi$ when $|x| \leq \pi/2$.

Definition 4.17 (Bohr set) Let $S \subset Z$ be a set of frequencies, and let $\rho > 0$. We define the *Bohr set* $\mathrm{Bohr}(S, \rho) = \mathrm{Bohr}_Z(S, \rho)$ as

$$\mathrm{Bohr}(S, \rho) := \left\{x \in Z : \sup_{\xi \in S} \|\xi \cdot x\|_{\mathbf{R}/\mathbf{Z}} < \rho\right\}.$$

We refer to S as the *frequency set* of the Bohr set, and ρ as the *radius*. The quantity $|S|$ is known as the *rank of the Bohr set.*

Remark 4.18 Note that if Z is a vector space over a finite field F, then every subspace of Z can be viewed as a Bohr set (with radius $O(1/|F|)$, and rank equal to the codimension). Thus Bohr sets can be viewed as a generalization of subspaces. Note that most finite groups Z tend to have very few actual subgroups (the extreme case being the cyclic groups $\mathbf{Z}_p$ of prime order), so it is convenient to be able to rely on the much larger class of Bohr sets as a substitute.

Remark 4.19 One way to think of Bohr sets is to consider the embedding of Z into the complex vector space $\mathbf{C}^S$ (and in particular to the standard unit torus inside $\mathbf{C}^S$) by the multiplicative map $x \mapsto (e(\xi \cdot x))_{\xi \in S}$. A Bohr set is thus the inverse image of a cube.

Observe that the $\| \|_{\mathbf{R}/\mathbf{Z}}$ norm is symmetric and subadditive; $\|-x\|_{\mathbf{R}/\mathbf{Z}} = \|x\|_{\mathbf{R}/\mathbf{Z}}$ and $\|x + y\|_{\mathbf{R}/\mathbf{Z}} \leq \|x\|_{\mathbf{R}/\mathbf{Z}} + \|y\|_{\mathbf{R}/\mathbf{Z}}$. From this we see that the Bohr sets $\mathrm{Bohr}(S, \rho)$ are symmetric, decreasing in S, and increasing in ρ (and fill out the whole space Z once $\rho > 1/2$); they are always unions of cosets of $S^\perp$, and if ρ is sufficiently small they consist entirely of $S^\perp$. One can also easily verify the intersection property

$$\mathrm{Bohr}(S, \rho) \cap \mathrm{Bohr}(S', \rho) = \mathrm{Bohr}(S \cup S', \rho)$$

and the addition property

$$\mathrm{Bohr}(S, \rho) + \mathrm{Bohr}(S, \rho') \subseteq \mathrm{Bohr}(S, \rho + \rho').$$

In particular we have

$$k\mathrm{Bohr}(S, \rho) \subseteq \mathrm{Bohr}(S, k\rho)$$

for any $k \geq 1$.

Next, we establish some bounds for the size of Bohr sets.

Lemma 4.20 (Size bounds) *If $S \subset Z$ and $\rho > 0$, then we have the lower bound*

$$\mathbf{P}_Z(\mathrm{Bohr}(S, \rho)) \geq \rho^{|S|} \tag{4.25}$$

and we have the doubling estimate

$$\mathbf{P}_Z(\mathrm{Bohr}(S, 2\rho)) \le 4^{|S|}\mathbf{P}_Z(\mathrm{Bohr}(S, \rho)). \tag{4.26}$$

This lemma should be compared with the Kronecker approximation theorem (Corollary 3.25); indeed the two results are very closely related.

Proof For each $\xi \in S$ let θ_ξ be an element of $\mathbf{R}/\mathbf{Z}$ chosen independently and uniformly at random. For any $x \in Z$, one can easily verify that

$$\mathbf{P}_Z(\|\xi \cdot x - \theta_\xi\|_{\mathbf{R}/\mathbf{Z}} < \rho/2 \text{ for all } \xi \in S) = \rho^{|S|}.$$

Summing this over all $x \in Z$ using linearity of expectation (1.4), we conclude

$$\mathbf{E}|\{x \in Z : \|\xi \cdot x - \theta_\xi\|_{\mathbf{R}/\mathbf{Z}} < \rho/2 \text{ for all } \xi \in S\}| \ge \rho^{|S|}|Z|$$

and thus there exists a choice of θ_ξ such that

$$|\{x \in Z : \|\xi \cdot x - \theta_\xi\|_{\mathbf{R}/\mathbf{Z}} < \rho/2 \text{ for all } \xi \in S\}| \ge \rho^{|S|}|Z|. \tag{4.27}$$

Now observe from the triangle inequality that if x, x' lie in the above set, then $x - x'$ lies in $\mathrm{Bohr}(S, \rho)$. The claim (4.25) follows.

Now we prove (4.26). By a limiting argument we may replace 2ρ by $2\rho - \varepsilon$ on the left-hand side for some small $\varepsilon > 0$. Observe that we can cover the interval $\{\theta \in \mathbf{R}/\mathbf{Z} : \|\theta\|_{\mathbf{R}/\mathbf{Z}} < 2\rho - \varepsilon\}$ by four intervals of the form $\{\theta \in \mathbf{R}/\mathbf{Z} : \|\theta - \theta_0\|_{\mathbf{R}/\mathbf{Z}} < \rho/2\}$. We can thus can cover $\mathrm{Bohr}(S, 2\rho)$ by $4^{|S|}$ sets of the type appearing in the left-hand side of (4.27). The claim follows by arguing as before. □

We have already mentioned that subspaces of a vector space are one example of a Bohr set. Progressions can form another example; for instance intervals such as $(-N, N)$ in a cyclic group $\mathbf{Z}_M$ can easily be seen to be a Bohr set of rank 1. We can combine these two examples by introducing the concept of a *coset progression*.

Definition 4.21 (Coset progressions) [157] A *coset progression* in an additive group Z is any set of the form $P + H$ where P is a progression and H is a finite subgroup of Z. We say that the coset progression $P + H$ is *proper* if P is proper and $|P + H| = |P||H|$ (i.e. all the sums in $P + H$ are distinct). We say that a coset progression $P + H$ has rank d if the component P has rank d. We say that $P + H$ is symmetric if P has the form $P = (-N, N) \cdot v$.

Of course, Corollary 3.8 shows that every coset progression can also be viewed as an ordinary progression, but possibly of much larger rank. If however Z is a cyclic group of prime order, then H will either be trivial or equal to the whole space, and will thus increase the rank by at most 1. Indeed we can view vector spaces over small finite fields on the one hand, and cyclic groups of prime order on the other, as the two extremes of additive behavior for finite groups Z.

Now we relate Bohr sets of rank d with coset progressions of rank d.

Lemma 4.22 (Bohr sets contain large coset progressions) *[160] Let* $\mathrm{Bohr}(S, \rho)$ *be a Bohr set of rank d in Z with $0 < \rho < \frac{1}{2}$. Then there exists a proper symmetric coset progression $P + H$ of rank $0 \leq d' \leq d$, obeying the inclusions*

$$\mathrm{Bohr}(S, (d')^{-2d'} \rho) \subseteq P + H \subseteq \mathrm{Bohr}(S, \rho). \tag{4.28}$$

In particular, from Lemma 4.20 we have

$$\mathbf{P}_Z(P + H) \geq \rho^d d^{-4d^2}. \tag{4.29}$$

Furthermore we have $H = S^{\perp}$.

Proof Let $\phi : Z \to (\mathbf{R}/\mathbf{Z})^S$ be the group homomorphism $\phi(x) := (\xi \cdot x)_{\xi \in S}$. Observe that $\phi(Z)$ is a finite subgroup of the torus $(\mathbf{R}/\mathbf{Z})^S$, and that $\mathrm{Bohr}(S, \rho)$ contains the inverse image of the cube $Q := \{(y_\xi)_{\xi \in S} \in \mathbf{R}^S : |y_\xi| \leq \rho\} \subset \mathbf{R}^S$ (which we identify with its projection in $(\mathbf{R}/\mathbf{Z})^S$) under ϕ.

Let $\Gamma \subseteq \mathbf{R}^S$ be the lattice $\phi(Z) + \mathbf{Z}^S$. Though it is a slight abuse of notation, we consider $\phi(Z) \cap Q$ to be the same as $\Gamma \cap Q$. Applying Lemma 3.36, we can find a progression $\tilde{P} := (-L, L) \cdot w$ for some linearly independent $w_1, \ldots, w_{d'} \subseteq \Gamma$ with $0 \leq d' \leq d$ such that

$$\Gamma \cap d'^{-2d'} \cdot Q \subseteq \tilde{P} \subseteq \Gamma \cap Q.$$

Since the w_j are independent, $\tilde{P}$ is necessarily proper. The claim now follows by setting v_j to be an arbitrary element of $\phi^{-1}(w_j)$ for each $1 \leq j \leq d'$, and setting H equal to the kernel of ϕ, which is of course just $S^{\perp}$. □

In the case of a cyclic group, we can dispense with the group H and sharpen the constants somewhat (though at the cost of losing the first inclusion in (4.28)):

Proposition 4.23 *Let $Z = \mathbf{Z}_N$ be a cyclic group, and let* $\mathrm{Bohr}(S, \rho)$ *be a Bohr set of rank d with $0 < \rho < \frac{1}{2}$. Then* $\mathrm{Bohr}(S, \rho)$ *contains a symmetric proper progression P of rank d and cardinality*

$$|P| \geq \frac{\rho^d}{d^d} N.$$

Furthermore we may choose P to be symmetric (i.e. $P = -P$).

Proof The main tool here will be Minkowski's second theorem. We use the standard bilinear form $\xi \cdot x = \xi x / N$, and write $S = (\xi_1, \ldots, \xi_d)$. Let $\alpha \in \mathbf{R}^d$ be the vector $\alpha := (\frac{\xi_1}{N}, \ldots, \frac{\xi_d}{N})$, and let Γ be the lattice $\mathbf{Z} \cdot \alpha + \mathbf{Z}^d$; this clearly has full rank, and by (3.12)

$$\mathrm{mes}(\mathbf{R}^d / \Gamma) = \mathrm{mes}(\mathbf{R}^d / \mathbf{Z}^d)/|\Gamma/\mathbf{Z}^d| \geq 1/N.$$

Let Q be the cube

$$Q := \{(x_1, \ldots, x_d) \in \mathbf{R}^d : |x_j| < \rho \text{ for all } 1 \leq j \leq n\},$$

and let $0 < \lambda_1 \leq \cdots \leq \lambda_d$ be the succesive minima of Q with respect to Γ, with a corresponding directional basis $v_1, \ldots, v_d \in \Gamma$ as given by Theorem 3.30. In particular we see that every coordinate of v_j has magnitude at most $\lambda_j \rho$.

Let $1 \leq j \leq d$ be arbitrary. Since $v_j \in \Gamma$, we see from the definition of Γ that there exists $w_j \in \mathbf{Z}_N$ such that $v_j \in \alpha w_j + \mathbf{Z}^d$. In particular we see that $\|\xi_i \cdot w_j\|_{\mathbf{R}/\mathbf{Z}} \leq \lambda_j \rho$ for all $1 \leq i, j \leq d$. Set $w := (w_1, \ldots, w_d)$. Now we let $M_j := \lfloor \frac{1}{d\lambda_j} \rfloor$, and let $M := (M_1, \ldots, M_d)$; we now claim that the progression $P := (-M, M) \cdot w$ is proper and lies in $\mathrm{Bohr}(S, \rho)$ (it is clearly symmetric). Let us first verify that $P \subseteq \mathrm{Bohr}(S, \rho)$. If $n = (n_1, \ldots, n_d) \in (-M, M)$, then for any $1 \leq j \leq d$ we have

$$\|\xi_j \cdot (n \cdot w)\|_{\mathbf{R}/\mathbf{Z}} \leq \sum_{j=1}^{d} |n_j| \|\xi_j \cdot w_j\|_{\mathbf{R}/\mathbf{Z}} < \sum_{j=1}^{d} \frac{1}{d\lambda_j} \lambda_j \rho = \rho$$

and hence $n_1 w_1 + \cdots + n_d w_d \in \mathrm{Bohr}(S, \rho)$. This proves the inclusion $P \subseteq \mathrm{Bohr}(S, \rho)$.

Now we show that P is proper. Suppose for contradiction that there exist distinct $n, n' \in (-M, M)$ such that $n \cdot w = n' \cdot w$; setting $\tilde{n} := n - n' \in (-2M, 2M)$, we thus see that $\tilde{n} \cdot w = 0$. In particular, $(\tilde{n} \cdot v)_i$ is an integer for each i. On the other hand, by arguing as before, we see that

$$|(\tilde{n} \cdot v)_i| \leq \sum_{j=1}^{d} |\tilde{n}_j| |\xi_i w_j / N| < \sum_{j=1}^{d} \frac{2}{d\lambda_j} \lambda_j \rho = 2\rho.$$

Since $\rho < 1/2$, we conclude that $(\tilde{n} \cdot v)_i = 0$ for all i, and thus $\sum_j \tilde{n}_j v_j = 0$. But this contradicts the linear independence of the directional basis $v_1, \ldots, v_d$. Thus P is proper.

Finally, the cardinality of the proper progression P is

$$|P| = \prod_{j=1}^{d} (2M_j - 1) \geq \prod_{j=1}^{d} \frac{1}{d\lambda_j}$$

and the claim follows from Minkowski's second theorem. □

One undesirable feature of Bohr sets of large rank d is that they have large doubling constant: (4.26) suggests that $\mathrm{Bohr}(S, \rho) + \mathrm{Bohr}(S, \rho)$ can be 4^d times larger than $\mathrm{Bohr}(S, \rho)$. A useful observation of Bourgain [39] is that if one considers an imbalanced sum $\mathrm{Bohr}(S, \rho) + \mathrm{Bohr}(S, \rho')$, with ρ' much smaller than ρ, then it is still possible for $\mathrm{Bohr}(S, \rho) + \mathrm{Bohr}(S, \rho')$ to be close to $\mathrm{Bohr}(S, \rho)$. This intuition is formalized by the notion of a *regular Bohr set.*

Definition 4.24 (Regular Bohr sets) A Bohr set $\mathrm{Bohr}(S, \rho)$ of rank d is said to be *regular* if one has the estimate

$$\begin{aligned}(1 - 100d|\kappa|)\mathbf{P}_Z(\mathrm{Bohr}(S, \rho)) &\leq \mathbf{P}_Z(\mathrm{Bohr}(S, (1 + \kappa)\rho)) \\ &\leq (1 + 100d|\kappa|)\mathbf{P}_Z(\mathrm{Bohr}(S, \rho))\end{aligned}$$

whenever $|\kappa| \leq \frac{1}{100d}$.

Not all Bohr sets are regular. However, it turns out that every Bohr set is "close" to a regular one:

Lemma 4.25 (Regular Bohr sets are ubiquitous) *[39] Let S be a non-empty additive set and let $0 < \varepsilon < 1$. Then there exists $\rho \in [\varepsilon, 2\varepsilon]$ such that* $\mathrm{Bohr}(S, \rho)$ *is regular.*

Proof Let $f : [0, 1] \to \mathbf{R}$ be the function $f(a) := \frac{1}{d} \log_2 \mathbf{P}_Z(\mathrm{Bohr}(S, 2^a \varepsilon))$. Observe that f is non-decreasing in a, and from Lemma 4.20 we have $f(1) - f(0) \leq 2$.

Suppose we could find $0.1 \leq a \leq 0.9$ such that $|f(a') - f(a)| \leq 20|a - a'|$ for all $|a - a'| \leq 0.1$. Then it is easy to see that $\mathrm{Bohr}(S, 2^a \varepsilon)$ is regular. Thus, it suffices to obtain an a with this property. This can be done directly from the Hardy–Littlewood maximal inequality (applied to the Lebesgue–Stieltjes measure df), or as follows. If no such a exists, then for every $0.1 \leq a \leq 0.9$ there exists a real interval I of length at most 0.1 and with one endpoint equal to a, such that $\int_I df > \int_I 20\,dx$. These intervals cover $\{a : 0.1 \leq a \leq 0.9\}$, which has measure 0.8. By the Vitali covering lemma (see exercises), one can find thus find a finite subcollection of disjoint intervals $I_1, \ldots, I_n$ of total length $|I_1| + \cdots + |I_n| \geq 0.8/5$ (say). But then we have

$$\log_2 5 \geq \int_0^1 df \geq \sum_{i=1}^{n} \int_{I_i} df \geq \sum_{i=1}^{n} \int_{I_i} 20dx \geq \frac{0.8}{5} \times 20,$$

a contradiction. □

We shall make a crucial use of this lemma in proving Bourgain's quantitative version of Roth's theorem in Section 10.4.

Exercises

4.4.1 Show that if $0 < \rho \le 1/6$ and $|S| \ge 1$, then $|\widehat{1_{\mathrm{Bohr}(S,\rho)}}(\xi)| \ge \frac{1}{2}\mathbf{P}_Z(\mathrm{Bohr}(S,\rho))$ for all $\xi \in S$. In particular Bohr sets are extremely non-uniform: $\|\mathrm{Bohr}(S,\rho)\|_u \ge \frac{1}{2}\mathbf{P}_Z(\mathrm{Bohr}(S,\rho))$. By applying Plancherel's theorem, conclude the additional bound $\mathbf{P}_Z(\mathrm{Bohr}(S,\rho)) \le \frac{4}{|S|}$.

4.4.2 Give examples to show that the density $\mathbf{P}_Z(\mathrm{Bohr}(S,\rho))$ of a Bohr set can be as low as $\Theta(\rho)^{|S|}$, and as large as $\Theta(1/|S|)$, even when ρ is small and $|S|$ is large. Thus the bounds in (4.25) and the preceding exercise cannot be significantly improved.

4.4.3 Establish the bound $\mathbf{P}_Z(\mathrm{Bohr}(S,k\rho)) \le O(k)^{|S|}\mathbf{P}_Z(\mathrm{Bohr}(S,\rho))$ for any $k \ge 1$. Using the Ruzsa covering lemma (Lemma 2.14), conclude that one can cover $\mathrm{Bohr}(S,k\rho)$ by $O(k)^{|S|}$ translates of $\mathrm{Bohr}(S,\rho)$. In particular, in the notation of Definition 2.25, $\mathrm{Bohr}(S,\rho)$ is a $O(1)^{|S|}$-approximate group.

4.4.4 In the setting of Lemma 4.22, show that $\mathrm{Bohr}(S,\rho)$ can be covered by $O(d)^{d^2}$ translates of $P + H$.

4.4.5 Show that a Bohr set $\mathrm{Bohr}(S,\rho)$ of rank d always contains an arithmetic progression of length $\Theta(|\mathrm{Bohr}(S,\rho)|^{1/d})$ and non-zero step size. (Hint: if $|\mathrm{Bohr}(S,\rho)|^{1/d}$ is large, use the preceding exercise to show that $\mathrm{Bohr}(S,\rho/k)$ contains a non-zero element for some integer $k = \Theta(|\mathrm{Bohr}(S,\rho)|^{1/d})$.)

4.4.6 [160] Let A be an additive set in Z that contains 0. Show that there exists a set S of frequencies with $|S| \le 1 + \log_2|A|$ such that $A \cap \mathrm{Bohr}(S,1/4) = \{0\}$. (Hint: choose $1 + \lfloor \log_2|A| \rfloor$ frequencies randomly and independently (allowing for collisions) and use the first moment method.)

4.4.7 (Vitali-type covering lemma) Let $\mathcal{I}$ be a finite collection of intervals in the real line. Show that there exist a subcollection $I_1, \ldots, I_n$ of these intervals whose interiors are disjoint, and such that $\sum_{i=1}^n |I_i| \ge \frac{1}{5}\mathrm{mes}(\bigcup_{I\in\mathcal{I}} I)$. (Hint: use a greedy algorithm, picking the largest intervals first.) By being more sophisticated in the argument, lower $\frac{1}{5}$ to $\frac{1}{2}$. (Hint: eliminate nested intervals, and then move greedily from left to right to cover $\bigcup_{I\in\mathcal{I}} I$ by two families of interior-disjoint intervals.)

4.4.8 (Hardy–Littlewood maximal inequality) Let μ be a non-negative finite measure on the real line, and let $M\mu$ denote the *Hardy–Littlewood maximal function* $M\mu(x) := \sup_{r>0} \frac{1}{2r}\mu\{y : x - r < y < x + r\}$. (It can be verified that $M\mu$ is a measurable function.) Using the Vitali-type covering lemma, establish the distributional inequality

$$\mathrm{mes}(\{x : M\mu(x) \ge \lambda\}) \le \frac{2}{\lambda}\mu(\mathbf{R}).$$

4.5 $\Lambda(p)$ constants, $B_h[g]$ sets, and dissociated sets

In Section 4.3 we discussed one Fourier-analytic characteristic of an additive set A in a finite additive group Z, namely its linear bias. In this section we discuss a rather different characteristic, namely the $\Lambda(p)$ constants of a set S of frequencies. These constants measure how "dissociated" or "Sidon-like" a set[1] S is; in more practical terms, the $\Lambda(p)$ constants quantify the independence of the characters associated to S in a certain $L^p(Z)$ sense. These constants can be used to obtain precise control on the arithmetic structure of S, for instance in controlling iterated sum sets of S. One feature of these constants is that they are stable under passage to subsets, thus $\Lambda(p)$ constants will also control iterated sum sets of subsets S' of S. This stability (which is not present in the Fourier bias, unless one takes random subsets as in Lemma 4.16) is useful for a number of applications.

We begin with the formal definition of the $\Lambda(p)$ constants.

Definition 4.26 (**$\Lambda(p)$ constants**) Let S be an additive set in a finite[2] additive group Z, and let $2 \leq p \leq \infty$. We define the $\Lambda(p)$ *constant* of S, denoted $\|S\|_{\Lambda(p)}$, to be the best constant such that the inequality

$$\left\| \sum_{\xi \in S} c(\xi) e(\xi \cdot x) \right\|_{L^p(Z)} \leq \|S\|_{\Lambda(p)} \|c\|_{l^2(S)} \tag{4.30}$$

holds for all sequences $c : S \to \mathbf{C}$ of complex numbers.

One can easily establish the bound

$$\|S\|_{\Lambda(p)} \leq |S|^{1/2-1/p}, \tag{4.31}$$

for $2 \leq p \leq \infty$, with equality at the endpoints $p = 2, \infty$; see Exercise 4.5.6. This exercise indicates that largeness of $\Lambda(p)$ constants is correlated to strong additive structure of S. At the other extreme, we now show that smallness of $\Lambda(p)$ constants is correlated to strong lack of additive structure of S.

Definition 4.27 (**B_h sets**) Let $h \geq 2$. A non-empty subset S of an additive group Z is a B_h *set* if for any $\xi_1, \ldots, \xi_h, \eta_1, \ldots, \eta_h \in S$, one has $\xi_1 + \cdots + \xi_h = \eta_1 + \cdots + \eta_h$ if and only if $(\xi_1, \ldots, \xi_h)$ is a permutation of $(\eta_1, \ldots, \eta_h)$. We say S is a *Sidon set* if it is a B_2 set.

These sets are the $g = 1$ version of the $B_h[g]$ sets, encountered in Section 1.7.1; Sidon sets were also briefly mentioned in Section 2.2. Note that we do not bother with the notion of a B_1 set, since every set is trivially a B_1 set.

[1] Here, we use "Sidon set" to denote a set whose pairwise sums are all disjoint. There is another, more Fourier-analytic, notion of a Sidon set related to $\Lambda(p)$ constants which we will not discuss here.

[2] One can also define the concept of a $\Lambda(p)$ constant for subsets of the integers, or more general additive groups, but we will not need to do so in this book.

Example 4.28 For any $M > 1$, the set $S := \{0\} \cup (M^\wedge\mathbf{N}) = \{0, 1, M, M^2, \ldots\}$ is a B_h set in $\mathbf{Z}$ if and only if $h < M$. In particular, the powers of 2 form a Sidon set. One can of course truncate these examples to finite additive groups such as $\mathbf{Z}_N$; note that any non-empty subset of a B_h set is also a B_h set.

Proposition 4.29 *Let S be a non-empty subset of a finite additive group Z. Then we have*

$$\|S\|_{\Lambda(4)} \geq \left(2 - \frac{1}{|S|}\right)^{1/4}, \tag{4.32}$$

with equality holding if and only if S is a Sidon set. More generally, if $h \geq 1$, then there exists a number $1 \leq \alpha(h, |S|) < (h!)^{1/2h}$ depending on h and $|S|$ such that $\|S\|_{\Lambda(2h)} = \alpha(h, |S|)$ when S is a B_h set, and $\|S\|_{\Lambda(2h)} > \alpha(h, |S|)$ otherwise.

Proof We first prove (4.32). By testing (4.30) with c_ξ identically equal to 1, it will suffice to show that

$$\left\| \sum_{\xi \in S} e(\xi \cdot x) \right\|_{L^4(Z)}^4 \geq \left(2 - \frac{1}{|S|}\right) |S|^2.$$

The left-hand side can be expanded as

$$\sum_{\xi_1, \xi_2, \eta_1, \eta_2 \in S} \mathbf{E}_{x \in Z} e((\xi_1 + \xi_2 - \eta_1 - \eta_2) \cdot x).$$

By Lemma 4.5 this simplifies to

$$|\{\xi_1, \xi_2, \eta_1, \eta_2 \in S : \xi_1 + \xi_2 = \eta_1 + \eta_2\}|.$$

Clearly $\xi_1 + \xi_2$ will equal $\eta_1 + \eta_2$ when (ξ_1, ξ_2) is a permutation of (η_1, η_2), so this expression is at least as large as

$$\sum_{\xi_1, \xi_2, \eta_1, \eta_2 \in S : \{\xi_1, \xi_2\} = \{\eta_1, \eta_2\}} 1 = |S|(|S| - 1)2 + |S| = \left(2 - \frac{1}{|S|}\right) |S|^2$$

as claimed. Note that this argument also shows that the inequality in (4.32) is strict if S is not a Sidon set, since then we have additional terms coming from pairs (ξ_1, ξ_2) and (η_1, η_2) which are not permutations of each other.

Now suppose that S is a Sidon set. To prove equality in (4.32) it suffices to show that

$$\left\| \sum_{\xi \in S} c_\xi e(\xi \cdot x) \right\|_{L^4(Z)}^4 \leq 2 - \frac{1}{|S|}$$

assuming the normalization $\sum_{\xi \in S} |c_\xi|^2 = 1$. The left-hand side can be expanded as

$$\sum_{\xi_1,\xi_2,\eta_1,\eta_2 \in S} c_{\xi_1} c_{\xi_2} \overline{c_{\eta_1} c_{\eta_2}} \mathbf{E}_{x \in Z} e((\xi_1 + \xi_2 - \eta_1 - \eta_2) \cdot x)$$

which as before simplifies to

$$\sum_{\xi_1,\xi_2,\eta_1,\eta_2 \in S : \xi_1+\xi_2=\eta_1+\eta_2} c_{\xi_1} c_{\xi_2} \overline{c_{\eta_1} c_{\eta_2}}.$$

Since S is a Sidon set, (η_1, η_2) must be a permutation of (η_1, η_2). Splitting into the cases $\xi_1 = \xi_2$ and $\xi_1 \neq \xi_2$, we can thus rewrite the previous expression as

$$\sum_{\xi \in S} |c_\xi|^4 + 2 \sum_{\xi_1,\xi_2 \in S : \xi_1 \neq \xi_2} |c_{\xi_1}|^2 |c_{\xi_2}|^2$$

which by the normalization $\sum_{\xi \in S} |c_\xi|^2 = 1$ can be written as

$$2 - \sum_{\xi \in S} |c_\xi|^4.$$

But from Cauchy–Schwarz and the normalization $\sum_{\xi \in S} |c_\xi|^2 = 1$ we have $\sum_{\xi \in S} |c_\xi|^4 \geq 1/|S|$, and the claim follows.

The general case $h \geq 2$ is similar but is left to Exercise 4.5.9. □

Another quantification of the heuristic that large $\Lambda(p)$ constants corresponds to strong additive structure is given by

Lemma 4.30 *Let S be a non-empty subset of a finite additive group Z, and let $h \geq 1$. Then we have*

$$|h_1 S - h_2 S| \geq \frac{|S|^h}{\|S\|_{\Lambda(2h)}^{2h}}$$

whenever $h_1, h_2 \geq 0$ are such that $h_1 + h_2 = h$. In particular we have

$$|hS| \geq \frac{|S|^h}{\|S\|_{\Lambda(2h)}^{2h}}.$$

Remark 4.31 This lemma shows that if S has a small $\Lambda(2h)$ constant, then not only do the sum sets hS become very large, but so do the sum sets hS' of all subsets S' of S, thanks to the monotonicity of $\Lambda(p)$ constants. The converse statement is also true up to logarithmic factors; see exercises. Thus $\Lambda(2h)$ constants measure the failure of S, or any of its subsets, to have good closure properties under h-fold sums.

Proof From (4.30) with $p := 2h$, and c_ξ set identically equal to 1, we have

$$\left\| \sum_{\xi \in S} e(\xi \cdot x) \right\|_{L^{2h}(Z)}^{2h} \leq \|S\|_{\Lambda(2h)}^{2h} |S|^h.$$

The left-hand side is equal to

$$\left\| \left(\sum_{\xi \in S} e(\xi \cdot x) \right)^{h_1} \left(\sum_{\xi \in -S} e(\xi \cdot x) \right)^{h_2} \right\|_{L^2(Z)}^2$$

since $e(-\xi \cdot x)$ is the conjugate of $e(\xi \cdot x)$. We can expand

$$\left(\sum_{\xi \in S} e(\xi \cdot x) \right)^{h_1} \left(\sum_{\xi \in -S} e(\xi \cdot x) \right)^{h_2} = \sum_{\xi \in Z} r_{h_1,h_2}(\xi) e(\xi \cdot x)$$

where r_{h_1,h_2} is the counting function

$$r_{h_1,h_2}(\xi) := |\{(\xi_1, \ldots, \xi_{h_1}, \xi'_1, \ldots, \xi'_{h_2}) \in S^{h_1+h_2} : \xi = \xi_1 + \cdots + \xi_{h_1} - \xi'_1 - \cdots - \xi'_h\}|.$$

By (4.2) we thus have

$$\sum_{\xi \in Z} r_{h_1,h_2}(\xi)^2 \leq \|S\|_{\Lambda(2h)}^{2h} |S|^h.$$

On the other hand, the function r_{h_1,h_2} is supported in $h_1 S - h_2 S$, so by Cauchy–Schwarz

$$\sum_{\xi \in Z} r_{h_1,h_2}(\xi) \leq |h_1 S - h_2 S|^{1/2} \|S\|_{\Lambda(2h)}^h |S|^{h/2}.$$

But from the definition of r_{h_1,h_2} we have

$$\sum_{\xi \in Z} r_{h_1,h_2}(\xi) = |S^{h_1+h_2}| = |S|^{h_1+h_2}$$

The claim follows. □

We now investigate the $\Lambda(p)$ constants of Sidon-like sets as $p \to \infty$.

Definition 4.32 An additive set S with cardinality $|S| = d$ is said to be *dissociated* if the cube $[0, 1]^d \cdot S$ is proper, or in other words, the 2^d subset sums

$$FS(S) := \left\{ \sum_{\xi \in S'} \xi : S' \subseteq S \right\}$$

are all distinct.

This should be compared with the concept of a Sidon set, which is a set S of cardinality d whose $\frac{d(d+1)}{2}$ pairwise sums $\{\xi_1 + \xi_2 : \xi_1, \xi_2 \in S\}$ are all distinct (except for the trivial identification $\xi_1 + \xi_2 = \xi_2 + \xi_1$). A good example of a dissociated set is the set of powers of 2: $S = \{1, 2, \ldots, 2^n\}$ in any cyclic group $\mathbf{Z}/N\mathbf{Z}$ with $N \geq 2^{n+1}$. Observe that if S is a dissociated set of cardinality d, and v is a non-zero element of $[-1, 1]^d$, then $v \cdot S \neq 0$ (since otherwise we could find two disjoint sets S_1, S_2 in S, corresponding to where the components of v are $+1$ or -1, such that $\sum_{\xi \in S_1} \xi = \sum_{\xi \in S_2} \xi$).

Dissociativity is the Fourier analog of joint independence. It leads to the following Fourier-analytic analog of Chernoff's inequality:

Lemma 4.33 (Rudin's inequality) *If S is dissociated, then we have*

$$\mathbf{E}_{x \in Z} \exp\left(\sigma \mathrm{Re} \sum_{\xi \in S} c(\xi) e(\xi \cdot x)\right) \leq e^{\sigma^2/2} \tag{4.33}$$

whenever $\|c\|_{l^2(S)} \leq 1$ and $\sigma \geq 0$. We also have the distributional estimates

$$\mathbf{P}_{x \in Z}\left\{\left|\sum_{\xi \in S} c(\xi) e(\xi \cdot x)\right| \geq \lambda\right\} = O_\varepsilon\left(e^{-\lambda^2/(4+\varepsilon)}\right) \tag{4.34}$$

for every $\varepsilon > 0$, and the $\Lambda(p)$ estimate

$$\|S\|_{\Lambda(p)} = O(\sqrt{p}) \tag{4.35}$$

for all $2 \leq p < \infty$.

Note that when $p = 2h$ then $(h!)^{1/2h}$ is comparable to $\sqrt{p}$ by Stirling's formula (1.52), and hence so (4.35) and shows that dissociated sets are comparable in $\Lambda(2h)$ constant to B_{2h} sets for any given h (if S is sufficiently large). This also shows that the bounds in the above lemma cannot be significantly improved except in the constants, even if one imposes even more additive independence conditions on S.

Proof Write $c(\xi) = |c(\xi)| e(\theta_\xi)$ for some phase $\theta_\xi \in \mathbf{R}/\mathbf{Z}$. We begin by observing the inequality

$$e^{tx} \leq \cosh(x) + t \sinh(x)$$

for all $x \geq 0$ and $-1 \leq t \leq 1$, which is simply a consequence of the convexity of e^{tx} as a function of t. In particular we see that

$$\exp(\sigma \mathrm{Re} c(\xi) e(\xi \cdot x)) \leq \cosh(\sigma |c(\xi)|) + \sinh(\sigma |c(\xi)|) \mathrm{Re}\, e(\xi \cdot x + \theta_\xi),$$

which upon multiplying and taking expectations becomes

$$\mathbf{E}_{x\in Z}\exp\left(\sigma\sum_{\xi\in S}\mathrm{Re} c(\xi)e(\xi\cdot x)\right)$$
$$\leq \mathbf{E}_{x\in Z}\prod_{\xi\in S}\Big((\cosh(\sigma|c(\xi)|)+\frac{1}{2}\sinh(\sigma|c(\xi)|)e(\xi\cdot x+\theta_\xi)$$
$$+\frac{1}{2}\sinh(\sigma|c(\xi)|)e(-\xi\cdot x-\theta_\xi)\Big).$$

Now we multiply the product out and inspect its behavior in x. We obtain a large number of terms ($3^{|S|}$, to be exact) that are of the form $e((v\cdot S)\cdot\xi)$, for some $v\in[-1,1]^{|S|}$, times some constant independent of x, where we select some enumeration $S=(\xi_1,\ldots,\xi_{|S|})$ of S. There is one constant term, namely $\prod_{\xi\in S}\cosh(\sigma|c(\xi)|)$, but all the others have a non-zero frequency vector $v\cdot S$ because S is dissociated, and thus integrate out to zero by the Fourier inversion formula. Thus we have

$$\mathbf{E}_{x\in Z}\exp\left(\sigma\sum_{\xi\in S}\mathrm{Re}\, c(\xi)e(\xi\cdot x)\right)\leq\prod_{\xi\in S}\cosh(\sigma|c(\xi)|),$$

and the claim (4.33) then follows from the elementary inequality $\cosh(x)\leq e^{x^2/2}$ (which follows by comparing Taylor series). From Markov's inequality we thus obtain

$$\mathbf{P}_{x\in Z}\left(\mathrm{Re}\sum_{\xi\in S}c(\xi)e(\xi\cdot x)\geq\lambda\right)\leq e^{\sigma^2/2}e^{-\sigma\lambda}$$

for every $\lambda\geq 0$; choosing $\sigma:=\lambda/2$, we obtain

$$\mathbf{P}_{x\in Z}\left(\mathrm{Re}\sum_{\xi\in S}c(\xi)e(\xi\cdot x)\geq\lambda\right)\leq e^{-\lambda^2/4}.$$

Replacing λ by $(1-\varepsilon)\lambda$ and rotating $c(\xi)$ by an arbitrary angle $e(\theta)$, we obtain

$$\mathbf{P}_{x\in Z}\left(\mathrm{Re}\, e(\theta)\sum_{\xi\in S}c(\xi)e(\xi\cdot x)\geq(1-\varepsilon)\lambda\right)\leq e^{-\lambda^2(1-\varepsilon)^2/4}.$$

Taking the union of these estimates as $e^{i\theta}$ varies over a finite number of angles (depending on ε) we obtain (4.34).

To obtain (4.35), we observe from the identity

$$\left\|\sum_{\xi\in S}c(\xi)e(\xi\cdot x)\right\|_{L^p(Z)}^p=p\int_0^\infty\lambda^{p-1}\mathbf{P}_{x\in Z}\left(\left|\sum_{\xi\in S}c(\xi)e(\xi\cdot x)\right|\geq\lambda\right)d\lambda$$

and (4.34) (with $\varepsilon = 1$, say) that

$$\left\| \sum_{\xi \in S} c(\xi) e(\xi \cdot x) \right\|_{L^p(Z)}^p = O\left(p \int_0^\infty \lambda^{p-1} e^{-\lambda^2/5} \, d\lambda \right).$$

To estimate the integral, we observe from elementary calculus that the integrand $\lambda^{p-1} e^{-\lambda^2/5}$ is bounded by $O(p)^{p/2}$ for $\lambda = O(\sqrt{p})$, and then decays exponentially for $\lambda \gg \sqrt{p}$. From this we can easily bound the integrand by $p^{O(1)} O(p)^{p/2}$, and the claim follows (note that $p^{1/p}$ is bounded by e). □

In the next few sections we shall use Rudin's inequality to obtain structural control on various sets of frequencies.

Exercises

4.5.1 Show that the $\Lambda(p)$ constant of a set S does not depend on the choice of bilinear form used to define the Fourier transform, and is also invariant under translations or isomorphisms of the set S.

4.5.2 For any $2 \leq p \leq \infty$ and any disjoint S_1, S_2, show the triangle inequality $\|S\|_{\Lambda(p)} \leq \|S_1\|_{\Lambda(p)} + \|S_2\|_{\Lambda(p)}$ whenever $S \subseteq S_1 \cup S_2$.

4.5.3 Let ε be the uniform distribution on $\{-1, 1\}$, and let $\varepsilon_1, \ldots, \varepsilon_N$ be independent trials of ε. If $c_1, \ldots, c_N$ are arbitrary complex numbers and $2 \leq p < \infty$, prove *Bernstein's inequality* [25]

$$\left(\sum_{j=1}^N |c_j|^2 \right)^{1/2} \leq \mathbf{E} \left(\left| \sum_{j=1}^N \varepsilon_j c_j \right|^p \right)^{1/p}$$
$$\leq O\left(\sqrt{p} \left(\sum_{j=1}^N |c_j|^2 \right)^{1/2} \right).$$

(Hint: for the lower bound, compute the $p = 2$ moment. For the upper bound, modify the proof of Lemma 4.33; alternatively, apply Lemma 4.33 to the group $Z = \mathbf{Z}_2^N$, where S is the standard basis for $\mathbf{Z}_2^N$.) Conclude that if $f_1, \ldots, f_N$ are any complex-valued functions on Z, then we have *Khintchine's inequality*

$$\left\| \left(\sum_{j=1}^N |f_j|^2 \right)^{1/2} \right\|_{L^p(Z)} \leq \mathbf{E} \left(\left\| \sum_{j=1}^N \varepsilon_j f_j \right\|_{L^p(Z)}^p \right)^{1/p}$$
$$\leq O\left(\sqrt{p} \left\| \left(\sum_{j=1}^N |f_j|^2 \right)^{1/2} \right\|_{L^p(Z)} \right).$$

4.5.4 Let $f : Z_1 \times Z_2 \to \mathbf{C}$ be a function on two variables in two non-empty finite sets Z_1, Z_2, and let $2 \leq p < \infty$. Establish the Minkowski inequality

$$\left(\mathbf{E}_{y\in Z_2}(\mathbf{E}_{x\in Z_1}|f(x,y)|^2)^{p/2}\right)^{1/p} \leq \left(\mathbf{E}_{x\in Z_1}(\mathbf{E}_{y\in Z_2}|f(x,y)|^p)^{2/p}\right)^{1/2} \tag{4.36}$$

(Hint: use the triangle inequality for the $L^{p/2}$ norm.) Conclude that $\|S\|_{\Lambda(p)}$ is the best constant such that

$$\left\|\left\|\sum_{\xi\in S} c(\xi)e(\xi\cdot x)\right\|_H\right\|_{L^p(Z)} \leq \|S\|_{\Lambda(p)}\left(\sum_{\xi\in S}\|c(\xi)\|_H^2\right)^{1/2}$$

for all finite-dimensional Hilbert spaces H and all sequences $(c(\xi))_{\xi\in S}$ taking values in H. Using this, conclude that $\|S_1 \times S_2\|_{\Lambda(p)} = \|S_1\|_{\Lambda(p)}\|S_2\|_{\Lambda(p)}$ whenever S_1, S_2 are additive sets in finite additive groups Z_1, Z_2 and $2 \leq p \leq \infty$.

4.5.5 [33], [20] Let $n \geq 1$ be an integer, let $Z := \mathbf{Z}_2^n$. For $\xi = (\xi_1, \ldots, \xi_n) \in \mathbf{Z}_2^n$, let $|\xi|$ denote the number of coefficients $\xi_1, \ldots, \xi_n$ which are equal to one. Establish the *Bonami–Beckner inequality*

$$\left\|\sum_{\xi\in Z}\varepsilon^{|\xi|}c(\xi)\right\|_{L^{1+1/\varepsilon^2}(Z)} \leq \|c\|_{l^2(Z)}$$

for all $0 < \varepsilon < 1$ and all $c \in l^2(Z)$. (Hint: first establish this by hand for $n = 1$, and then exploit (4.36) to obtain the general case.) Conclude in particular that if $S_k := \{\xi \in \mathbf{Z}_2^n : |\xi| = k\}$, then $\|S_k\|_{\Lambda(p)} \leq (p-1)^{k/2}$ for all $2 < p < \infty$.

4.5.6 Let $2 \leq p \leq \infty$, and let S be a non-empty subset of Z. Prove (4.31). (Hint: use the Hausdorff–Young inequality.) If $2 < p < \infty$, show that equality occurs if and only if S is a translate of a subgroup of Z. (You may need Exercise 4.2.9.)

4.5.7 Let S be an additive set in a finite additive group. Show that

$$\|S\|_{\Lambda(p)} \geq \min\left(1, |Z|^{-1/p}|S|^{1/2}\right)$$

for all $2 \leq p < \infty$. It turns out that these bounds are essentially sharp for randomly chosen sets S in Z of a fixed cardinality: see [35].

4.5.8 Let S be a B_h set in a finite additive group Z. Show that $|S| \leq |Z|^{1/h}$.

4.5.9 Complete the proof of Proposition 4.29.

4.5.10 Let S be an additive subset of Z. Show that $E(S, S) \leq \|S\|_{\Lambda(4)}^4|S|^2$; thus the additive energy of an additive set is controlled by its $\Lambda(4)$ constant.

4.5.11 Let S be an additive set, and let $h \geq 1$. Suppose that $A > 0$ is a constant such that

$$|hS'| \geq \frac{|S'|^h}{A^{2h}}$$

for all non-empty subsets S' of S. Show that

$$\|S\|_{\Lambda(2h)} = O(A(1 + \log |S|));$$

thus Lemma 4.30 can be reversed after conceding a factor of a logarithm. (Hint: first verify the estimate (4.30) when c is a characteristic function by reversing the proof of Lemma 4.30. For general c, decompose c into at most $O(1 + \log |S|)$ functions which are comparable to constant multiples of characteristic functions, by partitioning the range of c using powers of 2, and discarding those values of c smaller than (say) $|S|^{-100}\|c\|_{l^2}$.)

4.5.12 [251] Show that $\|S\|_{\Lambda(p)}$ is the best constant such that

$$\|\hat{f}\|_{l^2(S)} \leq \|S\|_{\Lambda(p)} \|f\|_{L^{p'}(Z)}$$

for all random variables f, where p' is the dual exponent to p, thus $1/p + 1/p' = 1$. Next, write

$$\|\hat{f}\|^2_{l^2(Z)} = \frac{|S|}{|Z|}\|f\|^2_{L^2(Z)} + \mathbf{E}_{x,y \in Z} f(x)\overline{f(y)}\mathbf{I}(x \neq y) \sum_{\xi \in S} e(\xi \cdot (x - y))$$

and observe the inequalities

$$\left| \mathbf{E}_{x,y \in Z} f(x) g(y) \mathbf{I}(x \neq y) \sum_{\xi \in S} e(\xi \cdot (x - y)) \right| \leq \|f\|_{L^2(Z)} \|g\|_{L^2(Z)}$$

and

$$\left| \mathbf{E}_{x,y \in Z} f(x) g(y) \mathbf{I}(x \neq y) \sum_{\xi \in S} e(\xi \cdot (x - y)) \right| \leq |Z| \|S\|_u \|f\|_{L^1(Z)} \|g\|_{L^1(Z)}.$$

Using Riesz–Thorin interpolation (or arguing as in Exercise 4.2.3) conclude that

$$\left| \mathbf{E}_{x,y \in Z} f(x) g(y) \mathbf{I}(x \neq y) \sum_{\xi \in S} e(\xi \cdot (x - y)) \right|$$
$$\leq (|Z| \|S\|_u)^{1-2/p} \|f\|_{L^{p'}(Z)} \|g\|_{L^{p'}(Z)}.$$

From this, conclude the *Tomas–Stein inequality*

$$\|S\|^2_{\Lambda(p)} \leq |S||Z|^{-\frac{2}{p}} + (\|S\|_u |Z|)^{1-\frac{2}{p}}$$

(compare with (4.31)). Thus, Fourier-uniform sets tend to have fairly small $\Lambda(p)$ constants. See also Lemma 10.22.

4.6 The spectrum of an additive set

We now use Fourier analysis to investigate the spectral properties of additive sets A which have high additive energy $E(A, A)$; examples of such sets include sets with small sum set $|A + A|$ or small difference set $|A - A|$ (cf. (2.8)). One can already conclude from estimates such as (4.23) that such sets must be highly non-uniform, i.e. 1_A contains non-trivial Fourier coefficients. However, this by itself is not the strongest Fourier-analytic statement one can say about such sets. In order to proceed further it is convenient to introduce the notion of the *α-spectrum* of a set.

Definition 4.34 (Spectrum) Let A be an additive set in a finite additive group Z with a non-degenerate symmetric bilinear form $\cdot$ and let $\alpha \in \mathbf{R}$ be a parameter. We define the *α-spectrum* $\mathrm{Spec}_\alpha(A) \subseteq Z$ to be the set

$$\mathrm{Spec}_\alpha(A) := \{\xi \in Z : |\widehat{1_A}(\xi)| \geq \alpha \mathbf{P}_Z(A)\}.$$

One could define this spectrum without the assistance of the bilinear form $\cdot$, but then it would be a subset of the Pontryagin dual group $\hat{Z}$ rather than Z.

From Lemma 4.9 we see that the sets $\mathrm{Spec}_\alpha(A)$ are symmetric, decreasing in α, empty for $\alpha > 1$, contain the origin for $\alpha \leq 1$, and are the whole space Z whenever $\alpha \leq 0$. Thus the spectrum is really only an interesting concept when $0 < \alpha \leq 1$. In the extreme case $\alpha = 1$ the spectrum becomes a group, see Exercise 4.6.2.

From (4.16) (and Markov's inequality) we observe the upper bound

$$|\mathrm{Spec}_\alpha(A)| \leq \alpha^{-2}/\mathbf{P}_Z(A) \tag{4.37}$$

on the cardinality of the α-spectrum. In fact we can use Rudin's inequality to obtain a more precise structural statement, in which the polynomial loss in $\mathbf{P}_Z(A)$ is replaced with a logarithmic loss. To prove this statement, we first need an easy lemma (cf. Corollary 1.42).

Lemma 4.35 (Cube covering lemma) *[36] Let S be an additive set in an ambient group Z, and let $d \geq 1$ be an integer. Then we can partition $S = D_1 \cup \cdots \cup D_k \cup R$ where $D_1, \ldots, D_k$ are disjoint dissociated subsets of S of cardinality $d + 1$, and the remainder set R is contained in a cube $[-1, 1]^d \cdot (\eta_1, \ldots, \eta_d)$ for some $\eta_1, \ldots, \eta_d \in Z$.*

Proof We use the greedy algorithm. We initially set $k = 0$. If we can find a dissociated subset D of S of cardinality $d + 1$, we remove it from S and add it to the collection $D_1, \ldots, D_k$, thus incrementing k. We continue in this manner until we are left with a remainder R where all dissociated subsets of S have cardinality d or less. Let $\{\eta_1, \ldots, \eta_{d'}\}$ be a dissociated subset of R with maximal cardinality; thus $d' \leq d$. Observe that if R contained an element ξ which was not contained in $[-1, 1]^{d'} \cdot (\eta_1, \ldots, \eta_{d'})$, then $\{\eta_1, \ldots, \eta_{d'}, \xi\}$ would be dissociated, so contradicting maximality of d'. Thus we have $R \subseteq [-1,1]^{d'} \cdot (\eta_1, \ldots, \eta_{d'})$, and the claim follows (padding out the progression with some dummy elements $\eta_{d'+1}, \ldots, \eta_d$ if necessary). □

Lemma 4.36 (Fourier concentration lemma) *[48] Let A be an additive set in a finite additive group Z, and let $0 < \alpha \leq 1$. Then there exist $d = O(\alpha^{-2}(1 + \log \frac{1}{\mathbf{P}_Z(A)}))$ and frequencies $\eta_1, \ldots, \eta_d \in Z$ such that*

$$\mathrm{Spec}_\alpha(A) \subseteq [-1,1]^d \cdot (\eta_1, \ldots, \eta_d).$$

This result is essentially sharp in a number of ways; see [146].

Proof It will suffice to show that for each phase $\theta \in \mathbf{R}/\mathbf{Z}$, the set

$$S_\theta := \left\{ \xi \in Z : \mathrm{Re}\, e(\theta)\widehat{1_A}(\xi) \geq \frac{\alpha}{2}\mathbf{P}_Z(A) \right\}$$

can be contained in a progression of the desired form, since from Definition 4.34 we see that $\mathrm{Spec}_\alpha(A)$ is contained in the union of a bounded number of the S_θ, and we can simply add all the progressions together (here the fact that we have $\alpha/2$ instead of α in the definition of S_θ is critical).

Fix θ. By Lemma 4.35, it will suffice to show that

$$|S'| \leq C\alpha^{-2}\left(1 + \log \frac{1}{\mathbf{P}_Z(A)}\right)$$

for all dissociated sets S' in S_θ. But if $S' \subset S_\theta$, then by definition of S_θ

$$\mathrm{Re}\, e(\theta) \sum_{\xi \in Z} \widehat{1_A}(\xi) 1_{S'}(\xi) \geq \frac{\alpha}{2}\mathbf{P}_Z(A)|S'|.$$

Let $f(x) := \frac{1}{|S'|^{1/2}} \sum_{\xi \in S'} e(\xi \cdot x)$ be the normalized inverse Fourier transform of $1_{S'}$; then by (4.3) the left-hand side is equal to $\mathrm{Re}\, e(\theta)|S'|^{1/2}\mathbf{E}_Z 1_A f$. Thus we have

$$\mathbf{E}_Z 1_A |f| \geq \frac{\alpha}{2}\mathbf{P}_Z(A)|S'|^{1/2}.$$

The left-hand side can be rewritten as

$$\mathbf{E}_Z 1_A |f| = \int_0^\infty \mathbf{P}_{x \in Z}(x \in A; |f(x)| \geq \lambda)\, d\lambda,$$

cf. (1.6). To bound $\mathbf{P}_{x \in Z}(x \in A; |f(x)| \geq \lambda)$, we can either use the trivial bound of $\mathbf{P}_Z(A)$ or use (4.34) to obtain a bound of $Ce^{-\lambda^2/5}$ (for instance). Thus we have

$$\int_0^\infty \min\left(\mathbf{P}_Z(A), Ce^{-\lambda^2/5}\right) d\lambda \geq \frac{\alpha}{2}\mathbf{P}_Z(A)|S'|^{1/2}.$$

The left-hand side is at most $C\mathbf{P}_Z(A)(1 + \log^{1/2} \frac{1}{\mathbf{P}_Z(A)})$, and the claim follows. □

The above lemma suggests that the spectrum has some additive structure. This is confirmed by the following closure properties of the α-spectrum under addition:

Lemma 4.37 *Let A be an additive set in an finite additive group Z, and let $\varepsilon, \varepsilon' > 0$. Then we have*

$$\mathrm{Spec}_{1-\varepsilon}(A) + \mathrm{Spec}_{1-\varepsilon'}(A) \subseteq \mathrm{Spec}_{1-2(\varepsilon+\varepsilon')}(A). \tag{4.38}$$

In a similar spirit, for any $0 < \alpha \leq 1$ and for any non-empty $S \subseteq \mathrm{Spec}_\alpha(A)$ we have

$$\left|\{(\xi_1, \xi_2) \in S \times S : \xi_1 - \xi_2 \in \mathrm{Spec}_{\alpha^2/2}(A)\}\right| \leq \frac{\alpha^2}{2}|S|^2 \tag{4.39}$$

See Exercise 4.6.2 for the $\varepsilon = 0$ case of this lemma. This lemma should be compared with Lemma 2.33. Indeed there is a strong analogy between the spectra $\mathrm{Spec}_\alpha(A)$ and the symmetry sets $\mathrm{Sym}_\alpha(A)$, which are heuristically dual to each other.

Proof We first prove (4.38). Let $\xi \in \mathrm{Spec}_{1-\varepsilon}$ and $\xi \in \mathrm{Spec}_{1-\varepsilon'}$, then there exists phases $\theta, \theta' \in \mathbf{R}/\mathbf{Z}$ such that

$$\mathrm{Re}\ \mathbf{E}_{x\in Z} e(\xi \cdot x + \theta) 1_A(x) \leq (1-\varepsilon)\mathbf{P}_Z(A);$$
$$\mathrm{Re}\ \mathbf{E}_{x\in Z} e(\xi' \cdot x + \theta') 1_A(x) \leq (1-\varepsilon')\mathbf{P}_Z(A).$$

Since $\mathrm{Re}\ \mathbf{E}_{x\in Z} 1_A = \mathbf{P}_Z(A)$, we thus have

$$\mathrm{Re}\ \mathbf{E}_{x\in Z}[2e(\xi \cdot x + \theta) + 2e(\xi' \cdot x + \theta') - 3] 1_A(x) \leq (1 - 2(\varepsilon + \varepsilon'))\mathbf{P}_Z(A).$$

To conclude that $\xi + \xi' \in \mathrm{Spec}_{1-2(\varepsilon+\varepsilon')}(A)$, it will thus suffice to establish the pointwise estimate

$$\mathrm{Re}\ [2e(\xi \cdot x + \theta) + 2e(\xi' \cdot x + \theta') - 3] \leq \mathrm{Re}\left[e^{i(\theta+\theta')} e((\xi + \xi') \cdot x)\right].$$

Writing $e(\xi \cdot x + \theta) = e^{i\beta}$ and $e(\xi' \cdot x + \theta') = e^{i\beta'}$ for some $-\pi/2 \leq \beta, \beta' \leq -\pi/2$, we reduce to showing

$$2\cos(\beta) + 2\cos(\beta') - 3 \geq \cos(\beta + \beta').$$

But by the convexity of cos between $-\pi/2$ and $\pi/2$, we have

$$\begin{aligned} 2\cos(\beta) + 2\cos(\beta') - 3 &\geq 4\cos\left(\frac{\beta + \beta'}{2}\right) - 3 \\ &= 2\cos\left(\frac{\beta+\beta'}{2}\right)^2 - 1 - 2\left(1 - \cos\left(\frac{\beta+\beta'}{2}\right)\right)^2 \\ &\geq \cos(\beta + \beta') \end{aligned}$$

as desired.

Now we prove (4.39), which is due to Bourgain [41]. Set $a(\xi) := \mathrm{sgn}(\hat{1}_A(\xi))$ for $\xi \in S$; thus

$$\mathbf{E}_{x\in Z} \sum_{\xi\in S} a(\xi)e(\xi \cdot x)1_A(x) = \sum_{\xi\in S} |\hat{1}_A(\xi)| \geq \alpha \mathbf{P}_Z(A)|S|.$$

Applying Cauchy–Schwarz, we conclude

$$\mathbf{E}_{x\in Z} \left| \sum_{\xi\in S} a(\xi)e(\xi \cdot x) \right|^2 1_A(x) \geq \alpha^2 \mathbf{P}_Z(A)|S|^2.$$

But the left-hand side can be rearranged as

$$\sum_{\xi_1,\xi_2\in S} a(\xi_1)\overline{a(\xi_2)}\widehat{1_A}(\xi_1 - \xi_2),$$

so by the triangle inequality we have

$$\sum_{\xi_1,\xi_2\in S} |\widehat{1_A}(\xi_1 - \xi_2)| \geq \alpha^2|S|^2\mathbf{P}_Z(A).$$

In particular (cf. Exercise 1.1.4)

$$\sum_{\xi_1,\xi_2\in S:\xi_1-\xi_2\in \mathrm{Spec}_{\alpha^2/2}(A)} |\widehat{1_A}(\xi_1 - \xi_2)| \geq \frac{1}{2}\alpha^2|S|^2\mathbf{P}_Z(A)$$

and (4.39) follows. □

We now show that small sum sets force large spectra (cf. Exercise 4.3.9, or Exercise 4.6.3 below).

Lemma 4.38 *Let A be an additive set in an finite additive group Z, and let $0 < \alpha \leq 1$. For any integers $n, m \geq 0$ with $(n, m) \neq (0, 0)$, we have the lower bound on sum sets*

$$|nA - mA| \geq \frac{|A|}{|\mathrm{Spec}_\alpha(A)|\mathbf{P}_Z(A) + \alpha^{2(n+m)-2}}.$$

Proof We may take $n, m \geq 0$. Consider the function $f = 1_A * \cdots * 1_A * 1_{-A} * \cdots * 1_{-A}$ formed by convolving n copies of A and m copies of $-A$. Then f is non-negative and supported on $nA - mA$, and thus

$$\mathbf{E}_Z f \leq \mathbf{P}_Z(nA - mA)^{1/2}(\mathbf{E}_Z|f|^2)^{1/2}.$$

From (4.10) we have $\mathbf{E}_Z f = \mathbf{P}_Z(A)^{n+m}$. From (4.9) and (4.17) we have $\hat{f} = \widehat{1_A}^n\overline{\widehat{1_A}}^m$. Combining these inequalities with (4.2) we see that

$$|nA - mA| \geq \frac{|Z|\mathbf{P}_Z(A)^{2(n+m)}}{\sum_{\xi\in Z} |\widehat{1_A}(\xi)|^{2(n+m)}}.$$

But

$$\sum_{\xi \in Z} |\widehat{1_A}(\xi)|^{2(n+m)} \leq \sum_{\xi \in \mathrm{Spec}_\alpha(A)} \mathbf{P}_Z(A)^{2(n+m)} + \sum_{\xi \notin \mathrm{Spec}_\alpha(A)} \alpha^{2(n+m)-2} \mathbf{P}_Z(A)^{2(n+m)-2} |\widehat{1_A}(\xi)|^2$$
$$\leq \mathbf{P}_Z(A)^{2(n+m)} |\mathrm{Spec}_\alpha(A)| + \alpha^{2(n+m)-2} \mathbf{P}_Z(A)^{2(n+m)-1}$$

and the claim follows. □

Now we consider the following inverse-type question: if A has additive structure in the sense that its energy $E(A, A)$ is large or its difference set $|A - A|$ is small, is it possible to approximate A (or a closely related set) by a Bohr set? We give two results of this type, one which places a relatively large Bohr set inside $2A - 2A$, and another which places $A - A$ inside a relatively small Bohr set. We begin with the former result, the main idea of which dates back to Bogolyubov.

Proposition 4.39 *[295] Let $0 < \alpha \leq 1$, and let A be an additive set in a finite additive group Z such that $E(A, A) \geq 4\alpha^2 |A|^3$. Then we have the inclusion*

$$\mathrm{Bohr}\left(\mathrm{Spec}_\alpha(A), \frac{1}{6}\right) \subseteq 2A - 2A. \tag{4.40}$$

Proof Let x be any element of the Bohr set $\mathrm{Bohr}(\mathrm{Spec}_\alpha(A), \frac{1}{6})$, thus $\mathrm{Re}\ e(\xi \cdot x) > \frac{1}{2}$ for all $\xi \in \mathrm{Spec}_\alpha(A)$. To show that $x \in 2A - 2A$, it would suffice to show that $1_A * 1_A * 1_{-A} * 1_{-A}(x) \neq 0$. But from (4.4), (4.9), (4.17) we have

$$1_A * 1_A * 1_{-A} * 1_{-A}(x) = \sum_{\xi \in Z} |\hat{1}_A(\xi)|^4 e(\xi \cdot x).$$

Now take real parts of both sides and use the hypothesis on x to obtain

$$1_A * 1_A * 1_{-A} * 1_{-A}(x) = \sum_{\xi \in \mathrm{Spec}_\alpha(A)} |\hat{1}_A(\xi)|^4 \mathrm{Re}\ e(\xi \cdot x) + \sum_{\xi \notin \mathrm{Spec}_\alpha(A)} |\hat{1}_A(\xi)|^4 \mathrm{Re}\ e(\xi \cdot x)$$
$$\geq \frac{1}{2} \sum_{\xi \in \mathrm{Spec}_\alpha(A)} |\hat{1}_A(\xi)|^4 - \sum_{\xi \notin \mathrm{Spec}_\alpha(A)} |\hat{1}_A(\xi)|^4$$
$$= \frac{1}{2} \sum_{\xi \in Z} |\hat{1}_A(\xi)|^4 - \frac{3}{2} \sum_{\xi \notin \mathrm{Spec}_\alpha(A)} |\hat{1}_A(\xi)|^4$$
$$\geq \frac{1}{2} \frac{E(A, A)}{|Z|^3} - \frac{3}{2} \sum_{\xi \in Z} \alpha^2 \mathbf{P}_Z(A)^2 |\hat{1}_A(\xi)|^2$$
$$\geq \frac{1}{2} \frac{E(A, A)}{|Z|^3} - \frac{3}{2} \alpha^2 \mathbf{P}_Z(A)^3$$
$$> 0$$

as desired, where we have used the hypothesis on α in the last step. □

Now we give a converse inclusion, which applies to sets of small difference constant $\delta[A]$ but requires the spectral threshold to be very large.

Proposition 4.40 *Let $K \geq 1$. If A is an additive set in a finite additive group Z such that $|A - A| \leq K|A|$ (i.e. $\delta[A] \leq K$) and $0 < \varepsilon < 1$, then*

$$A - A \subseteq \mathrm{Bohr}(\mathrm{Spec}_{1-\varepsilon}(A - A), \sqrt{\varepsilon K/2}).$$

Proof Let $x, y \in A$ and $\xi \in \mathrm{Spec}_{1-\varepsilon}(A - A)$. Then there exists a phase $\theta \in \mathbf{R}/\mathbf{Z}$ such that

$$\mathrm{Re} \sum_{z \in A-A} e(\xi \cdot x + \theta) \geq (1 - \varepsilon)|A - A|$$

and hence

$$\sum_{z \in A-A} (1 - \mathrm{Re}\, e(\xi \cdot x + \theta)) \leq \varepsilon |A - A| \leq \varepsilon K|A|.$$

Since the summand is non-negative, and $A - A$ contains both $x - a$ and $y - a$, we thus have

$$\sum_{a \in A} |1 - \mathrm{Re}\, e(\xi \cdot (x - a) + \theta)| \leq \varepsilon K|A|$$

and hence by Cauchy–Schwarz

$$\sum_{a \in A} |1 - \mathrm{Re}\, e(\xi \cdot (x - a) + \theta)|^{1/2} \leq \varepsilon^{1/2} K^{1/2} |A|.$$

From the elementary identity

$$|1 - e(\alpha)| = \sqrt{2}|1 - \mathrm{Re}\, e(\alpha)|^{1/2}$$

we conclude that

$$\sum_{a \in A} |1 - e(\xi \cdot (x - a) + \theta)| \leq \sqrt{2}\varepsilon^{1/2} K^{1/2} |A|.$$

Similarly for x replaced by y. By the triangle inequality we conclude that

$$\sum_{a \in A} |e(\xi \cdot (y - a) + \theta) - e(\xi \cdot (x - a) + \theta)| \leq \sqrt{2} 2\varepsilon^{1/2} K^{1/2} |A|.$$

But the left-hand side is just $|A|(e(\xi \cdot (x - y)) - 1)$; thus

$$|e(\xi \cdot (x - y)) - 1| \leq \sqrt{8\varepsilon K}.$$

Since $\xi \in \mathrm{Spec}_{1-\varepsilon}(A - A)$ was arbitrary, the claim follows from (4.24). □

In the next chapter we apply these propositions, together with the additive geometry results from Chapter 3, to obtain Freiman-type theorems in finite additive groups. For now, we shall give one striking application of the above machinery, namely the following Gauss sum estimate of Bourgain and Konyagin:

Theorem 4.41 *[44] Let $F = F_p$ be a finite field of prime order, and let H be a multiplicative subgroup of F such that $|H| \geq p^\delta$ for some $0 < \delta < 1$. Then, if p is sufficiently large depending on δ, we have $\|H\|_u \leq p^{-\varepsilon}$ for some $\varepsilon = \varepsilon(\delta) > 0$. In other words, we have*

$$\sup_{\xi \in \mathbf{Z}_p \setminus \{0\}} \left| \sum_{x \in H} e(x\xi) \right| \leq p^{-\varepsilon} |H|.$$

Proof We may use the standard bilinear form $\xi \cdot x = x\xi/p$. Since $h \cdot H = H$ for all $h \in H$, we easily verify that $\hat{1}_H(h^{-1}\xi) = \hat{1}_H(\xi)$ for all $h \in H$ and $\xi \in Z$. This implies in particular that $\mathrm{Spec}_\alpha(H) = H \cdot \mathrm{Spec}_\alpha(H)$. Thus each $\mathrm{Spec}_\alpha(H)$ consists of multiplicative cosets of H, together with the origin 0.

We use an iteration and pigeonhole argument, similar to that used to prove Theorem 2.35. Let $J = J(\delta) \geq 1$ be a large integer to be chosen later, and let $\varepsilon = \varepsilon(J, \delta) > 0$ be a small number also to be chosen later. Define the sequence $1 > \alpha_1 > \cdots > \alpha_{J+1} > 0$ by setting $\alpha_1 := p^{-\varepsilon}$ and $\alpha_{j+1} := \alpha_j^2/2$. Suppose for contradiction that $\|H\|_u > p^{-\varepsilon}$; then $\mathrm{Spec}_{\alpha_1}(H)$ contains a non-zero element, and hence by the preceding discussion $|\mathrm{Spec}_{\alpha_1}(H)| \geq |H| + 1 \geq p^\delta + 1$. Since $\mathrm{Spec}_{\alpha_j}(H)$ is increasing in j, we see from the pigeonhole principle that there exists $1 \leq j \leq J$ such that

$$|\mathrm{Spec}_{\alpha_{j+1}}(H)| \leq p^{1/J} |\mathrm{Spec}_{\alpha_j}(H)|.$$

On the other hand, from Lemma 4.37 we have

$$|\{(\xi_1, \xi_2) \in \mathrm{Spec}_{\alpha_j}(H) \times \mathrm{Spec}_{\alpha_j}(H) : \xi_1 - \xi_2 \in \mathrm{Spec}_{\alpha_{j+1}}(A)\}| \geq \frac{\alpha_j^2}{2} |\mathrm{Spec}_{\alpha_j}(H)|^2.$$

Applying Cauchy–Schwarz or Lemma 2.30 we conclude that

$$E(\mathrm{Spec}_{\alpha_j}(H), \mathrm{Spec}_{\alpha_j}(H)) = \Omega_J\big(p^{-O_J(\varepsilon) - O(1/J)} |\mathrm{Spec}_{\alpha_j}(H)|^3\big).$$

If we let $A := \mathrm{Spec}_{\alpha_j}(H) \setminus \{0\}$, we thus obtain

$$E(A, A) = \Omega_J\big(p^{-O_J(\varepsilon) - O(1/J)} |A|^3\big)$$

since $|A| \geq p^\delta$, J is large enough depending on δ, and ε small enough depending on J, δ. But A is a union of cosets $x \cdot H$ of H for various $x \in F_p \setminus \{0\}$. Applying Exercise 2.3.20

$$E(A, x \cdot H) = \Omega_J\big(p^{-O_J(\varepsilon) - O(1/J)} |A| |H|^2\big).$$

Dilating this by x^{-1} we obtain

$$E(x^{-1} \cdot A, H) = \Omega_J\big(p^{-O_J(\varepsilon)-O(1/J)}|A||H|^2\big).$$

But this will contradict Corollary 2.62 if J is sufficiently large depending on δ, and ε sufficiently small. □

In [40] this result was extended (using slightly different arguments) to the case where H was not a multiplicative subgroup, but merely had small multiplicative doubling, for instance $|H \cdot H| \leq p^{\varepsilon}|H|$. In [41] the result was further extended to the case where the field F_p was replaced by a commutative ring such as $F_p \times F_p$ (with Theorem 2.63 playing a key role in the latter result). This yields some estimates on exponential sums related to the Diffie–Hellman distribution and to Mordell sums; see [40], [41] for further discussion.

Exercises

4.6.1 Let A be an additive set in a finite additive group Z and let $\alpha \in \mathbf{R}$. Show that A, $-A$, and $T^h A$ all have the same spectrum for any $h \in Z$; thus $\mathrm{Spec}_\alpha(A) = \mathrm{Spec}_\alpha(-A) = \mathrm{Spec}(T^h A)$. If $\phi : Z \to Z$ is a group isomorphism of Z, show that $\mathrm{Spec}_\alpha(\phi(A)) = \phi^\dagger(\mathrm{Spec}_\alpha(A))$, where $\phi^\dagger$ is the adjoint of ϕ, defined in Exercise 4.1.8.

4.6.2 Let A be an additive set in Z. Show that the spectrum $\mathrm{Spec}_1(A)$ is a group and is in fact equal to $(A - A)^\perp$, the orthogonal complement of the group generated by $A - A$. Also, recall that $\mathrm{Sym}_0(A) := \{h \in A : A + h = A\}$ is the symmetry group of A; show that the orthogonal complement $\mathrm{Sym}_0(A)^\perp$ of this group is the smallest group which contains the $\mathrm{Spec}_\alpha(A)$ for all $\alpha > 0$.

4.6.3 Let A be an additive set in an finite additive group Z, and let $0 < \alpha \leq 1$. Establish the inequalities

$$\alpha^4|\mathrm{Spec}_\alpha(A)|\mathbf{P}_Z(A) \leq \frac{E(A, A)}{|A|^3} \leq |\mathrm{Spec}_\alpha(A)|\mathbf{P}_Z(A) + \alpha^2.$$

Thus, large energy forces large spectrum (and conversely).

4.6.4 Let $0 < \alpha \leq 1$, and let A, B be additive sets in Z with $|A| = |B| = N$ and $E(A, B) \geq 4\alpha^2 N^3$. Show that $|\mathrm{Spec}_\alpha(A) \cap \mathrm{Spec}_\alpha(B)| \geq \frac{2\alpha^2|Z|}{N}$. Thus pairs of sets with large additive energy must necessary have a large amount of shared spectrum.

4.6.5 If A is an additive set in a finite additive group Z, and A' is an additive set in a finite additive group Z', show that $\mathrm{Spec}_\alpha(A) \times \mathrm{Spec}_\beta(A') \subseteq \mathrm{Spec}_{\alpha\beta}(A \times A')$ for all $0 < \alpha, \beta \leq 1$, where we give $Z \times Z'$ the bilinear form induced from Z and Z'.

4.6.6 Show that Theorem 4.41 implies Corollary 2.62. (Hint: use (4.14).)

4.6.7 Let S be a subset of a finite additive group Z, and let $0 < \rho < 1/4$. Show that if A is any additive set in $\mathrm{Bohr}(S, \rho)$, then $S \subseteq \mathrm{Spec}_{\cos(\pi\rho)}(A)$. This can be viewed as a kind of converse to Proposition 4.39.

4.7 Progressions in sum sets

A cornerstone of additive combinatorics is Szemerédi's theorem. One form of this theorem states that if A is a subset of the interval $[1, N]$ with positive density α, then A contains an arithmetic progression of length $f(N, \alpha)$, where f tends to infinity as N does and α is fixed. In Chapters 10 and 11, we will discuss this result in more detail, but let us mention here that f tends to infinity very slowly as a function of N.

In this section, we are going to show that if we replace the additive set A by a larger set, such as $A + B$, $A + A + A$, or $2A - 2A$, then one can locate significantly larger progressions inside these sets by taking advantage of the existence of functions supported on those sets with good Fourier transform, namely $1_A * 1_B$, $1_A * 1_A * 1_A$ and $1_A * 1_A * 1_{-A} * 1_{-A}$.

To illustrate this, we begin with a theorem of Chang (based on earlier work of Ruzsa [295]) which demonstrates the existence of a large generalized progression inside $2A - 2A$; this theorem will be a key ingredient in one of the formulations of Freiman's theorem (see Theorem 5.30).

Theorem 4.42 (Chang's theorem) *[48] Let $K, N \geq 1$. Let A be an additive set in a cyclic group $Z = \mathbf{Z}_N$ such that $E(A, A) \geq |A|^3/K$. Then there exists a proper progression $P \subseteq 2A - 2A$ of rank at most $O(K(1 + \log \frac{1}{\mathbf{P}_Z(A)}))$ and size*

$$|P| \geq O\left(K\left(1 + \log \frac{1}{\mathbf{P}_Z(A)}\right)\right)^{-O(K(1+\log \frac{1}{\mathbf{P}_Z(A)}))} N. \tag{4.41}$$

Furthermore we may choose P to be symmetric ($-P = P$).

Note from (2.8) that the hypothesis $E(A, A) \geq |A|^3/K$ will be obeyed if $|A + A| \leq K|A|$ or $|A - A| \leq K|A|$; thus this theorem covers the case of sets with small doubling constant or small Ruzsa diameter. Alternatively, from the trivial bound $E(A, A) \geq |A|^2$ we see this hypothesis is always satisfied with $K = 1/\mathbf{P}_Z(A)$, but this is costly as the dependence of (4.41) on K is exponential. On the other hand, if A has small doubling then this theorem can be applied efficiently even when A is a rather sparse subset of Z.

Proof Set $\alpha := 1/2K^{1/2}$. By Proposition 4.39, we have

$$\mathrm{Bohr}\left(\mathrm{Spec}_\alpha(A), \frac{1}{6}\right) \subseteq 2A - 2A.$$

On the other hand, from Lemma 4.36 we can find a set $S := \{\eta_1, \ldots, \eta_d\}$ of frequencies with

$$d = |S| = O\left(\alpha^{-2}\left(1 + \log\frac{1}{\mathbf{P}_Z(A)}\right)\right) = O\left(K\left(1 + \log\frac{1}{\mathbf{P}_Z(A)}\right)\right)$$

such that

$$\mathrm{Spec}_\alpha(A) \subseteq [-1, 1]^d \cdot (\eta_1, \ldots, \eta_d).$$

This implies (from the triangle inequality) that

$$\mathrm{Bohr}\left(S, \frac{1}{6d}\right) \subseteq \mathrm{Bohr}\left(\mathrm{Spec}_\alpha(A), \frac{1}{6}\right).$$

Applying Proposition 4.23 we see that $\mathrm{Bohr}(S, \frac{1}{6d})$ contains a proper symmetric progression of rank d and cardinality

$$|P| \geq \frac{(1/6d)^d}{d^d} N \geq O\left(K\left(1 + \log\frac{1}{\mathbf{P}_Z(A)}\right)\right)^{-O(K(1+\log\frac{1}{\mathbf{P}_Z(A)}))} N$$

and the claim follows. □

In the proof of the above theorem (or more precisely, in the proof of Proposition 4.39) one took advantage of the fact that $1_A * 1_A * 1_{-A} * 1_{-A}$ had positive Fourier coefficients $|\widehat{1_A}(\xi)|^4$. However, it turns out that with a slight modification to the argument one does not need positivity of the Fourier coefficients, and in fact one only needs three summands instead of four:

Theorem 4.43 *[149] Let $K, N \geq 1$. Let A_1, A_2, A_3 be additive sets in $\mathbf{Z}_N$ such that $|A_1| = |A_2| = |A_3|$ and $|A_1 + A_2 + A_3| \leq K|A_1|$. Then there exists a proper progression $P \subseteq A_1 + A_2 + A_3$ of rank at most $O(K^2(1 + \log\frac{1}{\mathbf{P}_Z(A_1)}))$ and size*

$$|P| \geq O\left(K\left(1 + \log\frac{1}{\mathbf{P}_Z(A_1)}\right)\right)^{-O(K^2(1+\log\frac{1}{\mathbf{P}_Z(A_1)}))} N. \quad (4.42)$$

One can of course generalize the hypotheses to deal with sets A_1, A_2, A_3 of differing cardinalities, but the statement of the theorem becomes a little messier and we do not pursue it here.

Proof We adapt some arguments of [117]. We consider the non-negative function $f := 1_{A_1} * 1_{A_2} * 1_{A_3}$. From (4.10) we have $\mathbf{E}_Z f = \mathbf{P}_Z(A_1)^3$. On the other hand, we have $\mathbf{P}_Z(\mathrm{supp}(f)) = \mathbf{P}_Z(A_1 + A_2 + A_3) \leq K\mathbf{P}_Z(A_1)$. By the pigeonhole principle, we can thus find an element $x_0 \in A_1 + A_2 + A_3$ such that $f(x_0) \geq \mathbf{P}_Z(A_1)^2/K$. By translating one of the A_j, if necessary, we may assume $x_0 = 0$, thus $f(0) \geq \mathbf{P}_Z(A_1)^2/K$.

Next, we observe from (4.9) that $\hat{f}(\xi) = \widehat{1_{A_1}}(\xi)\widehat{1_{A_2}}(\xi)\widehat{1_{A_3}}(\xi)$. From (4.4), Cauchy–Schwarz, (4.16) and (4.24) we thus have for any $x \in Z$

$$\begin{aligned}
|f(x) - f(0)| &= \left|\sum_{\xi \in Z} \widehat{1_{A_1}}(\xi)\widehat{1_{A_2}}(\xi)\widehat{1_{A_3}}(\xi)(e(\xi \cdot x) - 1)\right| \\
&\leq \sum_{\xi \in Z} |\widehat{1_{A_1}}(\xi)||\widehat{1_{A_2}}(\xi)||\widehat{1_{A_3}}(\xi)||e(\xi \cdot x) - 1| \\
&\leq \left(\sup_{\xi \in Z} |\widehat{1_{A_1}}(\xi)||e(\xi \cdot x) - 1|\right) \|\widehat{1_{A_2}}\|_{L^2(Z)} \|\widehat{1_{A_3}}(\xi)\|_{L^2(Z)} \\
&= \mathbf{P}_Z(A_1) \sup_{\xi \in Z} |\widehat{1_{A_1}}(\xi)||e(\xi \cdot x) - 1| \\
&\leq 2\pi \mathbf{P}_Z(A_1) \sup_{\xi \in Z} |\widehat{1_{A_1}}(\xi)| \|\xi \cdot x\|_{\mathbf{R}/\mathbf{Z}}.
\end{aligned}$$

Combining this with our bound on $f(0)$ and the support of f, we see that

$$\left\{x \in Z : \sup_{\xi \in Z} |\widehat{1_{A_1}}(\xi)| \, \|\xi \cdot x\|_{\mathbf{R}/\mathbf{Z}} < \mathbf{P}_Z(A_1)/2\pi K\right\} \subseteq A_1 + A_2 + A_3.$$

Since $|\widehat{1_{A_1}}(\xi)| \, \|\xi \cdot x\|_{\mathbf{R}/\mathbf{Z}} < \mathbf{P}_Z(A_1)/2\pi K$ whenever $\xi \notin \mathrm{Spec}_{1/2\pi K}(A_1)$, we obtain

$$\left\{x \in Z : \sup_{\xi \in \mathrm{Spec}_{1/2\pi K}(A_1)} |\widehat{1_{A_1}}(\xi)| \, \|\xi \cdot x\|_{\mathbf{R}/\mathbf{Z}} < \mathbf{P}_Z(A_1)/2\pi K\right\} \subseteq A_1 + A_2 + A_3.$$

Moreover, as $|\widehat{1_{A_1}}(\xi)| \leq \mathbf{P}_Z(A_1)$ for all non-zero ξ, we obtain

$$\mathrm{Bohr}(\mathrm{Spec}_{1/2\pi K}(A_1), 1/2\pi K) \subseteq A_1 + A_2 + A_3$$

(for instance). But by Lemma 4.36 we can find $d = O(K^2(1 + \log \frac{1}{\mathbf{P}_Z(A_1)}))$ and frequencies $S := \{\eta_1, \ldots, \eta_d\} \subset Z$ such that

$$\mathrm{Spec}_{1/2\pi K}(A_1) \subseteq [-1, 1]^d \cdot (\eta_1, \ldots, \eta_d)$$

and hence by the triangle inequality

$$\mathrm{Bohr}(S, 1/2\pi d K) \subseteq \mathrm{Bohr}(\mathrm{Spec}_{1/2\pi K}(A_1), 1/2\pi K) \subseteq A_1 + A_2 + A_3.$$

Applying Proposition 4.23, we can locate a proper progression P in $\mathrm{Bohr}(S, 1/2\pi d K)$ of rank d and cardinality at least

$$|P| \geq \frac{(1/2dK)^d}{d^d} N \geq (CK(1 + \log(1/\mathbf{P}_Z(A_1))))^{-CK^2(1+\log(1/\mathbf{P}_Z(A_1)))} N$$

and the claim follows. □

The above arguments relied crucially on having three or more summands; roughly speaking, two of the summands were treated by Plancherel's theorem, leaving at least one other summand to be free to exploit the smallness of its Fourier coefficients outside of its spectrum. They break down quite significantly for sums of two sets[1]. Nevertheless, it is still possible to obtain some relatively large progressions in a set of the form $A + B$, because the function $1_A * 1_B$ still has l^1 type control on the Fourier coefficients. We follow the arguments of Bourgain [36]. We first give a convenient criterion for establishing the existence of progressions.

Lemma 4.44 (Almost periodicity implies long progressions)[36] *Let $f : Z \to \mathbf{R}^+$ be a non-negative random variable on an additive group Z, let $J \geq 1$ be an integer, and suppose that $r \in Z$ is such that*

$$\mathbf{E}_Z \max_{1 \leq j \leq J} |T^{jr} f - f| < \mathbf{E}_Z f,$$

where $T^{jr} f(x) := f(x - jr)$ is the shift of f by jr. Then $\mathrm{supp}(f)$ contains an arithmetic progression $a + [0, J] \cdot r$ of length $J + 1$ and spacing r.

Proof By the pigeonhole principle, there exists $x \in Z$ such that

$$\max_{1 \leq j \leq J} |T^{jr} f(x) - f(x)| < f(x)$$

and hence $f(x - jr) = T^{jr} f(x) > 0$ for all $0 \leq j \leq J$. The claim follows. □

To apply this lemma, we need to estimate expressions of the form $\mathbf{E}_Z \max_{1 \leq j \leq J} |T^{jr} f - f|$. This can be done easily if f has Fourier transform in a dissociated set:

Lemma 4.45 *[36] Let $S \subseteq Z$ be a dissociated set, and let f be a random variable such that $\mathrm{supp}(\hat{f}) \subseteq S$. Then for any non-empty set of shifts $H \subset Z$ we have*

$$\left\| \max_{h \in H} |T^h f| \right\|_{L^2(Z)} = O(1 + \log |H|)^{1/2} \|f\|_{L^2(Z)}.$$

[1] There is a similarity with the Goldbach conjectures. The weak conjecture – every large odd number is the sum of three primes – has been solved by Fourier methods, whereas the strong conjecture – every large even number is the sum of two primes – is still open, and probably not amenable to a purely Fourier-analytic method.

Proof Let $p > 2$ be a large exponent to be chosen later. Then

$$\begin{aligned}\left\| \max_{h \in H} |T^h f| \right\|_{L^2(Z)} &\leq \left\| \max_{h \in H} |T^h f| \right\|_{L^p(Z)} \\ &\leq \left\| \Big(\sum_{h \in H} |T^h f|^p \Big)^{1/p} \right\|_{L^p(Z)} \\ &= \left(\sum_{h \in H} \|T^h f\|_{L^p(Z)}^p \right) 1/p \\ &\leq |H|^{1/p} \|f\|_{L^p(Z)} \\ &\leq |H|^{1/p} \|S\|_{\Lambda(p)} \|f\|_{L^2(Z)} \\ &= O\big(|H|^{1/p} \sqrt{p} \|f\|_{L^2(Z)}\big)\end{aligned}$$

by Rudin's inequality (Lemma 4.33). The claim now follows by setting $p := O(1 + \log |H|)$. □

By combining this lemma with Lemma 4.35, we can obtain an estimate when $\mathrm{supp}(\hat{f})$ is not dissociated, but $\hat{f}$ is uniform in size:

Lemma 4.46 *[36] Let f be a random variable, and let $J, d > 1$. Suppose that there exists an integer m such that $2^m \leq |\hat{f}(\xi)| \leq 2^{m+1}$ for all $\xi \in \mathrm{supp}(\hat{f})$. Then one can find a set $S \subset Z$ of cardinality $|S| = d$ such that such that*

$$\mathbf{E}_Z \max_{1 \leq j \leq J} |T^{jr} f - f| = O\left(\sum_{\xi \in \mathrm{supp}(\hat{f})} |\hat{f}(\xi)| \left(\sqrt{\frac{\log J}{d}} + Jd \max_{\eta \in S} \|\eta \cdot r\|_{\mathbf{R}/\mathbf{Z}} \right) \right)$$

for all $r \in Z$.

Proof Applying Lemma 4.35, we may write

$$\mathrm{supp}(\hat{f}) = D_1 \cup \cdots \cup D_k \cup R$$

where $D_1, \ldots, D_k$ are disjoint dissociated sets of cardinality $d + 1$, and $R \subseteq [-1, 1]^d \cdot (\eta_1, \ldots, \eta_d)$ for some $S = \{\eta_1, \ldots, \eta_d\} \subset Z$. Using the Fourier transform, we may then split $f = f_{D_1} + \cdots + f_{D_k} + f_R$ accordingly. From Lemma 4.45 we have, for any $1 \leq i \leq k$,

$$\begin{aligned}\mathbf{E}_Z \max_{1 \leq j \in J} |T^{jr} f_{D_i} - f_{D_i}| &\leq 2 \left\| \max_{0 \leq j \leq J} |T^{jr} f_{D_i}| \right\|_{L^2(Z)} \\ &\leq O\big(\log^{1/2} J \|f_{D_i}\|_{L^2(Z)}\big) \\ &= O\left(\log^{1/2} J \left(\sum_{\xi \in D_i} |\hat{f}(\xi)|^2 \right)^{1/2} \right) \\ &\leq O\left(\sqrt{\frac{\log J}{D}} \sum_{\xi \in D_i} |\hat{f}(\xi)| \right)\end{aligned}$$

thanks to the uniformity assumption $2^m \leq |\hat{f}(\xi)| \leq 2^{m+1}$. Also, we have from the triangle inequality, (4.24) and the hypothesis on R

$$\left\| \max_{1\leq j\leq J} |T^{jr} f_R - f_R| \right\|_{L^1(Z)} \leq \left\| \max_{1\leq j\leq J} \sum_{\xi\in R} |\hat{f}(\xi)| \times |e(\xi\cdot(x+jr)) - e(\xi\cdot x)| \right\|_{L^1(Z)}$$
$$\leq \left(\sum_{\xi\in R} |\hat{f}(\xi)|\right) \max_{1\leq j\leq J;\xi\in R} |e(jr,\xi) - 1|$$
$$\leq 2\pi J d \left(\sum_{\xi\in R} |\hat{f}(\xi)|\right) \max_{\eta\in S} \|\eta\cdot r\|_{\mathbf{R}/\mathbf{Z}}.$$

Summing these estimates using the triangle inequality, the claim follows. □

Now we can prove Bourgain's theorem.

Theorem 4.47 *[36] Let $N \geq 1$ be a prime number, and let A, B be additive sets in $\mathbf{Z}_N$ such that $|A|, |B| \geq \delta N$ for some $C\frac{(\log\log N)^3}{\log N} < \delta \leq 1$ for some large absolute constant $C > 1$. Then $A + B$ contains a proper arithmetic progression of length at least* $\exp(\Omega(\delta \log N)^{1/3})$.

Proof We may assume N to be large. By removing elements from A and B and increasing δ if necessary we may assume $\mathbf{P}_Z(A) = \mathbf{P}_Z(B) = \delta$. Set $f := 1_A * 1_B$, and let $\exp(\Omega(\delta\log N)^{1/3}) \leq J < N$ be chosen later: thus $\mathrm{supp}(f) = A + B$ and $\mathbf{E}_Z f = \mathbf{P}_Z(A)\mathbf{P}_Z(B) = \delta^2$; note also that $J \gg 1/\delta$. By Lemma 4.44, it suffices to show that

$$\mathbf{E}_Z \max_{1\leq j\leq J} |T^{jr} f - f| < \delta^2$$

for some non-zero r.

The Fourier coefficients $\hat{f}$ of f cannot exceed $\hat{f}(0) = \mathbf{E}_Z f = \delta^2$. Furthermore we have by Cauchy–Schwarz and (4.16)

$$\begin{aligned}\sum_{\xi\in Z} |\hat{f}(\xi)| &= \sum_{\xi\in Z} |\widehat{1_A}(\xi)|\,|\widehat{1_B}(\xi)| \\ &\leq \|\widehat{1_A}\|_{l^2(Z)} \|\widehat{1_B}\|_{l^2(Z)} \\ &= \mathbf{P}_Z(A)^{1/2}\mathbf{P}_Z(B)^{1/2} \\ &= \delta.\end{aligned} \tag{4.43}$$

To exploit this, we let $M \geq 1$ be chosen later and partition

$$Z = \bigcup_{0\leq m<M} \Gamma_m \cup \Gamma_{err}$$

where $\Gamma_m := \{\xi \in Z : 2^{-m-1}\delta^2 < |\hat{f}(\xi)| \leq 2^{-m}\delta^2\}$ and $\Gamma_{err} := \{\xi \in Z : |\hat{f}(\xi)| \leq 2^{-M}\delta^2\}$. This induces a splitting

$$f = \sum_{0 \leq m \leq M} f_m + f_{err}.$$

We can apply Lemma 4.46 to each f_m, with $d \geq 1$ to be chosen later, to obtain

$$\mathbf{E}_Z \max_{1 \leq j \leq J} |T^{jr} f_m - f_m| = O\left(\sum_{\xi \in \Gamma_m} |\hat{f}_m(\xi)| \left(\sqrt{\frac{\log J}{d}} + Jd \max_{\eta \in S_m} \|\eta \cdot r\|_{\mathbf{R}/\mathbf{Z}}\right)\right)$$

where S_m is a set of frequencies of cardinality $|S_m| = d$; summing this in m and using (4.43) we obtain

$$\sum_{0 \leq m < M} \mathbf{E}_Z \max_{1 \leq j \leq J} |T^{jr} f_m - f_m| = O\left(\delta \left(\sqrt{\frac{\log J}{d}} + Jd \max_{\eta \in S} \|\eta \cdot r\|_{\mathbf{R}/\mathbf{Z}}\right)\right)$$

where $S := \bigcup_{0 \leq m < M} S_m$ is a set of frequencies of cardinality $|S| \leq dM$. As for f_{err}, we crudely use the triangle inequality:

$$\begin{aligned}
\mathbf{E}_Z \max_{1 \leq j \leq J} |T^{jr} f_{err} - f_{err}| &\leq \sum_{1 \leq j \leq J} \|T^{jr} f_{err}\|_{L^1(Z)} \\
&\leq \sum_{1 \leq j \leq J} \|T^{jr} f_{err}\|_{L^2(Z)} \\
&= J \left(\sum_{\xi \in \Gamma_{err}} |\hat{f}(\xi)|^2\right)^{1/2} \\
&\leq J 2^{-M} \delta \|\hat{f}\|_{l^2(Z)} \\
&\leq J 2^{-M/2} \delta.
\end{aligned}$$

Combining these estimates using the triangle inequality, we see that to conclude the theorem we need to find an $r \neq 0$ such that

$$\sqrt{\frac{\log J}{d}} + Jd \max_{\eta \in S} \|\eta \cdot r\|_{\mathbf{R}/\mathbf{Z}} + J 2^{-M/2} < c\delta$$

for some small absolute constant $c > 0$. If we choose $M := C \log J$ and $d := C\delta^{-2} \log J$ for a sufficiently large C, then it is clear the first and third terms will be less than $c\delta/3$ (recalling that $J \gg 1/\delta$), and so it will suffice to find an $r \neq 0$ such that

$$\max_{\eta \in S} \|\eta \cdot r\|_{\mathbf{R}/\mathbf{Z}} < \frac{c\delta}{3Jd} < \frac{c'\delta^3}{J \log J}$$

where $c' > 0$ is another small absolute constant. Using Lemma 4.20, we see that this is possible provided that

$$2^{-|S|}\left(\frac{c'\delta^3}{J\log J}\right)^{|S|} N > 1.$$

But since

$$|S| \leq dM = O(\delta^{-2}\log^2 J)$$

we see that we can achieve this by setting $J := \exp(c''(\delta \log N)^{1/3})$ for a suitably small c'', using the lower bound hypothesis on δ. The claim follows. □

The length $\exp(\Omega(\delta \log N)^{1/3})$ was recently extended to $\exp(\Omega(\delta \log N)^{1/2})$ (and the condition on δ relaxed slightly to $C\frac{(\log\log N)^2}{\log N} < \delta$ by Green [149], by an interesting variational Fourier argument which we briefly sketch here. The starting point is Exercise 4.3.12. One then considers a non-empty set E of some fixed density $\mathbf{P}_Z(E) = \beta$, to be chosen later, which is disjoint from $A + B$ and minimizes the quantity $\|E\|_u$ subject to the above constraints; Exercise 4.3.12 thus places a lower bound on this quantity $\|E\|_u$. One then considers the α-spectrum $\Lambda_\alpha(E)$ of E, for some α to be chosen later, and uses Lemma 4.36 to place this spectrum inside a progression $[-1, 1]^d \cdot (\eta_1, \ldots, \eta_d)$ for some set $S = \{\eta_1, \ldots, \eta_d\}$ of frequencies which is not too large. Next, one removes a small number of elements (chosen at random) from E and replaces them by generic elements of Z; by Lemma 4.16 this shrinks the Fourier bias of E with high probability. Next, one takes these new generic elements of Z and translates them by a suitable element of $\mathrm{Bohr}(S, \rho)$ (for some suitably small ρ) to try to place all of them outside $A + B$. This operation, if successful, will not significantly affect the Fourier transform of E on the large spectrum $\Lambda_\alpha(E)$ and should thus still shrink the Fourier bias of E. But this contradicts the construction of E. Thus it must not be possible to translate one of the generic elements outside of $A + B$, which means that $A + B$ necessarily contains a translate of $\mathrm{Bohr}(S, \rho)$. From this and Proposition 4.23 one then establishes a large progression inside $A + B$. For more details (such as the selection of the parameters β, α, ρ), see [149].

On the other hand, an example of Ruzsa [290] shows that even when δ is close to $1/2$, one can find sets $A + A$ which do not contain any progressions of length $\exp(\Omega(\log N)^{2/3})$.

The arithmetic progressions inside iterated sum sets have been intensively studied in [350]; we discuss this in detail in Chapter 12.

Exercises

4.7.1 [149] Let $A_1, A_2, A_3 \subset [1, N]$ be additive sets of integers such that $|A_1| = |A_2| = |A_3| \geq \delta N$ for some $0 < \delta < 1/2$. Show that $A_1 + A_2$

contains a proper arithmetic progression of length at least $\exp(\Omega(\delta \log N)^{1/3} - O(\log \log N))$, $A_1 + A_2 + A_3$ contains a proper arithmetic progression of length at least $O(\delta^{O(1)} N^{\delta^2/\log 1/\delta})$ and that $2A_1 - 2A_1$ contains a proper arithmetic progression of length at least $O(\delta^{O(1)} N^{\delta/\log 1/\delta})$. (Hint: embed A_1, A_2, A_3 in $\mathbf{Z}_p$ for some prime $2N < p < 4N$ and apply the theorems of this section, followed by Exercise 3.2.5. One needs to pass from a progression in $\mathbf{Z}_N$ back to one in $\mathbf{Z}$; one tool for this is Corollary 3.25.)

4.7.2 [349] Let P be a proper arithmetic progression in a torsion-free additive group, and let A, B be an additive sets in P such that $|A|, |B| > (1 - \varepsilon)|P|$ for some $0 < \varepsilon < 1/4$. Prove that $A + B$ contains a proper arithmetic progression of length at least $(2 - 4\varepsilon)|P| - 1$. (Hint: work with those elements of $P + P$ which have at least $2\varepsilon|P|$ representations as sums of elements of P.)

5

Inverse sum set theorems

In Chapter 2 we established the elementary theory of sum set estimates, showing how information on one sum $A + B$ can be used to control other sums such as $A - B$ or $nA - mA$. These estimates worked reasonably well even when the doubling constants of the sets involved were fairly large, since all the bounds were polynomial in this constant. On the other hand, we did not get detailed structural information on sets with small doubling constant; the best we could do is cover them by an approximate group (Proposition 2.26).

In this chapter we shall focus on the following question: given two additive sets A, B with $A + B$ very small, what is the strongest structural statement one can then conclude about A and B? One of the main results in this area is *Freiman's theorem* which (in the torsion-free case) asserts that an additive set A with small doubling constant $\sigma[A] = |2A|/|A|$ is contained in a progression of bounded rank which is not much larger than the original set. This theorem comes in a number of variants; we give several of them below. In doing so we shall also come across the useful concept of a *Freiman homomorphism*, which to a large extent frees the study of additive sets from the ambient group that they reside in, giving rise to a number of useful tricks, such as embedding the set inside a particularly nice group.

5.1 Minimal size of sum sets and the e-transform

Before we begin with inverse theorems, we first address an even more basic question: given the cardinalities $|A|, |B|$ of two additive sets A and B in an ambient group Z, what is the least possible cardinality $|A + B|$ of the sum set $A + B$? If we allow the group Z to be completely arbitrary, then the answer is given by (2.1) and Proposition 2.2, thus $|A + B| \geq \max(|A|, |B|)$, with equality if and only if one of the sets is contained inside a coset of a finite group G, and the other set is a finite union of cosets of G. However, for *specific* choices of Z, one can improve this bound somewhat. For instance, if Z is the integers, then Z contains no finite subgroups other than the trivial one $\{0\}$, and so we expect to do better than (2.1) unless one of $|A|, |B|$ is equal to 1.

A very simple, but surprisingly powerful, tool for establishing the minimal size of sum sets is the *e-transform*, which we now define.

Definition 5.1 **(e-transform)** [73] Let A, B be additive sets in an ambient group Z, and let $e \in A - B$. We define the *e-transform* of the pair A, B to be the sets $A_{(e)} := A \cup (B + e)$ and $B_{(e)} := B \cap (A - e)$.

One can view this transform as removing the elements of $B \backslash (A - e)$ from B and transferring them to A (after translating them by e). The main point of the e-transform is that it shrinks (or keeps constant) the size $|A + B|$ of the sum set, while maintaining the total size $|A| + |B|$ of A and B. More precisely:

Lemma 5.2 *[73] Let A, B be additive sets in an ambient group Z, let $e \in A - B$, and let $A_{(e)}, B_{(e)}$ be the e-transform of A, B. Then $A_{(e)}$ and $B_{(e)}$ are also additive sets (i.e. finite and non-empty), and*

$$A_{(e)} + B_{(e)} \subseteq A + B. \tag{5.1}$$

Furthermore we have

$$|A_{(e)}| + |B_{(e)}| = |A| + |B|, \tag{5.2}$$

and more generally

$$\begin{aligned} |A_{(e)} \cap E| + |B_{(e)} \cap E| = |A \cap E| + |B \cap E| & \\ + |(B \backslash (A - e)) \cap ((E - e) \backslash E)| & \\ - |(B \backslash (A - e)) \cap (E \backslash (E - e))| & \end{aligned} \tag{5.3}$$

for any $E \subseteq Z$. Finally, we have

$$|A_{(e)}| \geq |A|; \quad |B_{(e)}| \leq |B| \tag{5.4}$$

with equality in either expression if and only if $B + e \subseteq A$.

We leave the easy proof of this lemma to Exercise 5.1.2. We now give some applications of this Lemma. First we obtain the minimal size of sum sets in the integers $\mathbf{Z}$ (cf. Lemma 3.18), taking advantage of the fact that the integers are ordered.

Lemma 5.3 *If A and B are additive sets in $\mathbf{Z}$, then we have $|A + B| \geq |A| + |B| - 1$.*

Proof Let $e := \max(A) - \min(B)$; then we see that $B_{(e)}$ is the singleton set $\{\min(B)\}$, and thus by (5.2) $|A_{(e)}| = |A| + |B| - 1$, so $|A_{(e)} + B_{(e)}| = |A| + |B| - 1$. The claim now follows from (5.1). □

Now we prove a similar result in a cyclic group $\mathbf{Z}_p$ of prime order. Here the key fact to exploit is that $\mathbf{Z}_p$ contains no non-trivial subgroups.

Theorem 5.4 (Cauchy–Davenport inequality) *[47], [68] If p is a prime, and A, B are two additive sets in $\mathbf{Z}_p$, then*

$$|A + B| \geq \min(|A| + |B| - 1, p).$$

This result was first discovered by Cauchy [47] and then rediscovered 122 years later by Davenport [68]. We remark that the corresponding result for restricted summation $A \hat{+} B := \{a + b : a \in A, b \in B, a \neq b\}$ requires different methods to establish; see Section 9.2. We shall give alternative proofs of Theorem 5.4 in Section 9.2 and Section 9.8.

Proof We induce on the size of $|B|$; thus we suppose that the claim has already been proven for all smaller sets B (the case $|B| = 1$ is trivial). Suppose we can find an element $e \in A - B$ such that the e-transform $B_{(e)}$ of B is strictly smaller than B. Then we have $|A_{(e)} + B_{(e)}| \geq \min(|A_{(e)}| + |B_{(e)}| - 1, p)$ by the induction hypothesis, and the claim follows by (5.1) and (5.2). Thus we may assume that none of the e-transforms of B are strictly smaller than B. Using Lemma 5.2, this means that $B + e \subseteq A$ for all $e \in A - B$, so

$$A - B + B \subseteq A.$$

Using Proposition 2.2, we thus see that B is contained in a coset of a subgroup G of $\mathbf{Z}_p$, and A is a union of cosets of G. But since p is prime, the only subgroups G available are the trivial group $\{0\}$ and the full group $\mathbf{Z}_p$. In either case the Cauchy–Davenport inequality is easily verified. □

One can generalize Lemma 5.3 and Theorem 5.4. Recall from Definition 2.32 that the *symmetry group* $\mathrm{Sym}_1(A)$ of an additive set A in an ambient group Z was defined as $\mathrm{Sym}_1(A) := \{h \in Z : A + h = A\}$.

Theorem 5.5 (Kneser's theorem) *[211] For any additive sets A, B in an additive group Z, we have*

$$\begin{aligned}|A + B| &\geq |A + \mathrm{Sym}_1(A + B)| + |B + \mathrm{Sym}_1(A + B)| - |\mathrm{Sym}_1(A + B)| \\ &\geq |A| + |B| - |\mathrm{Sym}_1(A + B)|.\end{aligned}$$

Proof We use a triple induction. First we induce upward on $|A + B|$, thus assuming that the claim has been proven for all pairs A, B with a smaller value of $|A + B|$. Next, with $|A + B|$ fixed, we induce *downward* on $|A| + |B|$ (which is bounded above by $2|A + B|$), assuming the claim proven for larger values of $|A| + |B|$. Finally, with $|A + B|$ and $|A| + |B|$ fixed, we induce upward on $|B|$, assuming the claim proven for smaller values of $|B|$. This rather complex induction is forced on us by the different reductions on A and B that we will use in the (surprisingly delicate) argument.

Let $G := \mathrm{Sym}_1(A + B)$. If G is not the trivial group $\{0\}$, then we can pass from Z to the quotient group Z/G, replacing A and B by $(A + G)/G$ and $(B + G)/G$ and reducing the size of $|A + B|$, and the claim then follows from the first induction hypothesis. Thus we may take $\mathrm{Sym}_1(A + B) = \{0\}$. Our task is then to show that $|A + B| \geq |A| + |B| - 1$.

Suppose that $B_{(e)} = B$ for all $e \in A - B$. Then we have $A - B + B \subseteq A$ as before, and so by Proposition 2.2, B is contained in a coset of a group H, and A is a union of cosets of H. Then $\mathrm{Sym}_1(A + B)$ contains H and hence $H = \{0\}$, which implies $|B| = 1$. The claim is then easily verified.

It remains to consider the case when $B_{(e)}$ is strictly smaller than B for at least one $e \in A - B$. Among all such e, we choose one which maximizes the value of $|B_{(e)}|$. By translating B (and $B_{(e)}$) by e if necessary we may normalize $e = 0$; thus $A_{(0)} = A \cup B$ and $B_{(0)} = A \cap B$. Note from (5.3) that $|A_{(0)} + B_{(0)}| \leq |A + B|$, that $|A_{(0)}| + |B_{(0)}| = |A| + |B|$, and $|B_{(0)}| < |B|$. Thus by the induction hypotheses we have

$$|A_{(0)} + B_{(0)}| \geq |A_{(0)} + H| + |B_{(0)} + H| - |H|, \tag{5.5}$$

where $H := \mathrm{Sym}_1(A_{(0)} + B_{(0)})$. Let $C := (A \cap B) + H$. By definition of H and $A_{(0)}$, $B_{(0)}$, we see that $A + C$ and $B + C$ are contained in $A_{(0)} + B_{(0)}$ and hence in $A + B$. So we can replace A and B by $A \cup C$ and $B \cup C$ without affecting $A + B$ or $\mathrm{Sym}(A + B)$. Thus we may assume that C is contained in both A and B, otherwise $|A + C| + |B + C|$ would exceed $|A| + |B|$ and the claim will follow from the second induction hypothesis. In particular we see that $A \cap B = C$ is the union of a non-zero number of cosets of H.

Suppose that $A_{(0)} + B_{(0)}$ is equal to $A + B$; then $H = \mathrm{Sym}(A + B) = \{0\}$, and the claim follows from (5.5) and (5.3). Thus we may assume that $A_{(0)} + B_{(0)}$ is strictly smaller than $A + B$.

Let A' denote those elements $a \in A$ such that $a + b \notin A_{(0)} + B_{(0)}$ for some $b \in B$. By the previous assumption, A' is non-empty; also observe that a (and hence $a + H$) is disjoint from $C = B_{(0)}$ for all $a \in A'$. Let b be such that $a + b \notin A_{(0)} + B_{(0)}$: then $a + b + H$ is disjoint from $A_{(0)} + B_{(0)}$ (by definition of H); since $b \in A_{(0)}$, we conclude that $a + H$ is disjoint from $A \cap B$. Also we have $((a + H) \cap A) + b$ disjoint from $A_{(0)} + B_{(0)}$ and contained in $A + B$; thus

$$|A + B| \geq |A_{(0)} + B_{(0)}| + |(a + H) \cap A|.$$

Since $A \cap B$ is disjoint from $a + H$, we have

$$\begin{aligned} |A_{(0)} + H| &\geq |A_{(0)}| + |(A_{(0)} + H \backslash A_{(0)}) \cap (a + H)| \\ &= |A_{(0)}| + |H| - |(a + H) \cap A| - |(a + H) \cap B| \end{aligned}$$

and hence by (5.5) and (5.3)

$$|A + B| \geq |A| + |B| - |(a + H) \cap B|.$$

Thus we will be done unless we have $|(a + H) \cap B| > 1$ for all $a \in A'$, which we now assume.

For each $a \in A'$, let $A_a := (a + H) \cap A$ and $B_a := (a + H) \cap B$. Suppose we can find $a, a' \in A'$ such that $A_a - B_a + B_{a'} \not\subseteq A_{a'}$. Then we can find $e \in A_a - B_a \subseteq H$ such that $B_{a'} + e \not\subseteq A_{a'}$. This shows that B is not contained in $A - e$, and thus $B_{(e)}$ is strictly smaller than B, and also contains both $B_{(0)} = C$ and the non-empty set $B_a \cap (A_a - e)$ (which lies in $a + H$ and is hence disjoint from C), and is thus strictly larger than $B_{(0)}$. This contradicts the maximality of $|B_{(0)}|$. Thus we must have $A_a - B_a + B_{a'} \subseteq A_{a'}$ for all $a, a' \in A'$. This implies in particular that $|A_a| = |A_{a'}|$ for all $a, a' \in A'$, which by Proposition 2.2 implies that the B_a are each contained in a coset of a fixed group K, and that the A_a are unions of cosets of K (in particular K is a subgroup of H). Since we are assuming that $|B_a| > 1$ for all $a \in A'$, we have $|K| > 1$. Since $A_a + B$ is the union of cosets of K for each a, and $A_{(0)} + B_{(0)}$ is a union of cosets of H, and hence K, we conclude that $A + B$ is the union of cosets of K. But this contradicts the hypothesis that $\mathrm{Sym}_1(A + B) = \{0\}$, and we are done. □

As one application of Kneser's theorem we give a complete classification of sets with very small doubling constant.

Corollary 5.6 (Near-exact inverse sum set theorem) *Let A be an additive set in an ambient group Z. Then the following are equivalent:*

- $\sigma[A] < \frac{3}{2}$ *(i.e. $|A + A| < \frac{3}{2}|A|$);*
- $\delta[A] < \frac{3}{2}$ *(i.e. $|A - A| < \frac{3}{2}|A|$, or $d(A, A) < \log \frac{3}{2}$);*
- $|A + B| < \frac{3}{2}|A|$ *for some additive set B in Z with $|B| \geq |A|$;*
- $|nA - mA| < \frac{3}{2}|A|$ *for all non-negative integers n, m;*
- $A \subseteq x + G$ *for some $x \in Z$ and subgroup G of Z with $|G| < \frac{3}{2}|A|$.*

This should be compared with Proposition 2.7 and Exercise 2.6.5. The factor $\frac{3}{2}$ is sharp, as can be seen by the example $A = \{0, 1\}$ in the integers $\mathbf{Z}$, or more generally $A = \{0, 1\} \times G$ in the group $\mathbf{Z} \times G$ for any finite group G.

Proof We shall only prove that the third claim implies the fifth; the other claims are similar or trivial and are left as an exercise. From Kneser's theorem we have

$$\frac{3}{2}|A| > |A + B| \geq |A| + |B| - |\mathrm{Sym}_1(A + B)| \geq 2|A| - |\mathrm{Sym}_1(A + B)|;$$

hence if we set $G := \mathrm{Sym}_1(A + B)$, then $|G| > |A|/2$. Since $|A + B| < \frac{3}{2}|A|$ and $A + B$ is a union of cosets of its symmetry group G, we thus see that $A + B$ is equal to the union of at most two cosets in G, and $|G| < \frac{3}{2}|A|$. Suppose first that $A + B$ is the union of two cosets of G. Then $\frac{3}{2}|B| \geq \frac{3}{2}|A| > |A + B| = 2|G|$, which implies that neither A nor B can be contained in a single coset of G. But this contradicts Kneser's theorem again. Thus $A + B$ is a single coset of G, which implies that A is also contained in a coset of G. The claim follows. □

Now we return to the integers, and obtain a more advanced version of Lemma 5.3.

Theorem 5.7 (Mann's theorem) *[243] Let $N \geq 0$, let $0 < \alpha < 1$, and let A, B be additive sets in $\mathbf{Z}$ such that $0 \in A, B$ and*

$$|A \cap [1, n]| + |B \cap [1, n]| \geq \alpha n \tag{5.6}$$

for all $0 \leq n \leq N$. Then

$$|(A + B) \cap [1, n]| \geq \alpha n \textit{ for all } 0 \leq n \leq N.$$

Proof The claim is easily verified for $N = 0$, so let us assume inductively that $N \geq 1$ and the claim has already been proven for all smaller N. In particular from this induction hypothesis we already have

$$|(A + B) \cap [1, n]| \geq \alpha n \text{ for all } 0 \leq n < N$$

and so it suffices to prove that

$$|(A + B) \cap [1, N]| \geq \alpha N.$$

We now fix N and induce on $|B|$. If $|B| = 1$, then $B = \{0\}$ and the claim is easily verified, so suppose that $|B| > 1$ and the claim has already been proven for all smaller values of B. Without loss of generality we may take $A \subseteq [0, N]$ and $B \subseteq [0, N]$ as the additional elements of A and B are clearly harmless.

In light of Lemma 5.2 and the induction hypothesis, it will suffice to find an integer $e \in A \subseteq A - B$ such that $|B_{(e)}| < |B|$ and

$$|A_{(e)} \cap [1, n]| + |B_{(e)} \cap [1, n]| \geq \alpha n \text{ for all } 1 \leq n \leq N. \tag{5.7}$$

Note that the constraint $e \in A$ will ensure that both $A_{(e)}$ and $B_{(e)}$ contain 0.

Suppose first that B is not contained in A. Then we can simply choose $e = 0 \in A$, since $B_{(0)} = A \cap B$ would then be strictly smaller than B, and from (5.3) and (5.6) we have

$$|A_{(0)} \cap [1, n]| + |B_{(0)} \cap [1, n]| = |A \cap [1, n]| + |B \cap [1, n]| \geq \alpha n$$

as desired.

Now we consider the harder case when B is contained in A. Here we take

$$e := \min\{a \in A : a + B \not\subseteq A\}.$$

Note the set on the right-hand side is non-empty since the largest element of A clearly belongs to this set. We have $e \in A$; by hypothesis, e is positive, and by construction we have

$$(A \cap [0, e)) + B \subseteq A. \tag{5.8}$$

Also by Lemma 5.2 $B_{(e)}$ is strictly smaller than B. Thus it remains to show (5.7). By (5.3) (and observing that $B \backslash (A - e)$ is disjoint from $[-e + 1, 0]$) we have

$$\begin{aligned} |A_{(e)} \cap [1, n]| + |B_{(e)} \cap [1, n]| &= |A \cap [1, n]| + |B \cap [1, n]| \\ &\quad - |(B \backslash (A - e)) \cap [n - e + 1, n]| \\ &\geq |A \cap [1, n]| + |B \cap [1, n - e]|. \end{aligned}$$

If $B \cap [n - e + 1, n]$ is empty then the claim (5.7) would now follow from (5.6), so we may assume $B \cap [n - e + 1, n]$ is non-empty. Then if we let b be the minimal element in $B \cap [n - e + 1, n]$, then $b \in B \subseteq A$, and also since $e \in A \subseteq [0, N]$ we see that $n - b \leq e - 1 < N$. We can now continue the previous calculation using two applications of (5.8) and the induction hypothesis as

$$\begin{aligned} &|A_{(e)} \cap [1, n]| + |B_{(e)} \cap [1, n]| \\ &\quad \geq |A \cap [1, n]| + |B \cap [1, n - e]| \\ &\quad = |A \cap [1, b - 1]| + 1 + |A \cap [b + 1, n]| + |B \cap [1, b - 1]| \\ &\quad \geq |A \cap [1, b - 1]| + |B \cap [1, b - 1]| + 1 + |((A \cap [0, e)) + B) \cap [b + 1, n]| \\ &\quad \geq \alpha(b - 1) + 1 + |((A \cap [0, e)) + b) \cap [b + 1, n]| \\ &\quad \geq \alpha b + |A \cap [1, n - b]| \\ &\quad \geq \alpha b + |((A \cap [0, e)) + B) \cap [1, n - b]| \\ &\quad \geq \alpha b + |(A + B) \cap [1, n - b]| \\ &\quad \geq \alpha b + \alpha(n - b) \\ &\quad = \alpha n \end{aligned}$$

as desired. □

For further discussion of Mann's theorem and several variants, see [168].

The e-transform method also allows one to characterize when the above inequalities are sharp. We begin with an inverse theorem for Lemma 5.3.

Proposition 5.8 *Let A and B be additive sets in $\mathbf{Z}$ such that $|A|, |B| \geq 2$. Then $|A + B| = |A| + |B| - 1$ if and only if A, B are arithmetic progressions of the same step.*

Proof The "if" part is clear, so we prove the "only if" part. Let $e := \max(A) - \min(B)$. From the proof of Lemma 5.3 we see that we must have

$$A + B = A_{(e)} + B_{(e)} = (A \cup (B + e)) + \min(B) = (A + \min(B)) \cup (B + \max(A)).$$

Now let $\min(B) + v$ be the second smallest element of B, after $\min(B)$; then $v > 0$ and for any $a \in A \backslash \{\max(A)\}$ we have

$$a + \min(B) + v \subseteq A + B = (A + \min(B)) \cup (B + \max(A))$$
$$= (A + \min(B)) \cup (B \backslash \{\min(B)\} + \max(A)).$$

Note that since $a < \max(A)$ and $\min(B) + v$ is the minimal value of $B \backslash \{\min(B)\}$, than $a + \min(B) + v$ cannot lie in $(B \backslash \{\min(B)\} + \max(A))$. We conclude that

$$a + v \in A \text{ for all } a \notin \max(A).$$

From this it is easy to see that A is an arithmetic progression of step v. In particular $\max(A) - v$ is the second largest value of A after $\max(A)$, and by adapting the previous argument we see that B is also an arithmetic progression of step v, and we are done. □

Now we give an inverse theorem for the Cauchy–Davenport inequality.

Theorem 5.9 (Vosper's theorem) *[375] Let p be a prime, and let A, B be additive sets in $\mathbf{Z}_p$ such that $|A|, |B| \geq 2$ and $|A + B| \leq p - 2$. Then $|A + B| = |A| + |B| - 1$ if and only if A and B are arithmetic progressions with the same step.*

A similar theorem has recently been proven [174] in the case when $|A + B| = |A| + |B|$. A version of Vosper's theorem exists for arbitrary groups Z but is more complicated to state; see [201], [231]. See also Exercise 5.1.11.

Proof The "if" part is easy, so we prove the "only if" part. We first prove this claim when A is an arithmetic progression $\{a, a + v, \ldots, a + nv\}$ for some $n \geq 1$. Then by Cauchy–Davenport

$$\begin{aligned} |B| + n &= |A| + |B| - 1 \\ &= |A + B| \\ &= |\{a, a + v, \ldots, a + (n - 1)v\} + \{0, v\} + B| \\ &\geq |B + \{0, v\}| + n - 1, \end{aligned}$$

and hence (by Cauchy–Davenport again) we have $|B + \{0, v\}| = |B| + 1$. Thus B and $B + v$ only differ by at most one element, which implies that B is a progression of length v (see Exercise 3.2.7). By symmetry we have the same claim when the roles of A and B are reversed.

Now we use a duality trick to claim the following variant: if the *sum set* $A + B$ is a proper arithmetic progression, and $|A + B| = |A| + |B| - 1$, then so is A and B, and all three progressions have the same step. To see this, set $C := -(\mathbf{Z}_p \backslash (A + B))$. Then C is also an arithmetic progression with the same step as $A + B$ and with cardinality $|C| = p - |A + B| = p + 1 - |A| - |B| \geq 2$. Observe also that $C + B \subseteq -(\mathbf{Z}_p \backslash A)$, because if any element $-a$ of $-A$ was contained in $C + B$, then

C would intersect $-a - B \subset -(A + B)$, a contradiction. Thus $|C + B| \le p - |A| = |C| + |B| - 1$, and hence by Cauchy–Davenport $|C + B| = |C| + |B| - 1$. Since C was an arithmetic progression of length at least 2, we see from the previous discussion that B is also, and has the same step as C. Similarly for A.

To summarize, we have now proven Vosper's theorem in the cases when at least one of A, B, or $A + B$ is an arithmetic progression. Now we handle the general case. We induce on the size of B. If $|B| = 2$ then B is an arithmetic progression already, and the claim has already been proved. Now suppose that $|B| > 2$ and the claim has already been proven for smaller B. Suppose first that we can find an $e \in A - B$ such that the e-transform $B_{(e)}$ of B has size $1 < |B_{(e)}| < |B|$. Since $|A + B| = |A| + |B| - 1$; by hypothesis, we see from (5.1), (5.2) and the Cauchy–Davenport inequality that we must have $A_{(e)} + B_{(e)} = A + B$ and

$$|A_{(e)} + B_{(e)}| = |A_{(e)}| + |B_{(e)}| - 1.$$

Using the induction hypothesis, we thus see that $A_{(e)}$ and $B_{(e)}$ are arithmetic progressions with the same step v, and hence $A + B = A_{(e)} + B_{(e)}$ is also an arithmetic progression, and the claim follows by the preceding discussion.

The only remaining case is if we have $|B_{(e)}| = 1$ or $|B_{(e)}| = |B|$ for all $e \in A - B$. But if $E \subseteq A - B$ denotes all the $e \in A - B$ such that $|B_{(e)}| = |B|$, then by Lemma 5.2 we have $B + E \subseteq A$, and hence $|E| \le |A| - |B| + 1$ by Cauchy–Davenport. Since $|A - B| \ge |A| + |B| - 1$ by Cauchy–Davenport, we thus see that $|B_{(e)}| = 1$ for at least $2|B| - 2$ values of e. Since $B_{(e)}$ is a singleton subset of B, we thus see from the pigeonhole principle that there exists $e, e' \in A - B$ and $b \in B$ such that $B_{(e)} = B_{(e')} = \{b\}$. Since $|A + B| = |A| + |B| - 1$ by hypothesis, we see from (5.1), (5.2) that

$$A + B = A_{(e)} + b = A_{(e')} + b$$

and hence

$$A \cup (B + e) = A \cup (B + e').$$

Since A intersects $B + e$ only in $b + e$, and A intersects $B + e'$ only in $b + e'$, we thus see that $B + e$ and $B + e'$ differ by at most one element. But this forces B to be a progression (of step $e' - e$), and the claim follows. □

We now develop an inverse theorem for sets A, B of integers with fairly small sum set. We need a preliminary lemma.

Lemma 5.10 *Let A be an additive set in $\mathbf{Z}$ such that $0 \in A$, let $N \ge 1$ be an integer, and let $\phi_N : \mathbf{Z} \to \mathbf{Z}_N$ be the canonical quotient map. For each $x \in \phi_N(A)$, let $\mu_x := |\{a \in A : \phi_N(a) = x\}|$ denote the multiplicity of ϕ_N at x, and denote $m := \min_{x \in \phi_N(A) \setminus \{0\}} \mu_x$. Then*

$$|2A| \ge |A| + |\phi_N(A)|(\mu_0 - 2m) + |2\phi_N(A)|(2m - 1)$$

Proof We split (using Lemma 5.3 and the observation $\sum_{x \in \phi_N(A)} \mu_x = |A|$)

$$\begin{aligned}
|2A| &= \sum_{x \in \phi_N(2A)} \left|2A \cap \phi_N^{-1}(\{x\})\right| \\
&\geq \sum_{x \in \phi_N(2A)} \sup_{y,z \in \phi_N(A): y+z=x} \left|\left(A \cap \phi_N^{-1}(\{y\})\right) + \left(A \cap \phi_N^{-1}(\{z\})\right)\right| \\
&\geq \sum_{x \in \phi_N(2A)} \sup_{y,z \in \phi_N(A): y+z=x} \left(\left|A \cap \phi_N^{-1}(\{y\})\right| + \left|A \cap \phi_N^{-1}(\{z\})\right| - 1\right) \\
&= \left(\sum_{x \in \phi_N(2A)} \sup_{y,z \in \phi_N(A): y+z=x} \mu_y + \mu_z\right) - |\phi_N(2A)| \\
&\geq \left(\sum_{x \in \phi_N(A)} \mu_0 + \mu_x\right) + \left(\sum_{x \in \phi_N(2A) \setminus \phi_N(A)} m + m\right) - |\phi_N(2A)| \\
&= \mu_0 \phi_N(A) + |A| + (|\phi_N(2A)| - |\phi_N(A)|)2m - |\phi_N(2A)|
\end{aligned}$$

as desired (noting that $2\phi_N(A) = \phi_N(2A)$). □

Now we give the inverse theorem.

Theorem 5.11 (**$3k - 3$ theorem**) *[116] Let A be an additive set in $\mathbf{Z}$ such that $|2A| < 3|A| - 3$. Then there exists a proper arithmetic progression $P = a + [0, |2A| - |A|] \cdot v$ of length $|2A| - |A| + 1$ that contains A.*

Proof We use an argument from [233]. By translating A we may assume that $\min(A) = 0$. We may also assume that the set A has no common divisor $d > 1$, since otherwise we could replace A by $\frac{1}{d} \cdot A$. We will assume that $|A| \geq 3$ as the cases $|A| = 1, 2$ can be verified directly.

Write $N := \max(A)$, thus $A \subseteq [0, N]$ and $0, N \in A$. It will suffice to show that $N \leq |2A| - |A|$. Suppose for contradiction that $N > |2A| - |A|$. We now apply Lemma 5.10. Observe in this case that $\mu_0 = 2$ and $m = 1$, and hence

$$|2A| \geq |A| + |2\phi_N(A)|. \tag{5.9}$$

Since we are assuming $N > |2A| - |A|$, we conclude that

$$|2\phi_N(A)| < N. \tag{5.10}$$

By Exercise 2.1.6 and the hypothesis $|2A| < 3|A| - 3$ we have

$$|2\phi_N(A)| < 2|A| - 3 = 2|\phi_N(A)| - 1.$$

If N were prime then we could apply the Cauchy–Davenport inequality to conclude the desired contradiction. But in general we must rely instead on Kneser's theorem. Let $H := \mathrm{Sym}_1(2\phi_N(A))$, then by Kneser's theorem we have

$$|2\phi_N(A)| \geq 2|\phi_N(A) + H| - |H|$$

and hence if we set $k := |\phi_N(A) + H| - |\phi_N(A)|$, then

$$0 \le k \le \frac{|H| - 2}{2}. \tag{5.11}$$

In particular $|H| \ge 2$. Also from (5.10) we have $|H| < N$. Since H is a subgroup of $\mathbf{Z}_N$, we see that $H = (h \cdot Z)/(N \cdot Z)$ for some $2 \le h < N$ which is a factor of N.

Note that $\phi_N(A)$ contains zero, but cannot be contained entirely inside H as this would mean that A has a common divisor of h, contradicting our hypothesis. So we know that $\phi_N(A)$ contains at least two cosets of H, or equivalently that $|\phi_h(A)| \ge 2$.

Now we apply Lemma 5.10 again, but with N replaced by h. From (5.11) we see that if $x + H \subseteq \mathbf{Z}_N$ is any non-trivial coset of H, then $H \cup (x + H)$ intersects $\phi_N(A)$ in at least $2|H| - k$ points; since $\phi_N(0) = \phi_N(N) = 0 \in H \cup (x + H)$, this implies that $\phi_N^{-1}(H \cup (x + H)) = \phi_h^{-1}(\{0, x \bmod h\})$ intersects A in at least $2|H| - k + 1$ points. In other words we have

$$\mu_0 + m \ge 2|H| - k + 1.$$

A similar argument gives

$$m \ge |H| - k.$$

But since H was the symmetry group of $2\phi_N(A)$, we see that $2\phi_h(A)$ has trivial symmetry group; furthermore from (5.10) we see that $|2\phi_h(A)| < h$. Thus by Kneser's theorem we have $|2\phi_h(A)| \ge 2|\phi_h(A)| - 1$. Inserting all these facts into Lemma 5.10, we obtain

$$\begin{aligned}
|2A| &\ge |A| + |\phi_h(A)|(\mu_0 - 2m) + (2|\phi_h(A)| - 1)(2m - 1)\\
&\ge |A| + |\phi_h(A)|(2|H| - k - 3m + 1) + (2|\phi_h(A)| - 1)(2m - 1)\\
&= |A| + |\phi_h(A)|(2|H| - k - 1) + (|\phi_h(A)| - 2)m + 1\\
&\ge |A| + |\phi_h(A)|(2|H| - k - 1) + (|\phi_h(A)| - 2)(|H| - k) + 1\\
&= |A| + 3|\phi_h(A)||H| - 2k|\phi_h(A)| - |\phi_h(A)| - 2|H| + 2k + 1\\
&\ge |A| + 3|\phi_h(A)||H| - (|H| - 2)|\phi_h(A)| - |\phi_h(A)| - 2|H| + (|H| - 2) + 1\\
&= |A| + 2|\phi_h(A)||H| + |\phi_h(A)| - 2|H| + (|H| - 2) + 1\\
&= |A| + 2(|A| + k) + |\phi_h(A)| - 2|H| + (|H| - 2) + 1\\
&= 3|A| + |\phi_h(A)| - 2|H| + 2(|H| - 2) - 1\\
&\ge 3|A| - 3
\end{aligned}$$

which contradicts the hypothesis $2|A| < 3|A| - 3$. □

Note that we have used a result on torsion groups to imply a result in the torsion-free case; this phenomenon will also come up in later proofs of Freiman's theorem. The original proof of Freiman was somewhat different; see [116], [257]. A treatment of the case $|2A| = 3|A| - 3$ appears in [113], [28]. For some partial progress in the case $|2A| = 3|A| + o(|A|)$, see [193]. There has also been much work on generalizing the $3k - 3$ theorem to pairs of sets [111], [336], [333], [233]. For instance one has the following result.

Theorem 5.12 *[233] Let A, B be additive sets in $\mathbf{Z}$ such that $|A + B| < |A| + |B| + \min(|A|, |B|) - 3$. Then A is contained in an arithmetic progression of length at most $|A + B| - |B| + 1$ and B is contained in an arithmetic progression of length at most $|A + B| - |A| + 1$, where both progressions have the same difference.*

For some further refinements to this theorem, see [233].

Exercises

5.1.1 Prove the remaining claims in Corollary 5.6.

5.1.2 Prove Lemma 5.2.

5.1.3 Show that Kneser's theorem implies Lemma 5.3 and the Cauchy–Davenport inequality.

5.1.4 [211] Let A, B be additive sets in an ambient group. Show that if $|A + B| < |A| + |B|$ then

$$|A+B| = |A+\mathrm{Sym}_1(A+B)|+|B+\mathrm{Sym}_1(A+B)| - |\mathrm{Sym}_1(A+B)|.$$

5.1.5 [244] Let A, B be additive sets in an ambient group Z such that $|A + B| < |A| + |B| - 1$. Show that $|(A + \mathrm{Sym}_1(A + B))\backslash A| < |\mathrm{Sym}_1(A + B)| - 1$; thus A is rather close to being a union of cosets of $\mathrm{Sym}_1(A + B)$.

5.1.6 [243] If A is a (possibly infinite) set of integers, define the *Schnirelmann density* $\sigma(A)$ of A to be the quantity

$$\sigma(A) := \inf_{N>0} \mathbf{E}_{x\in[1,N]}(x \in A) = \inf_{N>0} \frac{|A \cap [1, N]|}{|[1, N]|}.$$

(Note that this is distinct from the lower density $\underline{\sigma}(A)$ defined in Definition 1.21, due to the use of the inf rather than the lim inf.) Show that if A and B are any sets of integers with $0 \in A, B$, then $\sigma(A + B) \geq \min(\sigma(A) + \sigma(B), 1)$. (Hint: use Theorem 5.7.) Conclude that if $0 \in A$ and $\sigma(A) \geq 1/k$ for some integer $k > 0$, then $kA \subset \mathbf{Z}^+$. Thus every set of integers of positive Schnirelmann density that contains 0 is a basis for the positive integers.

5.1.7 [312] Let A, B be sets of integers such that $1 \in A$ and $0 \in B$. Show that $\sigma(A+B) \geq \sigma(A)+\sigma(B)-\sigma(A)\sigma(B)$, where $\sigma()$ is the Schnirelmann density from Exercise 5.1.6. (Hint: order the positive elements of A as $a_1 < a_2 < \cdots$, and observe that $|(A+B) \cap [a_n, a_{n+1})| \geq 1 + |B \cap [1, a_{n+1} - a_n - 1]|$.)

5.1.8 [311], [201], [202] Let A and B be additive sets in an ambient group Z. Prove that

$$|A+B| \geq |A|+|B| - \min_{c \in A+B} |\{(a,b) \in A+B : a+b=c\}|.$$

(This can be done either by Kneser's theorem, or more directly via the e-transform method.)

5.1.9 Let p be a prime, let $N \geq 1$, and let $A_1, \ldots, A_N$ be additive sets in $\mathbf{Z}_p$ such that $|A_1|+\cdots+|A_N| = p+N-1$. Use the Cauchy–Davenport inequality to show that $A_1+\cdots+A_N = Z_p$. Conversely, show that this statement can be used to imply the Cauchy–Davenport inequality.

5.1.10 What happens if one extends Theorem 5.9 to cover the cases $|A|=1$, $|B|=1$, or $|A+B|=p-1$? (The case $|A+B|=p$ is much more difficult to analyze and does not have as simple a characterization.)

5.1.11 [201] Let A, B be additive sets in ambient group Z such that $|A|, |B| > 1$, $|\mathrm{Sym}_1(A+B)| = 1$, and $|A+B| < |A|+|B|$. By analyzing the proof of Kneser's theorem (and Vosper's theorem) carefully, show that $A+B$ is either equal to an arithmetic progression, or there exists a finite subgroup G of Z such that $A+B$ consists of one or more cosets of G, and possibly a subset of one other coset of G. (Compare with Exercise 5.1.5 and Exercise 3.2.7.)

5.1.12 [242] Let A, B be open subsets of the torus $(\mathbf{R}/\mathbf{Z})^d$. Prove the *Mann–Kneser–Macbeath inequality* $\mathrm{mes}(A+B) \geq \min(\mathrm{mes}(A)+\mathrm{mes}(B), 1)$, where $\mathrm{mes}()$ denotes the usual Haar measure on the torus. (Hint: discretize the torus to $(\mathbf{Z}/p\mathbf{Z})^d$ for some large prime p, apply Kneser's theorem, and then take limits.) Give examples to show that this inequality cannot be improved. One can extend this result to arbitrary measurable subsets of the torus with some additional analytic arguments. See [27] for some recent developments concerning this inequality. This inequality should be contrasted with the Brunn–Minkowski inequality (Theorem 3.16), and shows that sum sets in $(\mathbf{R}/\mathbf{Z})^d$ and sum sets in $\mathbf{R}^d$ behave slightly differently.

5.1.13 [116] Let $N \geq 0$ be an integer, and let A, B be non-empty subsets of $[0, N]$ such that $0, N \in A$ and $|A|+|B| \geq N+3$. Prove that $|A+B| \geq |B|+N$.

5.1.14 Show that Theorem 5.11 fails when $|2A| = 3|A| - 3$, by considering a progression of rank 2. Also show that the quantity $2|A| - |A|$ in that theorem cannot be replaced by any smaller quantity.

5.1.15 Let A, B be additive sets of integers. If $A \hat{+} B := \{a + b : a, b \in A, a \neq b\}$ denotes the restricted sum set of A and B, show that $|A \hat{+} B| \geq |A| + |B| - 3$. (Hint: a direct application of the e-transform will not work, but if one deconstructs the proof of Lemma 5.3 one can modify it to deal with restricted sum sets.) If $|A| \neq |B|$, improve the preceding bound to $|A \hat{+} B| \geq |A| + |B| - 2$. (Hint: one needs to adapt some ideas from Proposition 5.8.) An analogous result for $\mathbf{Z}_p$ is known, but requires more non-elementary methods; see Section 9.2.

5.2 Sum sets in vector spaces

We now study the minimal size of sum sets in a real finite-dimensional vector space V, exploiting such concepts as convexity which are not readily available in other groups. Of course, since V contains a copy of $\mathbf{Z}$, we know from Lemma 5.3 that $|A + B|$ can be as small as $|A| + |B| - 1$. However, one can do better than this if one knows that $A + B$ is high-dimensional, or in other words that it is not contained in a low-dimensional affine vector space (a translate of a linear vector space).

We begin with the case $A = B$, which is somewhat easier. Define the *rank* $\mathrm{rank}(A)$ of a subset of V to be the smallest d such that A is contained in an affine space of dimension d.

Lemma 5.13 (Freiman's lemma) *[116] Let A be an additive set in a finite-dimensional vector space V, and let suppose that* $\mathrm{rank}(A) \geq d$ *for some* $d \geq 1$. *Then we have*

$$|A + A| \geq (d+1)|A| - \frac{d(d+1)}{2}.$$

Proof We induce on d. If $d = 1$ then the claim follows from Theorem 5.5, so let us assume $d \geq 2$ and that the claim is already proven for $d - 1$. Now we fix d and induce on $|A|$. The claim is vacuously true if say $|A| = 1$, so assume $|A| \geq 2$ and that the claim is already been proven for smaller sets A. Let $a \in A$ be any extreme point of A; thus a is a vertex on the convex hull of A. Let $A' := A - \{a\}$. We divide into two cases. If $\mathrm{rank}(A') \geq d$, then by induction hypothesis

$$|A' + A'| \geq (d+1)|A'| - \frac{d(d+1)}{2}.$$

Since a lies outside of the convex hull of A' and $\mathrm{rank}(A') \geq d$, there must exist (by the greedy algorithm) at least d extreme points $x_1, \ldots, x_d$ of A' which are visible from a in the sense that the line segments joining a to $x_1, \ldots, x_d$ lie outside the convex hull of A'. In particular we see that the $d+1$ points $a, \frac{a+x_1}{2}, \ldots, \frac{a+x_d}{2}$ lie outside the convex hull of A', and in particular outside of $\frac{1}{2} \cdot (A' + A')$. Dilating this by 2 we see that $a+a, a+x_1, \ldots, a+x_d$ are disjoint from $A' + A'$. Thus

$$|A + A| \geq d + 1 + |A' + A'| \geq (d+1)|A| - \frac{d(d+1)}{2}$$

thus closing the induction.

It remains to consider the case when $\operatorname{rank}(A') < d$, thus A is contained in a $d-1$-dimensional affine space W. Since $\operatorname{rank}(A) \geq d$, we have $a \notin W$. This means that $2a$, $a + W$, and $2W$ are all disjoint; thus $a + a$, $a + A'$, and $A' + A'$ are all disjoint; thus

$$|A + A| \geq 1 + |A| - 1 + |A' + A'|.$$

But since $\operatorname{rank}(A) \geq d$, we have $\operatorname{rank}(A') = \operatorname{rank}(A \backslash \{a\}) \geq d - 1$, and hence by induction

$$|A' + A'| \geq d|A'| - \frac{d(d-1)}{2} = d|A| - \frac{d(d+1)}{2}$$

and the claim again follows by induction. □

Now we consider the problem of sums of two sets A, B in V. To make this problem more precise, let us temporarily define the quantity $S(d, n, t)$ for any $n \geq 1$, $t \geq 0$, and $d \geq 0$, to be the least value of $|A + B|$, where A, B ranges over all additive sets in a finite-dimensional vector space V, such that $|A| \geq n$, $|B| \geq n - t$, and $\operatorname{rank}(A + B) \geq d$. Since $|A + B| \geq |A|$ we have the trivial bound

$$S(d, n, t) \geq n. \tag{5.12}$$

This bound is however not sharp in general, and we shall improve it presently. We first need a lemma analyzing the behavior of $A + B$ near an extreme point of A and B, similar to that used in the proof of Lemma 5.13.

Lemma 5.14 *[296] Let A, B be additive sets in a finite-dimensional vector space V such that A and B both contain* 0, *and suppose that* 0 *is a vertex on the convex hull of $A \cup B$. Let $A' := A - \{0\}$ and $B' := B - \{0\}$, and $C := (A' \cup B') \backslash (A' + B')$. Then $A + B$ lies in the subspace of V spanned by C.*

Proof Without loss of generality we may take $V = \mathbf{R}^n$. By the Hahn–Banach theorem, there exists a linear functional $\phi : V \to \mathbf{R}$ such that $\phi(x) > 0$ for all $x \in (A \cup B) \backslash \{0\}$. We need to show that every element x of $A + B$ lies in the span of C. We shall prove this by induction on $\phi(x)$, which is a non-negative integer. If $\phi(x) = 0$, then $x = 0$ and there is nothing to prove. Now suppose that $\phi(x) > 0$ and the claim has already been shown for all smaller values of $\phi(x)$. If $x \in A' + B'$, then we can write $x = a + b$ where $a \in A'$ and $b \in B'$. But since $\phi(x) = \phi(a) + \phi(b)$ and $\phi(a), \phi(b) > 0$, we see that $\phi(a), \phi(b)$ are strictly less than $\phi(x)$, and the claim follows from induction. The only remaining case is when $\phi(x) > 0$ and $x \notin (A' + B')$. But since $x \in A + B$, this implies that $x \in C$, and we are done. □

We can now obtain the following recursive inequality on $S(d, n, t)$.

Proposition 5.15 *[296] Let $d \geq 1$, $n \geq 2$, and $t \leq n - 2$. Then*

$$S(d, n, t) \geq \min(S(d, n-1, t) + d + 1, S(d-1, n-1, t) + n, \\ S(d-1, n-1, t-1) + n - t).$$

Proof Let A, B be as in the definition of $S(d, n, t)$; note that A and B contain at least two elements. Since A and B are finite, we can find a linear functional $\phi : V \to \mathbf{R}$ which is injective on $A \cup B$ (indeed one could select ϕ randomly). Since ϕ is injective, we see that there is a unique element $a_0 \in A$ which minimizes ϕ on A, i.e. $\phi(a) > \phi(a_0)$ for all $a \in A \backslash \{a_0\}$. Similarly we can find a $b_0 \in B$ which minimizes ϕ on B, so that $\phi(b) > \phi(b_0)$ for all $b \in B \backslash \{b_0\}$. By translating A and B if necessary we may assume $a_0 = b_0 = 0$. Thus A and B now both contain 0, and if we define $A' := A \backslash \{0\}$ and $B' := B \backslash \{0\}$, then ϕ is strictly positive on both A' and B'. In particular ϕ is strictly positive on $A' + B'$, which therefore does not contain 0.

From Lemma 5.14 we have

$$|(A' \cup B') \backslash (A' + B')| \geq d$$

and hence (since 0 is contained in $A + B$ but not A', B', or $A' + B'$)

$$|A + B| \geq |A' + B'| + d + 1.$$

Let $c + W$ denote the affine span of $A' + B'$, where $c \in V$ and W is a linear subspace of W. If we knew that $\operatorname{rank}(A' + B') = \dim(W) \geq d$, we could then conclude that $|A' + B'| \geq S(d, n-1, t)$, and we would be done. Thus we may assume that $\dim(W) \leq d - 1$. Thus if we pick $a_1 \in A'$ and $b_1 \in B'$ arbitrarily, then we have $A' \in a_1 + W$ and $B' \in b_1 + W$. Thus $A + B$ is contained in the span of W, a_1, and b_1. By hypothesis, this means that at least one of a_1, b_1 must lie outside of W.

We now divide into a number of cases depending of the relative position of a_1 and b_1 with respect to W. Suppose first that a_1 and b_1 are linearly independent modulo W. Then $A = \{0\} \cup A'$ lies in $\{0, a_1\} + W$, and is thus disjoint from $A + B'$, which lies in $\{b_1, a_1 + b_1\} + W$; so

$$|A + B| \geq |A + B'| + |A| \geq |A + B'| + n.$$

On the other hand, $\operatorname{rank}(A + B') \geq \operatorname{rank}(A + B) - 1 \geq d - 1$, which implies $|A + B'| \geq S(d - 1, n - 1, t)$. The claim thus follows in this case.

Now suppose that a_1, b_1 are linearly dependent modulo W and $b_1 \notin W$. Then $A' \subset a_1 + W$ and $A' + B' \subset a_1 + b_1 + W$ are disjoint, while 0 is disjoint from A' (by definition) and $A' + B'$ (by previous remarks). Thus

$$|A + B| \geq 1 + |A'| + |A' + B'| \geq n + |A' + B'|.$$

On the other hand, since $A + B$ is contained in the span of W and b_1, we have $\operatorname{rank}(A' + B') = \dim(W) \geq \operatorname{rank}(A + B) - 1 \geq d - 1$, hence $|A' + B'| \geq S(d - 1, n - 1, t)$. The claim again follows.

The only remaining case is when $b_1 \in W$, which forces $a_1 \notin W$ by previous discussion. Then $A' + B$ and B are disjoint, thus

$$|A + B| \geq |B| + |A' + B| \geq n - t + |A' + B|.$$

But since $\operatorname{rank}(A' + B) \geq \operatorname{rank}(A + B) - 1 \geq d - 1$, we have $|A' + B| \geq S(d - 1, n - 1, t - 1)$, and the claim again follows. □

Corollary 5.16 *[296] For any $n \geq 1$, $t \geq 0$, $d \geq 0$ we have*

$$S(d, n, t) \geq \sum_{n-d \leq r \leq n} r - \sum_{1 \leq s \leq t} \min(s, d).$$

Proof The cases $d = 0$, $n = 1$, or $r \geq n - 1$ can be easily verified from (5.12), so we may restrict ourselves to the case $d \geq 1$, $n \geq 2$, and $t \leq n - 2$. We shall induce on the positive quantity $n + d + t$, assuming inductively that the claim has already been proven for all smaller values of $n + d + t$. But then we have

$$\begin{aligned}
S(d, n-1, t) + d + 1 &\geq \sum_{n-d-1 \leq r \leq n-1} r - \sum_{1 \leq s \leq t} \min(s, d) + d + 1 \\
&= \sum_{n-d \leq r \leq n} r - \sum_{1 \leq s \leq t} \min(s, d) \\
S(d-1, n-1, t) + n &\geq \sum_{n-d \leq r \leq n-1} r + n - \sum_{1 \leq s \leq t} \min(s, d-1) \\
&\geq \sum_{n-d \leq r \leq n} r - \sum_{1 \leq s \leq t} \min(s, d) \\
S(d-1, n-1, t-1) + n - t &\geq \sum_{n-d \leq r \leq n-1} r - \sum_{1 \leq s \leq t-1} \min(s, d) + n - t \\
&\geq \sum_{n-d \leq r \leq n} r - \sum_{1 \leq s \leq t} \min(s, d)
\end{aligned}$$

and the claim follows from Proposition 5.15. □

This inequality is sharp in many cases, although there have been some refinements using techniques relating to the Brunn–Minkowski inequality (Theorem 3.16); see [128], [129]. As a consequence of the inequality we obtain the following generalization of Theorem 5.13:

Theorem 5.17 *[296] Let V a finite-dimensional vector space and $d \geq 0$, and let A, B be additive sets in V such that* $\mathrm{rank}(A+B) \geq d$*, then have* $|A+B| \geq |A| + d|B| - \frac{d(d+1)}{2}$.

Proof Apply Corollary 5.16 with $n := |A|$, $t := |A| - |B|$ and use the trivial bound $\sum_{1 \leq s \leq t} \min(s, d) \geq t$ to obtain

$$|A+B| \geq (d+1)\left(n - \frac{d}{2}\right) - t = n + d(n-t) - \frac{d(d+1)}{2}$$

as desired. □

We now return to additive sets in a vector space with small doubling. Define a *d-parallelepiped P* in a vector space V to be any set of the form

$$P = a + \overline{I} \cdot v_1 + \cdots + \overline{I} \cdot v_d$$

where $v_1, \ldots, v_d$ are vectors in V (not necessarily linearly independent), $a \in V$, and $\overline{I} = \{x \in \mathbf{R} : -1 \leq x \leq 1\}$ is the closed unit ball. The 2^d points $a + \{-1, 1\} \cdot v_1 + \cdots + \{-1, 1\} \cdot v_d$ (which may possibly have multiplicity) are called the *corners* of this d-parallelepiped, while a is the *center*; note that the corners form a progression of rank d and dimensions $(2, \ldots, 2)$, which may or may not be proper. A remarkable fact, known as the *Freiman cube lemma*, is that if an additive set A in a d-dimensional vector space has small doubling, then there is a d-parallelepiped which contains a large fraction of A and whose corners lie in the set A. This is certainly not true for general sets A, as can be seen for instance by considering the set $\{(n, n^2) : -N \leq n \leq N\}$ in $\mathbf{Z}^2 \subset \mathbf{R}^2$. To prove the Freiman cube lemma we first prove an auxiliary lemma which is useful for inductive purposes:

Lemma 5.18 *[28] Let V be an d-dimensional vector space, and let W be a $d-r$-dimensional linear subspace of V for some $0 < r \leq d$. Let A be a symmetric additive set in V (thus $-A = A$) and let $K = \sigma[A] = |A+A|/|A|$ be the doubling constant. Then there exists a r-paralleopiped P with corners in A and center 0 such that*

$$|A \cap (P + W)| \geq (9K)^{-2^{r-1}+1}|A|.$$

Proof We induce on the codimension r. First suppose that $r = 1$. Without loss of generality we may take V to be a Euclidean space $\mathbf{R}^d$. We let v_1 be an element of A which maximizes the quantity $\mathrm{dist}(v_1, W)$; then it is easily seen that the 1-parallelepiped $P = 0 + \overline{I} \cdot v_1$ will obey the desired properties (here we exploit the symmetry of A to place both corners of P in A).

Now suppose that $r \geq 2$ and that the claim has already been proven for all smaller values of r. We place W inside a $d-1$-dimensional hyperplane $H \subset V$, which divides V into the hyperplane H and into two open half-spaces H_- and H_+. By the pigeonhole principle, one of the three sets $A \cap H$, $A \cap H_-$, and $A \cap H_+$ has cardinality at least $|A|/3$.

Suppose first that $|A \cap H| \geq |A|/3$. Then by applying the induction hypothesis (with V replaced by H and d replaced by $d-1$) we can find an $r-1$-parallelepiped $P \subset H \subset V$ with corners in $A \cap H \subseteq H$ and center 0 such that

$$\begin{aligned} |A \cap (P+W)| &\geq |(A \cap H) \cap (P+W)| \geq (9K)^{-2^{r-2}+1}|A|/3 \\ &\geq (9K)^{-2^{r-1}+1}|A|. \end{aligned}$$

The claim then follows by adding a dummy vector $v_r = 0$ to P to make it a r-parallelepiped.

Without loss of generality, it remains to consider the case when $|A \cap H_+| \geq |A|/3$. Since $|2(A \cap H_+)| \leq |2A| \leq K|A|$, we conclude that $\sigma[A \cap H_+] \leq 3K$. By Exercise 2.3.14, we see that $A \cap H_+$ contains a set F symmetric around some origin $a = x/2$ (since $F = x - F$) with $|F| \geq |A|/9K$ and $\sigma[F] \leq 9K^2$. Since F is contained entirely in the half-space H_+, we see that $a \in H_+$ also. In particular, $a \notin W$. Now let W' be the $d-r+1$-dimensional linear space spanned by W and a, and apply the induction hypothesis with A replaced by $F-a$, K replaced by $9K^2$, W replaced by W' and r replaced by $r-1$. This allows us to find a $r-1$-parallelepiped $P' = a + \overline{I} \cdot v_1 + \cdots + \overline{I} \cdot v_{r-1}$ with center a and corners in F such that

$$|F \cap (P' + W')| \geq (81K^2)^{-2^{-r-2}+1}|F| \geq (9K)^{-2^{-r-1}+1}|A|.$$

Now we let P be the r-parallelepiped $\overline{I} \cdot a + \overline{I} \cdot v_1 + \cdots + \overline{I} \cdot v_{r-1}$; since F and $-F$ are both contained in A (by the symmetry of A) we see that the corners of P lie in A, and P is certainly centered at the origin. To conclude the proof we need to show that

$$|A \cap (P+W)| \geq |F \cap (P' + W')|.$$

To prove this, we use a sliding argument taking advantage of the symmetries of A and F. Let us split $W' = W_{>0} \cup W_{\leq 0}$, where $W_{>0}$ is the open half-space in W' with boundary W which contains a, and $W_{\leq 0}$ is the closed half-space in W' with boundary W which excludes a. Then

$$\begin{aligned}|F \cap (P' + W')| &= |F \cap (P' + W_{>0})| + |F \cap (P' + W_{\leq 0})| \\ &= |(F - 2a) \cap (P' + W_{>0} - 2a)| + |F \cap (P' + W_{\leq 0})| \\ &= |(-F) \cap (P' + W_{>0} - 2a)| + |F \cap (P' + W_{\leq 0})| \\ &= |[(-F) \cap (P' + W_{>0} - 2a)] \cup [F \cap (P' + W_{\leq 0})]|\end{aligned}$$

since F is symmetric around a, and F and $-F$ are disjoint (one lies in H_+ and the other lies in H_-). It thus suffices to show that the sets $-F \cap (P' + W_{>0} - 2a)$ and $F \cap (P' + W_{\leq 0})$ lie in $A \cap (P + W)$. That these sets lie in A is clear, since A contains both F and $-F$. Also observe that $(P' + W_{>0} - 2a)$ is contained in $-(P' + W_{\leq 0})$ since P' is symmetric around a. Thus it only remains to show that $F \cap (P' + W_{\leq 0}) \subseteq P + W$. But since $F = 2a - F$ lies in H_+, and the corners of P' lie in F, and W lies in H, we see that both F and $P' + W$ lie in the slab between H and $2a + H$. Thus $F \cap (P' + W_{\leq 0})$ lies in the set $P' - \{ta : 0 \leq t \leq 2\} + W = P + W$, and the claim follows. □

As a corollary we obtain

Corollary 5.19 (Freiman cube lemma) *Let A be an additive set in a d-dimensional vector space V, and let $K = \sigma[A]$ be the doubling constant. Then there exists a d-parallelepiped with corners in A such that $|A \cap P| \geq (3K)^{-2^d}|A|$.*

Proof [28] Applying Exercise 2.3.14 we have $\sigma[F] \leq K^2$. Now apply Lemma 5.18 with $W = \{0\}$ and $r = d$. □

Lemma 5.13 shows, roughly speaking, that if A is an additive set in a vector space then $\mathrm{rank}(A)$ is controlled by a linear function of the doubling constant $\sigma[A]$. The following remarkable theorem shows that if one is willing to pass from A to a significant subset of A, then one can in fact control the rank by a *logarithmic* function of the doubling constant.

Theorem 5.20 (Freiman 2^n theorem) *Let $d \geq 1$, and let A be an additive set in a vector space V with doubling constant $K = \sigma[A] < 2^d$. Then there exists a subset A' of A with $\mathrm{rank}(A') < d$ such that $\sigma[A'] \leq K$ and $|A'| = \Theta_{d,K}(|A|)$.*

See [28] for further discussion, including the dependence of constants in the $\Theta_{d,K}()$ notation.

Proof [28] We fix d and induce on K. For $K \leq 1$ the claim is vacuously true. Now suppose that $K > 1$ and that the claim has already been proven for values of $K \leq K - \varepsilon(d, K)$ for some $\varepsilon(d, K) > 0$ which is bounded from below for K in any compact interval $\{1 \leq K \leq 2^d - \delta\}$; if we can prove the claim under such a hypothesis, then the claim follows unconditionally by a standard continuity argument (the set of K obeying the theorem is open, closed, and contains 1).

Fix A, V, K, and let $\varepsilon = \varepsilon(d, K)$ be chosen later. If there exists a set $A'' \subset A$ with $|A''| \geq \frac{\varepsilon}{K}|A|$ and $\sigma[A''] \leq K - \varepsilon$, then the claim would follow by applying the induction hypothesis with A replaced by A'' and K by $K - \varepsilon$. Thus we may assume that $\sigma[A''] \geq K - \varepsilon$ whenever $|A''| \geq \frac{\varepsilon}{K}|A|$. In particular we see that

$$|2A''| \geq K|A''| - \varepsilon|A| \text{ for all non-empty } A'' \subseteq A \tag{5.13}$$

(treating the case of small A'' and large A'' separately). Note that this also holds with $A'' = \emptyset$ if we adopt the convention that $2A'' = \emptyset$ in this case.

Let $r = \mathrm{rank}(A)$. Without loss of generality we may assume that V is r-dimensional, since otherwise we can restrict V to the affine span of A (and translate to the origin). If A is small, say $|A| \leq 10K^2$, then the claim follows just by setting A' to be a single point, so assume $|A| > 10K^2$. By Lemma 5.13 we conclude $r \leq K$. We will in fact show that the hypotheses on A force $r \leq d$, at which point we can take $A' := A$ and be done.

We now claim that (5.13) implies the bound

$$|A \cap W| = O(\varepsilon|A|) \tag{5.14}$$

for all affine hyperplanes W in V. To see this, observe that W divides V into the hyperplane W and two open half-spaces W_-, W_+. Since A has full rank, at least one of $A \cup W_+$, $A \cup W_-$ is non-empty. Let us say that $A \cup W_+$ is non-empty. Let a be a point in $A \cup W_+$ that minimizes the distance to W. One then observes from the convexity and disjointness of W, W_-, W_+ that the midpoint sets $\frac{1}{2} \cdot 2(A \cap W)$, $\frac{1}{2} \cdot 2(A \cup W_+)$, $\frac{1}{2} \cdot (a + (A \cap W))$, and $\frac{1}{2} \cdot (2(A \cup W_-))$ are all disjoint. Since all these sets are contained in $\frac{1}{2} \cdot 2A$, we see that

$$|2(A \cap W)| + |2(A \cup W_+)| + |A \cap W| + |2(A \cup W_-)| \leq |2A| = K|A|.$$

Applying (5.13) we conclude (5.14).

Next, we apply the Freiman cube lemma to obtain a r-parallelepiped P with corners in A such that

$$|A \cap P| = \Omega_K(|A|). \tag{5.15}$$

Comparing this with (5.14) we see that P cannot be contained in a affine hyperplane (if ε is chosen sufficiently small). Since the parallelepiped P has $2r \leq 2K$ faces, each of which lies on an affine hyperplane, we thus see that, with $\mathrm{int}(P)$ denoting the interior of P, then

$$|A \cap \mathrm{int}(P)| \geq |A \cap P| - O(K\varepsilon|A|).$$

If Q denotes the 2^r corners of P, we observe that the sets $\{x + \mathrm{int}(P) : x \in Q\}$ are all disjoint; thus

$$|2(A \cap P)| \geq 2^r |A \cap \mathrm{int}(P)| \geq 2^r |A \cap P| - O(2^r K\varepsilon|A|). \tag{5.16}$$

The complement $V \backslash P$ of P in V can be partitioned into at $2r$ (unbounded) convex regions $B_1 \cup \cdots \cup B_{2r}$ (Exercise 5.2.4). Observe from convexity and disjointness that the midpoint sets $\frac{1}{2} \cdot 2(A \cap B_j)$ are disjoint from each other and from $2(A \cap P)$. Thus

$$|2A| \geq \sum_{j=1}^{2r} |2(A \cap B_j)| + |2(A \cap P)|.$$

Applying (5.13) we conclude

$$|2(A \cap P)| \leq K|A \cap P| + 2r\varepsilon|A|.$$

Combining this with (5.16), (5.15) and using the bound $r \leq K$ we see that

$$2^r \leq K + O_K(\varepsilon).$$

By choosing ε sufficiently small depending on $K < 2^d$ and d we obtain $r < d$ as desired. □

Exercises

5.2.1 [118] Show that Lemma 5.13 is still true if $A + A$ is replaced by $A - A$.

5.2.2 Let $d \geq 1$, $B := \{0, 1\}^d \subseteq \mathbf{R}^d$, and A be an additive subset of the convex hull of B (i.e. A lies in the solid unit cube $\{(x_1, \ldots, x_d) : 0 \leq x_1, \ldots, x_d \leq 1\}$. Show that

$$|A + B| \geq (\sqrt{2} - o_{d\to\infty}(1))^d |A|.$$

(Hint: reduce to the case where A is a subset of B, and then reduce further to the case where A consists of elements $(n_1, \ldots, n_d) \in \{0, 1\}^d$ where $n_1 + \cdots + n_d$ is fixed. Then restrict the elements of B in a similar manner and apply the covering principle and Stirling's formula (1.52). You may find working out the counterexample in the next exercise to be helpful.)

5.2.3 Show that the quantity $\sqrt{2}$ in Exercise 5.2.2 cannot be improved, by setting A equal to those elements $(n_1, \ldots, n_d) \in B$ such that $n_1 + \cdots + n_d = \lfloor \frac{d}{2} \rfloor$.

5.2.4 Let V be an r-dimensional vector space, and let P be a r-parallelepiped in V which is not contained in any hyperplane. Show that $V \backslash P$ is the union of $2r$ unbounded convex regions (not necessarily open).

5.3 Freiman homomorphisms

We now introduce the fundamental concept of a *Freiman homomorphism*, that allows us to transfer an additive problem in one group Z to another group Z' in a way which is more flexible than the usual algebraic notion of group homomorphism. Roughly speaking, the role of Freiman homomorphisms is to additive sets as group homomorphisms are to additive groups. To avoid confusion we shall often write additive sets A more fully as (A, Z), where Z is the ambient group of A.

Definition 5.21 (Freiman homomorphisms) Let $k \geq 1$, and let A, B be additive sets with ambient groups Z and W respectively. A *Freiman homomorphism* of order k ϕ from (A, Z) to (B, W) (or more succinctly from A to B) is a map $\phi : A \to B$ with the property that

$$a_1 + \cdots + a_k = a'_1 + \ldots + a'_k \implies \phi(a_1) + \cdots + \phi(a_k) = \phi(a'_1) + \cdots + \phi(a'_k)$$

for all $a_1, \ldots, a_k, a'_1, \ldots, a'_k$. If in addition there is an inverse map $\phi^{-1} : B \to A$ which is a Freiman homomorphism of order k from (B, W) to (A, Z), then we say that ϕ is a *Freiman isomorphism of order* k, and that (A, Z) and (B, W) are *Freiman isomorphic of order* k.

For an equivalent characterization of a Freiman isomorphism, see Exercise 5.3.1.

It is easy to verify that a Freiman homomorphism of order k will also be Freiman homomorphic of all orders $k' < k$. Of course it is the $k \geq 2$ cases that are interesting; any map from A to B will be Freiman homomorphic of order 1, and any bijection will be Freiman isomorphic of order 1. Also, the identity map id from (A, Z) to (A, Z) is always a Freiman isomorphism of any order, and the composition of two Freiman homomorphisms (resp. isomorphisms) of order k is another Freiman homomorphism (resp. isomorphism) of order k; in particular, the relation of being Freiman isomorphic of order k is an equivalence relation. Thus the class of additive sets, and the Freiman homomorphisms of a fixed order k between them, form a category.

Remark 5.22 We digress to give an analogy with the differential geometry of manifolds. Manifolds can either be viewed extrinsically (embedded inside an ambient space such as a Euclidean space $\mathbf{R}^d$) or intrinsically (as a set endowed with certain structures such as a topology, Riemannian metric, etc.). One can easily get from the former viewpoint to the latter by restricting certain structures of the ambient space to the embedded set; reversing this procedure and embedding an intrinsic manifold inside a given ambient space is often much harder. Throughout this book we have taken the extrinsic approach, embedding the additive set A inside an ambient group Z. However one could also take a purely intrinsic viewpoint, fixing the order k of the Freiman homomorphism and viewing the additive set as $(A, \sim_k)$, where A is now thought of an abstract set (rather than a subset of an additive group) and $\sim_k$ is the equivalence relation on A^k defined (extrinsically) by setting $(a_1, \ldots, a_k) \sim_k (a'_1, \ldots, a'_k)$ if and only if $a_1 + \cdots + a_k = a'_1 + \cdots + a'_k$. This is still enough to develop the theory of Freiman homomorphism and isomorphisms, and one can define notions such as sum sets, additive energy, etc. in this intrinsic setting. However there do not appear to be any major advantages with this approach, especially since the embedding problem turns out to be relatively easy to solve (in contrast with the situation for, say, Riemannian manifolds). See Exercise 5.5.6 below.

We now give some examples of Freiman homomorphisms.

- If $\phi : Z \to Z'$ is a group homomorphism (resp. isomorphism) from one group Z to another Z', then it induces a Freiman homomorphism (resp. isomorphism) from (A, Z) to $(\phi(A), Z')$ of arbitrary order. In particular, the reflection map $\phi : Z \to Z$ defined by $\phi(x) := -x$ is a Freiman isomorphism from (A, Z) to $(-A, Z)$ of arbitrary order.
- If (A, Z) and (B, W) are two additive sets such that $Z \subseteq W$ and $A \subseteq B$, then the inclusion map $\iota : A \to B$ is a (rather trivial) Freiman homomorphism of arbitrary order. Thus, if $\phi : (B, W) \to (B', W')$ is any Freiman homomorphism, then the restriction $\phi|_A : (A, Z) \to (B', W')$ will be a Freiman homomorphism of the same order.
- If $x \in Z$, then the translation map $\phi : Z \to Z$ defined by $\phi(y) := y + x$ is a Freiman isomorphism from (A, Z) to $(A + x, Z)$ of any order.
- Let $N, M \geq 1$ be integers. Let $\phi : \mathbf{Z} \to \mathbf{Z}_M$ be the canonical quotient homomorphism, and let $\psi : [0, N) \to \phi([0, N))$ be the restriction of ϕ to $[0, N)$. Then ψ is a Freiman homomorphism of any order. But ψ is only a Freiman isomorphism of order k when $M \geq kN$, in which case ψ^{-1} is also a Freiman isomorphism. Thus it is possible to have a Freiman isomorphism between a set in a torsion-free group and a set in a torsion group, which would be impossible if one were only considering group homomorphisms.
- Let a, r be elements of an additive group Z, and let $P := a + [0, N) \cdot r$ be the arithmetic progression $P = \{a, a + r, \ldots, a + (N - 1)r\}$. Then the map $\phi : [0, N) \to P$ defined by $\phi(n) := a + nr$ is a Freiman homomorphism from $([0, N), \mathbf{Z})$ to (P, Z) of any order. It is a Freiman isomorphism of order k if and only if $\mathrm{ord}(r) \geq kN$. In particular, if r is non-zero and Z is torsion-free, then ϕ is a Freiman isomorphism of all orders.
- Let $N, M, d \geq 1$ be integers, and let $\phi : \mathbf{Z}^d \to \mathbf{Z}$ be the map $\phi(a_1, \ldots, a_d) := \sum_{j=1}^{d} a_j M^{j-1}$. Then the map ϕ is a Freiman homomorphism from $[0, N)^d$ to

$\phi([0, N)^d)$ of any order, and is a Freiman isomorphism of order k when $M \geq kN$.

- The sets $\{0, 1, 10, 11\}$ and $\{0, 1, 100, 101\}$ in $\mathbf{Z}$ are Freiman isomorphic of order k for any $k < 10$, but are not Freiman isomorphic of order k for any $k \geq 10$.

The relevance of Freiman homomorphisms to the theory of sum sets lies in the following lemma:

Lemma 5.23 *Let (A, G) be an additive set, and let $\phi : (A, G) \to (\phi(A), H)$ be a surjective Freiman homomorphism of order k. Then we have*

$$|\varepsilon_1\phi(A_1) + \cdots + \varepsilon_k\phi(A_k)| \leq |\varepsilon_1 A_1 + \cdots + \varepsilon_k A_k|$$

whenever $A_1, \ldots, A_k$ are non-empty subsets of A and $\varepsilon_1, \ldots, \varepsilon_k = \pm 1$. If ϕ is in fact a Freiman isomorphism of order k, then we may replace inequality with equality. In particular, if A and B are Freiman isomorphic of order k, then

$$|lB - mB| = |lA - mA| \text{ whenever } l, m \geq 0 \text{ and } l + m \leq k.$$

Proof Define an equivalence relation $\sim$ on $A_1 \times \cdots \times A_k$ by by declaring

$$(a_1, \ldots, a_k) \sim (a'_1, \ldots, a'_k) \iff \varepsilon_1 a_1 + \cdots + \varepsilon_k a_k = \varepsilon_1 a'_1 + \cdots + \varepsilon_k a'_k.$$

Observe that the number of equivalence classes in $A_1 \times \cdots \times A_k$ is precisely $|\varepsilon_1 A_1 + \cdots + \varepsilon_k A_k|$. Also observe that we can rewrite the above condition

$$\varepsilon_1 a_1 + \cdots + \varepsilon_k a_k = \varepsilon_1 a'_1 + \cdots + \varepsilon_k a'_k$$

in a positive form as

$$\sum_{j:\varepsilon_j=1} a_j + \sum_{j:\varepsilon_j=-1} a'_j = \sum_{j:\varepsilon_j=1} a'_j + \sum_{j:\varepsilon_j=-1} a_j.$$

From this it is clear that the equivalence relation is respected by any Freiman homomorphism of order k. Combining these observations yields the lemma. □

Thus Freiman isomorphisms will preserve the cardinality of iterated sum and difference sets (as well as related quantities such as the doubling constant, difference constant, and energy); see Exercise 5.3.5. Of course, in many applications one wants to take sum sets involving *two* additive sets A, B in an ambient group Z rather than one. One way to resolve this is to work with the union $A \cup B$, since Lemma 5.23 then shows that Freiman isomorphisms of $A \cup B$ will preserve the cardinality of sets such as $A + B$ or $A - B$ (if the order of the isomorphism is at least 2). But this has the slight drawback that one loses the freedom to translate A and B independently. One way to get around this is to define the *disjoint union* $A \uplus B$ of A and B, defined in the ambient group $Z \times \mathbf{Z}$ as

$$A \uplus B := (A \times \{0\}) \cup (B \times \{1\}).$$

Then any Freiman isomorphism of the disjoint union will preserve sum sets (see Exercise 5.3.7). Note that the obvious projection map from $A \uplus B$ to $A \cup B$ is a Freiman homomorphism of any order.

Freiman homomorphisms also preserve the property of being a progression:

Proposition 5.24 *Let $\phi : A \to B$ be a Freiman homomorphism of order at least 2, and let $P = a + [0, N] \cdot v$ be a progression in A. Then $\phi(P)$ is a progression in B with the same rank, dimensions, and volume as P. Furthermore, if ϕ is in fact a Freiman* isomorphism *of order at least 2, then $\phi(P)$ is proper if and only if P is proper.*

Proof We may assume that the components N_j of N are all strictly positive, since if one of the components N_j is zero then we can simply remove it and lower the rank by 1. By translation invariance we may suppose that the base point a is equal to 0, and that $\phi(0)$ is also zero. In particular P, and thus A, contains all the basis vectors $v_1, \ldots, v_d$.

Since ϕ is a Freiman homomorphism of order 2 and $\phi(0) = 0$, we see that $\phi(x + v_j) = \phi(x) + \phi(v_j)$ whenever x and $x + v_j$ both lie in A and $1 \leq j \leq d$. Iterating this we see from induction that $\phi(n \cdot v) = n \cdot \phi(v)$ for any $n \in [0, N]$, where $\phi(v) \in B^{\oplus d}$ is the d-tuple $\phi(v) := (\phi(v_1), \ldots, \phi(v_d))$. Thus $\phi(P) = [0, N] \cdot \phi(v)$ and is thus a progression with the same rank, dimensions, and volume as P. To prove the last part of the proposition, observe that if ϕ is a Freiman isomorphism then $|P| = |\phi(P)|$, and hence $|P| = |[0, N]|$ if and only if $|\phi(P)| = |[0, N]|$. □

We now show that torsion-free additive groups are no richer than the integers, for the purposes of understanding sums and differences of finite sets.

Lemma 5.25 *Let A be a finite subset of a torsion-free additive group Z. Then for any integer k, there is a Freiman isomorphism $\phi : A \to \phi(A)$ of order k to some finite subset $\phi(A)$ of the integers $\mathbf{Z}$. The same is true if we replace $\mathbf{Z}$ by $\mathbf{Z}_N$, if N is sufficiently large depending on A.*

Note that the converse is trivial: one can always embed the integers in any other torsion-free additive group, and hence any additive set in the integers can be embedded in any other torsion-free additive group such as $\mathbf{R}^d$. However, many of these embeddings are trivial, living in some subspace of $\mathbf{R}^d$. The question of the largest dimension one can "non-trivially" embed an additive set in will lead to the concept of *Freiman dimension*, which we shall study in Section 5.5.

Proof By Corollary 3.6 we may take $Z = \mathbf{Z}^n$ for some $n \geq 0$. By translating A we may assume that A in fact lives in $(\mathbf{Z}^+)^n$, i.e. all the coordinates are non-negative. Since A is finite, we see that A is a subset of $[0, M/k)^n$ for some large integer M (a multiple of k). Now define the map $\phi : A \to \mathbf{Z}$ by

$$\phi(a_1, \ldots, a_n) := a_1 + a_2 M + a_3 M^2 + \cdots + a_n M^{n-1}.$$

In other words, we view elements of A as digit strings of integers base M. This is a Freiman isomorphism of order k (with ϕ_k being defined the same way as ϕ, but restricted to kA); the point is that if M is large enough we never have to "carry" a digit. This shows that we can map A to the integers via a Freiman isomorphism; the same argument shows that we can map to $\mathbf{Z}/(N \cdot \mathbf{Z})$ if $N \geq M^n$. □

As we shall see later, the machinery of Freiman homomorphisms and Freiman isomorphisms will also be very useful when dealing with torsion groups, for instance we can use it to pass from a problem on the integers to a problem on a cyclic group or vice versa. If one is willing to only work with a fixed fraction of an additive set A, then the following compression lemma allows one to work in a cyclic group whose order is only a little bit larger than that of A itself.

Lemma 5.26 *[295] Let A be an additive set whose ambient group Z is either torsion-free or a cyclic group of prime order, and let $n \geq 1$ be a positive integer. Let N be an integer such that*

$$2n|nA - nA| < N < |Z|$$

(note the condition $N < |Z|$ is vacuous if Z is torsion-free). Then there exists a subset $A' \subseteq A$ of cardinality $|A'| \geq |A|/n$ and a Freiman isomorphism $\pi : A' \to B$ from A' to a subset $B \subseteq \mathbf{Z}_N$ of order n.

Proof By Lemma 5.25 it suffices to consider the case where Z is a cyclic group $\mathbf{Z}_p$ of prime order.

We shall use the first moment method. Let $\lambda \in \mathbf{Z}_p \backslash \{0\}$ be an invertible element of $\mathbf{Z}_p$ chosen uniformly at random. The map $x \mapsto \lambda \cdot x$ is thus an additive group isomorphism on $\mathbf{Z}_p$, and is in particular a Freiman isomorphism on $\mathbf{Z}_p$ of all orders. This freedom to dilate A by an arbitrary amount will be needed to avoid a certain "collision" problem which will become apparent shortly.

We now define the projection $\pi : \mathbf{Z}_p \to \mathbf{Z}_N$ by setting

$$\pi(m) := \iota(m) \bmod N,$$

where $\iota : \mathbf{Z}_p \to [0, p)$ is the obvious map that sends the residue class $m + (p \cdot \mathbf{Z})$ to m for $m = 0, \ldots, p - 1$.

The map π is not quite an additive homomorphism; however note, for $j = 0, 1, \ldots, n - 1$, that π is a Freiman homomorphism of order n when restricted to the set $Z_j := (jp/n, (j + 1)p/n]$, which is a set that occupies roughly $\frac{1}{n}$ of the original field $\mathbf{Z}_p$. By the pigeonhole principle, for each λ, there exists a $0 \leq j =$

$j(\lambda) < n$ such that the set $A' := \lambda \cdot A \cap Z_j$ has cardinality $|A'| \geq |A|/n$. Thus if we set $B := \pi(A') \subseteq \mathbf{Z}_N$, then the map $\pi : A' \to B$ is a surjective Freiman homomorphism of order n.

We are almost done; however we have not established that π is a Freiman *iso*morphism. The only possible obstruction is that there may be collisions in nA', in the sense that

$$\pi(x_1) + \cdots + \pi(x_n) = \pi(x_1') + \cdots + \pi(x_n')$$

while $x_1 + \cdots + x_n \neq x_1' + \cdots + x_n'$, for some $x_1, \ldots, x_n, x_1', \ldots, x_n' \in A'$. Fortunately, this type of collision rarely occurs, if N is large enough and λ is chosen randomly. Indeed, if we do have the above collision, then we see that

$$\iota(x_1) + \cdots + \iota(x_n) - (\iota(x_1') + \cdots + \iota(x_n'))$$

must be a non-zero multiple of N. Since $x_1, \ldots, x_n, x_1', \ldots, x_n'$ lie in A', and hence in λA, we thus see that a collision can only occur if $n\iota(\lambda A) - n\iota(\lambda A)$ contains a non-zero multiple of N. However, we can compute the probability that this occurs:

$$\begin{aligned}
&\mathbf{P}(\exists k \in \mathbf{Z}\backslash 0 : kN \in n\iota(\lambda A) - n\iota(\lambda A)) \\
&\quad \leq \sum_{|k| \leq np/N; k \neq 0} \mathbf{P}(kN \in n\iota(\lambda A) - n\iota(\lambda A)) \\
&\quad \leq \sum_{|k| \leq np/N; k \neq 0} \mathbf{P}(kN + p \cdot \mathbf{Z} \in n\lambda A - n\lambda A) \\
&\quad = \sum_{|k| \leq np/N; k \neq 0} \sum_{x \in nA - nA} \mathbf{P}(kN = \lambda x \bmod p) \\
&\quad = \sum_{|k| \leq np/N; k \neq 0} \sum_{x \in nA - nA} \mathbf{P}(\lambda = (kN)^{-1} x \bmod p) \\
&\quad \leq \sum_{|k| \leq np/N; k \neq 0} \sum_{x \in nA - nA} \frac{1}{p-1} \\
&\quad \leq \frac{2np}{N} |nA - nA| \frac{1}{p-1},
\end{aligned}$$

where we have used the fact that p is prime (to invert kN modulo p). By our hypotheses on N we thus see that this probability is strictly less than 1. Thus we may choose λ so that $\pi : A' \to B$ will be a Freiman isomorphism of order n as claimed. $\square$

The above argument should be compared with the proof of Theorem 1.3.

Exercises

5.3.1 Let $\phi : A \to B$ be a map between two additive sets, and let $k \geq 1$. Show that ϕ is a Freiman isomorphism of order k if and only if ϕ is surjective

and

$$a_1 + \cdots + a_k = a_1' + \cdots + a_k' \iff \phi(a_1) + \cdots + \phi(a_k) = \phi(a_1') + \cdots + \phi(a_k')$$

for all $a_1, \ldots, a_k, a_1', \ldots, a_k' \in A$.

5.3.2 [257] Let $n > 1$. Show that $\{0, 1, n+1\}$ is Freiman isomorphic to $\{0, 1, n\}$ of order n but not $n+1$.

5.3.3 Show that given any $k \geq 1$ and any additive set A, that A is Freiman isomorphic of order k to some subset of a finite abelian group.

5.3.4 Let (A, Z) and (B, W) be additive sets, and let $\phi : A \to B$ be a map which is a Freiman homomorphism of any order k. Suppose also that Z is the group generated by A. Show that there exists a unique group homomorphism $\psi : Z \to W$ and an element $c \in Z'$ such that $\phi(x) = \psi(x) + c$ for all $x \in A$.

5.3.5 Let (A, Z) and (B, W) be Freiman isomorphic of order at least 2. Show that $\sigma[A] = \sigma[B]$, that $\delta[A] = \delta[B]$, and that $E(A, A) = E(B, B)$. For any $\alpha \in \mathbf{R}$, show that $|\mathrm{Sym}_\alpha(A)| = |\mathrm{Sym}_\alpha(B)|$. (See Definitions 2.4, 2.8, 2.32 for the meanings of these terms.)

5.3.6 Let (A, Z) and (B, W) be additive sets which contain the origin 0, and let $\phi : (A, Z) \to (B, W)$ be a Freiman isomorphism of order at least 3 which fixes the origin, thus $\phi(0) = 0$. Show that for any $K \geq 1$, that A is a K-approximate group if and only if B is. Show that if one replaces "K-approximate group" by "translate of a K-approximate group" then one can drop the requirement that $\phi(0) = 0$ and that A, B contain 0.

5.3.7 Let (A, Z), (B, Z), (A', Z'), (B', Z') be additive sets, and suppose that $\phi : A \uplus B \to A' \uplus B'$ is a Freiman isomorphism of order k which maps A to A' and B to B'. Show that $|n_1 A - n_2 A + n_3 B - n_4 B| = |n_1 A' - n_2 A' + n_3 B' - n_4 B'|$ whenever $|n_1| + |n_2| + |n_3| + |n_4| \leq k$. If $k \geq 2$, show that $d(A, B) = d(A', B')$ and $E(A, B) = E(A', B')$. Also, show that A can be covered by K translates of B if and only if A' can be covered by K' translates of B'.

5.3.8 Suppose that two additive sets A and B are Freiman isomorphic of order k. If $n, m, k' \geq 0$ are such that $k'(n+m) \leq k$, show that $nA - mA$ and $nB - mB$ are Freiman isomorphic of order k'.

5.3.9 Show that all Sidon sets of a fixed cardinality N are Freiman isomorphic of order 2 to each other. More generally, for any $h \geq 2$, show that all B_h sets of cardinality N are Freiman isomorphic to each other of order h, and that the image of a B_h set under a Freiman isomorphism is still a B_h set. Thus one could work with a "standard" B_h set of order N, such as the basis $e_1, \ldots, e_N$ of $\mathbf{Z}^N$, and many additive results concerning that standard set would automatically transfer over to an arbitrary B_h set.

5.3.10 Let (A, Z) and (A', Z') be additive sets in finite additive groups Z, Z' which are Freiman isomorphic of order h for some $h \geq 1$. Show that $\|A\|_{\Lambda(2h)} = \|A'\|_{\Lambda(2h)}$, where the $\Lambda(p)$ constants are as in Definition 4.26.

5.3.11 [29] Let p be a prime, let $k \geq 1$, and let $(A, \mathbf{Z}_p)$ be an additive set in $\mathbf{Z}_p$ such that $|A| \leq \log_{2k} p$. Show that there exists an additive set $(A', \mathbf{Z})$ such that the canonical projection map from $\mathbf{Z}$ to $\mathbf{Z}_p$ is a Freiman isomorphism of order k from A' to A. (Hint: the claim is obvious if A is contained in the arithmetic progression $[-p/2k, p/2k] \cdot 1$ in $\mathbf{Z}_p$. For the general case, use the Kronecker approximation theorem (Corollary 3.25) to locate an integer n coprime to p such that $n \cdot A$ lies in this progression $[-p/2k, p/2k] \cdot 1$, and then find an integer m with $nm = 1 \pmod{p}$ to "invert" the dilation $x \mapsto n \cdot x$.)

5.3.12 [29] Let p be a prime, written in binary as $p = 2^{n_1} + \cdots + 2^{n_r}$ where $n_1 < \cdots < n_r$. Let $(A, \mathbf{Z}_p)$ be the additive set

$$A := \{0\} \cup \{1, 2^1, \ldots, 2^{n_r+1}\} \cup \{2^{n_1} + \cdots + 2^{n_j} : 1 \leq j \leq r\}.$$

Show that $|A| \leq 2\log_2 p + 1$, but there does not exist any set of integers A' which is Freiman isomorphic of order 2 to A. This shows that the estimate $|A| \leq \log_{2k} p$ in Exercise 5.3.11 is very close to being sharp.

5.3.13 Let (A, Z), (B, Z) be additive sets such that $A + B$ can be covered by K translates of A for some $K \geq 1$, and let $\phi : A \uplus B \to C$ be a Freiman homomorphism of order 4. Show that $\phi(A) + \phi(B)$ can be covered by K translates of $\phi(A)$.

5.3.14 Let Q be a progression of rank d, let $k \geq 1$, and let $N \geq k^d|Q|$. Show that there exists an additive set $(Q', \mathbf{Z}_N)$ in the cyclic group $\mathbf{Z}_N$ and a surjective Freiman homomorphism $\phi : Q' \to Q$ of order k. If Q is proper, one can also ensure that ϕ is injective. This fact is useful for viewing progressions as dense subsets of cyclic groups.

5.4 Torsion and torsion-free inverse theorems

We can now use all the machinery developed thus far to prove two inverse sum set theorems, one in the setting of r-torsion groups and one in the setting of torsion-free groups. The two arguments are quite different, but they will be combined to obtain an inverse sum set theorem for an arbitrary group in Section 5.6.

We begin with the r-torsion case.

Theorem 5.27 (Freiman theorem for r-torsion groups) *[300], [154] Suppose A is an additive set in an r-torsion group Z such that $|A + A| \leq K|A|$ or*

$|A - A| \leq K|A|$. *Then there exists a subgroup* H *of* Z *of cardinality* $|A| \leq |H| \leq r^{K^{O(1)}}|A|$ *such that* A *is contained in a translate of* H.

Proof By Proposition 2.26 we can find a $K^{O(1)}$-approximate group H such that A is contained in a translate of H. But then $H \pm H \subseteq H + X$ for some additive set X of cardinality at most $K^{O(1)}$. We conclude that the set $G := H + \langle X \rangle$ is a genuine group, where $\langle X \rangle$ is the group generated by X. But from the r-torsion hypothesis we have $|\langle X \rangle| \leq r^{|X|} \leq r^{K^{O(1)}}$, and the claim follows. □

Remark 5.28 The upper bound on $|G|$ has been improved to r^{2K^2-1} in [154], using the Green–Ruzsa covering lemma and the Plünnecke inequalities; see Exercise 5.4.1. The exponential dependence in K here is necessary, as the example $Z = \mathbf{Z}_r^K$, $A = \{e_1, \ldots, e_K\}$ shows. However if one relaxes the claim that A is completely contained in a translate of H then one should do better. For instance, it is conjectured by Marton [300] that in the above setting we can in fact find a group $H \subseteq Z$ of cardinality at most $|A|$ such that A can be covered by $O(K^{O_r(1)})$ translates of H. This would be sharp up to polynomial losses, since in that case one can easily verify that $|A + A|, |A - A| = O(K^{O_r(1)}|A|)$.

As a corollary we can also obtain a Chang-type theorem in the r-torsion case.

Corollary 5.29 (Chang theorem for r-torsion groups) *Suppose* A *is an additive set in an* r*-torsion group* Z *such that* $E(A, A) \geq |A|^3/K$. *Then* $2A - 2A$ *contains a subgroup of* Z *of cardinality at least* $r^{-O(K^{O(1)})}|A|$.

Proof We may take $r \geq 2$ as the case $r = 1$ is trivial. Using the Balog–Szemerédi–Gowers theorem (Theorem 2.31) and translating A if necessary, we may find a subset A' of A with $|A'| = \Omega(K^{-O(1)}|A|)$ which is contained in a $K^{O(1)}$-approximate group G of size $|G| = O(K^{O(1)}|A|)$. Using Theorem 5.27 we may place the approximate group G inside a genuine group H of cardinality at most $r^{K^{O(1)}}|A|$; thus $\mathbf{P}_H(A') \geq r^{-K^{O(1)}}$. By Proposition 4.39, we thus see that $2A' - 2A'$ contains a Bohr set $\mathrm{Bohr}_H(\mathrm{Spec}_\alpha(A'), \frac{1}{6})$ for some $\alpha = \Omega(K^{-O(1)})$. Using Lemma 4.36 as in the proof of Theorem 4.42, we conclude that $2A' - 2A'$ (and hence $2A - 2A$) contains a Bohr set $\mathrm{Bohr}_H(S, \frac{1}{6|S|})$ for some set of frequencies $S \subset H$ with $|S| = O(K^{O(1)})$. In particular, it contains the subgroup $\mathrm{Bohr}_H(S, 0)$. But as H is an r-torsion group, $\mathrm{Bohr}_H(S, 0) = \mathrm{Bohr}_H(S, 1/r)$, and so from (4.25) we see that

$$\begin{aligned}|\mathrm{Bohr}_H(S, 0)| &\geq r^{-O(K^{O(1)})}|H| \\ &\geq r^{-O(K^{O(1)})}|A'| \\ &= \Omega\big(r^{-O(K^{O(1)})}K^{-O(1)}|A|\big)\end{aligned}$$

and the claim follows (using the hypothesis $r \geq 2$ to absorb the lower order terms). □

We now turn to the torsion-free case. We begin with two preliminary results of interest in their own right. The first exploits all the above machinery of Freiman homomorphisms, as well as the powerful techniques of harmonic analysis from Chapter 4 and the additive geometry results in Chapter 3 (as encapsulated in Theorem 4.42), to show that if A has small doubling, then $2A - 2A$ contains a large proper progression.

Theorem 5.30 (Ruzsa–Chang theorem) *[295], [48] Let A be an additive set in a torsion-free additive group Z such that $|A + A| \leq K|A|$ for some $K \geq 1$. Then $2A - 2A$ contains a proper symmetric progression P of rank $O(K(1 + \log K))$ such that $|P| \geq e^{-O(K(1+\log^2 K))}|A|$.*

Proof Let p be the first prime number larger than $16|8A - 8A|$. By Lemma 5.26 and Bertrand's postulate (Exercise 1.10.3) one can then find a subset A' of A of cardinality $|A'| \geq |A|/8$, which is Freiman isomorphic of order 8 to an additive set B in $\mathbf{Z}_p$. Observe that

$$|B + B| = |A' + A'| \leq |A + A| \leq K|A| \leq 8K|B|$$

so B has doubling constant at most $8K$. Applying Theorem 4.42 we then obtain a proper symmetric progression Q inside $2B - 2B$ of rank at most $O(K(1 + \log K))$ and cardinality at least $O(K(1 + \log K))^{-O(K(1+\log K))}|B|$. In particular we have

$$|Q| \geq e^{-O(K \log^2 K)}|B|.$$

Since A' is Freiman isomorphic to B of order 8, $2A' - 2A'$ is Freiman isomorphic to $2B - 2B$ of order 2 (see Exercise 5.3.8). $2A - 2A$, contains a symmetric progression P which is Freiman isomorphic to Q, and the claim follows. □

The second result is a variant of the Ruzsa covering lemma which gives good constants when the doubling constant is small.

Lemma 5.31 (Chang's covering lemma) *[48] Let $K, K' \geq 1$, and let A, B be additive sets in an ambient group Z such that $|nA| \leq K^n|A|$ for all $n \geq 1$, and such that $|A + B| \leq K'|B|$. Then, for any $a_0 \in A$, there exists elements $v_1, \ldots, v_d$ in $A - A$ with $d = 4K(1 + \log_2(KK'))$ such that $A \subseteq B - B + [0, 1]^d \cdot (v_1, \ldots, v_d) + a_0$.*

Proof Without loss of generality we may take K to be an integer. By translation we may take $a_0 = 0$. We construct a sequence of enlargements $B = B_0 \subseteq B_1 \subseteq \cdots \subseteq B_N$ by iterating the argument of Lemma 2.14 as follows. Set $B_0 := B$. Now suppose inductively that $n \geq 0$ and B_n has already been constructed. Consider the collection $\{a + B_n : a \in A\}$ of translates of B_n by elements of A. If we can find at least $2K$ such translates which are disjoint, we set B_{n+1} to be the union of these $2K$

translates; thus $B_{n+1} = B_n + A_n$ for some subset A_n of A of cardinality $2K$, and $|B_{n+1}| = 2K|B_n|$, and then continue the algorithm. If we cannot find $2K$ disjoint translates, we select a family of disjoint translates of maximal cardinality, set B_{n+1} to be the union of these translates, and then halt the algorithm setting $N := n + 1$. Thus in the terminating case we have $B_{n+1} = B_n + A_n$, where A_n is a subset of A of cardinality less than $2K$.

Let us first see why this algorithm even terminates. By induction we see that $B_n \subseteq B + nA$ for all $0 \leq n < N$, but we also have $|B_n| = (2K)^n|B|$. On the other hand, from Lemma 2.6, we have

$$|B + nA| \leq \frac{|B - A||A + nA|}{|A|} \leq K'K^{n+1}|B|.$$

Thus the algorithm must terminate by the time $(2K)^n$ exceeds $K'K^{n+1}$, and we therefore have the bound $N \leq 1 + \log_2(KK')$.

Now let a be any element of A. Observe that $B_{N-1} + a$ cannot be disjoint from B_N, since otherwise we could have added it to the collection of disjoint translates comprising B_N. Thus $a \in B_N - B_{N-1}$ for all $a \in A$, and hence

$$A \subseteq B_N - B_{N-1} = B - B + A_0 - A_0 + A_1 - A_1 + \cdots + A_{N-1} - A_{N-1} + A_N.$$

By Lemma 3.11, we see that each of the A_j (or $-A_j$) can be contained in a progression of the form $[0, 1]^{d_j} \cdot v$ for some $d_j \leq 2K$, where the components of v lie in A_j and hence in $A - A$ (since $0 \in A$ and $A_j \subseteq A$). The claim then follows from several applications of (3.2). □

As a consequence of these two results we obtain an inverse theorem in the torsion-free case.

Theorem 5.32 (Freiman's theorem for torsion-free groups) *[116], [295], [48] Let A be an additive set in a torsion-free group Z such that $|A + A| \leq K|A|$. Let $a_0 \in A$. Then there exists a proper progression P contained in $2A - 2A$ of rank at most $O(K(1 + \log K))$ and cardinality at most $|P| \leq |2A - 2A| \leq K^{O(1)}|A|$, and vectors $v_1, \ldots, v_d$ in $4A - 4A$ with $d = O(K^{O(1)})$, such that $A \subseteq P + [0, 1]^d \cdot (v_1, \ldots, v_d) + a_0$.*

Proof By translation we may assume that $a = 0$, so $0 \in A$. Applying Theorem 5.30 we see that $2A - 2A$ contains a proper progression P of rank at most $CK(1 + \log K)$ and cardinality at least $e^{-O(K(1+\log^2 K))}|A|$. Note from Corollary 2.23 that $|P| \leq |2A - 2A| \leq K^{O(1)}|A|$. Now we use Lemma 5.31 to cover A by $P - P$. First from Corollary 2.23 note that

$$|A + P| \leq |3A - 2A| \leq K^{O(1)}|A| \leq e^{O(K(1+\log^2 K))}|P|$$

and that $|nA| \leq K^{O(n)}|A|$ for all $n \geq 1$. Thus by Lemma 5.31 (and the remarks immediately following that lemma) we have

$$A \subseteq P - P + [0, 1]^d \cdot (v_1, \ldots, v_d)$$

for some $v_1, \ldots, v_d \in A - A$ and $d = O(K^{O(1)})$. Also, from Lemma 3.10 we have $P - P \subseteq P + [0, 1]^{d'} \cdot (w_1, \ldots, w_{d'})$ where $d' = O(K(1 + \log K))$ is the rank of P and $w_1, \ldots, w_{d'} \in P - P \subseteq 4A - 4A$. Combining these facts using (3.2) we obtain the result. □

One can reduce the rank of the containing progression to $K - 1$, at the cost of worsening the size of $|P|$:

Theorem 5.33 *[48] Let A be an additive set in a torsion-free group Z such that $|A + A| \leq K|A|$ and $|A| \geq 100K^2$. Then there exists a proper progression P of rank at most $K - 1$ which contains A such that $|P| \leq \exp(O(K^{O(1)}))|A|$.*

Proof Without loss of generality we may assume that A contains the origin, and then we may assume that Z is generated by A otherwise we could pass from Z to the group $\langle A \rangle$ generated by A. From Theorem 5.32 and (3.2) we can contain A inside a progression Q of rank $d = O(K^{O(1)})$ and cardinality at most $\exp(O(K^{O(1)}))|A|$. Now consider the progression $2Q - 2Q$, which has the same rank as Q and essentially the same bounds on the cardinality. By Theorem 3.40 we can find a symmetric proper progression $R = [-N, N] \cdot v$ of some rank $d' \leq d$ containing $2Q - 2Q$ such that $|R| \leq \exp(O(K^{O(1)}))|A|$. In particular, the set A (which is contained inside $Q - Q$) is Freiman isomorphic of order 2 to a subset $\tilde{A}$ of $[-N, N] \subset \mathbf{Z}^{d'}$; thus $\tilde{A}$ has doubling constant at most K. By Freiman's lemma (Lemma 5.13) we may place $\tilde{A}$ in a subspace V of $\mathbf{Z}^{d'}$ of dimension at most $K - 1$.

We now use the "rank reduction argument". If $d' \leq K - 1$ then we are done (by setting $P = R$), so suppose $d' > K - 1$. The intersection of $[-N, N] \subset \mathbf{Z}^{d'}$ with V is the intersection of a convex subset with a lattice of rank strictly less than d' with cardinality at most $\exp(O(K^{O(1)}))|A|$, so by Lemma 3.36 we may contain it in a progression of rank strictly less than d' and cardinality at most $\exp(O(K^{O(1)}))|A|$, with steps inside $[-N, N]$. Using the Freiman isomorphism, this allows us to contain A in a progression Q' of rank strictly less than d and cardinality at most $\exp(O(K^{O(1)}))|A|$. We then iterate the above argument (replacing Q by Q') at most d times until one can contain A in a progression P of length $K - 1$. As the rank decreases at each stage it is easy to see that the final progression P will have size at most $\exp(O(K^{O(1)}))$. □

The exponential factors in Theorem 5.33 cannot be removed directly, as can be seen by considering the additive set $Z = \{e_1, \ldots, e_K\}$ in $\mathbf{Z}^K$. However it is conjectured that if one weakens the containment $A \subseteq P$ then one can do better, for instance

Conjecture 5.34 (Polynomial Freiman–Ruzsa conjecture) *Let A be an additive set in a torsion-free group Z such that $|A + A| \leq K|A|$. Then there exists a progression P of rank at most $O(K^{O(1)})$ such that $|P| = O(K^{O(1)}|A|)$ and $|A \cap P| = \Omega(K^{-O(1)}|A|)$.*

This would be the analog of Marton's conjecture mentioned earlier in this section. Such a conjecture, if true, would allow one to obtain substantially better bounds on many results whose proof involves Freiman's theorem. See [151], [152] for further discussion.

By combining Theorem 5.33 with Theorem 5.20 one can show

Proposition 5.35 *[28] Let A be an additive set in a torsion-free group Z such that $|A + A| \leq K|A|$ for some $K < 2^d$. Then there exists a proper progression P of rank at most d and size $|P| = \Theta_{K,d}(|A|)$ such that $|A \cap P| = \Theta_{K,d}(|A|)$.*

We leave the deduction of this proposition from the previous results to Exercise 5.4.5. Recently, a more quantitative version of this proposition was obtained:

Proposition 5.36 *[162] Let A be an additive set in a torsion-free group Z such that $|A + A| \leq K|A|$. Then for any $0 < \varepsilon \leq 1$ there exists a proper progression P of rank at most $\lfloor \log_2 K + \varepsilon \rfloor$ and size at most $|A|$ such that A is covered by $\exp(O(K^3 \log^3 K))/\varepsilon^{O(K)}$ translates of P.*

Exercises

5.4.1 [154] Using Lemma 2.17 and Corollary 6.28, improve the factor of $r^{K^{O(1)}}$ in Theorem 5.27 to r^{2K^2-1}.

5.4.2 Show that the term $(d+1)|A| - \frac{d(d+1)}{2}$ in Corollary 5.13 cannot be replaced by any smaller quantity.

5.4.3 Using Corollary 6.28, improve the bounds in Theorem 5.32 and Theorem 5.33 as much as you can.

5.4.4 [300], [151] Let Z, Z' be two r-torsion groups, let $K \geq 1$, and let $f : Z \to Z'$ be a function which is a "K-almost homomorphism" in the sense that the set $\{f(x+y) - f(x) - f(y) : x, y \in Z\}$ has cardinality at most K. Show that there exists a genuine group homomorphism $g : Z \to Z'$ such at $\{f(x) - g(x) : x \in Z\}$ has cardinality at most r^K. It is conjectured that one can improve r^K to $O_r(K^{O_r(1)})$; this would essentially imply Marton's conjecture. See [151], [152] for further discussion.

5.4.5 Prove Proposition 5.35. (In addition to Theorem 5.33 and Theorem 5.20, you may use the rank reduction argument as in the proof of Theorem 5.33.)

5.4.6 Let A be a bounded non-empty open set in $\mathbf{R}^d$ such that $\text{mes}(A + A) \leq K\text{mes}(A)$. Show that $K \geq 2^d$, and that one has the containment $A \subseteq B + P$, where B is a ball and P is a progression of rank $O(K^{O(1)})$ and volume $O(\exp(K^{O(1)})\text{mes}(A)/\text{mes}(B))$. (Hint: take B to be a ball contained in A. Now replace $\mathbf{R}^d$ with a lattice adapted to the scale of B.)

5.5 Universal ambient groups

In this section we fix the order k of Freiman homomorphisms and isomorphisms, and shall frequently omit the phrase "of order k".

It is possible for two additive sets to be Freiman isomorphic even though their ambient groups are very different. For instance, the additive sets $(\{1, 2, 3\}, \mathbf{Z}_6)$, $(\{1, 2, 3\}, \mathbf{Z}_7)$, and $(\{1, 2, 3\}, \mathbf{Z})$ are all Freiman isomorphic of order 2, despite the groups $\mathbf{Z}_6, \mathbf{Z}_7, \mathbf{Z}$ being different. On the other hand, the additive set $(\{1, 2, 3\}, \mathbf{Z}_3)$ is not Freiman isomorphic of order 2 to any of the above sets and has a quite different additive structure. It is natural to ask whether there is some universal ambient group that one can place an additive set in, after Freiman isomorphism. To phrase this more precisely, we introduce

Definition 5.37 (Universal ambient group) Let (A, Z) be an additive set, and let the order k of the Freiman homomorphisms be fixed. We say that Z is a *universal ambient group* (of order k) for the additive set A if, every Freiman homomorphism $\phi : (A, Z) \to (B, W)$ has a unique extension to a group homomorphism $\phi_{\text{ext}} : Z \to W$ (thus $\phi_{\text{ext}}(x) = \phi(x)$ for all $x \in A$). More generally, we say that an additive group Z' is a universal ambient group for (A, Z) if there exists an additive set (A', Z') which is Freiman isomorphic to (A, Z) such that Z' is a universal ambient group for A'; we then call (A', Z') an *embedding* of (A, Z) inside the ambient group Z'.

Examples 5.38 Let $k = 2$, and consider the additive set $(A, Z) = (\{1, 2, 3\}, \mathbf{Z}_7)$. The group $\mathbf{Z}_7$ is not a universal ambient group for $A = \{1, 2, 3\}$, as can be seen for instance by considering the Freiman homomorphism $\phi : A \to \mathbf{Z}$ defined by $\phi(1) = 1$, $\phi(2) = 2$, $\phi(3) = 3$. This homomorphism cannot extend to a group homomorphism on $\mathbf{Z}_7$, since 1 has order 7 in $\mathbf{Z}_7$ but has infinite order in $\mathbf{Z}$. Even if one replaces the ambient group $\mathbf{Z}_7$ with $\mathbf{Z}$, the additive set $(\{1, 2, 3\}, \mathbf{Z})$ is still not placed inside a universal ambient group, because the translation map $\phi(x) := x + 1$ is a Freiman homomorphism on $\{1, 2, 3\}$ but does not extend to a group homomorphism on $\mathbf{Z}$. On the other hand, the additive set $(\{(1, 1), (2, 1), (3, 1)\}, \mathbf{Z}^2)$ *is* placed inside a universal ambient group, as one can easily verify. But the additive set $(\{(1,1,0),(2,1,0),(3,1,0)\}, \mathbf{Z}^3)$ is not placed inside a universal ambient group for a different reason, namely that the extension of Freiman homomorphisms to group homomorphisms is not unique (one has too much freedom to decide what to do with the third coordinate).

As stated, the definition of a universal ambient group is invariant under Freiman isomorphism. Also, if an additive set A has two universal ambient groups Z and Z', then they are necessarily group isomorphic (as can be seen by extending the obvious Freiman isomorphism between the two associated embeddings of A). Thus universal ambient groups, if they exist, are unique up to group isomorphism (for fixed k).

Lev and Konyagin [232] observed that universal ambient groups always exist:

Theorem 5.39 (Existence of universal ambient groups) *[232] Fix $k \geq 2$, and let (A, Z) be an additive set. Then there exists a universal ambient group Z' for A. Furthermore, if A' is an embedding of A inside this ambient group Z', then Z' is generated as a group by A'. In particular, Z' is finitely generated.*

Proof Let $\mathbf{Z}^A$ be a group of rank $|A|$ which is freely generated by some basis $\{e_a : a \in A\}$. Let $\langle X \rangle$ be the subgroup of $\mathbf{Z}^A$ generated by the elements

$$X := \{e_{a_1} + \cdots + e_{a_k} - e_{a'_1} - \cdots - e_{a'_k} : a_1, \ldots, a_k, a'_1, \ldots, a'_k \in A, a_1 + \cdots + a_k = a'_1 + \cdots + a'_k\}.$$

We then define $Z' := \mathbf{Z}^A / \langle X \rangle$, and let A' be the image of the basis $\{e_a : a \in A\}$ under the canonical quotient map $\pi : \mathbf{Z}^A \to \mathbf{Z}^A/\langle X \rangle$. It is clear that Z' is generated by A'. We now show that the map $\iota : A \to A'$ defined by $\iota(a) := \pi(e_a)$ is a Freiman isomorphism. Since this map is surjective, it suffices by Exercise 5.3.1 to show that

$$a_1 + \cdots + a_k = a'_1 + \cdots + a'_k \iff \iota(a_1) + \cdots + \iota(a_k) = \iota(a'_1) + \cdots + \iota(a'_k).$$

But this is clear from the construction of $\mathbf{Z}^A/\langle X \rangle$.

Next, let $\phi : (A', Z') \to (B, W)$ be a Freiman homomorphism. Let $\psi : \mathbf{Z}^A \to W$ be the unique group homomorphism such that $\psi(e_a) = \phi(\iota(a))$ for all $a \in A$; this is uniquely defined since the basis $\{e_a : a \in A\}$ freely generates $\mathbf{Z}^A$. Also it is clear that ψ annihilates X, and hence $\langle X \rangle$. Thus ψ descends to a group homomorphism $\phi_{\mathrm{ext}} : \mathbf{Z}^A/\langle X \rangle \to W$, and it is easily verified that ϕ_{ext} extends ϕ. This proves existence of extensions. To prove uniqueness, it suffices to show that any two group homomorphisms from Z' to W which agree on A' will agree on all of Z'. But this follows since Z' is generated by A'. □

For an alternative construction of the universal ambient group, see Exercise 5.5.1. For some examples of universal ambient groups, see Exercise 5.5.1 and Exercise 5.5.17.

If (A, Z) is an additive set with universal ambient group Z, then we can define a *degree map* $\deg : Z \to \mathbf{Z}$ to be the group homomorphism extending the trivial Freiman homomorphism $a \mapsto 1$. Thus deg equals 1 on A, equals 2 on $2A$, and more generally equals $l - m$ on $lA - mA$. Thus in the universal ambient group the sets nA for $n \in \mathbf{Z}$ are all disjoint. Also observe that deg must annihilate the torsion group $\mathrm{Tor}(Z) := \{x \in Z : nx = 0 \text{ for some } n \in \mathbf{Z}^+\}$ of Z, since the range $\mathbf{Z}$ of deg is torsion-free. This shows that $Z/\mathrm{Tor}(Z)$ is a non-trivial torsion-free additive group, and hence by Corollary 3.6 is group isomorphic to $\mathbf{Z}^{d+1}$ for some $d \geq 0$. Since all universal ambient groups are group isomorphic, this quantity d depends only on the additive set A, and we give it a name:

Definition 5.40 (Freiman dimension) Let A be an additive set. We define the *Freiman dimension* of A to be the unique non-negative integer $\dim(A) = d$ such that $Z/\mathrm{Tor}(Z)$ is group isomorphic to $\mathbf{Z}^{d+1}$ for every universal ambient group Z of A.

Note that the Freiman dimension depends on the choice k of the order of Freiman homomorphism; see Exercise 5.5.11. Traditionally one works with the Freiman dimension corresponding to the case $k = 2$. We caution that Freiman dimension is not monotone; again, see Exercise 5.5.11. The Freiman dimension can be interpreted as the largest rank that is attainable by a Freiman isomorphic copy of A in a vector space; see Exercise 5.5.10.

Let (A, Z) be an additive set with a universal ambient group Z, and let d be the Freiman dimension of A, and let Z be a universal ambient group for A. Then by Definition 5.40 we may identify $Z \equiv \mathbf{Z}^d \times \mathbf{Z} \times \mathrm{Tor}(Z)$; by applying a group isomorphism if necessary we may assume that the degree map $\deg : Z \to \mathbf{Z}$ corresponds to the $\mathbf{Z}$ coordinate of this identification, thus $\deg(n, m, x) = m$ for all $n \in \mathbf{Z}^d$, $m \in \mathbf{Z}$, $x \in \mathrm{Tor}(Z)$. Now let $\pi : Z \to \mathbf{Z}^d$ be the projection to the first factor. We call the additive set $[A] := (\pi(A), \mathbf{Z}^d)$ a *torsion-free universal representation* of A. It is easy to see that the torsion-free universal representation $[A]$ of an additive set A is unique up to affine group isomorphisms on $\mathbf{Z}^d$ (i.e. up to translations and elements of $SL_d(\mathbf{Z})$). Also, since A generates Z, we see that $\pi(A)$ must generate $\mathbf{Z}^{d+1}$, which implies that $\mathbf{Z}^d$ lies in the affine span of $[A]$. In other words, $\mathrm{rank}([A]) = d$.

Note that π induces a surjective Freiman homomorphism from A to $[A]$. If Z has no torsion group then this is in fact a Freiman isomorphism, but in general if A contains enough "torsion" then A and $[A]$ will not be Freiman isomorphic;

see Exercise 5.5.9. Nevertheless, $[A]$ remains a universal embedding of A in the category of embeddings into *torsion-free groups*. More precisely:

Proposition 5.41 *Let A be an additive set with Freiman dimension d, and let $[A] \subset \mathbf{Z}^d$ be a torsion-free universal representation of A. Let $\pi : A \to [A]$ be the associated Freiman homomorphism, and let $\phi : A \to (A', Z')$ be any Frieman homomorphism into a torsion-free additive group Z'. Then there exists a unique vector $v = (v_1, \ldots, v_d) \in (Z')^d$ and $a \in Z'$ such that $\phi(b) = a + \pi(b) \cdot v$ for all $b \in A$.*

Proof We may assume that A is embedded inside a universal ambient group $Z = \mathbf{R}^d \times \mathbf{R} \times \mathrm{Tor}(Z)$, and that $[A] = \pi(A)$ where $\pi : Z \to \mathbf{R}^d$ is the projection to the first factor. On the other hand, ϕ extends to a group homomorphism $\phi_{\mathrm{ext}} : \mathbf{R}^d \times \mathbf{R} \times \mathrm{Tor}(Z) \to Z'$. Since Z' is torsion-free, ϕ_{ext} must annihilate $\mathrm{Tor}(Z)$, and thus ϕ_{ext} must take the form $\phi_{\mathrm{ext}}(n, m, x) = n \cdot v + m \cdot a$ for all $n \in \mathbf{R}^d$, $m \in \mathbf{R}$, $x \in \mathrm{Tor}(Z)$, where $v \in (Z')^d$ and $a \in \mathbf{Z}$. Since A is a subset of $\mathbf{R}^d \times \{1\} \times \mathbf{Z}$ and $\pi(n, m, x) = 1$, we thus have $\phi(b) = \phi_{\mathrm{ext}}(b) = \pi(b) \cdot v + a$ for all $b \in A$, as desired. □

From this and Freiman's lemma we can obtain

Corollary 5.42 *Let $k \geq 2$, and let A be an additive set in a torsion-free additive group Z such that* $\min(|A + A|, |A - A|) \leq (d + 1)|A| - \frac{d(d+1)}{2}$ *for some integer $K \geq 1$. Then* $\dim(A) < d$.

Proof Let $[A] = \pi(A)$ be a torsion-free universal representation of A. By Proposition 5.41 we have a Freiman homomorphism from $[A]$ back to A, and hence A and $[A]$ are Freiman isomorphic. Hence we may without loss of generality work with $[A]$ instead of A. But then the claim follows from Lemma 5.13 (or Exercise 5.2.1), since $\mathrm{rank}([A]) = d$. □

Thus, in the torsion-free case at least, sets with small doubling necessarily have small Freiman dimension. A slightly weaker statement is true when A is not a torsion-free additive group:

Corollary 5.43 *Let $k \geq 2$, and let A be an additive set. Then* $\dim(A) < \sigma[A]^{O(1)}$.

Proof Let $K := \sigma[A]$ and $d := \dim(A)$. If $K < \frac{3}{2}$ then $d = 0$ (Exercise 5.5.13). Hence we may assume $K \geq \frac{3}{2}$, and it will now suffice to show $d = O(K^{O(1)})$. Without loss of generality we may assume that A is embedded in a universal ambient group Z. From Proposition 2.26 we see that $A + A$ can be covered by $O(K^{O(1)})$ translates of A. Applying the quotient map $\pi : Z \to Z/\mathrm{Tor}(Z) \equiv \mathbf{Z}^{d+1}$, we then see that $\pi(A) + \pi(A)$ can be covered by $O(K^{O(1)})$ copies of $\pi(A)$, and thus $|2\pi(A)| \leq K^{O(1)}|\pi(A)|$. But $\pi(A)$ is Freiman isomorphic to a torsion-free universal representation $[A]$ of A; thus $|2[A]| \leq K^{O(1)}|[A]|$. On the other hand, since $\text{rank}([A]) = d$, we see from Lemma 5.13 that $|2[A]| > (d+1)|[A]| - \frac{d(d+1)}{2}$. Since $|[A]| \geq \text{rank}(A) + 1 = d + 1$ (for instance), we thus have $|2[A]| > \frac{d}{2}|[A]|$ (for instance). Combining this with the upper bound on $|2[A]|$ we obtain the result. □

For a refinement of the bounds in this corollary, see Exercise 6.5.18.

Exercises

5.5.1 For any additive sets A, B, let $\text{Hom}_k(A \to B)$ denote the space of Freiman homomorphisms (of order k) from A to B. Since A is an additive set, observe that $\text{Hom}_k(A \to \mathbf{R}/\mathbf{Z})$ is an additive group which can be viewed as a compact subgroup of a torus. In particular it has a Pontryagin dual $Z' := \widehat{\text{Hom}_k(A \to \mathbf{R}/\mathbf{Z})}$, defined as the space of all continuous group homomorphisms from $\text{Hom}_k(A \to \mathbf{R}/\mathbf{Z})$ to the circle group $\mathbf{R}/\mathbf{Z}$. For any $a \in A$, define the *Gelfand transform* $\hat{a} \in Z'$ of a by the formula

$$\hat{a}(\chi) := \chi(a) \text{ for all } \chi \in \text{Hom}_k(A \to \mathbf{R}/\mathbf{Z}),$$

and let $A' := \{\hat{a} : a \in A\}$. Show that (A', Z') is Freiman isomorphic to (A, Z), and that Z' is a universal ambient group for A.

5.5.2 Let A be an additive set. Show that $\mathbf{Z}^{|A|+1}$ is a universal ambient group for A if and only if A is a B_k set (see Definition 4.27), in which case the additive set $(\{e_j + e_{|A|+1} : 1 \leq j \leq |A|\}, \mathbf{Z}^{|A|+1})$ is an embedding of A into $\mathbf{Z}^{|A|+1}$. Here of course $e_1, \ldots, e_{|A|+1}$ is the standard basis for $\mathbf{Z}^{|A|+1}$.

5.5.3 Let A be an additive set, and let $\chi : A \to \mathbf{R}/\mathbf{Z}$ be a Freiman homomorphism. Let us say that χ is *infinitely divisible* if for every integer n there exists a Freiman homomorphism $\chi/n : A \to \mathbf{R}/\mathbf{Z}$ which, when multiplied by n, yields χ. Show that χ is infinitely divisible if and only if there exists a Freiman homomorphism $\phi : A \to \mathbf{R}$ such that $\phi \bmod 1 = \chi$. Conclude that the tangent space of the compact group $\text{Hom}_k(A \to \mathbf{R}/\mathbf{Z})$ at the origin is canonically identifiable with $\text{Hom}_k(A \to \mathbf{R})$.

5.5.4 Let $\phi : A \to B$ be a Freiman homomorphism (resp. isomorphism). Show that the map $\phi^\dagger : \text{Hom}_k(B \to \mathbf{R}/\mathbf{Z}) \to \text{Hom}_k(A \to \mathbf{R}/\mathbf{Z})$ defined by $\phi^\dagger(\chi) := \chi \circ \phi$ is a group homomorphism (resp. isomorphism). Also, if $\phi : A \to B$ and $\psi : B \to C$ are Freiman homomorphisms, show that $(\phi \circ \psi)^\dagger = \psi^\dagger \circ \phi^\dagger$. Show that the *adjoint functor* $\phi \mapsto \phi^\dagger$ is a bijection between Freiman homomorphisms from A to B, and group homomorphisms from $\text{Hom}_k(B \to \mathbf{R}/\mathbf{Z})$ to $\text{Hom}_k(A \to \mathbf{R}/\mathbf{Z})$.

5.5.5 Let G be an additive set which is also an additive group (i.e. $G + G = G$). Show that $\mathrm{Hom}_k(G \to \mathbf{R}/\mathbf{Z})$ is canonically identifiable with $\hat{G} \times (\mathbf{R}/\mathbf{Z})$, where $\hat{G}$ is the Pontryagin dual of G, i.e. the space of group homomorphisms from G to $\mathbf{R}/\mathbf{Z}$. If A is an additive set contained in G, give examples to show that $\mathrm{Hom}_k(A \to \mathbf{R}/\mathbf{Z})$ can be much larger or much smaller than $\mathrm{Hom}_k(G \to \mathbf{R}/\mathbf{Z})$, although Freiman duality will convert the inclusion map from A to G to a group homomorphism from $\mathrm{Hom}_k(G \to \mathbf{R}/\mathbf{Z})$ to $\mathrm{Hom}_k(A \to \mathbf{R}/\mathbf{Z})$.

5.5.6 Let $k = 2$. Show that the universal ambient group of $A = (\{1, 2, 3\}, \mathbf{Z}_6)$ (or $(\{1, 2, 3\}, \mathbf{Z}_7)$, or $(\{1, 2, 3\}, \mathbf{Z})$) is canonically identifiable with $\mathbf{Z}^2$, with $\hat{A}$ being identified with $\{(1, 1), (2, 1), (3, 1)\}$. Show on the other hand that the universal ambient group of $A = (\{1, 2, 3\}, \mathbf{Z}_3)$ is canonically identified with $\mathbf{Z}_3 \times \mathbf{Z}$, with $\hat{A}$ identified with $\{(1, 1), (2, 1), (3, 1)\}$. Show that the universal ambient group of $A = (\{1, 2, 4, 5\}, \mathbf{Z})$ is canonically identifiable with $\mathbf{Z}^3$, with $\hat{A}$ being identified with $\{(0, 0, 1), (0, 1, 1), (1, 0, 1), (1, 1, 1)\}$.

5.5.7 Let (A, Z) be an additive set embedded inside a universal ambient group Z, let (B, W) be another additive set, let $\phi : A \to B$ be a Freiman homomorphism, and let $\phi^{\mathrm{ext}} : Z \to W$ be the group homomorphism extension. Show that ϕ is a Freiman isomorphism if and only if the kernel $\ker(\phi^{\mathrm{ext}}) := \{x \in Z : \phi^{\mathrm{ext}}(x) = 0\}$ of ϕ^{ext} is disjoint from $(kA - kA)\backslash\{0\}$, or equivalently if ϕ^{ext} is injective on kA.

5.5.8 Let (A, Z) be an additive set embedded inside a universal ambient group Z, and let G be an additive group. Show that G contains a subset A' that is Freiman isomorphic to A if and only if G contains a subgroup H that is group isomorphic to Z/Γ for some subgroup Γ of Z which is disjoint from $(kA - kA)\backslash\{0\}$.

5.5.9 Let (A, Z) be an additive set embedded inside a universal ambient group Z. Show that A and $[A]$ are Freiman isomorphic if and only if $(kA - kA) \cap \mathrm{Tor}(Z) = \{0\}$. Note from Proposition 5.41 that A can be embedded into a torsion-free additive group if and only if A and $[A]$ are Freiman isomorphic.

5.5.10 Let A be an additive set in a torsion-free additive group Z. Show that there exists a Freiman-isomorphic copy (A', V') of (A, Z) inside a vector space V' such that $\mathrm{rank}(A') = \dim(A)$. Furthermore, we have $\mathrm{rank}(A'') \leq \dim(A)$ for any other Freiman isomorphic copy (A'', V'') of (A, Z) in a vector space.

5.5.11 Let (A, Z) be the additive set $(\{1, 2, 4, 5\}, \mathbf{Z})$. Show that $\dim(A) = 4$ if $k = 1$, that $\dim(A) = 2$ if $k = 2$, and $\dim(A) = 1$ for $k \geq 3$. In particular, when $k = 2$, conclude that $\dim(\{1, 2, 4, 5\}) > \dim(\{1, 2, 3, 4, 5\})$, thus demonstrating that Freiman dimension is not monotone.

5.5.12 Show that the Freiman dimension $\dim(A) = \dim_k(A)$ of an additive set is a non-increasing function of k, thus $\dim_{k+1}(A) \leq \dim_k(A)$.

5.5.13 Let $k \geq 2$, and let A be an additive set such that $\sigma[A] < \frac{3}{2}$. Show that $\dim(A) = 1$. (Hint: embed A in a universal ambient group and apply Corollary 5.6.)

5.5.14 Let (A, Z) and (A', Z') be additive sets. Show that $\dim(A \oplus A') = \dim(A) + \dim(A')$.

5.5.15 Let $\phi : A \to A'$ be a surjective Freiman homomorphism. Show that $\dim(A') \leq \dim(A)$.

5.5.16 Let A be an additive set, and let Z be a universal ambient group for A. Show that $\mathrm{Tor}(Z) = \{0\}$ if and only if the group homomorphism $\pi : \mathrm{Hom}_k(A \to \mathbf{R}) \to \mathrm{Hom}_k(A \to \mathbf{R}/\mathbf{Z})$ defined by $\pi(\phi) := \phi \bmod 1$ is surjective, or in other words every Freiman homomorphism from A to $\mathbf{R}/\mathbf{Z}$ lifts up to a Freiman homomorphism from A to $\mathbf{R}$.

5.5.17 Let $k = 2$ and consider the set $A := \{2e_1, e_1 + e_2, 2e_2, e_2 + e_3, 2e_3, e_3 + e_4, 2e_4, e_4 + e_1\}$ in $\mathbf{Z}^4$, where e_1, e_2, e_3, e_4 is the standard basis; one can view this as a generic skew quadrilateral together with the midpoints. Show that $(A, \mathbf{Z}^4)$ has $\mathbf{Z}^4 \times (\mathbf{Z}/2\mathbf{Z})$ as a universal ambient group. Thus it is possible for the universal ambient group to contain some torsion even when the additive set can be embedded in a torsion-free additive group. Write down an embedding of A in the universal ambient group $\mathbf{Z}^2 \times (\mathbf{Z}/2\mathbf{Z})$, and compare it with a torsion-free representation $[A]$ of A; are they Freiman isomorphic to each other?

5.5.18 Generalize Theorem 5.11 to handle additive sets A in any torsion-free additive group.

5.6 Freiman's theorem in an arbitrary group

Now we use the universal group, combined with Fourier analysis and additive geometry, to obtain Freiman's theorem in an arbitrary additive group. This result was first obtained by Green and Ruzsa [157]; the approach here is inspired by their argument but is arranged somewhat differently, relying in particular on volume bounds on polar bodies instead of the Ruzsa–Chang theorem (Theorem 5.30), and working in the universal ambient group rather than by introducing a sequence of successively smaller ambient groups to contain the additive set A.

Observe that in some inverse sum set theorems (Corollary 5.6, Theorem 5.27) a set with small doubling was contained inside a finite group (or a coset of such a group), whereas in other inverse sum set theorems (Theorem 5.11, Theorem 5.32, and to a lesser extent Corollary 5.19) a set with small doubling was contained inside a progression. In general, it is convenient to place a set of small doubling inside a *coset progression* $P + H$, which was defined in Definition 4.21.

Theorem 5.44 (Freiman's theorem in an arbitrary group) *[157] Let $K \geq 1$, and let (A, Z) be an additive set in an arbitrary group Z such that $|A + A| \leq K|A|$. Then there is a coset progression $P + H$ of rank at most* $\dim(A)$ *such that $A \subseteq P + H$ and $|P||H| \leq \exp(O(K^{O(1)}))|A|$. If Z is the universal ambient group of A, then we can take $H = \mathrm{Tor}(Z)$.*

One can make the constants in $\exp(O(K^{O(1)}))$ more explicit; see [157].

Proof Here we shall fix the order k of the Freiman homomorphisms under consideration to be $k = 2$. Without loss of generality we may assume Z is the universal ambient group; the general case then follows from Definition 5.37 (and the observation that the image of a group or progression under a group homomorphism is still a group or progression). We write $d := \dim(A)$; from Corollary 5.43 we have $d = O(K^{O(1)})$.

We know that Z is isomorphic to $\mathbf{Z}^d \times \mathbf{Z} \times \mathrm{Tor}(Z)$; we shall abuse notation and *identify* Z with $\mathbf{Z}^d \times \mathbf{Z} \times \mathrm{Tor}(Z)$, in particular identifying $\mathrm{Tor}(Z)$ with $\{(0, 0)\} \times \mathrm{Tor}(Z)$. We can also arrange matters so that the $\mathbf{Z}$ component of Z is given by the degree map, thus $\deg((n, m, x)) = m$ for all $n \in \mathbf{Z}^d$, $m \in \mathbf{Z}$, $x \in \mathrm{Tor}(Z)$, and A lives entirely in $\mathbf{Z}^d \times \{1\} \times \mathrm{Tor}(Z)$. By using a group isomorphism to translate A in the $\mathbf{Z}^d \times \mathrm{Tor}(Z)$ direction if necessary, we may assume that $(0, 1, 0) \in A$.

At present, Z is not a finite group and so we cannot directly apply the Fourier analytic techniques from Chapter 4. Thus we shall truncate Z to a finite group (cf. the use of Lemma 5.26 to prove Theorem 5.30); an alternative approach (which we do not pursue here, due to some minor measure-theoretic and analytic issues which arise) is to extend the theory of the Fourier transform and of Chapter 4 to infinite additive groups. We choose an extremely large prime number p depending on A (much larger than any of the $d + 1$ coefficients of elements of A in the $\mathbf{Z}^{d+1}$ component of Z), and let $\pi_p : Z \to Z_p$ be the canonical projection from $Z = \mathbf{Z}^d \times \mathbf{Z} \times \mathrm{Tor}(Z)$ to the finite additive group $Z_p := \mathbf{Z}_p^d \times \mathrm{Tor}(Z)$. If p is sufficiently large, then π_p is a Freiman isomorphism from A to the additive set $A_p := \pi_p(A)$. We endow Z_p with the symmetric non-degenerate bilinear form

$$(\xi, \eta) \cdot (x, y) = \frac{x\xi}{p} + \eta \cdot y$$

for all $x, \xi \in \mathbf{Z}_p^d$ and $y, \eta \in \mathrm{Tor}(Z)$, where $\eta \cdot y$ is some symmetric non-degenerate bilinear form on $\mathrm{Tor}(Z)$ (the exact choice of which will be irrelevant).

Let $\alpha := 1 - \frac{1}{10^5 K^2}$. Now we establish some lower bounds on the spectrum $\mathrm{Spec}_\alpha(A_p - A_p)$ of $A_p - A_p$, as defined in Definition 4.34. □

Lemma 5.45 *We have* $|\mathrm{Spec}_\alpha(A_p - A_p)| \geq \exp(-O(K^{O(1)}))|Z_p|/|A_p|$.

Proof We first control the size of sum sets nA for very large n. Since A_p is Freiman isomorphic to A, we have $\sigma[A_p] \leq K$. By Proposition 2.26 we can thus contain A_p inside a translate of a K^C-approximate group H of size $|H| \leq K^C|A_p|$; thus $2H \subseteq H + X$ for some X of cardinality $O(K^{O(1)})$. Iterating this we see that $nH \subseteq H + (n-1)X$, and thus

$$\begin{aligned} |n(A_p - A_p)| &\leq |2nH| \\ &\leq |H||(2n-1)X| \\ &\leq K^{O(1)}|A_p|\binom{|X| + 2n - 2}{|X|} \\ &\leq K^{O(1)}|A_p|(|X| + 2n - 2)|X|. \end{aligned}$$

If we then set $n := CK^C$ for a sufficiently large constant C, we can ensure that

$$|n(A_p - A_p)| \leq \frac{1}{2}\alpha^{2-2n}|A_p - A_p|.$$

We then apply Lemma 4.38 to obtain

$$|\mathrm{Spec}_\alpha(A_p - A_p)|\,\mathbf{P}_Z(A_p - A_p) \geq \frac{1}{2}\alpha^{2-2n}$$

and the claim follows (recall $|A_p - A_p| \leq K^2|A_p|$ from Ruzsa's triangle inequality). □

Now we can use the theory of Freiman homomorphisms and the universal ambient group to eliminate the role of the torsion group. Let $\Pi : Z \to \mathbf{Z}^d \subset \mathbf{R}^d$ be the canonical projection from $Z = \mathbf{Z}^d \times \mathbf{Z} \times \mathrm{Tor}(Z)$ to $\mathbf{Z}^{d+1}$, thus $\Pi(A)$ is a subset of $\mathbf{Z}^{d+1}$ and hence of $\mathbf{R}^{d+1}$.

Lemma 5.46 *We have* $\mathrm{Spec}_\alpha(A_p - A_p) \subseteq \mathbf{Z}_p^d \times \{0\}$. *Furthermore, if* $\xi' \in \mathbf{Z}_p^d$ *is such that* $(\xi', 0) \in \mathrm{Spec}_\alpha(A_p - A_p)$, *then there exists* $\tilde{\xi} \in \frac{1}{p} \cdot \mathbf{Z}^d \subset \mathbf{R}^d$ *with* $\tilde{\xi} = \xi'/p \pmod 1$ *such that* $|\langle x, \tilde{\xi}\rangle| \leq \frac{1}{5}$ *for all* $x \in \Pi(A) - \Pi(A)$.

Proof From Ruzsa's triangle inequality we have $|A_p - A_p| \leq K^2|A_p|$. From Proposition 4.40 we thus see that $A_p - A_p \subseteq \mathrm{Bohr}_Z(\mathrm{Spec}_\alpha(A_p - A_p), \frac{1}{50})$. Thus if $\xi \in \mathrm{Spec}_\alpha(A_p - A_p)$, then $|e(\xi \cdot x) - 1| \leq \frac{1}{50}$ for all $x \in A_p - A_p$. In particular we can find a phase $e^{2\pi i\theta}$ for some $\theta \in \mathbf{R}$ such that $|e(\xi \cdot x) - e^{2\pi i\theta}| \leq \frac{1}{50}$ for all

$x \in A_p$. We can thus find a function $\chi : A_p \to \mathbf{R}$ such that $e(\xi \cdot x) = e(\chi(x))$ and $\theta - \frac{1}{10} < \chi(x) < \theta + \frac{1}{10}$ for all $x \in A_p$. It is then easy to see that $\chi : A_p \to \mathbf{R}$ is a Freiman homomorphism, and hence $\chi \circ \pi_p : A \to \mathbf{R}$ is a Freiman homomorphism. Since Z is a universal ambient group for A, we thus see that we can extend $\chi \circ \pi_p$ to a group homomorphism $(\chi \circ \pi)_{\mathrm{ext}} : Z \to \mathbf{R}$. But since $\mathbf{R}$ is torsion-free, this group homomorphism must annihilate the torsion group $\mathrm{Tor}(Z)$. In particular, the map $\phi : x \mapsto (\chi \circ \pi_p)_{\mathrm{ext}}(x) \bmod 1$ is a group homomorphism from Z to $\mathbf{R}/\mathbf{Z}$ which annihilates $\mathrm{Tor}(Z)$. On the other hand, the map $\tilde{\phi} : x \mapsto \xi \cdot \pi_p(x)$ is another group homomorphism from Z to $\mathbf{R}/\mathbf{Z}$ which agrees with ϕ on A. Since Z is a universal ambient group for A, this means that $\phi = \tilde{\phi}$, and thus $\tilde{\phi}$ must also annihilate $\mathrm{Tor}(Z)$. In other words we see that $\xi \cdot x = 0$ whenever $x \in \mathrm{Tor}(Z)$, which means that $\xi \in \mathbf{Z}_p^d \times \{0\}$, and the first claim follows.

Now let $\xi' \in \mathbf{Z}_p^d$ be such that $(\xi', 0) \in \mathrm{Spec}_\alpha(A_p - A_p)$. Then as before we can find a Freiman homomorphism $\chi : A_p \to \mathbf{R}$ such that

$$(\xi', 0) \cdot x = \chi(x) \bmod 1 \text{ for all } x \in A_p \tag{5.17}$$

and a $\theta \in \mathbf{R}$ such that

$$\theta - \frac{1}{10} < \chi(x) < \theta + \frac{1}{10} \text{ for all } x \in A_p - A_p, \tag{5.18}$$

and we have a group homomorphism $(\chi \circ \pi)_{\mathrm{ext}} : Z \to \mathbf{R}$ which extends $\chi \circ \pi$ and annihilates $\mathrm{Tor}(Z)$. Since $Z = \mathbf{Z}^d \times \mathbf{Z} \times \mathrm{Tor}(Z)$, we thus see that there exist $\tilde{\xi} \in \mathbf{R}^d$ and $\eta \in \mathbf{R}$ such that

$$(\chi \circ \pi)_{\mathrm{ext}}(n, m, x) = n \cdot \tilde{\xi} + m\eta \text{ for all } n \in \mathbf{Z}^d, m \in \mathbf{Z}, x \in \mathrm{Tor}(Z).$$

Restricting this to elements of A (which lie in $\mathbf{Z}^d \times \{1\} \times \mathrm{Tor}(Z)$, we obtain

$$\chi((n \bmod p, x)) = \chi(\pi(n, 1, x)) = n \cdot \tilde{\xi} + \eta \text{ whenever } (n, 1, x) \in A. \tag{5.19}$$

Applying (5.17) we obtain

$$n \cdot \xi'/p = n \cdot \tilde{\xi} + \eta \pmod 1 \text{ whenever } (n, 1, x) \in A.$$

Since $(0, 1, 0) \in A$, we conclude that $\eta = 0 \pmod 1$. Since A generates all of $Z = \mathbf{Z}^d \times \mathbf{Z} \times \mathrm{Tor}(Z)$, we infer that $\tilde{\xi} = \xi'/p \pmod 1$ as desired; in particular $\tilde{\xi} \in \frac{1}{p} \cdot \mathbf{Z}^d$. Next, we apply (5.18) to deduce that

$$\theta - \frac{1}{10} < n \cdot \tilde{\xi} + \eta < \theta + \frac{1}{10} \text{ whenever } (n, 1, x) \in A$$

and thus

$$|(n - n') \cdot \tilde{\xi}| < \frac{1}{5} \text{ whenever } n, n' \in \Pi(A),$$

and the claim follows (note that the dot product $n \cdot x$ and the inner product $\langle n, x \rangle$ agree when $n \in \mathbf{Z}^d$ and $x \in \mathbf{R}^d$).

Since $\Pi(A) - \Pi(A)$ is a subset of $\mathbf{Z}^d$, it is also a subset of $\mathbf{R}^d$. Let B be the convex body generated by the open convex hull of $\Pi(A) - \Pi(A)$; note that B is open and non-empty because A generates Z, and hence $\Pi(A)$ generates $\mathbf{Z}^d$. Introducing the polar body

$$B^\circ := \{x \in \mathbf{R}^d : |x \cdot y| < 1 \text{ for all } y \in B\}$$

of B, we can rewrite the conclusion of Lemma 5.46 as

$$\tilde{\xi} \in \frac{1}{5} \cdot B^\circ.$$

Combining this with Lemma 5.45, we thus see that

$$\left| \left(\frac{1}{5} \cdot B^\circ \right) \cap \left(\frac{1}{p} \cdot \mathbf{Z}^d \right) \right| \geq \frac{\exp(-CK^C)|Z_p|}{|A_p|} = \frac{\exp(-O(K^{O(1)}))p^d |\mathrm{Tor}(Z)|}{|A|}$$

and thus

$$p^{-d} \left| B^\circ \cap \left(\frac{1}{p} \cdot \mathbf{Z}^d \right) \right| \geq \frac{\exp(-CK^C)|\mathrm{Tor}(Z)|}{|A|}.$$

Now we take limits as $p \to \infty$. Since B° is open and bounded, the left-hand side is just the Riemann sum for $\mathrm{mes}(B^\circ)$, and thus

$$\mathrm{mes}(B^\circ) \geq \exp\left(-O\left(K^{O(1)}\right)\right)|\mathrm{Tor}(Z)|/|A|.$$

Now we use the machinery from Chapter 3. Using the rather crude bound

$$\mathrm{mes}(B^\circ)\mathrm{mes}(B) \leq O(1)^d = O(1)^{K^{O(1)}} \tag{5.20}$$

(see Exercise 5.6.1), we can convert this lower bound on B° to an upper bound for B:

$$\mathrm{mes}(B) \leq \exp\left(O\left(K^{O(1)}\right)\right)|A|/|\mathrm{Tor}(Z)|.$$

Note that $B \cap \mathbf{Z}^d$ contains $\Pi(A) - \Pi(A)$; since $\Pi(A)$ generates $\mathbf{Z}^d$, we thus conclude that $B \cap \mathbf{Z}^d$ linearly spans $\mathbf{R}^d$. From this and Lemma 3.26 we see that

$$|B \cap \mathbf{Z}^d| \leq \exp\left(O\left(K^{O(1)}\right)\right)|A|/|\mathrm{Tor}(Z)|$$

where we have used the earlier observation $d = O(K^{O(1)})$ to absorb the $3^d d!/2^d$ factor from that Lemma. Applying the discrete John theorem (Lemma 3.36) we can thus place $B \cap \mathbf{Z}^d$ inside a progression $Q \subseteq \mathbf{Z}^d$ of rank at most d and volume

$$|Q| \leq \exp\left(O\left(K^{O(1)}\right)\right)|A|/|\mathrm{Tor}(Z)|,$$

again using the observation $d = O(K^{O(1)})$, this time to absorb the factors of $(d^{2d})^d$ that will appear. Since A was normalized to contain $(0, 1, 0)$, we have the inclusions $\Pi(A) \subseteq \Pi(A) - \Pi(A) \subseteq B \cap \mathbf{Z}^d \subseteq Q$, and hence $A \subseteq \Pi^{-1}(Q)$. But we may write $\Pi^{-1}(Q) = P + G$ where P is an isomorphic copy of Q, and $G := \mathrm{Tor}(Z)$. Theorem 5.44 follows. □

Remark 5.47 It seems of interest to improve the exponential losses $\exp(O(K^{O(1)}))$ in the above argument. Many of these losses are really exponential in the Freiman dimension d rather than in the doubling constant K, so one expects to gain somewhat when the Freiman dimension is small. However, the main step where the exponential losses are largest lies in the proof of Lemma 5.45, where one is forced to control extremely large sum sets of A_p in order to obtain a lower bound on the size of the spectrum. It may be that one will have to use a non-Fourier-analytic approach in order to avoid this type of loss. On the other hand, the asymptotic behavior of iterated sum sets is certainly relevant to the task of containing A inside a convex body or arithmetic progression (see Exercise 5.6.4). However, it may well be that this type of argument can at least be pushed to improve $\exp(O(K^{O(1)}))$ to a factor like $\exp(O(K \log^{O(1)} K))$ or perhaps even $\exp(O(K))$.

We now comment briefly on the slightly different argument of Green and Ruzsa [157] in establishing the above theorem. Instead of working in a universal ambient group, which could be infinite, they proceed by first using a Freiman isomorphism (of order at least 16, say) to embed A inside a very large finite group (similar to the group Z_p used in the analysis here), and then to use an estimate similar to Lemmas 5.45 and 5.46 to reduce the size of this ambient group Z iteratively until $|Z| \leq \exp(CK^C)|A|$ (the point being that if $|Z| > \exp(CK^C)|A|$, then the arguments of Lemmas 5.45 and 5.46 can be used to locate a narrow Bohr set that contains A, which is then Freiman isomorphic to a subset of a smaller group than Z. At this point one can apply an extension of Theorem 4.42 (for arbitrary finite additive groups, not necessarily cyclic) to show that $2A - 2A$ contains the sum of a large progression and a large group, at which point one can conclude a Ruzsa–Chang type theorem for arbitrary groups, which then implies the above theorem by an argument similar to how Theorem 5.30 implies Theorem 5.32. In particular, they establish

Theorem 5.48 (Ruzsa–Chang theorem in arbitrary groups) *[157] Let A be an additive set in an arbitrary additive group Z such that $|A + A| \leq K|A|$ for some $K \geq 1$. Then $2A - 2A$ contains a set of the form $P + G$ where P is a proper symmetric progression of rank at most $CK(1 + \log K)$ and G is a finite subgroup of Z such that $|P + G| = |P||G| \geq e^{-CK(1+\log^2 K)}|A|$.*

Exercises

5.6.1 Let B be a symmetric convex body, and consider the Euclidean Fourier transform

$$\widehat{1_B}(\xi) := \int_{\mathbf{R}^d} 1_B(x) e(-\xi \cdot x)\, d\xi.$$

Show that this Fourier transform is large on a large subset of the polar body B°, and use this and the Plancherel theorem on $\mathbf{R}^d$ to establish (5.20). (A much sharper inequality than (5.6.1) is available, namely *Santalo's inequality* [306], but we will not need this inequality here.)

5.6.2 [157] Let A be an additive set with $|A + A| \leq K|A|$. Show that there exists a finite group Z of order $|Z| \leq \exp(O(K^{O(1)}))|A|$ such that A is Freiman isomorphic of order 2 (say) to a subset of Z. (Hint: combine the analysis of this section with Exercise 5.5.8.)

5.6.3 [154] Suppose p is a prime number, and A is an additive set in $\mathbf{Z}_p$ such that $|A + A| \leq K|A|$ for some $K \geq 1$. Suppose also that $|A| \leq \exp(-O(K^{O(1)}))p$ for some sufficiently large absolute constant $C > 1$. Show that A is Freiman isomorphic of order 2 to a subset of the integers $\mathbf{Z}$. This is known as the *Freiman rectification principle*; see [29], [154] for further discussion.

5.6.4 Let A be an additive set in $\mathbf{Z}^d$ which generates $\mathbf{Z}^d$, and let B be the convex hull of A. Show that $|nA| = (1 + o_{n\to\infty}(1))n^d \mathrm{mes}(B)$ as $n \to \infty$. (See [261] for more precise results of this type.)

6

Graph-theoretic methods

Additive combinatorics is a subfield of combinatorics, and so it is no surprise that graph theory plays an important role in this theory. Graph theory has already made an implicit appearance in previous chapters, most notably in the proof of the Balog–Szemerédi–Gowers theorem (Theorem 2.29). However there are several further ways in which graph theoretical tools can be utilized in additive combinatorics. We will only discuss a representative sample of these applications here. First we discuss *Turán's theorem*, which shows that sparse graphs contain large independent sets, and which is useful for constructing sum-free sets. Next we give a very brief tour of *Ramsey theory*, which allows one to find monochromatic structures in colored graphs (or other colored objects), in particular allowing one to find monochromatic progressions in any coloring of the integers (*van der Waerden's theorem*). Then we use some results about connectivity of dense graphs to establish the *Balog–Szemerédi–Gowers theorem*, which relates partial sum sets to complete sum sets and which has already been exploited in Chapter 2. Finally, we use the theory of commutative directed graphs to establish the *Plünnecke inequalities*, which are perhaps the sharpest inequalities known for sum sets and which strengthen several of the results already established in Chapter 2.

In Chapter 10 and Chapter 11 we shall discuss one final graph-theoretical tool, the *Szemerédi regularity lemma*, which has had many applications in several areas of discrete mathematics, but which in additive combinatorics has had an especially crucial role in the study of arithmetic progressions in dense sets.

Graph-theoretic tools are especially useful when combined with the *probabilistic method*, which we already saw in Chapter 1, and indeed many of our arguments here will be probabilistic in nature.

6.1 Basic notions

A graph $G = G(V, E)$ consists of a finite set V of *vertices (points, nodes)* and a finite set E of *edges*, where each edge is an unordered pair $\{a, b\}$ of distinct vertices (thus we do not allow loops).

If $\{a, b\} \in E$, we say that the two vertices a and b are *adjacent* or *neighbors*. The collection of all the neighbors of a shall be denoted $N(a)$. The cardinality of $N(a)$ is called the *degree* of a and is denoted deg(a).

Consider a subset V' of V. We refer to the graph $G' = G'(V', E')$ where $E' := \{e \in E : e \subseteq V'\}$ as the *induced* subgraph of G which is *spanned* by V'. A set $V' \subset V$ is *independent* if it spans an empty graph, i.e., there is no edge with both endpoints in V'.

We say that the vertices $a_0, \ldots, a_k$ form a *path* of length k if $\{a_i, a_{i+1}\}$ is an edge for all $0 \leq i \leq k-1$. If $a_k = a_0$, we refer to the path as a *cycle*. Three vertices a, b, c form a *triangle* if they form a cycle of length 3, i.e. $\{a, b\}$, $\{b, c\}$ and $\{c, a\}$ are edges.

A graph is *bipartite* if one can partition its vertex set into two disjoint sets A and B so that every edge has one end point in A and another in B; A and B are called the *color classes* of G. Bipartite graphs play an important role in what follows and when dealing with them, we prefer to use the notation $G(A, B, E)$ instead of $G(V, E)$. Note that in a bipartite graph $G = G(A, B, E)$, two vertices in the same color class can only be connected by paths of even length, while vertices in opposite color classes can only be connected by paths of odd length. In particular all cycles must be of even length.

Exercises

6.1.1 Prove that a graph G is bipartite if and only if all cycles are of even length.

6.1.2 Let A be a symmetric additive set (so $A = -A$) in a finite additive group Z and $0 \notin A$. The *Cayley graph* of A is defined to be the graph with vertex set Z, and two vertices x, y connected by an edge if and only if $x - y \in A$. Show that $\deg(v) = |A|$ for all $v \in Z$, and that two points $v, w \in Z$ are connected by a path of length n if and only if $v - w \in n(A)$. Show that G is connected if and only if A spans Z.

6.1.3 (Popularity principle for bipartite graphs) Let $G(V_1, V_2, E)$ be a bipartite graph with V_2 non-empty. Show that there exists a bipartite subgraph $G'(V_1, V_2', E')$ of $G(V_1, V_2, E)$ with $|E'| \geq |E|/2$ and $\deg_{G'}(v_2) \geq |E|/2|V_2|$ for all $v_2 \in V_2'$.

6.1.4 (Cauchy–Schwarz for bipartite graphs) Let $G(V_1, V_2, E)$ be a bipartite graph with V_1, V_2 non-empty. Show that G contains at least $|E|^2/|V_2|$

paths of length two with both endpoints in V_1, including degenerate paths. Show, G also contains at least $|E|^4/|V_1|^2|V_2|^2$ cycles of length four.

6.1.5 [198] Let $G(V_1, V_2, E)$ be a bipartite graph with V_1, V_2 non-empty. Show, for any $k \geq 1$, that G contains at least $|E|^{2k}/|V_1|^{k-1}|V_2|^k$ paths of length $2k$ with both endpoints in V_1, including degenerate paths, and also that G contains at least $|E|^{2k+1}/|V_1|^k|V_2|^k$ paths of length $2k+1$ from V_1 to V_2. (Hint: using the popularity principle, one can obtain lower bounds like this but losing an absolute constant depending on k. Then use the tensor power trick (as in Corollary 2.19) to remove this constant.)

6.1.6 Let $G = G(V, E)$ be a graph. Using the first moment method, show that G contains a bipartite subgraph $G'(A, B, E')$ with $|E'| \geq \frac{1}{2}|E|$. Give an example to show that the number $\frac{1}{2}$ cannot be replaced by any larger constant.

6.2 Independent sets, sum-free subsets, and Sidon sets

Intuitively one expects graphs with small degrees to have large independent sets. The following theorem, due to Turán, quantifies this intuition.

Theorem 6.1 (Turán's theorem) *Let $G = G(V, E)$ be a graph on n vertices. Then G contains an independent set of size at least $\sum_{v\in V} \frac{1}{\deg(v)+1}$. In particular, if G has maximal degree d, then G has an independent set of size at least $n/(d+1)$.*

Proof We shall use the probablistic method, or more precisely the first moment method. Let $\pi : V \to [1, n]$ be a bijection chosen uniformly at random. Let us call a vertex $v \in V$ *good* if it is larger than all its neighbors, in the sense that $\pi(w) < \pi(v)$ whenever $w \in N(v)$, and let S be the set of all good vertices. It is clear that S is an independent set. Also, for any $v \in V$, the probability that v is good can be easily verified to be $\frac{1}{\deg(v)+1}$. Thus by linearity of expectation (1.4) we have

$$\mathbf{E}(|S|) = \sum_{v\in V} \mathbf{P}(v \in S) = \sum_{v\in V} \frac{1}{\deg(v)+1}$$

and so $|S| \geq \sum_{v\in V} \frac{1}{\deg(v)+1}$ with positive probability. The claim follows. □

6.2.1 Sum-free subsets

In 1965, Erdős and Moser [86] (see also [166], Problem C14) posed the following question. If $B \subset A$ are two additive sets, let us say that B is *sum-free* with respect to A if no element of A can be represented as the sum of two distinct elements of B. Given any additive set A, let $\phi(A)$ be the cardinality of the largest subset of A

which is sum-free with respect to A. Let $\phi(n)$ be the smallest value of $\phi(A)$ among all sets A of size n; thus $\phi(n)$ is the largest number such that every set A of n reals contains a subset of cardinality $\phi(n)$ which is sum-free with respect to A.

Note that it is important that we require the elements of B be distinct in order for this problem to be interesting. To see this, consider the set $A := 2^\wedge[1, n] = \{2, 2^2, \ldots, 2^n\}$. Clearly, if B is any subset of A of two or more elements, then there exists an element of A which is the sum of two (equal) elements in B.

It was remarked by Klarner (unpublished) and mentioned by Erdős in [86] that $\phi(n) = \Omega(\log n)$ for large n. The first published proof of this bound appeared in Choi's paper [55] about ten years later:

Theorem 6.2 *Let n be a large integer. Any set A of n real numbers contains a subset B of cardinality $\log n - O(1)$ which is sum-free with respect to A. In other words, $\phi(n) \geq \log n - O(1)$.*

Proof Let us first prove the claim for sets A of positive reals. Let us order the elements of A as $a_1 > a_2 > \cdots > a_n > 0$. Consider the graph G with vertices A, with two distinct elements $a, b \in A$ connected by an edge if and only if $a + b \in A$. By Theorem 6.1, this graph contains an independent vertex set B of size

$$|B| \geq \sum_{i=1}^{n} \frac{1}{\deg(a_i) + 1}.$$

Since B is independent in G, we see that B is sum-free with respect to A. Also, since $a_i + a_j > a_i$, and there are only $n - i$ elements of A larger than a_i, we see that $\deg(a_i) \leq n - i$ for all i. Since $\sum_{i=1}^{n} \frac{1}{n-i+1} = \log n - O(1)$, the claim follows.

To prove the general case, observe from the pigeonhole principle that any set of n reals either contains a subset of $n/2 - O(1)$ positive reals or $n/2 - O(1)$ negative reals, and the claim then follows (for large n) from the preceding paragraph. □

Let us now discuss the upper bound. Thus, we are interested in constructing sets A which do not contain large sum-free subsets. Erdős and Moser [86] proved that $\phi(n) \leq n/3$ and suggested that it probably has order $o(n)$. The first improvement over the Erdős and Moser result was due to Selfridge, who showed $\phi(n) \leq n/4$. Choi [55], using sieve methods, proved that $\phi(n) \leq O_\epsilon\left(n^{2/5+\epsilon}\right)$ for all $\epsilon > 0$. He also noted that in this problem it suffices to consider the special case when A is a set of positive integers. Choi's result was slightly improved by Baltz, Schoen and Srivastav [17], who showed that $\phi(n) \leq O(n^{2/5} \log^{2/5} n)$. A significant improvement of the upper bound was very recently obtained by Ruzsa [303] who proved that

$$\phi(n) = e^{O(\sqrt{\log n})}.$$

In the following we describe Ruzsa's construction, which, besides being very clever, is short and instructive. A key trick is to use a Freiman isomorphism to embed the problem in a very large-dimensional space (see also Exercise 10.1.4).

We shall need a dimension $d = \Theta(\sqrt{\log n})$. Using a Freiman isomorphism (see Lemma 5.25) it is enough to construct a set $A \subset \mathbf{Z}^d$ such that $|A| > n$ and $\phi(A) \leq e^{O(\sqrt{\log n})}$. For any $r > 0$, let $D_r \subset \mathbf{Z}^d$ be the set of integral lattice points in the ball of radius $\sqrt{r}$ centered at the origin, thus

$$D_r := \left\{ (x_1, \ldots, x_d) \in \mathbf{Z}^d \mid \sum_{i=1}^{d} x_i^2 \leq r \right\}.$$

We then set

$$A := \bigcup_{i=0}^{r-1} 2^i \cdot D_{r-i}$$

where $r = e^{O(\sqrt{\log n})}$. For an appropriate choice of d and r one can make $|A| > n$, and we claim that

$$\phi(A) \leq 2^d r = e^{O(\sqrt{\log n})}$$

Indeed, let $S \subset A$ have cardinality greater than $2^d r$. Then by the pigeonhole principle there exists $0 \leq i < r$ such that $|S \cap (2^i \cdot D_{r-i})| > 2^d$. Since $|D_1| = 2d + 1 < 2^d$, we see that $i < r - 1$. By the pigeonhole principle again, we can then find two vectors $s', s'' \in S \cap (2^i \cdot D_{r-i})$ which are congruent modulo $2 \cdot \mathbf{Z}^d$ (i.e. they have the same parity in each coordinate). Then one easily verifies that $s' + s'' \in 2^{i+1} \cdot D_{r-i-1} \subseteq A$, and so S is not sum-free with respect to A.

Remark 6.3 We return to the lower bound. In the same paper which established the upper bound, Ruzsa [297] improved Choi's result slightly by showing $\phi(n) > 2\log_3 n - 1$. Given the fact that Ruzsa's upper bound is sub-polynomial, one may suspect that $\phi(n) = \Theta(\log n)$, i.e., the right order of magnitude of $\phi(n)$ is $\log n$. It is, however, not the case. In a recent paper, Sudakov, Szemerédi and Vu [340] proved that $\phi(n)$ is super-logarithmic: thus in Landau notation

$$\phi(n) = \omega(n) \log n.$$

While this result improves Choi's result only slightly, its proof requires heavy machinery that involves the Balog–Szemerédi–Gowers theorem, Freiman's theorem, and Szemerédi's theorem. In this paper [340], the authors also proved a hypergraph version of the Balog–Szemerédi–Gowers theorem (see Section 6.4).

6.2.2 Turán's theorem and triangle-free graphs

Let G be a graph of n vertices and maximum degree d. The lower bound of $n/(d+1)$ for the size of an independent set that is given by Theorem 6.1 cannot be improved for general graphs G. Thus it was a stunning discovery when Ajtai, Komlós and Szemerédi [2] discovered that one can improve this bound by a factor $\Omega(\log d)$, provided that the graph is *triangle-free* (i.e. it contains no cycles of length three):

Theorem 6.4 *[2] Let $G = G(V, E)$ be a triangle-free graph on n vertices with maximum degree $d \geq 1$. Then G contains an independent set of size $\Omega(\frac{n}{d}\log d)$.*

Proof The original proof of Ajtai, Komlós and Szemerédi is one of the most important proofs in probabilistic combinatorics, as it inspired the development of the so-called semi-random method, which is one of the key achievements in discrete mathematics in the last twenty five years (see for example the introduction of [204]). That proof, however, is complicated and we choose to present a simpler one, found later by Shearer [316]. Shearer's proof also gives the specific lower bound $\frac{n \log_2 d}{8d}$.

Let X be the set of all independent sets in G; X is clearly non-empty. Let I be an element of X chosen uniformly at random; thus I is an independent set of G. It suffices to show that $\mathbf{E}(|I|) \geq \frac{n \log_2 d}{8d}$.

For each vertex $v \in V$ define a random variable

$$Y_v := d|I \cap \{v\}| + |N(v) \cap I|,$$

where we recall $N(v)$ is the set of neighbors of v. Since I is independent, we see that $Y_v = d$ when $v \in I$, and $Y_v = |\{w \in I : \{v, w\} \in E\}|$ otherwise. Since each vertex w in I can be in the neighborhoods of at most d other vertices, a simple counting argument then yields that

$$\sum_{v \in V} Y_v \leq 2d|I|.$$

Taking expectations of both sides and using linearity of expectation, we conclude that

$$\mathbf{E}(|I|) \geq \frac{1}{2d}\sum_{v \in V} \mathbf{E}(Y_v).$$

Thus to prove the desired lower bound on $\mathbf{E}(|I|)$, it will suffice to show that

$$\mathbf{E}(Y_v) \geq \frac{\log_2 d}{4}$$

for all $v \in V$.

Fix $v \in V$, and consider the induced subgraph $G'(V', E')$ spanned by $V' = V \backslash \{N(v) \cup \{v\}\}$. The set $I \cap V'$ is an independent set in V'. To prove the lower bound on $\mathbf{E}(Y_v)$ it will suffice to establish the conditional expectation bound

$$\mathbf{E}(Y_v | I \cap V' = I') \geq \frac{\log_2 d}{4},$$

for all independent sets I' in V'.

Fix this independent set I'. Let $J \subseteq N(v)$ be the set of vertices in V that are adjacent to v but not adjacent to any vertex in I'. Now we make a critical use of the triangle-free hypothesis. Since G is triangle-free, J is an independent set. Therefore, once I' is fixed, we can construct I by either adding v or adding a subset of J to I'. If $|J| = m$, then J has 2^m subsets and so there are exactly $2^m + 1$ choices for I. If v is added to I', $v \in I$ so $Y_v = d$. If a subset J' of J is added to I', then Y_v equals the cardinality of J'. Since the average cardinality of J' is $m/2$, and all choices of I are equally likely by construction, we obtain

$$\mathbf{E}(Y_v | I \cap V' = I') = \frac{d}{2^m + 1} + \frac{m}{2} \frac{2^m}{2^m + 1}.$$

A routine calculation shows that for any integers $d \geq 16$ and $m \geq 1$,

$$\frac{d}{2^m + 1} + \frac{m}{2} \frac{2^m}{2^m + 1} \geq \frac{\log_2 d}{4},$$

concluding the proof. □

Remark 6.5 Ajtai et al. conjectured that the bound $\Omega(\frac{n}{d} \log d)$ can be sharpened to $(1 + o_{n \to \infty; d}(1)) \frac{n}{d} \log d$; this has been confirmed by Shearer [317].

6.2.3 Sidon sets

Recall that an additive set S is called a *Sidon set* (also known as Sidon sequence) if the pairwise sums are all different (except for the trivial equalities $a + b = b + a$). This notion was introduced by Sidon [319] in 1932, motivated by problems in functional analysis.

It is well known, from the work of Erdős & Turán [100] and Singer [320] that the maximum cardinality of a finite Sidon sequence of integers contained in $[1, n]$ is asymptotically $\sqrt{n}$; see Exercises 2.2.6 and 2.2.7. In [320] Singer showed that for $n = p^2 + p + 1$, where p is a prime, there is a Sidon set consisting of $p + 1$ integers between 1 and n. Because of this property, any two translates of this set modulo $p^2 + p + 1$ intersect in exactly one residue class. Thus, the collection of all $p^2 + p + 1$ translates can be identified with the set of lines of a projective plane PF_p^2 of order p.

Estimates concerning infinite Sidon sequences are less satisfying. Erdős & Turán and Stöhr [339] proved that if S is a Sidon sequence, then

$$\limsup_{n\to\infty} \frac{|S \cap [1,n]|}{\sqrt{n}} = 0.$$

Using a greedy algorithm, it is easy to show that there is an infinite Sidon sequence S such that $|S \cap [1,n]| = \Omega(n^{1/3})$ (see Exercise 6.2.5).

It is quite hard to improve upon this trivial bound. The first break-through was due to Ajtai, Komlós and Szemerédi:

Theorem 6.6 *[2] There is an infinite Sidon sequence $S \subset \mathbf{Z}^+$ such that $|S \cap [1,n]| = \Omega(n^{1/3} \log^{1/3} n)$ for all sufficiently large n.*

The proof of this theorem used Theorem 6.4. In fact, this theorem was first developed as a lemma for the proof of Theorem 6.6. Recently, Ruzsa [298] has significantly improved the above result by constructing a Sidon sequence where $|S \cap [1,n]| \geq n^{\sqrt{2}-1+o(1)} = \Omega(n^{.4142})$, using a different method.

Remark 6.7 One can generalize the definition of Sidon sequences by considering sequences where the sums of any two h-tuples are different. Such sequences are called B_h sets and have been studied by various authors. For instance, in [53, 192], it was shown that if h is even, then a B_h set consisting of integers contained in $[1,n]$ cannot have more than $(h/2)^{1/h}((h/2)!)^{2/h} n^{1/h} + O(n^{1/2h})$ elements. These papers also study B_h sets modulo a prime.

Remark 6.8 Let S be a subset of $[1,n]$. We say that S is a *maximal Sidon set* (with respect to $[1,n]$) if S is Sidon and is maximal with respect to inclusion (i.e., adding any element from $[n]\backslash S$ to S would destroy the Sidon property). It is reasonable to ask what is the minimum size of a maximal Sidon set. It is easy to prove that any maximal Sidon set should have at least $n^{1/3}$ elements. Ruzsa [299], using Singer's construction [320], showed that there is a maximal Sidon set with at most $cn^{1/3} \log^{1/3} n$ elements.

Exercises

6.2.1 Without using Theorem 6.1, give an elementary proof of the fact that a graph G on n vertices with maximum degree d must contain an independent set of size $n/(d+1)$. (Hint: use the greedy algorithm.) Give examples that show that this $n/(d+1)$ bound cannot be improved.

6.2.2 Let $G = G(V, E)$ be a graph. Show that G contains an independent set of size at least $\Theta(\frac{|V|^2}{|E|+|V|})$.

6.2.3 Generalize Theorem 6.2 to the case to the case when A takes values in an arbitrary torsion-free additive group. (The torsion-free condition is absolutely necessary, as can be seen by considering the case when $A = \mathbf{Z}_N$).

6.2.4 Let $S \subset [1, n]$ be a maximal Sidon set in $[1, n]$. Show that $2S - S$ contains $[1, n]$, and conclude that $|S| = \Omega(n^{1/3})$.

6.2.5 [339] Let $S = \{1, 2, 4, 8, 13, 21, 31, \ldots\}$ be the Sidon set of positive integers constructed by the greedy algorithm (this set is sometimes known as the *Mian–Chowla sequence*). Show that the kth element of S does not exceed $(k-1)^3 + 1$, and hence $|S \cap [1, n]| = \Omega(n^{1/3})$ as $n \to \infty$.

6.2.6 (Minkowski's bound for sphere packing) A *sphere packing* P in $\mathbf{R}^n$ is a collection of non-intersecting open spheres with equal radii, and its density $\Delta(P)$ is the fraction of space covered by their interior. Define Δ_n to be the supremum of $\Delta(P)$ taken over all packings in $\mathbf{R}^n$. Prove that $\Delta_n = \Omega(2^{-n})$. (This is a special case of the Hlawka–Minkowski problem of packing convex sets in $\mathbf{R}^n$.)

6.2.7 [218] Let the notation be as in the previous exercise. Prove that $\Delta_n = \Omega(n2^{-n})$. (Hint: Discretize the problem, convert the sphere packing problem to one of finding a large independent set, and apply Ajtai et al.'s theorem.) Up to a constant this is the best bound known for sphere packing.

6.2.8 Prove the following extension of Theorem 6.4. Let $G = G(V, E)$ be a triangle-free graph on n vertices with maximum degree d and T triangles. Then G contains an independent set of size $\Omega(\frac{n}{d} \log \frac{dn^{1/2}}{n^{1/2}+T^{1/2}})$. (Hint: Apply Theorem 6.4 to a properly defined random subgraph of G.)

6.3 Ramsey theory

We now briefly consider another application of graph theory, or more precisely *Ramsey theory*, to additive combinatorics. This theory typically can produce results of the following form: if an explicit set (such as $[1, N]$) is colored into finitely many colors, then at least one of the color classes contains a specific arithmetic structure (e.g. an arithmetic progression). The simplest example of this is the *pigeonhole principle*: if we color an n-element set by fewer than n colors, then there exists two elements with the same color. Indeed one can view Ramsey theory as the study of generalizations and repeated applications of the pigeonhole principle. We will focus on only two results in this field, namely Schur's theorem and the Hales–Jewett theorem (a generalization of van der Waerden's theorem); for a more thorough treatment of these topics, see [143].

We say that a graph G is *complete* if every pair of distinct vertices $v, w \in G$ is connected by exactly one edge. A *edge k-coloring* of a graph $G(V, E)$ is a partition of the edge set E into k classes $E_1, \dots, E_k$. We say that a subgraph G' of G is E_j*-monochromatic* if all of its edges lie in E_j.

Theorem 6.9 (Ramsey's theorem for two colors) *[276] Let $n, m \geq 1$ be integers, and let $G = (V, E)$ be a complete graph with at least $\binom{n+m-2}{n-1}$ vertices. Then for any edge 2-coloring $E = E_{\text{blue}} \cup E_{\text{red}}$, there either exists a blue-monochromatic complete subgraph G_{blue} with n vertices, or a red-monochromatic complete subgraph G_{red} with m vertices.*

Example 6.10 Any two-coloring of a complete graph with six or more vertices into red and blue edges will contain either a blue triangle or a red triangle.

Proof We shall induce on the quantity $n + m$. When $n + m = 2$ (i.e. $n = m = 1$) the claim is vacuously true. Now suppose that $n + m > 2$ and the claim has already been proven for all smaller values of $n + m$. If $n = 1$ then the claim is again vacuous, and similarly when $m = 1$. Thus we shall assume $n, m \geq 2$.

Let $G = (V, E)$ be a complete graph with at least $\binom{n+m-2}{n-1}$ vertices, and let $v \in V$ be an arbitrary vertex. This vertex is adjacent to at least

$$\binom{n+m-2}{n-1} - 1 = \binom{n+m-3}{n-2} + \binom{n+m-3}{n-1} - 1$$

many edges, each of which is either blue or red. Thus by the pigeonhole principle, either v is adjacent to at least $\binom{n+m-3}{n-2}$ blue edges, or is adjacent to at least $\binom{n+m-3}{n-1}$ red edges. Suppose first that we are in the former case. Then we can find a complete subgraph G' of G with at least $\binom{n+m-3}{n-2}$ edges such that every vertex of G' is connected to v by a blue edge. By the induction hypothesis (with (n, m) replaced by $(n-1, m)$), G' either contains a blue-monochromatic complete subgraph G'_{blue} with $n-1$ vertices, or a red-monochromatic complete subgraph G'_{red} with m vertices. In the latter case we are already done by taking $G_{\text{red}} := G'_{\text{red}}$, and in the former case we can find a blue-monochromatic complete subgraph G_{blue} of G with n vertices by adjoining v to G'_{blue} (and adding in all the edges connecting v and G'_{blue}, which are all blue by construction). This disposes of the case when v is adjacent to at least $\binom{n+m-3}{n-2}$ blue edges; the case when v is connected to at least $\binom{n+m-3}{n-1}$ red edges is proven similarly (now using the inductive hypothesis at $(n, m-1)$ instead of $(n-1, m)$). $\square$

Remark 6.11 The bound $\binom{n+m-2}{n-1}$ is sharp for very small values of n and m, but can be improved for larger values of n and m, although computing the precise constants is very difficult (for instance, when $n = m = 5$ the best constant is only known to be somewhere between 43 and 49 inclusive). On the other hand, lower bounds are known (see Exercise 6.3.6).

One can iterate this theorem to arbitrary number of colors:

Corollary 6.12 (Ramsey's theorem for many colors) *[276] Given any positive integers $n_1, \ldots, n_m$, there exists a number $R(n_1, \ldots, n_m; m)$ such that given any complete graph $G = (V, E)$ with at least $R(n_1, \ldots, n_m; m)$ vertices, and any edge m-coloring $E = E_1 \cup \cdots \cup E_m$, there exists a $1 \leq j \leq m$ and a E_j-monochromatic complete subgraph G_j of G with n_j vertices.*

Proof We induct on m. The case $m = 1$ is trivial, and the case $m = 2$ is just Theorem 6.9. Now suppose inductively that $m > 2$ and the claim has already been proven for all smaller values of m. We set

$$R(n_1, \ldots, n_m; m) := R(R(n_1, \ldots, n_{m-1}; m-1), n_m; 2).$$

Suppose we color the edges of $K_{R(n_1,\ldots,n_m;m)}$ into m color classes $E_1, \ldots, E_m$. We coarsen this edge m-coloring into an edge 2-coloring $E_1 \cup \cdots \cup E_{m-1}, E_m$. By the induction hypothesis, we see that with respect to the coarsened coloring, either G contains an E_m-monochromatic complete subgraph G_m with n_m elements, or G contains an $E_1 \cup \cdots \cup E_{m-1}$-monochromatic complete subgraph $G_{1,\ldots,m-1}$ with $R(n_1, \ldots, n_{m-1}; m-1)$ elements. In the first case we are done; in the second case we are done by applying the induction hypothesis once again, this time to the complete graph $G_{1,\ldots,m-1}$. This closes the induction and completes the proof. □

We now give an immediate application of Ramsey's theorem to an arithmetic setting.

Theorem 6.13 (Schur's theorem) *[315] If m, k are positive integers, there exists a positive integer $N = N(m, k)$ such that, given any partition of $[1, N]$ into m sets $[1, N] = A_1 \cup \cdots \cup A_m$, at least one of the A_j contains a subset of the form $\{x_1, \ldots, x_k, x_1 + \cdots + x_k\}$. In fact we can choose $N := R(k+1, \ldots, k+1; m) - 1$, using the notation of Corollary 6.12.*

Remarks 6.14 Schur's theorem (in the $k = 2$ case) is equivalent to the assertion that the set $[1, N]$ cannot be covered by m sum-free sets if N is sufficiently large depending on m; in particular, the integers cannot be partitioned into any finite number of sum-free sets. Even when $k = 2$, the value of N given by the above arguments grows double-exponentially in m (Exercise 6.3.4); this is not best possible. For instance, it is known that given any 2-coloring of $[1, N]$, there exist at least $\frac{1}{22}N^2 - \frac{7}{22}N$ monochromatic triples of the form $(x, y, x+y)$, and that this bound is sharp [280], [313] (see also [142]).

Proof Let $G = G(V, E)$ be the complete graph on the $N+1$ vertices $V := [1, N+1]$, and let us edge m-color this graph as $E = E_1 \cup \cdots \cup E_m$ where E_j is the set of those edges (a, b) for which $|a - b| \in A_j$. By Corollary 6.12, the graph G

must contain a complete subgraph G' of $k+1$ vertices which is E_r-monochromatic for some r. If we list the vertices of G' in order as $v_0 < v_1 < \cdots < v_k$, then the quantities $\mathbf{c}(v_i - v_j)$ for $i > j$ are all equal to each other. The claim then follows by setting $x_j := v_j - v_{j-1} \in A_r$. □

We now give the Hales–Jewett theorem, which we state in an "arithmetic" format. While not strictly a theorem about graphs, it is certainly close in spirit to Ramsey's theorem.

Theorem 6.15 (Hales–Jewett theorem) *[169] Let $m \geq 1$ and $n \geq 1$. Then there exists an integer $d = d(n, m) \geq 1$ such that if $[0, n-1]^d \subset \mathbf{Z}^d$ is partitioned into m non-empty sets $[0, n-1]^d = E_1 \cup \cdots \cup E_m$, then at least one of the sets E_j contains a proper arithmetic progression $a + [0, n-1] \cdot v$ of length n, for some $a \in [0, n-1]^d$ and $v \in [0, 1]^d$.*

This theorem can be proven by a double induction. It is a special case of the following more technical proposition, in which one either locates a single monochromatic progression of length n, or several linked monochromatic progressions of length $n-1$ (with each progression being monochromatic with a different color).

Proposition 6.16 *Let $m \geq 1$, $n \geq 1$, and $1 \leq s \leq m$. Then there exists an integer $\tilde{d} = \tilde{d}(n, m, s) \geq 1$ such that if $[0, n-1]^{\tilde{d}} \subset \mathbf{Z}^{\tilde{d}}$ is partitioned into m non-empty sets $[0, n-1]^{\tilde{d}} = E_1 \cup \cdots \cup E_m$, then either at least one of the sets E_j contains a proper arithmetic progression $a + [0, n-1] \cdot v$, or there exist distinct classes $E_{j_1}, \ldots, E_{j_s}$ and $a \in [0, n-1]^{\tilde{d}}$ and non-zero $v_1, \ldots, v_s \in [0, 1]^{\tilde{d}}$ such that $a + [1, n-1] \cdot v_i \subseteq E_{j_i}$ for all $1 \leq i \leq s$.*

Indeed, applying Proposition 6.16 with $s := m$ one can conclude Theorem 6.15, since if one has m distinct monochromatic progressions $a + [1, n-1] \cdot v_i$, then one of the progressions $a + [0, n-1] \cdot v_i$ must also be monochromatic by the pigeonhole principle.

Proof of Proposition 6.16 To abbreviate notation, we shall use "arithmetic progression" in this proof to denote any proper arithmetic progression $a + [0, n-1] \cdot v$ or $a + [1, n-1] \cdot v$ in a lattice $\mathbf{Z}^d$ where $a \in [0, n-1]^d$ and $v \in [0, 1]^d$.

We use two induction loops. For the outer loop, we induce on n. The claim is trivial when $n = 1$, so we assume that $n > 1$ and the claim has already been proven for $n-1$ (and for arbitrary m, s). In particular, by the above discussion we see that we may assume Theorem 6.15 for $n-1$.

Now we begin our inner loop, inducing on s. When $s = 1$ the claim follows from Theorem 6.15 for $n-1$ (shifting $[1, n-1]$ to $[0, n-2]$), so assume that $2 \leq s \leq m$ and the claim has already been proven for $s-1$ (and the same value of n, but with arbitrary m). We set $\tilde{d} := \tilde{d}(n, m, s) := d_1 + d_2$, where $d_1 := \tilde{d}(n, m, s-1)$

and $d_2 := d(n-1, m^s n^{sd_1})$. Let $[0,n-1]^{\tilde{d}} = E_1 \cup \cdots \cup E_m$ be a partition of $[0,n-1]^{\tilde{d}}$ into m distinct color classes. Suppose that none of the E_j contain any arithmetic progressions of length n. Our task is then to show that there are s distinct classes $E_{j_1}, \ldots, E_{j_s}$, $a \in [0,n-1]^{\tilde{d}}$, and $v_1, \ldots, v_s \in [0,1]^{\tilde{d}}$ such that $a + [1, n-1] \cdot v_i \subseteq E_{j_i}$ for all $1 \leq i \leq s$.

We write $[0,n-1]^{\tilde{d}} = [0,n-1]^{d_1} \times [0,n-1]^{d_2}$, and for each $x \in [0,n-1]^{d_2}$ we consider the partition $[0,n-1]^{d_1} = E_{1,x} \cup \cdots \cup E_{m,x}$, where $E_{j,x} := \{y \in [0,1]^{d_1} : (y,x) \in E_j\}$. Since none of the E_j contain an arithmetic progression of length n, neither do the $E_{j,x}$. By definition of d_1 and the inner induction hypothesis, we conclude that for each x there exist distinct color classes $j_{1,x}, \ldots, j_{s-1,x}$, $a_x \in [0,n-1]^{d_1}$ and $v_{1,x}, \ldots, v_{s-1,x} \in [0,1]^{d_1}$ such that

$$a_x + [1, n-1] \cdot v_{i,x} \in E_{j_{i,x}} \tag{6.1}$$

for all $1 \leq i \leq s-1$. Note that a_x itself must then belong to another color class $j_{s,x}$ distinct from $j_{1,x}, \ldots, j_{s-1,x}$, otherwise one of the classes $E_{j,x}$ would contain an arithmetic progression of length n. If we set $v_{s,x} := 0$ then we see that (6.1) now holds for $i = s$ also, although in that case the progression $a_x + [1, n-1] \cdot v_{i,x}$ is not proper. This will however be rectified by means of the d_2 coordinates.

The map $x \mapsto (j_{1,x}, \ldots, j_{s,x}, a_x, v_{1,x}, \ldots, v_{s-1,x})$ is a map from $[0,n-1]^{d_2}$ to a set of cardinality at most $m^s n^{sd_1}$. Thus it induces a partition $[0,n-1]^{d_2} = F_1 \cup \cdots \cup F_{m^s n^{sd_1}}$ into $m^s n^{sd_1}$ color classes (some of which may be empty). By definition of d_2 and the outer induction hypothesis (again shifting $[1, n-1]$ to $[0, n-2]$), we conclude that one of the color classes F_t contains an arithmetic progression $a_* + [1, n-1] \cdot v_*$ with $a_* \in [0,n-1]^{d_2}$ and $v_* \in [0,1]^{d_2}$. This means that there exist distinct $j_{1,(t)}, \ldots, j_{s,(t)} \in [1,m]$, $a_{(t)} \in [0,n-1]^{d_1}$, and $v_{1,(t)}, \ldots, v_{s,(t)} \in [0,1]^{d_1}$ (with $v_{s,(t)} = 0$) such that $a_{(t)} + [1, n-1] \cdot v_{i,(t)} \in E_{j_{i,x}}$ for all $x \in a_* + [1, n-1] \cdot v_*$ and $1 \leq i \leq s$. But if we now set $a := (a_{(t)}, a_*) \in [0,n-1]^{\tilde{d}}$ and $v_i := (v_{i,(t)}, v_*) \in [0,1]^{\tilde{d}}$, we see that $a + [1,n] \cdot v_i \in E_{j_i}$ for all $1 \leq j \leq s$, and that each of the $a + [1,n] \cdot v_i$ are *proper* arithmetic progressions of length $n-1$. This closes the induction loop, and the claim follows. □

This theorem has a number of consequences, the most notable being perhaps van der Waerden's theorem.

Theorem 6.17 *(van der Waerden, [371]) Let $k, m \geq 1$ be integers. Then there exists an integer $N = N(k,m) \geq 1$ such that given any proper arithmetic progression P of length at least N (in an arbitrary additive group Z), and any partition $P = E_1 \cup \cdots \cup E_m$ of P into m color classes, at least one of these classes E_j contains a monochromatic proper arithmetic sub-progression P' of P of length $|P'| = k$.*

We leave the proof as an exercise. Let us, however, remark that if we fix k then the bound on m which follows from Hales–Jewett's theorem is very poor, growing as fast as the infamous Ackermann function. One can use Gowers' theorem [138] and the pigeonhole principle to deduce a much better bound.

Remark 6.18 In the case of $k = 3$, Solymosi observed (private communication) that one can obtain a rather good bound (which is comparable to the bound one gets from Roth's theorem) by a simple argument which does not involve Fourier analysis. For simplicity, let us assume that we color a group Z of cardinality N by k colors. We now show that there is a monochromatic arithmetic progression of length 3, assuming that k is sufficiently small compared with N. Let C_1 be the most popular color and let $a_1, \ldots, a_{m_1}$ be the elements colored by C_1. Clearly $m_1 \geq n/k$. By the pigeonhole principle, there is an element $x \in Z$ such that there are at least $\binom{m_1}{2}/n$ pairs (a_i, a_j), $i < j$ such that $a_j - a_i = x$. If there is no monochromatic arithmetic progression of length 3, then $b_j = a_j + x$ is not colored by C_1. Thus we end up with a set S_1 of at least

$$\binom{m_1}{2}\Big/n \geq n/3k^2 = n_1$$

elements which are not colored by C_1. Now repeat the argument with the set S_1; we end up with a set S_2 of size at least $\binom{n_1}{2}/n \geq n/27k^4 = n_2$ elements which are not colored by either C_1 or C_2 (Exercise 6.3.8). Iterating this argument k times, we end up with a set of $n_k = n/3^{2^k-1}k^{2^k}$ elements which cannot be colored by any color. This is a contradiction if $n \geq 3^{2^k-1}k^{2^k}$.

Exercises

6.3.1 Using Schur's theorem, show that if the positive integers $\mathbf{Z}^+$ are finitely colored and $k \geq 1$ is arbitrary, then there exist infinitely many monochromatic sets in $\mathbf{Z}^+$ of the form $\{x_1, \ldots, x_k, x_1 + \cdots + x_k\}$. (Hint: Schur's theorem can easily produce *one* such set; now color all the elements of that set by new colors and repeat.) Conversely, show that if the previous claim is true, then it implies Schur's theorem.

6.3.2 Show that if the positive integers $\mathbf{Z}^+$ are finitely colored then there exist infinitely many *distinct* integers x and y such that $\{x, y, x + y\}$ are monochromatic. (Hint: refine the coloring so that x and $2x$ always have different colors.) A more challenging problem is to establish a similar result for general k, i.e. to find infinitely many distinct $x_1, \ldots, x_k$ such that $\{x_1, \ldots, x_k, x_1 + \cdots + x_k\}$ is monochromatic.

6.3.3 Show that if the positive integers $\mathbf{Z}^+$ are finitely colored and $k \geq 1$ is arbitrary, then there exist infinitely many monochromatic sets of the form $\{x_1, \ldots, x_k, x_1 \ldots x_k\}$. Thus Schur's theorem can be adapted to products instead of sums. However, nothing is known about the situation when one has both sums *and* products; for instance, it is not even known that if one finitely colors the positive integers that one can find even a single monochromatic set of the form $\{x, y, x+y, xy\}$ for some positive integers x, y (not both equal to 1).

6.3.4 Show that the quantity $N(m, k)$ in Schur's theorem can be taken to be $O(1)^{k^m}$.

6.3.5 Let k be an integer, and let A be an additive set in an ambient group Z such that $|A| \geq \binom{2k-2}{k-1}$, and let C be an arbitrary subset of Z. Show that there exists a set $B \subseteq A$ of cardinality $|B| = k$ such that either $B + B \subseteq C$ or $B + B$ is disjoint from C.

6.3.6 [84] Show that if $n \geq 3$ and $N \leq 2^{n/2}$ then there exists a two-coloring of the edges of the complete graph on N vertices which does not contain a monochromatic complete subgraph of n vertices. (Hint: color the graph randomly.)

6.3.7 Prove van der Waerden's theorem. (Hint: set $N = k^d$ for a large d, and identify P with $[0, k-1]^d$. Then apply Theorem 6.15.)

6.3.8 Consider Remark 6.18. Show that if after the ith step we get an element y which is colored by C_j for some $j < i$, then $y, y - (d_i + \cdots + d_j), y - 2(d_i + \cdots + d_j)$ are all of color C_j, where d_l is the "popular" difference in step l.

6.3.9 Let Z be an arbitrary finite additive group, partitioned into m color classes $E_1 \cup \cdots \cup E_m$. Show that for any $k \geq 1$ there exists a color class E_j such that

$$\mathbf{P}_{a,r \in Z}(a, a+r, \ldots, a+(k-1)r \in E_j) = \Omega_{k,m}(1).$$

(Hint: apply Theorem 6.17 to a random progression in Z of a suitable length $N(k, m)$ and use the first moment method.) This is a weak form of Varnavides' version of Szemerédi's theorem, see Theorem 11.1.

6.3.10 Let A be an additive set, and let $P(n)$ be a statement pertaining to an element $n \in A$. Let us say that the property P is *k-choosable* for some $k \geq 1$ if, given every proper arithmetic progression of length k in A, at least one element n of that progression obeys the property $P(n)$. Show that if the properties $P_1(n), \ldots, P_m(n)$ are k-choosable, then the joint property $P_1(n) \wedge \cdots \wedge P_m(n)$ is $O_{k,m}(1)$-choosable. (This statement is in fact equivalent to van der Waerden's theorem, and plays a key role in the original proof [345] of Szemerédi's theorem.)

6.3.11 (Multi-dimensional Hales–Jewett theorem) [169] Let $n, m, r \geq 1$. Show that there exists an integer $d = d(n, m, r) \geq 1$ such that, given any partition of $[0, n-1]^d$ into m color classes $E_1, \ldots, E_m$, then at least one of the color classes contains a proper generalized arithmetic progression $a + [0, n-1]^r \cdot (v_1, \ldots, v_r)$, where $a \in [0, n-1]^d$ and $v_1, \ldots, v_r \in [0, 1]^d$. (Hint: apply Theorem 6.15 with n replaced by n^r.)

6.3.12 (Gallai's theorem) Let $k \geq 1$, $d \geq 1$, $m \geq 1$, and let $v_1, \ldots, v_k$ be elements of $\mathbf{Z}^d$. Show that there exists an $N = N(k, d, m, v_1, \ldots, v_k)$ such that for partition of the cube $[1, N]^d \subset \mathbf{Z}^d$ into m color classes $E_1, \ldots, E_m$, then at least one of the color classes contains a set of the form $\{x + rv_1, \ldots, x + rv_k\}$ for some $x \in \mathbf{Z}^d$ and some non-zero integer r.

6.4 Proof of the Balog–Szemerédi–Gowers theorem

Let A and B be two additive sets with common ambient group. Let $G = G(A, B, E)$ be a bipartite graph whose color classes are A and B and whose edge set is E (an edge is a pair (a, b) where $a \in A$ and $b \in B$). Recall that the *partial sum set* $A \overset{G}{+} B$ is defined as the collection of the sums $a + b$ where $a \in A, b \in B$ and $(a, b) \in E$.

Balog and Szemerédi [16] proved that if A and B are two sets of cardinality N and $|E| \geq n^2/K$ and $|A \overset{G}{+} B| \leq K'n$ for some K, K', then one can find $A' \subset A$ and $B' \subset B$ such that $|A'|, |B'|, |A' + B'| = \Theta_{K,K'}(n)$.

As stated, the above theorem is only useful if K and K' are independent of n (or extremely slowly growing in n). With a new proof, Gowers [138] has recently strengthened this statement by showing that the implicit constants in the $\Theta_{K,K'}()$ notation can be taken to be polynomial in K and K', and hence the theorem remains effective even when K and K' are as large as n^ε for some absolute constant $\varepsilon > 0$; we have already stated this result in Theorem 2.29. This has proven to be immensely valuable in a number of applications in which polynomial-type bounds are desired, for instance in Gowers' proof of Szemerédi's theorem (see in particular Section 11.3). The polynomials in Gowers' proof were implicit, but by following his ideas, one can work out the explicit version given in Theorem 2.29. Our treatment here is based on that in [340].

As it turns out, one can view the Balog–Szemerédi–Gowers theorem as a statement about dense bipartite graphs. Clearly, if a bipartite graph $G(A, B, E)$ has many edges, then there will be many pairs of vertices $a \in A$, $b \in B$ which are connected by paths of length 1. One then expects there to be many pairs $a, a' \in A$ which are connected by paths of length two, and many pairs $a \in A, b \in B$ which are connected by paths of length three. Furthermore, this connectivity becomes increasingly more "uniform" as the length of the path increases; compare with the results on arithmetic progressions in sum sets in Section 4.7. It is this uniformity which is essential to the proof of the Balog–Szemerédi–Gowers theorem.

We begin by formalizing the above principle for paths of length two and length three.

Lemma 6.19 (Paths of length two) *Let $G(A, B, E)$ be a bipartite graph with $|E| \geq |A||B|/K$ for some $K \geq 1$. Then, for any $0 < \varepsilon < 1$, there exists a subset $A' \subseteq A$ such that*

$$|A'| \geq \frac{|A|}{\sqrt{2}K}$$

and such that at least $(1 - \varepsilon)$ of the pairs of vertices $a, a' \in A'$ are connected by at least $\frac{\varepsilon}{2K^2}|B|$ paths of length two in G.

Proof By decreasing K if necessary we may assume $|E| = |A||B|/K$. Observe the combinatorial identities

$$\mathbf{E}_{b \in B} \frac{|N(b)|}{|A|} = \mathbf{E}_{a \in A} \frac{|N(a)|}{|B|} = \frac{|E|}{|A||B|} = \frac{1}{K}$$

and

$$\mathbf{E}_{b \in B} \frac{|N(b)|^2}{|A|^2} = \mathbf{E}_{a,a' \in A} \frac{|N(a) \cap N(a')|}{|B|}.$$

Applying Cauchy–Schwarz we conclude that

$$\mathbf{E}_{a,a' \in A} \frac{|N(a) \cap N(a')|}{|B|} \geq \frac{1}{K^2}.$$

Let Ω be the set of all pairs (a, a') such that $|N(a) \cap N(a')| < \frac{\varepsilon}{2K^2}|B|$; in other words, $(a, a') \in \Omega$ if a, a' are *not* connected by at least $\frac{\varepsilon}{2K^2}$ paths of length two. Clearly we have

$$\mathbf{E}_{a,a' \in A} \mathbf{I}((a, a') \in \Omega) \frac{|N(a) \cap N(a')|}{|B|} < \frac{\varepsilon}{2K^2}$$

and hence

$$\mathbf{E}_{a,a' \in A} \left(1 - \frac{1}{\varepsilon} \mathbf{I}((a, a') \in \Omega)\right) \frac{|N(a) \cap N(a')|}{|B|} \geq \frac{1}{2K^2}.$$

The left-hand side can be rearranged as

$$\mathbf{E}_{b \in B} \frac{1}{|A|^2} \sum_{a,a' \in N(b)} \left(1 - \frac{1}{\varepsilon} \mathbf{I}((a, a') \in \Omega)\right)$$

and hence by the pigeonhole principle there exists $b \in B$ such that

$$\frac{1}{|A|^2} \sum_{a,a' \in N(b)} \left(1 - \frac{1}{\varepsilon}\mathbf{I}((a,a') \in \Omega)\right) \geq \frac{1}{2K^2}.$$

In particular this implies that $|N(b)| \geq \frac{|A|}{\sqrt{2K}}$ and that $|\{a, a' \in N(b) : (a, a') \in \Omega)\}| \leq \varepsilon |N(b)|^2$. The claim then follows by setting $A' := N(b)$. □

We now obtain an analogous result for paths of length three.

Corollary 6.20 (Paths of length three) *Let $G(A, B, E)$ be a bipartite graph with $|E| \geq |A||B|/K$ for some $K \geq 1$. Then there exists $A' \subseteq A$, $B' \subseteq B$ with $|A'| \geq \frac{|A|}{4\sqrt{2K}}$ and $|B'| \geq \frac{|B|}{4K}$, such that every $a \in A'$ and $b \in B'$ is connected by at least $\frac{|A||B|}{2^{12}K^5}$ paths of length three.*

Proof Before we apply Lemma 6.19 it is convenient to prepare the graph G a little bit. Let $\tilde{A}$ be the set of vertices in A that have degree at least $|B|/2K$, and let $\tilde{G} = \tilde{G}(\tilde{A}, B, \tilde{E})$ be the induced subgraph. Since at most $|A||B|/2K$ edges are removed when passing from G to $\tilde{G}$, we see that $\tilde{G}$ has at least $|A||B|/2K$ edges. Writing $|A| = L|\tilde{A}|$ for some $L \geq 1$ and applying Lemma 6.19 to $\tilde{G}$ (with K replaced by $2K/L$ and $\varepsilon := \frac{1}{16K}$) we can find a subset $\tilde{A}'$ of A' of size

$$|\tilde{A}'| \geq \frac{|\tilde{A}|}{\sqrt{2}(2K/L)} = \frac{|A|}{2\sqrt{2}K}$$

and such that $1 - \frac{1}{16K}$ of the pairs $a, a' \in \tilde{A}'$ are connected by at least $L^2|B|/128K^3$ paths of length two.

Let us call a pair $(a, a') \in \tilde{A}' \times \tilde{A}'$ *bad* if they are not connected by at least $\frac{L^2|B|}{128K^2}$ paths of length two; thus there are at most $\frac{1}{16K}|\tilde{A}'|^2$ bad pairs. Let A' be the set of all $a \in \tilde{A}'$ such that at most $\frac{1}{8K}|\tilde{A}'|$ pairs (a, a') are bad. Then $|\tilde{A}' \backslash A'| \leq \frac{|\tilde{A}'|}{2}$, and thus

$$|A'| \geq \frac{1}{2}|\tilde{A}'| \geq \frac{|A|}{4\sqrt{2}K}.$$

Having constructed A', we turn now to B'. Since every element in $\tilde{A}$ (and hence in $\tilde{A}'$) has degree at least $|B|/2K$, we have

$$\sum_{b \in B} |\{a \in \tilde{A}' : (a, b) \in E)\}| = |\{(a, b) \in E : a \in A'\}| \geq |\tilde{A}'|\frac{|B|}{2K},$$

so if we let

$$B' := \left\{b \in B : |\{a \in \tilde{A}' : (a, b) \in E)\}| \geq \frac{|\tilde{A}'|}{4K}\right\}$$

then we have

$$|\tilde{A}'||B'| \geq \sum_{b \in B'} |\{a \in \tilde{A}' : (a, b) \in E\}| \geq |\tilde{A}'|\frac{|B|}{2K} - \frac{|\tilde{A}'|}{4K}|B| = \frac{|\tilde{A}'||B|}{4K}.$$

In particular we have $|B'| \geq |B|/4K$.

Finally, let $a \in A'$ and $b \in B'$ be arbitrary. By the construction of B', then b is adjacent to at least $|\tilde{A}'|/4K$ elements a' of $\tilde{A}'$. By construction of A', at most $|\tilde{A}'|/8K$ of the pairs (a, a') are bad. Thus there are at least $|\tilde{A}'|/8K \geq |A|/16\sqrt{2}K^2$ vertices a' which are simultaneously adjacent to b, and are connected to a by at least $\frac{L^2|B|}{128K^2}$ paths of length two. Thus a and b are connected by at least

$$\frac{|A|}{16\sqrt{2}K^2}\frac{L^2|B|}{128K^3} \geq \frac{|A||B|}{2^{12}K^5}$$

paths of length three. □

We can now derive as a consequence the Balog–Szemerédi–Gowers theorem, Theorem 2.29.

Proof of Theorem 2.29 First observe that we may ensure that A and B are disjoint, by the artificial trick of replacing the ambient group Z with $Z \times \mathbf{Z}$, replacing A with $A \times \{0\}$, and B with $B \times \{1\}$. Let us view the set $G \subset A \times B$ in the theorem as a bipartite graph on A and B. Applying Corollary 6.20, we can find A', B' obeying (2.18), (2.19), and such that every pair $a \in A'$, $b \in B'$ is connected by at least $|A||B|/2^{12}K^5$ paths of length three:

$$|\{(a', b') \in A \times B : (a, b'), (a', b'), (a', b) \in G\}| \geq \frac{|A||B|}{2^{12}K^5}.$$

Exploiting the obvious identity

$$a + b = (a + b') - (a' + b') + (a' + b)$$

and writing $x := a + b'$, $y := a' + b'$, $z := a' + b$, we conclude that

$$|\{(x, y, z) \in (A \overset{G}{+} B)^3 : x - y + z = a + b\}| \geq \frac{|A||B|}{2^{12}K^5}.$$

Since the total number of triples (x, y, z) is at most

$$|A \overset{G}{+} B|^3 \leq (K')^3|A|^{3/2}|B|^{3/2},$$

we conclude that the total number of possible values for $a + b$ is at most $2^{12}K^5(K')^3|A|^{1/2}|B|^{1/2}$, and the claim follows. □

Note that in this proof it is not critical that the group is abelian. For a multiplicative group, we can replace $a + b = (a + b') - (a' + b') + (a' + b)$ by $ab = (ab')(a'b')^{-1}(a'b)$, and the rest of the proof is the same.

To conclude this section, let us mention a generalization of Balog–Szemerédi–Gowers result for hypergraphs. Let $A_1, \ldots, A_k$ be additive sets with common ambient group (which we may take to be disjoint, by the trick used above) and let E be some family of ordered k-tuples $(a_1, \ldots, a_k)$ such that $a_i \in A_i$, $1 \leq i \leq k$. The sets $A_1, \ldots, A_k$ together with E are known as a *k-uniform k-partite hypergraph* which we shall call H; the set E is then known as the *edge set* of H (notice that a bipartite graph is a special case when $k = 2$). We denote by $\bigoplus_{H\,i=1}^{k} A_i$ the collection of the sums $a_1 + \cdots + a_k$ where $(a_1, \ldots, a_k) \in E$. For the case $k = 2$, we are talking about bipartite graphs.

Theorem 6.21 *[340] Let $k \geq 1$, and let n, K be positive numbers. If $A_1, \ldots, A_k$ are additive sets in a group Z of cardinality at most n, and $H(A_1, \ldots, A_k, E)$ is a k-partite k-uniform hypergraph with at least n^k/K edges and $\left|\bigoplus_{H\,i=1}^{k} A_i\right| \leq Kn$, then one can find subsets $A_i' \subset A_i$ such that*

- $|A_i'| = \Omega_k(n/K^{O_k(1)})$ *for all* $1 \leq i \leq k$.
- $|A_1' + \cdots + A_k'| = \Omega_k(K^{O_k(1)}n)$.

The heart of the proof is the following claim.

Claim 6.22 *Let $A_1, \ldots, A_k$ and n, K be as in the theorem above. Set $X = \bigoplus_{H\,i=1}^{k} A_i$. There are subsets $A_i' \subset A_i, i = 1, \ldots, k$ of cardinality at least $\Omega_k(n/K^{O_k(1)})$ and sets $Y_j \subseteq Z, 1 \leq j \leq 2k-2$ of cardinality at most $O_k(K^{O_k(1)}n)$, such that every element in $A_1' + \cdots + A_k'$ can be written in the form $x + \sum_{j=1}^{2k-2} y_j$ where $x \in X$, $y_j \in Y_j$ in at least $\Omega_k(n^{2k-2}/K^{O_k(1)})$ ways.*

It is easy to deduce Theorem 6.21 from this claim. For the sets $A_1', \ldots, A_k'$ as in the claim, we have

$$|A_1' + \cdots + A_k'| \leq \frac{|X| \prod_{j=1}^{2k-2} |Y_j|}{\Omega_k(n^{2k-2}/K^{O_k(1)})} = \Omega_k(K^{O_k(1)}n)$$

as desired. The proof of Claim 6.22 is left as an exercise.

Exercises

6.4.1 Let $G = G(A, B, E)$ be a bipartite graph such that $|E| \geq |A||B|/K$. Show that there exists a subset A' of A of cardinality $|A'| \geq |A|/K$ such that any two elements in A' are connected by at least one path of length 2 in G. Show that $|A|/K$ cannot be improved to $|A|/K + 1$, even when A, B, and K are large.

6.4.2 [210] Let d be a large integer. Let $V = \{0, 1\}^d$, be the d-dimensional discrete cube, and let $G = G(V, E)$ be the bipartite graph formed by joining an edge between $x, y \in V$ if x and y differ in at most $d/2$ coordinates (i.e. if the Hamming distance between x and y is at most $d/2$). Show that $|E| = (\frac{1}{4} + o_{d\to\infty}(1))|V|^2$, but if V' is any subset of V with size $|V'| \geq c|V|$ then there exist x, x' in V' that are connected by fewer than $o_{d\to\infty}(|V|)$ paths of length 2 in G. (Hint: use a volume-packing argument to find two points x, x' in V' which are almost antipodal in the sense that their Hamming distance is $d - O(1)$.) Convert this example into a bipartite example and show that one cannot expect to eliminate the $(1-\varepsilon)$ factor in Lemma 6.19 even if one lets ε be sufficiently small depending on K.

6.4.3 (Benny Sudakov, private communication) Let G be a bipartite graph $G = G(A, B, E)$ with $|A| = |B| = N$ and $|E| = \Theta(N^2)$ where N is sufficiently large. Show that G contains a complete bipartite graph with $\Omega(\log N)$ vertices in each color class. Show that the bound $\Omega(\log N)$ is best possible.

6.4.4 Let Z be the finite additive group $Z = \mathbf{Z}_2^d$ for some integer d, and let $\hat{Z}$ be the Pontryagin dual. Let $G = G(Z, \hat{Z}, E)$ be the bipartite graph formed by connecting $x \in Z$ to $\chi \in Z$ whenever $\chi(x) = 0$. Show that $|E| = |A||B|/2$. Using (4.2), show that one has $|A||B| \leq |Z|$ whenever $A \subseteq Z, B \subseteq Z'$ is a bipartite clique in G. Conversely, whenever N_1 and N_2 are positive integers such that $N_1 N_2 = |Z|$, show that there exists a bipartite clique $A \subseteq Z$, $B \subseteq Z'$ in G with $|A| = N_1$ and $|B| = N_2$. Compare this result with Exercise 6.4.3.

6.4.5 (Dyadic pigeonhole principle) Let $G = G(A, B, E)$ be a bipartite graph with $|E| \geq |A||B|/K$ for some $K \geq 1$. Show that there exists some $1 \leq K' \leq K$ and some induced subgraph $G' = G(A', B, E')$ of $G(A, B, E)$ with

$$|E|/(C + C\log K) \leq |E'| \leq |E|; \quad |A|/(C + C\log K) \leq |A'| \leq |A|$$

such that $|B|/2K' \leq \deg_{G'}(a) \leq |B|/K'$ for all $a \in A'$.

6.4.6 (Simultaneous popularity principle) Let $G = G(A, B, E)$ be a bipartite graph with $|E| \geq |A||B|/K$ for some $K \geq 1$. Show that there exists an induced subgraph $G' = G(A', B', E')$ with the bounds

$$|A'||B'| \geq |E'| \geq \frac{|A||B|}{2K^2}$$
$$|A'| \geq \frac{|A|}{K^2}$$
$$|B'| \geq \frac{|B|}{K^2}$$

such that $\deg_{G'}(a) \geq |B|/2K$ and $\deg_{G'}(b) \geq |A|/2K$ for all $a \in A'$ and $b \in B'$. (Hint: choose A', B' to maximize the quantity $\frac{|E \cap (A' \times B')|}{|A'|^{1/2}|B'|^{1/2}}$.)

6.4.7 Prove Claim 6.22. (Hint: use induction.)

6.4.8 Using the same hypotheses as Theorem 2.29, show that for any $\varepsilon > 0$ there exists a set $G' \subseteq A' \times A'$ such that $|G'| \geq (1-\varepsilon)|A'|^2$ and $|A' \overset{G'}{-} A'| \leq \frac{2(KK')^2}{\varepsilon}|A|$.

6.4.9 Improve the 2^{12} factor in Theorem 2.29 to 2^{10} by exploiting the fact that all of the paths of length three constructed in Corollary 6.20 pass through $\tilde{A}'$, which is a slightly smaller set than A.

6.4.10 [38] Let A, B be additive sets in an ambient group Z, and let $G \subset A \times B$ be such that $|G| \geq |A||B|/K$ and $|A \overset{G}{+} B| \leq K|A|^{1/2}|B|^{1/2}$ for some $K \geq 1$. Show that there exist subsets A', B' of A, B such that $|A'| = \Omega(K^{-O(1)}|A|)$, $|B'| \geq \Omega(K^{-O(1)}|B|)$, $d(A', B') = O(1 + \log K)$, and $|G \cap (A' \times B')| = \Omega(K^{-O(1)}|A||B|)$. (Hint: the novelty here is that we still wish the refinement $A' \times B'$ to capture a large portion of G. This requires that one revisit the arguments in Lemma 6.19 and Corollary 6.20 and perform some additional "popularity" refinements to ensure that every time one reduces the size of A or B, one still keeps a significant fraction of elements from G. One may also need to use Lemma 2.30 at times to ensure that one also keeps a large number of "popular differences" between various refinements of A and B.) For an earlier result of this type, see [223].

6.5 Plünnecke's theorem

One of the most useful tools for the study of sum sets is Plünnecke's theorem. In order to state this theorem, we first need some notation.

Definition 6.23 (Magnification ratio) A *directed bipartite graph* is a triple $G(A, B, E)$, where A, B are finite sets (not necessarily disjoint) and $E \subset A \times B$ is a collection of pairs (a, b) from A and B. We write $G : A \to B$ to emphasize the directed nature of this graph, and also write $a \mapsto_G b$ to denote the statement that $(a, b) \in E$. If $X \subset A$, we use $G(X) := \{b \in B : a \mapsto b \text{ for some } a \in X\}$ to denote the image of X, and then define the *magnification ratio* $\|G\|$ of G to be the quantity

$$\|G\| = \min_{X \subseteq A : X \neq \emptyset} \frac{|G(X)|}{|X|}.$$

Equivalently, $\|G\|$ is the largest number such that $|G(X)| \geq \|G\||X|$ for all sets $X \subseteq A$.

If $G : A \to B$ and $H : B \to C$ are two directed bipartite graphs, with A, B, C disjoint, we define the *composition* $H \circ G : A \to C$ to be the directed graph defined by setting $a \mapsto_{H \circ G} c$ in $H \circ G$ if and only if there exists $b \in B$ such that $a \mapsto_G b \mapsto_H c$.

One can also view a directed bipartite graph $G : A \to B$ as a multiply-valued function from A to B, and the magnification ratio is then a measure of the multiplicity of this function.

Example 6.24 Let A, B be additive sets with common ambient group. Then we can form the directed bipartite graph $G_{A,B} : A \to A + B$ by setting $a \mapsto_{G_{A,B}} a + b$ if and only if $a \in A$ and $b \in B$. Observe that

$$\|G_{A,B}\| := \min_{X \subseteq A : X \neq \emptyset} \frac{|X + B|}{|X|} \leq \frac{|A + B|}{|A|}.$$

Also, observe that if A, B, C are additive sets with $A, A + B, A + B + C$ disjoint, then $G_{A+B,C} \circ G_{A,B} = G_{A,B+C}$.

For general directed bipartite graphs one has the inequality $\|H \circ G\| \geq \|G\|\|H\|$. However there is a deeper inequality available for certain families of directed bipartite graphs known as *Plünnecke graphs*. While this concept can be given for abstract graphs, it is easiest to describe for graphs whose vertices lie in an additive group (which is always the case for our applications).

Definition 6.25 (Plünnecke graphs) Let A_0, A_1, A_2 be three additive sets in an additive group Z. Two directed bipartite graphs $G_1 : A_0 \to A_1$ and $G_2 : A_1 \to A_2$ are said to be *commutative* if, whenever $a, b, c \in Z$ are such that $a \mapsto_{G_1} a + b \mapsto_{G_2} a + b + c$ in G_2, then one also has $a \mapsto_{G_1} a + c \mapsto_{G_2} a + b + c$. More generally, if $k \geq 2$, and $A_0, \ldots, A_k$ are additive sets in Z, we define a *Plünnecke graph* of order k to be a k-tuple $(G_1, \ldots, G_k)$ of bipartite graphs $G_j : A_{j-1} \to A_j$ such that each adjacent pair G_j, G_{j+1} for $1 \leq j < k$ is commutative.

Here is a more informal way to describe commutativity: if two adjacent edges of a parallelogram lie in $G_1 \cup G_2$, then so do the other two edges of the parallelogram.

Example 6.26 Let A, B be additive sets. Then the k-tuple

$$(G_{A,B}, G_{A+B,B}, \ldots, G_{A+(k-1)B,B})$$

of directed bipartite graphs (as defined in Example 6.24) forms a Plünnecke graph.

We are now ready to state Plünnecke's theorem.

Theorem 6.27 (Plünnecke's theorem) *[273] Let $(G_1, \ldots, G_k)$ be a Plünnecke graph of order k. Then the sequence of magnification ratios $\|G_i \circ \cdots \circ G_1\|^{1/i}$, $i = 1, \ldots, k$ is non-increasing in i. In particular, we have*

$$\|G_k \circ \cdots \circ G_1\| \leq \|G_1\|^k.$$

Applying this theorem to Example 6.26, we immediately obtain

Corollary 6.28 (Plünnecke's inequality) *If A and B are two additive sets in an ambient group Z and $|A + B| \leq K|A|$, then for any positive integer k there is a subset X of A such that*

$$|X + kB| \leq K^k |X|.$$

In particular we have

$$|kB| \leq K^k |A|.$$

This inequality has a number of applications to sum set estimates. For instance, from this inequality and the Ruzsa triangle inequality we obtain

Corollary 6.29 (Plünnecke–Ruzsa estimates) *Suppose that A, B are two additive sets in an ambient group Z such that $|A + B| \leq K|A|$. Then we have $|nB - mB| \leq K^{n+m}|A|$ for all $n, m \geq 1$.*

In particular, this implies that if $|A \pm A| \leq K|A|$, then $|nA - nA| \leq K^{2n}|A|$ for all $n \geq 1$; thus sets which are approximately closed under addition or subtraction are also approximately closed under repeated additions and subtractions.

6.5.1 Main ideas of the proof

To prove Plünnecke's theorem, it suffices to prove that

$$\|G_k \circ \cdots \circ G_1\|^{1/k} \leq \|G_i \circ \cdots \circ G_1\|^{1/i} \tag{6.2}$$

for all $1 \leq i < k$, since the claim then follows by truncating k to equal $i + 1$. In fact, it will suffice to show a special "normalized" case of this inequality:

Proposition 6.30 (Normalized Plünnecke inequality) *Let $(G_1, \ldots, G_k)$ be a Plünnecke graph of order k such that $\|G_k \circ \cdots G_1\| \geq 1$. Then we have $\|G_i \circ \cdots \circ G_1\| \geq 1$ for all $1 \leq i < k$.*

Our proof consists of two steps. In the first, we show that Proposition 6.30 implies the theorem. In the second step we prove this proposition.

The main tool for the first step is the so-called "tensor product" trick. We first show that Proposition 6.30 implies an inequality somewhat weaker than what we want to prove. Applying this inequality to a high power of the graph under consideration and taking limits will enable us to obtain the full version.

The second step is a pure graph-theoretical argument, whose main ingredient is the classical theorem of Menger about the number of disjoint paths in a graph. The reader may sense a connection here as the assumption $\|G_k \circ \cdots \circ G_1\|^{1/k} \geq 1$ simply means that the number of vertices in A_k which can be reached from a subset X of A_0 by a directed path of length k is at least $|X|$.

6.5.2 The first step

If $G : A \to B$ and $G' : A' \to B'$ are bipartite graphs, we define the *direct sum* $G \oplus G' : A \oplus A' \to B \oplus B'$ by requiring $(a, a') \mapsto_{G \oplus G'} (b, b')$ if and only if $a \mapsto_G b$ and $a' \mapsto_{G'} b'$. It turns out that the notion of direct sum interacts well with those of magnification ratio and composition.

Claim 6.31

$$\|G \oplus H\| = \|G\|\|H\|. \tag{6.3}$$

$$(G_k \circ \cdots \circ G_1) \oplus (H_k \circ \cdots \circ H_1) = (G_k \oplus H_k) \circ \cdots \circ (G_1 \oplus H_1). \tag{6.4}$$

The proofs are left as exercises.

To prove (6.2), it now suffices to prove the apparently weaker inequality

$$\|G_k \circ \cdots \circ G_1\|^{1/k} \leq O_{i,k}\big(\|G_i \circ \cdots \circ G_1\|^{1/i}\big) \tag{6.5}$$

for some constant $C_{i,k} > 0$ depending on i and k. For, if we could prove (6.5) for all Plünnecke graphs G, we could in particular apply it to higher powers $G^{\oplus M}$ for any large M. Using the above claim, it follows that

$$\|G_k \circ \cdots \circ G_1\|^{M/k} \leq O_{i,k}\big(\|G_i \circ \cdots \circ G_1\|^{M/i}\big)$$

for all $M \geq 1$. Taking Mth roots and then letting $M \to \infty$ we obtain (6.2).

We next deduce (6.5) from Proposition 6.30. First we deal with the case $\|G_k \circ \cdots \circ G_1\|^{1/k} \leq 1$. Let N be the smallest positive integer such that

$$\|G_k \circ \cdots \circ G_1\|^{1/k} \geq \frac{k}{N}.$$

As $\|G_k \circ \cdots \circ G_1\|^{1/k} \leq 1$, then $N \geq k \geq 2$ and the definition of N implies that

$$\|G_k \circ \cdots \circ G_1\|^{1/k} < \frac{k}{N-1} \leq \frac{2k}{N}.$$

We introduce an auxiliary Plünnecke graph $(H_1, \ldots, H_k)$ of order k, constructed as follows. Let $E := \{e_1, \ldots, e_N\}$ be the basis vectors of $\mathbf{Z}^N$, and set

$$(H_1, \ldots, H_k) = (G_{0,E}, G_{E,E}, G_{2E,E}, \ldots, G_{(k-1)E,E}),$$

where we use the notation of Example 6.24. In other words, we have $u \mapsto_{H_i} u + e_j$ whenever u is the sum of $i-1$ basis vectors and $1 \le j \le n$. It is easy to show that the ith vertex set iE has cardinality $\frac{(N+i-1)!}{(N-1)!i!}$. Since

$$\frac{N^i}{i^k} < \frac{N^i}{i!} \le \frac{(N+i-1)!}{(N-1)!i!} \le N^i,$$

we have that

$$\frac{1}{i}N \le \|H_i \circ \cdots \circ H_1\|^{1/i} \le N.$$

Consider the graph $G' = G \oplus H$. Using the claim, we have

$$\|G'\|^{1/k} = \|G\|^{1/k}\|H\|^{1/k} \ge \frac{k}{N}\frac{N}{k} = 1,$$

which guarantees the assumption of Proposition 6.30 for G'. Applying this proposition to G', we obtain for every $1 \le i \le k$

$$\|G'_i \circ \cdots \circ G'_1\|^{1/i} = \|G_i \circ \cdots \circ G_1\|^{1/i}\|H_i \circ \cdots \circ H_1\|^{1/i} \ge 1.$$

Since $\|H_i \circ \cdots \circ H_1\|^{1/i} \le N$, it follows that

$$\|G_i \circ \cdots \circ G_1\|^{1/i} \ge \frac{1}{N} \ge \frac{1}{2k}\|G_k \circ \cdots \circ G_1\|^{1/k},$$

completing the proof.

To deal with the case when $\|G_k \circ \cdots \circ G_1\|_{1/k} > 1$, we define N to be the largest positive integer such that $\|G_k \circ \cdots \circ G_1\|_{1/k} \ge N$. Replacing the Plünnecke graph $(H_1, \ldots, H_k)$ by its transpose $(H_k^*, \ldots, H_1^*)$, formed by reversing all the arrows, one can easily verify that

$$\frac{1}{i}N^{-1} \le \|H_k^* \circ \cdots \circ H_1^*\|^{1/i} \le N^{-1}.$$

The rest of the proof is similar.

6.5.3 The second step

The key ingredient of this step is a classical theorem due to Menger. Consider a directed graph G and let A and B be two disjoint sets of vertices. We say that a set C of vertices is a *cut* separating A and B if by removing C we destroy all directed paths from A to B (a path is from A to B if it starts in A and ends in B). Let Γ be a collection of (mutually) vertex disjoint paths from A to B with maximum cardinality N. It is trivial that any cut C has cardinality at least N, as C should contain at least one vertex from each path in Γ. It turns out that this bound is always sharp:

Theorem 6.32 (Menger's theorem) *Let G, A, B, N be as above. Then there is a cut C with cardinality N separating A and B.*

For a proof of this classical theorem, see Section 6 of [238], or the exercises below.

Now consider a Plünnecke graph consisting of directed bipartite graphs $G_1 : A_0 \to A_1, \ldots, G_k : A_{k-1} \to A_k$. By the trick of replacing the ambient group Z with $Z \times \mathbf{Z}$, and A_j with $A_j \times \{j\}$, we can ensure that the A_j are disjoint. Now let G be the union of all the graphs $G_1, \ldots, G_k$; thus G is a directed graph on $A_0 \cup \cdots \cup A_k$. Set $A = A_0$ and $B = A_k$ and let $\Gamma = \{\gamma_1, \ldots, \gamma_N\}$ be a maximum collection of vertex disjoint paths as above. By Theorem 6.32 we can find a vertex cut $C = \{c_1, \ldots, c_N\}$ in G separating A_0 from A_k such that $c_j \in V(\gamma_j)$ for all $1 \leq j \leq N$.

Since all the paths $\gamma_1, \ldots, \gamma_N$ start in A_0 and are vertex-disjoint, it is clear that $N \leq |A_0|$. The core of the proof is the following lemma.

Lemma 6.33 *Under the assumption of Proposition 6.30, we have $N = |A_0|$.*

Assuming this lemma, the rest of the proof is straightforward. If $N = |A_0|$, then every vertex v in A_0 must be the initial vertex of exactly one path in Γ. Since these paths are vertex-disjoint, we thus see that $|G_i \circ \cdots \circ G_1(X)| \geq |X|$ for all $X \subseteq A_0$ and the claim follows.

In order to prove Lemma 6.33 we partition the cut C as $C = C_0 \cup \cdots \cup C_k$ where $C_j := C \cap A_j$. The heart of the matter is the following lemma.

Lemma 6.34 *For any $1 \leq i \leq k-1$, $C' := (C \backslash C_i) \cup C_i^-$ is also a cut in G separating A_0 from A_k.*

Applying Lemma 6.34 iteratively, we can conclude that there is a cut which concentrates on A_0 and A_k. The union $C_0 \cup C_k$ (where $C_0 \subset A_0$, $C_k \subset A_k$) is a cut if and only if all paths starting from a point in $X = A_0 \backslash C_0$ end in C_k. The definition of the magnification ratio implies that

$$\|G_k \circ \cdots \circ G_1\| \leq \frac{|C_k|}{|X|}.$$

On the other hand, $|C_0| + |C_k| = N$ and $|X| = |A_0| - |C_0|$. Since

$$\|G_k \circ \cdots \circ G_1\| \geq 1,$$

it follows that $N \geq |A_0|$, proving Lemma 6.33.

It remains to prove the critical Lemma 6.34. This proof is actually the only place where one needs to utilize the commutativity property of the consecutive pairs G_i, G_{i+1}. Consider C_i as in the lemma. We can assume that C_i is not empty (otherwise there is nothing to prove). Let $C_i = \{c_1, \ldots, c_m\}$ for some $1 \leq m \leq N$.

Fix a maximum collection of mutually disjoint paths. For each $1 \leq j \leq m$, c_j is a vertex of exactly one path γ_j from this collection. Thus, there exist unique $c_j^- \in A_{i-1}$ and $c_j^+ \in A_{i+1}$ such that the edges $(c_j^- \to c_j)$ and $(c_j \to c_j^+)$ lie in γ_j. Let $C_i^\pm \in A_{i\pm1}$ denote the sets $C_i^\pm := \{c_1^\pm, \ldots, c_m^\pm\}$. Since the paths γ_j are vertex-disjoint, we have $|C_i^-| = |C_i| = |C_i^+|$. Also, $C_i^\pm$ must be disjoint from C, since each path γ_j in the collection contains exactly one cut point.

Suppose for contradiction that C' was not a cut, i.e., there was a path γ from A_0 to A_k which did not intersect C'. But since C is a cut, γ must intersect C. This forces γ to intersect A_{i-1} at a vertex $v \in A_{i-1}$ which does not lie in either C_{i-1} or C_i^-. Furthermore, the intersection of γ with C is a point in C_i. Let us define s_1 to be the number of edges from C_i^- to C_i, s_2 to be the number of edges from $C_i^- \cup \{v\}$ to C_i and s_3 to be the number of edges from C_i to C_i^+. In order to obtain a contradiction, we are going to prove the following three mutually inconsistent inequalities

$$s_1 < s_2, \quad s_2 \leq s_3, \quad s_3 \leq s_1.$$

The first (strict) inequality $s_1 < s_2$ is trivial, as v does not belong to C_i^- and there is an edge from v to C_i along the path γ. To prove $s_3 \leq s_1$, we are going to construct an injective map between the edges from C_i to C_i^+ and the edges from $C_i^- \cup \{v\}$ to C_i. Take any edge $c_j \to c_{j'}^+$ from C_i to C_i^+, for some $1 \leq j, j' \leq m$. Since G_i and G_{i+1} are commutative and $(c_j^- \to c_j) \in G_i$, $(c_{j'} \to c_{j'}^+) \in G_{i+1}$), we see that $(c_j^- \to c') \in G_i$ and $(c' \to c_{j'}^+) \in G_{i+1}$), where $c' := c_j^- + c_{j'}^+ - c_j$. Furthermore, c' must lie in C_i, otherwise we could find a path from A_0 to A_k avoiding the cut C by using γ_j to travel to c_j^-, then passing through c' to $c_{j'}^+$, and then using $\gamma_{j'}$ to travel to A_k. Thus we obtain an edge $(c_j^- \to c')$ from C_i^- to C_i. One can easily verify that this map is injective.

The proof of the remaining inequality is similar. When dealing with an edge from v, we, naturally, construct an avoiding path by using γ up to v.

Exercises

6.5.1 Show that one can take the set X in Corollary 6.28 to be as large as $(1-\varepsilon)|A|$ for any $\varepsilon > 0$, at the cost of replacing the factor K^k with $(K/\varepsilon)^k$. (Hint: apply Corollary 6.28 repeatedly, removing X from A at each iteration.)

6.5.2 Prove Claim 6.31 and Claim 6.4.

6.5.3 By induction on the number of edges in a graph $G(V, E)$, show that if the minimal cut needed to disconnect A and B has size N, then there exist N disjoint paths from A to B. (Hint: if there exists a minimal cut C that spans at least one edge $\{x, y\}$, then remove this edge and construct N disjoint paths from A to C and from C to B. If instead every minimal cut is independent, take an edge $\{x, y\}$ and contract it by identifying x with y (and removing the resulting loop). Show that the resulting "quotient graph" still has minimal cut N and apply the induction hypothesis.) Deduce Menger's theorem as a corollary.

6.5.4 Let A be an additive set. Show that the sequence of real numbers $|nA|^{1/n}$ is non-increasing in n, i.e. $|mA|^{1/m} \geq |nA|^{1/n}$ for all $n \geq m \geq 1$.

6.5.5 [297] Let N be a large integer, and let $A, B \subseteq \mathbf{Z}^3$ be the sets $A := ([1, N] \times [1, N] \times \{0\}) \cup (\{(0, 0)\} \times [1, N])$ and $B := ([1, N] \times \{0\} \times \{0\}) \cup (\{0\} \times [1, N] \times \{0\})$. Show that $|A| = \Theta(N^2)$, $|B| = \Theta(N)$, and $|A + B| = \Theta(N^2)$ but $|A + 2B| = \Theta(N^3)$.

6.5.6 [297] Let A, B be additive sets in an ambient group Z. Show that $|A + 2B| \leq \frac{|A+B|^2}{|A|^{1/2}}$. (Hint: use Exercise 6.5.1 to estimate $|A' + 2B|$ for a large $A' \subseteq A$, and use the crude bound $|(A \backslash A') + 2B| \leq |A \backslash A'||2B|$ and Corollary 6.29 to estimate the remainder. Use the tensor power trick as in Corollary 2.19 to eliminate any constants you encounter.) Compare this with Exercise 6.5.5.

6.5.7 Let $0 < \delta < 1$. Show that there exists additive sets A, B in an ambient group Z such that $|A + B| = \Theta(|A|)$ but such that for every subset A' of A for which $|A'| \geq (1 - \delta)|A|$, we have $|A' + B + B| = \Omega(|A|/\delta)$. (Hint: adapt the example in Exercise 6.5.5.)

6.5.8 Prove Corollary 6.29.

6.5.9 Suppose that A, B are additive sets with common ambient group such that $|A + B| \leq K|A|$ and $|2B| \leq K|B|$. Show that $|A + nB - mB| \leq K^{2\max(n,m)+3}|A|$ for all $n, m \geq 0$. (Hint: use Ruzsa's covering lemma, Lemma 2.14.) Compare this with Exercise 6.5.5.

6.5.10 [297] Let d be a large integer, let M be the nearest integer to $(7/6)^d$, and in the ambient group $\mathbf{Z}_7^N \times \mathbf{Z}$ let $A := (\mathbf{Z}_7^d \times \{0\}) \cup (\{0, 1, 3\}^d \times [1, M])$ and $B := \{0, 1, 3\}^d \times \{0\}$. Show that $|A| = \Theta(7^d)$, $|B| = \Theta(3^d)$, $|A + B| = \Theta(7^d)$, but $|A - B| = \Theta((49/6)^N)$. Thus even if $|A + B|$ is comparable to $|A|$, $|A - B|$ can be as large as $|A|^{2-\log_7 6}$.

6.5.11 [297] Let d be a large integer, let $B = \{e_1, \ldots, e_{2d}\}$ be the standard basis of $\mathbf{Z}^{2d}$, and let $A = dB$. Show that $|A + B| = \Theta(|A|)$ but $|A - B| = \Theta(|A| \log |A|)$. More generally, show that $|A' - B| = \Theta(|A'| \log |A|)$ for any non-empty subset A' of A. This shows that there is no analog of Corollary 6.29 for $n = -1$ unless one is willing to lose a logarithmic factor. On the other hand, see Exercise 6.5.12 below.

6.5.12 Let A, B be additive sets with common additive group such that $|A + B| \leq K|A|$, and let $N > 1$. Show that there exists an additive set A' in A with $|A'| \geq \frac{1}{2}|A|$ and $|A' - B| \leq (4K)^{2^N/N}|A|^{1+1/N}$. Compare this with Exercise 6.5.11. (Hint: first use Exercise 6.5.1 to locate a large set A'

such that $|A' + 2^N B| \leq (4K)^{2^N}|A'|$. Then use the pigeonhole principle to find $0 \leq j < N$ such that $|2^{j+1}B| \leq ((4K)^{2^N}|A|)^{1/N}|2^j B|$. Then control $|A' - B|$ by $|A' - 2^j B|$ and use Ruzsa's triangle inequality.)

6.5.13 Let A, B be non-empty subsets of F_p such that $p^\delta \leq |A|, |B| \leq p^{1-\delta}$ for some $0 < \delta \leq 1$. Show that there exists an $\varepsilon = \varepsilon(\delta) > 0$ depending only on δ such that either $|A + B| \geq p^\varepsilon|A|$ or $|B \cdot B| \geq p^\varepsilon|B|$. (You will of course need the results from Section 2.8.)

6.5.14 Let $G_1 : A_0 \to A_1$ and $G_2 : A_1 \to A_2$ be abstract directed graphs (not necessarily living in an additive group). We say that G_1 and G_2 are *abstractly commutative* if for every edge $a_1 \mapsto_{G_2} a_2$ and any collection of edges $a_1^1 \mapsto_{G_1} a_1, \ldots, a_1^n \mapsto_{G_1} a_1$, it is possible to find n forward paths from $a_1^1 \mapsto_{G_1} b^1 \mapsto_{G_2} a_2, \ldots, a_1^n \mapsto_{G_1} b^n \mapsto_{G_2} a_2$ with $b^1, \ldots, b^n$ all disjoint, and similarly if G_1, G_2 are replaced by their transposes G_2^*, G_1^*. Show that the commutative property implies the abstract commutative property, and furthermore the Plünnecke inequalities still hold if the commutative property is replaced with the abstract commutative property. Thus while the Plünnecke inequalities do require some additive structure on the underlying graph (and in particular the commutativity of the underlying group), the amount of structure needed is fairly minimal.

6.5.15 Improve the upper bound in (2.11) to $\sigma[A] \leq e^{2d(A,A)}$, or equivalently that $|A + A| \leq \frac{|A-A|^2}{|A|}$. Note that this gives another proof of the inequality $|A + A| \leq |A - A|^{3/2}$ (Exercise 2.3.13).

6.5.16 Obtain improvements to Corollary 2.23 and Corollary 2.24. Obtain as sharp a value of the constants as you can.

6.5.17 [Ben Green and Imre Ruzsa, private communication] Let $\pi : Z \to Z'$ be a group homomorphism, and let A be an additive set in Z. Show that $\sigma[\pi(A)] \leq \sigma[A]^2$ (compare with Exercises 2.2.10 and 2.3.8). Hint: use Plünnecke's theorem to find a subset $X \subset A$ with $|X + 2A| \leq \sigma[A]^2|X|$ small. Let M be the largest multiplicity of π on X. Establish the bounds $|X + 2A| \geq M|2\pi(A)|$ and $M|\pi(A)| \geq |X|$.

6.5.18 Use the preceding exercise to obtain sharper bounds in Corollary 5.43.

6.5.19 [162] Let $A \subset \mathbf{R}^d$ be an additive set containing the cube $\{0, 1\}^d$. Show that $|A + A| \geq 2^{d/2}|A|$. (Hint: from Exercise 3.4.8 we know that $|B + A + \{0, 1\}^n| \geq 2^d|B|$ for all subsets B of A. Now use the Plünnecke inequality.) In the converse direction, show that there exist arbitrarily large sets A containing $\{0, 1\}^d$ with doubling constant comparable to $(3/2)^d$.

7

The Littlewood–Offord problem

Let $v_1, \ldots, v_d$ be d elements of an additive group Z (which we refer to as the *steps*). Consider the 2^d sums $\epsilon_1 v_1 + \cdots + \epsilon_d v_d$ with $\epsilon_1, \ldots, \epsilon_d \in \{-1, 1\}$. In this chapter we investigate the largest possible repetitions among these sums.

We are going to consider two, opposite, problems:

- The *Littlewood–Offord problem*, which is to determine, given suitable non-degeneracy conditions on $v_1, \ldots, v_d$ and Z (e.g. excluding the trivial case when all of the steps are zero), what the largest possible repetition or concentration is among these sums.
- The *inverse Littlewood–Offord problem*, which supposes as a hypothesis that the $v_1, \ldots, v_d$ have a large number of repeated sums, or sums concentrating in a small set, and asks what one can then deduce as a consequence on the steps $v_1, \ldots, v_d$.

These two problems have a similar flavor to that of sum set estimates and inverse sum set estimates respectively, and occur naturally in certain problems of additive combinatorics, in particular in considering the set of subset sums $FS(A) = \{\sum_{a \in B} a : B \subset A\}$ of a given set A, or in the determinant and singularity properties of random matrices with entries ± 1. These problems have also arisen in several other contexts, ranging from the zeroes of complex polynomials (which was the original motivation of Littlewood and Offord [237]), to database security (see [163]). Note that the problem of determining which elements are representable as a sum $\epsilon_1 v_1 + \cdots + \epsilon_d v_d$ is essentially the notorious *subset-sum problem*, which is known to be NP-complete in general. Furthermore, by thinking of the sum $\sum_{i=1}^{n} \epsilon_i v_i$ as a random variable depending on the atom variables ϵ_i, we can view the Littlewood–Offord problem as a special case of the problem of computing the probability distribution of a random variable, which is a well-developed topic in probability theory.

In this chapter, we present two different approaches. The first is the combinatorial approach of Erdős and later authors, which phrases the problem in the theory of *set systems* (collections of subsets of a given set), thus allowing one to apply the theory of extremal set systems. This approach is very elegant and gives sharp results, but it is difficult to extend it to cases in which one has more complicated constraints on the steps v_j. The second, and rather different approach, is the Fourier-analytic one introduced by Halász. The bounds obtained by this approach are usually off by an absolute constant from the best possible results, but the arguments are more flexible.

A general theme will be that strong concentration or repetition of the above sums is closely related to strong additive structure among the steps $v_1, \ldots, v_n$. At one extreme, if the group Z has no 2-torsion, then all the sums are distinct if and only if the $v_1, \ldots, v_n$ are dissociated (see Definition 4.32). At another extreme, if the $v_1, \ldots, v_n$ are contained inside an arithmetic progression of small rank and volume, then one expects plenty of repetitions among the sums. The situation is thus somewhat analogous to the theory of sum set estimates and inverse sum set theorems studied in previous chapters, and indeed there will be strong similarities in our treatment of the two (in particular, the parallel use of combinatorial and Fourier-analytic methods).

7.1 The combinatorial approach

The fundamental concept in this approach is that of an *anti-chain*.

Definition 7.1 (Anti-chains) A collection $\mathcal{A}$ of sets is known as an *anti-chain* if none of the sets is contained in any other; thus $A \not\subseteq B$ for any distinct $A, B \in \mathcal{A}$.

Anti-chains are sometimes also referred to as *Sperner systems*, especially in older literature.

Lemma 7.2 (LYM inequality) *[240], [246], [385] Let $\mathcal{A}$ be an anti-chain of subsets of a finite set X. Then we have*

$$\sum_{A \in \mathcal{A}} \frac{1}{\binom{|X|}{|A|}} \leq 1.$$

Proof We give a probabilistic proof of Bollobás, using Katona's method of random maps. Let $\phi : X \to [1, |X|]$ be a random bijection from X to $[1, |X|]$, chosen uniformly at random among all $|X|!$ such bijections. A simple combinatorial argument shows that

$$\mathbf{P}(\phi(A) = [1, |A|]) = \frac{1}{\binom{|X|}{|A|}}$$

for each $A \in \mathcal{A}$. On the other hand, since none of the A are contained in each other, the events $\phi(A) = [1, |A|]$ are disjoint. Thus, the sum of their probabilities is bounded by 1, which implies the claim. □

From the obvious inequality $\binom{|X|}{|A|} \leq \binom{|X|}{\lfloor |X|/2 \rfloor}$ we immediately conclude

Corollary 7.3 (Sperner's lemma) *[332] Let $\mathcal{A}$ be an anti-chain of subsets of a finite set X. Then $|\mathcal{A}| \leq \binom{|X|}{\lfloor |X|/2 \rfloor}$.*

Note that the bound is clearly optimal, as can be seen by taking $\mathcal{A}$ to be the anti-chain consisting of all subsets of X of cardinality $\lfloor |X|/2 \rfloor$.

We can apply Sperner's lemma to the Littlewood–Offord problem as follows.

Corollary 7.4 *[82] Let $v_1, \ldots, v_n$ be real numbers with $|v_i| \geq 1$ for all i. Let $I = \{x : x_0 - 1 < x < x_0 + 1\}$ be an open interval of length 2. Then the total number of n-tuples $(\epsilon_1, \ldots, \epsilon_n) \in \{-1, 1\}^n$ with $\epsilon_1 v_1 + \cdots + \epsilon_n v_n \in I$ is at most $\binom{n}{\lfloor n/2 \rfloor}$.*

Proof By reversing the signs of some of the v_i if necessary, we may assume that $v_i \geq 1$ for all i. Now let $\mathcal{A}$ be the set of all subsets A of $[1, n]$ such that $\sum_{i \in A} v_i - \sum_{i \notin A} v_i \in I$. One can easily verify that $\mathcal{A}$ is an anti-chain, and hence by Sperner's lemma $|\mathcal{A}| \leq \binom{n}{\lfloor n/2 \rfloor}$. The claim follows. □

Now let us give a different proof of Sperner's lemma. We need to complement the notion of an anti-chain with that of a chain.

Definition 7.5 (Chains) A *chain* is a sequence of sets $A_1, \ldots, A_m$ such that $A_i \subseteq A_{i+1}$ for all $1 \leq i < m$; we refer to m as the *length* of the chain. We say a chain is *connected* if $|A_{i+1} \backslash A_i| = 1$ for all $1 \leq i < m$. A connected chain in a finite set X is said to be *centered* if $|A_1| + |A_m| = |X|$, or equivalently if $|A_i| = \frac{|X|-m-1}{2} + i$ for all $1 \leq i \leq m$. Note that the length of a centered connected chain has to have the opposite parity as $|X|$.

Lemma 7.6 (Chain decomposition lemma) *[206] Let X be a finite set, and let $2^X = \{A : A \subseteq X\}$ be the power set of X. Then 2^X can be partitioned into disjoint non-empty centered connected chains.*

Proof We induce on $|X|$. The cases $|X| = 0, 1$ are trivial. Now suppose that $|X| > 1$ and the claim has already been proven for all smaller X. Write $X = X' \cup \{x_0\}$ where $|X'| = |X| - 1$. By hypothesis, $2^{X'}$ can be partitioned into disjoint non-empty centered connected chains in X'. For each such chain $A_1, \ldots, A_m$, observe that the chains

$$A_1, \ldots, A_m, A_m \cup \{x_0\}$$

and

$$A_1 \cup \{x_0\}, \ldots, A_{m-1} \cup \{x_0\}$$

are connected centered chains in 2^X, and can be easily be seen to partition 2^X. Note that the chains of the second type may be empty, but they can of course be omitted from the partition without difficulty. The claim follows. $\square$

Every centered connected chain in X has to contain exactly one subset of cardinality $\lfloor X/2 \rfloor$. Thus the total number of chains in Lemma 7.6 is exactly $\binom{|X|}{\lfloor |X|/2 \rfloor}$. More generally, we see the number of centered connected chains of length m given by this lemma is exactly $\binom{|X|}{(|X|-m+1)/2} - \binom{|X|}{(|X|-m-1)/2}$ if m has the opposite parity of $|X|$, and 0 otherwise.

Since an anti-chain can contain at most one element of every chain, we obtain a new proof of Sperner's lemma (compare also with Menger's theorem, Theorem 6.32). In fact, the same argument gives the following generalization.

Proposition 7.7 *[82] Let $\mathcal{A}_1, \ldots, \mathcal{A}_k$ be k disjoint anti-chains of subsets of a finite set X. Then*

$$|\mathcal{A}_1| + \cdots + |\mathcal{A}_k| \leq \sum_{i=-\lfloor k/2 \rfloor}^{\lfloor \frac{k-1}{2} \rfloor} \binom{|X|}{\lfloor (|X|+i)/2 \rfloor}.$$

We leave the proof of this proposition as an exercise. We can then extend Corollary 7.4 without difficulty:

Corollary 7.8 (Erdős's Littlewood–Offord inequality) *[82] Let $v_1, \ldots, v_n$ be real numbers with $|v_i| \geq 1$ for all i. Let $I = \{x : x_0 - k < x < x_0 + k\}$ be an open interval of length $2k$ for some integer $k \geq 1$. Then the total number of n-tuples $(\epsilon_1, \ldots, \epsilon_n) \in \{-1, 1\}^n$ with $\epsilon_1 v_1 + \cdots + \epsilon_n v_n \in I$ is at most $\sum_{i=-\lfloor k/2 \rfloor}^{\lfloor k/2 \rfloor} \binom{n}{\lfloor (n+i)/2 \rfloor}$.*

One can replace the real numbers **R** by higher-dimensional spaces, such as the complex numbers **C**. To do this, we need a product form of Sperner's lemma, as follows.

Lemma 7.9 (Product Sperner lemma) *[206] Let X and Y be finite sets, and let $\mathcal{A}$ be a collection of pairs (A, B) of subsets of X, Y, which are a* product anti-chain *in the sense that there are no distinct pairs (A, B), (A', B') in $\mathcal{A}$ with either $A = A'$ and $B \subsetneq B'$, or $A \subsetneq A'$ and $B = B'$. (To put it another way, for each fixed B, the collection of A for which $(A, B) \in \mathcal{A}$ forms an anti-chain, and vice versa.) Then $|\mathcal{A}| \leq \binom{|X|+|Y|}{\lfloor (|X|+|Y|)/2 \rfloor}$.*

We leave the proof of this lemma as an exercise. As a consequence we have the complex version of Corollary 7.4.

Corollary 7.10 *[206] Let $v_1, \ldots, v_n$ be complex numbers with $|v_i| \geq 1$ for all i. Let $B = \{z : |z - z_0| < 1\}$ be a ball of radius* 1. *Then the total number of n-tuples $(\epsilon_1, \ldots, \epsilon_n) \in \{-1, 1\}^n$ with $\epsilon_1 v_1 + \cdots + \epsilon_n v_n \in B$ is at most $\binom{n}{\lfloor n/2 \rfloor}$.*

Proof By randomly rotating the complex plane we may assume that none of the v_i are purely real or purely imaginary. By reversing the signs of some of the v_i if necessary we may assume that $\operatorname{Im} v_i > 0$ for all i. Let X be the set of all i with $\operatorname{Re} v_i > 0$, and Y be the set of all i with $\operatorname{Re} v_i < 0$; thus $X \cup Y = [1, n]$. Now let $\mathcal{A}$ be the set of all pairs (A, B) of sets $A \subset X$, $B \subset Y$ such that $\sum_{i \in A \cup B} v_i - \sum_{i \notin A \cup B} v_i \in I$. One can easily verify that $\mathcal{A}$ is a product anti-chain in the sense of Lemma 7.9, and the claim follows. □

In fact one has the analogous claim in general dimension, by a more sophisticated version of this argument; see [207].

This is only the tip of the iceberg concerning extremal combinatorics results of this type; see for instance [32] for a much more detailed treatment of these topics. Variants of this approach have also been successfully applied in cyclic groups; see [163].

Exercises

7.1.1 (Set-pair estimate)[31] Let $A_1, \ldots, A_m, B_1, \ldots, B_m$ be finite sets such that $A_i \cap B_j = \emptyset$ if and only if $i = j$. Show that

$$\sum_{i=1}^{m} \frac{1}{\binom{|A_i|+|B_i|}{|A_i|}} \leq 1.$$

Note that this includes Lemma 7.3 as a special case (where $B_i := X \backslash A_i$).

7.1.2 (Erdős–Ko–Rado theorem) [94] Let $A_1, \ldots, A_m$ be an anti-chain in $\mathbf{Z}_N$ such that any two A_i, A_j intersect (thus $A_i \cap A_j \neq \emptyset$ for all i, j), and $|A_i| \leq k$ for all i and some $k \leq N/2$. Show that $m \leq \binom{N-1}{k-1}$, and show that this bound is sharp. (Hint: first show that for any bijection $\phi : \mathbf{Z}_N \to \mathbf{Z}_N$, at most k of the sets $\phi(A_i)$ can be an interval of the form $[a + 1, a + |A_i|]$ for some $a \in \mathbf{Z}_N$; this elegant argument is due to Katona [196].)

7.1.3 Prove Proposition 7.7. (Hint: for any chain of length m, observe that at most $\min(m, k)$ elements of this chain can lie in $\mathcal{A}_1 \cup \cdots \cup \mathcal{A}_k$. Now count how many chains there are of a given length in Lemma 7.6.)

7.1.4 Prove Lemma 7.9. (Hint: if $A_1, \ldots, A_m$ is a connected chain in X, and $B_1, \ldots, B_n$ is a connected chain in Y, show that there are at most $\min(m, n)$ pairs of the form (A_i, B_j) in $\mathcal{A}$. Alternatively, decompose 2^Y into chains $B_1, \ldots, B_n$, and for each such chain apply Proposition 7.7.)

7.2 The Fourier-analytic approach

Now we present the Fourier-analytic approach of Halász. It is convenient to use the language of probability theory. For any n-tuple $\mathbf{v} = (v_1, \ldots, v_n)$ of steps in an additive group Z, we use the notation $X_\mathbf{v}$ to denote the random variable

$$X_\mathbf{v} := \epsilon_1 v_1 + \cdots + \epsilon_n v_n$$

where $\epsilon_1, \ldots, \epsilon_n$ are independent random variables taking values in $\{-1, +1\}$ with probability $1/2$ for each value. Clearly $\mathbf{P}(X_\mathbf{v} = x)$ equals the number of representations of x as $\epsilon_1 v_1 + \cdots + \epsilon_n v_n$ with $\epsilon_1, \ldots, \epsilon_n \in \{-1, 1\}$, divided by 2^n. Note that $X_\mathbf{v}$ is invariant under permutations of the n-tuple $\mathbf{v}$. We use $\mathbf{vw}$ to denote the concatenation of $\mathbf{v}$ and $\mathbf{w}$. The Littlewood–Offord problem then asks to control the distribution of $X_\mathbf{v}$ for a given $\mathbf{v}$, while the inverse Littlewood–Offord problem asks for some structural information on $\mathbf{v}$ given some unexpected distributional property of $X_\mathbf{v}$.

It will be useful to consider the more general random variables $X_\mathbf{v}^{(\mu)}$ for any $0 \leq \mu \leq 1$, defined as

$$X_\mathbf{v} := \epsilon_1^{(\mu)} v_1 + \cdots + \epsilon_n^{(\mu)} v_n,$$

where $\epsilon_1^{(\mu)}, \ldots, \epsilon_n^{(\mu)}$ are independent random variables which take the values $+1$ and -1 with probability $\mu/2$, and 0 with probability $1 - \mu$. Thus $X_\mathbf{v}^{(\mu)}$ is the same as $X_\mathbf{v}$ when $\mu = 1$, and at the other extreme $\mu = 0$ becomes the constant 0. The intermediate cases correspond to "lazy random walks" with step sizes $v_1, \ldots, v_n$. As ϵ_i can be 0 with considerable probability, one expects $X_\mathbf{v}^{(\mu)}$ to be more concentrated than $X_\mathbf{v}$, and this will indeed be the case. In practice, the cases $\mu \leq 1/2$ are more amenable to Fourier analysis than the $\mu = 1$ case due to a certain "positivity" property which we shall come to shortly.

In this section we shall consider the discrete problem of understanding the probabilities $\mathbf{P}(X_\mathbf{v}^{(\mu)} = x)$ that a random variable $X_\mathbf{v}^{(\mu)}$ concentrates at a single point. In the next section we briefly discuss the analogous probability $\mathbf{P}(X_\mathbf{v}^{(\mu)} \in Q)$ for concentration in a cube.

Let us first make some technical reductions to the problem. Firstly, we can reduce to the case when the ambient group Z is finite. This can be achieved by applying a suitable Freiman isomorphism of order n to the steps $v_1, \ldots, v_n$ (see Exercise 5.3.3) while noting that this does not affect the distribution of $X_\mathbf{v}$. Secondly, we can reduce further to the case that Z is odd. To see this, observe from Corollary 3.8 that any finite additive group can be written as the product of a 2-torsion group and a group of odd order. The behavior of the random variable $X_\mathbf{v}$, when projected down to the 2-torsion group is trivial (since $+v_j = -v_j$ in this group), so we may, without loss of generality, project onto the other factor. Note that if the original elements $v_1, \ldots, v_n$ lived in some torsion-free group such as $\mathbf{Z}^d$, then by Lemma 5.25 we could now place the vectors in a *cyclic* group of odd prime order. (In doing so we may temporarily obscure some of the "dimensional" structure of the elements $v_1, \ldots, v_d$, so in some cases it is convenient to revert back to the original ambient group at certain stages of the argument.)

With these reductions we can now express the distribution of $X_{\mathbf{v}}$ in terms of the Fourier transform. As usual we fix a symmetric non-degenerate bilinear form $\xi \cdot x$ on Z.

Lemma 7.11 (Fourier representation of $X_{\mathbf{v}}$) *Let Z be a finite group of odd order. If $\mathbf{v} = (v_1 \ldots v_n)$ is an n-tuple of elements of Z, then for any $0 \leq \mu \leq 1$ and $x \in Z$ we have*

$$\mathbf{P}\big(X_{\mathbf{v}}^{(\mu)} = x\big) = \mathbf{E}_{\xi \in Z} \cos(2\pi\xi \cdot x) \prod_{j=1}^{n} (1 - \mu + \mu \cos(2\pi\xi \cdot v_j)).$$

Proof Since the quantity $\prod_{j=1}^{n}(1 - \mu + \mu\cos(2\pi\xi \cdot v_j))$ is an even function of ξ, we can write the right-hand side as

$$\mathbf{E}_{\xi \in Z} e(-\xi \cdot x) \prod_{j=1}^{n} (1 - \mu + \mu \cos(2\pi\xi \cdot v_j)).$$

Observing that $1 - \mu + \mu \cos(2\pi\xi \cdot v_j) = \mathbf{E}(e(\xi \cdot \epsilon_j^{(\mu)} v_j))$ and using the independence of the $\epsilon_j^{(\mu)}$, we can rewrite this as

$$\mathbf{E}\mathbf{E}_{\xi \in Z} e\big(\xi \cdot (X_{\mathbf{v}}^{(\mu)} - x)\big).$$

But the claim now follows from Lemma 4.5. □

This lemma already highlights the special role of the case $0 \leq \mu \leq \frac{1}{2}$, as in this case $1 - \mu + \mu\cos(2\pi\xi \cdot v_j)$ becomes non-negative. In the further case $0 \leq \mu \leq \frac{1}{4}$, we have the elementary but very useful estimate

$$1 - \mu + \mu \cos(2\pi\xi \cdot v_j) = \exp\big(-\Theta\big(\mu\|\xi \cdot v_j\|_{\mathbf{R}/\mathbf{Z}}^2\big)\big) \qquad (7.1)$$

where we recall that $\|x\|_{\mathbf{R}/\mathbf{Z}}$ denotes the distance to the nearest integer.

From Lemma 7.11 we can immediately establish a number of useful bounds on how one distribution $X_{\mathbf{v}}^{(\mu)}$ controls another.

Corollary 7.12 *Let $\mathbf{v} = (v_1, \ldots, v_n)$, $\mathbf{w} = (w_1, \ldots, w_m)$ be tuples in an additive group Z which is torsion-free or is finite of odd order. Let $x \in Z$.*

- *(Domination) If $0 \leq \mu \leq \mu' \leq 1$, and at least one of $\mu' \leq 1/2$ or $\mu \leq \mu'/4$ hold, then*

$$\mathbf{P}\big(X_{\mathbf{vw}}^{(\mu')} = x\big) \leq \mathbf{P}\big(X_{\mathbf{v}}^{(\mu)} = 0\big) = \mathbf{E}_{\xi \in Z} \prod_{j=1}^{n} (1 - \mu + \mu \cos(2\pi\xi \cdot v_j)).$$

In particular, if $\mu \le 1/2$, *then* $X_{\mathbf{v}}^{(\mu)}$ *concentrates more at the origin than anywhere else.*

- *(Duplication) If* $0 \le \mu \le 1/2$, *then*

$$\mathbf{P}\big(X_{\mathbf{vw}}^{(\mu)} = x\big) \le \mathbf{P}\big(X_{\mathbf{v}^k}^{(\mu/k)} = 0\big)$$

for all integers $k \ge 1$, *where we use* $\mathbf{v}^k$ *to denote the concatenation of* k *copies of* $\mathbf{v}$.

- *(Hölder) If* $\mathbf{w}_1, \ldots, \mathbf{w}_k$ *are tuples in* Z *(possibly of different length) and* $0 \le \mu \le 1/2$, *then*

$$\mathbf{P}\big(X_{\mathbf{v}\mathbf{w}\mathbf{w}_1\ldots\mathbf{w}_k}^{(\mu)} = x\big) \le \prod_{i=1}^{k} \mathbf{P}\big(X_{\mathbf{v}\mathbf{w}_i^k}^{(\mu)} = 0\big)^{1/k}.$$

Proof As discussed earlier we may take Z to be finite of odd order. In all cases we rewrite the probabilities using Lemma 7.11. The Hölder formula is clear, as is the domination formula when $\mu' \le 1/2$. In the case $\mu \le \mu'/4$, one observes the elementary inequality

$$|\cos(\pi\theta)| \le \frac{3}{4} + \frac{1}{4}\cos(2\pi\theta)$$

and hence (by the triangle inequality)

$$|(1-\mu') + \mu'\cos(\pi\theta)| \le \left(1 - \frac{\mu'}{4}\right) + \frac{\mu'}{4}\cos(2\pi\theta).$$

The claim then follows from the change of variables $\xi \to 2\xi$ (which is invertible when Z has odd order).

The duplication formula similarly follows from the elementary inequality

$$(1-\mu) + \mu\cos(2\pi\theta) \le \left(\left(1 - \frac{\mu}{k}\right) + \frac{\mu}{k}\cos(2\pi\theta)\right)^k,$$

which can be seen by taking logarithms and exploiting the concavity of $\log(1-t)$ in the region $0 < t < 1$. □

The above corollary allows one to show that the quantity $\mathbf{P}(X_{\mathbf{v}}^{(\mu)} = 0)$ is fairly stable when one tinkers with the tuple $\mathbf{v}$ (for instance, by adding or removing duplicates) and the parameter μ, at least when $\mu \le 1/2$. As an application, let us give a Fourier-analytic analog of Corollary 7.4.

Corollary 7.13 *Let* $\mathbf{v} = (v_1, \ldots, v_n)$ *be an* n*-tuple in a torsion-free group* Z *such that at least* k *of the* v_j *are non-zero. Then for all* $0 < \mu \le 1$ *and* $x \in Z$ *we have*

$$\mathbf{P}\big(X_{\mathbf{v}}^{(\mu)} = x\big) = O\left(\frac{1}{\sqrt{\mu k}}\right).$$

Proof Using the domination property we may take $\mu \leq 1/2$. Without loss of generality we may take $v_1, \ldots, v_k$ to be non-zero. Applying Corollary 7.12 repeatedly we have

$$\begin{aligned}\mathbf{P}(X_{\mathbf{v}}^{(\mu)} = x) &\leq \mathbf{P}(X_{\mathbf{vv}}^{(\mu/2)} = 0)\\ &\leq \mathbf{P}(X_{\mathbf{v}v_j^k}^{(\mu/2)} = 0)\\ &\leq \mathbf{P}(X_{v_j^k}^{(\mu/2)} = 0)\end{aligned}$$

for some $1 \leq j \leq k$. The latter quantity is a standard quantity in the theory of random walks[1] and can be computed combinatorially using Stirling's formula (1.52), but we present here a Fourier-analytic approach. We can map v_j^k via a Freiman isomorphism to the identity 1 in a large cyclic group $\mathbf{Z}_N$, and use Lemma 7.11 to conclude

$$\mathbf{P}(X_{v_j^k}^{(\mu/2)} = 0) = \mathbf{E}_{\xi \in \mathbf{Z}_N} \left(1 - \frac{\mu}{2} + \frac{\mu}{2}\cos(2\pi\xi/N)\right)^k$$

and thus, on taking limits as $N \to \infty$,

$$\mathbf{P}(X_{v_j^k}^{(\mu/2)} = 0) = \int_0^1 \left(1 - \frac{\mu}{2} + \frac{\mu}{2}\cos(2\pi\xi)\right)^k d\xi.$$

Using (7.1), it suffices to bound $\int_0^1 \exp(-\Theta(k\mu^2\xi))d\xi$. It is easy to show that most of the weight of this integral is in the interval $(0, C/\sqrt{\mu k})$ for some large constant C. The claim follows. □

We remark that in the case $\mu = 1$, Corollary 7.4 gives the sharp bound

$$\mathbf{P}(X_{\mathbf{v}}^{(1)} = x) \leq \frac{\binom{k}{\lfloor k/2 \rfloor}}{2^k} = \Theta\left(\frac{1}{\sqrt{k}}\right)$$

thanks to Stirling's formula (1.52). This shows that the Fourier-analytic method can give bounds which are sharp up to absolute constants.

If the steps $v_1, \ldots, v_n$ are sufficiently "high-dimensional" one can do better than this $O(1/\sqrt{k})$ type bound; see Exercise 7.2.3.

Now let us give a deeper distributional inequality which relies in particular on the Cauchy–Davenport inequality (Theorem 5.4).

Lemma 7.14 (Halász relative concentration inequality) *[195] Let Z be either torsion-free or cyclic of odd prime order. Let $\mathbf{v}$ be a tuple in Z. Then for any*

[1] Indeed, a useful heuristic is to think of $X_{v^k}^{(\mu)}$ as behaving (up to constants) similarly to the uniform distribution on the progression $[-\sqrt{\mu k}, \sqrt{\mu k}] \cdot v$; note that this heuristic is supported by the Chernoff inequality.

$0 < \mu \leq \mu' \leq 1$ with $\mu \leq 1/4$, we have

$$\mathbf{P}\big(X_{\mathbf{v}}^{(\mu')} = x\big) \leq O\left(\sqrt{\frac{\mu}{\mu'}}\mathbf{P}\big(X_{\mathbf{v}}^{(\mu)} = 0\big)\right) + O\big(\mathbf{P}\big(X_{\mathbf{v}}^{(\mu)} = 0\big)^{\Theta(\mu'/\mu)}\big)$$

for all $x \in Z$.

Note that the domination inequality only gives $\mathbf{P}(X_{\mathbf{v}}^{(\mu')} = x) \leq \mathbf{P}(X_{\mathbf{v}}^{(\mu)} = 0)$. Thus Halász's inequality becomes superior when μ is significantly smaller than μ', in which case it asserts that $X_{\mathbf{v}}^{(\mu)}$ concentrates at the origin substantially more often than $X_{\mathbf{v}}^{(\mu')}$ does. For some further discussion and more quantitative versions of this inequality, see [195], [364], [365].

Proof Using the domination inequality we may assume that $\mu' \leq 1/2$ and $x = 0$. We may also take μ'/μ to be large. By Corollary 5.25 we may take $Z = \mathbf{Z}_p$ for some odd prime p. Introduce the functions $F, G : Z \to \mathbf{R}^+$ by

$$F(\xi) := \prod_{j=1}^{n}(1 - \mu' + \mu' \cos(2\pi\xi \cdot v_j)); \quad G(\xi) := \prod_{j=1}^{n}(1 - \mu + \mu\cos(2\pi\xi \cdot v_j));$$

then by Lemma 7.11 our task is to show that

$$\mathbf{E}_{\mathbf{Z}_p}(F) = O\left(\sqrt{\frac{\mu}{\mu'}}\mathbf{E}_{\mathbf{Z}_p}(G)\right) + O\big(\mathbf{E}_{\mathbf{Z}_p}(G)^{\Omega(\mu'/\mu)}\big).$$

Now let $0 < \alpha \leq 1$ be arbitrary. Observe from (7.1) that if $\xi \in \mathbf{Z}_p$ is such that $F(\xi) > \alpha$, then

$$\left(\sum_{j=1}^{n} \|\xi \cdot v_j\|_{\mathbf{R}/\mathbf{Z}}^2\right)^{1/2} = O\left(\frac{\sqrt{\log\frac{1}{\alpha}}}{\sqrt{\mu'}}\right).$$

By the triangle inequality, we thus conclude that if $\xi_1, \ldots, \xi_m$ are arbitrary elements of the set $\{\xi \in \mathbf{Z}_p : F(\xi) \geq \alpha\}$, then

$$\left(\sum_{j=1}^{n} \|(\xi_1 + \cdots + \xi_m) \cdot v_j\|_{\mathbf{R}/\mathbf{Z}}^2\right)^{1/2} = O\left(m\frac{\sqrt{\log\frac{1}{\alpha}}}{\sqrt{\mu'}}\right).$$

If we take m to be $\lfloor c\sqrt{\frac{\mu'}{\mu}}\rfloor$ for some small absolute constant $c > 0$, another application of (7.1) then gives

$$G(\xi_1 + \cdots + \xi_m) > \alpha.$$

In other words we have established the sum set inclusion

$$m\{\xi \in \mathbf{Z}_p : F(\xi) > \alpha\} \subseteq \{\xi \in \mathbf{Z}_p : G(\xi) > \alpha\}.$$

Applying the Cauchy–Davenport inequality repeatedly, we have[1]

$$\mathbf{P}_{\mathbf{Z}_p}(\{\xi \in \mathbf{Z}_p : G(\xi) > \alpha\}) \geq \max(m\mathbf{P}_{\mathbf{Z}_p}(\{\xi \in \mathbf{Z}_p : F(\xi) > \alpha\}), 1).$$

If $\alpha \geq \mathbf{E}_{\mathbf{Z}_p}(G)$, then $\mathbf{P}_{\mathbf{Z}_p}(\{\xi \in \mathbf{Z}_p : G(\xi) > \alpha\}) < 1$ by Markov's inequality, and hence

$$\mathbf{P}_{\mathbf{Z}_p}(\{\xi \in \mathbf{Z}_p : F(\xi) > \alpha\}) \leq \frac{1}{m}\mathbf{P}_{\mathbf{Z}_p}(\{\xi \in \mathbf{Z}_p : G(\xi) > \alpha\}).$$

Integrating this in α, we conclude

$$\mathbf{E}_{\mathbf{Z}_p}(F\mathbf{I}(F \geq \mathbf{E}_{\mathbf{Z}_p}(G))) \leq \frac{1}{m}\mathbf{E}_{\mathbf{Z}_p}(G) = O\left(\sqrt{\frac{\mu}{\mu'}}\mathbf{E}_{\mathbf{Z}_p}(G)\right).$$

On the other hand, from (7.1) we have the pointwise bound

$$F(\xi) \leq G^{\Theta(\mu'/\mu)}(\xi)$$

and hence

$$\mathbf{E}_{\mathbf{Z}_p}(F\mathbf{I}(F < \mathbf{E}_{\mathbf{Z}_p}(G))) \leq \mathbf{E}_{\mathbf{Z}_p}(G)^{\Theta(\mu'/\mu)}.$$

Adding this to the preceding inequality, we obtain the claim. □

A modification of the above argument gives a more direct bound on $\mathbf{P}(X_{\mathbf{v}}^{(\mu)} = x)$.

Lemma 7.15 (Halász concentration inequality) *[167] Let Z be a cyclic group of prime odd order, and let $\mathbf{v} = (v_1, \ldots, v_n)$ be a tuple in Z with all the v_j non-zero. Then for any $0 < \mu \leq 1$ and $x \in Z$ we have*

$$\mathbf{P}(X_{\mathbf{v}}^{(\mu)} = x) \leq O\left(\frac{1}{\sqrt{\mu n}}\mathbf{P}_{\xi\in Z}\left(\sum_{j=1}^{n}\cos(\xi \cdot v_j) \geq \frac{n}{2}\right)\right) + \exp(-\Omega(\mu n)). \tag{7.2}$$

Proof Using the domination property we may take $\mu \leq 1/2$. By Lemma 7.11 and (7.1) we have

$$\mathbf{P}(X_{\mathbf{v}}^{(\mu)} = x) \leq \mathbf{E}_Z F \leq \mathbf{E}_{\xi\in Z}\exp\left(-\Theta\left(\mu\sum_{j=1}^{n}\|\xi \cdot v_j\|_{\mathbf{R}/\mathbf{Z}}^2\right)\right).$$

[1] To be absolutely precise here, we should have written

$$\mathbf{P}_{\mathbf{Z}_p}(m\{\xi \in \mathbf{Z}_p : G(\xi) > \alpha\}) \geq \max(m\mathbf{P}_{\mathbf{Z}_p}(\{\xi \in \mathbf{Z}_p : F(\xi) > \alpha\}) - (m-1)/p, 1),$$

since Cauchy–Davenport inequality only implies $|A + B| \geq \min\{|A| + |B| - 1, p\}$, for any two subsets A, B of Z_p. However, the term $(m-1)/p$ is negligible as we can take p arbitrarily large.

We can subdivide the right-hand side based on the size of $(\sum_{j=1}^n \|\xi \cdot v_j\|_{\mathbf{R}/\mathbf{Z}}^2)^{1/2}$, and bound the above expression by

$$O\left(\sum_{1\leq m\leq c\mu n} \exp(-\Theta(m))\mathbf{P}_{\xi\in Z}\left(\left(\sum_{j=1}^n \|\xi\cdot v_j\|_{\mathbf{R}/\mathbf{Z}}^2\right)^{1/2} \leq \sqrt{m/\mu}\right)\right)+\exp(-\Omega(c\mu n))$$

where $c > 0$ is a small absolute constant. Now observe that

$$\|\xi \cdot v_j\|_{\mathbf{R}/\mathbf{Z}}^2 = \Theta(1-\cos(2\pi\xi\cdot v_j)) \tag{7.3}$$

which in conjunction with Lemma 4.5 gives

$$\mathbf{E}_{\xi\in Z}\|\xi\cdot v_j\|_{\mathbf{R}/\mathbf{Z}}^2 = \Theta(1).$$

By linearity of expectation we thus have

$$\mathbf{E}_{\xi\in Z}\sum_{j=1}^n \|\xi\cdot v_j\|_{\mathbf{R}/\mathbf{Z}}^2 = \Theta(n);$$

in particular, we see that $\mathbf{P}_{\xi\in Z}(((\sum_{j=1}^n \|\xi\cdot v_j\|_{\mathbf{R}/\mathbf{Z}}^2)^{1/2} \leq c\sqrt{n}))$ is strictly less than one if c is small enough. Applying the Cauchy–Davenport inequality as in the preceding proof, we conclude

$$\begin{aligned}&\mathbf{P}_{\xi\in Z}\left(\left(\sum_{j=1}^n \|\xi\cdot v_j\|_{\mathbf{R}/\mathbf{Z}}^2\right)^{1/2} \leq \sqrt{m/\mu}\right)\\ &\quad\leq O\left(\sqrt{\frac{m}{\mu n}}\right)\mathbf{P}_{\xi\in Z}\left(\left(\sum_{j=1}^n \|\xi\cdot v_j\|_{\mathbf{R}/\mathbf{Z}}^2\right)^{1/2} \leq c\sqrt{n}\right).\end{aligned}$$

Using (7.3) again, we conclude

$$\mathbf{P}_{\xi\in Z}\left(\left(\sum_{j=1}^n \|\xi\cdot v_j\|_{\mathbf{R}/\mathbf{Z}}^2\right)^{1/2} \leq \sqrt{m/\mu}\right) \leq O\left(\sqrt{\frac{m}{\mu n}}\right)\mathbf{P}_{\xi\in Z}\left(\sum_{j=1}^n \cos(\xi\cdot v_j) \geq \frac{n}{2}\right)$$

if c is sufficiently small. The claim then follows from the observation that

$$\sum_{1\leq m\leq\sqrt{\mu n}} \exp(-\Theta(m))\sqrt{\frac{m}{\mu n}} = O\left(\frac{1}{\sqrt{\mu n}}\right)$$

(the geometric decay of $\exp(-\Theta(m))$ being more than sufficient to counteract the polynomial growth of $\sqrt{m}$). □

This bound easily implies Corollary 7.13, and is in fact significantly stronger. For instance, we have

Corollary 7.16 *[167] Let $0 < \mu \leq 1$, and let n be sufficiently large depending on μ. Let $\mathbf{v} = (v_1, \ldots, v_n)$ be a tuple of positive integers. For each integer $j > 0$, let m_j denote the number of times j occurs in $\mathbf{v}$, thus $m_j := \{1 \leq i \leq n : v_i = j\}$. Then for any $x \in Z$ we have*

$$\mathbf{P}(X_{\mathbf{v}}^{(\mu)} = x) \leq O\left(\mu^{-1/2} n^{-5/2} \sum_{j>0} m_j^2\right).$$

In particular, if all the v_i are distinct, then

$$\mathbf{P}(X_{\mathbf{v}}^{(\mu)} = x) \leq O(\mu^{-1/2} n^{-3/2}).$$

We remark that in the $\mu = 1$ case, the second half of this Corollary was first established by combinatorial means in [310] (with the precise threshold given in [330]).

Proof We may use a Freiman isomorphism to place $v_1, \ldots, v_n$ inside $\mathbf{Z}_p$ for some very large prime p. A direct application of Parseval's theorem (Theorem 4.2) gives

$$\mathbf{E}_{\xi \in \mathbf{Z}_p} \left| \sum_{j=1}^{n} \cos(\xi \cdot v_j) \right|^2 = O\left(\sum_{j>0} m_j^2\right)$$

and hence by Markov's inequality

$$\mathbf{P}_{\xi \in \mathbf{Z}_p}\left(\sum_{j=1}^{n} \cos(\xi \cdot v_j) \geq \frac{n}{2}\right) = O\left(\frac{1}{n^2} \sum_{j>0} m_j^2\right).$$

The claim then follows from Lemma 7.15 (observing that $\exp(-\Theta(\mu n)) = O(\mu^{-1/2} n^{-5/2})$ when n is large). □

Exercises

7.2.1 Show that in the condition $\mu \leq \mu'/4$ in the domination inequality of Corollary 7.12, the constant 4 cannot be replaced by any smaller constant, even in the most important case $\mu = 1$.

7.2.2 If $\mathbf{v} = (v_1, \ldots, v_n)$ are a tuple of integers, show that

$$\mathbf{P}(X_{\mathbf{v}}^{(\mu)} = m) = \int_0^1 \cos(2\pi m \xi) \prod_{j=1}^{n} (1 - \mu + \mu \cos(2\pi v_j \xi))\, d\xi$$

for all integers m.

7.2.3 [167] Let $1 \leq k \leq n$ and $d \geq 1$, and let $\mathbf{v} = (v_1, \ldots, v_n)$, a tuple of vectors in $\mathbf{R}^d$, be "non-degenerate" in the sense that every proper subspace of $\mathbf{R}^d$ contains at most $n - k$ of the $v_1, \ldots, v_n$. Show that

$$\mathbf{P}(X_{\mathbf{v}}^{(\mu)} = x) = O_d((\mu k)^{-d/2})$$

for every $0 < \mu \leq 1$ and $x \in \mathbf{R}^d$. (Hint: argue as Corollary 7.13, starting with an expression such as $\mathbf{P}(X^{(\mu/d)}_{\mathbf{v}^d} = 0)$ and applying Hölder's inequality suitably to arrive at a quantity such as $\mathbf{P}(X^{(\mu/d)}_{w_1^k w_2^k \ldots w_d^k} = 0)$, where $w_1, \ldots, w_d \in \mathbf{R}^d$ are linearly independent.) Give examples that show this bound is best possible up to the implicit constants in the $O_d()$ notation.

7.2.4 [364] With the notation and assumptions of Lemma 7.14, establish the following quantitative special case of the Halász inequality:

$$\mathbf{P}\big(X^{(1)}_{\mathbf{v}} = x\big) \leq \frac{1}{2}\mathbf{P}\big(X^{(1/16)}_{\mathbf{v}} = 0\big) + \mathbf{P}\big(X^{(1/16)}_{\mathbf{v}} = 0\big)^4.$$

7.2.5 Show that Lemma 7.14 can fail when Z is a non-cyclic finite group. In particular, if $Z = F_3^d$, show that $\mathbf{P}(X^{(\mu)}_{\mathbf{v}} = 0)$ can be comparable to $1/3^d$ for a large range of μ if the tuple $\mathbf{v}$ is chosen appropriately. This shows the pivotal role played by the Cauchy–Davenport inequality in the Halász argument.

7.2.6 Show that if the m_j are decreasing in j, then the right-hand side of Corollary 7.16 cannot be improved except for the implicit constant. (Hint: compute the variance of $X^{(\mu)}_{\mathbf{v}}$.)

7.2.7 Let $0 < \mu \leq 1$, and suppose n is sufficiently large depending on μ. Let $\mathbf{v} = (v_1, \ldots, v_n)$ take values in an additive set S in $\mathbf{Z}_p$ for some odd prime p. Show that for any even integer $k \geq 2$ and $x \in \mathbf{Z}$ we have

$$\mathbf{P}\big(X^{(\mu)}_{\mathbf{v}} = x\big) \leq O_k\left(\mu^{-1/2} n^{-2k-\frac{1}{2}} \|S\|^{2k}_{\Lambda(2k)} \left(\sum_{j \in S} m_j^2\right)^k\right)$$

where m_j is the number of times j occurs in $\mathbf{v}$, and the $\Lambda(2k)$ constant is defined in Definition 4.26. In particular, if the v_j are all distinct, then

$$\mathbf{P}\big(X^{(\mu)}_{\mathbf{v}} = x\big) \leq O_k\big(\mu^{-1/2} n^{-(2k+1)/2} \|S\|^{2k}_{\Lambda(2k)}\big).$$

Thus $X^{(\mu)}_{\mathbf{v}}$ can only concentrate significantly when the $\Lambda(p)$ constants of the support of $\mathbf{v}$ are large.

7.2.8 [167] Let $0 < \mu \leq 1$, and let n be sufficiently large depending on μ. Let $v_1, \ldots, v_n$ be non-zero integers, and let $k \geq 2$ be an even integer. Generalize Corollary 7.16 to show that for any $x \in \mathbf{Z}$ we have

$$\mathbf{P}\big(X^{(\mu)}_{\mathbf{v}} = x\big) \leq O_k\big(\mu^{-1/2} n^{-2k-\frac{1}{2}} R_k\big)$$

where R_k is the number of solutions to the equation

$$\epsilon_1 v_{i_1} + \cdots + \epsilon_{2k} v_{i_{2k}} = 0$$

where $\epsilon_1, \ldots, \epsilon_{2k} \in \{-1, +1\}$ and $i_1, \ldots, i_{2k} \in [1, n]$. In particular, if the v_j are all distinct and take values in a set S, then we have

$$\mathbf{P}(X_{\mathbf{v}}^{(\mu)} = x) \leq O\big(\mu^{-1/2} n^{-9/2} \mathbf{E}(S, S)\big).$$

Thus $X_{\mathbf{v}}^{(\mu)}$ can only concentrate significantly when the support has substantial additive energy. Explain heuristically why this result is related to the $\mu = 2k/n$ case of Lemma 7.14.

7.3 The Esséen concentration inequality

In several applications, we are not interested in the probability that a random walk $X_{\mathbf{v}}^{(\mu)}$ ends up in a specified point, but rather in a region of space such as a cube. In some "discrete" cases (e.g. when the $v_1, \ldots, v_n$ live in a lattice) one can simply use the union bound to pass from the former to the latter, but this is not always the best approach. One useful tool for dealing with concentration in general is a simple concentration inequality of Esséen.

Lemma 7.17 (Esséen concentration inequality) *[101] Let X be a random variable taking a finite number of values in $\mathbf{R}^d$. Let $x_0 \in \mathbf{R}^d$, and let $R, \varepsilon > 0$. Then*

$$\sup_{x_0 \in \mathbf{R}^d} \mathbf{P}(|X - x_0| \leq R) = O\left(\frac{R}{\sqrt{d}} + \frac{\sqrt{d}}{\varepsilon}\right)^d \int_{\xi \in \mathbf{R}^d : |\xi| < \varepsilon} |\mathbf{E}(e(\xi \cdot X))| \, d\xi.$$

Here $e(x) := \exp(2\pi i x)$, $\xi \cdot X$ denotes the usual inner product on $\mathbf{R}^d$, and $|\xi|$ denotes the usual magnitude.

Proof By rescaling X and R by ε we may take $\varepsilon = \sqrt{d}$. A simple covering argument (using for instance Corollary 3.15) then shows that it suffices to show that

$$\mathbf{P}(|X - x_0| \leq c\sqrt{d}) \leq O(1)^d \int_{\xi \in \mathbf{R}^d : |\xi| < \sqrt{d}} |\mathbf{E}(e(\xi \cdot X))| \, d\xi$$

for all $x_0 \in \mathbf{R}$ and some small absolute constant $c > 0$. By translating X by x_0 (which does not affect the right-hand side) we may take $x_0 = 0$. Now from the standard Gaussian integral identity

$$\int_{\xi \in \mathbf{R}^d} e^{-\pi C |\xi|^2} e(\xi \cdot X) \, d\xi = C^{-d/2} e^{-\pi |X|^2 / C}$$

for any $C > 0$, we see that

$$\left| \int_{\xi \in \mathbf{R}^d : |\xi| < \sqrt{d}/2} e^{-\pi C |\xi|^2} e(\xi \cdot X) \, d\xi \right| = \Omega(1)^d \tag{7.4}$$

whenever $|X| \leq c\sqrt{d}$, if c is chosen sufficiently small and C chosen sufficiently large. Squaring this we obtain

$$\int_{\xi\in\mathbf{R}^d:|\xi|<\sqrt{d}} e(\xi\cdot X)w(\xi)\,d\xi = \Omega(1)^d\mathbf{I}(|X|\leq c\sqrt{d})$$

where $w(\xi) := \int_{|\xi_1|,|\xi-\xi_1|<\sqrt{d}/2} e^{-\pi C|\xi_1|^2}e^{-\pi C|\xi-\xi_1|^2}$. Taking expectations of both sides we obtain

$$\int_{\xi\in\mathbf{R}^d:|\xi|<\sqrt{d}} |\mathbf{E}(e(\xi\cdot X))|w(\xi)\,d\xi \geq \Omega(1)^d\mathbf{P}(|X|\leq c\sqrt{d}).$$

From (3.8) we see that $w(\xi) = O(1)^d$, and the claim follows. □

Applying this in particular to the random variable $X_{\mathbf{v}}^{(\mu)}$ for some $\mathbf{v} = (v_1, \ldots, v_n)$ and $0 \leq \mu \leq 1$ we obtain the following analog of Lemma 7.11:

$$\mathbf{P}\left(\left|X_{\mathbf{v}}^{(\mu)} - x_0\right| \leq R\right) = O\left(\frac{\sqrt{d}}{\varepsilon} + \frac{R}{\sqrt{d}}\right)^d \int_{\xi\in\mathbf{R}^d:|\xi|<\varepsilon} \prod_{j=1}^n |1-\mu+\mu\cos(2\pi\xi\cdot v_j)|\,d\xi. \tag{7.5}$$

As an application we present a higher-dimensional analog of Corollary 7.10, but with the loss of a dimension-dependent constant.

Proposition 7.18 *[207], [167] Let $0 < \mu \leq 1$, and suppose n is sufficiently large depending on μ. Let $v_1, \ldots, v_n$ be elements of $\mathbf{R}^d$ with $|v_i| \geq 1$ for all i. Then for any $x_0 \in \mathbf{R}^d$, we have*

$$\mathbf{P}\left(\left|X_{\mathbf{v}}^{(\mu)} - x_0\right| \leq k\right) \leq O(1)^d \frac{k}{\sqrt{\mu n}}$$

for all $k \geq 1$.

It is worth noting that the right-hand side grows only linearly in k, instead of the k^d type growth that one might naively expect. This is a reflection of the heuristic that the random variable $X_{\mathbf{v}}^{(\mu)}$ tends to concentrate the strongest on one-dimensional spaces (cf. Exercise 7.2.3).

Proof In view of (7.5) (with $R = k$ and $\varepsilon = 1/k$), it suffices to show that

$$\int_{\xi\in\mathbf{R}^d:|\xi|\leq 1/k} \prod_{j=1}^n |1-\mu+\mu\cos(2\pi\xi\cdot v_j)|\,d\xi = O\left(\frac{1}{k\sqrt{d}}\right)^d \frac{k}{\sqrt{\mu n}}.$$

Applying Hölder's inequality, we reduce to showing that

$$\int_{\xi\in\mathbf{R}^d:|\xi|\leq 1/k} |1-\mu+\mu\cos(2\pi\xi\cdot v_j)|^n\,d\xi = O\left(\frac{1}{k\sqrt{d}}\right)^d \frac{k}{\sqrt{\mu n}}$$

for each $1 \leq j \leq n$. We can estimate

$$|1 - \mu + \mu \cos(2\pi \xi \cdot v_j)| \leq \exp\big(-\Omega(\mu \|2\xi \cdot v_j\|^2_{\mathbf{R}/\mathbf{Z}})\big)$$

(cf. (7.1)) and then make the change of variables $t = 2\xi \cdot v_j$ (using (3.8) to estimate the volume of the $d-1$-dimensional balls that are integrated out) to reduce to showing the one-dimensional estimate

$$\frac{1}{|v_j|} \int_{|t| \leq 2|v_j|/k} \exp\big(-\Omega(\mu n \|t\|^2_{\mathbf{R}/\mathbf{Z}})\big)\, dt = O\left(\frac{1}{\sqrt{\mu n}}\right).$$

Subdividing the t variable into unit intervals and using the periodicity of $\|t\|_{\mathbf{R}/\mathbf{Z}}$ and the hypothesis $|v_j| \geq 1$, the claim then follows from the easily verified estimate

$$\int_{-\infty}^{\infty} \exp(-\Omega(\mu n |t|^2))\, dt = O\left(\frac{1}{\sqrt{\mu n}}\right).$$

□

One can similarly develop analogs of many of the results of the preceding section, though the analysis is a little more technical as the analogs of Corollary 7.12 are somewhat messier. See [167] for further development of this theory.

Exercises

7.3.1 Prove (7.4).

7.3.2 Establish the following dimension-independent analog of the Esséen concentration inequality:

$$\sup_{x_0 \in \mathbf{R}^d} \mathbf{E}\big(e^{-\pi|X - x_0|^2}\big) \leq \int_{\xi \in \mathbf{R}^d} |\mathbf{E}(e(\xi \cdot X))| e^{-\pi|\xi|^2}\, d\xi.$$

7.3.3 [367] Obtain an analog of Exercise 7.2.3 for the probability $\mathbf{P}(X^{\mu}_{\mathbf{v}} \in B)$ for some unit ball B, assuming that, for every proper subspace of $\mathbf{R}^n$, at most $n - k$ of the vectors lie within a unit distance of this subspace.

7.3.4 Use the previous exercise to develop an analog of Erdős, results in any dimension [108, 367].

7.4 Inverse Littlewood–Offord results

In the preceding sections we considered *direct* Littlewood–Offord results, in which some assumptions were made on the steps $\mathbf{v} = (v_1, \ldots, v_n)$, and as a conclusion some upper bounds were obtained for concentration probabilities such as $\mathbf{P}(X^{(\mu)}_{\mathbf{v}} = x)$. In many applications it is of more interest to establish *inverse* Littlewood–Offord results, in which a lower bound on a concentration probability is assumed, and some structural property of $\mathbf{v}$ is deduced as a consequence. Of course, every direct Littlewood–Offord result can be converted into an inverse by taking contrapositives. For instance, from Corollary 7.13 we know that if $v_1, \ldots, v_n$ live in a torsion-free group Z and

$$\mathbf{P}\big(X_{\mathbf{v}}^{(\mu)} = x\big) \geq \frac{1}{\sqrt{\mu k}}$$

for some $0 < \mu \leq 1$ and some $x \in Z$, then at most $O(k)$ of the steps $v_1, \ldots, v_n$ are non-zero. Similarly, from Corollary 7.16, we see that if $v_1, \ldots, v_n$ are positive integers and $\mathbf{P}(X_{\mathbf{v}}^{(\mu)} = x)$ is much larger than $\mu^{-1/2}n^{-3/2}$ for some $0 < \mu \leq 1$ and $x \in \mathbf{Z}$, then at least two of the v_j are equal (in fact one can easily establish that a large number of pairs (v_i, v_j) must be equal).

Now we consider inverse Littlewood–Offord theorems that give more structure on the steps $v_1, \ldots, v_n$. The results in this section can be viewed in analogy with inverse sum set estimates, in which one assumes that a certain set A has small doubling constant and concludes some structural information on A, for instance containing A inside a progression. For simplicity we shall focus on the case $\mu = 1$ (though one can use results such as Corollary 7.12 or Lemma 7.14 to then extend to more general μ).

Let us start with an example when $\max_x \mathbf{P}(X_{\mathbf{v}}^1 = x)$ is large. This example has been the main motivation of our results.

Example 7.19 Let P be a symmetric generalized arithmetic progression of (constant) rank d and volume V in Z. Let $v_1, \ldots, v_n$ be (not necessarily different) elements of V. Then the sum $\sum_{i=1}^n \epsilon_i v_i$ takes values in the generalized arithmetic progression nP which have volume $n^d V$. From the pigeonhole principle it follows that

$$\max_x \mathbf{P}\big(X_{\mathbf{v}}^1 = x\big) \geq n^{-d} V^{-1}. \tag{7.6}$$

The above example shows that if the elements of $\mathbf{v}$ belong to a generalized arithmetic progression with small rank and small volume then $\mathbf{P}_\mu(\mathbf{v})$ is large. One might hope that the inverse of this also holds, namely,

If $P_\mu(\mathbf{v})$ is large, then the elements of $\mathbf{v}$ belong to a generalized arithmetic progression with small rank and small volume.

We are going to present a few results which support this statement. Let us first give a simple, but rather weak, result.

Proposition 7.20 *Let $\mathbf{v} = (v_1, \ldots, v_n)$ be a tuple in an additive group Z which is either torsion-free or finite of odd order, such that $\mathbf{P}(X_{\mathbf{v}}^{(1)} = x) > 2^{-d-1}$ for some $x \in Z$ and $d \geq 0$. Then all the steps $v_1, \ldots, v_n$ are contained in a cube $[-1, 1]^d \cdot (w_1, \ldots, w_d)$ of dimension d.*

Proof Suppose the conclusion failed. Then from Lemma 4.35 we see that $\mathbf{v}$ must contain a dissociated subword $\mathbf{w} = (w_1, \ldots, w_{d+1})$ of length $d+1$. By conditioning on the variables not associated to $\mathbf{w}$, we observe that

$$2^{-d-1} < \mathbf{P}(X_{\mathbf{v}}^{(1)} = x) \leq \sup_{y \in Z} \mathbf{P}(X_{\mathbf{w}}^{(1)} = y).$$

On the other hand, since $\mathbf{w}$ is dissociated, and Z has no 2-torsion, all the sums in $X_{\mathbf{w}}^{(1)}$ are distinct and so $\mathbf{P}(X_{\mathbf{w}}^{(1)} = y) \leq 2^{-d-1}$, thus yielding the desired contradiction. □

In practice, this proposition is not very useful because the dimension d of the cube can be rather large (typically it is like $\log n$). However, one can lower dimension its by increasing the side lengths, and allowing some exceptional steps v_j to lie outside of the resulting progression.

Proposition 7.21 *Let Z be either torsion-free or finite of odd order. For any integer $d \geq 1$, there is a positive constant δ_d such that the following holds. Let $k \geq 2$ be an integer, let $x \in Z$, and let $\mathbf{v} = (v_1, \ldots, v_n)$ be a tuple in Z. Then either*

$$\mathbf{P}(X_{\mathbf{v}}^{(1)} = x) \leq \delta_d k^{-d}$$

or there exists a progression $P = [-k, k]^{d-1} \cdot (w_1, \ldots, w_{d-1})$ in Z such that for all but at most k^2 exceptional values of $j \in [1, n]$, there exists $a_0 \in [1, k]$ such that $a_0 v_j \in P$. Furthermore, each of the w_i lies in the set $\{v_1, \ldots, v_n\}$.

Note that Corollary 7.13 (with $\mu = 1$) can be thought of as the $d = 1$ case of this proposition, while Proposition 7.20 can be viewed as the limiting case $k = 1$. Of course one should take $k < \sqrt{n}$ to avoid the claim being vacuous.

Proof Call a tuple $(w_1, \ldots, w_r)$ *k-dissociated* if the progression $[-k, k]^r \cdot (w_1, \ldots, w_r)$ is proper. We now construct an k-dissociated tuple $(w_1, \ldots, w_r)$ for some $0 \leq r \leq d$ by the following algorithm.

- Step 0. Initialize $r = 0$. In particular, $(w_1, \ldots, w_r)$ is trivially k-dissociated, and from Corollary 7.12 we have

$$\mathbf{P}\left(X^{(1/4d)}_{\mathbf{v}^{d-r} w_1^{k^2} \ldots w_r^{k^2}} = 0\right) \geq \mathbf{P}(X_{\mathbf{v}}^{(1)} = x). \tag{7.7}$$

- Step 1. Count how many $1 \leq j \leq n$ there are such that $(w_1, \ldots, w_r, v_j)$ is k-dissociated. If this number is less than k^2, halt the algorithm. Otherwise, move on to Step 2.
- Step 2. Applying Corollary 7.12, we can locate a v_j such that $(w_1, \ldots, w_r, v_j)$ is k-dissociated, and

$$\mathbf{P}\left(X^{(1/4d)}_{\mathbf{v}^{d-r} w_1^{k^2} \ldots w_r^{k^2}} = 0\right) \leq \mathbf{P}\left(X^{(1/4d)}_{\mathbf{v}^{d-r-1} w_1^{k^2} \ldots w_r^{k^2} v_j^{k^2}} = 0\right).$$

We then set $w_{r+1} := v_j$ and increase r to $r+1$. Return to Step 1. Note that $(w_1, \ldots, w_r)$ remains k-dissociated, and (7.7) remains true, when doing so.

Suppose that we terminate at some step $r \leq d-1$. Then we have an r-tuple $(w_1, \ldots, w_r)$ which is k-dissociated, but such that $(w_1, \ldots, w_r, v_j)$ is k-dissociated for at most k^2 values of v_j. Unwinding the definitions, this shows that for all but at most k^2 values of v_j, there exists $a_0 \in [1, k]$ such that $a_0 v_j \in Q - Q$, where $Q := [0, k]^r \cdot (w_1, \ldots, w_r)$ and $r \leq d-1$. The claim then follows by adding some dummy vectors to the w_j.

Now we prove that we must indeed terminate at some step $r \leq d-1$. Assume (for a contradiction) that we have reached step d. Then we have an k-dissociated tuple $(w_1, \ldots, w_d)$ such that

$$\mathbf{P}(X_{\mathbf{v}}^{(1)} = x) \leq \mathbf{P}\left(X_{w_1^{k^2} \ldots w_d^{k^2}}^{(1/4d)} = 0\right).$$

Let $\Gamma \subset \mathbf{Z}^d$ be the lattice

$$\Gamma := \{(m_1, \ldots, m_d) \in \mathbf{Z}^d : m_1 w_1 + \cdots + m_d w_d = 0\},$$

then by using independence we can write

$$\mathbf{P}(X_{\mathbf{v}}^{(1)} = x) \leq \mathbf{P}\left(X_{w_1^{k^2} \ldots w_d^{k^2}}^{(1/4d)} = 0\right) = \sum_{(m_1, \ldots, m_d) \in \Gamma} \prod_{j=1}^{d} \mathbf{P}\left(X_{1^{k^2}}^{(1/4d)} = m_j\right) \qquad (7.8)$$

where $X_{1^{k^2}}^{(1/4d)} = \epsilon_1^{(1/4d)} + \cdots + \epsilon_{k^2}^{(1/4d)}$.

Now we use a volume-packing argument. A simple computation involving the binomial formula (or induction on the k^2 parameter) shows that the expression $\mathbf{P}(X_{1^{k^2}}^{(1/4d)} = m)$ is even in m, and decreasing for positive m. It is also $\Theta_d(1/k)$ when $|m| \leq k$ (this can be seen either from Stirling's formula (1.52), or from Corollary 7.13 and variance and monotonicity considerations). Thus we have

$$\mathbf{P}\left(X_{1^{k^2}}^{(1/4d)} = m\right) = O_d\left(\frac{1}{k} \sum_{m' \in m + (-k/2, k/2)} \mathbf{P}\left(X_{1^{k^2}}^{(1/4d)} = m\right)\right)$$

and hence from (7.8) we have

$$\mathbf{P}(X_{\mathbf{v}}^{(1)} = x) \leq O_d\left(k^{-d} \sum_{(m_1, \ldots, m_d) \in \Gamma} \sum_{(m'_1, \ldots, m'_d) \in (m_1, \ldots, m_d) + (-k/2, k/2)^d} \prod_{j=1}^{d} \mathbf{P}\left(X_{1^{k^2}}^{(1/4d)} = m_j\right)\right).$$

Since $(w_1, \ldots, w_d)$ is k-dissociated, all the $(m'_1, \ldots, m'_d)$ tuples in $\Gamma + (-k/2, k/2)^d$ are different. Thus, we conclude

$$\mathbf{P}(X_{\mathbf{v}}^{(1)} = x) \leq O_d \left(k^{-d} \sum_{(m_1,\ldots,m_d)\in\mathbf{Z}^d} \prod_{j=1}^{d} \mathbf{P}\left(X_{1^{k^2}}^{(1/4d)} = m_j\right)\right).$$

But from the union bound we have

$$\sum_{(m_1,\ldots,m_d)\in\mathbf{Z}^d} \prod_{j=1}^{d} \mathbf{P}\left(X_{1^{k^2}}^{(1/4d)} = m_j\right) = 1.$$

To complete the proof, set the constant δ_d in the proposition to be larger than the hidden constant in $O_d(k^{-d})$. □

The a_0 factor in the above proposition is somewhat undesirable. With some more effort, one can remove this factor, but at the cost of enlarging the progression somewhat.

Theorem 7.22 (Inverse Littlewood–Offord theorem) *[366] Let* $0 < \mu < 1$ *and let* α *and* A *be arbitrary positive constants. Then there is a constant* $B = B(\mu, \alpha, A)$ *such that the following holds. Assume that* $\mathbf{v} = (v_1, \ldots, v_n)$ *is a tuple of rational numbers satisfying* $\max_x \mathbf{P}(X_{\mathbf{v}}^{\mu} = x) \geq n^{-A}$. *Then there is a generalized arithmetic progression* P *of rational numbers of rank at most* B *and volume at most* n^B *which contains all but at most* Bn^{α} *elements of* $\mathbf{v}$.

The proof of Theorem 7.22 is somewhat lengthy but is a modification of that of Proposition 7.21. For details see [366].

An inverse theorem in a similar spirit for the relative Halász inequality, Lemma 7.14, was also obtained in [365]:

Theorem 7.23 (Inverse Halász inequality) *[365] Let* Z *be either torsion-free or cyclic of odd prime order. Let* $\mathbf{v} = (v_1, \ldots, v_n)$ *be a tuple in* Z, *and suppose that* $\varepsilon_0 > \varepsilon_1 > 0$ *are such that*

$$\mathbf{P}\left(X_{\mathbf{v}}^{(1)} = 0\right) \geq \varepsilon_1 \mathbf{P}\left(X_{\mathbf{v}}^{(1/4-\varepsilon_0/100)} = 0\right)$$

and

$$\mathbf{P}\left(X_{\mathbf{v}}^{(1)} = 0\right) \geq \left(\frac{3}{4} + 2\varepsilon_0\right)^n.$$

Then there exists a proper progression P *of rank* $O_{\varepsilon_0,\varepsilon_1}(1)$ *and volume* $O_{\varepsilon_0,\varepsilon_1}(\frac{1}{\mathbf{P}(X_{\mathbf{v}}^{(1)}=0)})$ *which contain the* $v_1, \ldots, v_n$.

In fact some additional structural information was obtained, namely that the $v_1, \ldots, v_n$ are mostly contained in the "core" of the progression P, and under certain "non-triviality" assumptions on $\mathbf{v}$ (basically, that the set of signs $(\eta_1, \ldots, \eta_n) \in \{-1, 1\}^n$ for which $\eta_1 v_1 + \cdots + \eta_n v_n = 0$ has to span the hyperplane) one can also place the v_i in an *arithmetic* progression of length $n^{o(n)}$. For more precise statements and proofs see [365]. The main point is to inspect the use of the Cauchy–Davenport inequality in the proof of Lemma 7.14, and observe that this inequality is only efficient when sets such as $\{\xi \in \mathbf{Z}_p : F(\xi) > \alpha\}$ have small doubling constant. This in turn can be used (via some duality arguments) to place the $v_1, \ldots, v_n$ in a "Bohr set" of small doubling constant, at which point one can apply a Freiman-type theorem (e.g. Theorem 5.44) to place the v_j in a progression. This result played an essential role in establishing the bound $\mathbf{P}(\det(M_n) = 0) = (\frac{3}{4} + o(1))^n$ for $n \times n$ random Bernoulli matrices; see Section 7.5 for further discussion.

Exercise

7.4.1 Let the notation and hypotheses be as in Proposition 7.21, and let $1 \leq m \leq k$. Show that either

$$\mathbf{P}\big(X_{\mathbf{v}}^{(1)} = x\big) = O_d\big(mk^{-d/2}\big)$$

or there exists a progression $P = [-k, k]^{d-1} \cdot (w_1, \ldots, w_{d-1})$ in Z such that for all but at most k^2 exceptional values of $j \in [1, n]$, there exist at least k/m values $a_0 \in [1, k]$ such that $a_0 v_j \in P$. (Hint: argue as in Proposition 7.21, but work with $k/2$-dissociated tuples instead of k-dissociated ones, and add one extra copy of $\mathbf{v}$ in (7.7). Then if the latter conclusion fails, use Corollary 7.12 one final time to exploit the sparseness of the a_0 for which $a_0 v_j \in P$ and thence obtain the former conclusion.)

7.5 Random Bernoulli matrices

Let M_n be the random $n \times n$ matrix whose entries are independent uniformly distributed signs ± 1 (M_n is often referred to as the *random Bernoulli matrix*). The distribution of several quantities relating to M_n, such as its determinant and singular values, is of interest to a number of fields, including theoretical physics, combinatorics and theoretical computer science. It turns out that the tools developed in earlier sections are very well adapted for the study of M_n.

In this section we focus on a specific problem, namely to understand the singularity probability $\mathbf{P}(\det(M_n) = 0)$. An equivalent formulation is: given n vectors $X_1, \ldots, X_n$ chosen uniformly at random from the unit cube $\{-1, 1\}^n \in \mathbf{R}^n$, what is the probability that these vectors are linearly independent?

This simple-sounding problem has turned out to be surprisingly non-trivial. It is easy enough to show that

$$\mathbf{P}(X_i = \pm X_j \text{ for some } 1 \leq i < j \leq n \text{ and sign } \pm) = (1 + o(1))n^2 2^{-n}. \quad (7.9)$$

A similar argument (taking into account both the rows and columns of M_n) gives

$$\mathbf{P}(\det(M_n) = 0) \geq (2 + o(1))n^2 2^{-n}. \quad (7.10)$$

It is conjectured that this is sharp; thus

Conjecture 7.24 $\mathbf{P}(\det(M_n) = 0) = (2 + o(1))n^2 2^{-n}$. *In particular,* $\mathbf{P}(\det(M_n) = 0) = (\frac{1}{2} + o(1))^n$.

This conjecture remains open, although we will discuss some progress on this problem in this section. Notice that M_n is singular if and only there is a non-zero vector $v \in \mathbf{R}^n$ such that $M_n v = 0$. By restricting v to some special sets of vectors, we can obtain the conjectured bound $(1/2 + o(1))^n$. The following result is due to Komlós.

Theorem 7.25 *Let $n \geq 3$, and let Ω_1 be the set of vectors in $\mathbf{R}^n$ with at least $3n/\log_2 n$ coordinates. The probability that $M_n v = 0$ for some non-zero $v \in \Omega_1$ is $(1 + o(1))n^2 2^{-n}$.*

By considering the transpose of M_n, one can see that this theorem is equivalent to the following lemma.

Lemma 7.26 *Let $n \geq 3$, and let E denote the event that $a_1 X_1 + \cdots + a_n X_n = 0$ for some non-zero $(a_1, \ldots, a_n) \in \Omega_1$. Then $\mathbf{P}(E) = (1 + o(1))n^2 2^{-n}$.*

Proof To establish the upper bound, we use the union bound to give

$$\mathbf{P}(E) = \sum_{2 \leq k \leq n - 3n/\log_2 n} \mathbf{P}(E_k \backslash E_{k-1})$$

where E_k is the event that $a_1 X_1 + \cdots + a_n X_n = 0$ for some $(a_1, \ldots, a_n) \in \mathbf{R}^n$ with exactly k of the a_j being non-zero. (Note that the event E_1 is vacuous.) From (7.9) we easily see that $\mathbf{P}(E_2) = (1 + o(1))n^2 2^{-n}$, so it will suffice to show that

$$\sum_{3 \leq k \leq n - 3n/\log_2 n} \mathbf{P}(E_k \backslash E_{k-1}) = o(n^2 2^{-n}).$$

From symmetry we have $\mathbf{P}(E_k \backslash E_{k-1}) \leq \binom{n}{k} \mathbf{P}(F_k \backslash E_{k-1})$, where F_k is the event that $a_1 X_1 + \cdots + a_k X_k = 0$ for some non-zero $a_1, \ldots, a_k$. If $F_k \backslash E_{k-1}$ occurs, then the $n \times k$ matrix whose columns are $X_1, \ldots, X_k$ has rank exactly $k - 1$, and so $(a_1, \ldots, a_k)$ is essentially the wedge product of $k - 1$ of the rows of this matrix. There are $\binom{n}{k-1}$ ways to choose these rows, and then, on fixing all the entries of those rows (and hence fixing $a_1, \ldots, a_k$), we see from Corollary 7.4 that each of the other $n-k+1$ rows will be consistent with the equation $a_1X_1 + \cdots + a_kX_k = 0$ with probability $\binom{k}{\lfloor k/2 \rfloor}/2^k$. We conclude that

$$\sum_{3 \le k \le n-3\log_2 n} \mathbf{P}(E_k \backslash E_{k-1}) \le \sum_{3 \le k \le n-3n/\log_2 n} \binom{n}{k}\binom{n}{k-1}\left(\binom{k}{\lfloor k/2 \rfloor}/2^k\right)^{n-k+1}.$$

The claim then follows by direct computation (estimating $\binom{k}{\lfloor k/2 \rfloor}/2^k$ by $O(1/\sqrt{n})$ when $k = \Theta(n)$). □

Let us consider another restricted class. Let Ω_2 be the set of integer vectors in $\mathbf{R}^n$ where the coordinates have absolute values at most n^C, for some positive constant C.

Theorem 7.27 *The probability that $M_n v = 0$ for some non-zero $v \in \Omega_2$ is $(1/2 + o(1))^n$. (The error term $o(1)$ depends of course on C.)*

Proof The lower bound is trivial so we focus on the upper. For each non-zero vector v, let $p(v)$ be the probability that $X \cdot v = 0$, where X is a random Bernoulli vector. It is trivial that $\mathbf{P}(M_n v = 0) = p(v)^n$. Since a hyperplane can contain at most 2^{n-1} ± 1 vectors, $p(v)$ is at most $1/2$. For $j = 1, 2, \ldots$ let S_j be the number of non-zero vectors v in Ω_2 such that $2^{-j-1} < p(v) \le 2^{-j}$. Then the probability that $M_n v = 0$ for some non-zero $v \in \Omega_2$ is at most

$$\sum_{j=1}^{n} (2^{-j})^n S_j.$$

Let us now restrict the range of j. Notice that if $p(v) \ge n^{-1/3}$, then by Corollary 7.4 most of the coordinates of v are zero and then by Theorem 7.25 the contribution from these v is at most $(1/2 + o(1))^n$. Next, since the number of vectors in Ω_2 is at most $(2n^C + 1)^n \le n^{(C+1)n}$, we can ignore those j where $2^{-j} \le n^{-C-2}$. Now it suffices to show

$$\sum_{n^{-C-2} \le 2^{-j} \le n^{-1/3}} (2^{-j})^n S_j = o((1/2)^n).$$

Let ϵ be a small positive constant (say .001). As we have $j = \Theta(\log n)$ for all relevant j, we can find an integer $d = O(1)$ such that

$$n^{-(d-1+1/3)\epsilon} > 2^{-j} \ge n^{-(d+1/3)\epsilon}.$$

(The value of d depends on j, but is bounded from above by a constant.) Set $k = n^\epsilon$. Thus $2^{-j} \gg k^{-d}$ and we can use Proposition 7.21 to estimate S_j. Indeed, by invoking this theorem, we see that there are at most $\binom{n}{k^2}(2n^C + 1)^{k^2} = n^{O(k^2)} = n^{o(n)}$ ways to choose the positions and values of exceptional coordinates of v. There are only $(2n^C + 1)^{d-1} = n^{O(1)}$ ways to fix the generalized progression P. Once P is fixed, the number of ways to set the rest of the coordinates of v is at most $|P|^n = (2k+1)^{(d-1)n}$. Putting these together,

$$S_j \leq O(1)^n n^{O(k^2)} k^{(d-1)n}.$$

Since $k = n^\varepsilon$ and $2^{-j} \leq n^{-(d-1+1/3)\epsilon}$, it follows that

$$2^{-jn} S_j \leq O(1)^n n^{o(n)} n^{-\epsilon n/3}.$$

As the number of js is only $O(\log n)$, and $n^{-\Omega(n)} \log n = o((1/2)^n)$, we are done. □

By combining Theorem 7.25 with Corollary 7.13, we have the following consequence.

Corollary 7.28 *[215] Let $n \geq 3$. Then for any $1 \leq i \leq n$ we have*

$$\mathbf{P}(X_i \text{ is a linear combination of } X_1, \ldots, X_{i-1}) \leq \min\left(2^{i-n-1}, O\left(\frac{1}{\sqrt{n}}\right)\right).$$

Proof Let us first prove the upper bound of 2^{i-n-1}. Note that $X_1, \ldots, X_{i-1}$ span a space of dimension at most $i-1$, and so there exist $i-1$ coordinates which determine all the other coordinates of the space. But if one fixes $i-1$ coordinates of X_i then X_i is still uniformly distributed among 2^{n-i+1} remaining points, and the claim follows. Now we prove the bound of $O(\frac{1}{\sqrt{n}})$. We may assume n is large and i is close to n (say $i > .9n$). The vectors $X_1, \ldots, X_{i-1}$ will be contained in at least one hyperplane $\{(x_1, \ldots, x_n) \in \mathbf{R}^n : a_1x_1 + \cdots + a_nx_n = 0\}$; choose one arbitrarily. By Corollary 7.25, we certainly will have $\Theta(n)$ of the coordinates non-zero with probability $1 - O(\frac{1}{\sqrt{n}})$ (in fact, we can have much higher probability here). By Corollary 7.13, the probability that $X_i \cdot (a_1, \ldots, a_n) = 0$ is at most $O(\frac{1}{\sqrt{n}})$. Since this event is necessary in order for X_i to be a linear combination of $X_1, \ldots, X_{i-1}$, the claim follows. □

From this corollary, Bayes' identity, and independence, one easily verifies that

$$\mathbf{P}(\det(M_n) = 0) \leq \sum_{i=2}^{n} \mathbf{P}(X_i \text{ is a linear combination of } X_1, \ldots, X_{i-1})$$
$$= O\left(\frac{\log n}{\sqrt{n}}\right)$$

for large n. This bound was sharpened slightly to $O(\frac{1}{\sqrt{n}})$ in [215], [216] by a variant of this method.

Using a refinement of this argument, one can in fact obtain the following estimate for the determinant [364]

$$\mathbf{P}\big(|\det(M_n)| = \sqrt{n!}\exp\big(O\big(n^{1/2}\log^{1/2} n\big)\big)\big) = 1 - o(1).$$

The right-hand side is nearly optimal (see Exercises 7.5.3 and 7.5.4). With the help of recent results from [366], one can have $o(1) = 1/n^C$ for any fix C, at the cost of changing the hidden constant in the O on the left-hand side. It is not clear, however, that one can have $o(1) = \exp(-\Omega(n))$.

Now let us present a breakthrough result of Kahn, Komlós, and Szemerédi [195], which established an exponential bound without any restriction.

Theorem 7.29 *[195] There is a positive constant ε such that* $\mathbf{P}(\det(M_n) = 0) \leq (1 - \varepsilon)^n$.

In fact the explicit value $\varepsilon = 0.001$ was obtained in [195]. This was improved to roughly $\varepsilon = 0.042$ in [364], and then to $\varepsilon = \frac{1}{4} + o(1)$ in [365]. Conjecture 7.24 asserts that one can take $\varepsilon = \frac{1}{2} + o(1)$, which would be best possible.

We now sketch the proof of Theorem 7.29. It is convenient to rephrase the problem using the following lemma:

Lemma 7.30 *[195],[374],[364] We have*

$$\mathbf{P}(\det(M_n) = 0) = 2^{o(n)}\mathbf{P}(X_1, \ldots, X_n \textit{ span a hyperplane}).$$

Proof We already know that

$$\mathbf{P}(\det(M_n) = 0) = \mathbf{P}(X_1, \ldots, X_n \text{ linearly dependent}).$$

Thus the lower bound is obvious, and we need only to establish the upper. If $X_1, \ldots, X_n$ are linearly dependent, then there must exist $0 \leq d \leq n-1$ such that $X_1, \ldots, X_{d+1}$ span a d-dimensional subspace. Fixing d and conditioning on this event, we see from repeated application of Corollary 7.28 that $X_1, \ldots, X_n$ will span a hyperplane with probability $2^{-o(n)}$. The claim follows. □

Using this lemma followed by the union bound, it thus suffices to show

$$\sum_V \mathbf{P}(X_1, \ldots, X_n \text{ span } V) \leq (1 - \varepsilon + o(1))^n$$

where V ranges over all hyperplanes. Note that we can restrict our attention to the hyperplanes V which are spanned by their intersection with $\{-1, 1\}^n$; it is easy to see that this is a finite set. Let us call such hyperplanes *non-trivial*. An important quantity associated to a non-trivial hyperplane is its *density*

$$\mathbf{P}(X \in V) = \frac{|V \cap \{-1, 1\}^n|}{|\{-1, 1\}|^n}$$

where we think of X as a random element of $\{-1,1\}^n$. Note that $\mathbf{P}(X \in V) = \mathbf{P}(X_{\mathbf{v}}^{(1)} = 0)$ whenever $\mathbf{v}$ is a normal vector to V. We can exclude the contribution of all the hyperplanes of low density by the following lemma:

Lemma 7.31 *[195] For any $0 < \alpha < 1$, we have*

$$\sum_{V:\mathbf{P}(X\in V)\leq\alpha} \mathbf{P}(X_1, \ldots, X_n \text{ span } V) \leq n\alpha.$$

Proof If $X_1, \ldots, X_n$ span the hyperplane V, then there exists $1 \leq i \leq n$ such that the $n-1$ vectors formed by omitting X_i from $X_1, \ldots, X_n$ still span V. Fixing i and conditioning on this event, we see that V is determined by all the vectors other than X_i, and then X_i has a probability of at most α of also lying in V. The claim follows. □

Thus to establish the claim, it suffices to consider only the high-density hyperplanes for which $\mathbf{P}(X \in V) \geq (1-\varepsilon)^n$. On the other hand, from Lemma 7.26 and Corollary 7.13 we can control the extremely high-density hyperplanes for which $\mathbf{P}(X \in V) \gg \frac{1}{\sqrt{n}}$. So in fact we only need to deal with the range where $(1-\varepsilon)^n \leq \mathbf{P}(X \in V) \leq O(\frac{1}{\sqrt{n}})$.

We now crucially exploit the relative Halász inequality, Lemma 7.14. Let $0 < \mu \ll 1$ be a small parameter (independent of n), and let $Y \in \{-1, 0, 1\}^n$ be the random variable $Y = (\eta_1^{(\mu)}, \ldots, \eta_n^{(\mu)})$. Lemma 7.14 implies (if n is large enough) that Y concentrates on the above hyperplanes V more strongly than X does, if μ is sufficiently small:

$$\mathbf{P}(Y \in V) = O(\sqrt{\mu})\mathbf{P}(X \in V). \tag{7.11}$$

If we use the informal heuristic

$$\mathbf{P}(X_1, \ldots, X_n \text{ span } V) \approx \mathbf{P}(X \in V)^n$$

then we thus expect

$$\mathbf{P}(X_1, \ldots, X_n \text{ span } V) \leq O(\sqrt{\mu})^n \mathbf{P}(Y_1, \ldots, Y_n \text{ span } V)$$

where $Y_1, \ldots, Y_n$ are identical independent copies of Y. Summing this in V, and using the trivial fact that each $Y_1, \ldots, Y_n$ can span at most one hyperplane V we thus expect

$$\mathbf{P}(M_n = 0) \leq O(\sqrt{\mu})^n$$

which certainly gives Theorem 7.29 by setting μ small enough.

The above strategy almost works, except for a slight problem in that the $Y_1, \ldots, Y_n$ may be so linearly dependent that they will only span a subspace of V rather than V itself. The simplest way to solve this problem is to use only a small number of Y, say $Y_1, \dots, Y_{\delta n}$ for some small[1] δ. If V is sufficiently high-density and δ is small enough, we can ensure that $Y_1, \dots, Y_{\delta n}$ will remain linearly independent in V. This reduces the potential gain in this argument from $O(\sqrt{\mu})^n$ to only $O(\sqrt{\mu})^{\delta n}$, but this is still enough to establish Theorem 7.29.

More rigorously, we introduce $Y_1, \dots, Y_{\delta n}$ independently of $X_1, \dots, X_n$. Fix a density $(1-\varepsilon)^n \leq \sigma \leq O(\frac{1}{\sqrt{n}})$, and let V be such that $\mathbf{P}(X \in V) = (1 + O(\frac{1}{n}))\sigma$:

$$\mathbf{P}(Y_1, \dots, Y_{\delta n} \in V) \geq \Omega\left(\frac{1}{\sqrt{\mu}}\right)^{\delta n} \sigma^{\delta n}.$$

If δ is sufficiently small depending on μ, and ε is sufficiently small depending on δ and μ, then one can modify Corollary 7.28 to refine this to

$$\mathbf{P}(Y_1, \dots, Y_{\delta n} \text{ linearly dependent in } V) \geq \Omega\left(\frac{1}{\sqrt{\mu}}\right)^{\delta n} \sigma^{\delta n}; \tag{7.12}$$

we leave this as an exercise. From independence we thus have

$$\mathbf{P}(X_1, \dots, X_n \text{ span } V) \leq O(\sqrt{\mu})^{\delta n} \sigma^{-\delta n} \mathbf{P}(E_V)$$

where E_V is the event that $X_1, \dots, X_n$ span V and $Y_1, \dots, Y_{\delta n}$ are linearly independent in V. But if this event occurs, then there exist $n - \delta n$ vectors in $X_1, \dots, X_n$ which, together with $Y_1, \dots, Y_{\delta n}$, span V. If we fix all these vectors then V is also fixed, and the remaining δn vectors in $X_1, \dots, X_n$ have a probability of $\Theta(\sigma^{\delta n})$ of lying in V. We thus conclude that

$$\sum_{V:\mathbf{P}(X \in V) = (1+O(\frac{1}{n}))\sigma} \mathbf{P}(E_V) \leq \binom{n}{\delta n} \Theta(\sigma^{\delta n})$$

which, when combined with the preceding estimates, give

$$\sum_{V:\mathbf{P}(X \in V) = (1+O(\frac{1}{n}))\sigma} \leq \mathbf{P}(X_1, \dots, X_n \text{ span } V) \leq O(\sqrt{\mu})^{\delta n} \binom{n}{\delta n}.$$

If we choose δ sufficiently small depending on μ, and ε sufficiently small depending on δ, μ, we can make the right-hand side $(1 - \varepsilon + o(1))^n$. Summing over all relevant σ (there are only about $O(n^2)$ such σ to sum over) we obtain Theorem 7.29 as desired.

By using Theorem 7.23 one can boost ε to be as large as $\frac{1}{4} + o(1)$. The basic point is that Theorem 7.23 allows one to improve (7.11) significantly unless the hyperplane V has an exceptional form (in particular, the coordinates of its normal

[1] Strictly speaking we should use $\lfloor \delta n \rfloor$ instead of δn but we shall omit this inessential detail for ease of exposition.

vector lie in a fairly small generalized progression). These exceptional hyperplanes however are rather rare and can be treated by a direct counting argument.

Let us conclude by a refinement of Theorem 7.29, which allows us to fix a few rows of M_n. Let Y be a set of l independent vectors $y_1, \ldots, y_l$ and denote by M_n^Y the random matrix with rows $X_1, \ldots, X_{n-l}, y_1, \ldots, y_l$, where X_i are i.i.d copies of the random Bernoulli vector X.

Theorem 7.32 *[366] For any non-negative integer l, there is a positive constant ε such that the probability that M_n^Y is singular is at most* $(1-\varepsilon)^n$.

Exercises

7.5.1 Prove (7.9) and (7.10).

7.5.2 Prove (7.12).

7.5.3 Show that $\det(M_n) \in 2^{n-1} \cdot \mathbf{Z}$ and $|\det(M_n)| \le n^{n/2}$ for all Bernoulli matrices M_n.

7.5.4 Show that $\det(M_n)$ has expectation zero and variance $n!$, and $|\det(M_n)|^2$ has expectation $n!$ and variance $n(n!)^2$. Derive a upper bound for $|\delta(M_n)|$. (For a matching lower bound, see [364].)

7.5.5 [195] Show that $\sup_{x\in\mathbf{R}} \mathbf{P}(\det(M_n) = x) = (1-\varepsilon+o(1))^n$ for some absolute $\varepsilon > 0$.

7.5.6 [195] Show that for any $\varepsilon > 0$ we have

$$\sum_{(1-\varepsilon)^n \le \mathbf{P}(X\in V) \le O(\frac{1}{\sqrt{n}})} \mathbf{P}(X_1, \ldots, X_n \text{ lie in } V) = (o_{\varepsilon\to 0}(1))^n.$$

7.5.7 [195] Show that there exists an absolute constant $C > 0$ such that $\mathbf{P}(X_1, \ldots, X_{n-C} \text{ dependent}) = (\frac{1}{2}+o(1))^n$ whenever n is sufficiently large depending on ε, C. Conclude in particular that the probability that M_n has rank $n - C$ or less is $(\frac{1}{2}+o(1))^n$.

7.6 The quadratic Littlewood–Offord problem

The preceding sections studied the concentration of linear combinations of random variables such as $\eta_1 v_1 + \cdots + \eta_n v_n$. It is also of interest to study more general polynomial combinations. For simplicity we shall restrict ourselves to the quadratic expression

$$Q(\eta_1, \ldots, \eta_n) = \sum_{1\le i<j\le n} c_{i,j}\eta_i\eta_j + \sum_{i=1}^n d_i\eta_i$$

where $c_{i,j}, d_i$ take values in an additive group Z, and $\eta_1, \ldots, \eta_n$ are independent uniformly distributed random ± 1 signs.

One can now ask under what conditions one can establish upper bounds on the concentration of the random variable Q. In the special case when the c_{ij} are identically zero, we know from Corollary 7.13 that Q will not concentrate at a single point as soon as many of the d_i are non-zero. One can then hope to establish a similar result for the quadratic component, namely that Q will not concentrate at a single point as soon as many of the c_{ij} are non-zero. We give a sample result of this form as follows:

Proposition 7.33 *[64] Let Z be either torsion-free or finite of odd order. Let the notation be as above, and suppose that for at least k values of i, we have $c_{i,j} \neq 0$ for at least l values of j. Then for any $x \in Z$ we have $\mathbf{P}(Q = x) = O(\min(k, l)^{-1/8})$.*

Proof Without loss of generality we may take $k \leq l$. A greedy algorithm argument shows that we can find a set $A \subset [1, n]$ of cardinality $\lfloor (k+1)/2 \rfloor$, such that for each $i \in A$ we have $c_{i,j} \neq 0$ for at least $\lfloor (l+1)/2 \rfloor$ values of $j \in [1, n] \backslash A$. The basic idea is to view the quadratic object Q as a linear expression $\sum_j X_j \eta_j$, where the X_j are themselves linear expressions of $\eta_1, \ldots, \eta_n$, so that one can obtain a quadratic non-concentration result from two applications of the linear non-concentration result. However there is a "coupling" problem, arising from the fact that the X_j and η_j do not behave independently. This however can be resolved via the following *decoupling inequality*

$$\begin{aligned} \mathbf{P}(E(X, Y)) &\leq \mathbf{P}(E(X, Y) \wedge E(X, Y'))^{1/2} \\ &\leq \mathbf{P}(E(X, Y) \wedge E(X, Y') \wedge E(X', Y) \wedge E(X', Y'))^{1/4} \end{aligned} \quad (7.13)$$

whenever X, Y, X', Y' are independent random variables taking finitely many values, with X, X' having the same distribution and Y, Y' having the same distribution, and $E(X, Y)$ is any event depending only on X and Y. The proof of this inequality follows from two applications of the Cauchy–Schwarz inequality and is left as an exercise. We apply this inequality with $X := (\eta_i)_{i \in A}$ and $Y := (\eta_j)_{j \in [1,n] \backslash A}$, writing Q as $Q(X, Y)$, to obtain

$$\mathbf{P}(Q(X, Y) = x) \leq \mathbf{P}(Q(X, Y) = Q(X, Y') = Q(X', Y) = Q(X', Y') = x)^{1/4}$$

where $X' = (\eta'_1, \ldots, \eta'_{n/2})$ and $Y' = (\eta'_{n/2+1}, \ldots, \eta'_n)$ are identical independent copies of X and Y. In particular we have

$$\mathbf{P}(Q(X, Y) = x) \leq \mathbf{P}(Q(X, Y) - Q(X, Y') - Q(X', Y) + Q(X', Y') = 0)^{1/4}.$$

On the other hand, we have the factorization

$$\begin{aligned} Q(X, Y) - Q(X, Y') - Q(X', Y) + Q(X', Y') &= \sum_{i \in A} \sum_{j \in B} c_{ij} (\eta_i - \eta'_i)(\eta_j - \eta'_j) \\ &= \sum_{i \in A} v_i \eta_i^{(1/2)} \end{aligned}$$

where $v_i := \sum_{j\in B} 4c_{ij}\eta_j^{(1/2)}$ and $\eta_i^{(1/2)} = (\eta_i - \eta_i')/2$. Observe that the $\eta_i^{(1/2)}$ are all independent and have the distribution of $\eta^{(1/2)}$ (i.e. they equal 0 with probability 1/2, and ± 1 with probability 1/4 each). Also we make the crucial observation that the $(v_i)_{i\in A}$ and $(\eta_i^{(1/2)})_{i\in A}$ are *independent*.

It now suffices to show that

$$\mathbf{P}\left(\sum_{i\in A} v_i\eta_i^{(1/2)} = 0\right) = O\left(\frac{1}{k^{1/2}}\right).$$

For each $i \in A$, we have the easy bound

$$\mathbf{E}(\mathbf{I}(v_i = 0)) = \mathbf{P}(v_i = 0) \le \frac{3}{4}$$

as can be seen by conditioning all the η_j except for a single j for which $c_{i,j} \neq 0$. From Corollary 7.13 we also have

$$\mathbf{E}(\mathbf{I}(v_i = 0)) = \mathbf{P}(v_i = 0) \le O\left(\frac{1}{\sqrt{l}}\right).$$

By linearity of expectation we thus have

$$\mathbf{E}\left(\sum_{i\in A} \mathbf{I}(v_i = 0)\right) \le |A| \min\left(\frac{3}{4}, O\left(\frac{1}{\sqrt{l}}\right)\right).$$

In particular by Markov's inequality we have

$$\mathbf{P}\left(\sum_{i\in A} \mathbf{I}(v_i = 0) \le \frac{7}{8}|A|\right) \le \min\left(\frac{6}{7}, O\left(\frac{1}{\sqrt{l}}\right)\right);$$

since $|A| = \Omega(k)$, we conclude

$$\mathbf{P}(|\{1 \le i \le n/2 : v_i \neq 0\}| = \Omega(k)\}|) \ge \max\left(\frac{1}{7}, 1 - O\left(\frac{1}{\sqrt{l}}\right)\right).$$

Now if we condition on the above event (call it E), then the distribution and independence of the $\eta_i^{(1/2)}$ remain unaffected. Thus we may apply Corollary 7.13 again to obtain

$$\mathbf{P}\left(\sum_{i\in A} v_i\eta_i^{(1/2)} = 0 | E\right) = O\left(\frac{1}{\sqrt{k}}\right);$$

we also have the crude upper bound of $\frac{3}{4}$ as before. Thus

$$\mathbf{P}\left(\sum_{i\in A} v_i\eta_i^{(1/2)} \neq 0 | E\right) = \max\left(\frac{1}{4}, 1 - O\left(\frac{1}{\sqrt{k}}\right)\right).$$

Combining this with the estimate on $\mathbf{P}(E)$ and Bayes' formula, we obtain the claim. □

In [64] this estimate was used, together with some techniques from the preceding section, to obtain

Theorem 7.34 *[64] Let M_n be a random symmetric $n \times n$ matrix whose entries are random uniformly distributed signs ± 1, and with the entries in the upper triangular half being independent. (The entries in the strictly lower triangular half are of course determined from the upper half by symmetry.) Then* $\mathbf{P}(\det(M_n) = 0) = O_\varepsilon(n^{-1/8+\varepsilon})$ *for any $\varepsilon > 0$.*

Exercises

7.6.1 Give examples that show that for arbitrary $k, l \geq 1$, there exists Q obeying the hypothesis in Proposition 7.33 with $\mathbf{P}(Q = 0) = \Omega(\min(k, l)^{-1/2})$. Thus, except for the exponent $1/8$ and for absolute constants, the conclusion in Proposition 7.33 is best possible.

7.6.2 Obtain a generalization of Proposition 7.33 to polynomials of degree d in $\eta_1, \ldots, \eta_n$, with $1/8$ replaced by an exponent depending on d.

7.6.3 Improve the constant $1/8$ in Proposition 7.33 to $1/4$.

7.6.4 (Meshulam, private communication) Find a quadratic form $Q = \sum_{1 \leq i,j \leq n} c_{ij}\xi_i\xi_j$, where $c_{ij} \neq 0$ for all i, j and ξ_i are i.i.d Bernoulli random variables, such that

$$\mathbf{P}(Q = 0) \geq (2 - o(1))\frac{\binom{n}{\lfloor n/2 \rfloor}}{2^n}.$$

Compare this to the linear case (Corollary 7.4).

8

Incidence geometry

Incidence geometry deals with the incidences among basic geometrical objects such as points, lines and spheres. One can obtain useful and non-trivial information on these incidences by the classical combinatorial technique of *double-counting* the number of a certain type of configuration of incidences in two different ways.

In many situations, tools from from incidence geometry, combined with a clever double counting argument provide a simple, yet powerful, approach to hard problems. The goal of this chapter is to demonstrate several such applications, including several in additive combinatorics.

The material is organized as follows. We start with a result on the *crossing number* of graphs, which has a topological flavor. Next, we use this result to give simple proof of the famous *Szemerédi–Trotter theorem* concerning point-line incidences. In the next two sections, we use this theorem to prove several bounds on the Erdős–Szemerédi sum-product problem and reprove Andrews' theorem on the number of lattice points in a convex polygon. Next, we introduce the method of *cell decomposition* and use it to treat Erdős distinct distances problem in $\mathbf{R}^d$. Finally, we discuss a variant of Erdős–Szemerédi sum-product problem for complex numbers.

8.1 The crossing number of a graph

In this chapter, a *point* refers to a point in the plane $\mathbf{R}^2$, and a *line* refers to a line in $\mathbf{R}^2$, unless otherwise specified. By a *curve*, we refer to the image of a continuous injective embedding[1] of a compact interval [0, 1] into $\mathbf{R}^2$.

[1] In applications one deals with very explicit curves such as circular arcs or straight lines, and so we could restrict the class of curves to these sorts of objects if desired. In this way one does not need to invoke any difficult results from topology such as the Jordan curve theorem (which is implicit in our application of Euler's formula).

Consider a graph $G = G(V, E)$; recall we assume our graphs G to be undirected and have no loops or repeated edges. A *drawing* of G is any representation of G in the plane $\mathbf{R}^2$ by identifying each vertex in V with a distinct point, and each edge (u, v) in E with a curve in $\mathbf{R}^2$ connecting u and v. The *crossing number* of such a drawing is the number of pairs of edges with no common endpoints, where the corresponding curves intersect each other. The *crossing number* of G is the minimum number of crossings in a drawing. Here and later, we denote this parameter by $\mathrm{cross}(G)$.

It is expected that if G has many edges, then its crossing number is large. The following theorem, which confirmed this intuition, was proved by Ajtai, Chvátal, Newborn and Szemerédi [1], and, independently, by Leighton [224].

Theorem 8.1 *Let $G = G(V, E)$ be a graph with $|E| \geq 4|V|$. Then* $\mathrm{cross}(G) \geq \frac{|E|^3}{64|V|^2}$.

Proof A *planar graph* is a graph whose crossing number is zero. It is well known (and can be easily proved using Euler's formula) that a planar graph $G = G(V, E)$ has at most $3|V|$ edges (in fact it has at most $3|V| - 6$ if $|V| \geq 3$). Now observe that any graph G can be made planar by removing at most $\mathrm{cross}(G)$ edges (one for each crossing that occurs in an optimal drawing of G). Combining these two facts we obtain the preliminary inequality

$$\mathrm{cross}(G) \geq |E| - 3|V| \tag{8.1}$$

for an arbitrary graph $G(V, E)$.

This bound is, of course, much weaker than what we want to prove. However it is possible to amplify (8.1) substantially via the first moment method as follows.

Fix $G = G(V, E)$ with $|E| \geq 4|V|$, and let $0 < p \leq 1$ be a parameter to be chosen later. Let V' be a random subset of V, chosen so that the events $v \in V'$ are independent with probability p. Let $G' = G'(V', E')$ be the induced subgraph of G spanned by V'. Applying (8.1) to G' and then taking expectations, we see from linearity of expectation that

$$\mathbf{E}(\mathrm{cross}(G')) \geq \mathbf{E}(|E'|) - 3\mathbf{E}(|V'|).$$

Further application of linearity of expectation shows that

$$\mathbf{E}(|V'|) = p|V|; \qquad \mathbf{E}(|E'|) = p^2|E|,$$

since each vertex has probability p of being included in V', and each edge has probability p^2 of being included in E'. Now consider a drawing of G with exactly $\mathrm{cross}(G)$ crossings. Each crossing involves four vertices of V and thus has a probability p^4 of surviving when we pass to G'. Using linearity of expectation one last time we conclude

$$\mathbf{E}(\text{cross}(G')) \le p^4 \text{cross}(G);$$

we have inequality rather than equality since the drawing of G' constructed here may not have the minimal number of crossings. Putting all this together we have

$$\text{cross}(G) \ge p^{-3}|E| - 3p^{-3}|V|.$$

The claim then follows by setting $p := 4|V|/|E|$. □

Remark 8.2 One can improve the bound on cross(G) slightly by optimizing p. To obtain a more significant improvement, one needs additional arguments. The current best bound is due to Pach and Tóth [271].

Exercises

8.1.1 Let $G(V, E)$ be a planar graph with no loops or multiple edges. Using Euler's formula $V - E + F = 2$, show that $|E| \le \max(3|V| - 6, 1)$. Show that this bound $\max(3|V| - 6, 1)$ is best possible.

8.1.2 Show that a planar graph has a vertex of degree at most 5. Use this fact and induction to prove that a planar graph is vertex-colorable by 6 colors, where of course we require adjacent vertices to have distinct colors. Without using the four-color theorem, refine the argument to show that in fact every planar graph is vertex-colorable by 5 colors. (Hint: given any two colors, say red and green, one can swap all the red and green colors in a single red-green connected component without difficulty. Now given four colors red, blue, green, white adjacent in that order around a single uncolored vertex v, it cannot simultaneously be true that the red and green vertices lie in the same red-green connected component, and the blue and white vertices lie in the same red-white connected component.)

8.1.3 Show that for any $n, e \ge 1$ with $e \ge 4n$ that there exists a graph $G = G(V, E)$ with n vertices and e edges such that $\text{cross}(G) = \Theta(e^3/n^2)$, and so the crossing number inequality cannot be improved except for constants. (Hint: There are many ways to generate an example. One is to connect adjacent and nearly-adjacent points on the unit circle. Another is to use Exercise 8.2.2.)

8.1.4 Show that for any graph $G = G(V, E)$, we have $|E| = O(|V| + |V|^{2/3}\text{cross}(G)^{1/3})$.

8.1.5 [342] Let $m \ge 1$ be an integer, and let $G = G(V, E)$ be a multi-graph with maximum edge multiplicity m, thus each pair of vertices are allowed to be connected by up to m edges. Define the crossing number of a multigraph in the obvious manner. Show that if

$|E| \geq 5m|V|$, then $\mathrm{cross}(G) = \Omega(\frac{|E|^3}{|V|^2 m})$. In particular we have $|E| = O(m|V| + m^{1/3}|V|^{2/3}\mathrm{cross}(G)^{1/3})$.

8.2 The Szemerédi–Trotter theorem

Given a finite collection of points P and lines L, a basic question is to bound the number

$$I(P, L) := |\{(p, l) \in P \times L : p \in l\}|$$

of incidences between P and L. Clearly we can make $I(P, L)$ as small as zero without any difficulty, so the interesting question is to maximize $I(P, L)$ for fixed cardinalities $|P|$ and $|L|$. One of course has the trivial bound $I(P, L) \leq |P||L|$, and one can improve this further without difficulty to

$$I(P, L) \leq \min(|P|^{1/2}|L| + |P|, |L|^{1/2}|P| + |L|); \tag{8.2}$$

see exercises. In [348], Szemerédi and Trotter proved the following stronger estimate, which is sharp up to constants.

Theorem 8.3 (Szemerédi–Trotter theorem) *Let P be a finite set of points and let L be a finite set of lines. Then we have*

$$I(P, L) \leq 4|P|^{2/3}|L|^{2/3} + 4|P| + |L|.$$

Proof We may remove those lines $l \in L$ which do not contain any points in P, as they contribute nothing to the left-hand side. Thus we may assume that every line in L contains at least one point in P. Now let $G = G(P, E)$ be the graph whose vertices are the points in P, and two points a and b are connected if and only if the open line segment from a to b lies in a line in L and contains no points in P.

We now apply the double counting method to $|E|$, the number of edges. Observe that if a line l in L contains $k \geq 1$ points in P, then l contributes $k - 1$ edges to E. Summing over $l \in L$, we conclude

$$|E| = I(P, L) - |L|.$$

On the other hand, observe that G has a tautological drawing, with the vertices in P mapping to themselves, and the edge $[a, b]$ mapping to the line segment from a to b. Since any two lines in L can intersect in at most one point, we conclude that $\mathrm{cross}(G) \leq |L|^2$. Applying the crossing number inequality, we conclude that either $|E| \leq 4|P|$ or $\mathrm{cross}(G) \geq |E|^3/64|P|^2$. Thus $|E| \leq \max(4|P|, 4|P|^{2/3}|L|^{2/3})$, and the claim follows. □

Remark 8.4 The above proof is due to Székely [342]; the original proof of Szemerédi and Trotter is quite different (see Exercise 8.4.7 for a proof closer in spirit to that). The symmetry between P and L can be explained by projective duality; if we embed the plane $\mathbf{R}^2$ into the projective space of $\mathbf{R}^3$, then points become associated to subspaces of $\mathbf{R}^3$ of dimension 1, while lines are associated to subspaces of codimension 1.

Let us now derive a few corollaries from the theorem. An immediate consequence, which we leave as a exercise, allows us to bound the number of lines which are "rich" in the sense that they contain many elements of a given set P of points.

Corollary 8.5 (Rich lines) *If P is any finite set of points and $k \geq 2$, then*

$$|\{l \text{ a line} : |l \cap P| \geq k\}| = O\left(\max\left(\frac{|P|^2}{k^3}, \frac{|P|}{k}\right)\right).$$

Dually, for any finite set L of lines, we have

$$|\{p \in \mathbf{R}^2 : |\{l \in L : p \in l\}| \geq k\}| = O\left(\max\left(\frac{|L|^2}{k^3}, \frac{|L|}{k}\right)\right).$$

Remark 8.6 In typical applications, such as those below, $k \leq |P|^{1/2}$ so the term $\frac{|P|^2}{k^3}$ is dominating. The case $k > |P|^{1/2}$ can be treated by the cruder estimate (8.2). Similarly for the second half of the corollary.

Next, we bound the number of pairs of points which are connected by a rich line.

Corollary 8.7 (Rich pairs) *If P is any finite set of points and $k \geq 1$, then*

$$|\{(p,q) \in P \times P : p \neq q; k \leq |l_{p,q} \cap P| \leq 2k\}| = O\left(\max\left(\frac{|P|^2}{k}, |P|k\right)\right)$$

where $l_{p,q}$ is the unique line connecting p and q. In particular, if $1 \leq k \leq |P|^{1/2}$, then

$$\left|\{(p,q) \in P \times P : p \neq q; k \leq |l_{p,q} \cap P| \leq |P|^{1/2}\}\right| = O\left(\frac{|P|^2}{k}\right).$$

Proof For the first bound, we observe that each line l with $k \leq |l \cap P| \leq 2k$ contributes at most $O(k^2)$ pairs to the left-hand side, so the claim follows from Corollary 8.5. The second bound follows from the first by a standard dyadic decomposition argument. □

An easy modification of this argument, which we leave as an exercise, allows us to also control collinear triples that are not on too rich of a line:

Corollary 8.8 (Collinear triples) *Let P be a finite set of points. Then the number of triples (u, v, w) where u, v, w are three collinear distinct points in P, whose line contains at most $|P|^{1/2}$ points in P, is at most $O(|P|^2 \log |P|)$.*

Applying this in particular to Cartesian products $P = A \times B$, where A, B are sets of real numbers with $|A| = |B| = m$, we observe that $|P| = m^2$ and no line intersects P in more than $|P|^{1/2} = m$ points. We conclude

Corollary 8.9 *Let A and B be sets of real numbers of cardinality m. Then $A \times B$ contains at most $O(m^4 \log m)$ collinear triples.*

It is an easy matter to extend the Szemerédi–Trotter theorem to more general curves than lines.

Theorem 8.10 (Generalized Szemerédi–Trotter theorem) *[342] Let P be a finite collection of points in $\mathbf{R}^2$, and let L be a finite collection of curves in $\mathbf{R}^2$. Suppose that any two curves in L intersect in at most α points, and any two points in P are simultaneously incident to at most β lines; then*

$$|\{(p, l) \in P \times L : p \in l\}| = O\big(\alpha^{1/3}\beta^{1/3}|P|^{2/3}|L|^{2/3} + |L| + \beta|P|\big).$$

As an application of this theorem we prove the following remarkable result of Andrews [13].

Theorem 8.11 *Let $\Gamma \subset \mathbf{R}^2$ be a lattice (e.g. $\Gamma = \mathbf{Z}^2$). If C is a convex n-gon with vertices in Γ, then the interior of C contains $\Omega(n^3)$ lattice points.*

Proof Let $\mathcal{C}$ be the boundary of C and F be collection of (piecewise linear) curves obtained by translating $\mathcal{C}$ by the lattice points inside C. Let P be the set of lattice points covered by the union of the curves in F and m be the number of lattice points inside C. We have $|F| = m$ and $|P| = \Theta(m)$ (cf. (3.10)).

We apply the double counting method to the number of incidences between P and F. On the one hand, the generalized Szemerédi–Trotter theorem gives an upper bound of $O(m^{4/3})$ for these incidences. On the other hand, each translate of $\mathcal{C}$ contains exactly n points, so the number of incidences is at least nm. Comparing these bounds we obtain $m = \Omega(n^3)$ as desired. □

Remark 8.12 The above theorem generalizes for $\mathbf{R}^d$. For any fixed d, Andrews proved that a convex polytope in $\mathbf{R}^d$ with n non-coplanar integral points on its boundary has volume $\Omega(n^{(d+1)/(d-1)})$. The above proof, however, does not generalize for higher dimensions.

An important open problem is to extend the Szemerédi–Trotter theorem to planes over other fields, for instance the complex plane $\mathbf{C}^2$ or the finite field planes F_p^2. The crude estimate (8.2) applies in all of these situations, but one would like to improve this bound. In the case of F_p^2 it was shown that $I(P, L) = O_\delta(\max(|P|, |L|)^{3/2-\varepsilon(\delta)})$ whenever $|P|, |L| \leq p^{2-\delta}$ for all $\delta > 0$ and some $\varepsilon(\delta) > 0$ depending on δ; see [43], [44]. The main ingredients in this argument was the sum-product estimate in Corollary 2.58 and the Balog–Szemerédi–Gowers theorem (Theorem 2.29).

Exercises

8.2.1 Using only the basic facts that two distinct points determine at most one line, and two distinct lines intersect in at most one point, together with the Cauchy–Schwarz inequality, prove (8.2). Observe that this argument works over any field, not just $\mathbf{R}$. In the case where the field is F_{p^2}, show that the bound can be sharp when $|P| = |L| = p^2$, or when $|P| = |L| = p^4$.

8.2.2 Let $n, m \geq 1$ be given. Find an example of a set of points P and a set of lines L such that $|P| = n$, $|L| = m$, and the number of incidences between P and L is $\Theta(n^{2/3}m^{2/3} + n + m)$, thus demonstrating that the Szemerédi–Trotter theorem is sharp up to constants. (Hint: consider sets P of the form $P = [1, a] \times [1, ab]$ for various parameters a, b.)

8.2.3 Prove Corollary 8.5.

8.2.4 Prove Corollary 8.8.

8.2.5 Let P be a finite set of points, and let $k \geq 2$. Show that

$$|\{(p, l) : p \in P; l \text{ a line}; p \in l; |l \cap P| \geq k\}| = O\left(\frac{|P|^2}{k^2} + |P| \log |P|\right).$$

8.2.6 (Beck's theorem) [19] Let P be a finite set of points. Show that either there exists a line that is incident to $\Theta(|P|)$ points in P, or there exist $\Theta(|P|^2)$ lines that are each incident to at least two points in P.

8.2.7 (Sylvester–Gallai theorem) Let P be a finite set of points, not all of which are collinear. Show that there exists a line that contains exactly two points in P. (Hint: minimize the quantity $\mathrm{dist}(p, l)$, where l is a line containing two or more points in P and $p \in P \backslash l$. Using elementary geometry, show that this quantity is minimized only when l contains exactly two points from P.)

8.2.8 Prove Theorem 8.10. (Hint: use Exercise 8.1.5.)

8.2.9 Let γ be a strictly convex curve in $\mathbf{R}^2$. Show that $|(R \cdot \gamma) \cap \Gamma| = O_\gamma(R^{2/3})$ for all $R \geq 1$ and all lattices Γ.

8.2.10 Let γ be a strictly convex curve in $\mathbf{R}^2$, and let A be a finite set in $\mathbf{R}^2$. Show that $|\{(a, a') \in A \times A : a - a' \in \gamma\}| = O(|A|^{4/3})$. Deduce from this that $|\{|x - y| : x, y \in A\}| = \Omega(|A|^{2/3})$.

8.3 The sum-product problem in R

In Section 2.8 we considered the *sum-product problem*, where one wished to establish lower bounds on either the sum set $A + A$ or the product set $A \cdot A$ when A was an arbitrary non-empty finite subset of a field or ring. For instance, it was shown there that if the ambient field contained no proper subfields, then one had $|A + A| + |A \cdot A| = \Omega(|A|^{1+\varepsilon})$ for some explicit $\varepsilon > 0$. In the case when A is a set of integers (or more generally of real numbers), Erdős and Szemerédi conjectured the following stronger result:

Conjecture 8.13 (Erdős–Szemerédi conjecture) *[91] Let A be a finite non-empty set of integers or reals. Then for any $\varepsilon > 0$ we have*

$$|A + A| + |A \cdot A| \geq \Omega_\varepsilon(|A|^{2-\varepsilon}).$$

The condition $\varepsilon > 0$ is sharp; see Exercise 8.3.6.

In support of this conjecture, Erdős and Szemerédi [91] proved the bound $|A + A| + |A \cdot A| \geq \Omega(|A|^{1+\delta})$ for some absolute constant $\delta > 0$, when A is a set of integers. Nathanson [258] showed that one can set $\delta = 1/31$. Ford [105] improved δ to $1/15$. These proofs relied on properties of factorizations.

In 1997, Elekes [76] improved δ to $1/4$ and extended to the case of real numbers, using the Szemerédi–Trotter theorem in an ingenious way.

Theorem 8.14 *Let A be a finite non-empty set of reals. Then*

$$|A + A| \times |A \cdot A| = \Omega\big(|A|^{5/2}\big).$$

In particular

$$|A + A| + |A \cdot A| = \Omega\big(|A|^{5/4}\big).$$

Proof Let $P = \{(a, b) | a \in A + A, b \in A \cdot A\}$; P is a subset of the plane and has cardinality $|A + A||A \cdot A|$.

Consider the set L of lines of the form $\{(x, y) : y = a(x - b)\}$ where a, b are elements of A. Clearly, L has $|A|^2$ elements. Moreover, each such line contains at least $|A|$ points in P, namely the points $(b + c, ac)$ with $c \in P$. Thus $I(P, L) \geq |A|^3$. Applying the Szemerédi–Trotter theorem we conclude

$$|A|^3 \leq O\big((|A + A||A \cdot A|)^{2/3}(|A|^2)^{2/3} + |A + A||A \cdot A| + |A|^2\big),$$

and the claim follows by elementary algebra. □

Very recently, Solymosi [324] added a new twist to Elekes' argument, essentially improving ε to $3/11$.

Theorem 8.15 *[324] Let A be a finite set of real numbers with $|A| \geq 2$. Then we have $|A+A|^8|A/A|^3 = \Omega(|A|^{14})$, and $|A+A|^8|A\cdot A|^3 = \Omega(\frac{|A|^{14}}{\log^3|A|})$. Consequently*

$$|A+A| + |A\cdot A| = \Omega\big(|A|^{14/11}/\log^{3/11}|A|\big). \tag{8.3}$$

Proof We may remove zero if necessary and assume that all elements of A are non-zero. We shall need a dyadic decomposition of A/A, in order to control the multiplicity of quotients in A/A from both above and below. Let $2 \leq d \leq |A|$ be a power of two to be chosen later, and let $D_d \subseteq A/A$ be the set

$D_d := \{m \in A/A : m = a_1/a_2$ for between d and $2d$ values of $(a_1, a_2) \in A \times A\}$.

Let $P := A \times A$, and let L denote all the lines $\{(x, y) : y = mx + b\}$ with slope m in D_d, and which contain at least one point in P. Observe that L is finite, and that each point $p \in P$ is incident to $|D_d|$ lines in L. Thus by Corollary 8.5 we have

$$|A|^2 = |P| = O\left(\frac{|L|}{|D_d|} + \frac{|L|^2}{|D_d|^3}\right);$$

since $|D_d| \leq |A/A| \leq |A|^2$, this implies a lower bound on $|L|$:

$$|L| = \Omega\big(|A||D_d|^{3/2}\big). \tag{8.4}$$

Now let $P' := (A+A) \times (A+A)$. Observe that if $l \in L$, then l has some slope $m \in D_d$ and contains a point (a_1, a_2) in P. In particular, $l \cap P'$ contains the set $\{(a_1+a_3, a_2+a_4) : a_3, a_4 \in A; a_3/a_4 = m\}$, which has cardinality at least d by definition of D_d. Thus each line in L contains at least d points in P'; by Corollary 8.5 again, we conclude that

$$|L| = O\left(\frac{|P'|}{d} + \frac{|P'|^2}{d^3}\right) = O\left(\frac{|P'|^2}{d^3}\right),$$

where the latter bound follows since $d \leq |A|$ and $|P'| \geq |A|^2$. Inserting (8.4) and $|P'| = |A+A|^2$ we obtain after some algebra

$$|D_d| = O\left(\frac{|A+A|^{8/3}}{|A|^{2/3}d^2}\right). \tag{8.5}$$

In particular, by definition of D_d,

$$|\{(a_1, a_2) \in A \times A : a_1/a_2 \in D_d\}| = O\left(\frac{|A+A|^{8/3}}{|A|^{2/3}d}\right).$$

Summing this over d equal to all powers of two greater than $C|A+A|^{8/3}/|A|^{18/3}$ for some large absolute constant C, we obtain

$$\big|\{(a_1, a_2) \in A \times A : a_1/a_2 \in D_d \text{ for some } d \geq C|A+A|^{8/3}/|A|^{8/3}\}\big| \leq \frac{1}{2}|A|^2$$

and hence

$$\left|\{(a_1, a_2) \in A \times A : a_1/a_2 \in D_d \text{ for some } d < C|A + A|^{8/3}/|A|^{8/3}\}\right| \geq \frac{1}{2}|A|^2.$$

But for d as above, each $m \in D_d$ has at most $O(d) = O(|A + A|^{8/3}/|A|^{14/3})$ representations of the form a_1/a_2, and so we can conclude that $|A/A| = \Omega(|A|^2/(|A + A|^{8/3}/|A|^{8/3}))$ which gives the first inequality.

To prove the second inequality, we observe from (8.5) that

$$|\{(a_1, a_2, a_3, a_4) \in A \times A : a_1/a_2 = a_3/a_4 \in D_d\}| = O\left(\frac{|A + A|^{8/3}}{|A|^{2/3}}\right);$$

note that while the above argument was only for $d \geq 2$, the estimate here also holds for $d = 1$ by crudely bounding the left-hand side by $|A|^2$ and bounding $|A + A|$ from below by $|A|$. Summing this over d equal to all powers of 2 between 1 and $|A|$, we obtain

$$|\{(a_1, a_2, a_3, a_4) \in A \times A : a_1/a_2 = a_3/a_4\}| = O\left(\frac{|A + A|^{8/3}}{|A|^{2/3}} \log |A|\right).$$

On the other hand, by a simple double counting argument (cf. (2.8)) we have

$$|\{(a_1, a_2, a_3, a_4) \in A \times A : a_1/a_2 = a_3/a_4\}| \geq |A|^4/|A \cdot A|,$$

and the claim follows. □

A special case which draws lots of attention is when either $|A + A|$ or $|A \cdot A|$ is small. Elekes and Ruzsa [80] proved the following theorem.

Theorem 8.16 *Let A be a finite set of real numbers with $|A| \geq 2$. Then*

$$|A + A|^4 |A \cdot A| = \Omega\left(\frac{|A|^6}{\log |A|}\right).$$

In particular, if $|A + A| = O(|A|)$, then $|A \cdot A| = \Omega(|A|^2/\log |A|)$.

The logarithmic factor is necessary; if one has $A := [1, n]$ then it is known that $|A \cdot A| = O(\frac{n^2}{\log^c n})$ for some positive constant c. (See also Exercise 8.3.6.)

Proof It is easy to reduce to the case when the elements of A are positive. Let $P := ((A + A) \cup A) \times ((A + A) \cup A)$; thus P is a collection of points of cardinality $O(|A + A|^2)$. We shall apply the double counting method to the number of collinear triples in P. On the one hand, Corollary 8.9 shows that the number of such triples is $O(|A + A|^2 \log |A|)$. On the other hand, a standard Cauchy–Schwarz argument (cf. (2.8)) shows that

$$|\{(a, b, c, d) \in A \times A \times A \times A : ab = cd\}| \geq \frac{|A|^4}{|A \cdot A|}.$$

We may assume $|A \cdot A| \leq \frac{1}{2}|A|^2$ since the claim is trivial otherwise. We can then remove the $a = d$ contribution from the right-hand side and conclude

$$|\{(a, b, c, d) \in A \times A \times A \times A : ab = cd; a \neq d\}| = \Omega\left(\frac{|A|^4}{|A \cdot A|}\right).$$

For any (a, b, c, d) in the above set and $e, f \in A$, observe that the three points $(e, f), (e + a, f + c), (e + b, f + d)$ form a collinear triple in P. The number of triples obtained in this manner is $\Omega(\frac{|A|^6}{|A \cdot A|})$. Combining this with the upper bound, the claim follows. □

The above results show that if $|A + A|$ is close to $|A|$, then $|A \cdot A|$ is close to $|A|^2$. In the other direction, the best known results are due to Chang [49], who has established that if $|A \cdot A| \leq K|A|$ then $|A + A| \geq 36^{-K}|A|^2$, and more generally $|hA| \geq (2h^2 - h)^{-hK}|A|^h$ for all $h \geq 2$. Those arguments are not as elementary as those presented here, relying instead on a result of Freiman (Theorem 5.13) and the machinery of $\Lambda(p)$ constants from Section 4.5 in order to get good lower bounds on $|hA|$. See [49] for further details and some history of the problem.

Exercises

8.3.1 Show that the Erdős–Szemerédi conjecture for sets of integers is equivalent to the corresponding conjecture for sets of rationals. Show that the conjecture for sets of reals is equivalent to the conjecture for sets of *algebraic* integers. It is not known whether the conjecture for reals is equivalent to the conjecture for (rational) integers.

8.3.2 Let A, B be additive sets of real numbers with $|A|, |B| \geq 2$. Show that $|\frac{A-A}{(B-B)\backslash 0}| = \Omega(|A||B|)$. (Hint: apply Beck's theorem to $P = A \times B$.) In particular, in the notation of Section 2.8 we have $|Q[A]| = \Omega(|A|^2)$; compare this with Corollary 2.51 and Corollary 2.52.

8.3.3 Let A, B, C be additive sets of real numbers. Show that $|A + B \cdot C| = \Omega(|A|^{1/2}|B|^{1/2}|C|^{1/2})$. (Hint: if $|B| \leq |C|$, apply the Szemerédi–Trotter theorem with $P := B \times (A + B \cdot C)$ and L equal to those lines with slope in C and y-intercept in A.) Conclude that $|h(B \cdot C)| = \Omega((|B||C|)^{1-1/2^h})$ for all $h \geq 1$.

8.3.4 Generalize Theorem 8.14 by demonstrating the inequality $|A + B||B \cdot C| = \Omega(\min(|A||B||C|, |A|^{1/2}|B|^{1/2}|C|^{3/2}))$.

8.3.5 [79] Let $f : \mathbf{R} \to \mathbf{R}$ be any strictly convex function, and let A be an additive set of reals. Show that $|A + A||f(A) + f(A)| = \Theta(|A|^{5/2})$. (Hint: note that Theorem 8.14 addresses the case when $f(x) = \log x$; this should suggest a proof for the general case.)

8.3.6 Let n be a large integer. Using Theorem 1.6, show that all but at most $o(n^2)$ elements of $[1, n] \cdot [1, n]$ have $(2 + o(1)) \log \log n$ prime divisors. (Note that the convergence of the sum $\sum_{m=1}^{\infty} \frac{1}{m^2}$ shows that one can neglect those elements which have a large square factor.) Conclude that $|[1, n] \cdot [1, n]| = o(n^2)$. For much more precise estimates, see [106].

8.4 Cell decompositions and the distinct distances problem

Given a finite point set $P \subset \mathbf{R}^d$, let $g(P) := \{|x - y| : x, y \in P\}$ denote the number of distinct distances between the elements of P. Define $g_d(n) = \min_{P \subset \mathbf{R}^d, |P|=n} g(P)$. The well-known *distinct distances problem* of Erdős, posed in 1946 [83], asks to determine the correct rate of growth of $g_d(n)$ in n for each fixed d; this question remains open even when $d = 2$. (Clearly we have $g_1(n) = n - 1$.)

By considering the progression $P = [1, n^{1/d}]^d$ it is easy to see that $g_d(n) = O(dn^{2/d})$. Erdős and many other researchers conjecture that $g_d(n)$ is close to this upper bound; in particular it is conjectured that $g_d(n) = \Omega_{\varepsilon,d}(n^{2/d-\varepsilon})$ for any $\varepsilon > 0$.

It is quite easy to establish the lower bound $g_d(n) = \Omega_d(n^{1/d})$: see Exercise 8.4.2. There is a series of improvements for the case $d = 2$, due to Moser [252], Chung [57], Chung–Szemerédi–Trotter [60], Székely [342], Solymosi–Tóth [328], Tardos [353], Katz and Tardos [197]. The most current bound is $g_2(n) = \Omega(n^{0.8635})$ [197], using the approach in [328] combined with clever entropy arguments. Here we will present a slightly weaker bound due Székely; this argument forms the base for all the subsequent bounds mentioned above.

Theorem 8.17 *[342] We have $g_2(n) = \Omega(n^{4/5})$.*

Proof Let P be a set of n points in $\mathbf{R}^2$. Define an *isosceles triangle* to be a triple (p, q, q') of distinct points in P such that $|p - q| = |p - q'|$. We say that the isosceles triangle is *narrow* if the circular arc from q to q' with center at p contains no other points in P. We refer to the pair (q, q') as the *base* of the isosceles triangle, and p as the *apex*. For any $k \geq 1$, we say that a pair (q, q') is *k-rich* if it is the base of at least k narrow isosceles triangles, and *k-poor* otherwise.

Let N be the number of narrow isosceles triangles (p, q, q'). We shall apply a double counting argument to N. We begin with the lower bound. There are $|P|$ choices for p. Given p, the remaining $|P| - 1$ points in p are contained in at most $g_2(|P|)$ circles centered at p. Let $\mathcal{C}$ be the collection of such circles, then we easily verify that the number of isosceles triangles with apex p is

$$\sum_{C \in \mathcal{C}: |C \cap P| \geq 2} 2|C \cap P|.$$

Since $\sum_{C \in \mathcal{C}} |C \cap P| = |P| - 1$, we can write the above quantity as $2|P| - O(g_2(P))$. Summing over all p we conclude that

$$N \geq 2|P|^2 - O(|P| g_2(|P|)).$$

Now we obtain the upper bound. We let $k \geq 1$ be a parameter to be chosen later, and split $N = N_{\text{rich}} + N_{\text{poor}}$, where N_{rich} (resp. N_{poor}) is the number of narrow isosceles triangles with a k-rich (resp. k-poor) base. Observe that if (p, q, q') is an isosceles triangle with a k-rich base, then the perpendicular bisector l of q, q' contains p and also contains at least k points from P. Conversely, for fixed l and p there are at most $4g_2(|P|)$ pairs (q, q') with perpendicular bisector l for which (p, q, q') is a narrow isosceles triangle; this can be seen by covering the points in $P \backslash \{p\}$ into at most $g_2(|P|)$ circles and observing that each circle contributes at most four such triangles. Applying Exercise 8.2.5 we conclude

$$N_{\text{rich}} = O(g_2(|P|) \left(\frac{|P|^2}{k^2} + |P| \log |P| \right).$$

As for the poor triangles, consider the multi-graph drawing G whose vertices are P and whose edges are the circular arcs corresponding to narrow isosceles triangles (p, q, q') with a k-poor base. This graph has $|P|$ vertices and N_{poor} edges, and has edge multiplicity at most k. Thus by Exercise 8.1.5, we have

$$N_{\text{poor}} = O\big(k|P| + k^{1/3}|P|^{2/3}\text{cross}(G)^{1/3}\big).$$

On the other hand, since the drawing of G is contained in at most $|P|g_2(|P|)$ circles (each center $p \in P$ contributing at most $g_2(|P|)$ circles), and any two circles cross in at most two points, we see that $\text{cross}(G) \leq 2(|P|g_2(|P|))^2$; thus

$$N_{\text{poor}} = O\big(k|P| + k^{1/3}|P|^{4/3} g_2(|P|)^{2/3}\big).$$

Combining our upper bounds for N_{poor} and N_{rich} with the lower bound for N, we obtain

$$|P|^2 \leq O(|P|g_2(|P|)) + O\left(g_2(|P|)\left(\frac{|P|^2}{k^2} + |P| \log |P|\right)\right) + O\big(k|P| + k^{\frac{1}{3}}|P|^{\frac{4}{3}} g_2(|P|)^{\frac{2}{3}}\big).$$

We optimize this by setting $k := c|P|^{2/5}$ for some small constant $c > 0$, and some elementary algebra then gives $g_2(|P|) = \Omega(|P|^{4/5})$ as desired. □

The above argument generalizes to many other metrics than the Euclidean metric; see [130]. To go beyond $n^{4/5}$, however, it seems that one needs to use the finer arithmetic structure of Euclidean geometry. Very roughly, the results of [328], [353], [197] proceed by analyzing the perpendicular bisectors of all of the narrow isosceles triangles (p, q, q') with a given apex p and a k-rich base; note these bisectors are k-rich in the sense that they contain at least k points in P. Using polar coordinates around p, one can parameterize these bisectors using the sum of the angles of q and q'. One can then use some bounds on partial sum sets to obtain non-trivial lower bounds on the number of k-rich lines through p, which can then be combined with Exercise 8.2.5 to obtain an improvement to Theorem 8.17; see [328]. The further refinements in [353], [197] proceed similarly, but with a slightly weaker notion of narrow isosceles triangle, allowing the circular arc connecting q with q' to contain $O(1)$ other points from P. This provides several further partial sum sets to yield slightly better lower bounds on the number of k-rich lines through p.

In the higher-dimensional case $d > 2$ much less is known. However there are some reasonable results if one imposes some uniform distribution on the points. Let Q^d be the standard unit cube in $\mathbf{R}^d$, centered at the origin. Let us call a finite set $P \subset \mathbf{R}^d$ *homogeneous* if $P \subset |P|^{1/d} \cdot Q$, and $|P \cap (x + Q)| = O_d(1)$ for all $x \in \mathbf{R}^d$. A good example of a homogeneous set can be obtained by starting with the progression $[1, |P|^{1/d}]^d$ and perturbing each element of this progression by an arbitrary bounded displacement.

A weakened version of Erdős' original problem asks for the number of distinct distances in a homogeneous set. Homogeneous sets are interesting for at least two reasons. First, the best known upper bounds for the distance problem are homogeneous. Second, homogeneous sets play an important role in analysis (see e.g. [189]). In this section we prove

Theorem 8.18 *[326] Let $P \subset \mathbf{R}^d$ be a homogeneous set. Then $g_d(P) = \Omega_d(|P|^{\frac{2}{d}-\frac{1}{d^2}})$.*

This should be compared with the (homogeneous) lattice example $P = [1, |P|^{1/d}]^d$, which gives $g_d(P) = O_d(|P|^{\frac{2}{d}})$. As in Theorem 8.17, the proof starts by a double counting argument applied to narrow isosceles triangles. However, crossing number and Szemerédi–Trotter type results are not available in higher dimensions, and one instead uses the more flexible technique of *cell decomposition*. Given a large, complex incidence system S, we try to break it into many pieces, each of which has only a small number of incidences. After the decomposition is achieved, an (often tricky) double counting argument concerning the number of a properly defined object yields fairly efficient bounds.

Proof We may of course assume $|P| \geq 2$. By hypothesis, P is contained in the cube $|P|^{1/d} \cdot Q$. Let $1 \leq r < |P|^{1/d}$ be an integer to be chosen later. By using hyperplanes parallel to the coordinate axes, we can partition $|P|^{1/d} \cdot Q = C_1 \cup \cdots \cup C_{r^d}$, where each C_i is a cube of side-length $|P|^{1/d}/r$; we assign the boundary points of these cubes arbitrarily to one of the cubes of the partition. We refer to the cubes C_i as *cells*.

For each $p \in P$, the set $P\backslash\{p\}$ is contained in the union of at most $g_d(P)$ spheres centered at p. We denote by $\mathcal{S}_p$ the set of these spheres. For each sphere $S \in \mathcal{S}_p$, let $\mathcal{C}_S$ denote all the cells C_i which intersect S and which contain at least one point of P; from elementary geometry we see that $|\mathcal{C}_S| = O_d(r^{d-1})$.

We now apply a double counting argument to the quantity

$$N := |\{(p, S, C, q, q') : p \in P; S \in \mathcal{S}_p; C \in \mathcal{C}_S; q, q' \in P \cap S \cap C; q \neq q'\}|;$$

informally, N counts the number of isosceles triangles in P where the base points lie in the same cell (cf. the proof of Theorem 8.17). We begin with an upper bound. Observe that there are r^d possible cells C. A cell has side-length $|P|^{1/d}/r > 1$, so by homogeneity it contains $O_d(|P|/r^d)$ points in P. Thus there are $O_d(|P|/r^d)^2$ possible pairs q, q' that can be associated to C. For each such pair, observed that p must lie on the hyperplane bisecting q and q' (since q, q' lie on a sphere centered at p). By homogeneity again, this hyperplane contains at most $O(|P|^{(d-1)/d})$ elements of P. Finally, once p, q, q' are fixed, S is completely determined. Putting this all together we obtain the upper bound

$$N \leq r^d O_d(|P|/r^d)^2 O\big(|P|^{(d-1)/d}\big) = O_d(|P|^{3-\frac{1}{d}} r^{-d}). \tag{8.6}$$

Now we obtain a lower bound. Observe the explicit formula

$$N = \sum_{p\in P} \sum_{S \in \mathcal{S}_p} \sum_{C \in \mathcal{C}_S} |P \cap S \cap C|^2 - |P \cap S \cap C|.$$

From Cauchy–Schwarz we have

$$\sum_{C \in \mathcal{C}_S} |P \cap S \cap C|^2 \geq \frac{|P \cap S|^2}{|\mathcal{C}_S|}$$

and

$$\sum_{S \in \mathcal{S}_p} |P \cap S|^2 \geq \frac{(|P| - 1)^2}{g_d(P)}$$

and hence

$$N \geq \sum_{p \in P} \frac{(|P| - 1)^2}{g_d(P)|\mathcal{C}_S|} - (|P| - 1).$$

Since $|\mathcal{C}_S| = O_d(r^{d-1})$, we conclude

$$N \geq \Omega_d\left(\frac{|P|^3}{g_d(P) r^{d-1}}\right) - |P|^2.$$

Combining this with (8.6) and rearranging, we conclude

$$g_d(P) = \Omega_d\left(\frac{|P|^3}{r^{d-1}\big(|P|^{3-\frac{1}{d}} r^{-d} + |P|^2\big)}\right).$$

We optimize this by selecting r to be the nearest integer to $|P|^{\frac{1}{d}-\frac{1}{d^2}}$, and the claim follows. □

For $d \geq 3$ and general (inhomogeneous) sets, little has been known for a long time, as the method based on the Szemérédi–Trotter theorem cannot be generalized to dimension larger than 2. Clarkson, Edelsbrunner, Gubias, Sharir and Welzl [63] proved that $g_3(n) = \Omega(n^{1/2})$. In 2002, Aronov, Pach, Sharir and Tardos [14] proved that $g_3(n) = \Omega_\epsilon(n^{77/141-\epsilon})$ for any $\epsilon > 0$. More generally, they proved that $g_d(n) = \Omega_{d,\epsilon}(n^{1/(d-90/77)-\epsilon})$ for any $d \geq 3$. This result gives a non-trivial improvement for small d, compared to the previous bound $n^{1/d}$. On the other hand, as $d \to \infty$, the exponent $1/(d-90/77)-\epsilon$ converges to $1/d$, rather than to the conjectured bound $2/d$.

Very recently, Solymosi and Vu [327] managed to show that the exponent $2/d$ is best possible to top order, in the sense that it cannot be replaced by $(2-\epsilon+o_{d\to\infty}(1))/d$ for any positive constant $\epsilon > 0$. More precisely, they showed that that

$$g_d(n) = \Omega_d\left(n^{\frac{2}{d}-\frac{2}{d(d+2)}}\right)$$

for all $d \geq 4$, and also $g_3(n) = \Omega(n^{.5643})$.

This result and the previous bound of Aronov et al. were proved using the decomposition method combined with other arguments. Unlike the homogeneous case, the decomposition used here is more sophisticated and was first developed by Chazelle and Friedman [52] (see also [245]), motivated by problems in geometric searching in computer science. Let us conclude this section by briefly discussing this result.

One of the main techniques for doing a search is divide-and-conquer. In many problems, the situation looks as follows: given a set B of hyperplanes (of co-dimension 1) in $\mathbf{R}^d$, one would like to partition $\mathbf{R}^d$ in not too many parts so that each part intersects only few hyperplanes.

Definition 8.19 A hyperplane H *strongly intersects* a set P if $H \cap P$ is not empty and P has a point on both side of H.

Lemma 8.20 *Let B be a set of k hyperplanes in $\mathbf{R}^d$. For any $1 \leq r \leq k$, one can partition $\mathbf{R}^d$ into r sets $P_1, \ldots, P_r$ such that for each $1 \leq i \leq r$, there are only $O(k/r^{1/d})$ planes which strongly intersect P_i.*

The bound $O(k/r^{1/d})$ is best possible; the hidden constants in O depend on d but not on r. One can also guarantee that the sets P_i are generalized simplices. Strong intersection actually means intersection with the interior (see [245]). Let us now consider a little bit more complex situation when beside B we also have a set A of n points. We can require, in addition, that each part contains not too many points.

Lemma 8.21 *Let A be a set of n points and B be a set of k hyperplanes in $\mathbf{R}^d$. For any $1 \leq r \leq k$, one can partition $\mathbf{R}^d$ into r sets $P_1, \ldots, P_r$ such that for each $1 \leq i \leq r$, $|P_i \cap A| \leq 2n/r$ and P_i strongly intersects $O(k/r^{1/d})$ planes.*

Lemma 8.21 is not restricted to hyperplanes. It still holds if we replace a family of hyperplanes by a family of surfaces satisfying certain topological conditions. In particular, the lemma holds if we replace hyperplanes by (full-dimensional) spheres (see Section 6.5 of [245]). As an analog of Lemma 8.21, we obtain the following lemma, which was actually used in [327].

Definition 8.22 A sphere S strongly intersects a set P if $S \cap P$ is not empty and P has a point on both sides of S.

Lemma 8.23 *Let A be a set of n points and B be a set of k spheres in $\mathbf{R}^d$. For any $1 \leq r \leq k$, one can partition $\mathbf{R}^d$ into r sets $P_1, \ldots, P_r$ such that for each $1 \leq i \leq r$, $|P_i \cap A| = O(n/r)$ and there are only $O(k/r^{1/d})$ spheres which strongly intersect P_i.*

It would be very desirable to have a finite field analog of the above lemmas. Here is the simplest form of the problem: given a set of lines (or simple curves) on a finite plane, we would like to partition the plane into a few parts so that each part intersects only a few lines. The main obstacle here is that one needs to find a proper replacement for the topological condition of strong intersection. This condition was used to rule out extremal cases such as when all the hyperplanes go through the same point.

Exercises

8.4.1 Let A be a finite non-empty set of reals. Show that $|k((A - A)\hat{}2)| \geq g_k(|A|^2)$, where $X\hat{}2 := \{x^2 : x \in X\}$ is the set of squares in X, and $kX := X + \cdots + X$ is the k-fold sum set of X. Thus progress on the Erdős distance problem is linked to progress on questions of sum-product type; see [43] for some further development of this idea.

8.4.2 [83] Let $x_1, \ldots, x_d$ be d points in general position in $\mathbf{R}^d$. Show that if $x \in \mathbf{R}^d$, then the d distances $|x - x_1|, \ldots, |x - x_d|$ determine x up to a multiplicity of $O_d(1)$. Use this to show that $g_d(n) = O_d(n^{1/d})$ for all n. (Note that the degenerate case in which many points lie in a lower-dimensional space can be dealt with by an induction argument.)

8.4.3 [326] (Rich lines in three dimensions) Let P be a homogeneous set in $\mathbf{R}^3$. Show that $\sum_{l:|l \cap P| \geq k} |l \cap P| = O(|P|^2/k^3)$ for all $k \geq 2$.

8.4.4 [326] Let A be a homogeneous set of cardinality n in $\mathbb{R}^3$ and $\mathcal{P}$ be a collection of D pairwise non-parallel planes. Then there is a plane $P \in \mathcal{P}$

such that the orthogonal projection of A on P has $\min(\Omega(D^{1/3}n^{2/3}), n/4)$ elements.

8.4.5 [326] (Beck's lemma for homogeneous sets in $\mathbf{R}^3$) There is a positive constant K such that the following holds. Let B be a homogeneous set of s points in $\mathbb{R}^3$ and F be a set of f pairs of points of B. At least $f/2$ pairs of F are on lines incident to at most $K\frac{s}{f^{1/2}}$ points of B.

8.4.6 Let n be a large number, and let $P \subset \mathbf{R}^2$ be the set $A := [1, \sqrt{n}] \times [1, \sqrt{n}]$. Show that $|P| = \Theta(n)$ and $g_2(P) = o(n)$. (Hint: for primes $p = 3 \pmod 4$, any number divisible by p but not by p^2 cannot be written as the sum of two integer squares. Use this fact for all small p (say $p < \log\log n$) and the Chinese remainder theorem to improve upon the trivial bound of $g_2(P) = O(n)$.) Conclude in particular that $g_2(n) = o(n)$.

8.4.7 The purpose of this exercise is to sketch an alternative proof of the Szemerédi–Trotter theorem via *cell decomposition*. Let P, L be collections of points and lines, and let $1 \leq r \leq |L|/2$. Choose r lines from L at random; show that this divides the plane into $O(r^2)$ regions (known as "cells"), and that all the other lines in L intersect at most $O(r)$ of these cells. Show that there are at most $O(r|L|)$ of incidences (p, l) with p lying on the boundary of one or more cells. By applying (8.2) to the points and lines incident to the interior of each cell, argue heuristically that there are at most $O(r|L| + r^{-1/2}|P||L|^{1/2})$ incidences (p, l) with p in the interior of one of the cells. Optimize this in r to conclude the Szemerédi–Trotter theorem up to an absolute constant. (See [63] for a rigorous version of this argument.)

8.5 The sum-product problem in other fields

A natural extension of the sum-product problem is to consider sets from fields and rings other than $\mathbf{R}$. One example (when $\mathbf{R}$ is replaced by $\mathbf{Z}_p$ for a prime p) was consider in an earlier chapter. In this section, we consider the case when $\mathbf{R}$ is replaced by the set of complex numbers.

One way to attack the problem is to prove a complex version of Szemerédi–Trotter theorem and then repeat the proofs of Theorems 8.14 and 8.15. While it is believed that the statement of Szemerédi–Trotter theorem holds for complex lines and points, proving it is not easy as the technique using the crossing number no longer applies (see however the recent announcement by Tóth [368]).

In the following, we show that using a clever double counting argument, one can extend Elekes's result for complex numbers. In fact, the argument, which is due to Solymosi [325], is effective for several other number fields as well. (See the remark at the end of the proof.)

Theorem 8.24 *[325] For any finite non-empty sets of complex numbers A, B, and Q,*

$$|A+B| \cdot |A \cdot Q| = \Omega\big(|A|^{3/2}|B|^{1/2}|Q|^{1/2}\big).$$

By setting $Q = B = A$, it follows immediately that

$$|A+A| \cdot |A \cdot A| = \Omega\big(|A|^{5/2}\big)$$

and

$$|A+A| + |A \cdot A| = \Omega\big(|A|^{5/4}\big),$$

thus this theorem generalizes Theorem 8.14.

Proof We may assume $|A| \geq 2$ and $0 \neq Q$. From elementary algebra we observe that the map

$$(a, a', b, q) \mapsto (a+b, a'+b, aq, a'q)$$

is one-to-one from $A \times A \times B \times Q$ to $(A+B) \times (A+B) \times (A \cdot Q) \times (A \cdot Q)$ provided that we exclude the diagonal $a = a'$. This observation by itself is only enough to obtain the trivial bound $|A+B| \cdot |A \cdot Q| = \Omega(|A||B|^{1/2}|Q|^{1/2})$. However we can do better by exploiting the intuitive observation that if a' is close to a, then $a' + b$ is close to $a + b$ and aq is close to $a'q$.

More precisely, for each $a \in A$, define the *nearest neighbor* a' of a to be an element of $A \backslash a$ which minimizes the distance $|a - a'|$. (If there is more than one candidate for nearest neighbor, choose arbitrarily.) We refer to (a, a') as a *neighboring pair*, thus there are $|A|$ neighboring pairs. We caution that if (a, a') is a neighboring pair then (a', a) is not necessarily a neighboring pair also.

Call a quadruple (a, a', b, q) *good* if (a, a') is a neighboring pair, $b \in B$ and $q \in Q$, and one has the closeness properties

$$|\{u \in A+B : |a+b-u| \leq |a-a'|\}| \leq \frac{28|A+B|}{|A|} \tag{8.7}$$

and

$$|\{v \in A \cdot Q : |aq - v| \leq |aq - a'q|\}| \leq \frac{28|A \cdot Q|}{|A|}. \tag{8.8}$$

Informally, (8.7) and (8.8) assert that $a' + b$ is a fairly close neighbor of $a + b$ in $A + B$, and similarly $a'q$ is a fairly close neighbor of aq in $A \cdot Q$. We will apply a double counting argument to N, the number of good quadruples.

First we establish a lower bound. For each $a \in A$ let $D_a := \{z \in \mathbf{C} : |z - a| \leq |a' - a|\}$ be the disk of radius $|a' - a|$ centered at a. A simple geometric argument (which we leave as an exercise) shows that any complex number z can be contained in at most seven of these disks. In particular for any $b \in B$ we have

$$\sum_{a \in A} |\{u \in A + B : |a + b - u| \leq |a - a'|\}| = \sum_{z \in A+B-b} |\{a \in A : z \in D_a\}| \leq 7|A + B|$$

and similarly for any $q \in Q$

$$\sum_{a \in A} |\{v \in A \cdot Q : |aq - v| \leq |aq - a'q|\}| = \sum_{z \in A \cdot Q/q} |\{a \in A : z \in D_a\}| \leq 7|A \cdot Q|.$$

If we thus fix b and q and choose $a \in A$ uniformly at random, a simple application of Markov's inequality then shows that (a, a', b, q) will be good with probability at least $1/2$. This shows that

$$N \geq |B||Q|\frac{|A|}{2}.$$

Now we establish an upper bound. Recall that the quadruple (a, a', b, q) is uniquely determined by the quadruple $(a + b, a' + b, aq, a'q)$. There are $|A + B|$ choices for $a + b$ and $|A \cdot Q|$ choices for aq. For fixed $a + b$, we see from (8.7) that there are at most $\frac{28|A+B|}{|A|}$ elements of $A + B$ which are closer to or equally distant from $a + b$ than $a' + b$, and thus there are at most $\frac{28|A+B|}{|A|}$ values of $a' + b$. Similarly there are at most $\frac{28|A \cdot Q|}{|A|}$ values of $a'q$. This gives the upper bound

$$N \leq |A + B|\frac{28|A + B|}{|A|}|A \cdot Q|\frac{28|A \cdot Q|}{|A|}.$$

Combining this with the lower bound, we obtain the claim. □

Remark 8.25 A similar argument works for quaternions and for other hypercomplex numbers. In general, if T and Q are sets of similarity transformations and A is a set of points in space such that, from any quadruple $(t(p_1), t(p_2), q(p_1), q(p_2))$, the elements $t \in T$, $q \in Q$, and $p_1 \neq p_2 \in A$ are uniquely determined, then $c|A|^{3/2}|T|^{1/2}|Q|^{1/2} \leq |T(A)| \cdot |Q(A)|$, where c depends on the dimension of the space only.

To conclude this section, let us describe a recent result of Chang, who investigates the sum-product problem for matrices [51].

Theorem 8.26 *There is a function $\Phi(n)$ tending to infinity with n such that the following holds. Let d be a fixed integer and A be a finite set of $d \times d$ real matrices such that for any two different elements M and M' of A, $\det(M - M') \neq 0$. Then*

$$|A + A| + |A \cdot A| \geq \Phi(|A|)|A|.$$

Theorem 8.27 *For every d there is a positive constant $\epsilon = \epsilon(d)$ such that the following holds. Let A be a finite set of $d \times d$ real, symmetric, matrices. Then*

$$|A + A| + |A \cdot A| \geq |A|^{1+\epsilon}.$$

The proofs of these theorems are more complicated than those presented here and we refer the readers to [51] for details.

Exercise

8.5.1 With the notation in the proof of Theorem 8.24, show that every complex number is contained in at most seven of the disks D_a. (Hint: show that if z is contained in both D_a and $D_{a'}$ with a, a', z distinct, then a, a' subtend an angle of at least 60° with respect to z.)

9

Algebraic methods

In most of this book we have studied additive combinatorics problems in an ambient group Z, relying primarily on the additive structure of Z (as manifested for instance in the Fourier transform). However, in many cases the ambient group is in fact a *field* F, and thus supports a number of special functions, in particular polynomials. One can then use tools from algebraic geometry to exploit these polynomial structures; this is known as the *polynomial method*. One of the primary ideas here is to interpret an additive set (e.g. a sum set $A + B$) as the zero locus of one or more polynomials, possibly in several variables. One can then hope to control the size of such sets using results from algebraic geometry about the number and distribution of zeroes of polynomials. The most familiar example of such a theorem is the statement that a polynomial $P(t)$ of one variable with degree d in a field F can have at most d zeroes; however for most applications we will need to study the zero locus of polynomial(s) in many variables. In this chapter we present four related tools and techniques from algebraic geometry which allow one to control such a zero locus. The first is the powerful *combinatorial Nullstellensatz* of Alon (Theorem 9.2), which asserts that the zero locus of a polynomial $P(t_1, \ldots, t_k)$ cannot contain a large box $S_1 \times \cdots \times S_k$ if a certain monomial coefficient of P is non-vanishing; this is particularly useful for obtaining lower bounds on the size of restricted sum sets and similar objects. The second is the *Chevalley–Warning theorem* (Theorem 9.24), which shows that under certain conditions the cardinality of a zero locus of multiple polynomials must be a multiple of char(F), the characteristic of the underlying field. This is useful for demonstrating the existence of non-trivial solutions to a set of polynomial equations in F. The third is *Stepanov's method* (see Section 9.7), which obtains upper bounds on a set by using linear algebra methods to locate a polynomial that vanishes to very high order at each of the elements of the set; this has proven to be particularly useful for controlling additive combinations of multiplicative subgroups of a finite field, and thus has application to sum-product estimates. Finally we discuss *divisibility criteria*, which show that a polynomial cannot have certain types of zeroes if some combination of its coefficients are divisible (or not divisible) by p in a certain manner; the most well known example of this is Eisenstein's criterion (Exercise 9.8.2), but the combinatorial Nullstellensatz can also be viewed as a statement of this type, and another example arises in cyclotomic fields (Lemma 9.49). As an application of these criteria we present an uncertainty principle for $\mathbf{Z}_p$ which gives a Fourier-analytic proof of the Cauchy–Davenport inequality (Theorem 5.4).

Much of the theory pertains to arbitrary fields F. However, we will at times need to focus on two special types of fields. The first are *finite fields*, of which the primary example are the fields $F_p = \mathbf{Z}_p$ of prime order. We shall review the theory of these fields in Section 9.4. The second are the *cyclotomic fields*, generated by pth roots of unity; we shall review the theory of those fields in Section 9.8.

It is easy to see that in a field F, all non-zero elements have the same torsion as the identity element 1. We refer to this torsion as the *characteristic* $\mathrm{char}(F)$ of F; it is either zero (if F is torsion-free) or a prime p (which is for instance the case when F is finite). Some of our results will only hold if the characteristic of F is sufficiently large (or equal to zero).

9.1 The combinatorial Nullstellensatz

As is well known, a polynomial $P \in F[t]$ of one variable over a field F can have at most $\deg(P)$ zeroes, where $\deg(P)$ denotes the degree of P. Let us rewrite this fact as

Lemma 9.1 *Let $P \in F[t]$ be a polynomial of one variable over a field F and degree d (thus the t^d coefficient of P is non-zero) and let S be a subset of F such that $|S| > \deg(P)$. Then there exists $x \in S$ such that $P(x) \neq 0$.*

We now present a powerful generalization of this fact to polynomials of several variables, namely the *combinatorial Nullstellensatz* of Alon [4].

Theorem 9.2 (Combinatorial Nullstellensatz) *[4] Let F be an arbitrary field, let $P \in F[t_1, \ldots, t_n]$ be a polynomial of degree d which contains a non-zero coefficient at $t_1^{d_1} \cdots t_n^{d_n}$ with $d_1 + \cdots + d_n = d$, and let $S_1, \ldots, S_n$ be subsets of F such that $|S_i| > d_i$ for all $1 \leq i \leq n$. Then there exists $x_1 \in S_1, \ldots, x_n \in S_n$ such that $P(x_1, \ldots, x_n) \neq 0$.*

Proof We induce on n. The case $n = 1$ is just Lemma 9.1. Now suppose that $n \geq 2$ and the claim has already been proven for $n - 1$.

Let $g_n(t_n)$ be the polynomial of one variable

$$g_n(t_n) = \prod_{s_n \in S_n} (t_n - s_n) = t_n^{|S_n|} + \text{ lower order terms.}$$

Thus g_n has degree $|S_n|$ and the leading term is monic (i.e. it has coefficient 1). By applying the long division algorithm to P, we may write

$$P(t_1, \dots, t_n) = q_n(t_1, \dots, t_n)g_n(t_n) + r_n(t_1, \dots, t_n)$$

where the quotient q_n is a polynomial of degree at most $d - |S_n|$, and the remainder r_n is a polynomial of degree at most d such that no monomial contains a factor of $t_n^{|S_n|}$, thus

$$r_n(t_1, \dots, t_n) = \sum_{j=0}^{|S_n|} r_{n,j}(t_1, \dots, t_{n-1})t_n^j.$$

We can expand $q_n g_n$ as $q_n t_n^{|S_n|}$ plus lower-order terms, of degree at most

$$\deg(q_n) + |S_n| - 1 \leq (d - |S_n|) + |S_n| - 1 < d = d_1 + \cdots + d_n.$$

Thus the lower-order terms have a vanishing $t_1^{d_1} \cdots t_n^{d_n}$ coefficient. Since $|S_n| > d_n$, we see that $q_n t_n^{|S_n|}$ also has a vanishing $t_1^{d_1} \cdots t_n^{d_n}$ coefficient. Thus by hypothesis on P, the remainder r_n must have a non-zero $t_1^{d_1} \cdots t_n^{d_n}$ coefficient. In particular, r_{n,d_n} contains a non-zero $t_1^{d_1} \cdots t_{n-1}^{d_{n-1}}$ coefficient. Applying the induction hypothesis, we can find $x_1 \in S_1, \dots, x_{n-1} \in S_{n-1}$ such that $r_{n,d_n}(x_1, \dots, x_{n-1})$ is non-zero. Applying Lemma 9.1, we can then find $x_n \in S_n$ such that

$$r_n(x_1, \dots, x_n) = \sum_{j=0}^{|S_n|} r_{n,j}(x_1, \dots, x_{n-1})x_n^j \neq 0.$$

Since $g_n(x_n) = 0$, we thus have $P(x_1, \dots, x_n) \neq 0$, as desired. □

For an explanation as to the terminology "combinatorial Nullstellensatz", see Exercise 9.1.3. Based on the combinatorial Nullstellensatz, Alon, Nathanson and Ruzsa developed the so-called *polynomial method*, which is a very powerful tool for proving bounds concerning cardinalities of sum sets. The next several sections contain various applications of this method.

Exercises

9.1.1 (Schwartz–Zippel lemma) Let F be a field, let $Q \in F[t_1, \dots, t_n]$ be a non-zero polynomial of $n \geq 1$ variables, and let S be a non-empty finite subset of F. Let $x_1, \dots, x_n$ be elements of S chosen independently at random. Then

$$\mathbf{P}(Q(x_1, \dots, x_n) = 0) \leq \frac{\deg(Q)}{|S|}.$$

(Hint: modify the induction argument used to prove the Nullstellensatz.)

9.1.2 Let F be a field, let $d_1, \ldots, d_n \geq 0$, and let $P \in F[t_1, \ldots, t_n]$ be a non-zero polynomial such that every monomial that occurs in P divides $t_1^{d_1} t_2^{d_2} \cdots t_n^{d_n}$. Show that there exist functions $f_{i,1}, \ldots, f_{i,d_i} : F^{i-1} \to F$ for each $1 \leq i \leq n$ such that

$$\{(x_1, \ldots, x_n) \in F^n : P(x_1, \ldots, x_n) = 0\}$$
$$\subseteq \bigcup_{i=1}^{n} \bigcup_{j=1}^{d_i} \{(x_1, \ldots, x_n) \in F^n : x_i = f_{i,j}(x_1, \ldots, x_{i-1})\};$$

thus the zero locus of P can be covered by a small number of graphs. Note that when $i = 1$ the functions $f_{1,j}$ are simply constants. Conclude in particular that the combinatorial Nullstellensatz holds for this choice of P and $d_1, \ldots, d_n$.

9.1.3 [4] Let F be an arbitrary field and $P \in F[t_1, \ldots, t_n]$ be a polynomial. Let $S_1, \ldots, S_n$ be non-empty subsets of F and let $g_1, \ldots, g_n \in F[t_1, \ldots, t_n]$ be the polynomials defined by $g_i(t_1, \ldots, t_n) := \prod_{s \in S_i}(t_i - s)$ for each $1 \leq i \leq n$. If P vanishes on $S_1 \times \cdots \times S_n$, show that there are polynomials $h_1, \ldots, h_n \in F[t_1, \ldots, t_n]$ satisfying $\deg h_i \leq \deg P - \deg g_i$ so that

$$P = \sum_{i=1}^{n} h_i g_i.$$

Moreover, the coefficients of $h_1, \ldots, h_n$ can be chosen to lie in the ring generated by the coefficients of P and $g_1, \ldots, g_n$. Use this and the previous exercise to provide an alternative proof of Theorem 9.2. This should be contrasted with the *Hilbert Nullstellensatz*, which asserts that given arbitrary polynomials $P, g_1, \ldots, g_n \in F[t_1, \ldots, t_n]$, with P vanishing on the algebraic variety determined by $g_1, \ldots, g_n$, then some power P^K of P can be written as a linear combination $P^K = \sum_{i=1}^{n} h_i g_i$ of $g_1, \ldots, g_n$.

9.1.4 Let $d_1, \ldots, d_n \geq 0$ be integers, and let F be a field whose characteristic is either zero or is greater than $\max(d_1, \ldots, d_n)$. Let $P \in F[t_1, \ldots, t_n]$ be such that the $t_1^{d_1} \cdots t_n^{d_n}$ coefficient is non-zero, but that no other non-zero monomial in P is divisible by $t_1^{d_1} \cdots t_n^{d_n}$. Let $S_1, \ldots, S_n \subset F$ be such that $|S_i| > d_i$ for all $1 \leq i \leq n$. Show that there exist $x_1 \in S_1, \ldots, x_n \in S_n$ such that $P(x_1, \ldots, x_n) \neq 0$. (Hint: for each $1 \leq i \leq n$, construct a function $g_i : S_i \to F$ such that $\sum_{x_i \in S_i} g_i(x_i) x_i^j = \mathbf{I}(j = d_i)$ for all $0 \leq j \leq d_i$. Then consider the quantity $\sum_{x_1 \in S_1, \ldots, x_n \in S_n} P(x_1, \ldots, x_n) g_1(x_1) \ldots g_n(x_n)$.)

9.1.5 Let F be a field and m a positive integer. Let $F^{\{0,1\}^m}$ be the ring of functions from $\{0, 1\}^m$ to F, and for each $i \in [1, m]$ let $x_i \in F^{\{0,1\}^m}$ be the coordinate functions $(x_1, \ldots, x_m) \mapsto x_i$. Show that the multilinear monomials $\prod_{i \in I} x_i$, $I \subset [1, m]$ constitute a basis of $F^{\{0,1\}^m}$, viewed as a vector space over F. (Hint: to establish linear independence, use Theorem 9.2.) In the case $F = \mathbf{C}$ or $F = \mathbf{R}$, show that this result also follows from (4.4) applied to the group $\mathbf{Z}_2^m$.

9.2 Restricted sum sets

We now apply the combinatorial Nullstellensatz to obtain lower bounds for sum sets, and restricted sum sets. We begin with a general lemma which gives a criterion for when such lower bounds on restricted sum sets can be attained.

Lemma 9.3 *[11] Let F be a field, let $n \geq 1$, and let $h \in F[t_1, \ldots, t_n]$ be a polynomial. Let $K \geq 0$, and let $A_1, \ldots, A_n$ be additive sets in F such that $\sum_{i=1}^n |A_i| = K + n + \deg(h)$. Suppose also that the polynomial $(t_1 + \cdots + t_n)^K h(t_1, \ldots, t_n)$ contains a non-zero coefficient at $t_1^{|A_1|-1} \cdots t_n^{|A_n|-1}$. Then*

$$|\{a_1 + \cdots + a_n : a_i \in A_i \text{ for all } 1 \leq i \leq n; h(a_1, \ldots, a_n) \neq 0\}| \geq K + 1. \quad (9.1)$$

Proof Suppose for contradiction that (9.1) failed; then one can find a set $B \subseteq F$ of cardinality $|B| = K$ which contains the set in (9.1). Let $P \in F[t_1, \ldots, t_n]$ be the polynomial

$$P(t_1, \ldots, t_n) := h(t_1, \ldots, t_n) \prod_{b \in B} (t_1 + \cdots + t_n - b).$$

Observe that $\deg(P) = K + \deg(h)$. On the other hand, by construction of B we see that P vanishes on contains $A_1 \times \cdots \times A_n$. But this contradicts the combinatorial Nullstellensatz. □

This powerful lemma allows one to reduce the task of establishing lower bounds on restricted sum sets to that of verifying that a single coefficient of an explicit polynomial is non-zero in the field F. As two quick applications of this lemma we reprove the Cauchy–Davenport inequality (Theorem 5.4) and then derive a variant, first conjectured by Erdős and Heilbronn, concerning the restricted sums $A \hat{+} B := \{a + b : a \in A, b \in B, a \neq b\}$.

Theorem 9.4 (Cauchy–Davenport inequality, again) *[47], [68] Let $F = F_p$ be a finite field of prime order. If A, B are two additive sets in F, then*

$$|A + B| \geq \min(|A| + |B| - 1, p).$$

We shall give a third proof of this theorem via the Fourier transform in Section 9.8.

Proof The claim is trivial when $|A| + |B| > p$ (see Exercise 2.1.6) so let us take $|A| + |B| \leq p$. We apply Lemma 9.3 with $n = 2$, $(A_1, A_2) = (A, B)$, $h \equiv 1$, and $K := |A| + |B| - 2$; we will be done as soon as we verify that $(t_1 + t_2)^K$ has a non-zero coefficient at $t_1^{|A|-1} t_2^{|B|-1}$ in F_p. But this coefficient is simply $\binom{K}{|A|-1} \bmod p$, which is non-zero since $K < p$. □

As a special case of the Cauchy–Davenport inequality we see that $|A + A| \geq \min(2|A| - 1, p)$ for any additive set A in F_p. The analogous result for restricted sums $A \hat{+} A$ took much longer to prove. It is easy to see that $|A \hat{+} A| = p$ when $2|A| - 3 \geq p$ (Exercise 9.2.1). In 1964, Erdős and Heilbronn (see [89]) conjectured that $|A \hat{+} A| \geq \min(2|A| - 3, p)$; this bound is easily seen to be optimal (Exercise 9.2.3). This innocuous-seeming variant of the Cauchy–Davenport inequality resisted attempts at solution for about thirty years; the e-transform methods in Section 5.1 do not appear to be able to prove the Erdős–Heilbronn conjecture. The conjecture was finally solved in 1994 by da Silva and Hamidoune [66] who confirmed it using a general result concerning Grassman spaces. We now give a short proof due to Alon, Nathanson, and Ruzsa [11] using the combinatorial Nullstellensatz, which demonstrates the power and simplicity of this method. Indeed one can prove slightly more:

Theorem 9.5 *[11] Let $F = F_p$ for some prime p, and let let A, B be two additive sets in F. Then*

$$|A \hat{+} B| \geq \min(|A| + |B| - 3, p).$$

Furthermore, if $|A| \neq |B|$, then we can improve the above bound to

$$|A \hat{+} B| \geq \min(|A| + |B| - 2, p).$$

Proof The case $|A| + |B| - 2 \geq p$ is easy (Exercise 9.2.1), so suppose $|A| + |B| - 2 < p$. The cases $|A| = 1$ or $|B| = 1$ are also trivial (Exercise 9.2.2), so assume $|A|, |B| \geq 2$. By deleting one element from A or B if necessary it suffices to obtain the latter bound in the case $|A| \neq |B|$.

We now apply Lemma 9.3 with $n = 2$, $(A_1, A_2) = (A, B)$, $h(t_1, t_2) := t_1 - t_2$, and $K = |A| + |B| - 3$. We will be done as soon as we verify that $(t_1 - t_2) \times (t_1 + t_2)^K$ contains a non-zero coefficient at $t_1^{|A|-1} t_2^{|B|-1}$. But this quantity can be computed as

$$\begin{aligned} &\binom{|A| + |B| - 3}{|A| - 2} - \binom{|A| + |B| - 3}{|A| - 1} \bmod p \\ &= \frac{(|A| + |B| - 3)!}{(|A| - 2)!(|B| - 2)!}(|B| - |A|) \bmod p. \end{aligned}$$

Since $|A| + |B| - 2 < p$, we see that this quantity is non-zero, and we are done. □

Clearly one can obtain further applications of Lemma 9.3; see for instance Exercise 9.2.4. But when one considers restricted sums of multiple sets one begins to need to study the coefficients of increasingly complicated polynomials, frequently involving such expressions as Vandermonde determinants. We shall therefore turn our attention next to the study of such polynomials and their coefficients. Our computations here shall be completely abstract, valid for indeterminates $x_1, \ldots, x_n$ taking values in any field F.

Definition 9.6 (Vandermonde determinant) If $n \geq 1$ and $x_1, \ldots, x_n$ are indeterminates, we define the *Vandermonde determinant* to be the expression

$$\Delta_n(x_1, \ldots, x_n) := \prod_{1 \leq i < j \leq n} (x_j - x_i) = (-1)^{\binom{n}{2}} \prod_{1 \leq i < j \leq n} (x_i - x_j).$$

It is easy to verify the symmetries

$$\begin{aligned} \Delta_n(x_1 + y, \ldots, x_n + y) &= \Delta_n(x_1, \ldots, x_n); \\ \Delta_n(\lambda x_1, \ldots, \lambda x_n) &= \lambda^{\binom{n}{2}} \Delta_n(x_1, \ldots, x_n); \\ \Delta_n(\pi(x)) &= \operatorname{sgn}(\pi) \Delta_n(x) \end{aligned} \tag{9.2}$$

for any variables λ, y and $x = (x_1, \ldots, x_n)$, and any permutation $\pi \in S_n$. In fact this effectively determines Δ_n up to constants, see Exercise 9.2.5. The quantity $\Delta_n(1, \ldots, n) = \prod_{i=1}^n (i-1)!$ is sometimes called the *superfactorial* of n.

The following well-known fact will be left as an exercise:

Lemma 9.7 *Let $n \geq 1$, and for each $1 \leq i \leq n$ let $P_i(x)$ be a monic polynomial of degree $i - 1$. Then for any variables $x_1, \ldots, x_n$ we have the identity*

$$\begin{aligned} \det(P_i(x_j))_{1 \leq i,j \leq n} &= \sum_{\pi \in S_n} \operatorname{sgn}(\pi) \prod_{i=1}^n P_{\pi(i)}(x_i) \\ &= \sum_{\pi \in S_n} \operatorname{sgn}(\pi) \prod_{i=1}^n P_i\left(x_{\pi(i)}\right) \\ &= \Delta_n(x_1, \ldots, x_n). \end{aligned}$$

In particular we have

$$\Delta_n(x_1, \ldots, x_n) = \sum_{\pi \in S_n} \operatorname{sgn}(\pi) \prod_{i=1}^n x_i^{\pi(i)-1}. \tag{9.3}$$

The formula (9.3) computes the coefficients of $\Delta_n(x_1, \ldots, x_n)$ exactly. Multiplying it with the multinomial formula

$$(x_1 + \cdots + x_n)^K = \sum_{c_1, \cdots, c_n \geq 0 : c_1 + \ldots + c_n = K} \frac{K!}{c_1! \cdots c_n!} \prod_{i=1}^n x_i^{c_i}$$

we obtain the formula

$$(x_1 + \cdots + x_n)^K \Delta_n(x_1^m, \ldots, x_n^m)$$
$$= \sum_{c_1,\ldots,c_n \geq 0: c_1+\cdots+c_n = K+m\binom{n}{2}} \sum_{\pi \in S_n} \operatorname{sgn}(\pi) \frac{K!}{\prod_{i=1}^n (c_i - \pi(i)m + m)!} \prod_{i=1}^n x_i^{c_i} \tag{9.4}$$

where we adopt the convention that $1/k! = 0$ when k is a negative integer.

In certain cases, the expression on the right-hand side of (9.4) can be simplified. For instance, in the $m = 1$ case we have

Lemma 9.8 *Let $n, K \geq 0$. Then we have*

$$(x_1 + \cdots + x_n)^K \Delta_n(x_1, \ldots, x_n)$$
$$= \sum_{c_1,\ldots,c_n \geq 0: c_1+\cdots+c_n = K+\binom{n}{2}} \frac{K!}{c_1! \cdots c_n!} \Delta_n(c_1, \ldots, c_n) x_1^{c_1} \cdots x_n^{c_n}. \tag{9.5}$$

Proof By (9.4), it suffices to establish the identity

$$\sum_{\pi \in S_n} \operatorname{sgn}(\pi) \frac{K!}{\prod_{i=1}^n (c_i - \pi(i) + 1)!} = \frac{K!}{c_1! \cdots c_n!} \Delta_n(c_1, \ldots, c_n).$$

If we introduce the *falling factorial*

$$(x)_n := x(x-1)\cdots(x-n+1), \tag{9.6}$$

then from Lemma 9.7 we have

$$\Delta_n(c_1, \ldots, c_n) = \sum_{\pi \in S_n} \operatorname{sgn}(\pi) \prod_{i=1}^n (c_i)_{\pi(i)-1} = \sum_{\pi \in S_n} \operatorname{sgn}(\pi) \prod_{i=1}^n \frac{c_i!}{(c_i - \pi(i) + 1)!}$$

and the claim follows. □

This Lemma already gives a generalization of the Erdős–Heilbronn conjecture; see Exercises 9.2.9 and 9.2.10.

In a similar spirit we have

Lemma 9.9 *Let $n, m, K, k \geq 0$ be such that*

$$(k-1) + \cdots + (k-n) = K + m\binom{n}{2}.$$

Then the coefficient of $x_1^{k-n} \cdots x_n^{k-1}$ in $(x_1 + \cdots + x_n)^K \Delta_n(x_1^m, \ldots, x_n^m)$ is

$$\frac{K!}{\prod_{i=1}^n (k-1-(i-1)m)!} m^{\binom{n}{2}} \Delta_n(1, \ldots, n).$$

Proof By (9.4), it suffices to establish the identity

$$\sum_{\pi \in S_n} \operatorname{sgn}(\pi) \frac{K!}{\prod_{i=1}^n (k-n-1+i-\pi(i)m+m)!}$$
$$= \frac{K!}{\prod_{i=1}^n (k-1-(i-1)m)!} m^{\binom{n}{2}} \Delta_n(1,\ldots,n).$$

Relabeling i by $\pi(i)$ and using the fact that $\operatorname{sgn}(\pi) = \operatorname{sgn}(\pi^{-1})$, the left-hand side can be rewritten as

$$\sum_{\pi \in S_n} \operatorname{sgn}(\pi) \frac{K!}{\prod_{i=1}^n (k-n-1+\pi(i)-(i-1)m)!}$$

which can be rewritten further using the falling factorial (9.6) as

$$\frac{K!}{\prod_{i=1}^n (k-1-(i-1)m)!} \sum_{\pi \in S_n} \operatorname{sgn}(\pi) \prod_{i=1}^n (k-1-(i-1)m)_{n-\pi(i)}.$$

Writing $n-\pi(i) = \alpha(i)-1$ and noting that $\operatorname{sgn}(\pi) = (-1)^{\binom{n}{2}} \operatorname{sgn}(\alpha)$, we rewrite this further as

$$\frac{K!}{\prod_{i=1}^n (k-1-(i-1)m)!} (-1)^{\binom{n}{2}} \sum_{\alpha \in S_n} \operatorname{sgn}(\alpha) \prod_{i=1}^n (k-1-(i-1)m)_{\alpha(i)-1}$$

which by Lemma 9.7 becomes

$$\frac{K!}{\prod_{i=1}^n (k-1-(i-1)m)!} (-1)^{\binom{n}{2}} \Delta_n(k-1, k-1-m, \ldots, k-1-(n-1)m).$$

The claim now follows from (9.2). □

As a consequence of this computation, we have the following additive combinatorial consequence concerning multiple restricted addition where the restrictions are of the form $P_i(a_i) \neq P_j(a_j)$ for polynomials P_i, P_j.

Theorem 9.10 *[234] Let k, m, n be positive integers such that the quantity $K := (k-1)n - (m+1)\binom{n}{2}$ is non-negative. Let F be a field whose characteristic is either zero or is a prime number greater than* $\max(K, m, n-1)$. *Let $A_1, \ldots, A_n$ be subsets of F for which $|A_i| \geq k-n+i$ for all $1 \leq i \leq n$. Let $P_1, \ldots, P_n \in F[t]$ be monic polynomials of degree m. Then*

$$|\{a_1 + \cdots + a_n | a_i \in A_i, P_i(a_i) \neq P_j(a_j) \text{ if } i \neq j\}| \geq K+1.$$

Proof Without loss of generality, we can assume that $|A_i| = k-n+i = k_i$. Let $f \in F[t_1, \ldots, t_n]$ be the polynomial

$$f(t_1, \ldots, t_n) := \prod_{1 \leq i < j \leq n} (P_j(t_j) - P_i(t_i)).$$

Since $k_i = k - n + i$ we have

$$\sum_{i=1}^{n}(k_i - 1) = (k-1)n - \binom{n}{2} = K + \deg(f).$$

Thus, the coefficient of $t_1^{k_1-1}\cdots t_n^{k_n-1}$ in the polynomial $(t_1 + \cdots + t_n)^K \times f(t_1, \ldots, t_n)$ is the same as the coefficient of $t_1^{k-n}\cdots t_n^{k-1}$ in

$$(t_1 + \cdots + t_n)^K \prod_{1\le i<j\le n}\left(t_j^m - t_i^m\right) = (t_1 + \cdots + t_n)^K \Delta_n\left(t_1^m, \ldots, t_n^m\right).$$

Applying Lemma 9.3 and Lemma 9.9, we reduce to showing that

$$\frac{K!}{\prod_{i=1}^{n}(k-1-(i-1)m)!}m^{\binom{n}{2}}\Delta_n(1, \ldots, n)\cdot 1 \neq 0.$$

But since the characteristic of F is either 0 or exceeds $\max(K, n-1, m)$, the claim is easily verified. □

Next we consider what happens if we raise the factors $(x_j - x_i)$ in $\Delta_n(x_1, \ldots, x_n)$ to arbitrary powers. A useful result in this regard is

Theorem 9.11 (Dyson's conjecture) *Let $a_1, \ldots, a_n$ be positive integers. The coefficient of $\prod_{i=1}^{n} x_i^{(n-1)a_i}$ in*

$$\prod_{i,j\in[1,n]:i\neq j}(x_j - x_i)^{a_j}$$

is

$$\frac{(a_1 + \cdots + a_n)!}{a_1!\cdots a_n!}.$$

This result was conjectured by Dyson [74] based on a problem in particle physics. It was verified by Gunson [165] and independently by Wilson [383] in 1962. We present a short and elegant proof due to Good [135].

Proof Let $x = (x_1, \ldots, x_n)$, $a = (a_1, \ldots, a_n)$ and

$$F(x, a) = \prod_{i,j\in[1,n]:i\neq j}\left(1 - \frac{x_i}{x_j}\right)^{a_j},$$

and let $F_0(a)$ denote the constant term in $F(x, a)$. It will suffice to prove that $F_0(a) = \frac{(a_1+\cdots+a_n)!}{a_1!\cdots a_n!}$ whenever the a_i are non-negative integers.

We induce on n. The claim is trivial when $n = 0$, so suppose $n \geq 1$ and the claim has already been proven for $n - 1$. We can then assume that none of the a_i are zero since we can simply eliminate that variable (noting that in that case the x_i variable only appears with a positive exponent) and apply the induction hypothesis. Thus $a_i \geq 1$ for all $i \in [1, n]$.

Let $e_1, \ldots, e_n$ be the standard basis vectors of $\mathbf{Z}^n$. It will suffice to prove the recursion

$$F_0(a) = \sum_{i=1}^{n} F_0(a - e_i)$$

whenever $a_i \geq 1$, since the claim then follows from the multinomial Pascal identity and an easy induction on $\sum_{i=1}^n a_i$.

By applying Lagrange's interpolation formula (Exercise 9.2.8) to the function $f(x) \equiv 1$ we have the identity

$$1 = \sum_{j=1}^{n} \prod_{i \in [1,n]: i \neq j} (x_j - x_i)^{-1}(y - x_i)$$

for all y. Setting $y = 0$ we have

$$1 = \sum_{j=1}^{n} \prod_{i \in [1,n]: i \neq j} \left(1 - \frac{x_i}{x_j}\right)^{-1}. \tag{9.7}$$

By multiplying both sides of (9.7) with $F(x, a)$, we see that if $a_j > 0$ for all $1 \leq j \leq n$ then we have the recursion

$$F(x, a) = \sum_{j=1}^{n} F(x, a - e_j)$$

and the claim follows by extracting the constant coefficient. □

As one particular consequence of Theorem 9.11, we see that the coefficient of $\prod_{i=1}^n x_i^{(n-1)m}$ in $\Delta_n(x_1, \ldots, x_n)^{2m}$ is $(nm)!/(m!)^n$. Using this fact and some additional arguments, Hou and Sun [186] proved the following generalization.

Lemma 9.12 *Let $n, m, k \geq 0$, and let $s := k + m(n-1)$. Then the coefficient of $x_1^s \cdots x_n^s$ in $(x_1 + \cdots + x_n)^{km} \Delta_n(x_1, \ldots, x_n)^{2m}$ is*

$$(-1)^{m\binom{n}{2}} \frac{(km!)}{(m!)^n} \prod_{j=1}^{n} \frac{(jm)!}{(s - (j-1)m)!}.$$

The proof of this theorem is somewhat technical and we refer the reader to [186] for details. As an additive combinatorial consequence, we can control restricted sum sets where the differences $a_i - a_j$ are required to avoid certain specified sets.

Theorem 9.13 *[186] Let k, m, n be positive integers and F be a field of characteristic p where p is zero or p is a prime satisfying*

$$p \geq n \max\{m, n + m - mk - 1\}.$$

Let $A_1, \ldots, A_k$ be subsets of F with cardinality at least n. For any $i, j \in \{1, \ldots, k\}, i \neq j$ let S_{ij} be a subset of F with cardinality at most m. Then the set

$$C := \{a_1 + \cdots + a_k | a_i \in A_i, a_i - a_j \notin S_{ij} \text{ if } i \neq j\}$$

has cardinality at least

$$|C| \geq (n + m - mk - 1)k + 1.$$

Proof We first need the following variant of Theorem 9.3, whose proof we leave to Exercise 9.2.11.

Lemma 9.14 *Let $A_1, \ldots, A_k$ be finite subsets of a field F. Assume that $|A_i| \geq n_i$. Let $\lambda, \mu \in F[t_1, \ldots, t_k]$ be such that $\deg(\mu) > 0$. Define*

$$C := \{\mu(a_1, \ldots, a_k) | a_i \in A_i, \lambda(a_1, \ldots, a_k) \neq 0\}.$$

Then there is no polynomial $\omega \in F[t_1, \ldots, t_n]$ such that the polynomial $\lambda\omega\mu^{|C|}$ has degree $\sum_{i=1}^k (n_i - 1)$ and the coefficient of $x_1^{n_1-1} \cdots x_k^{n_k-1}$ in this polynomial is non-zero.

To prove Theorem 9.13, we can assume, without loss of generality, that $|A_i| = n$ and $|S_{ij}| = m$ for all i, j. Let $l := n + m = mk - 1$. Assume, for contradiction, that $|C| < ln$. Let $\lambda, \mu, \omega \in F[t_1, \ldots, t_k]$ be the polynomials

$$\lambda(t_1, \ldots, t_k) := \prod_{1 \leq i \neq j \leq k} \prod_{c_{ij} \in S_{ij}} (t_i - t_j - c_{ij})$$

$$\mu(t_1, \ldots, t_k) := t_1 + \cdots + t_k,$$

$$\omega := \mu^{kl - |C|}.$$

The polynomial $\lambda\omega\mu^{|C|}$ has total degree $mk(k-1) + lk = k(n-1) = \sum_{i=1}^k (|A_i| - 1)$. Moreover, the coefficient of $t_1^{n-1} \cdots t_k^{n-1}$ in this polynomial is the same as that in

$$\prod_{1 \leq i < j \leq k} (t_i - t_j)^{2m} \mu^{kl} = \Delta_k(t_1, \ldots, t_k)^{2m} (t_1 + \cdots + t_k)^{kl}.$$

But this is non-zero thanks to Lemma 9.9 and the hypotheses on the characteristic p. This contradicts Lemma 9.14, completing the proof. □

Exercises

9.2.1 Let Z be any finite additive group of odd order, and let A, B be additive sets in Z. Show that if $|A| + |B| - 2 \geq |Z|$, then $A \hat{+} B = Z$. (Compare with Exercise 2.1.6.)

9.2.2 Verify Theorem 9.5 when $|A| = 1$ or $|B| = 1$.

9.2.3 Give examples to show that the bound $|A \hat{+} A| \geq \min(2|A| - 3, p)$ cannot be improved. What about the bound $|A \hat{+} B| \geq \min(|A| + |B| - 2, p)$ when $|A| \neq |B|$?

9.2.4 Let $F = F_p$ be a finite field of prime order, and let A, B be additive sets in F_p. Show that

$$|\{a + b | a \in A, b \in B, ab \neq 1\}| \geq \min(|A| + |B| - 3, p).$$

9.2.5 Verify the symmetries (9.2). Furthermore, show that if $P(x_1, \ldots, x_n)$ is any polynomial which obeys the same symmetries (9.2) as Δ_n, then P is a scalar multiple of Δ_n.

9.2.6 Prove Lemma 9.7. (Hint: one can use Gaussian elimination to reduce to the case $P_i(x) = x^{i-1}$. Then locate several linear factors of $\Delta_n(x_1, \ldots, x_n)$ and use the factor theorem. Alternatively, use Exercise 9.2.5.)

9.2.7 Show that if $x_1, \ldots, x_n$ are integers, then $\Delta_n(x_1, \ldots, x_n)$ is a multiple of $\Delta_n(1, \ldots, n) = \prod_{i=1}^n (i-1)!$.

9.2.8 (Lagrange interpolation formula) Let F be a field, let $n \geq 0$, let $a_0, \ldots, a_n$ be $n + 1$ distinct elements of F and let $b_0, \ldots, b_{n+1}$ be $n + 1$ arbitrary elements of F. Show that there is exactly one polynomial $f \in F[t]$ with coefficients in F of degree at most n such that $f(a_i) = b_i$, and that this polynomial is given by

$$f(x) = \sum_{i=0}^{n} b_i \prod_{0 \leq j \neq i \leq n} (a_i - a_j)^{-1}(x - a_j).$$

9.2.9 [11] Let $F = F_p$ be a finite field of prime order, and let $A_1, \ldots, A_k$ be additive sets in F_p with $|A_1|, \ldots, |A_k|$ all distinct and $\sum_{i=1}^k |A_i| \leq p + \binom{k+1}{2} - 1$. Let B be the restricted sum set

$$B := \{a_1 + \cdots + a_k : a_i \in A_i, a_i \neq a_j \text{ for all } 1 \leq i < j \leq k\}.$$

Using Theorem 9.3 and Lemma 9.8, establish the inequality $|B| \geq \{\sum_{i=1}^k |A_i| - \binom{k}{2} + 1, p\}$.

9.2.10 [66] (Generalized Erdős–Heilbronn conjecture) Let $F = F_p$ be a finite field of prime order, and let A be an additive set in F_p. Let $k^\wedge A := \{a_1 + \cdots + a_k : a_1, \ldots, a_k \in A, a_i \neq a_j \text{ for all } 1 \leq i < j \leq k\}$ be the set of k-fold sums of *distinct* elements of A. Show that $|k^\wedge A| \geq \min(p, k|A| - k^2 + 1)$. (Hint: use Exercise 9.2.9.)

9.2.11 Prove Lemma 9.14. (Hint: Apply the combinatorial Nullstellensatz to the polynomial $f := \lambda\omega \prod_{c \in C}(\mu - c)$.)

9.3 Snevily's conjecture

In [322], Snevily made the following conjecture.

Conjecture 9.15 (Snevily's conjecture) *[322] Let Z be an additive group of odd order and let A, B be two additive sets in Z with $|A| = |B|$. Then there is a bijection $\phi : A \to B$ such that the sums $\{a + \phi(a) : a \in A\}$ are all distinct.*

The general case of this conjecture remains open, but many special cases are known. For instance, using the combinatorial Nullstellensatz, Alon [5] showed that the conjecture holds for cyclic groups of prime order.

Theorem 9.16 *[5] Let $F = F_p$ where $p > 2$ is an odd prime and let A, B be two additive sets in F with $|A| = |B|$. Then there is a bijection $\phi : A \to B$ such that the sums $\{a + \phi(a) : a \in A\}$ are all distinct.*

Proof If $A = B$ then one can simply choose π to be the identity map, taking advantage of the fact that p is odd, so we may assume $A \neq B$. In particular we can take $|A| = |B| < p$. Enumerate $A = \{a_1, \ldots, a_k\}$, and let $P \in F[t_1, \ldots, t_k]$ be the polynomial

$$P(t_1, \ldots, t_k) = \prod_{1 \le i < j \le k} (t_j - t_i)(t_j - t_i + a_i - a_j).$$

Then $\deg(P) = k(k-1)$. Also, from Theorem 9.11, the coefficient of $x_1^{k-1} \cdots x_k^{k-1}$ in P is $k! \cdot 1$, which is non-zero in F_p since $k < p$. Applying the combinatorial Nullstellensatz, there is an $s_i \in B$ such that $P(s_1, \ldots, s_k) \neq 0$. This means $s_j - s_i \neq 0$ and $s_j - s_i + a_i - a_j \neq 0$ for all $1 \le i < j \le k$. If we then define $\phi : A \to B$ by setting $\phi(a_i) := s_i$ we thus see that ϕ is injective (hence surjective), and that the sums $a_i + \phi(a_i) = a_i + s_i$ are all distinct, as desired. □

Let us notice that in the case $k < p$, we never used the assumption that the elements of A are different, and in this case one can in fact generalize to arbitrary fields of characteristic p or 0; see Exercise 9.3.2. Also, observe that the proof only used a very special case of Dyson's conjecture. Using this conjecture in full generality and modifying the rest of the proof accordingly, we have the following more general result.

Theorem 9.17 *Let $F = F_p$ a field of prime order, let $k < p$, and let $R_1, \ldots, R_k$ be additive sets in F_p, such that $\sum_{i=1}^{k} |R_i| + 1 < p$. Let $a_1, \ldots, a_k \in F_p$, and let $B_1, \ldots, B_k$ be subsets of F_p with cardinality $|B_i| > (k-1)(|R_i| + 1)$. Then there are k pairwise distinct elements $\{b_1, \ldots, b_k\}$, where $b_i \in B_i$, such that the sums $a_i + b_i$ are pairwise distinct and for every $i \neq j$, $a_i + b_i - (a_j + b_j) \notin R_i$.*

Proof Let $P \in F[t_1, \ldots, t_k]$ be the polynomial

$$P(t_1, \ldots, t_k) := \prod_{i,j \in [1,k]: i \neq j} \prod_{r \in R_i} (a_i + t_i - a_j - t_j - r).$$

Then $\deg(P) = \sum_{i=1}^k (k-1)(|R_i|+1)$. Also, by Theorem 9.11 the coefficient of $\prod_{1 \leq i \leq k} t_i^{(k-1)(|R_i|+1)}$ in P is, up to sign,

$$\pm \frac{\left(\sum_{i=1}^k (|R_i|+1)\right)!}{\prod_{i=1}^k (|R_i|+1)!} \cdot 1$$

which is non-zero in F_p, since $\sum_{i=1}^k (|R_i|+1) < p$ by the assumption of the theorem. The claim now follows from the combinatorial Nullstellensatz. □

DasGupta, Károly, Serra and Szegedy [67] obtained a multiplicative version of Snevily's conjecture. Define the *Vandermonde permanent* $\mathrm{Per}_n(x_1, \ldots, x_n)$ of n variables to be the quantity

$$\mathrm{Per}_n(x_1, \ldots, x_n) = \sum_{\pi \in S_n} \prod_{i=1}^n x_i^{\pi(i)-1}$$

(cf. (9.3).)

Lemma 9.18 *[67] Let F be an arbitrary field and $a_1, \ldots, a_k$ be elements of F. Assume that the Vandermonde permanent* $\mathrm{Per}_k(a_1, \ldots, a_k)$ *is non-zero. Then for any subset $B = \{b_1, \ldots, b_k\}$ of F there is a permutation $\pi \in S_k$ such that the products $a_1 b_{\pi(1)}, \ldots, a_k b_{\pi(k)}$ are all distinct.*

Proof Let $f \in F[t_1, \ldots, t_k]$ be the polynomial

$$f(t_1, \ldots, t_k) := \Delta_k(t_1, \ldots, t_k)\Delta_k(a_1 t_1, \ldots, a_k t_k).$$

Then $\deg(f) \leq k(k-1)$. Set $S_1 = B, \ldots, S_k = B$. By the combinatorial Nullstellensatz, it suffices to show that the coefficient of $t_1^{k-1} \ldots t_k^{k-1}$ is not zero. Notice that

$$\begin{aligned} f(t_1, \ldots t_k) &= \left(\sum_{\pi \in S_k} (-1)^{\sigma(\pi)} \prod_{i=1}^k t_{\pi(i)}^{i-1}\right) \left(\sum_{\tau \in S_k} (-1)^{\sigma(\tau)} \prod_{i=1}^k (a_{\tau(i)} t_{\tau(i)})^{i-1}\right) \\ &= \left(\sum_{\pi \in S_k} (-1)^{\sigma(\pi)} \prod_{i=1}^k t_{\pi(i)}^{i-1}\right) \left(\sum_{\pi \in S_k} (-1)^{\binom{k}{2} - \sigma(\pi)} \prod_{i=1}^k (a_{\pi(i)} t_{\pi(i)})^{k-i}\right). \end{aligned}$$

Thus the coefficient in concern is exactly

$$\sum_{\pi \in S_k} (-1)^{\binom{k}{2}} \prod_{i=1}^{k} a_{\pi(i)}^{i-1} = (-1)^{\binom{k}{2}} \mathrm{Per}_k(a_1, \ldots, a_k) \cdot 1,$$

which is not zero due to the assumption of the lemma. The proof is thus complete. (For an alternative proof, see Exercise 9.3.3.) □

One can convert this multiplicative statement to an additive statement by embedding an additive group as a multiplicative subgroup of a suitable field. For instance, one can now show that Snevily's conjecture holds for cyclic groups of odd order:

Corollary 9.19 *[67] Let $n \geq 1$ be an odd number, and let A, B be two additive sets in $\mathbf{Z}_n$ such that $|A| = |B|$. Then there exists a bijection $\phi : A \to B$ such that the sums $\{a + \phi(a) : a \in A\}$ are all distinct.*

Proof We shall use the theory of finite fields, which we shall review in Section 9.4. Let $\mathbf{Z}_n^\times$ be the set of multiplicatively invertible elements of n, and let $\phi(n) := |\mathbf{Z}_n^\times|$ be the Euler totient function of n. By Cauchy's theorem (Exercise 3.1.2), we have $2^{\phi(n)} = 1 \pmod{n}$. Let F be a finite field of order $2^{\phi(n)}$ and characteristic 2 (the existence of such a field follows from Exercise 9.4.4). From Lemma 9.22, the multiplicative group $F^\times$ of F contains an element of order n, and hence contains a subgroup G isomorphic to the additive group $\mathbf{Z}_n$. It now suffices to verify the multiplicative form of Snevily's conjecture for G. But if $A = \{a_1, \ldots, a_k\}$ is a subset of G, then since F has characteristic 2 one can replace permanents with determinants and compute

$$\mathrm{Per}_k(a_1, \ldots, a_k) = \Delta_k(a_1, \ldots, a_k) = \prod_{1 \leq i < j \leq k} (a_j - a_i) \neq 0.$$

The claim now follows from Lemma 9.18. □

A variant of this argument gives a strengthened version of the above result when the cyclic group has order p^k.

Theorem 9.20 *[67] Let $p > 2$ be an odd prime, let $q = p^\alpha$ be a power of p for some $\alpha > 1$, let $1 \leq k < p$, and let $a_1, \ldots, a_k$ be elements of $\mathbf{Z}_q$. Then for any set $B = \{b_1, \ldots, b_k\} \subseteq \mathbf{Z}_q$ of cardinality k, there exists a permutation $\pi \in S_k$ such that the sums $a_i + b_{\pi(i)}$, $1 \leq i \leq k$ are all distinct.*

Proof We will need the machinery of cyclotomic fields, which we shall review in Section 9.8. Let ω be a primitive qth root of unity, and let $\mathbf{Q}(\omega)$ be the associated cyclotomic field. Observe that $\mathbf{Q}(\omega)$ contains the multiplicative subgroup $G := \{\xi^n : n \in \mathbf{Z}\} \subset \mathbf{Q}(\omega)$ which is group isomorphic to the additive group $\mathbf{Z}_q$. Thus it suffices to show that for any $a_1, \ldots, a_k \in G$ and any $B = \{b_1, \ldots, b_k\} \subseteq G$

of cardinality k, there exists a permutation $\pi \in S_k$ such that the products $a_i b_{\pi(i)}$ are all distinct. Applying Lemma 9.18, it suffices to verify that the Vandermonde permanent $\mathrm{Per}_k(a_1, \ldots, a_k) = \sum_{\pi \in S_k} \prod_{i=1}^k a_i^{\pi(i)-1}$ is non-vanishing in $\mathbf{Q}(\omega)$. Note that each of the summands in this permanent is a qth root of unity, and the number $|S_k| = k!$ of summands is not divisible by p. The claim then follows from Lemma 9.49. □

Exercises

9.3.1 Show that Conjecture 9.15 fails whenever the ambient group Z has even order. (Hint: first consider the case $Z = \mathbf{Z}_2$.)

9.3.2 [67] Let p be a prime, let $1 \leq k < p$, and let F be a field of characteristic equal to p or zero. and let $a_1, \ldots, a_k \in F$. Then for any subset $B = \{b_1, \ldots, b_k\}$ of G, there is a permutation $\pi \in S_k$ such that the sums $a_1 + b_{\pi(1)}, a_2 + b_{\pi(2)}, \ldots, a_k + b_{\pi(k)}$ are all different. (By Exercise 9.4.4, this implies that Snevily's conjecture is true whenever G is the group $\mathbf{Z}_p^\alpha$ for any $\alpha \geq 0$.)

9.3.3 [67] Let $R \ni 1$ be a commutative ring, and let $\pi \in S_k$ be a permutation. Set $P_\pi \in R[u_1, \ldots, u_k, v_1, \ldots, v_k]$ to be the polynomial

$$P_\pi(u_1, \ldots, u_k; v_1, \ldots, v_k) := \prod_{1 \leq i < j \leq n} \big(u_j v_{\pi(j)} - u_i v_{\pi(i)}\big).$$

Verify the identity

$$\sum_{\pi \in S_k} P_\pi = \Delta_k(u_1, \ldots, u_k)\mathrm{Per}_k(v_1, \ldots, v_k)$$

and use this to derive an alternative proof of Lemma 9.18.

9.4 Finite fields

We now pause to develop some of the theory of finite fields. We have already encountered the finite fields $F_p = \mathbf{Z}_p$ of prime order, but we now discuss more general finite fields of composite (prime power) order.

To avoid degeneracies we always assume that our fields have order at least 2 (so that $0 \neq 1$). Note that a finite field F is a finite additive group $(F, 0, +, -)$, but if one removes the 0 element one obtains a multiplicative group $(F^\times, 1, \times, \cdot^{-1})$, where $F^\times := F \backslash \{0\}$. Strictly speaking, a finite field has two multiplicative structures, the multiplicative group structure $x \times y$ for $x, y \in F$ and the $\mathbf{Z}$-module structure $n \cdot x$ for $n \in \mathbf{Z}$, $x \in F$ coming from iterated addition, but they are clearly related by the identity $n \cdot x = (n \cdot 1) \times x$; because of this, we shall abuse notation and identify n with $n \cdot 1$, and also identify the two multiplicative structures.

The most important examples of a finite field are the cyclic groups $F_p := \mathbf{Z}_p$ of prime order $|F_p| = p$. More generally, for any prime p and any integer $k \geq 1$, one can create a finite field F_{p^k} of order $|F_{p^k}| = p^k$ (Exercise 9.4.4). Such fields are unique up to field isomorphism (Exercise 9.4.6).

Because a finite field has both an additive and a multiplicative group structure, we will sometimes subscript certain group-theoretic concepts by addition or multiplication as appropriate. For instance, we use $\mathrm{ord}_+(x)$ to denote the additive order of $x \in F$ and $\mathrm{ord}_\times(x)$ to denote the multiplicative order. We now observe that all non-zero elements $x \in F^\times$ of a finite field have the same additive order $\mathrm{ord}_+(x)$.

Lemma 9.21 *Let F be a finite field, and let $p := \mathrm{ord}_+(1)$. Then p is prime, and $\mathrm{ord}_+(x) = p$ for all $x \in F^\times$.*

Proof If $\mathrm{ord}_+(1) = nm$ is composite for some $n, m > 1$, then $m \cdot 1, n \cdot 1 \neq 0$ but $(n \cdot 1) \times (m \cdot 1) = 0$, which contradicts the fact that $F^\times$ is a multiplicative group. Thus $\mathrm{ord}_+(1)$ is equal to a prime p. Since $p \cdot x = (p \cdot 1) \times x = 0 \times x = 0$, we see that $\mathrm{ord}_+(x)$ divides p for all $x \in F^\times$; since $\mathrm{ord}_+(x) \neq 1$, the claim follows. □

We call the prime $\mathrm{char}(F) := p = \mathrm{ord}_+(1)$ the *characteristic* of the finite field F. It is easy to see that F is now a vector space over F_p; in particular it has some dimension $k \geq 1$, and so $|F| = p^k$. From Cauchy's theorem (Exercise 3.1.2) applied to $F^\times$ we see that $\mathrm{ord}_\times(x)$ divides $|F^\times| = |F| - 1$ for all $x \in F^\times$. In other words,

$$x^{|F|-1} = 1 \text{ for all } x \in F^\times \tag{9.8}$$

and thus

$$x^{|F|} = x \text{ for all } x \in F. \tag{9.9}$$

This has the following consequence. For any positive integer n, define the *Euler totient function* $\phi(n)$ of n to be the number of elements in $[1, n]$ which are coprime to n (or equivalently, $\phi(n) = |\mathbf{Z}_n^\times|$).

Lemma 9.22 *Let F be a finite field, and let $n \geq 1$ be an integer dividing $|F^\times| = |F| - 1$. Then we have $|\{x \in F^\times : x^n = 1\}| = n$ and $|\{x \in F^\times : \mathrm{ord}_\times(x) = n\}| = \phi(n)$.*

Proof Since $x^n - 1$ has degree n, it has at most n zeroes, thus $|\{x \in F^\times : x^n = 1\}| \leq n$. On the other hand, if we write $|F| - 1 = nm$, we see from (9.8) that y^m lies in the set $\{x \in F^\times : x^n = 1\}$ for all $y \in F^\times$. Since the polynomial $y^m - c$ has at most m zeroes for each $c \in F$, we thus see that $|\{x \in F^\times : x^n = 1\}| \geq$

$|F^\times|/m = n$. This gives the first claim. This implies that

$$\sum_{d|n} |\{x \in F^\times : \mathrm{ord}_\times(x) = d\}| = |\{x \in F^\times : x^n = 1\}|$$
$$= n$$
$$= \sum_{d|n} \phi(d)$$

and the second claim now follows from an induction argument. □

Since $\phi(n) \neq 0$ for all $n \geq 1$, we thus see in particular that $F^\times$ contains an element of order $|F| - 1$; we call such elements *primitive elements* of $F^\times$. This implies in particular that $F^\times$ is a multiplicative cyclic group of order $|F| - 1$. Another consequence is

Lemma 9.23 *Let F be a finite field. Then for any $k \geq 1$ and any $h_1, \ldots, h_k \geq 0$ such that* $\min(h_1, \ldots, h_k) < |F| - 1$, *we have* $\sum_{x_1,\ldots,x_k \in F} x_1^{h_1} \cdots x_k^{h_k} = 0$.

Proof By factorizing the left-hand side, we see that it suffices to show that $\sum_{x\in F} x^h = 0$ for all $0 \leq h < |F| - 1$. When $h = 0$ we have $\sum_{x\in F} x^h = |F| \cdot 1 = 0$, since $|F|$ is a multiple of the characteristic $\mathrm{char}(F)$. Now suppose that $0 < h < |F| - 1$, and let ω be any primitive element of $F^\times$. Then $x \mapsto \omega x$ is a bijection on F, and so

$$\sum_{x\in F} x^h = \sum_{x\in F} (\omega x)^h = \omega^h \sum_{x\in F} x^h.$$

Since ω is primitive, $\omega^h \neq 1$, and hence $\sum_{x\in F} x^h = 0$ as claimed. □

We can now give the classical theorem of Chevalley and Warning on the number of solutions of a system of multi-variable polynomials over a finite field.

Theorem 9.24 (Chevalley–Warning theorem) *Let F be a finite field, let $n \geq 1$, and $P_1, \ldots, P_m \in F[t_1, \ldots, t_n]$ be polynomials such that* $\sum_{i=1}^m \deg(P_i) < n$. *Then the number of solutions $(x_1, \ldots, x_n) \in F^n$ to the equations*

$$P_1(x_1, \ldots, x_n) = \cdots = P_m(x_1, \ldots, x_n) = 0 \tag{9.10}$$

is a multiple of $\mathrm{char}(F)$.

Proof From (9.8) we have

$$\mathbf{I}(P_i(x_1, \ldots, x_n) = 0) = 1 - P_i(x_1, \ldots, x_n)^{|F|-1},$$

so the number of solutions to (9.10), thought of as an element of F, can be expressed as

$$\sum_{x_1,\ldots,x_n \in F} \prod_{i=1}^{m} \left(1 - P_i(x_1, \ldots, x_n)^{|F|-1}\right).$$

To prove the theorem, it thus suffices to show that

$$\sum_{x_1,\dots,x_n \in F} \prod_{i=1}^{m} \left(1 - P_i(x_1, \dots, x_n)^{|F|-1}\right) = 0. \tag{9.11}$$

By expanding the product $\prod_{i=1}^{m}(1 - P_i(x_1, \dots, x_n)^{|F|-1})$ we get a linear combination of monomials of the form $\prod_{j=1}^{m} x_j^{a_j}$, each of which has degree at most $\sum_{i=1}^{m} \deg(P_i)(|F|-1) < n(|F|-1)$. By the pigeonhole principle this means that $\min(a_1, \dots, a_m) < |F|-1$, and thus by Lemma 9.23, each monomial gives a zero contribution to (9.11). The claim follows. □

Since $\mathrm{char}(F) \geq 2$, we have the following corollary:

Corollary 9.25 *Let $P_1, \dots, P_m$ be as in Theorem 9.24. Then if there is one solution in F^n to (9.10), there must also exist at least one other solution.*

Next, we give a useful lemma which shows that the zeroes of sparse polynomials cannot have too high a multiplicity.

Lemma 9.26 *[180],[120] Let F be a finite field of prime order, and let $P \in F[t]$ be a non-zero polynomial of degree at most $|F|-1$ with at most k non-zero coefficients. Then all the zeroes of P in $F^\times$ are of order at most $k-1$; in other words, P does not contain any factors of the form $(x - x_0)^k$ for any $x_0 \in F^\times$.*

Proof We prove this by induction on k. The claim is trivial if $k = 1$, so suppose $k > 1$ and the claim has already been proven for $k-1$. Suppose that the x^j coefficient of P was non-zero. If P contained a zero of order at least k in $F^\times$, the (formal) derivative P' must then contain a zero of order at least $k-1$, and so $xP' - jP$ must also contain a zero of order at least $k-1$. But $xP' - jP$ is a non-trivial polynomial with at most $k-1$ non-zero coefficients, contradicting the induction hypothesis. Thus all the zeroes of P in $F^\times$ are of order at most $k-1$. □

Exercises

9.4.1 Let R be a commutative ring containing 1, and let $R[t]^{\mathrm{monic}}$ be the multiplicative semigroup of all monic polynomials in $R[t]$ (polynomials with leading coefficient 1). We say that a monic polynomial is irreducible if it has no proper monic factors. Using the Euclidean algorithm, show that every monic polynomial can be uniquely factored into monic irreducible factors, up to permutations. In particular this shows that $F[t]$ is a unique factorization domain whenever F is a field.

9.4.2 Let F be a finite field. Define the *von Mangoldt function* $\Lambda : F[t]^{\mathrm{monic}} \to \mathbf{R}$ by setting $\Lambda(f) := \deg(g)$ if $f = g^k$ for some irreducible g and some $k \geq 1$, and $\Lambda(f) := 0$ otherwise. Using Exercise 9.4.1, show that $\deg(f) := \sum_{g \in F[t]^{\text{monic}}: g|f} \Lambda(g)$ for all $f \in F[t]^{\text{monic}}$, where we use $g|f$ to denote that g is a factor of f. Conclude in particular

$$\sum_{f \in F[t]^{\text{monic}}} \frac{\deg(f)}{|F|^{s\deg(f)}} = \left(\sum_{f \in F[t]^{\text{monic}}} \frac{\Lambda(f)}{|F|^{s\deg(f)}} \right) \sum_{f \in F[t]^{\text{monic}}} \frac{1}{|F|^{s\deg(f)}}$$

for all $s > 1$. From this, conclude the *prime number theorem* for $F[t]$:

$$\sum_{f \in F[t]^{\text{monic}}:\deg(f)=k} \Lambda(f) = |F|^k \text{ for all } k \geq 1.$$

From this, conclude *Bertrand's postulate* for $F[t]$: for every $k \geq 1$ there exists at least one irreducible monic polynomial in $F[t]^{\text{monic}}$ of degree k. Also, establish the *Riemann hypothesis* for $F[t]$:

$$|\{f \in F[t]^{\text{monic}} : \deg(f) = k,\ f \text{ irreducible}\}| = |F|^k/k + O\big(|F|^{k/2}\big).$$

Note that this is considerably easier to establish than the corresponding Riemann hypothesis for $\mathbf{Z}$!

9.4.3 Let F be a finite field of order $|F| = p^k$ for some prime p and some $k \geq 1$. Let $f(t) \in F_p[t]$ be a polynomial over F_p such that $f(t)|t^{p^k} - t$. Show that $f(t)$ has exactly $\deg(f)$ distinct zeroes in F. (Hint: if $t^{p^k} - t = f(t)g(t)$, the zeroes of $t^{p^k} - t$ are the union of the zeroes of $f(t)$ and the zeroes of $g(t)$.) In the language of Galois theory, this means that every factor of $t^{p^k} - t$ splits completely over F.

9.4.4 Let F be a finite field and $k \geq 1$ be an integer. Let $f(t) \in F[t]$ be a monic irreducible polynomial of degree k (which exists by Exercise 9.4.2). Show that the quotient ring $F[t]/(f(t))$ is a finite field of order $|F|^k$. Show that this finite field is isomorphic *as an additive group only* to the vector space F^k. Note that this construction shows that there exists a field of order p^k for any prime p and any $k \geq 1$.

9.4.5 Let F be a finite field of order $|F| = p^k$ for some prime p and some $k \geq 1$. Let ω be a primitive element of $F^\times$. Let $f(t) \in F_p[t]$ be the minimal polynomial of ω over F_p, i.e. the monic polynomial in $F_p[t]$ of minimal degree such that $f(\omega) = 0$. Show that $\deg(f) = k$, and that the vectors $1, \omega, \ldots, \omega^{k-1}$ form a basis for F, viewed as a vector space over F_p.

9.4.6 Let F and G be two finite fields of the same order $|F| = |G| = p^k$. Prove that F and G are isomorphic. (Hint: let ω be a primitive element of $F^\times$, and let $f(t)$ be the minimal polynomial of ω. Use Exercise 9.4.3 to find

$\omega' \in G^\times$ such that $f(\omega') = 0$, and then find a field isomorphism between F and G which maps ω to ω'.)

9.4.7 Let $F = F_{p^k}$ be a finite field of characteristic p, and let $\phi : F \to F$ be the *Frobenius map* $\phi(x) := x^p$. Show that ϕ is a field isomorphism. Furthermore, show that the iterates $\phi^0, \phi^1, \ldots, \phi^{k-1}$ of this map are the only field isomorphisms of F to itself.

9.4.8 Let $F = F_{p^k}$ be a finite field, and let $1 \leq k' \leq k$. Show that the set $G := \{x \in F : x^{p^{k'}} = x\}$ is a subfield of F of order $|G| = |p^{k'}|$.

9.4.9 (Wilson's theorem) If p is a prime, show that $(p-1)! \cdot 1 = -1$ in F_p. (Hint: show that if $x \in F_p^\times$, then $x = x^{-1}$ if and only if $x = \pm 1$.)

9.4.10 Show that Lemma 9.26 fails when F is not of prime order. (Hint: if $|F| = p^k$, consider the polynomial $x^p - x$.)

9.4.11 Use Corollary 9.25 to give an alternative proof of Exercise 4.3.16 which does not use the Fourier transform.

9.5 Davenport's problem

For an finite additive group Z, define the *Davenport number* $s = s(Z)$ of Z to be the smallest integer such that whenever $a_1, \ldots, a_s$ are elements of Z (not necessarily distinct), there exists a partial sum $\sum_{i \in I} a_i$ of the a_i for some non-empty $I \subseteq [1, s]$ which sums to zero. The problem of determining $s(Z)$ for arbitrary groups Z was posed by Davenport in 1966. A simple estimate is

Lemma 9.27 *If Z is a finite additive group, then $s(Z) \leq |Z|$.*

Proof Let $a_1, \ldots, a_{|Z|}$ be elements of Z; it suffices to show that some non-trivial partial sum of these elements is zero. Consider the $|Z|$ partial sums $a_1, a_1 + a_2, \ldots, a_1 + \cdots + a_{|Z|}$. If one of them is zero, we are done. Otherwise, by the pigeonhole principle there exists two such partial sums which are equal. Subtracting the shorter partial sum from the longer, we obtain the result. □

In 1961, Erdős, Ginzburg and Ziv [88] proved the following remarkable variant.

Theorem 9.28 *[88] Let Z be a finite additive group, and $a_1, \ldots, a_{2|Z|-1}$ be elements of $|Z|$. Then there exists $I \subset [1, 2|Z| - 1]$ with $|I| = |Z|$ such that $\sum_{i \in I} a_i = 0$.*

Proof Let us start with the special case when $Z = \mathbf{Z}_p$ is a cyclic group of prime order. In this case we use Chavelley–Warning theorem to derive the claim. Let

$F = F_p = \mathbf{Z}_p$ and let $P_1, P_2 \in F[t_1, \ldots, t_{2p-1}]$ be the polynomials

$$P_1(t_1, \ldots, t_{2p-1}) = \sum_{i=1}^{2p-1} a_i t_i^{p-1}; \quad P_2(t_1, \ldots, t_{2p-1}) = \sum_{i=1}^{2p-1} x_i^{p-1}.$$

Observe that $\deg(P_1) + \deg(P_2) = 2(p-1) < 2p-1$, and that $(0, \ldots, 0)$ is a simultaneous root of P_1 and P_2, and hence by Corollary 9.25 we can find another simultaneous root $(y_1, \ldots, y_{2p-1}) \neq 0$ of P_1 and P_2. But by (9.8) we see that $\sum_{i=1}^{2p-1} y_i^{p-1} := |\{i \in [1, 2p-1] : y_i \neq 0\}| \cdot 1$. The claim then follows by setting $I := \{i \in [1, 2p-1] : y_i \neq 0\}$.

In the general case, we induce on $|Z|$. If $|Z|$ is prime then we are already done, so suppose that $|Z| = pm$ for some prime p and some $1 \leq m < |Z|$. Then (using Corollary 3.8 if necessary) we can find a surjective homomorphism $\phi : Z \to \mathbf{Z}_p$ whose kernel $G := \ker(\phi)$ is a subgroup of Z of order m. Since we have already proven the theorem for $\mathbf{Z}_p$, we see that for any sequence of $2p-1$ elements of Z, we can already obtain a subsequence of size p which lies in G. By the greedy algorithm, we can thus locate $2m-1$ disjoint subsets $I_1, \ldots, I_{2m-1}$ of cardinality p inside $[1, 2|Z|-1]$ such that $\sum_{i \in I_j} a_i \in G$ for each $1 \leq j \leq 2m-1$. Now write $\sum_{i \in I_j} a_i = b_j$. By induction hypothesis we can find a subset $J \subset [1, 2m-1]$ of cardinality m such that $\sum_{j \in J} b_j = 0$. The claim now follows by setting $I := \bigcup_{j \in J} I_j$. □

From considering the sequence $1, \ldots, 1$ and Lemma 9.27 we see that $s(\mathbf{Z}_p) = p$ for any prime p, and more generally that

$$s(\mathbf{Z}_{p^{k_1}} \oplus \cdots \oplus \mathbf{Z}_{p^{k_l}}) \geq 1 + \sum_{i=1}^{l} (p^{k_i} - 1) \tag{9.12}$$

for any prime p and any $k_1, \ldots, k_l \geq 1$ (see Exercise 9.5.1).

Olson [266] proved that this bound is sharp. Let us first see this in the case $k_1 = \cdots = k_l = 1$, by modifying the proof of Theorem 9.28.

Proposition 9.29 *For any $l \geq 1$ and any prime p, we have $s(\mathbf{Z}_p^l) = 1 + l(p-1)$.*

Proof By (9.12) it suffices to prove the upper bound. Write $F := \mathbf{Z}_p$. Consider a sequence $a_1, \ldots, a_n \in F^l$ where $n \geq 1 + l(p-1)$. Each a_i can be viewed as an l-dimensional vector and we write $a_i = (a_{i1}, \ldots, a_{il})$. Let $P_1, \ldots, P_l \in F[t_1, \ldots, t_n]$ be the polynomials $P_j(t_1, \ldots, t_n) := \sum_{i=1}^n a_{ij} t_i^{p-1}$ for $1 \leq j \leq l$; then $\sum_{j=1}^l \deg(P_j) = l(p-1) < n$. Since $(0, \ldots, 0)$ is a simultaneous zero of $P_1, \ldots, P_l$, we thus see from Corollary 9.25 that there must exist another simultaneous zero $(y_1, \ldots, y_n) \neq (0, \ldots, 0)$. Setting $I := \{i \in [1, n] : y_i \neq 0\}$, we conclude using (9.8) as before that $\sum_{i \in I} a_i = 0$, as desired. □

This simple argument does not directly extend to the general groups considered in (9.12); nevertheless, Olson was able to proceed by a different argument.

Theorem 9.30 *[266] Let p be a prime and $k_1, \dots, k_l \geq 1$. Then the inequality (9.12) in fact holds with equality.*

Proof Again it suffices to prove the upper bound. It is convenient to use multiplicative notation. Let G be an abelian multiplicative group which is isomorphic to the additive group $\mathbf{Z}_{p^{k_1}} \oplus \dots \oplus \mathbf{Z}_{p^{k_l}}$, let $n \geq 1 + \sum_{i=1}^{l}(p^{k_i} - 1)$, and let $g_1, \dots, g_n \in G$. It will suffice to find $I \in [1, n]$ such that $\prod_{i \in I} g_i = 1$.

Let R be the group ring of G over $\mathbf{Z}_p$ (i.e. R is the space of formal linear combinations of elements of G with coefficients in $\mathbf{Z}_p$). In this ring we claim that

$$(1 - g_1) \cdots (1 - g_n) = 0.$$

To see this, let $x_1, \dots, x_l$ be the standard basis for G, where x_i has order p^{k_i}. Each g_j can be written as the product of a few x_is. We use the identity $1 - xy = (1 - x) + x(1 - y)$ iteratively to replace $1 - g_j$ as a linear combination (with coefficient in R) of the elements $1 - x_i$. Thus, it follows that the product $(1 - g_1) \cdots (1 - g_n)$ is a linear combination of elements of the form $\prod_{i=1}^{l}(1 - x_i)^{n_i}$ where $\sum_{i=1}^{l} n_i = n > \sum_{i=1}^{l}(p^{k_i} - 1)$. There must be some j such that $n_j \geq p^{k_j}$. On the other hand, in R, $(1 - x_j)^{p^{k_j}} = 1 - x_j^{p^{k_j}} = 0$. It follows that $(1 - g_1) \cdots (1 - g_n) = 0$, as claimed. This implies that for some non-trivial subsequence of the g_i has product 1, because otherwise, the coefficient of 1 in the product $(1 - g_1) \cdots (1 - g_n)$ would be non-zero. This proves Theorem 9.30. □

This allows us to prove variants of Theorem 9.28 for product groups. For instance:

Lemma 9.31 *[266] Let $Z := \mathbf{Z}_p^2$ where p is a prime. For any sequence $a_1, \dots, a_{3p-2} \in Z$, one can find a subsequence of length at most p whose sum is zero.*

Proof Embed Z in $Z' := \mathbf{Z}_p^3$ and a sequence $x + a_1, \dots, x + a_{3p-2}$, where x is an element of $Z' \backslash Z$. By Theorem 9.30 (or Proposition 9.29) we have $s(Z') = 3p - 3$, and thus some subsequence of $x + a_1, \dots, x + a_{3p-3}$ has sum zero. Rearranging subscripts, we may assume that $(x + a_1) + \dots + (x + a_n) = 0$, where $1 \leq n \leq 3p - 3$. This implies that $nx = 0$ and $a_1 + \dots + a_n = 0$. It follows that $n = p$ or $n = 2p$. If $n = p$ then we are done. If $n = 2p$, we apply Theorem 9.30 or Proposition 9.29 again, this time to the group Z. As $s(Z) = 2(p - 1)$, the sequence $a_1, \dots, a_{n-1}$ contains a subsequence whose sum is zero. Again by rearranging subscripts, we may assume that $a_1 + \dots + a_m = 0$ where $m \leq n - 1$. If $m \leq p$

then we are done. If $m > p$, then the sequence $a_{m+1}, \dots, a_n$ has length less than p and its sum is also zero since $a_1 + \cdots + a_n = 0$. The proof is complete. $\square$

By this Lemma and an induction argument similar to that used to prove Theorem 9.28 one can then obtain the following estimate on the Davenport number of product groups:

Theorem 9.32 *[267] Let Z and W be additive groups such that $|W|$ divides $|Z|$. Then $s(Z \oplus W) \leq |Z| + |W| - 1$.*

We leave the proof of this theorem to Exercise 9.5.2.

Finally, let us briefly discuss the version of Davenport's problem when the elements in the sequence are different. Under this condition, the magnitude of the Davenport number changes dramatically. Szemerédi [347] proved

Theorem 9.33 *There is a constant c such that the following holds. Let $S = \{a_1, \dots, a_s\}$ be a sequence of s different elements of $\mathbf{Z}_p$, where p is a prime and $s > c\sqrt{p}$. Then there is a non-empty subsequence of S whose elements sum up to zero.*

A more recent result of Hamidoune and Zemor [175] showed that one can set $c = \sqrt{2} + o(1)$, which is asymptocially best possible.

Assume that $A \subset \mathbf{Z}_p$ does not contain 0 and view the elements of A as integers between 1 and $p - 1$. It is clear that if $\sum_{a \in A} a < p$ then no subset of A sums up to 0. In [349, 352], Szemerédi and Vu showed that if A has sufficiently many elements, then this is essentially the only reason.

Theorem 9.34 *Let A be a subset of $\mathbf{Z}_p$, where p is a large prime. Assume that no subset of A sums up to 0. Then there is a subset A' of A with at most $p^{0.49}$ elements and a non-zero element $x \in \mathbf{Z}_p$ such that the sum of the elements in $x \cdot (A \backslash A')$ (viewed as positive integers between 1 and $p - 1$) is less than p.*

For another classification result of this kind, see Theorem 12.20. The approach to these two results relies on inverse arguments, in spirit of those discussed in Chapter 12.

Exercises

9.5.1 Prove (9.12).

9.5.2 By modifying the inductive argument in the proof of Theorem 9.28, deduce Theorem 9.32 from Lemma 9.31.

9.5.3 Let n be a positive integer, and let $\mathbf{Z}[i]$ be the ring of Gaussian integers. Show that a sequence of $2n^2 - 1$ Gaussian integers contains a subsequence of length n whose sum is divisible by n.

9.5.4 Let Z be an additive group of order n and let k be a positive integer divisible by n. Prove that for any sequence of elements of Z of length $k+n-1$ there is a subsequence of length divisible by k whose sum is 0.

9.5.5 [6] Let p be a prime, and let $v_1, \ldots, v_{3p} \in \mathbf{Z}_p^2$ be such that $\sum_{i=1}^{3p} v_i = 0$. Then there is a subset $J \subset [1, 3p]$ such that $|J| = p$ and $\sum_{j \in J} v_j = 0$. (Hint: modify the argument in Lemma 9.31.)

9.6 Kemnitz's conjecture

Define a parameter $s(n, d)$ as the smallest integer s such that any sequence of s elements from $\mathbf{Z}_n^d$ contains a subsequence of length n whose sum is 0 in $\mathbf{Z}_n^d$. In this terminology, Theorem 9.28 states that $s(n, 1) = 2n - 1$. Harborth [176] considered the problem of controlling $s(n, d)$ for higher d. He first observed the easy estimates

$$(n-1)2^d + 1 \leq s(n, d) \leq (n-1)n^d + 1 \tag{9.13}$$

and also derived a recursive inequality

$$s(mk, d) \leq s(m, d) + m(s(k, d) - 1); \tag{9.14}$$

we leave the proofs as exercises.

Exact computation of $s(n, d)$ is a difficult task, especially for large d; for instance the quantity $s(3, d)$ is closely related to the still unsolved problem of obtaining sharp constants for Roth's theorem in $\mathbf{Z}_3^d$ (see Exercise 10.2.4). But in the case when d is fixed and n is large, more is known. Alon and Doubnier [6] proved that $s(n, d) = O_d(n)$. Kemnitz conjectured that the lower bound in (9.13) is sharp for $d = 2$:

Conjecture 9.35 *[200] For any $n \geq 1$, we have $s(n, 2) = 4n - 3$.*

In [200] the conjecture was verified when the prime factors of n are from the set $\{2, 3, 5, 7\}$. Alon and Doubnier [6] proved that $s(n, 2) \leq 6n - 5$. They also sketched an argument which gives $s(p, 2) \leq 5p - 2$ for all sufficiently large prime p.

Rónyai [286] made significant progress by proving

Theorem 9.36 *[286] For every prime p we have $s(p, 2) \leq 4p - 2$.*

Theorem 9.36 and (9.14) imply that $s(n, 2) \leq \frac{41}{10}n$; see Exercise 9.6.3.

Proof The case $p = 2$ is trivial so we can assume that p is odd. Set $m := 4p - 2$, and let $v_1 = (a_1, b_1), \ldots, v_m = (a_m, b_m)$. By Exercise 9.5.5, it suffices to show that there is a subset $J \subset \{1, \ldots, m\}$ with $|J| = p$ or $|J| = 3p$ such that $\sum_{j\in J} v_i = 0$. Assume, for contradiction, that there is no such J. Let $\sigma, P_1, P_2 \in F_p(t_1, \ldots, t_m)$ be the polynomials

$$\sigma(t_1, \ldots, t_m) := \sum_{I\subset[1,m],|I|=p} \prod_{i\in I} t_i$$

$$P_1(t_1, \ldots, t_m) := \left(1 - \left(\sum_{i=1}^{m} a_i t_i\right)^{p-1}\right)\left(1 - \sum_{i=1}^{m} b_i t_i\right)^{p-1} - 1$$

$$\times \left(1 - \left(\sum_{i=1}^{m} t_i\right)^{p-1}\right)(2 - \sigma(t_1, \ldots, t_m))$$

$$P_2(t_1, \ldots, t_m) := \prod_{i-1}^{m}(1 - t_i)$$

and set $P := P_1 - 2P_2$.

We now claim that $P(x_1, \ldots, x_m) = 0$ whenever $x_1, \ldots, x_m \in \{0, 1\} \subset F_p$. There are several cases to consider, depending on the size of the set $J := \{i \in [1, m] : x_i = 1\}$. When J is empty then it is easy to see that $P_1(x_1, \ldots, x_m) = 2$ and $P_2(x_1, \ldots, x_m) = 1$. When J is non-empty, then $P_2(x_1, \ldots, x_m)$ is zero. To see that $P_1(x_1, \ldots, x_m)$ is also zero, we observe from (9.8) that $1 - (\sum_{i=1}^{m} x_i)^{p-1} = 0$ when $|J|$ is not divisible by p, and from Wilson's theorem (Exercise 9.4.9) that $\sigma(x_1, \ldots, x_m) = 2$ when $|J| = 2p$. Finally, when $|J| = p$ or $|J| = 3p$, we have by hypothesis that $\sum_{i=1}^{m}(a_i, b_i)x_i = \sum_{i\in J}(a_i, b_i) \neq 0$, and hence $(1 - \sum_{i=1}^{m} a_i x_i)^{p-1}((1 - \sum_{i=1}^{m} b_i x_i)^{p-1} - 1) = 0$.

Thus P vanishes on $\{0, 1\} \times \cdots \times \{0, 1\}$. Also, $\deg(P_1) = 4p - 3$ and $\deg(P_2) = m = 4p - 2$, thus $\deg(P) = 4p - 2$. Moreover, the coefficient of the monomial $t_1 \cdots t_m$ in P is $2(-1)^m \cdot 1 \neq 0$. This contradicts the combinatorial Nullstellensatz and concludes the proof. □

Remark 9.37 In the above proof one only used a very special case of the combinatorial Nullstellensatz; indeed one could rely just on Exercise 9.1.5, which can be proven by more elementary means – in fact, this was the original approach in [286].

Remark 9.38 Very recently, Reiher [279] has proved Kemnitz's conjecture, using the Chavelley–Warning theorem combined with a clever combinatorial argument.

For a further survey of results in this area, see [81].

Exercises

9.6.1 [176] Prove (9.13). (Hints: for the upper bound, use the pigeonhole principle. For the lower bound, take $n-1$ copies of $\{0,1\}^d$.)

9.6.2 [176] Prove (9.14). (Hint: modify the inductive argument in the proof of Theorem 9.28.)

9.6.3 Using Theorem 9.36 and (9.14) to deduce that $s(n,2) \leq \frac{41}{10}n$. (Hint: first verify the claim when all the prime divisors of n are less than 11, and then induct on n.)

9.6.4 [6] Modify the proof of Lemma 9.31 to prove that $s(n,d) = O_d(n)$.

9.7 Stepanov's method

In this section we fix a finite field F, and fix a multiplicative subgroup G of $F^\times$. The multiplicative structure of G can be determined explicitly:

Lemma 9.39 *Let G be a subgroup of $F^\times$. Then $|G|$ divides $|F^\times|$; thus we have $|F^\times| = |F| - 1 = |G|h$ for some $h \geq 1$. Furthermore we have the explicit formulas*

$$G = \left\{x \in F^\times : x^{|G|} = 1\right\} = \{y^h : y \in F^\times\}, \tag{9.15}$$

and if $G^\perp \subseteq F^\times$ denotes the orthogonal complement group $G^\perp := \{\xi \in F^\times : \xi^h = 1\}$, then $G^\perp$ indexes the multiplicative cosets $x \cdot G$ of G. Indeed if we define $G_\xi := \{x \in F^\times : x^{|G|} = \xi\}$ for all $\xi \in G^\perp$, then the sets $\{G_\xi : \xi \in G^\perp\}$ partition $F^\times$, and one has $x \cdot G = G_{x^{|G|}}$ for all $x \in F^\times$.

We leave the easy verification of this lemma to Exercise 9.7.1. In this section however we shall be more concerned with understanding the *additive* structure of G. A convenient way of quantifying this structure is via the sets $\Lambda(\xi) \subset F$ defined for all $\xi \in G^\perp$ by

$$\Lambda(\xi) := \left\{x \in F : x^{|G|} = (x-1)^{|G|} = \xi\right\} = G_\xi \cap (G_\xi + 1).$$

It is clear that these sets are disjoint as ξ ranges over $G^\perp$. The relevance of these sets to the additive structure of G lies in the easily verified identity

$$|G \cap (G+x)| = |(G-g) \cap \pm G_\xi| = |\Lambda(\xi^{-1})| \text{ whenever } \xi \in G^\perp, x \in G_\xi, g \in G; \tag{9.16}$$

see Exercise 9.7.2. As a consequence of (9.16) we have the following identities, whose verification we leave to Exercise 9.7.3.

Lemma 9.40 *We have* $\sum_{\xi \in G^\perp} |\Lambda(\xi)| = |\bigcup_{\xi \in G^\perp} \Lambda(\xi)| = |G| - 1$ *and* $E(G, G) = |G|^2 + |G| \sum_{\xi \in G^\perp} |\Lambda(\xi)|^2$, *where* $E(G, G)$ *is the additive energy of* G. *If* $-1 \in G^\perp$, *then we have* $|\Lambda(-\xi)| = |\Lambda(\xi)|$ *for all* $\xi \in G^\perp$.

In [337] Stepanov introduced a method for controlling various additive expressions involving G and related objects such as $|\Lambda(\xi)|$. For simplicity we shall restrict our attention just to the task of obtaining upper bounds on $|\Lambda(\xi)|$, following [180]. The idea is to use elementary linear algebra to construct a sparse polynomial P which vanishes to high order on several of the sets $\Lambda(\xi)$. One then applies tools such as Lemma 9.26 to obtain a non-trivial bound. We illustrate this method with the following result of Heath-Brown and Konyagin, which gives distributional information on the sizes of the $|\Lambda(\xi)|$.

Theorem 9.41 *[180] Let* $F = F_p$ *be a finite field of prime order, and let* G *be a multiplicative subgroup of* $F^\times$. *Let* $G^\perp$ *and* Λ *be defined as above. Then for any set* $\Gamma \subseteq G^\perp$ *with* $|\Gamma| = O(|F|^3/|G|^4)$, *we have*

$$\sum_{\xi \in \Gamma} |\Lambda(\xi)| = O\big(\min\big(|G|, |G|^{2/3}|\Gamma|^{2/3}\big)\big).$$

Proof Let $0 < c \ll 1$ be a small absolute constant to be chosen later. We may assume that G is large, $|G| > c^{-100}$, since the claim is trivial otherwise. Similarly we may assume that Γ is non-empty and that $|\Gamma| \leq c^{100}|F|^3/|G|^4$, since the claim for $|\Gamma| = \Theta(|F|^3/|G|^4)$ then follows by partitioning Γ into $O(1)$ sets of size at most $c^{100}|F|^3/|G|^4$.

When $|\Gamma| = \Omega(|G|^{1/2})$ then the claim already follows from Lemma 9.40, so we may assume that $|\Gamma| < c^{100}|G|^{1/2}$. Let us define the normalized quantities

$$A := \lfloor c^{10}|G|^{2/3}|\Gamma|^{-1/3} \rfloor; \quad B := \lfloor c|G|^{1/3}|\Gamma|^{1/3} \rfloor;$$

observe from our hypotheses on $|\Gamma|$ that we have the bounds

$$1 \leq B \leq A; \quad AB < |G|; \quad A^2|\Gamma| \leq cAB^2; \quad A + 2|G|B < |F| \qquad (9.17)$$

if c is chosen suitably small. By the disjointness of the $\Lambda(\xi)$, it then suffices to show that

$$\left|\bigcup_{\xi \in \Gamma} \Lambda(\xi)\right| = O\left(1 + \frac{|G|B}{A}\right). \qquad (9.18)$$

We now let $V \subseteq F[t]$ be the linear subspace (over F) of $F[t]$ generated by the AB^2 polynomials $t^a t^{b|G|}(t-1)^{b'|G|}$ where $0 \leq a < A$ and $0 \leq b, b' < B$. We first observe that V has large dimension:

Lemma 9.42 V *has linear dimension exactly* AB^2 *over* F.

Proof Suppose for contradiction that V had dimension less than AB^2. Then we could find coefficients $c_{a,b,b'} \in F$, not all zero, such that

$$\sum_{0 \le a < A} \sum_{0 \le b < B} \sum_{0 \le b' < B} c_{a,b,b'} t^a t^{b|G|} (t-1)^{b'|G|} = 0.$$

We may assume that there is at least one non-zero coefficient $c_{a,b,0}$, otherwise we could divide out by $(t-1)^{|G|}$. But then the polynomial $\sum_{0 \le a < A} \sum_{0 \le b < B} c_{a,b,0} t^a t^{b|G|}$ would have a zero of order $|G|$ at $t = 1$. On the other hand, this polynomial is non-zero and its Newton diagram contains at most AB points, which contradicts Lemma 9.26 and (9.17). □

We then exploit this large dimension to locate a polynomial which vanishes to high order on $\bigcup_{\xi \in \Gamma} \Lambda(\xi)$.

Lemma 9.43 *V contains a non-zero polynomial P which vanishes to order A at all elements of $\bigcup_{\xi \in \Gamma} \Lambda(\xi)$.*

Proof It is convenient to use an algebraic geometry perspective and work via commutative rings. Let R be the commutative ring over F generated by indeterminates $t, t^{-1}, s, s^{-1}, r, \varepsilon$ subject to the constraints

$$tt^{-1} = ss^{-1} = 1; \quad s = t - 1; \quad t^{|G|} = s^{|G|} = r; \prod_{\xi \in \Gamma} (r - \xi) = 0; \quad \varepsilon^A = 0; \tag{9.19}$$

in other words, R is the polynomial ring $F[t, t^{-1}, s, s^{-1}, r, \varepsilon]$ quotiented out by the ideal generated by the polynomials $tt^{-1} - 1$, $ss^{-1} - 1$, $s - t + 1$, $t^{|G|} - r$, $s^{|G|} - r$, $\prod_{\xi \in \Gamma}(r - \xi)$, and ε^A. Let $\iota : F[t] \mapsto R$ be the ring homomorphism that maps t to $t + \varepsilon$. We shall show that the image $\iota(V)$ of V has linear dimension strictly less than AB^2. By Lemma 9.42, this will force the existence of a non-zero polynomial $P \in V$ such that $\iota(P) = 0$; in other words we can find $Q_1, \ldots, Q_7 \in F[t, t^{-1}, s, s^{-1}, r, \varepsilon]$ such that

$$\begin{aligned} P(t + \varepsilon) &= Q_1(tt^{-1} - 1) + Q_2(ss^{-1} - 1) + Q_3(s - t + 1) \\ &\quad + Q_4(t^{|G|} - r) + Q_5(s^{|G|} - r) + Q_6 \prod_{\xi \in \Gamma}(r - \xi) + Q_7 \varepsilon^A \end{aligned}$$

for any indeterminates $t, t^{-1}, s, s^{-1}, r, \varepsilon$. Restricting this to $r := \xi \in \Gamma$, $t := x \in \Lambda(\xi) \subset F^\times$, $s := x - 1 \in F^\times$, $t^{-1} := x^{-1} \in F^\times$, $s^{-1} := (x-1)^{-1} \in F^\times$, $\varepsilon \in F$, we obtain

$$P(x + \varepsilon) = Q_7(x, x^{-1}, x - 1, (x-1)^{-1}, \xi, \varepsilon)\varepsilon^A$$

which shows that P vanishes to order A at x, which is an arbitrary element of $\bigcup_{\xi \in \Gamma} \Lambda(\xi)$.

It remains to bound the linear dimension of $\iota(V)$. Observe that this space is generated by the polynomials $\iota(t^a t^{b|G|}(t-1)^{b'|G|}) = (t+\varepsilon)^a (t+\varepsilon)^{b|G|}(s+\varepsilon)^{b'|G|}$. But by the Taylor expansion of $(t+\varepsilon)^{b|G|}$ and using the constraints (9.19), we have

$$(t+\varepsilon)^{b|G|} = t^{b|G|}\left(1 + \binom{b|G|}{1} t^{-1}\varepsilon + \binom{b|G|}{2} t^{-2}\varepsilon^2 + \cdots\right)$$
$$= r^b \left(1 + \binom{b|G|}{1} t^{-1}\varepsilon + \cdots + \binom{b|G|}{A-1} t^{-A+1}\varepsilon^{A-1}\right).$$

In particular we see that $(t+\varepsilon)^{b|G|}$ is equal in R to a polynomial expression in $t, t^{-1}, s, s^{-1}, r, \varepsilon$ of degree $O(A)$. Similarly for $(t+\varepsilon)^a$ and $(s+\varepsilon)^{b'|G|}$. Thus $\iota(V)$ lies in the space of polynomials in $t, t^{-1}, s, s^{-1}, r, \varepsilon$ of degree at most $O(A)$. Taking out a common denominator of $(ts)^{-O(A)}$, we obtain a space of polynomials in t, s, r, ε of degree at most $O(A)$. The variable s can be eliminated since $s = t - 1$ from (9.19). The variable r is limited to have degree at most $|\Lambda|$, again by (9.19). This shows that the dimension of $\iota(V)$ is at most $O(|\Lambda| A^2)$, which (9.17) is indeed less than the dimension AB^2 of V, as desired. □

Let P be as in Lemma 9.43. Since $P \in V$, we have $\deg(P) \leq A + 2|G|B < |F|$ thanks to (9.17). Since P can have at most $\deg(P)$ zeroes (counting multiplicity) in F, we obtain

$$A\left|\bigcup_{\xi\in\Gamma} \Lambda(\xi)\right| \leq A + 2|G|B,$$

which gives (9.18) as desired. □

Theorem 9.41 can already be used to give non-trivial sum set bounds on G, for instance via controlling the additive energy $E(G, G)$. In fact we can also control the additive energy $E(A, A)$ of subsets of G:

Lemma 9.44 *[44] Let $F = F_p$ be a finite field of prime order, and let G be a multiplicative subgroup of $F^\times$ of order $|G| = O(|F|^{3/4})$. Let A be an additive set in G. Then we have*

$$E(A, A) = O\big(|G||A|^{3/2}\big). \tag{9.20}$$

Comparing this with (2.7) we see that this bound is non-trivial when $|A| \geq |G|^{2/3}$. See also Corollary 2.62.

Proof For every $\xi \in G^\perp$, we define the counting function $\alpha(\xi)$ by

$$\alpha(\xi) := |\{(a_1, a_2) \in A \times A : a_1 - a_2 \in G_\xi\}|.$$

We observe that

$$\begin{aligned}E(A,A) &= |\{(a_1,a_2,a_3,a_4)\in A\times A\times A\times A : a_1-a_2=a_3-a_4\}|\\ &= |A|^2+\sum_{\xi\in G^\perp}|\{(a_1,a_2,a_3,a_4)\in A\times A\times A\times A : a_1-a_2=a_3-a_4\in G_\xi\}|\\ &\le |A|^2+\sum_{\xi\in G^\perp}\alpha(\xi)\sup_{d\in G_\xi}|\{(g_1,g_2)\in G\times G : g_1-g_2=d\}|\\ &= |A|^2+\sum_{\xi\in G^\perp}\alpha(\xi)|\Lambda(\xi^{-1})|\end{aligned}$$

thanks to (9.16). Since $|A|^2=|A|^{1/2}|A|^{3/2}=O(|G||A|^{3/2})$, it thus suffices to show that

$$\sum_{\xi\in G^\perp}\alpha(\xi)|\Lambda(\xi^{-1})| = O\big(|G||A|^{3/2}\big).$$

From the identity

$$\sum_{\xi\in G^\perp}\alpha(\xi)=|A|^2$$

we see that it suffices to show that

$$\sum_{\xi\in G^\perp:\Lambda(\xi^{-1})\ge|G||A|^{-1/2}}\alpha(\xi)|\Lambda(\xi^{-1})| = O\big(|G||A|^{3/2}\big).$$

But from (9.16) we also have the trivial bound

$$\alpha(\xi)\le |A|\sup_{g_1\in G}|\{g_2\in G : g_1-g_2\in G_\xi\}| = |A||\Lambda(\xi^{-1})|$$

and so it suffices to show that

$$\sum_{\xi\in G^\perp:\Lambda(\xi^{-1})\ge|G||A|^{-1/2}}|\Lambda(\xi^{-1})|^2 = O\big(|G||A|^{1/2}\big).$$

But if we order $G^\perp=\{\xi_1,\dots,\xi_M\}$ in decreasing order of $\Lambda(\xi_j^{-1})$, then by Theorem 9.41 we then have

$$j\big|\Lambda(\xi_j^{-1})\big| = O\big(\min\big(|G|,|G|^{2/3}j^{2/3}\big)\big)\text{ for all }1\le j\le M,$$

which implies that

$$\sum_{\xi\in G^\perp:\Lambda(\xi)\ge|G||A|^{-1/2}}|\Lambda(\xi^{-1})|^2 = \sum_{j=O(|A|^{3/2}/|G|)} O\big(|G|^{2/3}j^{2/3}/j\big)^2 = O\big(|G||A|^{1/2}\big)$$

as desired. □

As a consequence we can now give a sum-product estimate which improves somewhat on the results in Section 2.8.

Theorem 9.45 *[44] Let $F = F_p$ be a finite field of prime order, and let A be an additive set in $F^\times$. Let $Q[A] = \frac{A-A}{(A-A)\backslash 0}$ be the quotient set of A, as defined in Definition 2.49. Then there exists $\xi \in Q[A]$ such that*

$$|A + \xi \cdot A| \geq c \min\left(|F|, \frac{|A|^{5/2}}{|A \pm A|}, \frac{|A|^3}{|A \cdot A|}\right)$$

for either choice of sign $\pm$.

Proof If $|A| \geq |F|^{1/2}$ then the claim follows from Corollary 2.51, so suppose $|A| < |F|^{1/2}$. Let D be the set of popular quotients,

$$D := \left\{d \in F^* : |\{(a', a'') \in A \times A : a'/a'' = d\}| \geq \frac{2|A|^2}{9|A \cdot A|}\right\},$$

and let G be the multiplicative group generated by D. Then by the multiplicative version of Exercise 2.6.10, there exists a coset $\xi_0 \cdot G$ of G for some $\xi_0 \in F^*$ such that $|A \cap (\xi_0 \cdot G)| \geq |A|/3$. By dividing A by ξ_0 we may assume that $\xi_0 = 1$.

Lemma 9.46 *Let $H \subseteq G$ be the set of those $\xi \in G$ such that*

$$|A + \xi \cdot A| \geq \min\left(\frac{|A|^2|G|}{|A|^2 + |G|}, \frac{2|A|^3}{9|A \cdot A|}\right).$$

Then $H \cap Q[A]$ is non-empty.

Proof Suppose for contradiction that H and $Q[A]$ are disjoint. From Exercise 2.8.4 there exists a $\xi \in G$ such that $|A + \xi \cdot A| \geq \frac{|A|^2|G|}{|A|^2+|G|}$, and hence H is non-empty. Thus, $G \backslash Q[A]$ is non-empty, and is also a proper subset of G (since $1 \in Q[A] \cap G$). Next, observe that if $\xi \in G \backslash Q[A]$ and $d \in D$, then by Lemma 2.50, all the sums in $A + \xi \cdot A$ are distinct, and hence

$$|A + (\xi d) \cdot A| \geq |A||A \cap (d \cdot A)| \geq \frac{2|A|^2}{9|A \cdot A|}.$$

This shows that $D \cdot (G \backslash Q[A]) \subseteq H$. Since $H \subseteq G$ and H and $Q[A]$ are disjoint, we conclude $D \cdot (G \backslash Q[A]) \subseteq G \backslash Q[A]$; since D generates G, this implies that $G \cdot (G \backslash Q[A]) \subseteq G \backslash Q[A]$. But this contradicts the previous observation that $G \backslash Q[A]$ was a proper non-empty subset of G. □

Let ξ be as in the above lemma; thus

$$|A + \xi \cdot A| \geq c \min\left(|G|, |A|^2, \frac{|A|^3}{|A \cdot A|}\right).$$

Note that since $|A \cdot A| \geq |A|$, we can drop the $|A|^2$ term from the right-hand side. We will now be done unless $|G| \leq c|A|^{5/2}/|A \pm A|$ for some small $c > 0$.

Since $|A \pm A| \geq |A|$ and $|A| \leq |F|^{1/2}$, we have $|G| \leq c\frac{|A|^{5/2}}{|A \pm A|} \leq c|F|^{3/4}$. But then, from Theorem 9.41, (2.8) and the fact that $|A \cap G| \geq |A|/3$ we see that if $|G| = O(|F|^{3/4})$, then

$$\begin{aligned} |A \pm A| &\geq |(A \cap G) \pm (A \cap G)| \\ &\geq c\frac{|A \cap G|^4}{E(A \cap G, A \cap G)} \\ &\geq c|A \cap G|^{5/2}/|G| \\ &\geq c|A|^{5/2}/|G|, \end{aligned}$$

a contradiction. □

Exercises

9.7.1 Prove Lemma 9.39. (Hint: in Section 9.4 it was demonstrated that $F^\times$ is a cyclic group of order $|F^\times| = |F| - 1$.)

9.7.2 Prove (9.16).

9.7.3 Prove Lemma 9.40. (Hint: use (9.16) and Lemma 2.9.)

9.7.4 [44] Let $F = F_p$ be a finite field of prime order, and let A be an additive set in $F^\times$ such that $|A| \leq |F|^{1/2}$. Using Theorem 9.45, prove that $|A \cdot (A - A) + A \cdot (A - A)| = \Omega(|A|^{5/4})$. Use this to derive another proof of Corollary 2.58.

9.8 Cyclotomic fields, and the uncertainty principle

We now recall some of the elementary theory of cyclotomic fields $\mathbf{Q}(\omega)$, and apply this to obtain an uncertainty principle for the Fourier transform on $\mathbf{Z}_p$.

Definition 9.47 (Cyclotomic field) Let $n \geq 1$ be any positive integer. An nth *root of unity* is any complex number $\omega \in \mathbf{C}$ such that $\omega^n = 1$. An nth root of unity ω is said to be *primitive* if ω is not an mth root of unity for any $1 \leq m < n$. We define the *cyclotomic field of order* n to be the field $\mathbf{Q}(\omega)$ obtained by adjoining a primitive nth root of unity to the rationals $\mathbf{Q}$. We define the nth *cyclotomic polynomial* $\Phi_n \in \mathbf{C}[z]$ to be the polynomial $\Phi_n(z) := \prod_\omega (z - \omega)$, where ω ranges over the primitive nth roots of unity.

It is easy to see that for each n, there are $\phi(n)$ primitive roots of unity, and they are all powers of each other. Thus there is only one cyclotomic field $\mathbf{Q}(\omega)$ for each order n. In particular we see that Φ_n is a monic polynomial of degree $\phi(n)$. Some further basic properties of Φ_n are as follows.

Lemma 9.48 *Φ_n has integer coefficients (thus $\Phi_n \in \mathbf{Z}[z]$), and is irreducible in $\mathbf{Z}[z]$. Furthermore we have $\Phi_n(1) = p$ when n is a prime power $n = p^k$, $\Phi_1(1) = 0$, and $\Phi_n(1) = 1$ otherwise.*

Proof We first observe from the factor theorem that

$$z^n - 1 = \prod_{\omega:\omega^n=1} (z - \omega) \text{ for any } n \geq 1.$$

Since every nth root of unity is a primitive dth root of unity for some d, we obtain

$$z^n - 1 = \prod_{d|n} \Phi_d(z). \tag{9.21}$$

Thus one can obtain $\Phi_n(z)$ by factoring out $\prod_{d|n;d<n} \Phi_d(z)$ from $z^n - 1$. By an easy induction on n this implies that Φ_n is a monic polynomial with integer coefficients. Since $(z^n - 1)/\Phi_1(z) = (z^n - 1)/(z - 1)$ approaches n as $z \to 1$, we obtain the formula

$$n = \prod_{d|n;d>1} \Phi_d(1).$$

Taking logarithms we conclude that

$$\sum_{d|n;d>1} \Lambda(d) = \sum_{d|n;d>1} \log \Phi_d(1)$$

for all $n \geq 1$, where $\Lambda(d) := \log p$ when d is a prime power $d = p^k$ for some $k \geq 1$, and $\Lambda(d) = 0$ otherwise (cf. Exercise 1.10.6). Another easy induction on n then shows that $\Phi_n(1) = e^{\Lambda(n)}$ for all $n > 1$, which gives the desired formula for $\Phi_n(1)$.

Now we prove the irreducibility. When n is prime this can be easily verified from Eisenstein's criterion (Exercise 9.8.3), but the general case is trickier. We use an argument of Gauss. Suppose for contradiction that Φ_n is reducible in $\mathbf{Z}[z]$, then we can partition the primitive nth roots of unity into two disjoint non-empty classes A and B such that the monic polynomials $f(z) := \prod_{\omega \in A}(z - \omega)$ and $g(z) := \prod_{\omega \in B}(z - \omega)$ lie in $\mathbf{Z}[z]$. Of course we have $\Phi_n = fg$. Since any two primitive nth roots are powers of each other, we can find an $\omega \in A$ such that $\omega^m \in B$ for some integer m. By decomposing m into primes and arguing by contradiction, we can in fact locate a prime p and an $\omega \in A$ such that $\omega^p \in B$. This implies that the polynomials $f(z)$ and $g(z^p)$ have a common root, and hence by the Euclidean algorithm we can find a non-trivial monic polynomial $h(z) \in \mathbf{Z}[z]$ which divides both $f(z)$ and $g(z^p)$. This implies that $\Phi_n(z^p) = f(z^p)g(z^p)$ contains a factor of $h(z^p)h(z)$; by (9.21) we see that $z^{np} - 1$ also contains a factor of $h(z^p)h(z)$.

Now we work in the finite field F_p. In that setting we have $h(z^p) = h(z)^p$ and $(z^n - 1)^p = z^{np} - 1$ (cf. Exercise 9.4.7) and hence $(z^n - 1)^p$ contains a factor of

$h(z)^{p+1}$; in particular $z^n - 1$ must contain a factor of $h(z)^2$ in F_p (cf. Exercise 9.4.1). Taking formal derivatives, this implies that $z^n - 1$ and nz^{n-1} have a common factor of $h(z)$; but from the Euclidean algorithm and the fact that $n \neq 0 \pmod p$ we see that these polynomials have a greatest common divisor of 1, contradiction. □

As a consequence of Lemma 9.48 we obtain a useful criterion for non-vanishing of polynomial expressions of roots of unity, which was already exploited in the proof of Theorem 9.20.

Lemma 9.49 *Let p be a prime and q be a power of p. Let $P \in \mathbf{Z}[t_1, \ldots, t_k]$ be a polynomial with integer coefficients such that $P(z_1, \ldots, z_k) = 0$ for some qth roots of unity $z_1, \ldots, z_k$. Then the integer $P(1, \ldots, 1)$ is divisible by p.*

Proof Let ω be a primitive qth root of unity, then $z_i = \omega^{n_i}$ for some integers n_i. If we let $Q(t) := P(t^{n_1}, \ldots, t^{n_k})$, then $Q(\omega) = 0$. Thus $Q(t)$ shares a root in common with the irreducible polynomial $\Phi_q(t)$, which must then be a factor of $Q(t)$. Thus $Q(1) = P(1, \ldots, 1)$ has $\Phi_q(1) = p$ as a factor. □

We apply this lemma to prove a non-vanishing result on generalized Vandermonde determinants. We first need a coefficient computation.

Proposition 9.50 *[355] Let $n_1, \ldots, n_k$ be non-negative integers, and let $P \in \mathbf{Z}[z_1, \ldots, z_k]$ be the polynomial*

$$P(z_1, \ldots, z_k) = \sum_{\pi \in S_k} \operatorname{sgn}(\pi) \prod_{i=1}^{k} z_i^{n_{\pi(i)}}$$

(cf. (9.3)). Then we can factor $P = \Delta_k Q$, where $Q \in \mathbf{Z}[z_1, \ldots, z_k]$ is such that

$$Q(1, \ldots, 1) = \Delta_k(n_1, \ldots, n_k)/\Delta_k(1, \ldots, k).$$

Proof The expression $P(z_1, \ldots, z_k)$ can also be interpreted as the determinant of the $k \times k$ matrix $(z_i^{n_j})_{1 \le i, j \le k}$. This shows in particular that P vanishes when any two of the z_i are equal. Dividing out the factors of $z_i - z_j$ using long division and applying Definition 9.6 we conclude the existence of a polynomial $Q \in \mathbf{Z}[z_1, \ldots, z_k]$ such that $P = \Delta_k Q$. It remains to compute $Q(1, \ldots, 1)$. To do this we introduce the normalized differentiation operators $D_i := z_i \frac{d}{dz_i}$, and consider the expression $D_1^0 D_2^1 \ldots D_k^{k-1} P(1, \ldots, 1)$. We split P into factors

$$P(z_1, \ldots, z_k) = \prod_{1 \le i < j \le k} (z_j - z_i) \times Q(z_1, \ldots, z_k)$$

and apply the Leibniz rule $D_i(fg) = (D_i f)g + f(D_i g)$ repeatedly. Observe that there are $\binom{k}{2}$ linear factors in the expression to be differentiated, all of which vanish at $(1, \ldots, 1)$. There are also $\binom{k}{2}$ derivatives to be applied. Thus the only terms in the Leibniz rule which do not vanish at $(1, \ldots, 1)$ are those in which all the derivatives land on the linear factors. Furthermore each derivative must land on a distinct linear factor to yield a non-zero term. But this means that each of the D_k derivatives must land on one of the $z_k - z_i$ factors with $i < k$ (and there are $(k-1)!$ ways this can happen); similarly the D_{k-1} derivatives must then land on one of the $z_{k-1} - z_i$ factors with $i < k-1$ (with $(k-2)!$ ways this can happen), and so forth. We conclude that

$$D_1^0 D_2^1 \ldots D_k^{k-1} P(1, \ldots, 1) = (k-1)! \cdots 1!0! Q(1, \ldots, 1) = \Delta_k(1, \ldots, k) Q(1, \ldots, 1).$$

On the other hand, since each monomial $z_1^{n_1} \cdots z_k^{n_k}$ is an eigenfunction of D_i with eigenvalue n_i, we see from definition of P that

$$D_1^0 D_2^1 \cdots D_k^{k-1} P(z_1, \ldots, z_k) = \sum_{\pi \in S_k} \operatorname{sgn}(\pi) \prod_{i=1}^{k} {n_{\pi(i)}}^{i-1} z_i^{n_{\pi(i)}}.$$

Substituting $z_1 = \cdots = z_k = 1$ and applying (9.3) we obtain

$$D_1^0 D_2^1 \cdots D_k^{k-1} P(z_1, \ldots, z_k) = \Delta_k(n_1, \ldots, n_k).$$

Combining this with the previous identity, the claim follows. □

Combining Proposition 9.50 with Lemma 9.49 we obtain

Lemma 9.51 (Chebotarev's lemma) *Let $q = p^\alpha$ be a prime power, let $1 \leq k < p$, and let $z_1, \ldots, z_k$ be distinct qth roots of unity. Let $n_1, \ldots, n_k$ be integers which are distinct modulo p. Then the $k \times k$ matrix $(z_i^{n_j})_{1 \leq i,j \leq k}$ has non-zero determinant.*

Indeed, Chebotarev's lemma follows since $\Delta_k(z_1, \ldots, z_k)$ is non-zero and $\Delta_k(n_1, \ldots, n_k)$ is not divisible by p. We note that while this result was proved by Chebotarev in 1926 (see [338]), it has been independently rediscovered and reproved a number of times [278], [71], [263], [102], [355], [120], [131]. As a consequence of this lemma, one easily establishes the following uncertainty principle for $\mathbf{Z}_p$:

Theorem 9.52 *[355] Let p be a prime number. Let $f : \mathbf{Z}_p \to \mathbf{C}$ be a non-zero random variable, and let $\hat{f} : \mathbf{Z}_p \to \mathbf{C}$ be its Fourier transform (using the standard bicharacter $e(x, \xi) = \exp(2\pi i x\xi/p)$). Then we have $|\operatorname{supp}(f)| + |\operatorname{supp}(\hat{f})| \geq p+1$ Conversely, if A and B are two non-empty subsets of $\mathbf{Z}/p\mathbf{Z}$ such that $|A| + |B| \geq p+1$, then there exists a function f such that $\operatorname{supp}(f) = A$ and $\operatorname{supp}(\hat{f}) = B$.*

We leave the deduction of Theorem 9.52 from Lemma 9.51 to Exercise 9.8.9. This result should be compared with (4.21). As an application of this theorem we give yet another proof of the Cauchy–Davenport inequality, this proof being Fourier-analytic (or more precisely Fourier-algebraic) in nature.

Theorem 9.53 (Cauchy–Davenport inequality, yet again) *Let $F = F_p$ be a finite field of prime order. If A, B are two additive sets in F, then*

$$|A + B| \geq \min(|A| + |B| - 1, p).$$

Proof ([355] and Robin Chapman, private communication) Since A and B are non-empty, we may find two subsets X and Y of $\mathbf{Z}/p\mathbf{Z}$ such that $|X| = p + 1 - |A|$, $|Y| = p + 1 - |B|$, and $|X \cap Y| = \max(|X| + |Y| - p, 1)$. By Theorem 9.52 we may find a function f such that $\mathrm{supp}(f) = A$ and $\mathrm{supp}(\hat{f}) = X$, and a function g such that $\mathrm{supp}(g) = B$ and $\mathrm{supp}(\hat{g}) = Y$. Then $f * g$ has support contained in $A + B$ and has Fourier support equal to $X \cap Y$ (in particular, $f * g$ is non-zero), and hence by Theorem 9.52 again we have $|A + B| + |X \cap Y| \geq p + 1$, which gives $|A + B| \geq \min(|A| + |B| - 1, p)$ as desired. □

One can iterate Theorem 9.52 to also apply to the group $\mathbf{Z}_p^n$ for any $n \geq 1$, which we endow with the standard bilinear form, as in Example 4.2.

Corollary 9.54 *[249] Let p be a prime, $n \geq 1$ be an integer, and $f : \mathbf{Z}_p^n \to \mathbf{C}$ be a non-zero random variable. Then we have*

$$p^k |\mathrm{supp}(f)| + p^{n-k-1} |\mathrm{supp}(\hat{f})| \geq p^n + p^{n-1}$$

for all $0 \leq k \leq n - 1$.

Remark 9.55 These bounds can be seen to be sharp in a large number of situations, by taking the Cartesian product of the examples in Theorem 9.52 with subgroups of $\mathbf{Z}_p$. It has a nice geometric interpretation: if one plots the point $(|\mathrm{supp}(f)|, |\mathrm{supp}(\hat{f})|)$ in $\mathbf{Z} \times \mathbf{Z}$, then this point lies on or above the convex hull of the points (p^j, p^{n-j}) for $0 \leq j \leq n$, which correspond to the cases where f is the indicator function of a subgroup of $\mathbf{Z}_p^n$; this convex hull should be contrasted with the hyperbola corresponding to (4.21). In [249], this result was generalized further to arbitrary finite additive groups Z, see Exercise 9.8.11.

Proof We prove this by induction on n. For $n = 1$ this is just Theorem 9.52. Now suppose that $n > 1$, and the Corollary has already been proven for all smaller values of n. Fix f. We parameterize $\mathbf{Z}_p^n$ as $x = (\underline{x}, x_n)$, where $\underline{x} \in \mathbf{Z}_p^{n-1}$ and $x_n \in \mathbf{Z}_p$. If $g(\underline{\xi}, x_n)$ is the Fourier transform of $f(\underline{x}, x_n)$ in the $\underline{x}$ variable (with x_n fixed), then $\hat{f}(\underline{\xi}, \xi_n)$ is the Fourier transform of $g(\underline{\xi}, x_n)$ in the x_n variable (keeping $\underline{\xi}$ fixed).

Let $A \subset \mathbf{Z}_p$ be the set of all x_n such that $f(\cdot, x_n)$ (and hence $g(\cdot, x_n)$) is not identically zero. Observe that $1 \leq |A| \leq p$ and

$$|\mathrm{supp}(f)| = \sum_{x_n \in A} |\mathrm{supp}(f(\cdot, x_n))|.$$

Thus by the pigeonhole principle there exists an x_n such that

$$|A||\mathrm{supp}(f(\cdot, x_n))| \leq |\mathrm{supp}(f)|. \tag{9.22}$$

Fix this x_n. By induction we have

$$p^{k'}|\mathrm{supp}(f(\cdot, x_n))| + p^{n-k'-1}|\mathrm{supp}(g(\cdot, x_n))| \geq p^{n-1} + p^{n-2} \tag{9.23}$$

for all $0 \leq k' \leq n-2$. Also, for any $\underline{\xi}$ in the support of $g(\cdot, x_n)$, we see that $g(\underline{\xi}, \cdot)$ is supported in A, so by Theorem 9.52

$$|\mathrm{supp}(\hat{f}(\underline{\xi}, \cdot))| \geq p + 1 - |A|.$$

Summing this over all $\underline{\xi}$ in the support of $g(\cdot, x_n)$ we obtain

$$|\mathrm{supp}(\hat{f})| \geq (p + 1 - |A|)|\mathrm{supp}(g(\cdot, x_n))|.$$

Combining this with (9.22) we obtain

$$\begin{aligned} p^k|\mathrm{supp}(f)| + p^{n-k-1}|\mathrm{supp}(\hat{f})| \geq\ & p^k|A||\mathrm{supp}(f(\cdot, x_n))| \\ & + (p + 1 - |A|)p^{n-k-1}|\mathrm{supp}(g(\cdot, x_n))|. \end{aligned}$$

When $|A|$ is equal to 1 or p then the right-hand side here is at least $p^n + p^{n-1}$ thanks to (9.23). Since the right-hand side is linear in $|A|$, the same is true for the intermediate cases $1 < |A| < p$. This completes the induction. □

Exercises

9.8.1 Let p be a prime and $k \geq 1$. Prove that $\Phi_p(z) = 1 + z + z^2 + \cdots + z^{p-1}$ and $\Phi_{p^k}(z) = \Phi_p(z^{p^{k-1}})$.

9.8.2 (Eisenstein's criterion) Let p be a prime, and let $P(t) = a_n t^n + \cdots + a_0 \in \mathbf{Z}[t]$ be such that a_n is not divisible by p, that $a_{n-1}, \ldots, a_0$ are divisible by p, and a_0 is not divisible by p^2. Show that P is irreducible in $\mathbf{Z}[t]$.

9.8.3 Let p be a prime. Compute the polynomial $\Phi_p(t+1)$ explicitly, and then use Eisenstein's criterion to give a proof that $\Phi_p(t+1)$, and hence Φ_p itself, is irreducible in $\mathbf{Z}[t]$, without using Lemma 9.48.

9.8.4 Let $n \geq 1$ be an integer, and suppose that $x \in F_p^\times$ is such that $\Phi_n(x) = 0$. Show that $\mathrm{ord}_\times(x) = n$, and in particular n divides $p - 1$.

9.8.5 Let n, m be integers. Using Exercise 9.8.4, show that all the prime factors of $\Phi_n(m)$ are equal to 1 mod n and are coprime to m. Using this (and modifying Euclid's proof of the infinitude of primes) show that there are infinitely many primes equal to 1 mod n; this is a special case of *Dirichlet's theorem*.

9.8.6 Let $n \geq 1$, and let ω be a primitive nth root of unity. Show that the cyclotomic field $\mathbf{Q}(\omega)$ is a $\phi(n)$-dimensional vector space over $\mathbf{Q}$, and that the complex numbers $1, \omega, \omega^2, \ldots, \omega^{\phi(n)-1}$ form a linear basis for $\mathbf{Q}(\omega)$.

9.8.7 Let p be a prime, and let ω be a primitive pth root of unity. Let $\mathbf{Z}[\omega]$ be the ring generated by ω. Show that the quotient ring $\mathbf{Z}[\omega]/((1-\omega)\cdot\mathbf{Z}[\omega])$ is isomorphic to the field F_p. (Hint: exploit the fact that $\Phi_p(1) = p$, and hence $\Phi_p(\omega) - p$ contains a factor of $(1-\omega)$.)

9.8.8 [120] Let p be a prime, let ω be a primitive pth root of unity, let $z_1, \ldots, z_k$ be distinct pth roots of unity. Suppose there exists a polynomial $P \in \mathbf{Z}[\omega][z]$ of degree at most $p-1$ which vanishes at $z_1, \ldots, z_k$ and has at most k non-zero coefficients. Using Exercise 9.8.7 and Lemma 9.26, show that P is a multiple of $(1-\omega)$. Using this and an infinite descent argument, obtain another proof of Lemma 9.51 (at least in the case $q = p$, which is all one needs for Theorem 9.52).

9.8.9 [355] Deduce Theorem 9.52 from Lemma 9.51. (Hint: Lemma 9.51 implies that all the minors of the Fourier matrix $(e^{2\pi ijk/p})_{1\leq j,k\leq p}$ are invertible.) Conversely, show that Theorem 9.52 implies the $q = p$ case of Lemma 9.51.

9.8.10 Let p be a prime, let $G := \{z \in \mathbf{C} : z^p = 1\}$ be the pth roots of unity, and let $P \in \mathbf{C}[z]$ be a non-zero polynomial with $\deg(P) < p$. Show that that the number of zeroes of P in G cannot exceed the number of non-zero coefficients in P.

9.8.11 [249] Given any finite additive group Z and any real number k, let $\theta(Z;k)$ denote the quantity

$$\theta(Z;k) := \inf\{|\mathrm{supp}(\hat{f})| : f \in L^2(Z); f \not\equiv 0; |\mathrm{supp}(f)| \leq k\}.$$

Show that for every subgroup G of Z and any $1 \leq k \leq |Z|$, we have the inequality

$$\theta(Z;k) \geq \inf_{st=k} \theta(G;s)\theta(Z/G;t)$$

by adapting the proof of Corollary 9.54. Conclude via an inductive argument that for any non-zero function f in $L^2(Z)$, the lattice point $(|\mathrm{supp}(f)|, |\mathrm{supp}(\hat{f})|)$ lies on or above the convex hull of the points $(|G|, |Z|/|G|)$ as G ranges over all subgroups of Z.

10

Szemerédi's theorem for $k = 3$

A surprisingly fruitful and deep problem in additive combinatorics is that of determining whether a given set A contains non-trivial (i.e. proper) arithmetic progressions of a given length. We have already seen some special cases of this problem; in Section 4.7 we saw that sum sets such as $A + A$, $A + A + A$, or $2A - 2A$ contained very long arithmetic progressions (and generalized arithmetic progressions), while in Section 6.3 we saw that if we colored a large finite group (or a large interval of integers) into a small number of color classes, then one of the color classes must necessarily contain a long arithmetic progression. In this chapter and the next we shall discuss perhaps one of the deepest theorems known to additive combinatorics, namely *Szemerédi's theorem*:

Theorem 10.1 (Szemerédi's theorem) *[345] Let A be a subset of the positive integers with positive upper density*[1] $\overline{\sigma}(A) > 0$. *Then A contains arbitrarily long arithmetic progressions.*

This theorem was originally proved by Szemerédi in 1975 by a sophisticated combinatorial argument, introducing for the first time the powerful *Szemerédi regularity lemma*, which we discuss in Section 10.6. There are several other deep and important proofs of this theorem, including the ergodic-theoretic proof of Furstenberg [125], the additive combinatorial proof of Gowers [138], and the hypergraph regularity proofs of Gowers [140] and Nagle, Rödl, Schacht, and Skokan [254], [282], [283], [284]. These proofs will be discussed in the next chapter.

One can formulate Szemerédi's theorem in a more quantitative manner, using the following definition.

Definition 10.2 (Erdős–Turán constant) [99] Let A be an additive set, and let $k \geq 1$. We let $r_k(A)$ denote the size of the largest subset of A which does not contain any proper arithmetic progressions of length k.

[1] Upper and lower density were defined in Definition 1.21.

Examples 10.3 We have $r_1(A) = 0$ and $r_2(A) = 1$ for any additive set A. Clearly $r_k(A)$ is non-decreasing in A, and we have the trivial bound $r_k(A) \leq |A|$ for any A. If A lives in a p-torsion group (e.g. $A \subseteq F_p^n$) then $r_k(A) = |A|$ for all $k > p$.

Theorem 10.1 is then easily shown to be equivalent to the following version, which was first conjectured by Erdős and Turán [99].

Theorem 10.4 (Szemerédi's theorem, second formulation) *Let $k \geq 1$ and $N \geq 1$. Then $r_k([1, N]) = o_{N\to\infty;k}(N)$ and $r_k(\mathbf{Z}_N) = o_{N\to\infty;k}(N)$.*

One in fact has the following generalization:

Theorem 10.5 (Szemerédi's theorem, in an arbitrary group) *Let $k \geq 1$ and let Z be a finite additive group with $|Z|$ coprime to $(k-1)!$. Then $r_k(Z) = o_{|Z|\to\infty;k}(|Z|)$.*

This generalization either follows from the density Hales–Jewett theorem [124] or from the hypergraph proofs of Szemerédi's theorem [140], [254], [282], [283], [284], and will be discussed in Section 11.6.

A further famous conjecture of Erdős and Turán remains open:

Conjecture 10.6 (Erdős–Turán conjecture) *[99] Let $A \subset \mathbf{Z}^+$ be such that $\sum_{n\in A} \frac{1}{n} = \infty$. Then A contains arbitrarily long proper arithmetic progressions.*

Up to very small factors, such as $\log^{o(1)} N$, this conjecture is essentially equivalent to asking for $r_k([1, N]) = O_k(N/\log N)$ for all k and N (Exercise 10.0.6). This conjecture remains unsolved even for progressions of length 3 (though see Theorem 10.30 below). However a special case of this conjecture, restricted to the prime numbers $P = \{2, 3, 5, \ldots\}$, has recently been proven by Green and Tao:

Theorem 10.7 (Green–Tao theorem) *[158] Let $k \geq 1$ and $N > 1$. Then $r_k(P \cap [1, N]) = o_{N\to\infty;k}(|P \cap [1, N]|)$. In particular, the primes contain arbitrarily long arithmetic progressions.*

Note from (1.48) that the sum $\sum_p \frac{1}{p}$ is divergent.

For general k, Szemerédi's theorem and the Green–Tao theorem are rather involved and will be treated in Chapter 11. However, the $k = 3$ case is amenable to Fourier-analytic methods, and we have the following famous theorem of Roth:

Theorem 10.8 (Roth's theorem) *[287] We have $r_3([1, N])$, $r_3(\mathbf{Z}_N) = o_{N\to\infty}(N)$ for all $N > 1$. More generally, for any finite additive group Z of odd order we have $r_3(Z) = o_{|Z|\to\infty}(|Z|)$.*

The generalization to arbitrary additive groups Z of odd order is due to Meshulam [248]. Note that the restriction that Z be odd is necessary, since for 2-torsion groups, there are no proper progressions of length three and hence $r_3(Z) = |Z|$ in that case.

Both Roth's theorem and Szemerédi's theorem have a surprising diversity of different proofs, using such techniques as harmonic analysis, ergodic theory, graph theory, hypergraph theory, inverse sum set theory, and Ramsey theory. However, they all revolve around a fundamental dichotomy, namely the dichotomy between *arithmetically structured sets* (e.g. arithmetic progressions, Bohr sets, sets of small doubling, sets of large additive energy, almost periodic sets) and *arithmetically unstructured sets* (e.g. random sets, pseudo-random sets, "mixing" sets). The point is that one needs very different arguments to deal with either of the two cases, and so any proof of the above theorems must first decompose a general set somehow into a structured component and an unstructured one. To make such a decomposition rigorous, one needs some powerful tools, for instance from harmonic analysis, ergodic theory, or graph theory.

The purpose of this chapter is to give several proofs of Roth's theorem, both for general Z and in special cases, and to also discuss some variants of this theorem. These proofs serve as models for the more difficult Szemerédi and Green–Tao theorems, to be discussed in the next chapter. It turns out that *linear* Fourier analysis (as developed in Chapter 4) is a particularly well adapted tool to detect progressions of length 3; as we shall see however in the next chapter, progressions of longer length will require a *quadratic* or higher-order Fourier analysis.

Exercises

10.0.1 Establish the inequalities

$$r_k([1, N/k)) \leq r_k(\mathbf{Z}_N) \leq r_k([1, N])$$

for any $N > k > 1$. This shows that the two forms $r_k(\mathbf{Z}_N) = o_{N\to\infty;k}(N)$ and $r_k([1, N]) = o_{N\to\infty;k}(N)$ of Theorem 10.4 are equivalent.

10.0.2 Show that Theorem 10.4 is equivalent to Theorem 10.1. (Hint: to deduce Theorem 10.1 from Theorem 10.4 is rather easy. For the converse direction, argue by contradiction, obtaining dense subsets of $[1, N]$ without any proper arithmetic progressions, and paste those subsets together in some suitable way to contradict Theorem 10.1.)

10.0.3 Show that Theorem 10.1 is equivalent to the statement that every subset of the integers of positive upper density contains infinitely many progressions of length k, for each $k \geq 1$.

10.0.4 Show that Szemerédi's theorem implies van der Waerden's theorem (Theorem 6.17)

10.0.5 Give an example to show that if the positive integers $\mathbf{Z}^+$ are partitioned into two color classes, then it is not necessarily the case that one of the color classes contains an *infinitely long* proper arithmetic progression

$a + \mathbf{Z}^+ \cdot r$. Thus the properties of containing arbitrarily long proper arithmetic progressions, and infinitely long proper arithmetic progressions, are distinct.

10.0.6 Show that the Erdős–Turán conjecture is equivalent to the absolute convergence of the sum

$$\sum_{n=1}^{\infty} \frac{r_k([1, 2^n])}{2^n}.$$

10.0.7 Show that if A and B are additive sets which are Freiman isomorphic of order 2, then $r_k(A) = r_k(B)$ for all k.

10.0.8 If A and B are additive sets (possibly in different groups), show that $r_k(A \times B) \geq r_k(A) r_k(B)$.

10.0.9 If Z, Z' are two finite additive groups, show that $r_k(Z \times Z') \leq r_k(Z)|Z'|$.

10.0.10 Show that to prove Theorem 10.5 for arbitrary groups Z, it suffices to verify it for cyclic groups $\mathbf{Z}_N$ and for vector spaces $\mathbf{Z}_p^n$ over fields of prime order. (Hint: use Corollary 3.8 and the previous exercise.) A similar claim applies of course to Roth's theorem.

10.0.11 Let $n \geq 1$. Define a *capset* of order n to be any subset of the vector space F_3^n over the finite field F_3 which contains no (affine) lines. Show that the largest possible cardinality of a capset of order n is $r_3(F_3^n)$. Using Exercise 10.0.8, show that $r_3(F_3^n) \geq 2^n$.

10.0.12 If Z is a finite additive group whose order is coprime to $k!$, show that $r_k(Z) \leq (1 - \frac{1}{k})|Z|$. (Hint: if $A \subset Z$ has cardinality greater than $(1 - \frac{1}{k})|Z|$, choose $a \in Z$, $r \in Z \backslash \{0\}$ randomly and consider the probability of the events $a + jr \notin A$ for $j = 0, 1, \ldots, k - 1$.)

10.1 General strategy

In this section we make some general observations concerning progressions of length 3, and describe in high-level terms the various strategies one could employ to prove Roth-like theorems.

Let us work in a fixed finite additive group Z of odd order, and let A be a subset of Z. We shall think of A as being rather dense, so that the density $0 \leq \mathbf{P}_Z(A) \leq 1$ is moderately large. Roth's theorem is then an assertion that if $|Z|$ is sufficiently large, then A must contain progressions of length three.

To explain why this should be the case, it is convenient to introduce the trilinear form

$$\Lambda_3(f, g, h) := \mathbf{E}_{x, r \in Z} f(x) g(x + r) h(x + 2r) \qquad (10.1)$$

for any $f, g, h : Z \to \mathbf{C}$. Note in particular that

$$\Lambda_3(1_A, 1_A, 1_A) = \mathbf{P}_{x,r\in Z}(x, x+r, x+2r \in A) \tag{10.2}$$

so the quantity $\Lambda_3(1_A, 1_A, 1_A)$ measures the proportion of arithmetic progressions $(x, x+r, x+2r)$ in Z which are completely contained in A. Intuitively, if A is "randomly" distributed, then the events $x \in A$, $x+r \in A$, $x+2r \in A$ should be "independent", and we then expect

$$\begin{aligned}\Lambda_3(1_A, 1_A, 1_A) &\approx \mathbf{P}_{x,r\in Z}(x \in A)\mathbf{P}_{x,r\in Z}(x+r \in A)\mathbf{P}_{x,r\in Z}(x+2r \in A)\\ &= \mathbf{P}_Z(A)^3.\end{aligned} \tag{10.3}$$

Thus if A is fairly dense in Z, we expect $\Lambda_3(1_A, 1_A, 1_A)$ to be large. On the other hand, if $|Z|$ is odd and A has no proper progressions of length 3, then the only progressions $(x, x+r, x+2r)$ which can lie in A are those for which $x \in A$ and $r = 0$, whence

$$\Lambda_3(1_A, 1_A, 1_A) = \mathbf{P}_Z(A)/|Z|. \tag{10.4}$$

If $|Z|$ is sufficiently large, this seems to be in conflict with the heuristic (10.3). Thus to prove Roth's theorem it will suffice to establish some rigorous analog of (10.3). In particular, Roth's theorem will be implied by the following result.

Theorem 10.9 (Varnavides' theorem) *[372] Let Z be a finite additive group of odd order. Then for any non-empty set $A \subseteq Z$ we have*

$$\Lambda_3(1_A, 1_A, 1_A) = \Omega_{\mathbf{P}_Z(A)}(1).$$

In other words, we have $\Lambda_3(1_A, 1_A, 1_A) \geq c(\mathbf{P}_Z(A))$ where $c(\mathbf{P}_Z(A)) > 0$ depends only on the density $\mathbf{P}_Z(A)$ of A and not on the group Z. More generally, if $f : Z \to \mathbf{R}^+$ is a non-negative function which is not identically zero, and obeying the bound $0 \leq f(x) \leq 1$ for all $x \in Z$, then

$$\Lambda_3(f, f, f) = \Omega_{\mathbf{E}_Z(f)}(1).$$

Note that Varnavides' theorem is in fact a bit stronger than Roth's theorem, as it implies that any subset of Z of density δ will contain $\Omega_\delta(|Z|^2)$ proper arithmetic progressions of length 3, if Z is sufficiently large depending on δ. This is in contrast with Roth's theorem which would only provide a single proper arithmetic progression of length 3. Nevertheless, a simple averaging argument shows that the two theorems are equivalent: see exercises.

It is still not clear how to convert the heuristic (10.3) into a rigorous statement such as Theorem 10.9. Indeed (10.3) can fail for certain special A, with $\Lambda_3(1_A, 1_A, 1_A)$ ranging as high as $\mathbf{P}_Z(A)^2$ if A is a subgroup of Z, and as low as $\mathbf{P}_Z(A)^{\Omega(\log \frac{1}{\mathbf{P}_Z(A)})}$ if A is given by the Behrend example (see exercises). However, it turns out that $\Lambda_3(1_A, 1_A, 1_A)$ will be very close to $\mathbf{P}_Z(A)^3$ (as predicted by (10.3)) as long as A has very little *linear bias*. Recall from Definition 4.12 that the linear bias (or Fourier bias) $\|A\|_u$ of an additive set A was defined as

$$\|A\|_u := \sup_{\xi \in Z\backslash 0} |\hat{1}_A(\xi)| = \sup_{\xi \in Z\backslash 0} |\mathbf{E}_{x \in Z} 1_A(x) e(-\xi \cdot x)|.$$

Proposition 10.10 (Lack of progressions implies non-uniformity) *[287] Let A be an additive set in a finite additive group Z of odd order. Then*

$$|\Lambda_3(1_A, 1_A, 1_A) - \mathbf{P}_Z(A)^3| \leq \|A\|_u \mathbf{P}_Z(A).$$

In particular, if A contains no proper arithmetic progressions of length 3*, then we have the linear bias estimate*

$$\|A\|_u \geq \mathbf{P}_Z(A)^2 - \frac{1}{|Z|}.$$

Proof From the identity $a - 2(a + r) + (a + 2r) = 0$, and the observation that the map $x \mapsto 2 \cdot x$ is bijective on Z when $|Z|$ is odd, we see that

$$\Lambda_3(1_A, 1_A, 1_A) = \frac{1}{|Z|^2} |\{(a_1, a_2, a_3) \in A \times (-2 \cdot A) \times A : 0 = a_1 + a_2 + a_3\}|.$$

Applying Lemma 4.13 we obtain the first inequality. The second claim then follows from (10.4). □

This shows that the only way the heuristic (10.3) can fail is if the function 1_A has a large correlation with a linear character $e(\xi \cdot x)$. This very important observation can be viewed as an *inverse theorem* for Λ_3; we will return to this perspective in the next chapter. There is an analog of the above proposition for functions. Define the *linear bias* $\|f\|_{u^2(Z)}$ of a function $f : Z \to \mathbf{C}$ to be the quantity

$$\|f\|_{u^2(Z)} := \sup_{\xi \in Z} |\hat{f}(\xi)|. \tag{10.5}$$

The reason for the notation $u^2(Z)$ will be made clearer in the next chapter. Note for instance that $\|A\|_u = \|1_A - \mathbf{P}_Z(A)\|_{u^2(Z)}$ for any $A \subseteq Z$.

Proposition 10.11 *Let Z have odd order. For any functions $f, g, h : Z \to \mathbf{C}$, we have the identity*

$$\Lambda_3(f, g, h) = \sum_{\xi \in Z} \hat{f}(\xi)\hat{g}(-2\xi)\hat{h}(\xi). \tag{10.6}$$

We can then conclude the estimate

$$|\Lambda_3(f, g, h)| \leq \|f\|_{u^2(Z)} \|g\|_{L^2(Z)} \|h\|_{L^2(Z)}$$

and similarly with f, g, h permuted on the right-hand side.

Proof From the Fourier inversion formula (4.4) we have

$$f = \sum_{\xi_1} \hat{f}(\xi_1) e_{\xi_1}; \quad g = \sum_{\xi_2} \hat{g}(\xi_2) e_{\xi_2}; \quad h = \sum_{\xi_3} \hat{h}(\xi_3) e_{\xi_3}$$

and hence

$$\Lambda_3(f, g, h) = \sum_{\xi_1, \xi_2, \xi_3 \in Z} \hat{f}(\xi_1)\hat{g}(\xi_2)\hat{h}(\xi_3)\Lambda_3(e_{\xi_1}, e_{\xi_2}, e_{\xi_3}).$$

On the other hand, a direct computation using Lemma 4.5 shows

$$\Lambda_3(e_{\xi_1}, e_{\xi_2}, e_{\xi_3}) = \mathbf{I}(\xi_2 = -2\xi_1; \xi_3 = \xi_1)$$

which gives (10.6). From Parseval's identity (4.2) and the hypothesis that Z has odd order, we have

$$\sum_{\xi \in Z} |\hat{g}(-2\xi)|^2 = \|g\|^2_{L^2(Z)}; \quad \sum_{\xi \in Z} |\hat{h}(\xi)|^2 = \|h\|^2_{L^2(Z)}$$

and the claim then follows from Hölder's inequality. Similarly if the roles of f, g, h are permuted. □

To exploit inverse results such as Proposition 10.10 or Proposition 10.11, there are two arguments available: the *density increment argument* of Roth, and the *energy increment argument* developed separately by Furstenberg and Szemerédi (in very different contexts). The density increment argument proceeds informally as follows. To prove Roth's theorem, suppose for contradiction that one can find a dense set A in a large group Z (or interval $[1, N]$) which contains no progressions of length three. Proposition 10.10 then implies that A has large linear bias, thus 1_A correlates with some linear phase function $e(\xi \cdot x)$. It then turns out that this linear bias can be converted into a *density increment*, or more precisely some structured subset Z' (such as a subgroup, a sub-progression, or a Bohr set) of the original space Z on which A has larger density, thus $\mathbf{P}_{Z'}(A) > \mathbf{P}_Z(A)$. (Recall that $\mathbf{P}_{Z'}(A) = |A \cap Z'|/|Z'|$ and $\mathbf{P}_Z(A) = |A|/|Z|$.) One then passes to this structured subset and repeats the argument. If the original space Z was large enough, we can run this argument for so many steps that the relative density of A eventually exceeds 1, a contradiction.

The energy increment argument proceeds differently, aiming to prove Varnavides' theorem instead of Roth's theorem (i.e. one seeks non-trivial lower bounds on $\Lambda_3(f, f, f)$). Instead of continually changing the ambient space Z, we now hold Z fixed, but instead construct certain *low complexity approximations* $f_{U^\perp}$ to the original function f. Initially, our approximation will just be the density, $f_{U^\perp} = \mathbf{P}_Z(A)$. We now consider the error $f_U := f - f_{U^\perp}$ between the indicator function and the approximation. If this error is very linearly uniform (in the sense that the Fourier bias $\|f_U\|_{u^2(Z)}$ is small), then Proposition 10.11 can be used to approximate $\Lambda_3(f, f, f)$ by $\Lambda_3(f_{U^\perp}, f_{U^\perp}, f_{U^\perp})$, and one can exploit the low complexity of $f_{U^\perp}$ to obtain a non-trivial lower bound on the latter quantity. If instead the error exhibits linear bias, one can exploit this by refining the approximation $f_{U^\perp}$ to absorb this bias; this will increase the *energy* $\|f_{U^\perp}\|^2_{L^2(Z)}$ of $f_{U^\perp}$ by a significant amount. One then repeats the argument until the error f_U contains no further bias; a key point will be that f (and hence $f_{U^\perp}$) remain bounded throughout the iteration and so the energy of $f_{U^\perp}$ cannot increase indefinitely.

Exercises

10.1.1 Let Z be a finite additive group of odd order, let $0 < \delta < 1$, and let A be a random subset of Z such that the events $x \in A$ are independent with probability $\mathbf{P}(x \in A) = \delta$. Show that with probability $1 - o_{|Z|\to\infty;\delta}(1)$, we have $\mathbf{P}_Z(A) = \delta + o_{|Z|\to\infty;\delta}(1)$ and $\Lambda_3(1_A, 1_A, 1_A) = \delta^3 + o_{|Z|\to\infty;\delta}(1)$, thus confirming (10.3) in the random case. (Hint: use Corollary 1.9.)

10.1.2 Let Z be a finite additive group of odd order. Show that $\Lambda_3(1_A, 1_A, 1_A) \leq \mathbf{P}_Z(A)^2$, with equality attained if and only if A is the translate of a subgroup of Z.

10.1.3 Let $N, d, r \geq 1$ be integers, and consider the set

$$A = \{(n_1, \ldots, n_d) \in [0, N/2)^d : n_1^2 + \cdots + n_d^2 = r\},$$

viewed as a subset of $\mathbf{Z}_N^d$. Show that this set has no proper arithmetic progressions of length 3, and can have cardinality as large as $(N/2)^d/(d^2N^2)$ for a suitable choice of r. Conclude in particular that $r_3(\mathbf{Z}_N^d) \geq N^d/(2^d d^2 N^2)$.

10.1.4 (Behrend's example) [21] Using the preceding exercise and a Freiman isomorphism, establish the bounds

$$r_3(\mathbf{Z}_N), r_3([1, N]) = \Omega\big(Ne^{-O(\sqrt{\log N})}\big)$$

for all large N. In particular, it is not the case that $r_3([1, N])$, $r_3(\mathbf{Z}_N) = O(N^{1-\varepsilon})$ for any fixed $\varepsilon > 0$. This rules out a number of elementary approaches to proving Roth's theorem or Szemerédi's theorem (e.g arguments based entirely on Cauchy–Schwarz and pigeonhole principle type arguments) as these tend to only give polynomial type bounds. We remark that the more general estimate

$$r_k(\mathbf{Z}_N), r_k([1, N]) = \Omega_k\big(N \exp\big(- O_k(\log N)^{1/(1+\lfloor \log_2(k-1)\rfloor)}\big)\big)$$

for all $k \geq 3$ has been established in [277], [221] by a similar argument.

10.1.5 Given any $0 < \delta < 1$, give an example of an additive set A in a cyclic group $\mathbf{Z}_N$ such that $\mathbf{P}_Z(A) \geq \delta$ but

$$\Lambda(1_A, 1_A, 1_A) = O\left(\delta^{\Omega(\log \frac{1}{\delta})}\right).$$

(Hint: use the Behrend example.) Thus it is not possible to establish any lower bound of the form $\Lambda(1_A, 1_A, 1_A) = \Omega(\mathbf{P}_Z(A)^C)$ for any absolute constant $C > 0$.

10.1.6 [253] Let N be a large number. Show that one can color $\mathbf{Z}_N$ into $\exp(O(\sqrt{\log N}))$ color classes, such that none of the color classes contains a proper arithmetic progression of length three. Hint: modify the Behrend example.

10.1.7 Show that Varnavides' theorem for sets A implies Varnavides' theorem for functions f. (Hint: either bound f from below by a constant multiple of an indicator function, or construct a set A probabilistically using $f(x)$ as the probability that $x \in A$ and use the first moment method.)

10.1.8 Show that the special case $r_3([1, N]) = o_{N\to\infty}(N)$ of Roth's theorem implies Varnavides' theorem for $\mathbf{Z}_N$. (Hint: take a set A in $\mathbf{Z}_N$ and intersect it with a randomly chosen progression $a + [1, M] \cdot r$ for some moderately large M, and apply Roth's theorem to the progression $a + [1, M] \cdot r$. Then use the first moment method.)

10.1.9 Let F be a finite field. Show that the special case $r_3(F^n) = o_{n\to\infty;F}(N)$ of Roth's theorem implies Varnavides' theorem for F^n. (Hint: take a set A in F^n and intersect it with a randomly chosen m-dimensional affine subspace of F^n for some moderately large m. Then argue as in the preceding exercise.)

10.1.10 Show that Roth's theorem for arbitrary Z implies Varnavides' theorem for arbitrary Z.

10.1.11 Use Proposition 10.11 and the decomposition $1_A = (1_A - \mathbf{P}_Z(A)) + \mathbf{P}_Z(A)$ to provide an alternative proof of Proposition 10.10.

10.1.12 Assume Theorem 10.9. Let $(X, \mathcal{B}, d\mu)$ be any probability space (so $\mu(X) = 1$), and let $T : X \to X$ be any measure-preserving bijection on X, so $\mu(T^n(E)) = \mu(E)$ for all $E \in \mathcal{B}$ and $n \in \mathbf{Z}$. Show that if $f : X \to \mathbf{R}^+$ is any function with $0 \leq f(x) \leq 1$ almost everywhere and $\int_X f = \delta > 0$, then

$$\liminf_{N\to\infty} \mathbf{E}_{n\in[-N,N]} \int_X f(x) T^n f(x) T^{2n} f(x)\, d\mu(x) = \Omega_\delta(1).$$

10.2 The small torsion case

We now use the above Fourier-analytic methods and the density increment argument to prove the following simple special case of Roth's theorem.

Proposition 10.12 (Roth's theorem for p-torsion groups) *[248] Let Z be a p-torsion group (thus $px = 0$ for all $x \in Z$) for some odd prime p. Then*

$$r_3(Z) < \frac{3}{\log_p |Z|}|Z|.$$

Remark 10.13 Define a *capset* to be a subset of the vector space $\mathbf{Z}_3^n$ which contains no lines. Then the above proposition implies that capsets have density less than $3/n$. Rather amazingly, this simple bound is essentially the best known (other than improving the constant 3); in the converse direction, the best lower bound known on the density of capsets in $\mathbf{Z}_3^n$ is $(0.724581\ldots + o(1))^n$; see [75]. Any improvement of the upper bound to $o(1/n)$, or the lower bound to $(1 - o(1))^n$, would be a significant advance in our understanding of the Erdős–Turán conjecture.

Remark 10.14 A useful heuristic is that the cyclic group $\mathbf{Z}_N$ (or the interval $[1, N]$) should behave roughly like the p-torsion group $\mathbf{Z}_p^n$ whenever $N \sim p^n$. Using this heuristic and the above proposition, one would expect that $r_3([1, N])$ and $r_3(\mathbf{Z}_N)$ should be $O(N/\log N)$. Such a bound would essentially be equivalent to the Erdős–Turán conjecture (Conjecture 10.6) in the $k = 3$ case. Unfortunately the direct analog of the above argument gives $r_3([1, N]), r_3(\mathbf{Z}_N) = O(N\sqrt{\frac{\log\log N}{\log N}})$, see Theorem 10.30. In general, the p-torsion groups are somewhat easier to analyze than general groups, due to their vector space structure over the field F_p. To extend the p-torsion arguments to more general settings, one needs some additional machinery, in particular the theory of Bohr sets.

We now begin the proof of Proposition 10.12. We may view Z as a vector space over F_p. Assume for contradiction that we can find a set $A \subset Z$ of density $\mathbf{P}_Z(A) \geq \frac{3}{\log_p |Z|}$ which has no proper progressions of length 3. From Proposition 10.10 we already know that A must exhibit linear bias, thus $\|A\|_u$ is large. To use this fact, we need to convert linear bias to a more useful structural property. This is achieved as follows.

Lemma 10.15 (Non-uniformity implies density increment) *Let Z be a vector space over a finite field F_p of prime order, and let $f : Z \to \mathbf{R}$ be a function with mean zero, $\mathbf{E}_Z(f) = 0$. Then there exists a subspace Z' of Z of codimension 1 over F_p, and a point $x_0 \in Z$, such that*

$$\mathbf{E}_{x \in x_0 + Z'} f(x) \geq \frac{1}{2}\|f\|_{u^2(Z)}.$$

Proof Without loss of generality we may take $Z = F_p^n$, and use the bilinear form in Example 4.2.

By definition of $\|f\|_{u^2(Z)}$ and the mean zero hypothesis, we can find a non-zero $\xi \in Z$ and a phase $\theta \in \mathbf{R}/\mathbf{Z}$ such that

$$\operatorname{Re} \mathbf{E}_{y \in Z} f(y) e(\xi \cdot y + \theta) = \|f\|_{u^2(Z)},$$

where e is the exponential map defined by equation (4.1). Applying the mean zero hypothesis again, we conclude

$$\operatorname{Re} \mathbf{E}_{y \in Z} f(y)(e(\xi \cdot y + \theta) + 1) = \|f\|_{u^2(Z)}$$

Let $Z' := \{\xi\}^\perp = \{x \in Z : \xi \cdot x = 0\}$ be the orthogonal complement of ξ; then Z' is a subspace of Z of codimension 1, and the function $y \mapsto e(\xi \cdot y + \theta) + 1$ is constant on every coset of Z'. Making the change of variables $y = x_0 + x$ for each $x \in Z'$, and then averaging over x, we conclude

$$\begin{aligned}\operatorname{Re} \mathbf{E}_{y \in Z} f(y)(e(\xi \cdot y + \theta) + 1) &= \mathbf{E}_{x \in Z'} \mathbf{E}_{x_0 \in Z} f(x_0 + x) \operatorname{Re}(e(\xi(x + x_0) + \theta) + 1) \\ &= \mathbf{E}_{x_0 \in Z} (\mathbf{E}_{x \in x_0 + Z'} f(x)) \operatorname{Re}(e(\xi \cdot x_0 + \theta) + 1).\end{aligned}$$

By the pigeonhole principle there must therefore exist a coset $x_0 + Z'$ such that

$$(\mathbf{E}_{x \in x_0 + Z'} f(x)) \operatorname{Re}(e(\xi \cdot x_0 + \theta) + 1) \geq \|f\|_{u^2(Z)}.$$

Since $\operatorname{Re}(e(\xi \cdot x_0 + \theta) + 1) \leq 2$, the claim follows. □

Remark 10.16 The reason to add 1 to $e(\xi \cdot y + \theta)$ is to make sure that $\operatorname{Re}(e(\xi \cdot y + \theta) + 1)$ is non-negative. We will use this trick repeatedly in this chapter.

We can now prove Proposition 10.12, by using the density increment argument of Roth.

Proof of Proposition 10.12 By Corollary 3.8 we may take $Z = F_p^n$, with the standard bilinear form in Example 4.2. We induce on n. The claim is trivial when $n \leq 3$, so suppose $n > 3$. Suppose for contradiction that $r_3(F_p^n) \geq 3|Z|/n$, then we can find a set $A \subset Z$ with density $\mathbf{P}_Z(A) \geq 3/n$ containing no proper progressions of length 3. Then by Lemma 10.15 (applied to $f := 1_A - \mathbf{P}_Z(A)$) we have a coset $x_0 + Z'$ of Z of codimension one such that

$$\mathbf{P}_{x_0 + Z'}(A) \geq \mathbf{P}_Z(A) + \frac{1}{2}\|A\|_u.$$

Applying Proposition 10.10 we conclude

$$\begin{aligned}\mathbf{P}_{x_0+Z'}(A) &\geq \frac{3}{n} + \frac{1}{2}\frac{9}{n^2} - \frac{1}{2|Z|}\\ &\geq \frac{3}{n} + \frac{4}{n^2}\\ &\geq \frac{3}{n-1}\end{aligned}$$

since $|Z| = p^n \geq n^2$ and $n > 3$. By the induction hypothesis, the set $(A - x_0) \cap Z'$ thus contains a proper arithmetic progression of length 3, and hence A does also, which gives the desired contradiction. □

A very similar argument also establishes Varnavides' theorem in this setting:

Proposition 10.17 (Varnavides's theorem for p-torsion groups) *Let Z be a p-torsion group for some odd prime p, and let $f : Z \to \mathbf{R}^+$ be such that $0 \leq f(x) \leq 1$ for all $x \in Z$. Then*

$$\Lambda_3(f, f, f) \geq p^{-6/\mathbf{E}_Z(f)}.$$

Proof We induce on $n := \lfloor 3/\mathbf{E}_Z(f) \rfloor$. When $n \leq 3$ the claim is trivial, so suppose $n > 3$ and the claim has already been proven for $n - 1$. We may again view Z as a vector space over F_p, with a standard bilinear form. Write $f = f_{U^\perp} + f_U$, where $f_{U^\perp} := \mathbf{E}_Z(f)$ and $f_U := f - f_{U^\perp}$. Observe that

$$\Lambda_3(f_{U^\perp}, f_{U^\perp}, f_{U^\perp}) = \mathbf{E}_Z(f)^3.$$

If we had

$$\Lambda_3(f, f, f) \geq \mathbf{E}_Z(f)^3/9$$

(say) then we would be done (since $\mathbf{E}_Z(f)^3/9 \geq p^{-6/\mathbf{E}_Z(f)}$), so let us assume instead that

$$|\Lambda_3(f, f, f) - \Lambda_3(f_{U^\perp}, f_{U^\perp}, f_{U^\perp})| \geq 8\mathbf{E}_Z(f)^3/9.$$

We can rewrite the left-hand side as the telescoping sum of three terms,

$$|\Lambda_3(f_U, f, f) + \Lambda_3(f_{U^\perp}, f_U, f) + \Lambda_3(f_{U^\perp}, f_{U^\perp}, f_U)|.$$

From their definitions, we see that f_U has mean zero, and $f_{U^\perp}$ is constant. Thus one can easily verify that the latter two terms vanish. Hence

$$|\Lambda_3(f_U, f, f)| \geq 8\mathbf{E}_Z(f)^3/9.$$

Since f is bounded by 1, we have

$$\|f\|_{L^2(Z)}^2 = \mathbf{E}_Z(f^2) \leq \mathbf{E}_Z(f)$$

and hence by Proposition 10.11 we have

$$\| f_U \|_{u^2(Z)} \geq 8\mathbf{E}_Z(f)^2/9.$$

Applying Lemma 10.15, we can find a subspace Z' of Z of codimension 1 and x_0 in Z, such that

$$\mathbf{E}_{x \in x_0 + Z'} f(x) \geq \mathbf{E}_Z(f) + 4\mathbf{E}_Z(f)^2/9.$$

If we let $g : Z' \to \mathbf{R}$ be the function $g(x) := f(x + x_0)$, then g ranges between 0 and 1 and we have

$$\mathbf{E}_{Z'}(g) \geq \mathbf{E}_Z(f) + 4\mathbf{E}_Z(f)^2/9;$$

this in particular forces $\mathbf{E}_Z(f) \leq 3/4$, and then from elementary algebra one concludes

$$\frac{6}{\mathbf{E}_{Z'}(g)} \leq \frac{6}{\mathbf{E}_Z(f)} - 2.$$

By the induction hypothesis we then have

$$\Lambda_3(g, g, g) \geq p^2 p^{-6/\mathbf{E}_Z(f)},$$

while from definition of g and positivity of f we have $\Lambda_3(f, f, f) \geq p^{-2}\Lambda_3(g, g, g)$. This completes the induction. □

A remarkable phenomenon is that lower bounds of the above type still persist when the boundedness condition $f \leq 1$ is replaced by a more general condition $f \leq \nu$, providing that the enveloping weight ν is sufficiently *pseudo-random*. This phenomenon (essentially first observed in [212], [147]) was made more explicit in [158], when a *transference principle* was formulated. This principle was aimed at studying progressions of arbitrary length k and was phrased in an ergodic theory language, but a parallel Fourier-analytic principle in $k = 3$ exists, and was developed in [159]. We give a simplified formulation of this result below, in the special contexts of random subsets of p-torsion groups. Specifically, we shall prove

Theorem 10.18 (Roth's theorem in random subsets of torsion groups) *Let Z be a finite p-torsion group for some odd prime p, let $|Z|^{-0.01} \leq \tau \leq 1$, and let B be a random subset of Z with the events $x \in B$ being independent with probability $\mathbf{P}(x \in B) = \tau$. Then with probability $1 - o_{|Z| \to \infty; p}(1)$ we have $r_3(B) = o_{|Z| \to \infty; p}(|B|)$.*

Remark 10.19 The point of this theorem is that it allows us to detect arithmetic progressions in subsets of Z of density as low as $|Z|^{-0.01}$, which is well beyond the reach of Proposition 10.12, provided that those sets have large *relative* density compared to a random set. A modification of the proof given below can be used to establish that any subset of the primes of positive relative density contains infinitely many arithmetic progressions of length 3; see [147], [159]; the point was that the primes were contained in a set of "almost primes" which was very uniform (or "pseudo-random") and thus behaved very much like a random set in a certain Fourier-analytic sense. By replacing the Fourier-analytic methods with ergodic theory methods (and replacing linear uniformity with the notion of Gowers uniformity, which could be obtained for the almost primes by some number-theoretic arguments of Goldston and Yildirim), this result was then extended to cover arithmetic progressions of arbitrary length; see [158]. Note that the original proof in [212] relied on the Szemerédi regularity lemma (Lemma 10.42 below) instead of Fourier-analytic methods (and has weaker bounds as a consequence); on the other hand, it works for an arbitrary finite additive group Z of odd order, and allows the density τ to approach $|Z|^{-1/2}$, which is the optimal value (Exercise 10.2.2).

We now begin the proof of Theorem 10.18. We shall need the following extension of Proposition 10.17, in which f is not bounded by 1, but is instead bounded by a "pseudo-random measure", and also enjoys some Fourier bounds.

Theorem 10.20 *[159] Let Z be a finite p-torsion group for some odd prime p, and let $f : Z \to \mathbf{R}_{\geq 0}$ be a non-negative function such that*

$$\|\hat{f}\|_{l^q(Z)} \leq M \tag{10.7}$$

for some $2 < q < 3$ and $0 < M < \infty$. Suppose also that we have the bound $f \leq \nu$ where $\nu : Z \to \mathbf{R}_{\geq 0}$ obeys the pseudo-randomness condition

$$|\hat{\nu}(\xi) - \mathbf{I}(\xi = 0)| \leq \eta \tag{10.8}$$

for some $0 < \eta < 1$. Then we have

$$\Lambda_3(f, f, f) \geq 8p^{-12/\mathbf{E}_Z(f)} - 7M^3 \log_p^{1-3/q} \frac{1}{\eta}.$$

Note that Proposition 10.17 corresponds to the case $\nu = 1$, in which case we can take $\eta = 0$ (and q, M are irrelevant). More generally, this theorem is useful when η is very small compared to M. The constants can be improved somewhat but this will not concern us here.

Proof We may assume Z is a vector space over F_p, with a bilinear form as in Example 4.2. Let $\alpha := M/\log_p^{1/q} \frac{1}{\eta}$. We recall the spectrum $\mathrm{Spec}_\alpha(f) \subseteq Z$, defined as

$$\mathrm{Spec}_\alpha(f) := \{\xi \in Z : |\hat{f}(\xi)| \geq \alpha\}.$$

From the hypothesis (10.7) and Chebyshev's inequality we have

$$|\mathrm{Spec}_\alpha(f)| \leq M^q/\alpha^q = \log_p \frac{1}{\eta}. \tag{10.9}$$

Thus if we let $V = \mathrm{Spec}_\alpha(f)^\perp$ be the orthogonal complement to $\mathrm{Spec}_\alpha(f)$, then V is a subspace of Z and[1]

$$|V^\perp| \leq p^{|\mathrm{Spec}_\alpha(f)|} \leq \frac{1}{\eta}. \tag{10.10}$$

We split $f = f_U + f_{U^\perp}$, where $f_U := f - f * \frac{1_V}{\mathbf{P}_Z(V)}$ is the "uniform" component of f and $f_{U^\perp} := f * \frac{1_V}{\mathbf{P}_Z(V)}$ is the "anti-uniform" component. This allows us to split $\Lambda_3(f, f, f)$ into eight terms,

$$\Lambda_3(f, f, f) = \Lambda_3(f_U, f_U, f_U) + \cdots + \Lambda_3(f_{U^\perp}, f_{U^\perp}, f_U) + \Lambda_3(f_{U^\perp}, f_{U^\perp}, f_{U^\perp})$$

The idea is to use Proposition 10.17 to obtain lower bounds on the last term, and (10.6) to obtain magnitude bounds on the remaining seven terms.

We begin by controlling $f_{U^\perp}$. Since f is bounded pointwise by ν, we can use the Poisson summation formula (Exercise 4.1.7) and (10.10), (10.8) to obtain

$$\begin{aligned} f * \frac{1_V}{\mathbf{P}_Z(V)}(x) &\leq \nu * \frac{1_V}{\mathbf{P}_Z(V)}(x) \\ &= \sum_{\xi \in V^\perp} \hat{\nu}(\xi) e(\xi \cdot x) \\ &\leq 1 + |V^\perp| \sup_{\xi \in V^\perp} |\hat{\nu}(\xi) - \mathbf{I}(\xi = 0)| \\ &\leq 1 + \frac{1}{\eta}\eta = 2. \end{aligned}$$

We thus see that $f_{U^\perp}$ is bounded above by 2. Also it is non-negative and $\mathbf{E}_Z(f_{U^\perp}) = \mathbf{E}_Z(f)$ thanks to (4.10). Thus by Proposition (10.17) (applied to $f_{U^\perp}/2$) we have

$$\Lambda_3(f_{U^\perp}, f_{U^\perp}, f_{U^\perp}) \geq 8p^{-12/\mathbf{E}_Z(f)}.$$

Now we consider the other terms. From the Poisson summation formula again we have

$$\hat{f}_{U^\perp} = \hat{f} 1_{V^\perp} \text{ and } \hat{f}_U = \hat{f}(1 - 1_{V^\perp}).$$

In particular we have

$$\|\hat{f}_U\|_{l^q(Z)}, \|\hat{f}_{U^\perp}\|_{l^q(Z)} \leq M.$$

Furthermore, since $V^\perp$ contains $\mathrm{Spec}_\alpha(f)$, we see that

$$\sup_{\xi \in Z} |\hat{f}_U(\xi)| \leq \alpha.$$

[1] This is extremely crude. It is likely that one can use the machinery of dissociated sets as in Lemma 4.36 to do better here.

Applying (10.6) and Hölder's inequality we obtain

$$|\Lambda_3(f_U, f_{U^\perp}, f_{U^\perp})| \leq M^q \alpha^{3-q} = M^3 \log_p^{1-3/q} \frac{1}{\eta}$$

and similarly for the other six $\Lambda_3()$ expressions to be estimated. The claim follows. □

Remark 10.21 The strategy of the above transference argument was to identify a fairly coarse partition of Z (in this case, into cosets of V) to average against in order to produce a well-behaved approximant $f_{U^\perp}$ to f, with the error f_U between f and $f_{U^\perp}$ being so uniform (in the Fourier sense) as to be negligible. This philosophy was developed in a quantitative manner in [150], in which an arithmetic version of the Szemerédi regularity lemma was obtained.

The hypothesis (10.7) in this Corollary may seem to be restrictive, but in many cases one can control the l^q norm of $\hat{f}$, or at least the spectrum $\mathrm{Spec}_\alpha(f)$ of f, by exploiting the pseudo-randomness properties of ν. For instance, one has

Lemma 10.22 (Tomas–Stein argument) *Let Z be a finite additive group, and let $\nu : Z \to \mathbf{R}^+$ and $f : Z \to \mathbf{C}$ be such that (10.8) holds for some $0 < \eta < 1$, and such that $|f(x)| \leq \nu(x)$ for all $x \in Z$. For any $\alpha > 0$ let $\mathrm{Spec}_\alpha(f) := \{\xi \in Z : |\hat{f}(\xi)| \geq \alpha\}$. Then we have*

$$|\mathrm{Spec}_\alpha(f)| \leq 4/\alpha^2$$

for all $\alpha \geq 2\eta^{1/2}$.

Remark 10.23 This estimate should be compared with (4.37); the point is that no L^2 bound on f is assumed, otherwise this type of estimate would follow from Plancherel's theorem. The orthogonality argument used here plays a fundamental role in the restriction theory of the Fourier transform, see for instance [356] for a survey. It is also closely related to the *large sieve inequality* in analytic number theory.

Proof For each $\xi \in \mathrm{Spec}_\alpha(f)$ let $c(\xi) := \mathrm{sgn}(\hat{f}(\xi))$. Then we have

$$\left| \sum_{\xi \in \mathrm{Spec}_\alpha(f)} \hat{f}(\xi)\overline{c(\xi)} \right| = \sum_{\xi \in \mathrm{Spec}_\alpha(f)} |\hat{f}(\xi)| \geq \alpha |\mathrm{Spec}_\alpha(f)|.$$

But the left-hand side can be rewritten as

$$\mathbf{E}_Z \left(f \overline{\sum_{\xi \in \mathrm{Spec}_\alpha(f)} c(\xi) e_\xi} \right).$$

Since $f \leq \nu$, we may use Cauchy–Schwarz and conclude that

$$\alpha|\mathrm{Spec}_\alpha(f)| \leq \mathbf{E}_Z(\nu)^{1/2}\mathbf{E}_Z\left(\nu\left|\sum_{\xi\in\mathrm{Spec}_\alpha(f)} c(\xi)e_\xi\right|^2\right)^{1/2}.$$

Since $\mathbf{E}_Z(\nu) = \hat{\nu}(0) \leq 1+\eta \leq 2$, we thus conclude that

$$\mathbf{E}_Z\left(\nu\left|\sum_{\xi\in\mathrm{Spec}_\alpha(f)} c(\xi)e_\xi\right|^2\right) \geq \frac{1}{2}\alpha^2|\mathrm{Spec}_\alpha(f)|^2.$$

We can expand the left-hand side as

$$\sum_{\xi,\xi'\in\mathrm{Spec}_\alpha(f)} c(\xi)\overline{c(\xi')}\mathbf{E}_Z(\nu e_\xi \overline{e_{\xi'}}) = \sum_{\xi,\xi'\in\mathrm{Spec}_\alpha(f)} c(\xi)\overline{c(\xi')}\hat{\nu}(\xi'-\xi).$$

But since $|c(\xi)| = 1$ and $|\hat{\nu}(\xi'-\xi)| \leq \eta + \mathbf{I}(\xi'-\xi=0)$, we conclude that

$$\begin{aligned}\frac{1}{2}\alpha^2|\mathrm{Spec}_\alpha(f)|^2 &\leq \sum_{\xi,\xi'\in\mathrm{Spec}_\alpha(f)} \big(\eta + \mathbf{I}(\xi'-\xi=0)\big)\\ &\leq \eta|\mathrm{Spec}_\alpha(f)|^2 + |\mathrm{Spec}_\alpha(f)|.\end{aligned}$$

Since $\alpha \geq 2\eta^{1/2}$, we have $\eta|\mathrm{Spec}_\alpha(f)|^2 \leq \frac{1}{4}\alpha^2|\mathrm{Spec}_\alpha(f)|^2$, and the claim follows. $\square$

We can now prove Theorem 10.18.

Proof of Theorem 10.18 We may assume that $|Z|$ is sufficiently large depending on δ, p since the claim is vacuous otherwise. We shall abbreviate $o_{|Z|\to\infty;p}(1)$ simply as $o(1)$. From Corollary 1.9 we have $\mathbf{P}_Z(B) = \tau + O(|Z|^{-1/5})$ (say) with probability $1-o(1)$; in particular B is non-empty. Also, if we set $\nu := 1_B/\tau$, then by Lemma 4.16 (with A replaced by Z) we have

$$\sup_{\xi\in Z\backslash 0} |\hat{\nu}(\xi)| = O\big(|Z|^{-1/5}\big)$$

again with probability $1-o(1)$. Combining this with our density bound on $\mathbf{P}_Z(B)$, we thus have

$$\sup_{\xi\in Z} |\hat{\nu}(\xi) - \mathbf{I}(\xi=0)| = O\big(|Z|^{-1/5}\big) \tag{10.11}$$

with probability $1-o(1)$. Henceforth we shall condition on these events.

Let $\delta = \delta(|Z|, p) < 1$ be a small quantity decaying to zero very slowly as $|Z| \to \infty$ (i.e. $\delta = o(1)$); it will suffice to show that for δ sufficiently slowly decaying, and conditioning on the previous events, every subset A of B with relative density $|A|/|B| \geq \delta$ will contain a proper arithmetic progression of length 3.

Set $f := 1_A/\tau$. Clearly f is non-negative and $f \leq \nu$. Also $\mathbf{E}_Z(f) \geq \delta \mathbf{P}_Z(B) = \Theta(\delta\tau)$. From Lemma 10.22 and (10.11) we have

$$|\mathrm{Spec}_\alpha(f)| \leq 4/\alpha^2 \text{ whenever } \alpha = o\big(|Z|^{-1/10}\big),$$

while from (4.2) we have the very crude bound $\|\hat{f}\|^2_{l^2(Z)} \leq \tau^{-2} \leq |Z|^{0.02}$. Combining these two estimates, we easily obtain

$$\|\hat{f}\|_{l^{5/2}(Z)} = O(1) \tag{10.12}$$

(for instance); see Exercise 10.2.1. Applying Theorem 10.20 (with $\eta := |Z|^{-1/5}$), we conclude

$$\Lambda_3(f, f, f) \geq 8p^{-12/\delta} - O_p\big(\log^{-1/5}|Z|\big).$$

On the other hand, if A contained no arithmetic progressions of length 3, then we would have

$$\Lambda_3(f, f, f) = \frac{|A|}{\tau^3|Z|^2} = O\left(\frac{1}{\tau^3|Z|}\right) = O(|Z|^{-0.97}),$$

which would lead to a contradiction if Z was large compared with δ, p, and the claim follows. □

We remark that the above argument is quite quantitative, and it is not difficult to use it to extract specific bounds for Theorem 10.18, but we will not do so here.

Exercises

10.2.1 Deduce (10.12) from the bounds on $\mathrm{Spec}_\alpha(f)$ and $\|\hat{f}\|_{l^2(Z)}$. (Hint: one can use an analog of (1.7).)

10.2.2 Show that Theorem 10.18 fails if $\tau = |Z|^{-1/2-\varepsilon}$ for any absolute constant ε. (Hint: count the number of proper progressions of length 3 in B, and remove them to create A.)

10.2.3 [248] Let $s(n, d)$ be the quantity defined in Section 9.6. Show that $s(3, d) = \Theta(r_3(F_3^n))$. In particular, we have $s(3, d) = O(3^d/d)$ for large d.

10.3 The integer case

We now sketch the proof of Roth's theorem for integers (which was the original setting for Roth's argument). We shall be somewhat brief here as the result will be superseded by the Roth–Bourgain theorem, Theorem 10.30.

As in the proof of Proposition 10.12, we need two ingredients; first, we need to show that lack of progressions in $[1, N]$ implies some linear bias, and second we need to convert this linear bias to a density increment on a sub-progression of $[1, N]$. Because $[1, N]$ is not quite a group, we cannot apply Proposition 10.10 directly. However we have the following substitute.

Proposition 10.24 (Lack of progressions implies non-uniformity) *[287] Let P be an arithmetic progression of integers, and let $A \subset P$ be such that $|A| = \delta|P|$ for some $0 < \delta \leq 1$. Assume also that $|P| \geq 100/\delta^2$, and that A contains no arithmetic progressions of length 3. Then there exists $\xi \in \mathbf{R}/\mathbf{Z}$ such that*

$$|\mathbf{E}_{n\in P}(1_A(n) - \delta)e(n\xi)| = \Omega(\delta^2).$$

Proof By a rescaling argument one can take $P = [1, N]$. By Bertrand's postulate (Exercise 1.10.3) we can find a prime p between $2N$ and $4N$. We identify A with a subset of $\mathbf{Z}_p$ in the usual manner (and give $\mathbf{Z}_p$ the standard bilinear form), and observe from (10.2) and the hypothesis on A that

$$\Lambda_3(1_A, 1_A, 1_A) = \frac{1}{p^2}|A| \leq \frac{\delta}{4N}.$$

Let us now split $1_A = f_U + f_{U^\perp}$, where $f_{U^\perp} := \delta 1_{[1,N]}$ and $f_U := 1_A - f_{U^\perp}$. A simple computation shows that

$$\Lambda_3(f_{U^\perp}, f_{U^\perp}, f_{U^\perp}) \geq \frac{\delta^3}{100}$$

(say). By hypothesis on N, we conclude

$$|\Lambda_3(f_U + f_{U^\perp}, f_U + f_{U^\perp}, f_U + f_{U^\perp}) - \Lambda_3(f_{U^\perp}, f_{U^\perp}, f_{U^\perp})| = \Omega(\delta^3).$$

The left-hand side can be split as the sum of seven terms, so at least one of them is $\Omega(\delta^3)$. For sake of discussion let us suppose that

$$|\Lambda_3(f_U, f_U, f_U)| = \Omega(\delta^3);$$

the other six cases are similar (the point being that all of them involve at least one copy of f_U). Using (10.6) and the triangle inequality, we conclude that

$$\sum_{\xi\in\mathbf{Z}_p} |\hat{f}_U(\xi)|^2|\hat{f}_U(-2\xi)| = \Omega(\delta^3).$$

On the other hand, from Plancherel's theorem we have

$$\sum_{\xi\in\mathbf{Z}_p} |\hat{f}_U(\xi)|^2 = \|f_U\|^2_{L^2(Z)} = O\big(\|1_A\|^2_{L^2(\mathbf{Z}_p)} + \|\delta 1_{[1,N]}\|^2_{L^2(\mathbf{Z}_p)}\big) = O(\delta).$$

We thus conclude that there exists $\xi \in \mathbf{Z}_p$ such that

$$|\hat{f}_U(-2\xi)| = \Omega(\delta^2),$$

thus

$$|\mathbf{E}_{n\in[1,N]}(1_A(n) - \delta)e(2n\xi/p)| = \Omega(\delta^2).$$

The claim follows. □

Similarly, we have the following analog of Lemma 10.15.

Lemma 10.25 (Non-uniformity implies density increment) *[287] Let $f : \mathbf{Z} \to \mathbf{R}$ be a function supported on an arithmetic progression P such that $|f(n)| \leq 1$ for all n, $\sum_n f(n) = 0$, and*

$$|\mathbf{E}_{n\in P} f(n)e(n\xi)| \geq \sigma$$

for some $\xi \in \mathbf{R}/\mathbf{Z}$ and $\sigma > 0$. Then there exists a proper arithmetic progression $P' \subset P$ with $|P'| = \Omega(\sigma^2|P|^{1/2})$ and.

$$|\mathbf{E}_{n\in P'} f(n)| \geq \sigma/4$$

Proof Again we may take $P = [1, N]$. Using the Kronecker approximation theorem (Corollary 3.25) we can find an integer $1 \leq r \leq N^{1/2}$ such that $\|r\xi\|_{\mathbf{R}/\mathbf{Z}} \leq N^{-1/2}$. Let P_0 denote the progression $[1, \sigma N^{1/2}/100] \cdot r$. Then we have

$$\left|\sum_n \mathbf{E}_{x\in P_0} f(n+x)e(n\xi)e(x\xi)\right| = \left|\sum_n f(n)e(n\xi)\right| \geq \sigma N,$$

where e is defined in equation (4.1). On the other hand, since $x \in P_0$, we see from (4.24) that $|e(x\xi) - 1| \leq \sigma/10$, and so

$$\left|\sum_n \mathbf{E}_{x\in P_0} f(n+x)e(n\xi)(e(x\xi) - 1)\right| \leq \sum_{n\in[-N,N]} \sigma/10 \leq \sigma N/2$$

(say), and so by the triangle inequality

$$\left|\sum_n \mathbf{E}_{x\in P_0} f(n+x)e(n\xi)\right| \geq \sigma N/2.$$

In particular there exists a phase $\theta \in \mathbf{R}/\mathbf{Z}$ such that

$$\mathrm{Re} \sum_n \mathbf{E}_{x\in P_0} f(n+x)e(n\xi + \theta) \geq \sigma N/2.$$

Since f sums to zero, we have $\sum_n \mathbf{E}_{x\in P_0} f(n+x) = 0$, and hence

$$\sum_n \mathbf{E}_{x\in P_0} f(n+x)\mathrm{Re}(1 + e(n\xi + \theta)) \geq \sigma N/2.$$

Note that the summand is only non-zero when $n \in (-N, N]$. By the pigeonhole principle, there thus exists an n such that

$$\mathbf{E}_{x\in n+P_0} f(x) = \mathbf{E}_{x\in P_0} f(n+x) \geq \sigma N/4.$$

Since f is bounded by 1 and supported in $[1, N]$, we conclude in particular that

$$|(n + P_0) \cap [1, N]| \geq \sigma |P_0|/4 = \Omega\big(\sigma^2 N^{1/2}\big).$$

The claim then follows by taking $P' = (n + P_0) \cap [1, N]$. □

Combining this with the preceding proposition, we conclude

Corollary 10.26 (Lack of progressions implies density increment) *Let $A \subset P$ be such that $|A| = \delta |P|$ for some $0 < \delta \leq 1$. Assume that $|P| \geq 100/\delta^2$. Suppose also that A contains no arithmetic progressions of length 3. Then there exists a proper arithmetic progression P' in P with $|P'| = \Omega(\delta^4 |P|^{1/2})$ such that we have the density increment*

$$\mathbf{P}_{P'}(A) \geq \mathbf{P}_P(A) + \Omega(\delta^2).$$

By iterating this Corollary, one can eventually show that $r_3([1, N]) = O(\frac{N}{\log \log N})$; we leave this as an exercise to the reader.

There has been some recent progress in understanding the structure of subsets of $\mathbf{Z}/N\mathbf{Z}$ which attain the minimal number of progressions of length 3 among all sets with a given density; see [65]. It may be that this will lead to an alternative proof of Roth's theorem.

Exercises

10.3.1 [287] By iterating Corollary 10.26, establish the bound $r_3(P) = O(\frac{N}{\log \log N})$ for any arithmetic progression P of integers of length N, and hence $r_3(\mathbf{Z}_N) = O(\frac{N}{\log \log N})$. (Hint. You may want to first establish the weaker but easier bound $r_3(P) = o(|P|)$.)

10.3.2 [372] Let $f : \mathbf{Z}_N \to \mathbf{R}^+$ be such that $0 \leq f(x) \leq 1$ for all $x \in \mathbf{Z}_N$. By using the previous exercise and arguing as in Proposition 10.17, show that

$$\Lambda_3(f, f, f) = \Omega(\exp(-\exp(O(1/\mathbf{E}_Z(f))))).$$

10.4 Quantitative bounds

In the preceding section we obtained a bound of $O(N/\log \log N)$ for the quantity $r_3([1, N])$. The main reason for this double logarithm lies in the use of Kronecker's theorem in Lemma 10.25, which reduces the size of the progression P by roughly a square root, while only increasing the density by a small amount $O(\delta^2)$. This step is so inefficient that it is worthwhile to make the other parts of the argument more complicated in order to reduce the number of times one invokes Kronecker's theorem. One such approach, due to Heath-Brown and Szemerédi, is to apply

Kronecker's theorem to a large batch of frequencies at once, rather than one at a time. It yields the following improvement[1]:

Theorem 10.27 *[177, 344] For all large N, we have $r_3(\mathbf{Z}_N), r_3([1, N]) = O(N/\log^c N)$ for some absolute constant $c > 0$.*

Proof It suffices to verify the claim for $r_3([1, N])$. We refine the arguments in the preceding section, again skipping some details. First we need the following variant of Proposition 10.24.

Proposition 10.28 (Lack of progressions implies non-uniformity) *Let $A \subset [1, N]$ be such that $|A| = \delta N$ for some $0 < \delta \leq 1$ and such that A has no proper arithmetic progressions of length 3. Suppose also that $N \geq 100/\delta^2$. Let p be a prime between N and $2N$, and identify $[1, N]$ with a subset of $\mathbf{Z}_p$. Let $f_U : \mathbf{Z}_p \to \mathbf{R}$ be the function $f_U := 1_A - \delta 1_{[1,N]}$. Then there exists a set $S \subset \mathbf{Z}_N$ such that $|S| = O(\delta^{-5})$ and*

$$\sum_{\xi \in S} |\hat{f}_U(\xi)|^2 = \Omega(\delta^2 |S|^{1/5}).$$

Proof Write $f_{U^\perp} := \delta 1_{[1,N]}$. Arguing as in Proposition 10.28, we conclude once again that

$$\sum_{\xi \in \mathbf{Z}_p} |\hat{f}_{U^\perp}(\xi)|^2 |\hat{f}_U(-2\xi)| = \Omega(\delta^3)$$

or something very similar to this. A direct calculation (which we leave as an exercise) also shows that

$$\sum_{\xi \in \mathbf{Z}_p} |\hat{f}_{U^\perp}(\xi)|^3 = O(\delta^3) \tag{10.13}$$

and hence by Hölder's inequality we have

$$\sum_{\xi \in \mathbf{Z}_p} |\hat{f}_U(\xi)|^3 = \Omega(\delta^3). \tag{10.14}$$

Now suppose for contradiction that

$$\sum_{\xi \in S} |\hat{f}_U(\xi)|^2 \leq c\delta^2 |S|^{1/5}$$

for all sets S and some small $c > 0$ to be chosen later. Applying this in particular to the set $S = \{\xi : \hat{f}_U(\xi) \geq \lambda\}$ for some arbitrary parameter λ, we see that

$$\lambda^2 |\{\xi : \hat{f}_U(\xi) \geq \lambda\}| \leq c\delta^2 |\{\xi : \hat{f}_U(\xi) \geq \lambda\}|^{1/5}$$

[1] We thank Ben Green for presenting these arguments to the authors.

and hence

$$|\{\xi : |\hat{f}_U(\xi)| \geq \lambda\}| \leq c^{5/4}\delta^{5/2}\lambda^{-5/2}.$$

Multiplying this by $3\lambda^2$ and integrating we obtain

$$3\int_0^\delta |\{\xi : |\hat{f}_U(\xi)| \geq \lambda\}|\lambda^2\, d\lambda = O\big(c^{5/4}\delta^3\big).$$

But one can easily verify (e.g. using (4.15)) that $|\hat{f}_U(\xi)| \leq \delta$, and so the left-hand side simplifies to $\sum_{\xi\in\mathbf{Z}_p} |\hat{f}_U(\xi)|^3$. But this will contradict (10.14) if c is sufficiently small. The claim follows. □

Now we need the following variant of Lemma 10.25.

Lemma 10.29 (Non-uniformity implies density increment) *[287] Let N and p be as in the preceding lemma. Let $f : \mathbf{Z}_p \to \mathbf{R}$ be a function supported on $[1, n]$ such that $|f(n)| \leq 1$ for all n, and such that*

$$\sum_{\xi\in S} |\hat{f}(\xi)|^2 \geq \sigma \tag{10.15}$$

for some set $S \subset \mathbf{Z}_p$ and some $\sigma > 0$. Then there exists a proper arithmetic progression $P' \subset [1, N]$ with $|P'| = \Omega(\sigma N^{1/(|S|+1)})$ and

$$|\mathbf{E}_{n\in P'} f(n)| = \Omega(\sigma).$$

Proof By Kronecker's theorem, we can find $1 \leq r \leq N^{1-\frac{1}{|S|+1}}$ such that $\|r\xi\|_{\mathbf{R}/\mathbf{Z}} \leq N^{-\frac{1}{|S|+1}}$ for all $\xi \in S$. Let Q be the progression $Q = [1, N^{\frac{1}{|S|+1}}/10] \cdot r$, then a simple computation shows that

$$\frac{1}{\mathbf{P}_{\mathbf{Z}_p}(Q)}|\hat{1}_Q(\xi)| = \Theta(1) \text{ for all } \xi \in S.$$

In particular, from (4.2), (4.9), (10.15) and the previous line we have

$$\left\| f * \frac{1}{\mathbf{P}_{\mathbf{Z}_p}(Q)} 1_Q \right\|^2_{L^2(Z)} = \sum_{\xi\in\mathbf{Z}_p} \frac{1}{\mathbf{P}_{\mathbf{Z}_p}(Q)^2} |\hat{1}_Q(\xi)|^2 |\hat{f}(\xi)|^2 = \Omega(\sigma).$$

On the other hand, from the boundedness of f we have

$$\left\| f * \frac{1}{\mathbf{P}_{\mathbf{Z}_p}(Q)} 1_Q \right\|_{L^1(Z)} \leq \|f\|_{L^1(Z)} \left\| \frac{1}{\mathbf{P}_{\mathbf{Z}_p}(Q)} 1_Q \right\|_{L^1(Z)} \leq 1.$$

Also, observe that the expression inside the norm has mean zero. Hence by Hölder's inequality we have

$$\sup_{x\in Z} f * \frac{1}{\mathbf{P}_{\mathbf{Z}_p}(Q)} 1_{Q(x)} = \Omega(\sigma).$$

Thus there exists $x \in Z$ such that

$$\mathbf{E}_{y \in x-Q} f(y) = \Omega(\sigma).$$

Setting $P' = [1, N] \cap (x - Q)$, the claim follows. □

The rest of the proof is similar to the arguments in the previous section and is left as an exercise. □

A further refinement was achieved by Bourgain [39], dispensing with the need for Kronecker's theorem altogether. The idea was to avoid using arithmetic progressions, but work entirely with Bohr sets, and in particular with *regular* Bohr sets. As a consequence, the following result was obtained, which seems to be very close to the limit of the Fourier-analytic method (it is in some sense the natural generalization of Proposition 10.12):

Theorem 10.30 (Roth–Bourgain theorem) *For additive groups Z of large finite odd order, we have $r_3(Z) = O(\sqrt{\frac{\log\log|Z|}{\log|Z|}}|Z|)$. In particular for all large N $r_3(\mathbf{Z}_N), r_3([1, N]) = O(\sqrt{\frac{\log\log N}{\log N}}N)$.*

This theorem follows easily from the following variant, which can be viewed as a generalization of Theorem 10.17:

Theorem 10.31 *For all additive groups Z of large finite odd order, and all $f : Z \to \mathbf{R}^+$ with $0 \le f(x) \le 1$, we have $\Lambda_3(f, f, f) = \Omega(\mathbf{E}_Z(f)^{-O(\mathbf{E}_Z(f)^2)})$.*

We leave the deduction of Theorem 10.30 from Theorem 10.31 as an exercise to the reader. To prove Theorem 10.31, the main tool shall be the following result, which is a substitute for Corollary 10.26.

Proposition 10.32 (Lack of progressions implies density increment) *Let Z be an additive group of large odd order, let* $\mathrm{Bohr}(S, \rho)$ *be a regular Bohr set of rank d, and let $f : Z \to \mathbf{R}^+$ be such that $0 \le f(x) \le 1$ and $\mathbf{E}_{x \in x_0 + \mathrm{Bohr}(S,\rho)} f(x) \ge \delta$ for some $x_0 \in Z$. Suppose also that*

$$\Lambda_3(f, f, f) \le \left(\left(\frac{\delta}{2d}\right)^{100} \rho\right)^d.$$

Then there exists a regular Bohr set $\mathrm{Bohr}(S', \rho')$ *of rank at most $d + 1$ and radius $\rho' \ge (\frac{\delta}{2d})^{31}\rho$ and an element $x_0' \in Z$ such that*

$$\mathbf{E}_{x \in x_0' + \mathrm{Bohr}(S',\rho')} f(x) \ge \delta + \delta^2/2^{10}.$$

The deduction of Theorem 10.31 from Proposition 10.32 is also straightforward, and is left as another exercise to the reader.

Proof By translation we may take $x_0 = 0$. By increasing δ if necessary we may assume $\mathbf{E}_{x \in \mathrm{Bohr}(S,\rho)} f(x) = \delta$. By reducing f to zero outside of $\mathrm{Bohr}(S, \rho)$, we may assume that f is supported on $\mathrm{Bohr}(S, \rho)$. Now suppose for sake of contradiction that

$$\mathbf{E}_{x \in x_0' + \mathrm{Bohr}(S',\rho')} f(x) < \delta + \delta^2/2^{10} \tag{10.16}$$

for all $x_0' \in Z$ and all Bohr sets $\mathrm{Bohr}(S', \rho')$ of rank at most $d + 1$ and radius at least $(\frac{\delta}{2d})^{31}\rho$.

By Lemma 4.25 we can find $0 < \rho_3 < \rho_2 < \rho_1 < \rho$ such that for each $j = 1, 2, 3$, we have[1]

$$\left(\frac{\delta}{2d}\right)^{10j+1} \rho \leq \rho_j \leq \left(\frac{\delta}{2d}\right)^{10j} \rho$$

and that $\mathrm{Bohr}(S, \rho_j)$ is regular. Note that the sets $\mathrm{Bohr}(S, \rho)$, $\mathrm{Bohr}(S, \rho_1)$, $\mathrm{Bohr}(S, \rho_2)$, $\mathrm{Bohr}(S, \rho_3)$ will differ in size by factors of $\delta^{O(d)}$, which will be too large for our application. Hence we shall have to keep careful track of the densities of each of these Bohr sets separately.

By hypothesis and a change of variable, we have

$$\mathbf{E}_{x,r \in Z} f(x - r) f(x) f(x + r) = \Lambda_3(f, f, f) \leq \left(\left(\frac{\delta}{2d}\right)^{100} \rho\right)^d;$$

in particular, from (4.25) we have

$$\mathbf{E}_{x,r \in Z} f(x - r) f(x) f(x + r) \leq \frac{\delta^3}{4} \mathbf{P}_Z(\mathrm{Bohr}(S, \rho)) \mathbf{P}_Z(\mathrm{Bohr}(S, \rho_1))$$

(say). Since f is non-negative, we can localize r to $\mathrm{Bohr}(S, \rho_1)$ and conclude

$$\mathbf{E}_{x \in Z; r \in \mathrm{Bohr}(S,\rho_1)} f(x - r) f(x) f(x + r) \leq \frac{\delta^3}{4} \mathbf{P}_Z(\mathrm{Bohr}(S, \rho)). \tag{10.17}$$

Write $f_{U^\perp} := \delta 1_{\mathrm{Bohr}(S,\rho)}$. From the symmetry of the above expression in r, one can verify the identity

$$\begin{aligned} &\mathbf{E}_{x \in Z; r \in \mathrm{Bohr}(S,\rho_1)} f(x - r) f(x) f(x + r) \\ &\quad = \mathbf{E}_{x \in Z; r \in \mathrm{Bohr}(S,\rho_1)} f_{U^\perp}(x - r) f(x) f_{U^\perp}(x + r) \\ &\qquad + \mathbf{E}_{x \in Z; r \in \mathrm{Bohr}(S,\rho_1)} (f - f_{U^\perp})(x - r) f(x) (f + f_{U^\perp})(x + r). \end{aligned} \tag{10.18}$$

[1] The reader should not take the numerical quantities (especially the powers of 2) too seriously in this argument; they are certainly not optimal.

Observe that if $x \in \mathrm{Bohr}(S, \rho - \rho_1)$ and $r \in \mathrm{Bohr}(S, \rho_1)$, then $x \pm r \in \mathrm{Bohr}(S, \rho)$, and therefore

$$\mathbf{E}_{r \in \mathrm{Bohr}(S,\rho_1)} f_{U^\perp}(x - r) f(x) f_{U^\perp}(x + r) = \delta^2 f(x).$$

Thus by positivity of f and $f_{U^\perp}$

$$\mathbf{E}_{x \in Z; r \in \mathrm{Bohr}(S,\rho_1)} f_{U^\perp}(x - r) f(x) f_{U^\perp}(x + r) \geq \delta^2 \mathbf{E}_{x \in Z} f(x) 1_{\mathrm{Bohr}(S,\rho-\rho_1)}(x).$$

By hypothesis we have

$$\mathbf{E}_{x \in Z} f(x) 1_{\mathrm{Bohr}(S,\rho)}(x) = \delta \mathbf{P}_Z(\mathrm{Bohr}(S, \rho))$$

while from regularity of $\mathrm{Bohr}(S, \rho)$ we have

$$\mathbf{E}_{x \in Z} f(x) 1_{\mathrm{Bohr}(S,\rho-\rho_1)}(x) \geq \frac{\delta}{2} \mathbf{P}_Z(\mathrm{Bohr}(S, \rho))$$

(say). Combining the above three estimates we obtain

$$\mathbf{E}_{x \in Z; r \in \mathrm{Bohr}(S,\rho_1)} f_{U^\perp}(x - r) f(x) f_{U^\perp}(x + r) 1_{\mathrm{Bohr}(S,\rho_1)}(r) \geq \frac{\delta^3}{2} \mathbf{P}_Z(\mathrm{Bohr}(S, \rho))$$

Combining this with (10.17), (10.18) we conclude that

$$|\mathbf{E}_{x \in Z; r \in \mathrm{Bohr}(S,\rho_1)} (f - f_{U^\perp})(x - r) f(x) (f + f_{U^\perp})(x + r)| \geq \frac{\delta^3}{4} \mathbf{P}_Z(\mathrm{Bohr}(S, \rho));$$

we shift this by r to obtain

$$|\mathbf{E}_{x \in Z; r \in \mathrm{Bohr}(S,\rho_1)} (f - f_{U^\perp})(x) f(x + r) (f + f_{U^\perp})(x + 2r)| \geq \frac{\delta^3}{4} \mathbf{P}_Z(\mathrm{Bohr}(S, \rho));$$

We would like to use this fact to deduce some linear bias in $f - f_{U^\perp}$. Unfortunately the constraint $r \in \mathrm{Bohr}(S, \rho_1)$ is not favorable (it localizes r to a smaller scale than x). To resolve this we need to localize the x variable to a smaller scale, namely ρ_2. To do this we write $x = y + z$ where z is restricted to $\mathrm{Bohr}(S, \rho_2)$, and conclude that

$$\begin{aligned}&|\mathbf{E}_{y \in Z; r \in \mathrm{Bohr}(S,\rho_1); z \in \mathrm{Bohr}(S,\rho_2)} (f - f_{U^\perp})(y + z) f(y + z + r) (f + f_{U^\perp})(y + z + 2r)| \\ &\quad \geq \frac{\delta^3}{4} \mathbf{P}_Z(\mathrm{Bohr}(S, \rho)).\end{aligned}$$

Observe that we may localize y to $\mathrm{Bohr}(S, \rho + \rho_2)$ since the expression inside the expectation vanishes otherwise. Since f is bounded and $\mathrm{Bohr}(S, \rho)$ is regular, the contribution of $\mathrm{Bohr}(S, \rho + \rho_2) \backslash \mathrm{Bohr}(S, \rho)$ can be crudely bounded by

$$\mathbf{P}_Z(\mathrm{Bohr}(S, \rho + \rho_2)) - \mathbf{P}_Z(\mathrm{Bohr}(S, \rho)) \leq \frac{\delta^3}{8} \mathbf{P}_Z(\mathrm{Bohr}(S, \rho - \rho_2))$$

(say). Thus we can restrict y to $\mathrm{Bohr}(S, \rho - \rho_2)$ and use the triangle inequality to obtain

$$\mathbf{E}_{y \in \mathrm{Bohr}(S,\rho-\rho_2)} F(y) \geq \frac{\delta^3}{8}$$

where

$$F(y) := |\mathbf{E}_{r \in \mathrm{Bohr}(S,\rho_1); z \in \mathrm{Bohr}(S,\rho_2)} (f - f_{U^\perp})(y+z) f(y+z+r)(f + f_{U^\perp})(y+z+2r)|.$$

Now that the position variable z is localized to a smaller scale than the shift variable r we may now remove the shift restriction $r \in \mathrm{Bohr}(S, \rho_1)$ as follows. We rewrite

$$F(y) = \frac{|\mathbf{E}_{z \in \mathrm{Bohr}(S,\rho_2)} \mathbf{E}_{r \in Z} 1_{\mathrm{Bohr}(S,\rho_1)}(r)(f - f_{U^\perp})(y+z) f(y+z+r)(f + f_{U^\perp})(y+z+2r)|}{\mathbf{P}_Z(\mathrm{Bohr}(S, \rho_1))}.$$

Now note that for each fixed y and each fixed $z \in \mathrm{Bohr}(S, \rho_2)$, the function

$$1_{\mathrm{Bohr}(S,\rho_1)}(r) - 1_{y+\mathrm{Bohr}(S,\rho_1)}(y+z+r) 1_{2\cdot\mathrm{Bohr}(S,\rho_1)}(y+z+2r)$$

has an $L^1(Z)$ norm in the r variable of at most $\mathbf{P}_Z(\mathrm{Bohr}(S, \rho_1 + 2\rho_2) \backslash \mathrm{Bohr}(S, \rho_1 - 2\rho_2))$, which by the regularity of $\mathrm{Bohr}(S, \rho_1)$ will be at most $\frac{\delta^3}{16} \mathbf{P}_Z(\mathrm{Bohr}(S, \rho_1))$. Using this and the boundedness of f, we see that if we write

$$\begin{aligned} \tilde{F}(y) &:= \frac{1}{\mathbf{P}_Z(\mathrm{Bohr}(S, \rho_1))} |\mathbf{E}_{z \in \mathrm{Bohr}(S,\rho_2)} \mathbf{E}_{r \in Z} \\ &\quad 1_{y+\mathrm{Bohr}(S,\rho_1)}(y+z+r) 1_{2\cdot\mathrm{Bohr}(S,\rho_1)}(y+z+2r) \\ &\quad (f - f_{U^\perp})(y+z) f(y+z+r)(f + f_{U^\perp})(y+z+2r)| \\ &= \frac{|\Lambda_3((f - f_{U^\perp}) 1_{y+\mathrm{Bohr}(S,\rho_2)}, f 1_{y+\mathrm{Bohr}(S,\rho_1)}, (f + f_{U^\perp}) 1_{y+2\cdot\mathrm{Bohr}(S,\rho_1)})|}{\mathbf{P}_Z(\mathrm{Bohr}(S, \rho_1) \mathbf{P}_Z(\mathrm{Bohr}(S, \rho_2))}, \end{aligned}$$

then $F(y)$ and $\tilde{F}(y)$ differ by at most $\delta^3/16$. In particular we have

$$\mathbf{E}_{y \in \mathrm{Bohr}(S,\rho-\rho_2)} \tilde{F}(y) \geq \frac{\delta^3}{16} \qquad (10.19)$$

At this point we need to pause to address a technical issue, namely that the function $(f - f_{U^\perp}) 1_{y+\mathrm{Bohr}(S,\rho_2)}$ may have non-zero mean. Fortunately this can be dealt with by the first moment method. Let $G(y)$ denote the function

$$G(y) := \mathbf{E}_{x \in y+\mathrm{Bohr}(S,\rho_2)} (f - f_{U^\perp}).$$

Since f and $f_{U^\perp}$ range between 0 and 1 and have the same mean, we see that G is bounded in magnitude by 1 and has mean zero. Also, $G(y)$ vanishes when $y \in \mathrm{Bohr}(S, \rho + \rho_2)$, while from (10.16) we see that $G(y)$ is bounded above by

$\delta^2/2^{10}$ when $y \in \mathrm{Bohr}(S, \rho - \rho_2)$. Since $\mathrm{Bohr}(S, \rho)$ is regular, we thus see that

$$\begin{aligned}\mathbf{E}_{x\in Z} \max(G(y), 0) &\le \frac{\delta^2}{2^{10}}\mathbf{P}_Z(\mathrm{Bohr}(S, \rho - \rho_2)) \\ &\quad + \mathbf{P}_Z(\mathrm{Bohr}(S, \rho + \rho_2)\backslash \mathrm{Bohr}(S, \rho - \rho_2)) \\ &\le \frac{\delta^2}{2^9}\mathbf{P}_Z(\mathrm{Bohr}(S, \rho - \rho_2)).\end{aligned}$$

Since $|G(y)| = G(y) + 2\max(G(y), 0)$, we thus have

$$\mathbf{E}_{x\in \mathrm{Bohr}(S,\rho-\rho_2)}|G(y)| \le \frac{1}{\mathbf{P}_Z(\mathrm{Bohr}(S, \rho - \rho_1))}\mathbf{E}_{x\in Z}|G(y)| \le \frac{\delta^2}{2^8};$$

we can combine this with (10.19) to obtain

$$\mathbf{E}_{y\in \mathrm{Bohr}(S,\rho-\rho_2)}\tilde{F}(y) - 8\delta|G(y)| \ge \frac{\delta^3}{32}$$

and thus there exists $y \in \mathrm{Bohr}(S, \rho - \rho_2)$ such that

$$\tilde{F}(y) \ge 8\delta|G(y)| + \frac{\delta^3}{32}.$$

We fix this y and return to the analysis of $\tilde{F}(y)$. From Proposition 10.11 we have

$$\begin{aligned}\tilde{F}(y) \le &\frac{1}{\mathbf{P}_Z(\mathrm{Bohr}(S, \rho_1)\mathbf{P}_Z(\mathrm{Bohr}(S, \rho_2))}\|(f - f_{U^\perp})1_{y+\mathrm{Bohr}(S,\rho_2)}\|_{u^2(Z)} \\ &\times \|f1_{y+\mathrm{Bohr}(S,\rho_1)}\|_{L^2(Z)}\|(f + f_{U^\perp})1_{y+2\cdot \mathrm{Bohr}(S,\rho_1)}\|_{L^2(Z)}.\end{aligned}$$

From (10.16) we have

$$\|f1_{y+\mathrm{Bohr}(S,\rho_1)}\|^2_{L^2(Z)} \le 2\delta\mathbf{P}_Z(\mathrm{Bohr}(S, \rho_1)))$$

and

$$\|(f + f_{U^\perp})1_{y+2\cdot \mathrm{Bohr}(S,\rho_1)}\|^2_{L^2(Z)} \le 8\delta\mathbf{P}_Z(\mathrm{Bohr}(S, \rho_1))).$$

Thus we have

$$\tilde{F}(y) \le \frac{4\delta}{\mathbf{P}_Z(\mathrm{Bohr}(S, \rho_2))}\sup_{\xi\in Z}|[(f - f_{U^\perp})1_{y+\mathrm{Bohr}(S,\rho_2)}]^\wedge(\xi)|.$$

Thus there exists $\xi \in Z$ such that

$$\frac{1}{\mathbf{P}_Z(\mathrm{Bohr}(S, \rho_2)}|[(f - f_{U^\perp})1_{y+\mathrm{Bohr}(S,\rho_2)}]^\wedge(\xi)| \ge 2|G(y)| + \frac{\delta^2}{128}.$$

Since $y \in \mathrm{Bohr}(S, \rho - \rho_2)$, we have $f_{U^\perp} = \delta$ on $y + \mathrm{Bohr}(S, \rho_2)$. We can therefore find a phase $\theta \in \mathbf{R}/\mathbf{Z}$ such that

$$\mathrm{Re}\mathbf{E}_{x\in y+\mathrm{Bohr}(S,\rho_2)}(f(x) - \delta)e(-\xi \cdot x + \theta) \ge 2|\mathbf{E}_{x\in y+\mathrm{Bohr}(S,\rho_2)}(f - \delta)| + \frac{\delta^2}{128}.$$

In particular, by the triangle inequality we have

$$\mathbf{E}_{x\in y+\mathrm{Bohr}(S,\rho_2)}(f(x)-\delta)[2+\mathrm{Re}\, e(-\xi\cdot x+\theta)] \geq \frac{\delta^2}{128}.$$

The only remaining task is to eradicate the multiplier $2+\mathrm{Re}\, e(-\xi\cdot x+\theta)$. This shall be done by replacing the Bohr set $\mathrm{Bohr}(S,\rho_2)$ with the narrower one $\mathrm{Bohr}(S',\rho_3)$, where $S' := S\cup\{\xi\}$. Writing $x = w+z$ where $z\in\mathrm{Bohr}(S',\rho_3)$, we see that

$$\mathbf{E}_{w\in Z;z\in\mathrm{Bohr}(S',\rho_3)}1_{\mathrm{Bohr}(S,\rho_2)}(w+z)(f(w+z)-\delta)$$
$$[2+\mathrm{Re}\, e(-\xi\cdot w+\theta)e(-\xi\cdot z)] \geq \frac{\delta^2}{128}\mathbf{P}_Z(\mathrm{Bohr}(S,\rho_2)).$$

Since $z\in\mathrm{Bohr}(S',\rho_3)$, we have $|e(-\xi\cdot z)-1|\leq 2\pi\rho_3$ by (4.24). It is then easy to replace $e(-\xi\cdot z)$ by 1 incurring an error of at most $\frac{\delta^2}{512}\mathbf{P}_Z(\mathrm{Bohr}(S,\rho_2))$ (say), concluding that

$$\mathbf{E}_{w\in Z;z\in\mathrm{Bohr}(S',\rho_3)}1_{\mathrm{Bohr}(S,\rho_2)}(w+z)(f(w+z)-\delta)$$
$$[2+\mathrm{Re}\, e(-\xi\cdot w+\theta)] \geq \frac{3\delta^2}{512}\mathbf{P}_Z(\mathrm{Bohr}(S,\rho_2)).$$

A similar argument (exploiting the regularity of $\mathrm{Bohr}(S,\rho_2)$) allows one to replace the cut-off $1_{\mathrm{Bohr}(S,\rho_2)}(w+z)$ by $1_{\mathrm{Bohr}(S,\rho_2)}(w)$, to obtain

$$\mathbf{E}_{w\in Z;z\in\mathrm{Bohr}(S',\rho_3)}1_{\mathrm{Bohr}(S,\rho_2)}(w)(f(w+z)-\delta)$$
$$[2+\mathrm{Re}\, e(-\xi\cdot w+\theta)] \geq \frac{\delta^2}{256}\mathbf{P}_Z(\mathrm{Bohr}(S,\rho_2))$$

which we rewrite as

$$\mathbf{E}_{w\in\mathrm{Bohr}(S,\rho_2)}[2+\mathrm{Re}\, e(-\xi\cdot w+\theta)](\mathbf{E}_{x\in w+\mathrm{Bohr}(S',\rho_3)}f(x)-\delta) \geq \frac{\delta^2}{256}.$$

On the other hand, from (10.16) and the bound $2+\mathrm{Re}\, e(-\xi\cdot w+\theta)\leq 3$ the left-hand side is bounded by $3\frac{\delta^2}{2^{10}}$, a contradiction. □

Exercises

10.4.1 Prove (10.13).

10.4.2 Complete the proof of Theorem 10.27 given Proposition 10.28 and Lemma 10.15.

10.4.3 Deduce Theorem 10.30 from Theorem 10.31.

10.4.4 Deduce Theorem 10.31 from Proposition 10.32. (Hint: use an iteration argument with about $O(1/\mathbf{E}_Z(f))$ steps, with parameter sizes $\delta = \Omega(\mathbf{E}_Z(f))$, $d = O(1/\mathbf{E}_Z(f))$ and $\rho = \Omega(\mathbf{E}_Z(f)^{O(1/\mathbf{E}_Z(f))})$ throughout the iteration.)

10.5 An ergodic argument

In 1977, Furstenberg [121] gave a spectacular new proof of Szemerédi's theorem (and hence Roth's theorem), using the methods of ergodic theory rather than Fourier analysis or combinatorics. The argument relies on very little arithmetic structure, being based almost entirely on an analysis of the mixing properties of the shift operator $TA := A + 1$ on a set A of integers. As such it is very flexible and has led to several wide-ranging generalizations of Szemerédi's theorem, some of which we will discuss in the next chapter.

The initial ergodic arguments of Furstenberg were infinitary in nature, working with the integers $\mathbf{Z}$, and in fact embedding these integers in an abstract measure-preserving system $(X, \mathcal{B}, T, \mu)$. In several versions of the argument, the axiom of choice (in the guise of Zorn's lemma) was used to obtain a suitable structural decomposition of this measure-preserving system. More recently, however, there has been progress in establishing *finitary* versions of this argument, in which one works in a concrete and finite measure-preserving system, such as the cyclic group $\mathbf{Z}_N$ with the standard shift $TA := A + 1$. These finitary arguments, which were inspired by the Szemerédi regularity lemma, to be introduced in the next section, are somewhat messier than the elegant infinitary arguments, but lead to explicit (albeit poor) quantitative bounds for $r_k(\mathbf{Z}_N)$. Also these finitary ergodic arguments played an essential role in the proof of the Green–Tao theorem concerning progressions in the primes.

In this section we give a finitary ergodic proof of Roth's theorem, using a formulation from [358]. The proof is not fully ergodic because we shall exploit the Fourier transform, but in the next chapter we will discuss how one can remove this dependence on the Fourier transform (and thus extend the argument to higher k). The precise result we shall establish is

Theorem 10.33 *For all finite groups Z, and all $f : Z \to \mathbf{R}^+$ with $0 \leq f(x) \leq 1$ and $\mathbf{E}_Z(f) \geq \delta$, we have $\Lambda_3(f, f, f) = \Theta_\delta(1)$.*

This is of course weaker than what one can obtain by purely Fourier-analytic methods such as Theorem 10.31, but the proof is somewhat different and is easier to extend to higher k. In particular, it replaces the density increment argument of previous sections by an *energy increment argument*. Whereas in the previous arguments one constructed a series of objects (progressions or Bohr sets) on which f had increasingly large density, here we construct a series of σ*-algebras* or *partitions* with respect to which f has increasingly large *energy*. This eventually leads to constructing a "low-complexity" approximation $f_{U^\perp}$ to f, where the error $f_U := f - f_{U^\perp}$ is linearly uniform and thus has negligible impact on $\Lambda_3(f, f, f)$.

The low-complexity approximation $f_{U^\perp}$ turns out to be *almost periodic*, which will lead to a lower bound on $\Lambda_3(f_{U^\perp}, f_{U^\perp}, f_{U^\perp})$.

We turn to the details, beginning with the definition of almost periodicity. For convenience we shall take advantage of the Fourier transform to define this notion, though it is not essential (see exercises).

Definition 10.34 (Almost periodicity) Let $K \geq 1$ be an integer and $\sigma > 0$. We say that a function $f : Z \to \mathbf{C}$ is K*-quasiperiodic* if there exist frequencies $\xi_1, \ldots, \xi_K \in Z$ (possibly repeated) and complex numbers $c_1, \ldots, c_k$ with $|c_1|, \ldots, |c_k| \leq 1$ such that $f = \sum_{j=1}^k c_j e_{\xi_j}$, or in other words

$$f(x) = \sum_{j=1}^{k} c_j e(x \cdot \xi_j).$$

We say that a function $f : Z \to \mathbf{C}$ is (K, σ)*-almost periodic* if there exists a K-quasiperiodic function g such that $\|f - g\|_{L^2(Z)} \leq \sigma$.

A key observation is that Theorem 10.33 is easy to prove for almost periodic functions, if K is not too large and σ is sufficiently small. More precisely, we have

Proposition 10.35 (Almost periodic functions are recurrent) *Let $f : Z \to \mathbf{R}^+$ be such that $0 \leq f(x) \leq 1$ and $\mathbf{E}_Z(f) \geq \delta$. If f is (K, σ)-almost periodic for some $K \geq 1$ and $0 < \sigma < \delta^3/8$, then*

$$\Lambda_3(f, f, f) = \Omega((\delta/K)^K \delta^3).$$

This proposition should be compared with Lemma 4.44. A key point here is that the smallness condition on σ does not involve K. This will be important for us as K will eventually be quite large compared with σ.

Proof By definition, we can find frequencies $\xi_1, \ldots, \xi_K$ and coefficients $c_1, \ldots, c_K$ of magnitude $O(1)$ such that

$$f(x) = \sum_{j=1}^{K} c_j e(x \cdot \xi_j) + g(x)$$

for all $x \in Z$, where g has an $L^2(Z)$ norm of at most σ. Now let $S := \{\xi_1, \ldots, \xi_K\}$ and let $\rho > 0$ be a radius to be chosen later. If h lies in the Bohr set $\mathrm{Bohr}_Z(S, \rho)$, then $e(h \cdot \xi_j) = 1 + O(\rho)$, and hence

$$T^{jh} f(x) = f(x) + O(K\rho) + T^{jh} g(x)$$

for $j = 1, 2$, where $T^h f(x) = f(x + h)$ denotes the shift by h. In particular we have

$$\|T^{jh} f - f\|_{L^2(Z)} \leq O(K\rho) + 2\sigma,$$

while from the boundedness of f we have $\|T^{jh} f\|_{L^\infty(Z)} \le 1$. After a few applications of the triangle inequality and Hölder's inequality, we then conclude that

$$\|f(T^h f)(T^{2h} f) - f^3\|_{L^1(Z)} \le O(K\rho) + 4\sigma$$

and hence by the triangle inequality again

$$\mathbf{E}_{x\in Z} f(x) T^h f(x) T^{2h} f(x) \ge \mathbf{E}_{x\in Z}(f(x)^3) - O(K\rho) - 4\sigma.$$

On the other hand, from Hölder's inequality we have

$$\mathbf{E}_{x\in Z} f(x)^3 \ge \mathbf{E}_{x\in Z}(f(x))^3 = \delta^3$$

so by hypothesis on σ

$$\mathbf{E}_{x\in Z} f(x) T^h f(x) T^{2h} f(x) \ge \frac{1}{2}\delta^3 - O(K\rho).$$

Applying (4.25) and the positivity of f, we conclude that

$$\Lambda_3(f, f, f) = \mathbf{E}_{x,h\in Z} f(x) T^h f(x) T^{2h} f(x) \ge \rho^K \max\left(\frac{1}{2}\delta^3 - O(K\rho), 0\right).$$

The claim then follows by taking ρ to be a sufficiently small multiple of δ/K. □

To establish Theorem 10.33 in the general case, one now needs to approximate an arbitrary function f by an almost periodic one. Indeed we will establish the following fundamental proposition:

Proposition 10.36 (Koopman–von Neumann decomposition) *Let $f : Z \to \mathbf{R}^+$ be such that $0 \le f(x) \le 1$, let $\sigma > 0$, and let $F : \mathbf{R}^+ \times \mathbf{R}^+ \to \mathbf{R}^+$ be an arbitrary function. Then there exists a quantity $K = O_{\sigma,F}(1)$ and a decomposition $f = f_{U^\perp} + f_U$ with the following properties:*

- *the "anti-uniform" component $f_{U^\perp}$ obeys the bounds $0 \le f_{U^\perp} \le 1$ and $\mathbf{E}_Z f_{U^\perp} = \mathbf{E}_Z f$, and is (K, σ)-almost periodic;*
- *the "uniform" component f_U obeys the Fourier uniformity estimate $\|f_U\|_{u^2(Z)} \le \frac{1}{F(\sigma,K)}$.*

A remarkable feature of this proposition is that one can make the uniformity control on f_U arbitrarily strong by making F grow arbitrarily quickly. The price one pays for this is that the upper bound on K then deteriorates substantially.

We shall prove Proposition 10.36 in the rest of this section. For now, let us see how the proposition implies Theorem 10.33. We apply the proposition with $\sigma := \delta^3/8$ and F to be chosen later. From Proposition 10.35 we have

$$\Lambda_3(f_{U^\perp}, f_{U^\perp}, f_{U^\perp}) = \Omega((\delta/K)^K \delta^3).$$

Since $f, f_{U^\perp}$ are bounded between 0 and 1, f_U is bounded in magnitude by 1. Applying the Fourier uniformity estimate and Proposition 10.11, we conclude that

$$\Lambda_3(f, f, f) - \Lambda_3(f_{U^\perp}, f_{U^\perp}, f_{U^\perp}) = O\left(\frac{1}{F(\sigma, K)}\right).$$

Thus if we choose F to be sufficiently quickly growing, we can absorb the error term into the main term and conclude that

$$\Lambda_3(f, f, f) = \Omega((\delta/K)^K \delta^3).$$

Since $K = O_{\sigma,F}(1) = O_\delta(1)$, the claim follows.

It remains to prove Proposition 10.36. One can prove this proposition by a direct application of the Fourier transform (this is essentially the approach in [34]); however we shall use a more ergodic approach which extends more easily to progressions of longer length. A crucial tool here is that of *conditional expectation*.

Definition 10.37 (Conditional expectation) Define a *σ-algebra* of Z to be any collection $\mathcal{B}$ of subsets of Z which contains $\emptyset$ and Z, and is closed under unions, intersections, and complements. (The σ-algebras are in one-to-one correspondence with partitions of Z, and can be viewed as such.) If $\mathcal{B}, \mathcal{B}'$ are two σ-algebras, we define $\mathcal{B} \vee \mathcal{B}'$ to be the smallest σ-algebra which contains both. We say that a function $f : Z \to \mathbf{C}$ is *measurable* with respect to $\mathcal{B}$ if it is constant on every atom of $\mathcal{B}$, where an *atom* is any minimal non-empty element of $\mathcal{B}$. Given any $f : Z \to \mathbf{C}$, we define the *conditional expectation* $\mathbf{E}(f|\mathcal{B}) : Z \to \mathbf{C}$ to be the function

$$\mathbf{E}(f|\mathcal{B})(x) := \mathbf{E}_{\mathcal{B}(x)} f = \frac{1}{|\mathcal{B}(x)|} \sum_{y \in \mathcal{B}(x)} f(y)$$

where $\mathcal{B}(x)$ is the unique atom of $\mathcal{B}$ which contains x; equivalently, $\mathbf{E}(f|\mathcal{B})$ is the orthogonal projection in $L^2(Z)$ to the space of $\mathcal{B}$-measurable functions.

It turns out that certain σ-algebras $\mathcal{B}$ are "compact" in the sense that conditional expectations such as $\mathbf{E}(f|\mathcal{B})$ are automatically almost periodic. One precise formulation of this is

Proposition 10.38 (Characters generate compact σ-algebras) *Let $\xi \in Z$ and $0 < \varepsilon < 1$. Then there exists a σ-algebra $\mathcal{B}_{\varepsilon,\xi}$ with $O_\varepsilon(1)$ atoms which approximately contains the character $e_\xi(x) := e(\xi \cdot x)$ in the sense that*

$$\|e_\xi - \mathbf{E}(e_\xi|\mathcal{B}_{\varepsilon,\xi})\|_{L^\infty(Z)} = O(\varepsilon), \tag{10.20}$$

and also has the property that every $\mathcal{B}_{\varepsilon,\xi}$-measurable function f with $\|f\|_{L^\infty(Z)} \le 1$ is $(O_{\varepsilon,\sigma}(1), O(\sigma))$-almost periodic for every $\sigma > 0$.

Proof We use the first moment method. Let α be a randomly selected element of the unit square $Q := \{z \in \mathbf{C} : 0 \leq \mathrm{Re}(z), \mathrm{Im}(z) < 1\}$, and let $\mathcal{B}_{\varepsilon,\xi}$ be the σ-algebra generated by the sets

$$A_{a,b,\varepsilon,\alpha} := \{x \in Z : e_\xi(x) \in \varepsilon(Q + a + bi + \alpha)\}; a, b \in \mathbf{Z}.$$

These sets, which partition Z, are essentially translates of the Bohr set $\mathrm{Bohr}_Z(\{\xi\}, \varepsilon)$; at most $O(1/\varepsilon)$ of them are non-empty. Since e_ξ fluctuates by at most $O(\varepsilon)$ on each such set, we obtain the property (10.20). Now we prove the latter property. Observe that f is a linear combination of at most $O(1/\varepsilon)$ indicator functions $1_{A_{a,b,\varepsilon,\alpha}}$, with bounded coefficients, so it suffices to prove the claim for the $O(1/\varepsilon)$ non-trivial indicator functions $1_{A_{a,b,\varepsilon,\alpha}}$. The claim is trivial for $\sigma \geq 1$, and by approximating σ by the nearest power of 2 we thus see that it suffices to verify the claim for $\sigma = 2^{-n}$ for integer $n \geq 0$. By the Borel–Cantelli lemma it will thus suffice to show that

$$\mathbf{P}(1_{A_{a,b,\varepsilon,\alpha}} \text{ is } (O_{\varepsilon,n}(1), O(2^{-n}))\text{-almost periodic}) = 1 - O(\varepsilon 2^{-n})$$

for each $n \geq 1$ and $a, b \in \mathbf{Z}$.

Fix n, a, b. We rewrite

$$1_{A_{a,b,\varepsilon,\alpha}}(x) = 1_Q\left(\frac{e(x \cdot \xi)}{\varepsilon} - a - bi - \alpha\right).$$

Let B be the $\varepsilon 2^{-3n}$-neighborhood of the boundary of the square Q. From Urysohn's lemma followed by the Weierstrass approximation theorem, we can write

$$1_Q(z) = P_{n,\varepsilon}(z) + O(1_B(z)) + O(2^{-n}),$$

where $P_{n,\varepsilon}(z)$ is a polynomial of z and $\overline{z}$ depending only on n. We conclude that

$$\begin{aligned} &1_{A_{a,b,\varepsilon,\alpha}}(x) \\ &= P_{n,\varepsilon}\left(\frac{e(x \cdot \xi)}{\varepsilon} - a - bi - \alpha\right) + O(\mathbf{I}(e(x \cdot \xi))/\varepsilon \in a + bi + \alpha + B) + O(2^{-n}). \end{aligned} \tag{10.21}$$

The first term on the right-hand side can be easily verified to be $O_{n,\varepsilon}(1)$-quasiperiodic. An application of the first moment method easily shows that

$$\mathbf{E}(\|\mathbf{I}(e(x \cdot \xi)/\varepsilon \in a + bi + \alpha + B\|^2_{L^2(Z)}) = O(\varepsilon 2^{-3n}),$$

so by Markov's inequality we see that the second term in (10.22) has an $L^2(Z)$ norm of $O(2^{-n})$ with probability $1 - O(\varepsilon 2^{-n})$. We thus see that $1_{A_{a,b,\varepsilon,\alpha}}$ is $(O_{n,\varepsilon}(1), O(2^{-n}))$-quasiperiodic with probability $1 - O(\varepsilon 2^{-n})$, as desired. □

One can extend this to the σ-algebra generated by multiple characters:

Corollary 10.39 *Let $\xi_1, \ldots, \xi_n \in Z$ and $\varepsilon_1, \ldots, \varepsilon_n > 0$. Let $\mathcal{B} := \mathcal{B}_{\varepsilon_1,\xi_1} \vee \cdots \vee \mathcal{B}_{\varepsilon_n,\xi_n}$, where $\mathcal{B}_{\varepsilon,\xi}$ was defined in the previous proposition. Then every $\mathcal{B}$-measurable function f with $\|f\|_{L^\infty(Z)} \leq 1$ is $(O_{\varepsilon_1,\ldots,\varepsilon_n,n,\sigma}(1), O_n(\sigma))$-almost periodic for every $\sigma > 0$.*

Proof Observe that $\mathcal{B}$ has at most $O_{n,\varepsilon_1,\ldots,\varepsilon_n}(1)$ atoms and so it suffices to verify the claim for an indicator $f = 1_A$, where A is an atom of $\mathcal{B}$. But 1_A is then a product of n indicators $1_{A_1} \ldots 1_{A_n}$, where A_j is an atom of $\mathcal{B}_{\varepsilon,\xi_j}$, and the claim then follows from the previous proposition and the observation that the product of bounded almost periodic functions remains bounded and almost periodic (but with slightly worse constants). □

The heart of the proof of Proposition 10.36 now lies in the following key lemma. We define the *energy* $\mathcal{E}_f(\mathcal{B})$ of $\mathcal{B}$ with respect to f to be the quantity

$$\mathcal{E}_f(\mathcal{B}) := \|\mathbf{E}(f|\mathcal{B})\|^2_{L^2(Z)} = \mathbf{E}_{x\in Z}|\mathbf{E}(f|\mathcal{B})(x)|^2.$$

Lemma 10.40 (Lack of uniformity implies energy increment) *Let $\varepsilon, \mu > 0$ be such that $\varepsilon \leq \mu/4$, and let $f : Z \to \mathbf{R}^+$ be such that $0 \leq f(x) \leq 1$, and let $\mathcal{B}$ be a σ-algebra such that*

$$\|f - \mathbf{E}(f|\mathcal{B})\|_{u^2(Z)} \geq \mu.$$

Then there exists a frequency $\xi \in Z$ such that we have the energy increment property

$$\mathcal{E}_f(\mathcal{B} \vee \mathcal{B}_{\varepsilon,\xi}) \geq \mathcal{E}_f(\mathcal{B}) + \mu^2/4.$$

Proof By definition of $u^2(Z)$, we can find $\xi \in Z$ such that

$$|\langle f - \mathbf{E}(f|\mathcal{B}), e_\xi\rangle_{L^2(Z)}| \geq \mu.$$

On the other hand, from (10.20) we have we see that e_ξ fluctuates by at most 2ε on each atom of $\mathcal{B}_{\varepsilon,\xi}$, and hence on each atom of $\mathcal{B} \vee \mathcal{B}_{\varepsilon,\xi}$. Thus

$$\|e_\xi - \mathbf{E}(e_\xi|\mathcal{B} \vee \mathcal{B}_{\varepsilon,\xi})\|_{L^\infty(Z)} \leq 2\varepsilon;$$

since $f - \mathbf{E}(f|\mathcal{B})$ is bounded in magnitude by 1, we conclude

$$|\langle f - \mathbf{E}(f|\mathcal{B}), e_\xi - \mathbf{E}(e_\xi|\mathcal{B} \vee \mathcal{B}_{\varepsilon,\xi})\rangle_{L^2(Z)}| \leq 2\varepsilon.$$

Since $\varepsilon < \mu/4$, we deduce

$$|\langle f - \mathbf{E}(f|\mathcal{B}), \mathbf{E}(e_\xi|\mathcal{B} \vee \mathcal{B}_{\varepsilon,\xi})\rangle_{L^2(Z)}| \geq \mu/2.$$

From the easily verified identity

$$\langle f - \mathbf{E}(f|\mathcal{B} \vee \mathcal{B}_{\varepsilon,\xi}), \mathbf{E}(e_\xi|\mathcal{B} \vee \mathcal{B}_{\varepsilon,\xi})\rangle_{L^2(Z)} = 0$$

we thus have

$$|\langle \mathbf{E}(f|\mathcal{B} \vee \mathcal{B}_{\varepsilon,\xi}) - \mathbf{E}(f|\mathcal{B}), \mathbf{E}(e_\xi|\mathcal{B} \vee \mathcal{B}_{\varepsilon,\xi})\rangle_{L^2(Z)}| \geq \mu/2$$

and hence by Cauchy–Schwarz

$$\|\mathbf{E}(f|\mathcal{B} \vee \mathcal{B}_{\varepsilon,\xi}) - \mathbf{E}(f|\mathcal{B})\|_{L^2(Z)}^2 \geq \mu^2/4.$$

The claim then follows from Pythagoras' theorem. □

We now have enough tools to prove Proposition 10.36.

Proof of Proposition 10.36 We construct a nested pair of σ-algebras $\mathcal{B} \subset \mathcal{B}'$ and an integer $K \geq 1$ by the following double-loop algorithm.

- Step 0. Initialize $\mathcal{B} = \{\emptyset, Z\}$.
- Step 1. Let K be the smallest integer such that $\mathbf{E}(f|\mathcal{B})$ is $(K, \sigma/2)$-almost periodic. (Note from the Fourier inversion formula that K is finite.) Set $\mathcal{B}' := \mathcal{B}$; thus we trivially have $\mathcal{E}_f(\mathcal{B}') \leq \mathcal{E}_f(\mathcal{B}) + \sigma^2/4$.
- Step 2. If

$$\|f - \mathbf{E}(f|\mathcal{B}')\|_{u^2(Z)} \leq \frac{1}{F(\sigma, K)}$$

then we terminate the algorithm. If not, then we can apply Lemma 10.40 with $\varepsilon := \frac{1}{4F(\sigma,K)}$ to obtain a new σ-algebra $\mathcal{B}'' := \mathcal{B}' \vee \mathcal{B}_{\varepsilon,\xi}$ for some $\xi \in Z$ such that

$$\mathcal{E}_f(\mathcal{B}'') \geq \mathcal{E}_f(\mathcal{B}') + \frac{1}{4F(\sigma, K)^2}.$$

- Step 3. If we have

$$\mathcal{E}_f(\mathcal{B}'') \leq \mathcal{E}_f(\mathcal{B}) + \sigma^2/4$$

then we set $\mathcal{B}' := \mathcal{B}''$ and return to Step 2. If instead we have

$$\mathcal{E}_f(\mathcal{B}'') > \mathcal{E}_f(\mathcal{B}) + \sigma^2/4$$

then we set $\mathcal{B} = \mathcal{B}''$ and return to Step 1.

Observe that every time we return from Step 3 to Step 2, the energy $\mathcal{E}_f(\mathcal{B}')$ increases by at least $\frac{1}{4F(\sigma,K)^2}$, while K does not change. On the other hand, since f is bounded, $\mathcal{E}_f(\mathcal{B}')$ varies between 0 and 1. Thus we can only return from Step 3 to Step 2 at most $4F(\sigma, K)^2$ times before either terminating or returning to Step 1. Now, every time one returns from Step 3 to Step 1, the energy $\mathcal{E}_f(\mathcal{B})$ increases by at least $4/\sigma^2$, so one can only return from Step 3 to Step 1 at most $4/\sigma^2$ times. Thus this algorithm terminates after a finite number of steps. If we then set $f_{U^\perp} := \mathbf{E}(f|\mathcal{B}')$ and $f_U := f - \mathbf{E}(f|\mathcal{B}')$ we have $f = f_U + f_{U^\perp}$, that $\|f_U\|_{u^2(Z)} \leq \frac{1}{F(\sigma,K)}$, that

$0 \leq f_{U^\perp} \leq 1$, and $\mathbf{E}_Z f_{U^\perp} = \mathbf{E}_Z f$. Finally, from construction we have $\mathcal{E}_f(\mathcal{B}') \leq \mathcal{E}_f(\mathcal{B}) + \sigma^2/4$ and hence by Pythagoras's theorem $\|f_{U^\perp} - \mathbf{E}(f|\mathcal{B})\|_{L^2(Z)} \leq \sigma/2$. Since $\mathbf{E}(f|\mathcal{B})$ is $(K, \sigma/2)$-almost periodic by construction, we conclude that $f_{U^\perp}$ is (K, σ)-almost periodic.

The only remaining thing to verify is that $K = O_{\sigma,F}(1)$. Observe that at every stage, $\mathcal{B}$ and $\mathcal{B}'$ are the join of a finite number of σ-algebras of the form $\mathcal{B}_{\varepsilon,\xi}$. In particular, Corollary 10.39 applies to these σ-algebras. An easy induction argument then shows that at every stage of the iteration, $\mathcal{B}$ and $\mathcal{B}'$ are the join of at most $O_{\sigma,F}(1)$ σ-algebras, that the parameters ε involved are bounded from below by $\Omega_{\sigma,F}(1)$, and the parameter K is always bounded above by $O_{\sigma,F}(1)$. The claim follows. □

Exercises

10.5.1 Let $f, g : Z \to \mathbf{C}$ be functions bounded in magnitude by 1 which are both (K, σ)-almost periodic for some $0 < \sigma \leq 1$. Show that $f + g$ is $(2K, 2\sigma)$-almost periodic, and that fg is $(K^2, 4\sigma)$-almost periodic.

10.5.2 Let $f : Z \to \mathbf{C}$ be (K, σ)-almost periodic. Show that one can cover the set $\{T^h f : h \in Z\} \subset L^2(Z)$ by at most $O_{K,\sigma}(1)$ balls of radius 2σ in the $L^2(Z)$ metric. Conclude that

$$\mathbf{P}_{h\in Z}\big(\|T^h f - f\|_{L^2(Z)} \leq 4\sigma\big) = \Theta_{K,\sigma}(1),$$

which may help explain the terminology "almost periodic". For a converse to this result, see Exercise 10.5.5 below.

10.5.3 Let $\xi_1, \ldots, \xi_n$ be a dissociated subset of Z. Using Rudin's inequality (Lemma 4.33), show that

$$\mathbf{P}_{h\in Z}\left(\sum_{j=1}^{n} |e(\xi_j \cdot h) - 1|^2 < n\right) \leq \exp(-\Omega(n)). \qquad (10.22)$$

10.5.4 Let $f : Z \to \mathbf{C}$ be such that $\|f\|_{L^2(Z)} = \|\hat{f}\|_{l^2(Z)} \geq 4\sigma$ and $\|\hat{f}\|_{l^\infty}(Z) \leq \delta\sigma$ for some $\sigma, \delta > 0$. Establish the bound

$$\mathbf{P}_{h\in Z}\left(\sum_{\xi\in Z} |e(\xi \cdot h) - 1|^2 |\hat{f}(\xi)|^2 < 4\sigma^2\right) = O(\delta^c)$$

for some absolute constant $c > 0$. (Hint: normalize $\|\hat{f}\|_{l^2(Z)} = 1$, so $\sigma \leq 1/4$, and then let $\xi_1, \ldots, \xi_n$ be independent identical random variables with probability distribution $|\hat{f}(\xi)|^2$. Show that $\xi_1, \ldots, \xi_n$ are dissociated with probability $1 - O(2^n\delta)$, and apply (10.22) combined with the first moment method. Then optimize in n.)

10.5.5 Let $f : Z \to \mathbf{C}$ be normalized so that $\|f\|_{L^2(Z)} = \|\hat{f}\|_{l^2(Z)} = 1$. Suppose that one can cover the set $\{T^h f : h \in Z\} \subset L^2(Z)$ by M balls of radius σ in the $L^2(Z)$ metric. Show that f is $(O_{M,\sigma}(1), 4\sigma)$-almost periodic. (Hint: use the pigeonhole principle and the Fourier transform to establish a lower bound for

$$\mathbf{P}_{h \in Z}\left(\sum_{\xi \in Z} |e(\xi \cdot h) - 1|^2 |\hat{f}(\xi)|^2 < 4\sigma^2\right).$$

Remove the K largest Fourier coefficients from f, for some $K = O_{M,\sigma}(1)$ to be chosen later, and apply the previous exercise to conclude an upper bound on the l^2 norm of the remaining Fourier coefficients.) This result, combined with Exercise 10.5.2, gives a way to define almost periodicity purely in terms of the precompactness of the orbit $\{T^h f : h \in Z\}$, without explicit mention of the Fourier transform.

10.6 The Szemerédi regularity lemma

In the original proof of Szemerédi's theorem (Theorem 10.1), Szemerédi introduced an important result in graph theory, the *Szemerédi regularity lemma*. This lemma has since become one of the main tools in discrete mathematics. It asserts, roughly speaking, that any dense large graph can be decomposed into a relatively small number of disjoint subgraphs, most of which behave pseudo-randomly. A more "ergodic" way of viewing the lemma is as an assertion that the indicator function of a graph can be decomposed into a "low-complexity" component and a "pseudo-random" component.

To state the lemma, we need some notation.

Definition 10.41 (**ε-regularity**) Let $G(V, E)$ be a graph. If X, Y are disjoint non-empty subsets of V, we define the *edge density* $d(X, Y)$ between X and Y to be the quantity

$$d(X, Y) := \mathbf{P}_{x \in X, y \in Y}(\{x, y\} \in E).$$

If $\epsilon > 0$, we say that the pair (X, Y) is *ϵ-regular* if we have

$$|d(X', Y') - d(X, Y)| \leq \epsilon$$

whenever $X' \subseteq X$, $Y' \subseteq Y$ are such that $|X'| \geq \epsilon|X|$ and $|Y'| \geq \epsilon|Y|$.

A partition $V = V_1 \cup V_2 \cup \cdots \cup V_k$ is *near-uniform* if $-1 \leq |V_i| - |V_j| \leq 1$. Szemerédi's Regularity Lemma asserts that given a positive constant ϵ and a graph G, one can find a near-uniform partition of V in not too many parts so that most of the pairs (V_i, V_j) are ϵ-regular.

Lemma 10.42 *(Regularity Lemma) Let ϵ be a positive constant, $m \geq 1$ an integer, and $G = G(V, E)$ a graph. If $|V|$ is sufficiently large depending on ε and m, then there exists a near-uniform partition $V = V_1 \cup \cdots \cup V_k$ for some $m \leq k \leq O_{\epsilon,m}(1)$. such that all but at most ϵk^2 of the pairs (V_i, V_j) are ϵ-regular.*

Remark 10.43 The Regularity Lemma does not assert that all pairs (V_i, V_j) are regular, only that $(1 - \varepsilon)$ of the pairs are. In fact, there are examples showing that one cannot expect regularity of all the pairs (Exercise 10.6.5).

Remark 10.44 The theorem requires $|V|$ to be large depending on ε and m, or to put it another way, one needs $\varepsilon = o_{|V|\to\infty;m}(1)$. The proof of the Regularity Lemma allows us to have $\epsilon = O_m(\frac{1}{(\log_* |V|)^{1/5}})$, where $\log_*$ is the inverse to the tower exponential $n \mapsto e \uparrow\uparrow n$, defined recursively by $e \uparrow\uparrow 1 = e$ and $e \uparrow\uparrow (n+1) := \exp(e \uparrow n)$. Quite amazingly, Gowers has shown that this bound is essentially tight, namely, for any sufficiently large $|V|$, there are graphs where one cannot find an ϵ-regular partition with ϵ larger than $\epsilon = \frac{1}{(\log_* |V|)}$.

The proof of the regularity lemma can be found in various textbooks on graph theory; in Section 11.6 we shall give a proof of this lemma using "ergodic" techniques similar to that of the previous section. See also [359] for an information-theoretic perspective on the lemma, and [239] for an analytic perspective.

The survey paper [208] contains a wide range of applications of the regularity lemma. In this section, we restrict ourself to a few applications in additive combinatorics, and in particular to Roth's theorem.

To prove Roth's theorem via the regularity lemma, it is convenient to first prove some graph-theoretic results. Let $G = G(V, E)$ be a graph. A set $\{e_1, \ldots, e_k\}$ in E forms a *matching* if $e_1, \ldots, e_k$ are mutually disjoint. A matching is *induced* if the subgraph spanned by its endpoint does not contain any edge other than those already in the matching.

Proposition 10.45 *[304] Let $G = G(V, E)$ be a graph whose edge set is the union of $|V|$ induced matchings. Then $|E| = o_{|V|\to\infty}(|V|^2)$.*

Proof The strategy will be to apply the regularity lemma, combined with the intuitive fact that a dense ε-regular graph cannot support any large induced matchings.

Assume that the proposition failed. Then one could find an integer $m \geq 1$ and arbitrarily large graphs $G(V, E)$ with $|E| \geq \frac{6}{m}|V|^2$ (say) such that each of the graphs G was the union of $|V|$ induced matchings.

Fix one of these large graphs. Applying the regularity lemma (with $\varepsilon := 1/m$) we obtain a partition $V = V_1 \cup \cdots \cup V_k$ with $m \leq k \leq O_m(1)$ with $|V_i| = \frac{1}{k}|V| + O(1)$ for all i, j, and such that all but at most $\frac{1}{m}k^2$ of the pairs (V_i, V_j) are $\frac{1}{m}$-regular.

Call an edge e of G *bad* if one of the following three events occurs:

- e is contained in one of the V_i;
- e connects V_i to V_j, where $d(V_i, V_j) \leq \frac{2}{m}$;
- e connects V_i to V_j, where (V_i, V_j) is not $\frac{1}{m}$-regular.

One can easily verify that the total number of bad edges is at most

$$(1 + o_{|V|\to\infty;m}(1)) \left(k\binom{|V|/k + O(1)}{2} + \frac{2}{m}\binom{k}{2}\frac{|V|^2}{k^2} + \frac{1}{m}k^2\frac{|V|^2}{k^2}\right) \leq \frac{3}{m}|V|^2,$$

if V is large enough depending on m. Thus if we let $E' \subset E$ be the edges of E that are not bad, we still have $|E'| \geq \frac{3}{m}|V|^2$. By the pigeonhole principle, we can thus find an induced matching F of G which contains at least $\frac{3}{m}|V|$ edges from E'.

Call a set V_i *poor* if it contains at most $\frac{2}{m}|V_i|$ vertices from F. If we delete all the poor sets V_i (and their associated edges) from F, we will have deleted at most $\frac{2}{m}|V|$ edges in all. Thus the remaining matching F' will still contain an edge from E'. By definition, this edge connects two distinct sets V_i, V_j which are not poor, which have edge density at least $\frac{2}{m}$, and is $\frac{1}{m}$-regular. If we let $V_{i,F}$ and $V_{j,F}$ be the vertices from F in V_i, V_j respectively, we thus have

$$d(V_{i,F}, V_{j,F}) \geq d(V_i, V_j) - \frac{1}{m} \geq \frac{1}{m}.$$

On the one hand, since F is an induced matching, the number of edges in $V_{i,F}$ and $V_{j,F}$ cannot exceed $|V_{i,F}|$, and so the edge density cannot exceed $1/|V_{j,F}|$. We conclude that

$$|V_{j,F}| \leq m.$$

On the other hand, we have $|V_{j,F}| \geq \frac{2}{m}|V_j|$ (since V_j is not poor) and $|V_j| = \frac{1}{k}|V| + O(1)$. We conclude that $|V| = O_{m,k}(1) = O_m(1)$, contradicting the hypothesis that V could be arbitrarily large depending on m. The claim follows. □

There are several equivalent formulations of the above theorem; see the exercises. A slightly stronger version of the theorem is as follows.

Lemma 10.46 (Triangle removal lemma) *[304] Let $G = G(V, E)$ be a graph which contains at most $\delta|V|^3$ triangles. Then it is possible to remove $o_{\delta\to 0}(|V|^2)$ edges from G to obtain a graph which is triangle-free (it contains no triangles whatsoever).*

Lemma 10.46 can be proven by the same method used to prove Proposition 10.45 and is left as an exercise. In fact one can easily use Lemma 10.46 to deduce Proposition 10.45.

Now we use Proposition 10.45 to give another proof of Roth's theorem, Theorem 10.8.

Proof Fix a finite additive group Z of odd order, and a subset A of Z which contains no arithmetic progressions. It suffices to show that $|A| = o_{|Z|\to\infty}(|Z|)$.

We define a bipartite graph G as follows. The color classes are the sets $Z \times \{1\}$ and $Z \times \{2\}$. We draw an edge between $(a + r, 1)$ and $(a + 2r, 2)$ for every $a \in Z$ and $r \in A$. For each $a \in Z$, the edges between $(a + r, 1)$, $(a + 2r, 2)$ for $r \in A$ form a matching. We claim that this matching is induced. For, if there was another edge connecting $(a + r, 1)$ with $(a + 2s, 2)$ for some distinct $r, s \in A$, then by construction we would have $2s - r \in A$. But then $r, s, 2s - r$ would be a proper progression of length three in A, a contradiction. Thus G is the union of $|Z|$ induced matchings, and hence has at most $o_{|Z|\to\infty}(|Z|^2)$ edges. Since the number of edges in G is clearly $|A||Z|$, the claim follows. □

In fact, the above methods yield the following stronger form of Roth's theorem.

Proposition 10.47 *[3] Let Z be a finite additive group, and let $A \subset Z \times Z$ be such that A contains no "right-angled triangles" $(a, b), (a, b + r), (a + r, b)$ with $a, b, r \in Z$ and $r \neq 0$. Then $|A| = o_{|Z|\to\infty}(|Z|^2)$.*

We leave the proof of Proposition 10.47 (and its connection to Roth's theorem) to the exercises.

It is of interest to obtain more quantitative bounds for the $o()$ terms in the above results. By using an explicitly quantitative formulation of the regularity lemma, one can sharpen the $o_{|V|\to\infty}(|V|^2)$ expression in Proposition 10.45 to $O(|V|^2/(\log_* |V|)^{1/5})$, and similarly for Lemma 10.46, Roth's theorem and Proposition 10.47. Thus the quantitative bounds achieved by this method compare poorly to that achieved by the Fourier method. (However, the graph-theoretical method is slightly easier to extend to the case of general k; see the next chapter.) Given that the bounds of Roth's theorem are significantly better than what is achieved by the regularity lemma, one is then naturally led to ask the following question:

Question 10.48 *[139] Prove Proposition 10.45 (or Lemma 10.46) without using the Regularity Lemma. Find a better quantitative bound.*

In the case of Proposition 10.47, there has been some recent progress on this question [381], [314]. In particular, the best known bound here is $A = O(\frac{|Z|}{\log\log|Z|})$, due to Shkredov [314].

Exercises

10.6.1 [304] Show that Proposition 10.45 is equivalent to the following statement: Let $G(V, E)$ be a graph such that each edge is contained in at most one triangle. Then $|E| = o_{|V|\to\infty}(|V|^2)$.

10.6.2 ((6, 3)-theorem) [304] Show that Proposition 10.45 is equivalent to the following statement: let $G = G(V, E)$ be a 3-uniform hypergraph (thus each "edge" in E is a collection $\{x, y, z\}$ of three vertices in V) such that there is no set of six vertices in V which contain three or more edges in E. Then $|E| = o_{|V|\to\infty}(|V|^2)$.

10.6.3 [304] Show that Lemma 10.46 implies Proposition 10.45. (Hint: first reduce to the case of a bipartite graph which is the union of induced matchings. Add $|V|$ additional vertices to the graph, one for each induced matching, and connect each new vertex to all the vertices in an induced matching. This creates a tripartite graph with rather few triangles, but which requires many edges to be removed in order to make it triangle-free.)

10.6.4 [304]Use the regularity lemma to prove Lemma 10.46.

10.6.5 [8] Let $V_1 = \{v_1, \ldots, v_n\}$, $V_2 = \{w_1, \ldots, w_n\}$ be disjoint collections of vertices, let $V := V_1 \cup V_2$, and let $G = G(V_1, V_2, E)$ be the bipartite graph formed from all those edges $\{v_i, w_j\}$ for which $i \leq j$. Use this to show that even for very simple graphs one must require an exceptional set of pairs (V_i, V_j) which is not regular.

10.6.6 By modifying the proof of Roth's theorem, use Proposition 10.45 to prove Proposition 10.47.

10.6.7 [323] Use Lemma 10.46 to prove Proposition 10.47, without going through Proposition 10.45. (Hint: consider a graph whose vertices are the vertical lines $\{(a, b) : a = \text{const}\}$, horizontal lines $\{(a, b) : b = \text{const}\}$ and diagonal lines $\{(a, b) : a + b = \text{const}\}$ in Z^2, and with two vertices connected by an edge if their associated lines have distinct orientations and intersect in a point in A.)

10.6.8 Show that Proposition 10.47 implies Roth's theorem. (Hint: if $A \subseteq Z$, consider sets of the form $\{(a, b) \in Z \times Z : a + 2b \in A\}$.)

10.6.9 [136] Let V_1, V_2 be disjoint finite sets, and let $f_1 : V_1 \to \{-1, +1\}$ and $f_2 : V_2 \to \{-1, +1\}$ be functions. Let $G = G(V_1, V_2, E)$ be the bipartite graph formed by creating an edge between $x_1 \in V_1$ and $x_2 \in V_2$ if and only if $f_1(x_1) = f_2(x_2)$. Let $X_1 \subset V_1$ and $X_2 \subset V_2$ be non-empty, and let $0 < \varepsilon < 1$. Show that if (X_1, X_2) is ε-regular, then

$$|\mathbf{E}_{x_1\in X_1} f_1(x_1)|, |\mathbf{E}_{x_2\in X_2} f_2(x_2)| \geq 1 - O(\varepsilon).$$

This shows that any partition of V_1 and V_2 into regular pairs will have to essentially be a refinement of the sets $\{x_1 \in V_1 : f_1(x_1) = \pm 1\}$ and $\{x_2 \in V_2 : f_2(x_2) = \pm 2\}$.

10.6.10 [136] Let V be a large finite set. Show that there exist n functions $f_1, \ldots, f_n : V \to \{-1, +1\}$ for some $n = \Omega(\log |V|)$ with the property that for any distinct $x, x' \in V$, we have $f_i(x) = f_i(x')$ for at most $3n/4$ values of i, or in other words $|\mathbf{E}_{i\in[1,n]} f_i(x) f_i(x')| \leq 1/2$. (Hint: use the probabilistic method. Alternatively, identify V with an error-correcting code in $\{-1,+1\}^n$, constructed for instance using the greedy algorithm.) If $\lambda : V \to \mathbf{R}^+$ is any function such that $\|\lambda\|_{l^1(V)} = 1$ and $\|\lambda\|_{l^\infty(V)} \leq 1 - \varepsilon$ for some $\varepsilon > 0$, show that

$$\mathbf{E}_{i\in[1,n]} \left| \sum_{x\in V} \lambda(x) f_i(x) \right|^2 \leq 1 - \Omega(\varepsilon).$$

Conclude in particular that $|\sum_{x\in V} \lambda(x) f_i(x)| \leq 1 - \Omega(\varepsilon)$ for at least $\Omega(\varepsilon n)$ values of i.

10.6.11 [136] Let V be a large finite set, and let $f_1, \ldots, f_n : V \to \{-1,+1\}$ be as in the preceding exercise. Let W be another large finite set, let G be the graph with vertex set $[1,n] \times V \times W$, with any two distinct vertices $(i, x, w), (j, y, z)$ being connected by an edge if and only if $f_i(y) = f_j(x)$. Let $\varepsilon > 0$, and suppose that $[1,n] \times V$ is partitioned into $[1,n] \times V \times W = V_1 \cup \cdots \cup V_k$ as in the regularity lemma. Suppose further that for all but $O(\varepsilon k)$ of the sets V_s, there exists an $i_s \in [1,n]$ such that $|V_s \cap (\{i_s\} \times V \times W)| \geq (1 - O(\varepsilon))|V_s|$; thus up to errors of $O(\varepsilon)$, most of the cells V_s of the partition are essentially contained in one of the $\{i\} \times V$. Conclude that for all but $O(\varepsilon k)$ of the sets V_s, there exists $i_s \in [1,n]$ and $x_s \in V$ such that $|V_s \cap (\{i_s\} \times \{x_s\} \times W)| \geq (1 - O(\varepsilon))|V_s|$; thus any regular partition which essentially refines the partition $\{\{i\} \times V \times W\}$, must automatically essentially refine the finer partition $\{\{i\} \times \{x\} \times W\}$. (This is a more complicated version of Exercise 10.6.9, and requires use of the previous exercise, with $\lambda(x)$ being equal to the relative density of $V_s \cap (\{i_s\} \times \{x_s\} \times W)$ in $V_s \cap (\{i_s\} \times V \times W)$.) An iteration of this fact can be used to establish a lower bound of tower type for the Szemerédi regularity lemma; see [136].

10.7 Szemerédi's argument

In this section we give another proof of Roth's theorem due to Szemerédi (see e.g. [143]). This argument gives slightly better bounds than that obtained from the regularity lemma, but still worse than that given from the Fourier-analytic argument. However, it has the advantage of being completely elementary and rather short. A more complex version of this argument was also used in [343] to establish Szemerédi's theorem for progressions of length 4, but the general k case requires a rather different (and even more complex) combinatorial argument which we will not discuss here; see [345].

Intuitively, the idea is as follows. If A is a dense set of an interval $[1, N]$, then it should contain a large cube $a + [0, 1]^d \cdot v$. If A also has no proper progressions of length three, then this implies that A must be disjoint from a sumset $2a + [0, 1]^d \cdot 2v - A_0$ of a large set and a cube. This disjointness "squeezes" A into a collection of moderately long progressions, and the density of A must increase on at least one of them. This creates a density increment that one can then iterate as in the Fourier-analytic proof of Roth's theorem. Thus one is using the disjointness from a sumset as a substitute for Fourier bias (cf. Exercise 4.3.12).

We now give the main steps of the argument, leaving the proofs as exercises. First we need to show that dense sets contain cubes.

Lemma 10.49 *Let $0 < \delta < 1$, let P be a large proper arithmetic progression, and let A be a subset of P with $|A| \geq \delta|P|$. Then A contains a proper cube $a + [0, 1]^d \cdot v$ with $d = \Omega_\delta(\log\log|P|)$. In particular all the steps $v_1, \ldots, v_d$ are non-zero.*

The main point here is that the quantity d goes to infinity (somewhat slowly) as $|P| \to \infty$. As a corollary of this lemma, we have

Corollary 10.50 *Let $0 < \delta < 1$, let N be a sufficiently large integer depending on δ, and let A be a subset of $[1, N]$ with $|A| \geq \delta N$ which contains no proper progressions of length three. Then at least one of the following statements is true:*

- *(Density increment) There exists a progression $P \subset [1, N]$ of length $|P| \geq N/4 + O(1)$ such that $|A \cap P| \geq 1.1\delta|P|$.*
- *(Disjointness from sumset) There exists a set $A_0 \subset [1, N/4]$ and a cube $a + [0, 1]^d \cdot v \subset (N/4, N/2]$ with $d = \Omega_\delta(\log\log N)$ and all steps $v_1, \ldots, v_d$ non-zero such that $|A_0| = \Omega(\delta N)$ the sumset $2a + [0, 1]^d \cdot 2v - A_0 \subset [1, N]$ is disjoint from A.*

Proof Without loss of generality we may assume that $|A \cap (iN/4, (i + 1)N/4]| \leq 1.1\delta N/4 + O(1)$ for all $i = 0, 1, 2, 3$ otherwise we have a density increment. In particular this implies that $|A \cap (iN/4, (i + 1)N/4]| = \Omega(\delta N)$ for all i. Applying Lemma 10.49 to the set $A \cap (N/4, N/2]$ we see that this set contains a cube $a + [0, 1]^d \cdot v$ with the desired properties. If we then set $A_0 := A \cap [1, N/4)$, the claim then follows by observing that whenever $x \in A_0$ and $y \in a + [0, 1]^d \cdot v$, the sequence $x, y, 2y - x$ is a proper arithmetic progression and hence $2y - x$ cannot lie in A. □

Suppose we are in the situation in the above corollary, and the "disjointness from sumset" statement holds. Let $E_0 \subset E_1 \subset \cdots \subset E_d \subset [1, N]$ be the sets

$$E_i := 2a + [0, 1]^i \cdot (2v_1, \ldots, 2v_i) - A_0.$$

By the pigeonhole principle we can find $1 \leq i \leq d$ such that

$$|E_i| \leq |E_{i-1}| + O\left(\frac{N}{d}\right).$$

On the other hand, we have $E_i = E_{i-1} + \{0, 2v_i\}$. This shows that we can partition the set $[1, N]\backslash E_i$ into $O(\frac{N}{d})$ proper arithmetic progressions $P_1, \ldots, P_k$ of step $2v_i$ (see Exercises). Observe that

$$\sum_{j=1}^{k} |P_j| = N - |E_i| \leq N - |A_0| = (1 - \Omega(\delta))N.$$

On the other hand, since A is disjoint from E_i, we have

$$\sum_{j=1}^{k} |A \cap P_j| = |A| \geq \delta N.$$

We thus have

$$\sum_{j=1}^{k} |A \cap P_j| - (\delta + c\delta^2)|P_j| - cd > 0$$

for some small absolute constant $c > 0$. Thus by the pigeonhole principle, there exists a progression P_j such that $|P_j| \geq cd = \Omega_\delta(\log \log N)$ and $|A \cap P_j| \geq (\delta + c\delta^2)|P_j|$. This establishes a density increment of A on a progression whose length goes to infinity as $N \to \infty$. This is essentially Corollary 10.26 (but with somewhat worse explicit constants) and one can now iterate this Corollary as before to establish Roth's theorem.

A careful accounting of the bounds here yields the bound $r_3([1, N]) = O(N/\log_* N)$, which is marginally better than the bounds obtained by the regularity lemma.

Exercises

10.7.1 Prove Lemma 10.49. (Hint: first prove the following preliminary statement: if $|A| \geq \delta|P|$, and $|P|$ is large depending on δ, then there are $\Omega(\delta^2|P|)$ values of v such that $|A \cap (A + v)| = \Omega(\delta^2|P|)$.)

10.7.2 Let $A \subset [1, N]$ and $r \neq 0$ be such that $A + r \subset [1, N]$ and $|A + \{0, r\}| \leq |A| + k$. Show that $[1, N]\backslash(A + \{0, r\})$ can be partitioned into $O(k)$ disjoint arithmetic progressions of step r.

11

Szemerédi's theorem for $k > 3$

In this chapter we continue the study of Szemerédi's theorem (Theorem 10.1), but now focus on the case of longer arithmetic progressions $k > 3$. While we have seen that the $k = 3$ case of this theorem can be treated by Fourier-analytic methods, it turns out that the higher-k case cannot be dealt with by (linear) Fourier-analytic tools, even when $k = 4$; we will see some justifications for this fact later. Indeed, whereas Roth's treatment [287] of the $k = 3$ case appeared in 1953, it was only in 1969 that Szemerédi [343] established the $k = 4$ case of Theorem 10.1, by combining the density increment argument of Roth with some impressively complicated combinatorial arguments (based on those discussed in Section 10.7). Unfortunately, this argument did not yield any new bound on van der Waerden's theorem (Exercise 6.3.7), as this theorem was used in the proof; note that one of the original motivation of Erdős and Turán in introducing this problem in [99] was to obtain a more effective bound on van der Waerden's theorem than the Ackermann-type bounds obtained by the usual proof methods. In 1972 Roth [288] obtained an alternative proof of the $k = 4$ case by combining the Fourier method with Szemerédi's arguments, but again van der Waerden's theorem was involved.

In 1975, Szemerédi [345] finally established Theorem 10.1 for all k. The argument is purely combinatorial. It uses the density increment argument, van der Waerden's theorem, and an induction on k, although to execute this induction properly, a number of new combinatorial tools needed to be introduced, most notably the very useful and influential *Szemerédi regularity lemma*, which has already been discussed in Section 10.6. This lemma in particular introduces an *energy increment argument* to complement the density increment argument. This proof was elementary, and a technical masterpiece, but was rather complicated; while it did give (in principle) explicit quantitative bounds for Theorem 10.1, they were extremely large (involving iterated Ackermann functions). For reasons of space we will not present the original argument here.

Since Szemerédi's proof, several other important and very different proofs of this theorem have appeared (though as we will discuss later, all proofs share something in common, namely an analysis of the dichotomy between randomness and structure). In 1977 Furstenberg [121], [125], [122] introduced an elegant ergodic approach which proved Szemerédi's theorem for arbitrary k. The first observation was that there was a *correspondence principle* which showed that Theorem 10.1 was equivalent to a recurrence in the ergodic theory of measure-preserving systems, now known as the *Furstenberg multiple recurrence theorem*. This theorem in $k = 2$ was the classical Poincaré recurrence theorem, while Furstenberg observed (see [122]) that the $k = 3$ case, i.e. Roth's theorem, could be deduced from some spectral theory of the shift operator and some measure-theoretic constructions. Again, in the $k \geq 4$ cases, spectral methods (which are the ergodic analog of Fourier-analytic methods) proved insufficient to deduce the recurrence theorem; however, by developing a structural theorem for measure-preserving systems (which can be viewed as an infinitary analog of the Szemerédi regularity lemma, and is proven by an infinitary version of the energy increment argument, requiring Zorn's lemma), and by establishing a recurrence theorem for generalized almost periodic functions (this being a measure-theoretic analog of the van der Waerden theorem, and which in fact could be deduced from that theorem and some additional arguments), Furstenberg was able to obtain a conceptually simple, but non-elementary, proof of Szemerédi's theorem for all k. This proof has since proven to be extremely flexible and has yielded many other recurrence theorems of a similar type, the majority of which have not yet been obtainable by other means. On the other hand, due to the infinitary nature of the argument, and the reliance of the axiom of choice, no quantitative bound could be extracted from these arguments until very recently (see [357]). We shall briefly discuss this infinitary ergodic approach in Section 11.5, though we will not go into details as they require a certain amount of machinery not used elsewhere in this book. We will discuss the finitary version, which has less machinery but is more complicated, in Section 11.4.

The next important observation, which first appeared in [304] in 1978, was that the Szemerédi regularity lemma could handle the $k = 3$ case (i.e. Roth's theorem) directly, without need for van der Waerden's theorem, see Section 10.6; indeed one could also obtain a generalization of this result to right-angled triangles which was previously obtainable by ergodic means [124]. However, it became increasingly apparent that in order to generalize this argument to higher k one would have to generalize Szemerédi's regularity lemma to hypergraphs. There were a number of attempts in this direction, most notably by Frankl and Rödl [109], who managed to handle 3-uniform hypergraphs (which implies Szemerédi theorem for arithemtic progression of length 4). However, in the general case, even stating a valid formulation of a hypergraph regularity lemma which would then easily imply the full strength of Szemerédi's theorem remained elusive until very recently, when Gowers [140] and Rödl and Skokan [283], [284] finally derived a sufficiently strong hypergraph regularity lemma from which Theorem 10.1 easily follows. We will discuss this approach in Section 11.6.

It is also worth mentioning that in 1988, Shelah [318] discovered the first quantitative bound on van der Waerden numbers which was not of Ackermann type (i.e. it was primitive recursive), although it was still extremely large, for instance larger than a tower exponential bound. As with the other direct proofs known for van der Waerden's theorem, this result, while ingeniously simple, did not seem to offer any new proof of Szemerédi's theorem, although it would improve the bound arising from Szemerédi's argument somewhat (given that that argument requires the van der Waerden theorem).

Another major breakthrough then occured with the work of Gowers [137], [138] in 1998 and 2001, who combined the Fourier-analytic methods of Roth with methods of additive combinatorics (such as Freiman's theorem, and the Balog–Szemerédi-Gowers theorem), and several new ideas (such as introducing the *Gowers uniformity norms*, and initiating a theory of *quadratic Fourier analysis* as well as higher-order generalizations) to obtain a new proof of Szemerédi's theorem first in $k = 4$ and then for general k. This proof gave for the first time bounds in Szemerédi's theorem which were comparable in strength to Roth's original argument (though the analog of Bourgain's bound for $k = 3$ has not yet been replicated for higher k). This argument also revealed for the first time the new structures (such as quadratic phase functions) which had to be understood in order to handle the cases $k \geq 4$ and which were not present in the $k = 3$ theory. Interestingly, very similar conclusions were also being drawn at the same time from the ergodic theory approach, most notably through the recent work of Host and Kra [185] and Ziegler [386] but also from earlier work of Furstenberg, Weiss, Conze, Lesigne, and others. There are thus encouraging signs of a "higher-order Fourier analysis" which may yet unify the Fourier and ergodic approaches, but while developments here are proceeding rapidly, there does not yet appear to be a definitive and satisfactory theory in this regard. We shall discuss Gowers' approach in Sections 11.1–11.3.

Finally, an extremely recent development has been the coming together of many of the independent directions of research mentioned above, most notably in the recent result of Green and Tao [158] which establishes arbitrarily long progressions in the primes, or in any subset of the primes of positive relative density. This argument requires that one extend Szemerédi's theorem to the setting of dense subsets of "pseudo-random" sets such as the almost primes. To achieve this, one uses a quantitative version of the ergodic theory approach, set in a finitary rather than infinitary setting (and with no use of the axiom of choice); to make this scheme work, one must borrow heavily from the tools developed from the other approaches to Szemerédi's theorem, notably the energy increment strategy used to prove the regularity lemma, the Gowers uniformity norms used in the Fourier-analytic approach, and the relativization to pseudo-random sets which first appears in the graph and hypergraph regularity approach; in addition, some number-theoretic arguments (involving some analysis of the Riemann zeta function) is necessary to ensure that the almost primes have the desired amount of pseudo-randomness. We will discuss this result in Section 11.7.

Unfortunately we will not have space to give a full proof of Szemerédi's theorem or the Green–Tao theorem in this chapter, as all of the known proofs are either quite lengthy or require a fair amount of supporting theory. We will however give several partial results and key lemmas, and describe the main steps of most of these proofs, referring the reader to the original references for the complete details. Thus this chapter should be viewed as an introduction to the very active current area of research surrounding Szemerédi's theorem, rather than a comprehensive treatment of the field. One theme we wish to emphasize here is that despite the extraordinary diversity of methods and techniques in all the various proofs of Szemerédi's theorem, that there is some very strong unifying themes among all these approaches, such as the exploitation of a dichotomy between randomness and structure, and this chapter intends to highlight such common themes between all the known proofs of Szemerédi's theorem and related results.

11.1 Gowers uniformity norms

As in the previous chapter, it is convenient to attack Szemerédi's theorem by studying the k-linear form

$$\Lambda_k(f_0, \ldots, f_{k-1}) := \mathbf{E}_{x,r\in Z} f_0(x) f_1(x+r) \cdots f_{k-1}(x+(k-1)r)$$

defined for any finite additive group Z and any functions $f_0, \ldots, f_{k-1} : Z \to \mathbf{C}$. Thus, for instance, $\Lambda_k(1_A, \ldots, 1_A)$ is at least as large as $\mathbf{P}(A)/|Z|$, and will be larger if and only if A contains an arithmetic progression of length k with non-zero step; note that such progressions are proper if $|Z|$ is coprime to $k!$. This form of course generalizes the form $\Lambda_3(f, g, h)$ which featured prominently in the previous chapter.

Just as Varnavides' theorem (Theorem 10.9) is equivalent to Roth's theorem (Theorem 10.8), Szemerédi's theorem is equivalent to the following.

Theorem 11.1 (Szemerédi's theorem again) *Let $k \geq 3$, let Z be a finite cyclic group of prime order $|Z| \geq k$, and let $f : Z \to \mathbf{R}^+$ be a non-negative function* *which is not identically zero, and obeys the bound* $0 \le f(x) \le 1$ *for all* $x \in Z$; *then*

$$\Lambda_k(f, \ldots, f) = \Omega_{k,\mathbf{E}_Z(f)}(1).$$

In fact, this theorem is valid for all finite abelian groups, not just the cyclic group, as we shall see in Section 11.6.

Thus one strategy to prove Szemerédi's theorem is to obtain good bounds for quantities of the form $\Lambda_k(f_0, \ldots, f_{k-1})$ for various choices of $f_0, \ldots, f_{k-1}$. This is the approach taken by both Gowers' Fourier-analytic proof and in the finitary ergodic proof (and variants of this strategy also are used in the infinitary ergodic proof and the hypergraph proof). In the previous chapter, the linear bias norm $\|f\|_{u^2(Z)}$ was used to control this quantity effectively when $k = 3$, but this norm turns out to not be appropriate for higher k (see exercises). There are higher-order generalizations $\|f\|_{u^{k-1}(Z)}$ of the linear bias norm which we will discuss later, but in the absence of any useful quadratic or higher-order analog of Plancherel's theorem, it is difficult (though not impossible, see below) to use this norm to control Λ_k. Instead, there is a related norm, the *Gowers uniformity norm* $\|f\|_{U^{k-1}(Z)}$, which is more combinatorial than Fourier-analytic in nature, but controls the form Λ_k very easily. It is defined as follows.

Definition 11.2 (Gowers uniformity norm) Let $f : Z \to \mathbf{C}$ and $d \ge 1$. Then the *Gowers uniformity norm* $\|f\|_{U^d(Z)}$ of order d is defined recursively by[1]

$$\|f\|_{U^1(Z)} := |\mathbf{E}_Z(f)|; \quad \|f\|_{U^{d+1}(Z)} := \left(\mathbf{E}_{h\in Z}\|T^h f\overline{f}\|_{U^d(Z)}^{2^d}\right)^{1/2^{d+1}}$$

for all $d \ge 1$, where $T^h f(x) := f(x+h)$ is the shift of f by h.

Thus for instance we have

$$\begin{aligned}\|f\|_{U^2(Z)} &= (\mathbf{E}_{h\in Z}|\mathbf{E}_Z(T^h f\overline{f})|^2)^{1/4}\\ &= (\mathbf{E}_{x,h_1,h_2\in Z} f(x+h_1+h_2)\overline{f(x+h_1)f(x+h_2)}f(x))^{1/4}\end{aligned}$$

and more generally

$$\|f\|_{U^d(Z)} = \left(\mathbf{E}_{x,h_1,\ldots,h_d\in Z}\prod_{\omega\in\{0,1\}^d} C^{|\omega|} f(x+\omega\cdot h)\right)^{1/2^d} \tag{11.1}$$

where $Cf := \overline{f}$ is the conjugation operator, $\omega = (\omega_1, \ldots, \omega_d)$, $h := (h_1, \ldots, h_d)$, and $|\omega| := \omega_1 + \cdots + \omega_d$. In the case of an indicator function $f = 1_A$, we have

$$\|1_A\|_{U^d(Z)}^{2^d} = \mathbf{P}_{x,h_1,\ldots,h_d\in Z}(x + [0,1]^d \cdot (h_1, \ldots, h_d) \subset A); \tag{11.2}$$

[1] It would also be consistent to define $\|f\|_{U^0(Z)} = \mathbf{E}_Z(f)$, but this quantity is signed and thus is too pathological to be called a norm.

thus $\|1_A\|_{U^d(Z)}$ is a normalized measure of how many d-dimensional cubes are contained in A. In particular, we have the identity

$$\|1_A\|_{U^2(Z)}^4 = \mathbf{E}(A, A)/|Z|^3$$

which relates the U^2 norm of 1_A to the additive energy of A, defined in Definition 2.8.

At first glance, the Gowers uniformity norm $\|f\|_{U^{k-1}(Z)}$ norm looks even more complicated than the expression $\Lambda_k(f, \ldots, f)$ which it is meant to control, but as we shall see it has a significantly better structure which makes it more amenable to analysis.

In the $d = 2$ case the Gowers uniformity norm is also related to the Fourier transform by the simple formula

$$\|f\|_{U^2(Z)} = \|\hat{f}\|_{l^4(Z)}; \tag{11.3}$$

we leave the verification of this identity as an exercise. (Compare also with (4.18).) In particular this shows that the $U^2(Z)$ norm is indeed a norm. It turns out that the higher $U^d(Z)$ norms are also norms as well. To see this it is convenient to introduce the *Gowers inner product* $\langle (f_\omega)_{\omega\in\{0,1\}^d}\rangle_{U^d(Z)}$ of 2^d functions f_ω by the formula

$$\left\langle (f_\omega)_{\omega\in\{0,1\}^d}\right\rangle_{U^d(Z)} := \mathbf{E}_{x,h_1,\ldots,h_d\in Z} \prod_{\omega\in\{0,1\}^d} \mathcal{C}^{|\omega|} f_\omega(x + \omega\cdot h).$$

Thus for instance

$$\begin{aligned}
\langle (f_0, f_1)\rangle_{U^1(Z)} &= \mathbf{E}_{x,h\in Z}\overline{f_1(x+h)}f_0(x)\\
&= \mathbf{E}_Z(f_0)\overline{\mathbf{E}_Z(f_1)}\\
\langle (f_{00}, f_{01}, f_{10}, f_{11})\rangle_{U^2(Z)} &= \mathbf{E}_{x,h_1,h_2\in Z} f_{11}(x+h_1+h_2)\overline{f_{10}(x+h_1)}\\
&\quad \overline{f_{01}(x+h_2)}f_{00}(x)\\
&= \sum_{\xi\in Z}\hat{f}_{11}(\xi)\overline{\hat{f}_{10}(\xi)\hat{f}_{01}(\xi)\hat{f}_{00}(\xi)}.
\end{aligned}$$

Furthermore we see that

$$\|f\|_{U^d(Z)} = \left\langle (f)_{\omega\in\{0,1\}^d}\right\rangle_{U^d(Z)}^{1/2^d}. \tag{11.4}$$

An application of the Cauchy–Schwarz inequality in the h_d variable gives the bound

$$\left|\left\langle (f_\omega)_{\omega\in\{0,1\}^d}\right\rangle_{U^d(Z)}\right| \leq \left\langle (f_{\omega',0})_{\omega\in\{0,1\}^d}\right\rangle_{U^d(Z)}^{1/2}\left\langle (f_{\omega',1})_{\omega\in\{0,1\}^d}\right\rangle_{U^d(Z)}^{1/2} \tag{11.5}$$

where $\omega' := (\omega_1, \ldots, \omega_{d-1}) \in \{0, 1\}^{d-1}$ is the first $d-1$ components of ω. Similarly for permutations. Applying this inequality d times one obtains

$$\left|\left\langle (f_\omega)_{\omega\in\{0,1\}^d}\right\rangle_{U^d(Z)}\right| \leq \prod_{\tilde\omega\in\{0,1\}^d}\left|\left\langle (f_{\tilde\omega})_{\omega\in\{0,1\}^d}\right\rangle_{U^d(Z)}\right|^{1/2^d}.$$

Applying (11.4), we conclude the *Gowers–Cauchy–Schwarz inequality*

$$\left|\langle (f_\omega)_{\omega\in\{0,1\}^d}\rangle_{U^d(Z)}\right| \leq \prod_{\omega\in\{0,1\}^d} \|f_\omega\|_{U^d(Z)}. \tag{11.6}$$

Applying (11.6) to the special case when $f_\omega = f$ for $\omega_d = 0$ and $f_\omega = 1$ otherwise, one easily verifies the useful *monotonicity formula*

$$\|f\|_{U^{d-1}(Z)} \leq \|f\|_{U^d(Z)} \tag{11.7}$$

for all $d \geq 1$. Next, from (11.4), multilinearity, and the Gowers–Cauchy–Schwarz inequality we have

$$\begin{aligned}
\|f_0 + f_1\|_{U^d(Z)}^{2^d} &= \langle (f_0 + f_1)_{\omega\in\{0,1\}^d}\rangle_{U^d(Z)} \\
&= \sum_{I\subseteq\{0,1\}^d} \langle (f_{\mathbf{I}(\omega\in I)})_{\omega\in\{0,1\}^d}\rangle_{U^d(Z)} \\
&\leq \sum_{I\subseteq\{0,1\}^d} \prod_{\omega\in\{0,1\}^d} \left\|f_{\mathbf{I}(\omega\in I)}\right\|_{U^d(Z)} \\
&= \prod_{\omega\in\{0,1\}^d} \left(\|f_0\|_{U^d(Z)} + \|f_1\|_{U^d(Z)}\right)
\end{aligned}$$

from which we deduce the *Gowers triangle inequality*

$$\|f_0 + f_1\|_{U^d(Z)} \leq \|f_0\|_{U^d(Z)} + \|f_1\|_{U^d(Z)}.$$

This argument should be compared with the standard derivation of the Hilbert space triangle inequality from the Hilbert space Cauchy–Schwarz inequality. Since the Gowers norm $\|f\|_{U^d(Z)}$ is clearly non-negative and homogeneous, it is at least a semi-norm. When $d = 1$ it is not necessarily a norm (because $\|f\|_{U^1(Z)} = |\mathbf{E}_Z(f)|$ can vanish without f being identically zero), but from (11.3) and the injectivity of the Fourier transform we see that the U^2 norm, at least, is a norm, and then (11.7) implies that the higher U^d are also norms.

We now relate the Gowers uniformity norms to the forms Λ_k which are relevant to Szemerédi's theorem. It is convenient to introduce the following notation: we use $\mathbf{b}(x_1, \ldots, x_n)$ to denote any function of n variables $x_1, \ldots, x_n$ that is bounded in magnitude by 1. As with the $O()$ notation, the exact function used in the $\mathbf{b}()$ notation will vary from case to case. This notation is useful whenever dealing with complicated multilinear expressions involving an interesting function f, and several other less interesting functions whose only important features are their boundedness and the precise set of variables that they depend on; the $\mathbf{b}()$ notation can then be used to conceal the uninteresting functions and focus attention on the important terms in the expression.

We begin with a simple but very useful lemma, which controls the correlations of several functions f_a with an arbitrary bounded function $\mathbf{b}(a)$, in terms of the correlations of f_a with (shifts of) itself.

Lemma 11.3 (Van der Corput lemma) *Let Z be a finite additive group, and let A be a non-empty set. For each $a \in A$ let $f_a : Z \to \mathbf{C}$ be a function. Then we have*

$$|\mathbf{E}_Z(\mathbf{E}_{a\in A}\mathbf{b}(a) f_a)| \leq |\mathbf{E}_{a\in A, h\in Z}\mathbf{E}_Z(T^h f_a \overline{f_a})|^{1/2}.$$

Proof From the triangle inequality followed by Cauchy–Schwarz we have

$$\begin{aligned} |\mathbf{E}_Z(\mathbf{E}_{a\in A}\mathbf{b}(a) f_a)| &\leq \mathbf{E}_{a\in A}|\mathbf{E}_Z(f_a)| \\ &\leq (\mathbf{E}_{a\in A}|\mathbf{E}_Z(f_a)|^2)^{1/2} \\ &= (\mathbf{E}_{a\in A, x, x'\in Z} f_a(x')\overline{f_a(x)})^{1/2}. \end{aligned}$$

The claim then follows by making the substitution $x' = x + h$. □

As a consequence we have

Lemma 11.4 (Generalized von Neumann theorem) *Let Z be a finite additive group, let $k \geq 2$, and let $c_0, \ldots, c_{k-1}$ be distinct integers such that $c_i - c_j$ is coprime to $|Z|$ for all distinct i, j. Then for any function $f : Z \to \mathbf{C}$ we have*

$$|\mathbf{E}_{x,r\in Z} f(x + c_0 r)\mathbf{b}(x + c_1 r)\cdots\mathbf{b}(x + c_{k-1} r)| \leq \|f\|_{U^{k-1}(Z)}.$$

As a particular corollary, we see that if $|Z|$ is coprime to $(k-1)!$ then we have

$$|\Lambda_k(f_0, \ldots, f_{k-1})| \leq \min_{0\leq j\leq k-1} \|f_j\|_{U^{k-1}(Z)} \tag{11.8}$$

whenever $f_0, \ldots, f_{k-1} : Z \to \mathbf{C}$ are bounded in magnitude by 1. This result should be compared with Proposition 10.11.

Proof We induce on k. When $k = 2$ we observe that the map $(x, r) \mapsto (x + c_0 r, x + c_1 r)$ is bijective on $Z \times Z$ (since $|Z|$ is coprime to $c_0 - c_1$) and hence

$$|\mathbf{E}_{x,r\in Z} f(x + c_0 r)\mathbf{b}(x + c_1 r)| = |\mathbf{E}_Z(f)\mathbf{E}_Z(\mathbf{b})| \leq |\mathbf{E}_Z(f)| = \|f\|_{U^1(Z)}$$

as desired. Now suppose that $k \geq 3$ and the claim has already been proven for $k - 1$. By shifting x by $c_{k-1} r$ if necessary (and replacing c_j by $c_j - c_{k-1}$) we may take $c_{k-1} = 0$, so we can write the left-hand side as

$$|\mathbf{E}_{x\in Z}\mathbf{b}(x)\mathbf{E}_{r\in Z} f(x + c_0 r)\mathbf{b}(x + c_1 r)\cdots\mathbf{b}(x + c_{k-2} r)|.$$

Applying Lemma 11.3, we can bound this by

$$|\mathbf{E}_{x,h\in Z}\mathbf{E}_{r\in Z}(T^{c_0 h} f\overline{f})(x + c_0 r)\mathbf{b}(x + c_1 r, h)\cdots\mathbf{b}(x + c_{k-2} r, h)|^{1/2}.$$

Applying the induction hypothesis, we can bound this by

$$\left|\mathbf{E}_{h\in Z}\|T^{c_0 h} f\overline{f}\|_{U^{k-2}(Z)}\right|^{1/2}.$$

Since $c_0 = c_0 - c_{k-1}$ is coprime to $|Z|$ we can change variables and replace $c_0 h$ by h. Applying Hölder we can the bound the previous expression by

$$\left(\mathbf{E}_{h\in Z}\|T^h f\overline{f}\|_{U^{k-2}(Z)}^{2^{k-2}}\right)^{1/2^{k-1}},$$

and the claim now follows from the recusive definition of the $U^{k-1}(Z)$ norm. □

Let us informally refer to a function f as *Gowers uniform of order* $k-2$ if the quantity $\|f\|_{U^{k-1}(Z)}$ is small. It is easy to verify the bounds

$$\|f\|_{u^2(Z)} \leq \|f\|_{U^2(Z)} \leq \|f\|_{u^2(Z)}^{1/2} \tag{11.9}$$

whenever f is bounded in magnitude by 1 (see exercises), thus Gowers uniformity of order 1 is the same as linear (or Fourier) uniformity. In analogy with this, we shall refer to Gowers uniformity of order 2 as *quadratic uniformity*, Gowers uniformity of order 3 as *cubic uniformity*, and so forth. A partial explanation for this terminology can be found in Exercise 11.1.12; see also the next section.

The estimate (11.8) shows that functions which are Gowers uniform of order $k-2$ are negligible for the purposes of counting progressions of length k. One is then naturally led to the strategy of approximating an arbitrary function f by a much more structured function f, up to errors which are Gowers uniform. For instance, if one is lucky enough that $f - \mathbf{E}(f)$ is Gowers uniform of order $k-2$, then one can use (11.8) to approximate $\Lambda_k(f, \ldots, f)$ by $\Lambda_k(\mathbf{E}(f), \ldots, \mathbf{E}(f)) = \mathbf{E}(f)^k$. Of course, it is not always the case that $f - \mathbf{E}(f)$ is Gowers uniform. In such an event, it is important to understand which functions are *not* Gowers uniform, and more precisely what the *obstructions* to Gowers uniformity are. This will be the focus of the next section.

Exercises

11.1.1 Show that Theorem 11.1 is equivalent to Theorem 10.1. Also show that to prove Theorem 11.1 it suffices to do so in the special case when f is an indicator function, $f = 1_A$.

11.1.2 Let $Z = \mathbf{Z}_N$ for some prime $N > 3$, let ξ be a non-zero element of Z, and define the functions

$$f_0(x) := e(\xi x^2/N);$$
$$f_1(x) := e(-3\xi x^2/N);$$
$$f_2(x) := e(3\xi x^2/N);$$
$$f_3(x) := e(-\xi x^2/N).$$

Show that $\Lambda_4(f_0, f_1, f_2, f_3) = 1$, but that $\|f_j\|_{u^2(Z)} = N^{-1/2}$ for $j = 0, 1, 2, 3$. This shows that there is no direct analog of Proposition 10.11.

Modify this example to show that there is no direct analog of Proposition 10.10 either. (Hint: it is simpler to construct an example in a vector space such as F_5^n, based on a quadratic hypersurface, than in a cyclic group such as $\mathbf{Z}_N$, which would require some sort of "quadratic Bohr set".)

11.1.3 Modify the proof of the Gowers triangle inequality to provide a proof of the triangle inequality for $l^{2^d}(Z)$ for $d = 1, 2, 3, \ldots$ based purely on the Cauchy–Schwarz inequality.

11.1.4 Prove (11.1) and (11.2).

11.1.5 Prove (11.3). Use (11.3) and Plancherel's theorem to prove (11.9).

11.1.6 Prove (11.5).

11.1.7 Prove (11.7).

11.1.8 For any finite additive group Z, any $f : Z \to \mathbf{C}$, and any $d \geq 1$, show that $\|\overline{f}\|_{U^d(Z)} = \|f\|_{U^d(Z)}$ and $\|\mathrm{Re}(f)\|_{U^d(Z)}, \|\mathrm{Im}(f)\|_{U^d(Z)} \leq \|f\|_{U^d(Z)}$.

11.1.9 Let $\phi : Z \to Z'$ be a Freiman isomorphism of order 2 from Z to Z'. Show that $\|f \circ \phi\|_{U^d(Z)} = \|f\|_{U^d(Z')}$ for any $d \geq 1$ and any $f : Z' \to \mathbf{C}$. In particular we have the translation invariance $\|T^h f\|_{U^d(Z)} = \|f\|_{U^d(Z)}$ for any $h \in Z$.

11.1.10 If $f : Z \to \mathbf{C}$ and $f' : \mathbf{Z}' \to \mathbf{C}$ are functions on two finite additive groups Z, Z', show that $\|f \otimes f'\|_{U^d(Z \otimes Z')} = \|f\|_{U^d(Z)} \|f'\|_{U^d(Z')}$.

11.1.11 Use (11.8) to give another proof that the $\|f\|_{U^k(Z)}$ norms are non-degenerate for $k \geq 2$, at least in the case when $|Z|$ is coprime to $(k-1)!$.

11.1.12 Let $d \geq 1$, let $F = F_p$ be a field of prime order $p > d$, and let $P : F \to F$ be a polynomial of degree exactly d with coefficients in F. Let $f : F \to \mathbf{C}$ be the function $f(x) := e(P(x)/p)$, where the map $x \mapsto x/p$ is defined from F to $\mathbf{R}/\mathbf{Z}$ in the obvious manner. Show that $\|f\|_{U^{d'}(F)} = 1$ for all $d' > d$, but that $\|f\|_{U^{d'}(F)} \leq ((d-1)/p)^{1/2^d}$ for all $1 \leq d' \leq d$; this shows that the $U^d(F)$ norms are genuinely different for each $1 \leq d < p$. Informally, we see that f is Gowers uniform of order $d-1$ or less, but is not Gowers uniform of order d or more. In particular establish the *Weyl exponential sum estimate*

$$\sum_{x \in F} e(P(x)/p) = O\left(p^{1-2^{-d}}\right).$$

Compare this with Lemma 4.14.

11.1.13 For any finite additive group Z and any $f : Z \to \mathbf{C}$, show that

$$\|f\|_{U^d(Z)} \leq \|f\|_{L^{2^d/(d+1)}(Z)}.$$

for all $d \geq 1$, and that $\lim_{d \to \infty} \|f\|_{U^d(Z)} = \|f\|_{L^\infty(Z)}$. Show that the exponent $2^d/(d+1)$ cannot be replaced by any smaller quantity. (Hint: consider a Dirac mass, or the characteristic function of a subgroup.)

11.1.14 Let G be a subgroup of a finite additive group Z, and let $f : Z \to \mathbf{C}$ and $d \geq 1$. For each coset $y + G$ of G, define $\|f\|_{U^d(y+G)}$ in the obvious translation-invariant manner. Show that

$$\|f\|_{U^d(Z)} \leq \left(\mathbf{E}_{y \in Z} \|f\|_{U^d(y+G)}^{2^d/(d+1)}\right)^{(d+1)/2^d}$$

thus generalizing the previous exercise and demonstrating that "local uniformity norms" control "global uniformity norms".

11.1.15 Let $f : Z \to \mathbf{C}$ be a function on a finite additive group Z. Establish the Parseval-type identity $\|f\|_{U^3(Z)} = |Z|^{1/2} \|\hat{f}\|_{U^3(Z)}$, which shows that the Fourier transform does not simplify the U^3 norm. (This phenomenon is related to the fact that the Fourier transform of a Gaussian is again a Gaussian.) Deduce a similar Plancherel-type identity for the U^3 inner product. For the higher U^d norms, $d \geq 4$, the situation is even worse; the Fourier representation is more complicated than the spatial representation.

11.1.16 Let A be a subset of a finite additive group Z, and let $d \geq 1$. Using (11.7), show that there are at least $\mathbf{P}_Z(A)^{2^d} |Z|^{d+1}$ $d + 1$-tuples $(x, h_1, \ldots, h_d) \in Z^{d+1}$ such that the cube $x + [0, 1]^d \cdot (h_1, \ldots, h_d)$ is contained in A. Use this to give another proof of Lemma 10.49.

11.1.17 Let $f : Z \to \mathbf{C}$ be a random function such that the random variables $f(x)$ for $x \in Z$ are jointly independent, have mean zero, and are bounded by 1. Show that $\mathbf{E}\|f\|_{U^d(Z)}^{2^d} = O_d(1/|Z|)$ for all $d \geq 1$. Thus random balanced functions tend to be Gowers uniform of very high order.

11.2 Hard obstructions to uniformity

In this section we consider the following *inverse problem*: suppose $f : Z \to \mathbf{C}$ is a function bounded in magnitude by one which fails to be Gowers uniform of some order $k - 2$, say $\|f\|_{U^{k-2}(Z)} \geq \delta$ for some $0 < \delta \leq 1$. What structural information can one then conclude about f? As it turns out, a sufficiently strong answer to this question will lead to a proof of Szemerédi's theorem for progressions of length k. This is the strategy employed by Gowers [137], [138] in his proof of Szemerédi's theorem, with the focus on obtaining as strong an inverse theorem for the $U^{k-2}(Z)$ norm as possible. In Section 11.4 we describe a slightly different approach in which one obtains a much weaker (and easier to prove) inverse theorem, but one which is still sufficient to obtain Szemerédi's theorem (but with much worse quantitative bounds).

A good model case is provided by the case $k = 3$. From (11.9) we see that if $\|f\|_{U^2(Z)} \geq \delta$, then $\|f\|_{u^2(Z)} \geq \delta^2$, and hence there exists a linear phase function $g(x) := e(\xi \cdot x)$ which has a large inner product with f: $|\langle f, g\rangle_{L^2(Z)}| \geq \delta^2$. This fact, combined with (11.8), can be used to give a variant of Proposition 10.10 or Proposition 10.11, which in turn can be employed in either a density increment argument or energy increment argument to prove the $k = 3$ case of Szemerédi's theorem, as was done in Sections 10.2, 10.3 and Section 10.5 respectively. One can view these linear phase functions as being the *obstructions* to Gowers uniformity of order 1; we have just seen that failure of Gowers uniformity of order 1 implies correlation with one of these linear phase functions, and conversely the other inequality in (11.9) implies that correlation with a linear phase function implies lack of Gowers uniformity of order 1.

This model case, combined with the observations in Exercise 11.1.12, suggest that, more generally, lack of Gowers uniformity of order $k - 2$ should be tied to correlation with a phase function which is somehow polynomial of degree $k - 2$. This can be made precise as follows.

Definition 11.5 (Polynomial bias) Let Z be a finite additive group, and let $\phi : Z \to \mathbf{R}/\mathbf{Z}$ be a phase function. Given any $h \in Z$, we define the *difference operator* $(h \cdot \nabla)$ applied to ϕ as

$$(h \cdot \nabla)\phi(x) := \phi(x + h) - \phi(x).$$

We will sometimes subscript ∇ by ∇_x to emphasize the variable being differenced over (in case ϕ also depends on some other variables). If $d \geq 1$, we say that ϕ is a *phase polynomial of degree less than d* if we have

$$(h_1 \cdot \nabla_x) \cdots (h_d \cdot \nabla_x)\phi(x) = 0 \text{ for all } x, h_1, \ldots, h_d \in Z.$$

Phase polynomials of degree less than 2 will be referred to as *linear*, phase polynomials of degree less than 3 will be referred to as *quadratic*, and so forth. If $f : Z \to \mathbf{C}$ is a function, then we define the *polynomial bias* of f of degree d to be the quantity

$$\|f\|_{u^d(Z)} := \sup_\phi \left|\langle f, e(\phi)\rangle_{L^2(Z)}\right| = \sup_\phi |\mathbf{E}_{x\in Z} f(x)e(-\phi(x))|$$

where ϕ ranges over all phase polynomials of degree less than d.

More generally, if $B \subset Z$ is non-empty, we say that $\phi : B \to \mathbf{R}/\mathbf{Z}$ is a *locally polynomial phase function of degree less than d* if we have

$$(h_1 \cdot \nabla_x) \ldots (h_d \cdot \nabla_x)\phi(x) = 0 \text{ whenever } x + [0, 1]^d \cdot (h_1, \ldots, h_d) \subseteq B,$$

and then define

$$\|f\|_{u^d(B)} := \sup_\phi \left|\langle f, e(\phi)\rangle_{L^2(B)}\right| = \sup_\phi |\mathbf{E}_{x\in B} f(x)e(-\phi(x))|$$

where ϕ ranges over all phase functions which are locally polynomial on B of degree less than d.

To illustrate this definition, first observe that the only phase polynomials of degree less than 1 are the constants $\phi(x) = c$, and hence

$$\|f\|_{u^1(Z)} = |\mathbf{E}(f)| = \|f\|_{U^1(Z)}. \tag{11.10}$$

Thus $\|\cdot\|_{u^1(Z)}$ is a seminorm. For $d > 1$, one easily verifies that $\|\cdot\|_{u^d Z}$ is a genuine norm. For instance, from Exercise 4.1.4, we see that the only phase polynomials of degree less than 2 are the linear phases $\phi(x) = \xi \cdot x + c$, and thus the definition of the u^2 norm matches the one given in (10.5). In particular we still have the relation (11.9).

More generally, the $u^d(Z)$ and $U^d(Z)$ norms are quite related, enjoying the same symmetries. For instance, if ϕ is a phase polynomial of degree less than d, then one can easily verify that

$$\|fe(-\phi)\|_{u^d(Z)} = \|f\|_{u^d(Z)}; \quad \|fe(-\phi)\|_{U^d(Z)} = \|f\|_{U^d(Z)} \tag{11.11}$$

In particular, from (11.7) we have

$$|\mathbf{E}_{x\in Z} f(x)e(-\phi(x))| = \|fe(-\phi)\|_{U^1(Z)} \leq \|fe(-\phi)\|_{U^d(Z)} = \|f\|_{U^d(Z)}$$

and hence on taking suprema we have

$$\|f\|_{u^d(Z)} \leq \|f\|_{U^d(Z)}.$$

Thus correlation with a phase function of degree less than d implies lack of Gowers uniformity of order $d-1$. In light of (11.10), (11.9), one may hope that a converse statement is true, namely that lack of Gowers uniformity of order $d-1$ implies a correlation with a phase of degree less than d. One hopeful sign in this direction is the identity

$$\|e(\phi)\|_{U^d(Z)}^{2^d} = \mathbf{E}_{x,h_1,\ldots,h_d\in Z}\, e((h_1 \cdot \nabla_x) \cdots (h_d \cdot \nabla_x)\phi(x)) \tag{11.12}$$

whose verification we leave as an exercise. This suggests, though does not quite prove, that a function has large $U^d(Z)$ norm if and only if its phase is approximately polynomial of degree less than d. The above statement would then be an assertion that a phase which is approximately polynomial of degree d, in fact correlates with a genuine polynomial of degree d. Such an assertion should remind one of the Balog–Szemerédi–Gowers theorem, Theorem 2.29, and in fact that theorem plays a key role in establishing facts such as these.

In the case when Z is a vector space over a finite field of small order and $k = 4$, we can formalize these conjectures affirmatively as follows.

Theorem 11.6 (Inverse theorem for $U^3(F^n)$) *[137], [160] Let Z be a vector space over a finite field F, let $f : Z \to \mathbf{C}$ have magnitude bounded by 1, such that $\|f\|_{U^3(Z)} \geq \eta$ for some $0 < \eta \leq 1$. Then there exists a subspace $W \subset Z$*

with

$$\dim_F(W) \geq \dim_F(Z) - O\big(\eta^{-O(1)}\big) \tag{11.13}$$

such that

$$\mathbf{E}_{y \in Z} \|f\|_{u^3(y+W)} = \Omega\big(\eta^{O(1)}\big). \tag{11.14}$$

In particular, there exists $y \in Z$ such that $\|f\|_{u^3(y+W)} \geq \Omega(\eta^{O(1)})$.

The proof of this inverse theorem is quite lengthy, using techniques from previous chapters as well as a heavy reliance on Fourier-analytic methods and the van der Corput lemma (Lemma 11.3), and will be deferred to the next section. We remark that the case when F has characteristic 2 was not quite dealt with in the above-cited papers, but requires an additional observation of Samorodnitsky (private communication). Assuming it for now, we can now prove Szemerédi's theorem for vector spaces Z and in the case $k = 4$. In fact the inverse theorem allows us to give both a density increment proof and an energy increment proof. The density increment proof is based on the following proposition, analogous to (though somewhat weaker than in some respects) Lemma 10.15.

Proposition 11.7 (Lack of uniformity implies density increment) *Let Z be a vector space over a finite field F of odd prime order, and let $f : Z \to \mathbf{C}$ have magnitude bounded by 1 be such that $\mathbf{E}_Z(f) = 0$ and $\|f\|_{U^3(Z)} \geq \eta$ for some $0 < \eta \leq 1$. Then there exists a subspace Z' of Z with*

$$\dim_F(Z') \geq \frac{1}{2}\dim(Z) - O\big(\eta^{-O(1)}\big)$$

and a point $x_0 \in Z$, such that

$$\mathbf{E}_{x \in x_0 + Z'} f(x) \geq \Omega\big(\eta^{O(1)}\big).$$

Proof From Theorem 11.6 we can find a subspace W obeying the dimension bound (11.13) and the correlation bound (11.14). Also note that if $\mathbf{E}_{y+W}(f) \geq \Omega(\eta^{O(1)})$ for even a single $y \in Z$ then we will be done, so we may take $\mathbf{E}_{y+W}(f) \leq c\eta^C$ for any given absolute constants $c, C > 0$. Since we also have

$$\mathbf{E}_{y \in Z}\mathbf{E}_{y+W}(f) = \mathbf{E}_Z(f) = 0$$

we conclude that

$$\mathbf{E}_{y \in Z}|\mathbf{E}_{y+W}(f)| = 2\mathbf{E}_{y \in Z} \max(\mathbf{E}_{y+W}(f), 0) \leq 2c\eta^C$$

and so from (11.14) we have (choosing the constants c, C appropriately)

$$\mathbf{E}_{y \in Z} \|f\|_{u^3(y+W)} - 2|\mathbf{E}_{y+W}(f)| = \Omega\big(\eta^{O(1)}\big).$$

In particular we can find $y \in Z$ such that

$$\|f\|_{u^3(y+W)} \geq 2|\mathbf{E}_{y+W}(f)| + \Omega(\eta^{O(1)}).$$

By translating f by y if necessary we may take $y = 0$. By definition of the u^3 norm and Exercise 11.2.6, we can thus find a self-adjoint linear operator $M : W \to W$ and $\xi \in W$ such that

$$|\mathbf{E}_{x\in W} f(x)e(-Mx \cdot x)e(-\xi \cdot x)| \geq 2|\mathbf{E}_W(f)| + \Omega(\eta^{O(1)}).$$

Observe that the quantity $Mx \cdot x + \xi \cdot x$ only takes at most $|F|$ values. If we thus partition W into $|F|$ level sets $S_1, S_2, \ldots, S_{|F|}$, each of the form $\{x \in W : Mx \cdot x + \xi \cdot x = \text{const}\}$, then we have from the triangle inequality that

$$\sum_{j=1}^{|F|} |\mathbf{E}_{x\in W} 1_{S_j}(x) f(x)| \geq 2\left|\sum_{j=1}^{|F|} \mathbf{E}_W(1_{S_j}(x) f(x))\right| + \Omega(\eta^{O(1)})$$

and hence, by the identity $\max(y, 0) = (|y| + y)/2$,

$$\sum_{j=1}^{|F|} \max(\mathbf{E}_{x\in W} 1_{S_j}(x) f(x), 0) > \Omega(\eta^{O(1)})$$

and so by the pigeonhole principle we can find j such that

$$\mathbf{E}_{x\in W} 1_{S_j}(x) f(x) > \Omega(\eta^{O(1)})\mathbf{P}_W(S_j).$$

Now we need to take the quadratic surface S_j and partition it into affine spaces. We first observe that there exists a subspace U of W with dimension

$$\dim_F(U) \geq \frac{1}{2}\dim_F(W) - \frac{3}{2} \geq \frac{1}{2}\dim_F(W) - O(\eta^{-O(1)})$$

which is null with respect to M: see Exercise 4.3.16. Splitting S_j into cosets of U, we see from the pigeonhole principle that there exists a coset $x_1 + U$ such that

$$\mathbf{E}_{x\in x_1+U} 1_{S_j}(x) f(x) > \Omega(\eta^{O(1)})\mathbf{P}_{x_1+U}(S_j),$$

so in particular $S_j \cap (x_1 + U)$ is non-empty and

$$\mathbf{E}_{x\in S_j\cap(x_1+U)} f(x) > \Omega(\eta^{O(1)}).$$

The point of working on a coset $x_1 + U$ of a null space is that the quantity $Mx \cdot x + \xi \cdot x$ becomes linear with respect to x. Thus the intersection of S_j with $x_1 + U$ is an affine subspace $x_0 + Z'$ of $x_1 + U$ of codimension at most 1. The claim follows. □

Iterating this proposition as in the proof of Roth's theorem, one can eventually deduce the bound

$$r_4(F^n) = O\left(\frac{|F|^n}{\log^c n}\right) \tag{11.15}$$

for all $n > 1$ and some absolute constant c. It is also possible to adapt the energy increment argument from Section 10.5, with the the concept of quasi-periodic being replaced with that of being determined by a bounded number of quadratic phase functions, however the bounds on $r_4(F^n)$ obtained this way are rather poor. One can do a bit better by adapting the argument in Theorem 10.27, obtaining the bound

$$r_4(F^n) = O\left(\frac{|F|^n}{n^c}\right);$$

see [161].

It is likely that the above inverse theory extends to higher values of k, but there are some technical difficulties in carrying this out, and this has not yet been achieved at this time of writing.

Given the success of the inverse U^3 approach to establish in the finite field case, one then is led to see whether a similar inverse theorem holds for other groups, such as cyclic groups $\mathbf{Z}_N$. Here one encounters an interesting phenomenon, which is that the quadratic phase functions on $\mathbf{Z}_N$ do *not* form a complete set of obstructions to Gowers uniformity of order 2. An example is given as follows.

Proposition 11.8 (Furstenberg–Weiss example) *Let N be a large integer, and let $M := \lfloor\sqrt{N}\rfloor$, and let α be an irrational number obeying the diophantine condition $\|n\alpha\|_{\mathbf{R}/\mathbf{Z}} = \Omega(n^{-C})$ for some constant $C > 0$. Define the function $f : \mathbf{Z}_N \to \mathbf{C}$ by $f(x) := e(\alpha\lfloor x/M\rfloor^2)$ when $x \in [0, M/10) + M \cdot [0, M/10)$, and $f(x) := 0$ otherwise. Then $\|f\|_{U^3(\mathbf{Z}_N)} = \Theta(1)$, but $\|f\|_{u^3(\mathbf{Z}_N)} = o_{N\to\infty;\alpha}(1)$.*

As the name implies, this example was essentially discovered by Furstenberg and Weiss[126], though in a substantially different language to that presented here (they constructed a characteristic factor for quadruple recurrence which was not given by quadratic eigenfunctions).

Proof (Sketch) We can write $f(x) = e(\phi(x))1_P(x)$, where P is the progression $P := [0, M/10) + M \cdot [0, M/10)$ and ϕ is the phase function $\phi(x) := \alpha\lfloor x/M\rfloor^2$. One can easily verify that ϕ is locally quadratic on ϕ, and hence by (11.7)

$$\|f\|_{U^3(\mathbf{Z}_N)} = \|1_P\|_{U^3(\mathbf{Z}_N)} \geq \|1_P\|_{U^1(\mathbf{Z}_N)} = \mathbf{P}_{\mathbf{Z}_N}(P) = \Theta(1).$$

On the other hand, since f is bounded by 1, we have $\|f\|_{U^3(\mathbf{Z}_N)} \leq 1$. Thus $\|f\|_{U^3(\mathbf{Z}_N)} = \Theta(1)$ as claimed.

To prove the second claim, we see from (11.2.2) that it suffices to show that

$$\mathbf{E}_{x\in\mathbf{Z}_N} e(\phi(x) + (c_2x^2 + c_1x + c_0)/N)1_P(x) = o_{N\to\infty;\alpha}(1)$$

for all integers c_0, c_1, c_2. Writing $x = yM + z$ for $y, z \in [0, M/10)$, it suffices to show that

$$\mathbf{E}_{y,z\in[0,M/10)} e(\alpha y^2 + (c_2(yM + z)^2 + c_1(yM + z) + c_0)/N) = o_{N\to\infty;\alpha}(1).$$

To estimate this sum one has two choices. Either one can apply van der Corput's lemma (Exercise 11.2.9) twice in the y variable (with $H_1 = M^{1-\varepsilon}$ and $H_2 = M^{1-2\varepsilon}$ for some small ε), and reduce to showing that

$$\mathbf{E}_{h_1\in[1,H_1],h_2\in[1,H_2]}e(2(\alpha + c_2M^2)h_1h_2) = o_{N\to\infty;\alpha}(1);$$

or one can apply van der Corput's lemma once in the y variable and once in the z variable to reduce to showing that

$$\mathbf{E}_{h_1\in[1,H_1],h_2\in[1,H_2]}e(2c_2Mh_1h_2) = o_{N\to\infty;\alpha}(1).$$

While neither of these two bounds holds uniformly in c_2, it turns out that one of the two bounds is always true, the latter in the "minor arc" case when c_2M is not within $M^{-2+O(\varepsilon)}$ to being a rational with denominator at most $M^{O(\varepsilon)}$, and the former in the complementary "major arc" case. The exact verification of the bounds requires some basic machinery from Diophantine approximation, but we omit it as it is somewhat messy. □

This example shows that in addition to the *globally quadratic* phase obstructions that appeared in the finite field case, we now must consider *locally quadratic* phase obstructions, which are only defined on a suitable progression in the group such as $[0, M/10) + M\cdot[0, M/10)$. One can alternatively replace progressions with Bohr sets, which are of course closely related (cf. Section 4.4). A typical inverse theorem in this setting is as follows.

Theorem 11.9 (Inverse theorem for $U^3(Z)$) *[160] Let Z be an finite additive group of odd order, let $f : Z\to\mathbf{C}$ be a function bounded in magnitude by 1, such that $\|f\|_{U^3(Z)}\geq\eta$. Then there exists a regular Bohr set $B := B(S,\rho)$ in G with $|S|\leq O(\eta^{-O(1)})$ and $\rho = \Omega\big(\eta^{O(1)}\big)$ such that*

$$\mathbf{E}_{y\in Z}\|f\|_{u^3(y+B)} = \Omega\big(\eta^{O(1)}\big). \tag{11.16}$$

In particular, there exists $y\in Z$ such that $\|f\|_{u^3(y+B)} = \Omega(\eta^{O(1)})$.

The proof of this theorem is similar to that of Theorem 11.6 which we give below, but is somewhat more complicated as we must work with (regular) Bohr sets instead of subspaces (which ultimately arises from the application of a version of Chang's theorem, Theorem 4.42, for arbitrary groups). It can then be used to prove

Proposition 11.10 (Lack of uniformity implies density increment) *[137], [138] Let $Z = \mathbf{Z}_N$ be a cyclic group of odd prime order, and let $f : Z\to\mathbf{C}$ have magnitude bounded by 1 be such that $\mathbf{E}_Z(f) = 0$ and $\|f\|_{U^3(Z)}\geq\eta$ for some $0<\eta\leq 1$. If $N\geq\exp(O(\eta^{-O(1)})$, then there exists a proper arithmetic* *progression P in Z of length $|P| = \Omega(N^c)$ for some absolute constant $0 < c < 1$ such that*

$$\mathbf{E}_{x \in P} f(x) \geq \Omega(\eta^{O(1)}).$$

This result was first established by Gowers[1] [137], [138] without directly proving an inverse theorem. However, the method of proof of Theorem 11.9 in [160] is based almost entirely the techniques used in [137] to establish Proposition 11.10. By the usual iteration arguments, this proposition can be used to establish the bound

$$r_4(\mathbf{Z}_N) = O\left(\frac{N}{(\log\log N)^c}\right)$$

for some absolute constant $0 < c < 1$ and all large N; this is the best bound on $r_4(\mathbf{Z}_N)$ known to date. See [137], [138], [160] for further discussion. In a similar spirit, Theorem 11.9 can eventually be used to establish the more general result

$$r_4(Z) = O\left(\frac{|Z|}{(\log\log|Z|)^c}\right)$$

for any large finite additive group Z; see [160]. It seems likely that this bound can be improved to $O(\frac{|Z|}{\log^c|Z|})$ by using the arguments in Theorem 10.27 or Theorem 10.30 but this will probably be quite messy.

Exercises

11.2.1 Prove (11.11).

11.2.2 Let $\mathbf{Z}_N$ be a cyclic group (and thus also a ring), and let $\phi : \mathbf{Z}_N \to \mathbf{R}/\mathbf{Z}$ be a phase polynomial of degree less than d. Show that there exist $c_0, c_1, \ldots, c_{d-1} \in \mathbf{Z}_N$ such that $\phi(x) = (c_{d-1}x^{d-1} + \cdots + c_1 x + c_0)/N$ for all $x \in \mathbf{Z}_N$, where the map $x \mapsto x/N$ is defined from $\mathbf{Z}_N$ to $\mathbf{R}/\mathbf{Z}$ in the obvious manner. Conversely, every function of this form is a phase polynomial of degree less than d. Thus in the cyclic case, the concept of a phase polynomial collapses to the usual definition of a polynomial.

11.2.3 Prove (11.12). (You may need to reflect some of the variables or take conjugates to eliminate a $(-1)^d$ factor.)

11.2.4 Let $f : Z \to \mathbf{C}$ be a function bounded in magnitude by 1, and let $d \geq 1$. Show that $\|f\|_{u^d(Z)}, \|f\|_{U^d(Z)} \leq 1$, and that $\|f\|_{u^d(Z)} = 1$ if and only if $\|f\|_{U^d(Z)} = 1$.

[1] The original argument in [137] had a exponential dependence on η rather than a polynomial one for $\mathbf{E}_{x \in P} f(x)$, leading ultimately to the weaker bound of $O(\frac{N}{(\log\log\log N)^c})$ for $r_4(\mathbf{Z}_N)$. This is due to a reliance on Freiman's theorem instead of a Chang–Bogulybov type theorem; the problem being that the Freiman theorem employed (essentially Theorem 5.32) suffers an exponential loss in an unfavorable location.

11.2.5 Show that the $u^d(Z)$ norm enjoys the same invariances that the $U^d(Z)$ norm did in Exercises 11.1.8, 11.1.9, 11.1.10, as well as an analog of (11.7). Show that the more general $u^d(B)$ norms also obey a suitable analog of Exercises 11.1.9, 11.1.10.

11.2.6 [160] Let $F = F_p$ be a finite field of odd prime order, and let Z be a finite-dimensional vector space over F, with the usual bilinear form. Show that if $\phi : Z \to \mathbf{R}/\mathbf{Z}$ is a quadratic phase function, then we have the representation $\phi(x) = Mx \cdot x + \xi \cdot x + c$ for some unique $c \in \mathbf{R}/\mathbf{Z}, \xi \in Z$, and a self-adjoint F-linear operator $M : Z \to Z$. Conversely, every function of this form is a quadratic phase function. What happens if $F = F_2$ has order 2?

11.2.7 [160] (Quadratic Hahn–Banach theorem) Let F and Z be as in the preceding exercise, and let Z' be a subspace of Z. Show that any quadratic phase function on Z' can be extended (possibly non-uniquely) to a quadratic phase function on Z. Conclude in particular that for any $f : Z \to \mathbf{C}$ we have $\|f\|_{U^3(Z)} \geq \|f\|_{u^3(Z)} \geq \mathbf{P}_Z(Z') \sup_{y \in Z} \|f\|_{u^3(y+Z')}$; this can be viewed as a kind of converse to Theorem 11.6.

11.2.8 Use Proposition 11.7 and (11.8) to establish (11.15).

11.2.9 (Van der Corput lemma) If $1 \leq H < M$ and $f : [0, M) \to \mathbf{C}$ is a function bounded in magnitude by 1, show that

$$|\mathbf{E}_{x \in [0,M)} f(x)| \leq O(\mathbf{E}_{1 \leq h \leq H} |\mathbf{E}_{x \in [0,M-H)} f(x+h)\overline{f(x)}|)^{1/2} + O\left(\frac{H^{1/2}}{M^{1/2}}\right) + O\left(\frac{1}{H^{1/2}}\right).$$

(Hint: extend f by zero to the integers $\mathbf{Z}$, and obtain a preliminary upper bound of $(\mathbf{E}_{1 \leq h \leq H} |\mathbf{E}_{x \in [0,M-H)} f(x+h)|^2)^{1/2} + O(\frac{H^{1/2}}{M^{1/2}})$.) Compare this with Lemma 11.3.

11.3 Proof of Theorem 11.6

In this section we give a proof of Theorem 11.6. Let us fix F, Z, f, η with the above properties. The proof proceeds in several stages.

11.3.1 Locating a somewhat linear phase derivative

The first step is to apply the inverse theorem (11.9) for the $U^2(Z)$ norm. From the recursive definition of the $U^3(Z)$ norm we have

$$\mathbf{E}_{h \in Z} \|T^h f \overline{f}\|_{U^2(Z)}^4 \geq \eta^8$$

and hence by (11.9)

$$\mathbf{E}_{h\in Z}\|T^h f\overline{f}\|^2_{u^2(Z)} \geq \eta^8.$$

If we let $H \subset Z$ be the set

$$\{h \in H : \|T^h f\overline{f}\|^2_{u^2(Z)} \geq \eta^8/2$$

then we have

$$\mathbf{E}_{h\in Z}\|T^h f\overline{f}\|^2_{u^2(Z)} \leq \eta^8/2 + \mathbf{P}_Z(H)$$

and hence

$$\mathbf{P}_Z(H) \geq \eta^8/2. \tag{11.17}$$

By definition of H, we can thus find a function $\xi : H \to Z$ such that

$$|\mathbf{E}_{x\in Z} T^h f(x)\overline{f(x)}e(-\xi(h)\cdot x)|^2 \geq \eta^8/2 \tag{11.18}$$

for all $h \in H$. Informally, if we use $\phi(x)$ to denote the phase of $f(x)$, this estimate is asserting that $\phi(x+h) - \phi(x) - \xi(h)\cdot x$ is in some sense approximately constant in x, so that $\phi(x+h) - \phi(x)$ is approximately linear. The challenge is thus to "integrate" this fact and conclude that ϕ is somehow approximately quadratic. To do this, the first task shall be to obtain some linearity of $\xi(h)$ (this reflects the fact that we expect the quantity $(h\cdot\nabla)\phi$ to somehow be linear in h). We sum the preceding expression over all $h \in H$ using (11.17) and conclude

$$\mathbf{E}_{h\in Z} 1_H(h)|\mathbf{E}_{x\in Z} T^h f(x)\overline{f(x)}e(-\xi(h)\cdot x)|^2 \geq \eta^{16}/4.$$

Expanding this out as in Lemma 11.3 we conclude

$$|\mathbf{E}_{x,h,k\in Z} 1_H(h) T^{h+k} f(x)\overline{T^h f(x) T^k f(x)} f(x) e(\xi(h)\cdot k)| \geq \eta^{16}/4.$$

In order to focus on ξ, we suppress the explicit mention of the functions f using the $\mathbf{b}()$ notation. After collecting some terms we obtain

$$|\mathbf{E}_{x,h,k\in Z}\mathbf{b}(x+h,k)\mathbf{b}(x,k)1_H(h)e(\xi(h)\cdot k)| \geq \eta^{16}/4.$$

We can eliminate the $\mathbf{b}(x,k)$ factor using Lemma 11.3, concluding that

$$\begin{aligned}|\mathbf{E}_{x,h,h_1,k\in Z}&\mathbf{b}(x+h,k)\mathbf{b}(x+h+h_1,k)1_H(h)1_H(h+h_1)e((h_1\cdot\nabla)\xi(h)\cdot k)| \\ &\geq \eta^{32}/16.\end{aligned}$$

Making the substitution $y = x+h$ and collecting some terms this becomes

$$|\mathbf{E}_{y,h,h_1,k\in Z}\mathbf{b}(y,k,h_1)1_H(h)1_H(h+h_1)e((h_1\cdot\nabla)\xi(h)\cdot k)| \geq \eta^{32}/16.$$

We can eliminate $\mathbf{b}(y, k, h_1)$ using Lemma 11.3, concluding that

$$|\mathbf{E}_{y,h,h_1,h_2,k\in Z} 1_H(h)1_H(h+h_1)1_H(h+h_2)1_H(h+h_1+h_2) \\ e((h_2 \cdot \nabla)(h_1 \cdot \nabla)\xi(h) \cdot k)| \geq \eta^{64}/256.$$

The point of eliminating all the $\mathbf{b}()$ factors now becomes clear, as the y averaging can be dropped, and we can sum the k sum using Lemma 4.5, to obtain

$$|\mathbf{E}_{h,h_1,h_2\in Z} 1_H(h)1_H(h+h_1)1_H(h+h_2)1_H(h+h_1+h_2) \\ \mathbf{I}((h_2 \cdot \nabla)(h_1 \cdot \nabla)\xi(h) = 0)| \geq \eta^{64}/256$$

or in other words

$$\mathbf{P}_{h,h_1,h_2\in Z}(h, h+h_1, h+h_2, h+h_1+h_2 \in H; \\ \xi(h+h_1+h_2) - \xi(h+h_1) - \xi(h+h_2) + \xi(h) = 0) \geq \eta^{64}/256.$$

This is as assertion that ξ behaves approximately like a Freiman homomorphism of order 2; observe that the unknown function f and the phase oscillations have completely disappeared from this estimate. This allows us to now employ the tools of additive combinatorics.

11.3.2 Obtaining a perfectly linear phase derivative

We now convert the somewhat linear phase function $\xi(h)$ into a genuinely linear phase function. This shall be done using the inverse sum set technology of previous chapters, though one needs to be a little careful to make sure that the density bounds one obtains are polynomial in the η parameter rather than exponential.

Let $\Gamma \subset Z \times Z$ denote the set

$$\Gamma := \{(h, \xi(h)) : h \in H\};$$

then the above statement can be rephrased as a lower bound on the additive energy of Γ (see Definition 2.8):

$$E(\Gamma, \Gamma) \geq \eta^{64}|Z|^3/256.$$

On the other hand, we have $|\Gamma| \leq |H| \leq |Z|$. We can thus apply the Balog–Szemerédi–Gowers theorem, Theorem 2.31, to conclude that there exists a $O(\eta^{-O(1)})$-approximate group $G \subset Z \times Z$ of cardinality $O(\eta^{-O(1)}|Z|)$, such that

$$|\Gamma \cap (G + (h_0, \xi_0))| = \Omega\big(\eta^{O(1)}|Z|\big)$$

for some $(h_0, \xi_0) \in Z \times Z$. In particular $|G| = \Omega(\eta^{O(1)}|Z|)$. We can analyze G further using a Freiman-type or Chang-type theorem. There are many ways to do this; we shall use Corollary 5.29. This shows that $2G - 2G$ contains a subspace V of $Z \times Z$ of size

$$|V| = \Omega\big(|F|^{-O(\eta^{-O(1)})}|G|\big) = \Omega\big(|F|^{-O(\eta^{-O(1)})}|Z|\big),$$

or in other words

$$\dim_F(V) \geq \dim_F(Z) - O\big(\eta^{-O(1)}\big).$$

Since G is a $O(\eta^{-O(1)})$-approximate group, we see that

$$|G + V| \leq |G + 2G - 2G| \leq O\big(\eta^{-O(1)}\big)|G| = O\big(\eta^{-O(1)}\big)|Z|$$

and thus

$$\begin{aligned} |\Gamma \cap (G + V + (h_0, \xi_0))| &\geq |\Gamma \cap (G + (h_0, \xi_0))| \\ &= \Omega\big(\eta^{O(1)}|Z|\big) \\ &= \Omega\big(\eta^{O(1)}|G + V|\big). \end{aligned}$$

Splitting $G + V$ into $|G + V|/|V|$ cosets of V (this is a very special case of the Ruzsa covering lemma, Lemma 2.14) and using the pigeonhole principle, we can thus find a coset $V + (h_1, \xi_1)$ of V such that

$$|\Gamma \cap (V + (h_1, \xi_1))| = \Omega\big(\eta^{O(1)}|V|\big). \tag{11.19}$$

Thus we have replaced the approximate group G with a genuine subspace V, though V is somewhat smaller than G.

Let $V_0 := V \cap (0 \times \{Z\})$ denote the vertical component of V. Since Γ is a (partial) graph, we see that all the sums in $\Gamma + V_0$ are distinct. In particular we have

$$|V_0||\Gamma \cap (V + (h_1, \xi_1))| = |\Gamma \cap (V + (h_1, \xi_1)) + V_0| \leq |V + (h_1, \xi_1)| = |V|,$$

which, when combined with (11.19), gives the bound $|V_0| = O(\eta^{-O(1)})$. Now from elementary linear algebra we can write $V = V_0 + V_1$, where $V_1 = \{(h, Mh) : h \in W_1\}$ is the graph of a linear transformation $M : W_1 \to Z$, and W_1 is a subspace of Z of dimension

$$\dim_F(W_1) = \dim_F(V_1) = \dim_F(V) - \dim_F(V_0) \geq \dim_F(Z) - O\big(\eta^{-O(1)}\big).$$

Covering V by translates of V_1 and applying the pigeonhole principle to (11.19), we can find a coset $V_1 + (h_2, \xi_2)$ of V_1 such that

$$|\Gamma \cap (V_1 + (h_2, \xi_2))| = \Omega\big(\eta^{O(1)}|V_2|\big).$$

Unfolding the definition of Γ, we conclude that

$$\mathbf{P}_{h \in W_1}(h + h_2 \in H; \xi(h + h_2) = Mh + \xi_2) = \Omega\big(\eta^{O(1)}\big).$$

Thus we have established that ξ exhibits exact linear behavior on a large fraction of a coset $h_2 + W_1$. Recalling the definition (11.18) of $\xi(h)$, we thus have

$$\mathbf{P}_{h\in W_1}(|\mathbf{E}_{x\in Z}T^{h+h_2}f(x)\overline{f(x)}e(-(Mh+\xi_2)\cdot x)|^2 \geq \eta^8/2) = \Omega\big(\eta^{O(1)}\big). \quad (11.20)$$

Ignoring the lower-order terms h_2 and ξ_2, and writing $\phi(x)$ for the phase of $f(x)$, this estimate is informally asserting that $(h\cdot\nabla_x)\phi(x) \approx Mh\cdot x$ for a large fraction of x and h. We would like to somehow "integrate" this and conclude that $\phi(x)$ behaves like $\frac{1}{2}Mx\cdot x$. However it turns out that to achieve this we need to ensure that M is somehow "symmetric". This is the purpose of the next stage of the argument.

11.3.3 Symmetrizing the derivative

We now show that (11.20) forces a certain symmetry property on M. Note that this estimate implies that

$$|\mathbf{E}_{h\in W_1}\mathbf{b}(h)\mathbf{E}_{x\in Z}T^{h+h_2}f(x)\overline{f(x)}e(-(Mh+\xi_2)\cdot x)| = \Omega\big(\eta^{O(1)}\big)$$

for some choice of bounded function $\mathbf{b}(h)$. We will focus on the term $e(-Mh\cdot x)$, and conceal all the other terms using the $\mathbf{b}()$ notation, thus obtaining

$$|\mathbf{E}_{h\in W_1;x\in Z}\mathbf{b}(h)\mathbf{b}(x)\mathbf{b}(x+h)e(-Mh\cdot x)| = \Omega\big(\eta^{O(1)}\big).$$

Splitting Z into cosets of W_1 and using the pigeonhole principle, we can find $x_1 \in Z$ such that

$$|\mathbf{E}_{x,h\in W_1}\mathbf{b}(h)\mathbf{b}(x+x_1)\mathbf{b}(x+h+x_1)e(-Mh\cdot(x+x_1))| = \Omega\big(\eta^{O(1)}\big);$$

absorbing the x_1 factors into the $\mathbf{b}()$ notation we conclude

$$|\mathbf{E}_{x,h\in W_1}\mathbf{b}(h)\mathbf{b}(x)\mathbf{b}(x+h)e(-Mh\cdot x)| = \Omega\big(\eta^{O(1)}\big).$$

Now we proceed as in previous steps, using Lemma 11.3 to eliminate the $\mathbf{b}()$ terms, though this time we get rid of the variables in a slightly different way. First we eliminate the $\mathbf{b}(h)$ factor using Lemma 11.3 to conclude

$$|\mathbf{E}_{x,y,h\in W_1}\mathbf{b}(x)\mathbf{b}(y)\mathbf{b}(x+h)\mathbf{b}(y)e(-Mh\cdot(y-x)\})| = \Omega\big(\eta^{O(1)}\big).$$

Normally we would make the substitution $y = x + h'$, but instead we make the substitution $z = x + y + h$ to obtain

$$|\mathbf{E}_{x,y,z\in W_1}\mathbf{b}(z,x)\mathbf{b}(z,y)e(-M(z-x-y)\cdot(y-x))| = \Omega\big(\eta^{O(1)}\big). \quad (11.21)$$

Since

$$\begin{aligned} &e(-M(z-x-y)\cdot(y-x)) \\ &\quad = e(Mx\cdot y - My\cdot x)e(-Mz\cdot y + My\cdot y)e(Mz\cdot x - Mx\cdot x) \end{aligned}$$

we conclude (after absorbing some factors into the $\mathbf{b}()$ terms) that

$$|\mathbf{E}_{x,y,z\in W_1}\mathbf{b}(z,x)\mathbf{b}(z,y)e(Mx\cdot y - My\cdot x)| = \Omega\big(\eta^{O(1)}\big).$$

Pigeonholing in z, we derive

$$|\mathbf{E}_{x,y\in W_1}\mathbf{b}(x)\mathbf{b}(y)e(Mx\cdot y - My\cdot x)| = \Omega\big(\eta^{O(1)}\big).$$

We eliminate $\mathbf{b}(x)$ using Lemma 11.3 to conclude

$$|\mathbf{E}_{x,h,y\in W_1}\mathbf{b}(y)\mathbf{b}(y+h)e(Mx\cdot h - Mh\cdot x)| = \Omega\big(\eta^{O(1)}\big)$$

Applying the triangle inequality to eliminate $\mathbf{b}(y)$, $\mathbf{b}(y+h)$ we deduce

$$\mathbf{E}_{h,y\in W_1}|\mathbf{E}_{x\in W_1}e(Mx\cdot h - Mh\cdot x)| = \Omega\big(\eta^{O(1)}\big).$$

Introduce the symmetry space

$$W := \{h \in W_1 : Mx\cdot h = Mh\cdot x \text{ for all } h \in W_1\}.$$

Then from Lemma 4.5 we have

$$\mathbf{E}_{x\in W_1}e(Mx\cdot h - Mh\cdot x) = \mathbf{I}(h\in W)$$

and thus

$$|W|/|W_1| = \Omega\big(\eta^{O(1)}\big).$$

In particular we have

$$\dim_F(W) \geq \dim_F(W) - O\left(\log\frac{1}{\eta}\right) \geq \dim_F(Z) - O\big(\eta^{-O(1)}\big).$$

Returning to (11.20), we see from covering W_1 by translates of W and using the pigeonhole principle that there exists $h_3 \in Z$ such that

$$\mathbf{P}_{h\in W}(|\mathbf{E}_{x\in Z}T^{h+h_2+h_3}f(x)\overline{f(x)}e(-(M(h+h_3)+\xi_2)\cdot x)|^2 \geq \eta^8/2) = \Omega\big(\eta^{O(1)}\big). \tag{11.22}$$

11.3.4 Eliminating the quadratic phase

We are now ready to finish the proof of Theorem 11.6. From (11.22) we see in particular that

$$|\mathbf{E}_{h\in W}\mathbf{b}(h)\mathbf{E}_{x\in Z}T^{h+h_2+h_3}f(x)\overline{f(x)}e(-(M(h+h_3)+\xi_2)\cdot x)| = \Omega\big(\eta^{O(1)}\big)$$

for some bounded $\mathbf{b}(h)$. We now focus on the $\overline{f(x)}$ term and conceal many of the other terms using the $\mathbf{b}()$ notation, obtaining

$$|\mathbf{E}_{h\in W;x\in Z}\mathbf{b}(h)\mathbf{b}(x+h)\overline{f(x)}e(-Mh\cdot x)| = \Omega\big(\eta^{O(1)}\big),$$

where we used the identity $e(\xi \cdot x) = e(\xi \cdot (x+h))e(-\xi \cdot h)$ to eliminate the phase terms which were linear in x. Splitting x into cosets of W and using the triangle inequality we conclude

$$|\mathbf{E}_{y\in Z}\mathbf{E}_{h,x\in W}\mathbf{b}(h)\mathbf{b}(x+y+h)\overline{f(x+y)}e(-Mh\cdot(x+y))| = \Omega(\eta^{O(1)}),$$

which we rewrite as

$$|\mathbf{E}_{y\in Z}\mathbf{E}_{h,x\in W}\mathbf{b}(h,y)\mathbf{b}(x+h,y)\overline{f(x+y)}e(-Mh\cdot x)| = \Omega(\eta^{O(1)}).$$

By construction of W, we know that $Mh \cdot x = Mx \cdot h$ for all $x, h \in W$. We now divide into two cases depending on whether F has characteristic 2 or not. If F has odd characteristic, then we have

$$e(-Mh\cdot x) = e\left(-\frac{1}{2}M(x+h)\cdot(x+h)\right)e\left(\frac{1}{2}Mx\cdot x\right)e\left(\frac{1}{2}Mh\cdot h\right).$$

If we then set $f_y : W \to \mathbf{C}$ to be the function

$$f_y(x) = \overline{f(x+y)}e\left(\frac{1}{2}Mx\cdot x\right)$$

we conclude that

$$|\mathbf{E}_{y\in Z}\mathbf{E}_{h,x\in W}\mathbf{b}(h,y)\mathbf{b}(x+h,y)f_y(x)| = \Omega(\eta^{O(1)}).$$

On the other hand, from Lemma 11.4 (after a linear change of variables) we see that

$$|\mathbf{E}_{h,x\in W}\mathbf{b}(h,y)\mathbf{b}(x+h,y)f_y(x)| \le \|f_y\|_{U^2(W)}$$

and thus by (11.9)

$$\left|\mathbf{E}_{y\in Z}\|f_y\|_{u^2(W)}^{1/2}\right| = \Omega(\eta^{O(1)}).$$

By Cauchy–Schwarz we conclude

$$|\mathbf{E}_{y\in Z}\|f_y\|_{u^2(W)}| = \Omega(\eta^{O(1)}).$$

Since the $u^3(W)$ norm controls the $u^2(W)$ norm, and the quadratic phase $e(\frac{1}{2}Mx \cdot x)$ does not affect the $u^3(W)$ norm, we have

$$\|f_y\|_{u^2(W)} \le \|f_y\|_{u^3(W)} = \|\overline{f(y+\cdot)}\|_{u^3(W)} = \|f\|_{u^3(y+W)}$$

and we obtain (11.14) as desired.

Now we argue for the case when F has characteristic 2, using an observation of Alex Samorodnitsky (private communication). Since M is symmetric on W, the function $x \mapsto Mx \cdot x$ is in fact linear on W (here we rely on the characteristic 2 hypothesis). Thus we can write $Mx \cdot x = \xi \cdot x$ for some $\xi \in W$. By passing to the orthogonal complement of ξ in W if necessary we may assume that $\xi = 0$, thus $Mx \cdot x = 0$ for all $x \in W$. This allows us to find a transformation $A : W \to W$ such that $Mh \cdot x = Ah \cdot x + Ax \cdot h$; for instance, one can write M as a matrix with coefficients in F, use the hypothesis $Mx \cdot x = 0$ to show that the matrix has zero diagonal, and then take A to be the upper triangular portion of M. We then have

$$e(-Mh \cdot x) = e(-A(x+h) \cdot (x+h))e(Ax \cdot x)e(Ah \cdot h)$$

and the rest of the argument proceeds as before. □

Remark 11.11 The fact that we have to pass from the original space Z to a subspace W of somewhat lower dimension is a defect of the argument. If one knew the polynomial Freiman–Ruzsa conjecture (Conjecture 5.34) one could set $W = Z$, which would lead to somewhat stronger results in applications.

We now comment briefly on extending these arguments to higher k, to obtain Szemerédi's theorem in general. At this time of writing the inverse U^3 theorem has not been extended to higher k, even in the simple case of a vector space over a finite field. However, Proposition 11.10 has been extended successfully to general k:

Proposition 11.12 (Lack of uniformity implies density increment) *[138] Let $Z = \mathbf{Z}_N$ be a cyclic group of odd prime order, let $k \geq 3$, and and let $f : Z \to \mathbf{C}$ have magnitude bounded by 1 be such that $\mathbf{E}_Z(f) = 0$ and $\|f\|_{U^{k-1}(Z)} \geq \eta$ for some $0 < \eta \leq 1$. If $N \geq \exp(O_k(\eta^{-O_k(1)})$, then there exists a proper arithmetic progression P in Z of length $|P| = \Omega(N^{c_k})$ for some absolute constant $0 < c_k < 1$ such that*

$$\mathbf{E}_{x \in P} f(x) \geq \Omega_k\left(\eta^{O_k(1)}\right).$$

This leads ultimately to the bound

$$r_k(\mathbf{Z}_N) = O\left(\frac{N}{(\log\log N)^{c_k}}\right) \qquad (11.23)$$

for all $k \geq 3$ and large N, where $c_k > 0$ depends only on k; in fact in [138] the explicit value $c_k = 1/2^{2^{k+9}}$ is attained. This is currently the best bound known for $r_k(\mathbf{Z}_N)$ for general $k \geq 4$ and large N. It is however likely that this can be improved to $O(\frac{N}{(\log N)^{c_k}})$ based on analogy with the $k = 3$ case.

The proof of Proposition 11.12 is quite lengthy and difficult. In principle, one wishes to induct on k, leveraging inverse theorems for U^{k-2} to obtain inverse theorems for U^{k-1}. This was the strategy employed at the start of the proof of Theorem 11.6, using the simple inverse theorem (11.9) for U^2 to create the partially defined derivative $\xi(h)$, which one then obtains arithmetic structure on. Unfortunately this strategy has not yet been made to work even for $k = 5$ and for the model case of a vector space over a finite field, mainly because the inverse theorem for U^3 is much weaker than that for U^2, in particular involving an unknown space W (or a Bohr set B), which will ultimately depend on a certain shift parameter h in an unpleasant way. To prove Proposition 11.12, Gowers employed a slightly different approach, starting with the original function f and taking $k-3$ "derivatives" $f \mapsto T^h f \overline{f}$ to reduce the U^{k-1} norm to the U^2 norm. Employing the U^2 inverse theorem, one then obtains a $k-3$-fold derivative function $\xi(h_1, \ldots, h_{k-3})$. The strategy is then to establish some multilinearity properties of this function ξ in order to execute a similar scheme to the one described above. This requires a substantial amount of new combinatorial technology, not least of which is a multilinear version of the Balog–Szemerédi–Gowers theorem, which cannot be established simply by applying the Balog–Szemerédi–Gowers theorem separately in each variable (again because of the issue that the structures obtained in this way for one variable will depend on the other variables). See [138] for details.

Exercises

11.3.1 (Alex Samorodnitsky, private communication) Let $f : Z \to \mathbf{C}$, and let $D : Z \times Z \to \mathbf{R}^+$ denote the quantity $D(h, \xi) := |\widehat{T^h f \overline{f}}(\xi)|^2$. Establish the identity

$$\sum_{\xi_1,\xi_2,\xi_3,\xi_4 \in Z:\xi_1+\xi_2=\xi_3+\xi_4} \mathbf{E}_{h_1,h_2,h_3,h_4 \in Z:h_1+h_2=h_3+h_4} \prod_{j=1}^{4} D(h_j, \xi_j) = \sum_{\xi \in Z} \mathbf{E}_{h \in Z} D(h, \xi)^4.$$

(Hint: first show that D is essentially its own Fourier transform.) This identity can be used as a substitute for the first part of the above argument.

11.4 Soft obstructions to uniformity

In the last two sections we described the approach of Gowers in proving Szemerédi's theorem. There were three main components to the argument. First, there was the generalized von Neumann theorem (11.8) which showed among other things that one could approximate $\Lambda_k(f, \ldots, f)$ by $\mathbf{E}_Z(f)^k$ as long as $f - \mathbf{E}_Z(f)$ was sufficiently Gowers uniform of order $k-2$. Second, there was the inverse theorem, which implied that if $f - \mathbf{E}_Z(f)$ was not Gowers uniform of order $k-2$ then there was enough structure on f to conclude a density increment for f on a subspace or sub-progression of Z. Finally there was the standard density incrementation argument that iterated the previous two observations to conclude the proof of Szemerédi's theorem.

Of the three components mentioned above, the second was by far the most difficult. The reason is that this approach requires a rather strong type of inverse theorem, and in particular requires one to give quite "concrete" or "hard" obstructions to Gowers uniformity, in order to conclude the desired density increment. There is however an alternative approach, similar to the finitary ergodic argument given in Section 10.5, which requires much "softer" obstructions to Gowers uniformity, in the sense that these obstructions are not presented in as explicit a form as, say, a polynomial phase function. This makes the second stage of the argument immensely simpler. However, one must now make the third stage of the argument more complicated, replacing the density incrementation argument by an energy incrementation argument, and then establishing some sort of recurrence result for the soft obstructions. This last step now becomes rather difficult, for instance involving van der Waerden's theorem. One consequence of this is that the quantitative bounds obtained by this method are extremely poor. Nevertheless, this approach is quite robust, requiring very little arithmetic structure as compared with Gowers' approach.

To describe this approach to Szemerédi's theorem, let us first review the ingredients used in the finitary ergodic proof of Roth's theorem in Section 10.5. The strategy was to approximate the original function f by some low complexity approximation $f_{U^\perp}$, such that the error $f_U = f - f_{U^\perp}$ was suitably uniform. One achieves this iteratively: if one has some preliminary approximation $f_{U^\perp}$ whose error f_U is not sufficiently uniform, then one concludes that f_U correlates with a certain obstruction to uniformity, which in this case was a character e_ξ. One then constructs a σ-algebra out of this obstruction e_ξ and uses that algebra to refine the approximation $f_{U^\perp}$ to f, increasing the energy (L^2 norm) of $f_{U^\perp}$ in the process. One repeats this procedure until the error finally becomes uniform (and hence negligible). The only remaining task is then to establish some recurrence property for the approximation $f_{U^\perp}$, namely a lower bound on $\Lambda_k(f_{U^\perp}, \ldots, f_{U^\perp})$. The key here was that the approximation $f_{U^\perp}$ was built out of the σ-algebras associated to characters, and was hence *almost periodic*; this led to a non-trivial recurrence property for $f_{U^\perp}$.

The above argument used Fourier analysis by involving the characters e_ξ. However, one could replace this family of functions by any other family of functions, provided that two properties hold: firstly, that there were enough functions to provide a complete set of obstructions to Gowers uniformity of order $k-2$, and secondly, that any function generated by these functions (or more precisely by their associated σ-algebras) had enough "almost periodicity" to lead to recurrence.

Using this observation, it becomes possible to dispense with Fourier analysis altogether by working with a somewhat different family of functions, replacing the characters with *dual functions of order* $k-1$ and almost periodic functions with *uniformly almost periodic functions of order* $k-2$.

We now discuss these concepts in more detail. We begin with the concept of a dual function.

Definition 11.13 (Dual function) If $f : Z \to \mathbf{C}$ and $d \geq 1$, we define the *dual function* $\mathcal{D}_d(f) : Z \to \mathbf{C}$ recursively by

$$\mathcal{D}_1(f)(x) = \mathbf{E}_Z(f); \quad \mathcal{D}_{d+1}(f)(x) = \mathbf{E}_{h\in Z} T^h f(x)\overline{\mathcal{D}_d(T^h f \overline{f})(x)}.$$

When $d = 2$ one can compute the dual function in terms of the Fourier transform:

$$\begin{aligned} \mathcal{D}_2(f)(x) &= \mathbf{E}_{h\in Z} T^h f(x)\overline{\mathbf{E}_Z(T^h f \overline{f})} \\ &= \mathbf{E}_{h,k\in Z} T^h f(x) T^k f(x) \overline{T^{h+k} f(x)} \\ &= \sum_{\xi\in Z} |\hat{f}(\xi)|^2 \hat{f}(\xi) e(\xi \cdot x). \end{aligned} \tag{11.24}$$

we leave this as an exercise. The formula for higher d is more complicated, for instance

$$\mathcal{D}_3(f)(x) = \mathbf{E}_{h,k,l\in Z} T^h f(x) T^k f(x) T^l f(x) \overline{T^{h+k} f(x) T^{h+l} f(x) T^{k+l} f(x)} T^{h+k+l}(x).$$

We observe the useful translation and conjugation invariance

$$\mathcal{D}_d(T^h f) = T^h \mathcal{D}_d(f); \quad \mathcal{D}_d(\overline{f}) = \overline{\mathcal{D}_d(f)} \tag{11.25}$$

which is easily established by induction.

Dual functions are intimately connected with the Gowers uniformity norm. An easy induction gives the identity

$$\|f\|_{U^d(Z)}^{2^d} = \langle f, \mathcal{D}_d(f)\rangle_{L^2(Z)} = \mathbf{E}_{x\in Z} f(x)\overline{\mathcal{D}_d(f)(x)} \tag{11.26}$$

while from the Gowers–Cauchy–Schwarz inequality (11.6) we have the inequality

$$\left|\langle g, \mathcal{D}_d(f)\rangle_{L^2(Z)}\right| \leq \|g\|_{U^d(Z)} \|f\|_{U^d(Z)}^{2^d-1} \tag{11.27}$$

for all $f, g : Z \to \mathbf{C}$. In particular we have the dual characterization of $U^d(Z)$:

$$\|g\|_{U^d(Z)} = \sup\left\{\left|\langle g, \mathcal{D}_d(f)\rangle_{L^2(Z)}\right| : \|f\|_{U^d(Z)} \leq 1\right\} \tag{11.28}$$

which explains the terminology "dual function". From (11.26) we immediately obtain an easy inverse theorem:

Lemma 11.14 (Soft inverse theorem) *Let $f : Z \to \mathbf{C}$ be a function bounded in magnitude by* 1, *and let $F = \mathcal{D}_d(f)$ be the dual function. If $\|f\|_{U^d(Z)} \geq \eta$, then $|\langle f, F\rangle| \geq \eta^{2^d}$.*

Thus dual functions are a complete set of obstructions to Gowers uniformity, and will play the role that the characters e_ξ played in Section 10.5. (To see the connection, observe that $\mathcal{D}_2(e_\xi) = e_\xi$ for any character e_ξ, thus characters are themselves a kind of dual function.) To use this inverse theorem effectively in the finitary ergodic argument, we need to show that functions that are generated out of σ-algebras of dual functions obey some sort of "almost periodicity" property. The actual definition is rather strange-looking and to motivate it we first give an informal discussion. For sake of concreteness we work in the group $\mathbf{Z}_N$. In this setting, all functions $f : \mathbf{Z}_N \to \mathbf{C}$ are of course periodic of order N, but we are interested in almost periodicity properties which occur for shifts much smaller than N, in the sense that the shifts $T^n f$ are somehow compressed into a space of "dimension" much smaller than N, whatever that means. As it turns out, there will be a different notion of almost periodicity for each order $d - 1$; roughly speaking, a function should be almost periodic of order $d - 1$ if its phase or phases behave like a polynomial of degree $d - 1$.

Let us quantify this intuition with examples. The function $f(x) = e(\xi x/N)$ is a model example of a function which we expect to be "almost periodic of order 1", as its shifts $T^n f$ are quite recurrent. Indeed we have the formula

$$T^n f = c_n f$$

where c_n are the constants $c_n = e(\xi n/N)$. If we instead take the function $f(x) = e(\xi_1 x/N) + e(\xi_2 x/N)$, then this function would still be considered almost periodic of order 1, since we have the formula

$$T^n f = c_{n,1} g_1 + c_{n,2} g_2$$

where $c_{n,j}$ are the constants $c_{n,j} = e(\xi_j n/N)$, and g_j are the bounded functions $g_j(x) = e(\xi_j x/N)$. Thus in this case the shifts $T^n f$ of f only vary in a two-dimensional space.

Next, we consider the function $f(x) = e(ax^2/N)$. This function would not be considered almost periodic in the usual sense, as the shifts seem to take values in a very high-dimensional space (as large as N). Indeed we have the shift formula

$$T^n f = c_n f$$

where the c_n are no longer constant, but are themselves linearly independent functions of x: $c_n(x) = e((2anx + n^2)/N)$. However, observe that while the c_n are not constant, they are still "simpler" than the original function f because they are almost periodic of order 1, whereas we expect the quadratic object f to be almost periodic of order 2.

One can of course continue these examples. They lead to the following recursive heuristic: a function f should be considered almost periodic of order $d - 1$ if one has some representation of the form $T^n f = c_{n,1}g_1 + c_{n,2}g_2 + \cdots$, where the $g_1, g_2, \ldots$ are bounded functions and the $c_{n,1}, c_{n,2}, \ldots$ are almost periodic of order $d-2$. Of course one should also provide some bound as to how many terms appear in this expansion, otherwise everything will be almost periodic of every order.

A convenient way to formalize the above intuition is as follows.

Definition 11.15 (Uniform almost periodicity norms) [357] If $f : Z \to \mathbf{C}$, we define $\|f\|_{UAP^0(Z)}$ to be infinite if f is non-constant, and equal to $|c|$ if f is equal to a constant c. If we now inductively assume that the $UAP^d(Z)$ norm has been defined for some d, we define the *$UAP^{d+1}(Z)$ norm* of f to be the infimum of all the constants $M > 0$ for which one has a representation formula of the form

$$T^n F = M\mathbf{E}(c_{n,h}g_h) \text{ for all } n \in Z, \tag{11.29}$$

where H is a finite non-empty set, $g = (g_h)_{h\in H}$ is a collection of functions from Z to $\mathbf{C}$ with $\|g_h\|_{L^\infty(Z)} \leq 1$, $c = (c_{n,h})_{n\in Z, h\in H}$ is a collection of functions from Z to $\mathbf{C}$ with $\|c_{n,h}\|_{UAP^d(Z)} \leq 1$, and h is a random variable taking values in H.

We informally refer to a function as *uniformly almost periodic of order $d-1$* if its $UAP^{d-1}(Z)$ norm is bounded.

One can easily check inductively that the $UAP^d(Z)$ norms are finite for $d \geq 1$, and are indeed norms, in particular obeying the triangle inequality

$$\|f + g\|_{UAP^d(Z)} \leq \|f\|_{UAP^d(Z)} + \|g\|_{UAP^d(Z)}. \tag{11.30}$$

Moreover, we have the important *Banach algebra property*

$$\|fg\|_{UAP^d(Z)} \leq \|f\|_{UAP^d(Z)}\|g\|_{UAP^d(Z)}. \tag{11.31}$$

We leave the easy verification of these facts as an exercise; the rather complicated construction in Definition 11.15 was designed primarily in order to obtain these nice properties (11.30), (11.31).

The UAP^{d-1} norms are a kind of dual to the U^d norms; see Exercise 11.4.8. The UAP^1 norm is the same as the Wiener algebra norm, see Exercise 11.4.10. They are also connected to dual functions:

Lemma 11.16 *Let $f : Z \to \mathbf{C}$ be a function bounded in magnitude by 1. Then $\|\mathcal{D}_d(f)\|_{UAP^{d-1}(Z)} \leq 1$ for all $d \geq 1$.*

Proof We induce on d. The case $d = 1$ is clear. Now suppose that $d \geq 2$ and the claim has already been proven for $d-1$. From the definition of $\mathcal{D}_d(f)$ and (11.25), and the change of variables $n + h = h'$, we have

$$T^n\mathcal{D}_d(f) = \mathbf{E}_{h\in Z}(T^{n+h} f\overline{\mathcal{D}_{d-1}(T^{n+h} f\overline{T^n f})}) = \mathbf{E}_{h'\in Z}(\overline{\mathcal{D}_{d-1}(T^{h'} f\overline{T^n f})}T^{h'} f).$$

The claim then follows by setting $M := 1$, $H := Z$, $c_{n,h} = \mathcal{D}_{d-1}(\overline{T^h f} T^n f)$, and $g_h := T^h f$. □

Combining this with Lemma 11.14 we see that the uniformly almost periodic functions of order $d-1$ form a complete set of obstructions for the Gowers uniformity norm of order d:

Corollary 11.17 (Soft inverse theorem, II) *Let $f : Z \to \mathbf{C}$ be a function bounded in magnitude by 1 with $\|f\|_{U^d(Z)} \geq \eta$. Then there exists a function $F : Z \to \mathbf{C}$ such that $\|F\|_{UAP^{d-1}} \leq 1$ and $|\langle f, F\rangle| \geq \eta^{2^d}$.*

One now has enough machinery to prove the following variant of Proposition 10.36.

Proposition 11.18 (Koopman–von Neumann decomposition) *[357] Let $k \geq 3$, let $f : Z \to \mathbf{R}^+$ be such that $0 \leq f(x) \leq 1$, let $\sigma > 0$, and let $F : \mathbf{R}^+ \times \mathbf{R}^+ \to \mathbf{R}^+$ be an arbitrary function. Then there exists a quantity $K = O_{\sigma,F,k}(1)$ and a decomposition $f = f_{U^\perp} + f_U$ with the following properties:*

- *the "anti-uniform" component $f_{U^\perp}$ obeys the bounds $0 \leq f_{U^\perp} \leq 1$ and $\mathbf{E}_Z f_{U^\perp} = \mathbf{E}_Z f$. Furthermore there exists an approximation f_{UAP} to $f_{U^\perp}$ with $0 \leq f_{UAP} \leq 1$, $\|f_{U^\perp} - f_{UAP}\|_{L^2(Z)} \leq \sigma$, and $\|f_{UAP}\|_{UAP^{k-2}(Z)} \leq K$;*
- *the "uniform" component f_U obeys the Gowers uniformity estimate $\|f_U\|_{U^{k-1}(Z)} \leq \frac{1}{F(\sigma,K)}$.*

This proposition is proven by almost identical means to Proposition 10.36 and we leave it as an exercise. The soft inverse theorem in Corollary 11.17 allows us to use uniformly almost periodic functions as a substitute for characters (and for quasi-periodic functions); the Banach algebra properties of such functions are the substitute for the fact that polynomial combinations of almost periodic functions are almost periodic. Otherwise the proof is much the same.

To conclude the proof of Szemerédi's theorem $r_k(\mathbf{Z}_N) = o_{N\to\infty;k}(N)$, one needs a recurrence theorem for the almost periodic component:

Proposition 11.19 (Uniformly almost periodic functions are recurrent) *[357] Let $k \geq 3$, let N be a large prime, let $f_{U^\perp}, f_{UAP} : \mathbf{Z}_N \to \mathbf{R}^+$ be such that $0 \leq f_{U^\perp}, f_{UAP} \leq 1$, $\mathbf{E}_{\mathbf{Z}_N} f_{U^\perp} \geq \delta$, $\|f_{U^\perp} - f_{UAP}\|_{L^2(\mathbf{Z}_N)} \leq \frac{\delta^2}{1024k}$, and $\|f_{UAP}\|_{UAP^{k-2}(\mathbf{Z}_N)} \leq K$. Then we have*

$$\Lambda_k(f_{U^\perp}, \ldots, f_{U^\perp}) = \Omega_{k,\delta,K}(1).$$

From Proposition 11.19, Proposition 11.18 and (11.8) one can conclude Szemerédi's theorem by the same argument as in Section 10.5. The proof of Proposition 11.19, however, is rather difficult, invoking an induction on k, the use of an energy increment argument to regularize certain σ-algebras which will appear, some Hilbert space arguments to locally compactify shift orbits such as $\{T^n f_{UAP} : f_{UAP} \in \mathbf{Z}_N\}$, and then van der Waerden's theorem to find monochromatic arithmetic progressions, where the coloring is determined by the local compactification. We will not prove it in full generality here, referring the reader to [357] for full details. However, we will sketch the somewhat simpler $k = 3$ version of the argument below. In this case one could instead rely on Exercise 11.4.10 and Proposition 10.35 to obtain a simpler proof with much more efficient bounds, but the argument we give below does not require the Fourier transform and can be extended (with additional arguments) to the higher k case.

Proof of Proposition 11.19 in the $k = 3$ case (Sketch) We consider the shifts $\{T^n f_{UAP} : n \in \mathbf{Z}_N\}$ as a subset of $L^2(\mathbf{Z}_N)$. Since $\|f_{UAP}\|_{UAP^1} \leq K$, we see that there exists a random variable h taking values in a finite set H and functions $g_h : \mathbf{Z} \to \mathbf{C}$ with $\|g_h\|_{L^\infty(\mathbf{Z}_N)} \leq 1$, such that all the shifts $T^n f_{UAP}$ are contained in the set

$$\Gamma := \{K\mathbf{E}_h(c_h g_h) : c_h \in \mathbf{C}, |c_h| \leq 1 \text{ for all } h \in H\} \tag{11.32}$$

which can be thought of as a kind of high-dimensional cube. It turns out that this set is "compact" in the sense that it can be covered by $O_{k,\delta,K}(1)$ balls in $L^2(\mathbf{Z}_N)$ of radius $\delta^2/1024$ (see Exercise 11.4.13). This induces a coloring of $\mathbf{Z}_N$ by $O_{\delta,K}(1)$ colors, by assigning to each $n \in \mathbf{Z}_N$ one of the balls that contains $T^n f$. By van der Waerden's theorem (Exercise 6.3.9), we conclude that for $\Omega_{\delta,K}(1)$ of the pairs $(a, r) \in \mathbf{Z}_N$, the triple $a, a + r, \ldots, a + 2r$ are monochromatic, so that the functions $T^a f_{UAP}, T^{a+r} f_{UAP}, T^{a+2r} f_{UAP}$ lie in the same $\delta^2/1024$-ball. This implies that the functions $T^a f_{U^\perp}, T^{a+r} f_{U^\perp}, T^{a+2r} f_{U^\perp}$ are distance at most $\delta^2/512$ apart. Since these functions are also bounded between 0 and 1 and have mean δ, an application of Markov's inequality then shows that these functions are simultaneously greater than $\delta/4$ (say) on a set of density at least $\delta/4$. Thus $\mathbf{E}(T^a f_{U^\perp} T^{a+r} f_{U^\perp} T^{a+2r} f_{U^\perp}) = \Omega_\delta(1)$ for all such pairs (a, r). Taking averages over all a, r we obtain the claim. □

Exercises

11.4.1 Prove (11.24) and (11.25).

11.4.2 Prove (11.26), (11.27), and (11.28).

11.4.3 Verify that $\|f\|_{UAP^d(Z)}$ is well-defined and finite for all $d \geq 1$, and obeys (11.30) and (11.31). In particular, verify that the $UAP^d(Z)$ norm is indeed a norm.

11.4.4 Establish the monotonicity property $\|f\|_{UAP^{d+1}(Z)} \leq \|f\|_{UAP^d(Z)}$ for all $f : Z \to \mathbf{C}$ and $d \geq 0$.

11.4.5 Let $\phi : Z \to Z'$ be a Freiman isomorphism of order 2. Show that $\|f \circ \phi\|_{UAP^{d-1}(Z)} = \|f\|_{UAP^{d-1}(Z')}$ for all $f : Z' \to \mathbf{C}$ and $d \geq 1$. In particular, the $UAP^{d-1}(Z)$ norms are translation-invariant.

11.4.6 Let $\phi : Z \to \mathbf{R}/\mathbf{Z}$ be a phase polynomial of degree less than d. Show that $\|e(\phi)f\|_{UAP^{d-1}(Z)} = \|f\|_{UAP^{d-1}(Z)}$ for all $f : Z \to \mathbf{C}$.

11.4.7 Let $\phi : Z \to \mathbf{R}/\mathbf{Z}$ be a phase polynomial of degree less than d. Show that $\mathcal{D}_d(e(\phi)) = e(\phi)$ and $\|e(\phi)\|_{UAP^{d-1}(Z)} = 1$, thus every polynomial phase function is a dual function.

11.4.8 [357] Obtain the inequality

$$\left|\langle f, g\rangle_{L^2(Z)}\right| \leq \|f\|_{U^d(Z)}\|g\|_{UAP^{d-1}(Z)}$$

for any $d \geq 1$ and $f, g : Z \to \mathbf{C}$. (Hint: use induction on d.) Thus functions which are uniformly almost periodic of order $d - 1$ are almost orthogonal to Gowers uniform functions of order d. This can be viewed as a partial converse to Corollary 11.17. Note in particular that we have

$$\|f\|^2_{L^2(Z)} \leq \|f\|_{U^d(Z)}\|f\|_{UAP^{d-1}(Z)}$$

thus a function cannot be simultaneously uniformly almost periodic and Gowers uniform without also being small.

11.4.9 Let $f, g : Z \to \mathbf{C}$ be functions bounded in magnitude by 1. Establish the inequality

$$\|fg\|^{2^d}_{U^d(Z)} \leq \|f\|_{UAP^{d-1}(Z)}\|g\|_{U^d(Z)}$$

for all $d \geq 1$. (Hint: use Lemma 11.16 applied to fg, together with the algebra property of $UAP^{d-1}(Z)$.)

11.4.10 (Ben Green, private communication) Show that $\|f\|_{UAP^1(Z)} = \|\hat{f}\|_{l^1(Z)}$ for all $f : Z \to \mathbf{C}$. (Hint: from Exercise 11.4.7 and the triangle inequality one can obtain the inequality $\|f\|_{UAP^1(Z)} \leq \|\hat{f}\|_{l^1(Z)}$. To obtain the other inequality, first use Plancherel's theorem to establish that $\mathbf{E}_{n,x\in Z}c_{n,h}g_h(x)b(x+n)| \leq 1$ whenever $c_{n,h}$ is a constant bounded by 1, g_h is a function with $\|g_h\|_{L^\infty(Z)} \leq 1$, and b is a function with $\|\hat{b}\|_{l^\infty(Z)} \leq 1$.)

11.4.11 [357] Prove Theorem 11.18.

11.4.12 Use Proposition 11.19, Proposition 11.18 and (11.8) to deduce that $r_k(\mathbf{Z}_N) = o_{N\to\infty;k}(N)$ for all $k \geq 1$ and all large N.

11.4.13 [357] Let Γ be the set defined in (11.32). Show that given any $\varepsilon > 0$, the set Γ can be covered by $O_{\varepsilon,K}(1)$ balls in $L^2(Z)$ of radius ε. (Hint: find a maximal orthonormal set $v_1, \ldots, v_J$ such that $\mathbf{E}|\langle g_h, v_j\rangle_{L^2(Z)}|^2 \geq \varepsilon^2/4$ for all $1 \leq j \leq J$, and use Bessel's inequality and linearity of expectation to obtain an upper bound on J. Show that the quantities Γ stay within $\varepsilon/2$ of the J-dimensional space spanned by $v_1, \ldots, v_J$.)

11.5 The infinitary ergodic approach

In this section we discuss some of the ideas underlying Furstenberg's infinitary ergodic approach to Szemerédi's theorem. These arguments are the shortest and most elegant way to prove the theorem, but also require a certain amount of machinery concerning infinite measure spaces. Also it is quite difficult to extract a quantitative bound from these methods. As the techniques here are rather disjoint from those in the rest of this book we shall not provide full details, referring the reader instead to [122]. However, the insights developed here were essential in developing several of the finitary arguments in this chapter, most notably the finitary ergodic proof of Szemerédi's theorem, and the Green–Tao theorem on arithmetic progressions in the primes.

Define a *measure-preserving system* to be a (possibly infinite) space X with a σ-algebra $\mathcal{B}$, a probability measure $\mathbf{P}_X$ on $\mathcal{B}$, and a bijection $T : X \to X$ such that all the powers T^n of T with $n \in \mathbf{Z}$ are measure-preserving, thus $\mathbf{P}_X(T^n A) = \mathbf{P}_X(A)$ for all $A \in \mathcal{B}$. In this infinite setting, a σ-algebra cannot be rigorously viewed as a partition; instead it is a collection of sets closed under countable unions, intersections, and complements, and containing $\emptyset$ and X. We define an expectation $\mathbf{E}_X$ on bounded measurable functions from X to $\mathbf{R}$ in the usual manner, and define a shift operator T^n on such functions by $T^n f(x) := f(T^{-n}x)$. To simplify the notation slightly we shall only work with real-valued functions in this section rather than complex-valued ones.

Example 11.20 **(Circle shift)** Let X be the unit circle $\mathbf{R}/\mathbf{Z}$ with Lebesgue measure, and let T be the shift $Tx = x + \alpha$ for some fixed $\alpha \in \mathbf{R}$. The dynamics of this system depend on whether α is rational or irrational; for instance, in the former case the shift T is periodic, but not in the latter case. However in both cases we have the following almost periodicity property: given any bounded measurable function on X, the shifts $\{T^n f : n \in \mathbf{Z}\}$ are pre-compact in $L^2(X)$. In particular given any ε we have $\|T^n f - f\|_{L^2(X)} \leq \varepsilon$ for infinitely many ε. Because of this property we say that this measure-preserving system is *compact*.

Example 11.21 **(Skew shift)** Let X be the torus $(\mathbf{R}/\mathbf{Z}) \times (\mathbf{R}/\mathbf{Z})$ with Lebesgue measure, and let T be the skew shift $T(x, y) := (x + \alpha, y + x)$ for some fixed $\alpha \in \mathbf{R}$. Note that the orbits $T^n(x, y)$ are linear in n in the x variable, but quadratic in n in the y variable. This system is not compact, but contains a non-trivial compact factor, namely the σ-algebra $\mathcal{B}_0$ consisting of all the sets of the form $A \times (\mathbf{R}/\mathbf{Z})$, where A is Borel measurable in $\mathbf{R}/\mathbf{Z}$. (To put this another way, the $\mathcal{B}_0$-measurable functions are precisely those functions which do not depend on the y variable.) This factor is isomorphic to the circle shift mentioned earlier. It turns out that the skew shift is a *relatively compact extension* of the circle shift, though we will not quantify precisely what this means here except to observe that if f is a smooth function on $(\mathbf{R}/\mathbf{Z}) \times (\mathbf{R}/\mathbf{Z})$, then the orbits $\{T^n f : n \in \mathbf{Z}\}$ form a precompact set on each fiber $\{x = \text{constant}\}$ of $\mathcal{B}_0$, endowed with the obvious one-dimensional measure.

Example 11.22 (Bernoulli shift) Now consider the infinite unit cube $X := [0, 1]^{\mathbf{Z}}$ of infinite binary sequences $(\omega_n)_{n \in \mathbf{Z}}$, with the usual product topology and Borel σ-algebra $\mathcal{B}$. Let $B \subset X$ denote the "cylinder" of sequences where $\omega_0 = 1$, and let T be the shift operator defined by $T^h(\omega_n)_{n\in\mathbf{Z}} := (\omega_{n+h})_{n\in\mathbf{Z}}$. Using the Kolmogorov extension theorem (or Caratheodory's extension theorem and Tychonoff's theorem) we can find a measure $\mathbf{P}$ on X such that

$$\mathbf{P}(T^{h_1} B \cap \cdots \cap T^{h_m} B) = 2^{-m}$$

whenever $h_1, \ldots, h_m$ are distinct integers. Informally, one can view this system as the probability space corresponding to an infinite number of coin tosses, one for each integer h; the event $T^h B$ is then the event that the hth coin turns up heads, and the shift operator corresponds to relabeling all of the coins up by 1. The behavior here is completely different from the compact case; indeed, if f is bounded and measurable, and has mean zero, one can show that $\langle T^n f, f\rangle_{L^2(X)} \to 0$ as $n \to \infty$. A system with this property is known as *strongly mixing*.

Furstenberg derived Szemerédi's theorem by proving the following equivalent formulation.

Theorem 11.23 (Furstenberg multiple recurrence theorem) *[121], [125], [122] Let $(X, \mathcal{B}, \mathbf{P}, T)$ be a measure-preserving system, and let $f : X \to \mathbf{R}^+$ be a non-negative bounded measurable function with $\mathbf{E}(f) > 0$. Then for all $k \geq 1$ we have*

$$\liminf_{N\to\infty} \mathbf{E}_{1\leq n\leq N} \mathbf{E}_X f T^n f \cdots T^{(k-1)n} f > 0.$$

It is fairly easy to deduce this theorem from Szemerédi's theorem; we leave this as an exercise. The converse deduction of Szemerédi's theorem from Furstenberg's theorem is a little trickier, requiring some measure-theoretic tools:

Proof of Theorem 10.1 assuming Theorem 11.23 (Sketch) Suppose for contradiction that we can find a set $A \subseteq \mathbf{Z}$ of positive upper progressions containing no progressions of length k. Thus we can find a sequence of integers $N_1, N_2, \ldots$ going to infinity such that $\liminf_{j\to\infty} \mathbf{P}_{[-N_j,N_j]}(A) > 0$. Now use the Hahn–Banach theorem to construct a linear functional λ on bounded real-valued sequences $(c_j)_{j=1}^{\infty}$ such that

$$\liminf_{j\to\infty} c_j \leq \lambda((c_j))_{j=1}^{\infty} \leq \limsup_{j\to\infty} c_j.$$

Now consider the infinite unit cube $X := [0,1]^{\mathbf{Z}}$ of infinite binary sequences $(\omega_n)_{n\in\mathbf{Z}}$, with the usual product topology and Borel σ-algebra $\mathcal{B}$. Let $B \subset X$ denote the "cylinder" of sequences where $\omega_0 = 1$, and let T be the shift operator defined by $T^h(\omega_n)_{n\in\mathbf{Z}} := (\omega_{n+h})_{n\in\mathbf{Z}}$. Using the Kolmogorov extension theorem (or Caratheodory's extension theorem and Tychonoff's theorem) we can find a measure $\mathbf{P}$ on X such that

$$\mathbf{P}(T^{h_1}B \cap \cdots \cap T^{h_m}B) = \lambda\big((\mathbf{P}_{[-N_j,N_j]}((A+h_1)\cap\cdots\cap(A+h_m)))_{j=1}^{\infty}\big)$$

for all $h_1,\ldots,h_m \in \mathbf{Z}$. In particular we see that $\mathbf{P}(B) > 0$. By Theorem 11.23 applied to $f = 1_B$ we conclude that $\mathbf{P}(B \cap T^n B \cap \cdots \cap T^{(k-1)n}B)$ for at least one non-zero B, which implies that A contains a progression of length k. □

One can prove the multiple recurrence theorem in a manner similar to that in the previous sections. For instance, there is an analog of the Gowers uniformity norm $\|f\|_{U^d(X)}$, defined inductively for bounded measurable f by $\|f\|_{U^0(X)} := \mathbf{E}_X(f)$ and

$$\|f\|_{U^d(X)} := \lim_{N\to\infty}\left(\mathbf{E}_{1\le n\le N}\|fT^hf\|_{U^{d-1}(X)}^{2^{d-1}}\right)^{1/2^d}.$$

(The existence of this limit is guaranteed by the von Neumann ergodic theorem; see [185].) One can verify that these U^d norms obey properties similar to their finitary counterparts; see [185], with a key distinction that it is now quite possible for a non-zero function f to have a vanishing U^d norm. We have an important analog of the generalized von Neumann theorem (11.8), namely that

$$\lim_{N\to\infty}\mathbf{E}_{1\le n\le N}\mathbf{E}_X f_0 T^n f_1 \cdots T^{(k-1)n} f_{k-1} = 0$$

whenever $f_0,\ldots,f_{k-1}$ are bounded measurable functions with at least one of the f_j having a vanishing U^{k-1} norm. Thus functions with vanishing U^{k-1} norm have a negligible impact on recurrence.

Again, attention now turns towards the obstructions to uniformity. It turns out that in the infinitary setting these obstructions have a rather nice description. Let $U^{k-1}(X)^*$ denote the space of all bounded functions f for which the expression

$$\|f\|_{U^{k-1}(X)^*} := \sup\{|\mathbf{E}_X(fg)| : \|g\|_{U^{k-1}(X)} \le 1\}$$

is finite. It turns out (see [185]) that there exists a unique σ-algebra $\mathbf{Z}_{k-2}$ such that the closure of $U^{k-1}(X)^*$ in the L^2 topology consists precisely of those square-integrable functions which are measurable with respect to $\mathbf{Z}_{k-2}$; the $\mathbf{Z}_{k-2}$ are thus the *universal characteristic factor* for the $U^{k-1}(X)$ norm. As a consequence one can precisely quantify which functions are Gowers uniform of order $k-1$:

$$\|f\|_{U^{k-1}(X)} = 0 \iff \mathbf{E}(f|\mathbf{Z}_{k-2}) = 0.$$

Here the conditional expectation $f \mapsto \mathbf{E}(f|\mathbf{Z}_{k-2})$ is defined as the L^2-orthogonal projection onto the space of $\mathbf{Z}_{k-2}$-measurable functions.

One consequence of the above discussion is that in order to prove the Furstenberg recurrence theorem, it suffices to do so under the additional assumption that f is $\mathbf{Z}_{k-2}$-measurable (because the error $f - \mathbf{E}(f|\mathbf{Z}_{k-2})$ has a vanishing $U^{k-1}(X)$ norm and is hence irrelevant). To do this, it is clearly of importance to understand the factors $\mathbf{Z}_{k-2}$ of $\mathcal{B}$ as much as possible.

The factor $\mathbf{Z}_0$ turns out to be the space of invariant sets in X, i.e. $\mathbf{Z}_0 := \{A \in \mathcal{B} : TA = A\}$. This is essentially the *von Neumann ergodic theorem*, which we leave to the exercises. The factor $\mathbf{Z}_1$ is known as the *Kronecker factor* and is generated by all the almost periodic functions, or equivalently by the eigenfunctions of the shift operator T. The higher factors are more difficult to describe explicitly. However, it can be shown without too much difficulty (see e.g. [121], [185], [236]); a closely related result is in [386]) that each factor $\mathbf{Z}_{d+1}$ is a *relatively compact extension* of the preceding factor $\mathbf{Z}_d$ (in fact, it is the maximal relatively compact extension). What this means is a little bit tricky to describe precisely, but it roughly means that for a dense set of f which are measurable in $\mathbf{Z}_{d+1}$, the orbits $\{T^n f : n \in \mathbf{Z}\}$ are precompact relative to $\mathbf{Z}_d$, which informally means that they are precompact when restricted to each "atom" or "fiber" of $\mathbf{Z}_d$. See [122] for a rigorous formulation of these assertions (which requires the theory of disintegration of measures). Using some tools from measure theory and analysis, as well as a combinatorial argument closely related to the van der Waerden theorem, it was shown in [121], [125] that if the Furstenberg recurrence theorem holds for any factor $\mathbf{Z}_d$, then it also holds for a relatively compact extension $\mathbf{Z}_{d+1}$; this is analogous to Proposition 11.19. This fact, combined with the preceding discussion, yields the Furstenberg recurrence theorem and thus Szemerédi's theorem.

Recently, there has been significant progress by Host–Kra [185] (and subsequently by Ziegler [386]) in understanding the factors $\mathbf{Z}_{k-2}$. (Strictly speaking, Ziegler treats a slight variant $\mathbf{Y}_{k-2}$ of the factors $\mathbf{Z}_{k-2}$; see [236] for a comparison between the two.) It turns out that the factors $\mathbf{Z}_{k-2}$ are isomorphic to the inverse limit of $k-2$*-step nilsystems*, or in other words a system $(G/\Gamma, \mathcal{B}, T, \mathbf{P})$, where G is a nilpotent Lie group of order $k-2$, Γ is a co-compact subgroup of G, $\mathcal{B}$ is the usual Borel algebra, T is a left shift operator $T : x\Gamma \mapsto gx\Gamma$ for some fixed group element $g \in G$, and $\mathbf{P}$ is normalized Haar measure. Thus for instance the circle shift in Example 11.20 is a 1-step nilsystem, whereas the skew shift turns out to be isomorphic to a 2-step nilsystem. These characterizations of $\mathbf{Z}_{k-2}$ are roughly analogous to the "hard" inverse theorems discussed in Section 11.2; see [160] for further discussion of this in the $k = 4$ case. Just as these hard inverse theorems lead to better quantitative results on Szemerédi's theorem, the characterizations of $\mathbf{Z}_{k-2}$ given here lead to stronger recurrence theorems; for instance, they can be used to replace the limit inferior in the Furstenberg recurrence theorem with a limit, and in fact obtain the stronger result that the averages $\mathbf{E}_{1 \le n \le N} T^n f \cdots T^{(k-1)n} f$ converge in L^2 norm to a non-zero (and somewhat explicitly describable) function. See [185], [386]. A current area of research is to develop and simplify these ergodic theory results (which are currently quite difficult and lengthy to prove) and clarify their connection with the analogous developments in the Fourier-analytic and combinatorial approaches.

The ergodic approach is well suited for establishing stronger combinatorial results than Szemerédi's theorem, several of which have not yet been proven by other means. We describe some of them here.

Theorem 11.24 (Multi-dimensional Szemerédi theorem) *[123] Let $d \ge 1$, and let $A \subset \mathbf{Z}^d$ be such that $\limsup_{N \to \infty} \mathbf{P}_{[-N,N]^d}(A) > 0$. Then for any $v_1, \ldots, v_k \in \mathbf{Z}^d$, there exist infinitely many pairs $(a, r) \in \mathbf{Z}^d \times \mathbf{Z}^+$ such that $a + rv_1, \ldots, a + rv_k \in A$.*

Theorem 11.25 (Polynomial Szemerédi theorem) *[23] Let $P_1, \ldots, P_k : \mathbf{Z} \to \mathbf{Z}$ be polynomials that map the integers to the integers such that $P_1(0) = \cdots = P_k(0) = 0$. Let $A \subset \mathbf{Z}$ have positive upper density. Then there exist infinitely many pairs $(a, r) \in \mathbf{Z}^d \times \mathbf{Z}^+$ such that $a + P_1(r), \ldots, a + P_k(r) \in A$.*

Theorem 11.26 (Density Hales–Jewett theorem) *[124] Let $n \ge 1$ and $0 < \delta \le 1$. Then there exists an integer $d = d(|A|, \delta) \ge 1$ such that if A is any subset of $[0, n-1]^d$ with cardinality $|A| \ge \delta n^d$, then A contains a proper arithmetic progression $a + [0, n-1] \cdot v$ of length n, for some $a \in [0, n-1]^d$ and $v \in [0, 1]^d$.*

Further refinements include additional structural information on the pairs (a, r) constructed by the above theorems, as well as convergence of various limits; in addition, there is much current work in extending the description of the characteristic factor for the U^k norm and for multiple recurrence to these more complex recurrence theorems. Unfortunately a complete survey of these exciting developments is well beyond the scope of this book.

Exercises

11.5.1 Show that Theorem 11.1 for a fixed k implies Theorem 11.23 for the same value of k.

11.5.2 (Poincaré recurrence theorem) Using only the pigeonhole principle and elementary measure theory, prove Theorem 11.23 in the $k = 2$ case.

11.5.3 (Von Neumann ergodic theorem) Let $(X, \mathcal{B}, \mathbf{P}, T)$ be a measure-preserving system. Show that the spaces $\{f \in L^2(X) : Tf = f\}$ and $\{Tf - f : f \in L^2(X)\}$ are complementary orthogonal subspaces of $L^2(X)$. Use this to conclude that if $\mathbf{Z}_0 := \{A \in \mathcal{B} : TA = A\}$, then $\mathbf{E}_{1 \leq n \leq N} T^n f$ converges in $L^2(X)$ to $\mathbf{E}(f|\mathbf{Z}_0)$ for any $f \in L^2(X)$, and that $\|f\|_{U^1(Z)} = \|\mathbf{E}(f|\mathbf{Z}_0)\|_{L^2(X)}$. Note that these results simplify in the case when the system is *ergodic* (which means that $\mathbf{Z}_0 = \{\emptyset, X\}$), since in that case $\mathbf{E}(f|\mathbf{Z}_0)$ is just $\mathbf{E}_X(f)$. In particular we have $\|f\|_{U^1(Z)} = |\mathbf{E}_X(f)|$ in this case, just as in the finitary case.

11.5.4 (Khintchine's recurrence theorem) Let A be a subset of a measure-preserving system $(X, \mathcal{B}, \mathbf{P}, T)$. Show that for every $\varepsilon > 0$ that there exist infinitely many $n \in \mathbf{Z}$ such that $\mathbf{P}(A \cap T^n A) \geq \mathbf{P}(A)^2 - \varepsilon$. (Hint: obtain lower and upper bounds for $\|\mathbf{E}_{1 \leq n \leq N} 1_{T^n A}\|_{L^2(Z)}$. Alternatively, use the von Neumann ergodic theorem.) Show that the theorem fails if $\mathbf{P}(A)^2 - \varepsilon$ is replaced by $\mathbf{P}(A)^2 + \varepsilon$, regardless of how small $\mathbf{P}(A)$ and ε are. It is natural to then conjecture that $\mathbf{P}(A \cap T^n A \cap \cdots \cap T^{(k-1)nA}) \geq \mathbf{P}(A)^k - \varepsilon$ for infinitely many n; this is true for $k = 1, 2, 3, 4$ under the additional assumption of ergodicity, but fails for $k > 4$, see [22].

11.5.5 Let $(X, \mathcal{B}, \mathbf{P}, T)$ be a compact measure-preserving system (so the orbits $\{T^n f : n \in \mathbf{Z}\}$ are precompact in L^2 whenever f is bounded and measurable). Prove the Furstenberg recurrence theorem in this special case. (Compare with Proposition 10.35 or the $k = 3$ proof of Proposition 11.19.)

11.5.6 Let $(X, \mathcal{B}, \mathbf{P}, T)$ be a weakly mixing measure-preserving system, which means that $\lim_{N \to \infty} \mathbf{E}_{1 \leq n \leq N} |\langle T^n f, f \rangle_{L^2(X)}|^2 = 0$ whenever f is bounded, measurable, and has expectation zero. (This is weaker than *strong mixing*, which demands that $\lim_{n \to \infty} \langle T^n f, f \rangle = 0$ under the same hypotheses.) Show that $\|f\|_{U^{k-1}(X)} = 0$ if and only if $\mathbf{E}_X(f) = 0$, and establish the Furstenberg recurrence theorem in this special case.

11.5.7 Let $(X, \mathcal{B}, \mathbf{P}, T)$ be measure-preserving system, and let f be bounded and measurable. Show that if f is almost periodic (thus the orbit $\{T^n f : n \in \mathbf{Z}\}$ is precompact in $L^2(X)$), then $\mathbf{E}_X(fg) = 0$ whenever g is bounded, measurable, and vanishing in $U^2(X)$ norm. Compare this with Exercise 11.4.8.

11.5.8 Let $(X, \mathcal{B}, \mathbf{P}, T)$ be measure-preserving system. Let $\mathbf{Z}_1$ be the smallest σ-algebra with respect to which all almost periodic functions are measurable. If f is bounded and measurable, show that $\|f\|_{U^2(X)} = 0$ if and only if $\mathbf{E}(f|\mathbf{Z}_1) = 0$. (Hint: the "only if" part follows from the preceding exercise. For the "if" part, construct the dual function $\mathcal{D}_2 f := \lim_{N \to \infty} \mathbf{E}_{-N \leq n \leq N} T^n f \mathbf{E}_X(f T^n f|\mathbf{Z}_0)$, and show that this function is almost periodic. You may need the fact that Volterra integral operators are compact.)

11.5.9 (Koopman–von Neumann theorem) Let $(X, \mathcal{B}, \mathbf{P}, T)$ be measure-preserving system, and let $f \in L^2(X)$. Show that there is a unique decomposition $f = f_{U^\perp} + f_U$, where $\|f_U\|_{U^2(X)} = 0$ and $f_{U^\perp}$ is the limit in $L^2(X)$ of almost periodic functions.

11.6 The hypergraph approach

In Section 10.6 we saw that the Szemerédi regularity lemma led to a result in graph theory, namely the *triangle removal lemma*, which in turn implied Roth's theorem (as well as a generalization to right-angled triangles). It is then natural to ask whether a similar approach can prove Szemerédi's theorem for more general k. This turns out to be the case, but requires one work with *hypergraphs* (also known as *set systems*) instead of graphs. We need some notation. If A is a finite set and $k \geq 0$, we let $\binom{A}{k}$ denote the collection of all the k-element subsets of A. Define a *k-uniform hypergraph* $H = H(V, E)$ to be any pair (V, E), where V is a finite set (the *vertex set*), and E is a subset of $\binom{V}{k}$ (the *edge set*). Thus a 2-uniform hypergraph is the same as an ordinary graph.

The triangle removal lemma, Lemma 10.46, can be generalized as follows. If $H = H(V, E)$ is a k-uniform hypergraph, we define a *k-simplex* in H to be any set $S = \{v_1, \ldots, v_{k+1}\} \subset V$ of $k+1$ vertices such that $\binom{S}{k} \subset e$, i.e. the $k+1$ edges $S\backslash\{v_1\}, \ldots, S\backslash\{v_{k+1}\}$ all lie in E. Note that a 2-simplex is the same as a triangle.

Theorem 11.27 (Simplex removal lemma) *[283],[284],[140] Let $k \geq 2$, and let $H = H(V, E)$ be a k-uniform hypergraph which contains at most $\delta|V|^{k+1}$ k-simplices. Then it is possible to remove $o_{\delta\to 0;k}(|V|^2)$ edges from H to obtain a hypergraph which is simplex-free (it contains no k-simplices whatsoever).*

This result was conjectured by Erdős, Frankl and Rödl [87] in 1986, but not proven in full until much later. The $k = 2$ case dates back of course to [304] in 1978, but even the $k = 3$ case did not appear until 2002 [110] (though unpublished versions of this result existed much earlier, see for instance [109]); see also [139]. The full result was proven independently and simultaneously by Rödl and Skokan [283], [284] (see also [282], [254]) and Gowers [140]. A slight strengthening of this result was later established in [360] for the purposes of establishing arbitrary constellations in the Gaussian primes. For a recent survey of developments, see [281].

Just as Lemma 10.46 implies Proposition 10.47, Theorem 11.27 implies the following higher-dimensional analog:

Proposition 11.28 *Let Z be a finite additive group, let $k \geq 2$, and let $A \subset Z^k$ be such that A contains no "right-angled simplices" $x, x + re_1, \dots, x + re_k$ with $x \in Z^k$ and $r \in Z \backslash 0$, where (by slight abuse of notation) we write re_i for $(0, \dots, 0, r, \dots, 0)$ with the ith position being the only non-zero one. Then $|A| = o_{|Z| \to \infty; k}(|Z|^k)$.*

This in turn can be used to deduce Theorem 11.24 as well as Szemerédi's theorem. In fact it yields Szemerédi's theorem for an arbitrary group: more precisely we have $r_k(Z) = o_{|Z| \to \infty; k}(|Z|)$ for any finite additive group Z with $|Z|$ coprime to $(k-1)!$.

Just as the triangle removal lemma can be proven by the Szemerédi regularity lemma (indeed, this is currently its only proof known), the simplex removal lemma can be proven by a hypergraph regularity lemma. It turns out, however, that unlike the situation with the regularity lemma, where there is essentially one formulation (up to equivalences), there are several choices of hypergraph regularity lemma to choose from. The first such regularity lemma, introduced by Chung and Graham [58], regularizes the k-edge set E in terms of a partition of the vertex sets V, and is proven very similarly to the regularity lemma for graphs. Unfortunately, this lemma seems to be too weak to easily deduce the simplex removal lemma, the problem being that the regularity properties conferred by this lemma are not sufficient to obtain an accurate count for the number of simplices in the hypergraph, even in the 3-uniform case. The situation is intriguingly similar to the phenomenon noted in earlier sections that Fourier uniformity is insufficient to count progressions of length 4 or greater, even though Fourier analysis does not make an appearance in the regularity lemma. The solution (again in the 3-uniform case for simplicity) is to regularize the 3-edges by a partition of the 2-*edge set* $\binom{V}{2}$, and then regularize the 2-edge partition further (essentially using the ordinary regularity lemma) using a partition of the vertex set V (or equivalently $\binom{V}{1}$). This however leads to some new issues not present in the ordinary graph case. First, it is possible for the secondary partition to somehow disrupt the regularity obtained by the primary partition. Second, one has to decide on the relative strength of regularity between the primary partition and the secondary partition; this is particularly important since there is an expensive (tower-exponential) trade-off between the amount of regularity conferred by a partition, and the number of cells needed in the partition, and one may need the regularity in one partition to dominate the number of cells in another. Finally, even after all the appropriate regularity has been attained, one still needs to accurately count the number of simplices in the hypergraph.

These problems can all be solved, but require a certain amount of technicality. We will not give the general details here, but we will do the $k = 2$ case (i.e. the usual regularity lemma) in detail, in a way which allows for a relatively easy extension to the higher k case. Our treatment here follows [360], [359] (see also [7], [282] for some closely related arguments). Namely, we will view the regularity lemma as being akin to the Koopman–von Neumann theorems of previous sections, decomposing a function into a "compact" component and a "uniform" component. Indeed we have the following analog of Proposition 11.18. Let us say that a function $f : V_1 \times \cdots \times V_d \to \mathbf{R}^+$ is *K-constant* if for each $1 \leq i \leq d$ there exists a partition of V_i into K cells $V_{i,1}, \ldots, V_{i,K}$ (some of the cells may be empty or otherwise unequal in size) such that f is constant on each of the products $V_{1,i} \times \cdots \times V_{d,j}$. This will be analogous to the concept of almost periodicity used in previous sections.

Proposition 11.29 (Preliminary regularity lemma) *[359] Let $V_1, \ldots, V_d$ be arbitrary finite non-empty sets, let $f : V_1 \times \cdots \times V_d \to \mathbf{R}^+$ be such that $0 \leq f(x_1, \ldots, x_d) \leq 1$, let $\sigma > 0$, and let $F : \mathbf{R}^+ \times \mathbf{R}^+ \to \mathbf{R}^+$ be an arbitrary function. Then there exists a quantity $K = O_{\sigma,F}(1)$ and a decomposition $f = f_{U^\perp} + f_U$ with the following properties:*

- *the "anti-uniform" component $f_{U^\perp}$ obeys the bounds $0 \leq f_{U^\perp} \leq 1$. Furthermore there exists a K-constant approximation f_C to $f_{U^\perp}$ with $0 \leq f_C \leq 1$ and $\|f_{U^\perp} - f_C\|_{L^2(V_1 \times \cdots \times V_d)} \leq \sigma$;*
- *the "uniform" component f_U obeys the regularity estimate*

$$|\mathbf{E}_{x_1 \in V_1, \ldots, x_d \in V_d} 1_{A_1 \times \cdots \times A_d} f_U(x_1, \ldots, x_d)| \leq \frac{1}{F(\sigma, K)}$$

for all $A_1 \subseteq V_1, \ldots, A_d \subseteq V_d$.

Remark 11.30 For application to graph regularity one only needs $d = 2$. For more general d, the above lemma is closely related to the hypergraph regularity lemma of Chung and Graham [58].

Let us assume this proposition for the moment, and establish the regularity lemma in the more traditional formulation of Lemma 10.42.

Proof of Lemma 10.42 assuming Proposition 11.29 Let ϵ, m, and $G = G(V, E)$ be as in the lemma. We set $V_1 = V_2 = V$, and let $f(x_1, x_2)$ be the incidence matrix of G, i.e. $f(x_1, x_2) = \mathbf{I}(\{x_1, x_2\} \in E)$. Let $\sigma > 0$ be a small number to be chosen later (it will eventually be a small multiple of ε^4), and let $F : \mathbf{R}^+ \times \mathbf{R}^+ \to \mathbf{R}^+$ be a growth function (depending on m) to be chosen later. We apply Proposition 11.29 with $d = 2$ to obtain a decomposition $f = f_{U^\perp} + f_U$, a K-constant approximation f_C to $f_{U^\perp}$ for some $K = O_{\sigma,F}(1)$, and partitions $V_{1,1}, \ldots, V_{1,K}$ and $V_{2,1}, \ldots, V_{2,K}$ of V with respect to which f_C is constant. We can take the common refinement $V'_{K(i-1)+j} := V_{1,i} \cap V_{2,j}$ for $i, j \in [1, K]$ to obtain a unified partition $V'_1, \ldots, V'_{K^2}$ such that f_C is constant on each product $V'_i \times V'_j$. Next, we let N be the largest integer less than $\sigma|V|/100mK^2$ (we will assume V large enough depending on m, K, σ that $N \geq 1$), and partition each cell V'_j arbitrarily into disjoint sub-cells of size N, plus an error of size at most N. Note that all the errors, when put together, will give an error set V_* of size at most $K^2 N \leq \sigma|V|/100m$, while the remaining sub-cells yield a partition $V''_1, \ldots, V''_k$ of the remaining set $V \backslash V_*$, with each V''_j having cardinality exactly N. Thus we have

$$|V| - \frac{\sigma|V|}{100m} \leq kN \leq |V|,$$

thus $k \geq m$ and $k = \Theta(|V|/N) = \Theta(mK^2/\sigma)$. The partition $V''_1, \ldots, V''_k, V_*$ is not quite near-uniform in the sense of Section 10.6, because of the exceptional set V_*. But we can break up V_* arbitrarily into k near-uniform pieces and distribute them to the V''_j, replacing each set V''_j by a slightly larger set V'''_j such that the $V'''_1, \ldots, V'''_k$ are a near-uniform partition of V and

$$|V'''_j \backslash V''_j| = O\left(\frac{|V_*|}{k} + 1\right) = O\left(\frac{K^2 N}{mK^2/\sigma} + 1\right) = O(\sigma N)$$

where we again assume V to be suitably large depending on m, K, σ. Thus V'''_j is only larger than V''_j by a factor of $1 + O(\sigma)$.

Let us now investigate the ϵ-regularity of a pair V'''_i, V'''_j. Let $X \subset V'''_i, Y \subset V'''_j$ be such that $|X| \geq \epsilon|V'''_i|$ and $|Y| \geq \epsilon|V'''_j|$, in particular $|X|, |Y| \geq \epsilon N$. We wish to see if

$$|d(X', Y') - d(V'''_i, V'''_j)| \leq \epsilon;$$

this will follow if we can show

$$\left|\mathbf{E}_{x_1 \in V'''_i, x_2 \in V'''_j} f(x_1, x_2)\left(1_X(x_1)1_Y(y_1) - \frac{|X|}{|V'''_i|}\frac{|Y|}{|V'''_j|}\right)\right| \leq \varepsilon^3. \qquad (11.33)$$

for arbitrary subsets $X \subset V'''_i, Y \subset V'''_j$ (with no lower bound on cardinality). We make some preliminary reductions. Observe that we may restrict X to V''_i and Y to Y''_i, and replace the range of x_1 and x_2 to V''_i and V''_j respectively, and only incur an error of $O(\sigma)$ on the left-hand side. Similarly we may replace $|X|/|V'''_i|$ and $|Y|/|V'''_j|$ by $|X|/N$ and $|Y|/N$ and again only accept an error of $O(\sigma)$, thus estimating the left-hand side of (11.33) by

$$\left|\mathbf{E}_{x_1 \in V''_i, x_2 \in V''_j} f(x_1, x_2)\left(1_X(x_1)1_Y(y_1) - \frac{|X|}{N}\frac{|Y|}{N}\right)\right| + O(\sigma).$$

Now since f_C is constant on $V''_i \times V''_j$ we have

$$\mathbf{E}_{x_1 \in V''_i, x_2 \in V''_j} f_C(x_1, x_2)\left(1_X(x_1)1_Y(y_1) - \frac{|X|}{N}\frac{|Y|}{N}\right) = 0.$$

Next, by the uniformity of f_U we have

$$\left|\mathbf{E}_{x_1\in V_i'',x_2\in V_j''} f_U(x_1,x_2)\left(1_X(x_1)1_Y(y_1)-\frac{|X|}{N}\frac{|Y|}{N}\right)\right| \le \frac{|V|^2}{N^2}\frac{2}{F(\sigma,K)}$$
$$= O\left(\frac{mK^2}{\sigma F(\sigma,K)}\right).$$

By choosing F suitably (e.g. $F(\sigma,K)=mK^2/\sigma^2$) we can ensure that the right-hand side is $O(\sigma)$. Putting this all together, we can bound the left-hand side of (11.33) by

$$\left|\mathbf{E}_{x_1\in V_i'',x_2\in V_j''}(f_{U^\perp}-f_C)(x_1,x_2)\left(1_X(x_1)1_Y(y_1)-\frac{|X|}{N}\frac{|Y|}{N}\right)\right| + O(\sigma).$$

By the triangle inequality, this is less than

$$2\mathbf{E}_{x_1\in V_i'',x_2\in V_j''}|(f_{U^\perp}-f_C)(x_1,x_2)| + O(\sigma).$$

Now from Proposition 11.29 and Cauchy–Schwarz we have

$$\mathbf{E}_{x_1\in V,x_2\in V}|(f_{U^\perp}-f_C)(x_1,x_2)| \le \sigma$$

which after trimming away the exceptional set V_* gives

$$\mathbf{E}_{x_1\in V_1''\cup\cdots\cup V_k'',x_2\in V_1''\cup\cdots\cup V_k''}|(f_{U^\perp}-f_C)(x_1,x_2)| = O(\sigma).$$

By Markov's inequality (and the uniform sizes of the V_j'') we conclude that

$$\mathbf{E}_{x_1\in V_i'',x_2\in V_j''}|(f_{U^\perp}-f_C)(x_1,x_2)| = O(\sigma/\varepsilon)$$

for all but at most ϵ of the pairs (i,j). In such a case we obtain a bound of $O(\sigma/\varepsilon)$ for the left-hand side of (11.33), which will be acceptable by choosing σ equal to a small multiple of ε. Finally, the bound $K=O_{\sigma,F}(1)$ now implies a bound $k=O_{\epsilon,m}(1)$ as required. This establishes the lemma (with the partition $V_1'',\ldots,V_k''$). □

Note that in the above proof only a very specific choice of function $F()$ was needed. However, the ability to set the function F *arbitrarily* becomes very important in the hypergraph theory, as it is the easiest way to reconcile the problem mentioned earlier of needing to have the regularity control given by one partition dominate the number of cells of another partition without totally losing control of all the error terms. Of course the price one pays for this is that the total number of cells at the end of the argument becomes extremely large.

We now begin the proof of Proposition 11.29. The reader may wish to focus on the $d=2$ case for sake of familiarity, although the general d case is no different. We will re-interpret the partitions $V_{i,1},\ldots,V_{i,K}$ of V_i as σ-algebras $\mathcal{B}_i$ on V_i for

$1 \leq i \leq d$, which induces a further σ-algebra $\mathcal{B}_1 \otimes \cdots \otimes \mathcal{B}_d$ on $V_1 \times \cdots \times V_d$, formed by the Cartesian products $V_{1,i_1} \times \cdots \times V_{d,i_d}$. Note in particular that the function $f_C := \mathbf{E}(f|\mathcal{B}_1 \otimes \cdots \otimes \mathcal{B}_d)$ will be a K-constant function between 0 and 1. The decomposition $f = f_{U^\perp} + f_U$ will be given by $f_{U^\perp} := \mathbf{E}(f|\mathcal{B}'_1 \otimes \cdots \otimes \mathcal{B}'_d)$ and $f_U := f - f_{U^\perp}$, where the $\mathcal{B}'_i$ are somewhat finer σ-algebras than the $\mathcal{B}_i$. The exact choice of $\mathcal{B}_i$, and $\mathcal{B}'_i$ will be determined by an energy increment algorithm very similar to that used to prove Proposition 10.36.

We turn to the details. We fix $V_1, \ldots, V_d$ and the function $f : V_1 \times \cdots \times V_d \to \mathbf{R}^+$. Given any σ-algebras $\mathcal{B}_1, \ldots, \mathcal{B}_d$ of V_1, V_2, we define the *energy* $\mathcal{E}_f(\mathcal{B}_1 \otimes \cdots \otimes \mathcal{B}_d)$ by

$$\mathcal{E}_f(\mathcal{B}_1 \otimes \cdots \otimes \mathcal{B}_d) := \|\mathbf{E}(f|\mathcal{B}_1 \otimes \cdots \otimes \mathcal{B}_d)\|^2_{L^2(V_1 \times \cdots \times V_d)},$$

thus the energy ranges between 0 and 1 and finer σ-algebras have higher energy. Just as Proposition 10.36 relied on Lemma 10.40, Proposition 11.29 will rely on the following analog.

Lemma 11.31 (Lack of uniformity implies energy increment) *Let $\mu > 0$ and $K' \geq 1$, and for each $1 \leq i \leq d$ let $\mathcal{B}'_i$ be a σ-algebra on V_i with at most K' atoms each such that*

$$|\mathbf{E}_{x_1 \in V_1, \ldots, x_d \in V_d} 1_{A_1 \times \cdots \times A_d}(f - \mathbf{E}(f|\mathcal{B}'_1 \otimes \cdots \otimes \mathcal{B}'_d))(x_1, \ldots, x_d)| \geq \mu$$

for some $A_1 \subseteq V_1, \ldots A_d \subseteq V_d$. Then for each $1 \leq i \leq d$ there exists finer σ-algebras $\mathcal{B}''_i$ than $\mathcal{B}'_i$ with at most $2K'$ atoms each such that

$$\mathcal{E}_f(\mathcal{B}''_1 \otimes \cdots \otimes \mathcal{B}''_d) \geq \mathcal{E}_f(\mathcal{B}'_1 \otimes \cdots \otimes \mathcal{B}'_d) + \mu^2.$$

Proof For $1 \leq i \leq d$, let $\mathcal{B}''_i$ be the σ-algebra generated by $\mathcal{B}'_i$ and A_i. Observe that

$$\mathbf{E}_{x_1 \in V_1, \ldots, x_d \in V_d} 1_{A_1 \times \cdots \times A_d}(f - \mathbf{E}(f|\mathcal{B}''_1 \otimes \cdots \otimes \mathcal{B}''_d))(x_1, \ldots, x_d) = 0$$

since $A_1 \times \cdots \times A_d$ is the union of atoms in $\mathcal{B}''_1 \otimes \cdots \otimes \mathcal{B}''_d$, on each of which $f - \mathbf{E}(f|\mathcal{B}''_1 \otimes \cdots \otimes \mathcal{B}''_d)$ has mean zero. Subtracting this from the hypothesis we conclude

$$\begin{aligned}|\mathbf{E}_{x_1 \in V_1, \ldots, x_d \in V_d} 1_{A_1 \times \cdots \times A_d}(&\mathbf{E}(f|\mathcal{B}''_1 \otimes \cdots \otimes \mathcal{B}''_d) \\ &- \mathbf{E}(f|\mathcal{B}'_1 \otimes \cdots \otimes \mathcal{B}'_d))(x_1, \ldots, x_d)| \geq \mu\end{aligned}$$

and hence by Cauchy–Schwarz

$$\|\mathbf{E}(f|\mathcal{B}''_1 \otimes \cdots \otimes \mathcal{B}''_d) - \mathbf{E}(f|\mathcal{B}'_1 \otimes \cdots \otimes \mathcal{B}'_d)\|^2_{L^2(V_1 \times \cdots \times V_d)} \geq \mu^2.$$

The claim then follows from Pythagoras' theorem. □

Proof of Proposition 11.29 This will be almost identical to Proposition 10.36. We construct a nested pair of σ-algebras $\mathcal{B}_i \subset \mathcal{B}'_i$ on V_i for each $1 \leq i \leq d$ and an integer $K \geq 1$ by the following double-loop algorithm.

- Step 0. Initialize $\mathcal{B}_i = \{\emptyset, V_i\}$ for each i.
- Step 1. Let K be the smallest integer such that each of the $\mathcal{B}_i$ have at most K atoms. Set $\mathcal{B}'_i := \mathcal{B}_i$ for each i; thus we trivially have $\mathcal{E}_f(\mathcal{B}'_1 \otimes \cdots \otimes \mathcal{B}'_d) \leq \mathcal{E}_f(\mathcal{B}_1 \otimes \cdots \otimes \mathcal{B}_d) + \sigma^2$.
- Step 2. If

$$|\mathbf{E}_{x_1 \in V_1, \ldots, x_d \in V_d} 1_{A_1 \times \cdots \times A_d}(f - \mathbf{E}(f|\mathcal{B}'_1 \otimes \cdots \otimes \mathcal{B}'_d))(x_1, \ldots, x_d)| \leq \frac{1}{F(\sigma, K)}$$

  for all $A_1 \subseteq V_1, \ldots, A_d \subseteq V_d$, then we terminate the algorithm. If not, then we can apply Lemma 11.31 to obtain for each $1 \leq i \leq d$ a new σ-algebra $\mathcal{B}''_i$ with at most twice as many atoms as $\mathcal{B}'_i$ such that

$$\mathcal{E}_f(\mathcal{B}''_1 \otimes \cdots \otimes \mathcal{B}''_d) \geq \mathcal{E}_f(\mathcal{B}'_1 \otimes \cdots \otimes \mathcal{B}'_d) + \frac{1}{F(\sigma, K)^2}.$$

- Step 3. If we have

$$\mathcal{E}_f(\mathcal{B}''_1 \otimes \cdots \otimes \mathcal{B}''_d) \leq \mathcal{E}_f(\mathcal{B}_1 \otimes \cdots \otimes \mathcal{B}_d) + \sigma^2$$

  then we set $\mathcal{B}' := \mathcal{B}''$ and return to Step 2. If instead we have

$$\mathcal{E}_f(\mathcal{B}''_1 \otimes \cdots \otimes \mathcal{B}''_d) > \mathcal{E}_f(\mathcal{B}_1 \otimes \cdots \otimes \mathcal{B}_d) + \sigma^2$$

  then we set $\mathcal{B} = \mathcal{B}''$ and return to Step 1.

Once the algorithm terminates we set $f_{U^\perp} := \mathbf{E}(f|\mathcal{B}'_1 \otimes \cdots \otimes \mathcal{B}'_d)$, $f_C := \mathbf{E}(f|\mathcal{B}_1 \otimes \cdots \otimes \mathcal{B}_d)$, and $f_U := f - f_{U^\perp}$. The verification that the algorithm does indeed terminate in finite time and gives the desired properties is almost identical to the analogous arguments in Proposition 10.36 and is left to the reader as an exercise. □

Remark 11.32 A closer inspection of the above argument shows that the number of atoms K in the σ-algebra $\mathcal{B}$ can increase from K to as much as $K2^{F(\sigma,K)^2}$ whenever we return from Step 3 to Step 1. Since the latter step can occur as often as $1/\sigma^2$ times, we see that the final complexity will most likely be a tower exponential in K or worse (unless we restrict F to have logarithmic growth or so, see [239] for some discussion of this type of lemma). As mentioned in Section 10.6, this tower-exponential behavior is unavoidable, see [136].

Now we discuss the extension of the regularity lemma to hypergraphs. To simplify the exposition we shall consider only the 3-uniform case. First it turns out that a minor modification of the proof of Proposition 11.29 yields the following variant, in which we obtain much stronger uniformity control (with respect to arbitrary 2-edge sets rather than vertex sets), but with a weaker and more complex notion of K-constancy. More precisely, let us say that a function $f : V_1 \times V_2 \times V_3 \to \mathbf{R}^+$ is $(K, 2)$-*constant* if there exist partitions $V_i \times V_j = E_{ij,1} \cup \cdots , E_{ij,K}$ for $ij = 12, 23, 31$ such that f is constant on each set

$$\{(x_1, x_2, x_3) \in V_1 \times V_2 \times V_3 : (x_i, x_j) \in E_{ij,a_{ij}} \text{ for all } ij = 12, 23, 31\}$$

for all $a_{12}, a_{23}, a_{31} \in [1, K]$.

Proposition 11.33 (Preliminary hypergraph regularity lemma) *[360] Let $V_1, \ldots, V_3$ be arbitrary finite non-empty sets, let $f : V_1 \times \cdots \times V_3 \to \mathbf{R}^+$ be such that $0 \leq f(x_1, x_2, x_3) \leq 1$, let $\sigma > 0$, and let $F : \mathbf{R}^+ \times \mathbf{R}^+ \to \mathbf{R}^+$ be an arbitrary function. Then there exists a quantity $K = O_{\sigma,F}(1)$ and a decomposition $f = f_{U^\perp} + f_U$ with the following properties:*

- *the "anti-uniform" component $f_{U^\perp}$ obeys the bounds $0 \leq f_{U^\perp} \leq 1$. Furthermore there exists a $(K, 2)$-constant approximation f_C to $f_{U^\perp}$ with $0 \leq f_C \leq 1$ and $\|f_{U^\perp} - f_C\|_{L^2(V_1 \times V_2 \times V_d)} \leq \sigma$;*
- *the "uniform" component f_U obeys the regularity estimate*

$$|\mathbf{E}_{x_1 \in V_1, x_2 \in V_2, x_3 \in V_3} 1_{A_{12}}(x_1, x_2) 1_{A_{23}}(x_2, x_3) 1_{A_{31}}(x_3, x_1) f_U(x_1, \ldots, x_d)| \leq \frac{1}{F(\sigma, K)}$$

for all $A_{12} \subseteq V_1 \times V_2$, $A_{23} \subseteq V_2 \times V_3$, $A_{31} \subseteq V_3 \times V_1$.

The proof of Proposition 11.33 is almost identical to that of Proposition 11.29 but with a somewhat heavier notational burden. We leave it as an exercise.

One can then deduce a full-strength regularity lemma for 3-uniform hypergraphs. The exact formulation of this lemma is rather messy and too unenlightening to be given here (see [139], [283], [284], [282], [360]), but we will describe the formulation indirectly by informally outlining the *proof* of the lemma, following [360]. Given a function $f : V_1 \times V_2 \times V_3 \to \mathbf{R}^+$, an initial error tolerance σ, and a growth function $F : \mathbf{R}^+ \times \mathbf{R}^+ \to \mathbf{R}^+$, we then decide upon a much faster growth function $F^{\text{fast}} : \mathbf{R}^+ \times \mathbf{R}^+ \to \mathbf{R}^+$, the exact choice of which will be chosen later. Applying Proposition 11.33 with this much faster growth function F^{fast} we obtain a primary decomposition $f = f_{U^\perp} + f_U$, where f_U is extremely regular with respect to 2-edge partitions (enjoying the fast function F^{fast} in the denominator), and $f_{U^\perp}$ is approximable by a $(K, 2)$-constant function f_C, where K has some (rather lousy) upper bound. The K-constant function can be described using $O(K)$ edge sets $E_{ij,a}$ in $V_i \times V_j$ for $ij = 12, 23, 31$. We then apply Proposition 11.29 to the indicators $1_{E_{ij,a}} : V_i \times V_j \to \mathbf{R}^+$ each of these edge sets $E_{ij,a}$, using the original function F, and replacing the error tolerance σ by something smaller, e.g. $1/F(\sigma, K)$. Strictly speaking, we need a "multiple function" or "vector-valued" version of Proposition 11.29 in which we regularize multiple functions simultaneously using a single partition, but this is not hard to set up. This gives us a new parameter $K' := O_{K,F,\sigma}(1)$, such that we have secondary decompositions of each of the indicator functions $1_{E_{ij,a}}$ into a K'-constant main term and some manageable errors. Finally, we choose F^{fast} so that $F^{\text{fast}}(\sigma, K)$ dominates any expression that will arise from K'; this basically means that F^{fast} is a tower-iterated version of F, and will ensure that the error term f_U in the primary decomposition is also manageable.

Thus to summarize (and glossing over the delicate issues regarding the relative sizes of various parameters), we start with a function $f(x_1, x_2, x_3)$ of three variables, and approximate it by a combination of $O(K)$ functions $1_{E_{ij,a}}(x_i, x_j)$ of two variables, plus manageable errors; we then approximate each of the $1_{E_{ij,a}}(x_i, x_j)$ by a combination of $O(K')$ functions of one variable (i.e. the indicators of the vertex classes), again plus manageable errors. With carefully chosen relative sizes of parameters as given above, this regularization of the original function f is suitable for such tasks as accurately counting the number of 3-simplices in a 3-uniform hypergraph, in a manner similar in spirit to (but somewhat lengthier than) the proof of Lemma 10.46. This in turn eventually leads to a proof of Theorem 11.27, which in turn implies Szemerédi's theorem and a number of other consequences.

Exercises

11.6.1 Deduce Proposition 11.28 from Lemma 10.46. (Hint: the vertex set V for the k-uniform hypergraph should consist of coordinate hyperplanes such as $\{(x_1, \ldots, x_k) : x_i = \text{const}\}$, as well as the diagonal hyperplanes $\{(x_1, \ldots, x_k) : x_1 + \cdots + x_k = \text{const}\}$.)

11.6.2 Use Proposition 11.28 to deduce Theorem 11.24.

11.6.3 Use Proposition 11.28 to deduce the claim $r_k(Z) = o_{|Z|\to\infty;k}(|Z|)$ whenever $|Z|$ is coprime to $(k-1)!$.

11.6.4 [139] Let V, W be disjoint sets of n vertices each. Let us color the vertices in V red or blue randomly and independently, with equal probability of each. Suppose we also color the *edges* in W (i.e. the elements of $\binom{W}{2}$) red or blue randomly and independently. Let $H = H(V \cup W, E)$ be the 3-uniform hypergraph with edge set consisting of all triples $\{v, w, w'\}$, where $v \in V$ and $\{w, w'\} \in \binom{W}{2}$ have the same color, together with all triples of the form $\{v, v', w\}$ with $\{v, v'\} \in \binom{V}{2}$ and $w \in w$. Let us also define the competing 3-uniform hypergraph $H' = H'(V \cup W, E')$, where E' consists of all the triples of the form $\{v, v', w\}$ with $\{v, v'\} \in \binom{V}{2}$ and $w \in W$, and with each triple of the form $\{v, w, w'\}$ with $v \in V$ and $\{w, w'\} \in \binom{W}{2}$ belonging to E' with independent probability 1/2. Show that with large probability, the number of 3-edges joining any three large subsets A, B, C of $V \cup W$ is about the same for H and H', but that H and H' have very different numbers of 3-simplices. (Of course, one should quantify these vague statements precisely, for instance using Chernoff's inequality.) This shows that regularization based entirely on vertex partition will not be sufficient to easily conclude the simplex removal lemma.

11.6.5 Prove Proposition 11.33.

11.7 Arithmetic progressions in the primes

We now discuss the Green–Tao theorem, Theorem 10.7. We will not give a complete proof of this theorem here, referring the reader to the original paper [158] and to the survey articles [358], [217], [184], [153], [361] for further details. Instead we shall give a somewhat informal discussion, in particular focusing on the connections with the other arguments discussed in this chapter.

We begin by a very brief history of the problem. This result has been conjectured for some time; indeed, long progressions of primes were already studied by Lagrange and Waring in 1770. The Erdős–Turan conjecture (Conjecture 10.6), formulated in 1936, was certainly motivated in part by this problem; it implies Theorem 10.7 but is much stronger (and still open). The first significant progress on the problem was in 1939, when Van der Corput [370] used Fourier-analytic methods (but not the density increment or energy increment arguments) to establish that the primes contained infinitely many progressions of length three. A key step of the argument is to obtain good bounds for exponential sums such as $\mathbf{E}_{1\leq n\leq N}\Lambda(n)e(\alpha n)$, where Λ is the von Mangoldt function and α is a real number (which may be close to a rational with small denominator, or far away from one). However, as discussed earlier, Fourier methods (also known as the *Hardy–Littlewood circle method* in analytic number theory) do not directly work for progressions of length 4 or higher. Progress on this problem thus became very slow. Szemerédi's theorem did not directly give any new results on the primes, as they had density zero, and even the powerful quantitative bounds of Bourgain (Theorem 10.30) for $k = 3$ and Gowers (11.23) were insufficient to attack the primes (which would require a bound roughly of the form $r_k(\mathbf{Z}_N) = o(N \log\log N/\log N)$).

Meanwhile, the methods of sieve theory were developed by analytic number theorists, in part to solve questions concerning the existence of patterns of primes such as arithmetic progressions. While these methods seem unable by themselves to count primes directly (due to the notorious *parity problem* in sieve theory, the discussion of which is beyond the scope of this book), they have proven to be enormously successful in counting *almost-primes* – products of very few primes.

For instance, it is not too hard to use sieve theory methods to show that for any given k, there are infinitely many progressions of length k, the elements of which are each the product of $O_k(1)$ prime factors. However to pass from the almost-primes to the primes remained difficult; one notable result is that of Heath-Brown [179] in 1981, who showed that there were infinitely many progressions of length 4 where three elements were prime and the fourth was the product of at most two primes. In another direction, Balog [15] in 1992 was able to find infinitely many k-tuples of primes $p_1, \ldots, p_k$ whose midpoints $(p_i + p_j)/2$ were also prime. Meanwhile, in 1996, Kohayakawa, Luczak, and Rödl [212] extended the Szemerédi regularity lemma to subgraphs of a certain type of random subgraph, and in so doing extended Roth's theorem to show that *relatively* dense subsets of a random set contained many progressions of length 3 (see Theorem 10.18). More recently, Green [147] used Fourier methods to obtain a Roth theorem for the primes, in other words showing that any subset of the primes of positive relative density contained infinitely many arithmetic progressions of length 3. This was then refined by Green and Tao [159], who showed (roughly speaking) that any dense subset of a set which was well controlled by a sieve would contain infinitely many progressions of length 3.

In [158] this type of result was extended to arbitrary k. The precise statement requires some notation.

Definition 11.34 (Pseudo-random measure) [158] A function $\nu : \mathbf{Z}_N \to \mathbf{R}^+$ is said to be *k-pseudo-random* if we have $\mathbf{E}_{\mathbf{Z}_N} \nu = 1 + o_{N\to\infty}(1)$, and more generally we have the *linear forms condition*

$$\mathbf{E}_{x_1,\ldots,x_t \in \mathbf{Z}_N} \prod_{i=1}^{m} \nu\left(\sum_{j=1}^{t} L_{ij} x_j + b_i\right) = 1 + o_{N\to\infty;k}(1)$$

whenever $0 \leq m \leq k2^{k-1}, t \leq 3k-4$, and $b_1, \ldots, b_m \in \mathbf{Z}_N$ are arbitrary, and L_{ij} are rational numbers with numerator and denominator of magnitude at most k, such that none of the m t-tuples $(L_{ij})_{j=1}^{t}$ are rational multiples of any other. Furthermore we assume the *correlation condition*

$$\mathbf{E}_{x\in\mathbf{Z}_N} \prod_{i=1}^{m} \nu(x + h_i) \leq \sum_{1\leq i<j\leq m} \tau(h_i - h_j)$$

for all $1 \leq m \leq 2^{k-1}$ and all $h_1, \ldots, h_m \in \mathbf{Z}_N$, where $\tau : \mathbf{Z}_N \to \mathbf{R}^+$ is a function obeying the moment conditions $\mathbf{E}\tau^q = O_{q,k}(1)$ for all $1 \leq q < \infty$.

The above definition is rather complicated, but one should view these conditions as an assertion that the weight function (or "measure") ν is very randomly distributed. If we have $\nu = \frac{1}{\mathbf{P}(A)} 1_A$ for some set $A \subset \mathbf{Z}_N$, these conditions are essentially asserting that the events $\sum_{i=1}^{m} L_{ij}x_j + b_i \in A$ are essentially independent of each other if the $(L_{ij})_{j=1}^{t}$ are not commensurate, and the events $x + h_i \in A$ are only mildly correlated to each other for generic choices of $h_1, \dots, h_m$.

The key result in [158] then takes the Szemerédi theorem, in the form of Theorem 11.1, and generalizes it to pseudo-random measures.

Theorem 11.35 (Relative Szemerédi theorem) *Let $k \geq 3$, let $\mathbf{Z}_N$ be a finite cyclic group of large prime order N, and let $f : Z \to \mathbf{R}^+$ is a non-negative function which is not identically zero, and obeys the bounds $0 \leq f(x) \leq \nu(x)$ and $\mathbf{E}_{\mathbf{Z}_N}(f) \geq \delta > 0$ for all $x \in \mathbf{Z}_N$ for some k-pseudo-random measure ν. Then*

$$\Lambda_k(f, \dots, f) = \Omega_{k,\delta}(1) - o_{N\to\infty;k,\delta}(1).$$

This strengthening of Szemerédi's theorem allows one to detect arithmetic progressions not just in sets of positive density, but now also in sets of positive *relative* density with respect to sufficiently "pseudo-random" sets, even if the latter sets have density zero. For instance, given any set $B \subset \mathbf{Z}_N$ for which $\frac{1}{\mathbf{P}(B)} 1_B$ is k-pseudo-random, the above theorem will guarantee that $r_k(B) = o_{N\to\infty;k}(|B|)$, provided one has a mild condition such as $\mathbf{P}(B) \geq N^{-1/k}$ in order to neglect the diagonal $r = 0$ term in $\Lambda_k(f, \dots, f)$. In particular, any subset A of B of large relative density $|A|/|B| \geq \delta$ will contain a proper arithmetic progression of length k as soon as N is sufficiently large depending on δ and k.

As it turns out, the primes P do not quite fall into the above framework, because they are unevenly distributed with respect to small residue classes (e.g. they are almost all odd), and any set B containing P for which P has positive relative density will also necessarily have some uneven distribution in small residue classes (this is ultimately due to the divergence of the Euler product $\prod_p (1 - \frac{1}{p})^{-1}$). On the other hand, pseudo-random measures are necessarily evenly distributed among such classes (see exercises). However, this can be easily fixed, by the simple trick of using the pigeonhole principle to pass to a single residue class among small divisors. More precisely, one defines $W := \prod_{p<w} p$ for some small w (e.g. $w = \log\log N$ will suffice), and replaces the primes P by the set $P_{W,b,N} = \{q \in [\varepsilon_k N, 2\varepsilon_k N] : Wp + b \in P\}$ for some b coprime to W (in fact one can use Dirichlet's theorem on distribution of primes in residue classes to take $b = 1$). Here $\varepsilon_k := 1/2^k(k+4)!$ is a small number needed for some minor technical reasons (related to the denominators of the L_{ij} in the k-pseudo-random condition). See [158], [361] for more details of this "W-trick".

It turns out that $P_{W,b,N}$ can be contained effectively in a k-pseudo-random measure. More precisely, there exists a k-pseudo-random measure $\nu : \mathbf{Z}_N \to \mathbf{R}^+$ such that $\mathbf{E}_{\mathbf{Z}_N} 1_{P_{W,b,N}} \nu = \Theta_k(1)$, and also one has the mild upper bound $\|\nu\|_{L^\infty(\mathbf{Z}_N)} = O(N^{1/k})$ (again needed to order to neglect the $r = 0$ diagonal term).

This fact, combined with Theorem 11.1, is enough to establish arithmetic progressions of length k in the primes, and even to establish the stronger result that $r_k(P \cap [1, N]) = o_{N\to\infty;k}(|P \cap [1, N]|) = o_{N\to\infty;k}(N/\log N)$. The construction of this measure relies on a version of the Selberg sieve used by Goldston and Yıldırım [134], [132], [133] (see also [363], [184], [361]); it is purely number-theoretical in nature and we do not reproduce it here. However, we do remark that ν can be thought of as being a (smoothed out) version of the normalized indicator function on the *almost-primes* $P_k = \{n : n \text{ is the product of } O_k(1) \text{ primes}\}$, or more precisely of the portion of P_k in the residue class $b \pmod W$. As mentioned earlier, modern sieve theory techniques such as the Selberg sieve are very accurate at counting correlations of almost-primes, and thus can verify the k-pseudo-randomness of ν by fairly standard arguments. In contrast, verifying the k-pseudo-randomness of a normalized counting function of the primes themselves (or of a related object such as $P_{W,b,N}$) is still beyond the reach of current technology, being roughly equivalent to the notorious *Hardy–Littlewood prime tuples conjecture*, which would imply not just the Green–Tao theorem but also the twin prime conjecture, Goldbach's conjecture, and many other difficult and unsolved problems in additive number theory. Thus one crucially needs a tool such as the relative Szemerédi theorem to bridge the gap between the almost-primes (which we understand quite well) and the primes (which are still very mysterious).

We briefly discuss the proof of Theorem 11.35. It turns out that this theorem is proven by a means very similar to that to the proof of Szemerédi's theorem outlined in Section 11.4, but now the functions involved are not bounded by 1, but are instead bounded by some k-pseudo-random measure ν. Nevertheless, it is still possible to adapt most of the arguments in that section (with the exception of the useful UAP^{k-2} norms, which do not seem to have a suitable analog in this setting). First of all one can generalize the generalized von Neumann theorem (11.8) to obtain the bound

$$|\Lambda_k(f_0, \dots, f_{k-1})| = O_k\left(\min_{0\le j\le k-1} \|f_j\|_{U^{k-1}(\mathbf{Z}_N)}\right) + o_{N\to\infty;k}(1) \qquad (11.34)$$

whenever $f_0, \dots, f_{k-1} : \mathbf{Z}_N \to \mathbf{R}^+$ are bounded in magnitude by $\nu + 1$. The original bound (11.8) was proven using multiple applications of the van der Corput lemma, which in turn is essentially just the Cauchy–Schwartz inequality; similarly, the bound (11.34) is also proven using several applications of the Cauchy–Schwarz inequality, the main task being to keep track of all the weights involving ν and to use the linear forms condition to ensure that after a certain point these weights can be replaced by 1 with only a negligible error. See [158] for full details.

The bound (11.34) tells us that even in the pseudo-random setting, functions which are Gowers uniform of order $k - 2$ can still be safely ignored. This opens the way to prove Theorem 11.35 by using a Koopman–von Neumann theorem. Here, the relevant theorem is as follows.

Proposition 11.36 (Generalized Koopman–von Neumann structure theorem) *[158] Let ν be a k-pseudo-random measure, and let $f : \mathbf{Z}_N \to \mathbf{R}^+$ be such that $0 \leq f(x) \leq \nu(x)$ for all $x \in \mathbf{Z}_N$. Let $0 < \varepsilon \ll 1$ be a small parameter, and assume $N > N_0(\varepsilon)$ is sufficiently large. Then there exists a σ-algebra $\mathcal{B}$ and an exceptional set $\Omega \in \mathcal{B}$ such that:*

- *(smallness condition)*

$$\mathbf{E}(\nu 1_\Omega) = o_{N\to\infty;\varepsilon,k}(1); \tag{11.35}$$

- *(ν is uniformly distributed outside of Ω)*

$$\|(1 - 1_\Omega)\mathbf{E}(\nu - 1|\mathcal{B})\|_{L^\infty(\mathbf{Z}_N)} = o_{N\to\infty;\varepsilon,k}(1); \tag{11.36}$$

and

- *(Gowers uniformity estimate)*

$$\|(1 - 1_\Omega)(f - \mathbf{E}(f|\mathcal{B}))\|_{U^{k-1}(\mathbf{Z}_N)} \leq \varepsilon^{1/2^k}. \tag{11.37}$$

Assuming this proposition, one can now write $(1 - 1_\Omega)f = f_U + f_{U^\perp}$, where $f_U := (1 - 1_\Omega)(f - \mathbf{E}(f|\mathcal{B}))$ is Gowers uniform of order $k - 2$, and $f_{U^\perp} := (1 - 1_\Omega)\mathbf{E}(f|\mathcal{B})$ is bounded by $1 + o_{N\to\infty;\varepsilon,k}(1)$ (since $\mathbf{E}(f|\mathcal{B}) \leq 1 + \mathbf{E}(\nu - 1|\mathcal{B})$) and non-negative. Furthermore by using (11.35) one can show that $f_{U^\perp}$ almost has the same mean as f: $\mathbf{E}_{\mathbf{Z}_N} f_{U^\perp} = \mathbf{E}_{\mathbf{Z}_N} f - o_{N\to\infty;\varepsilon,k}(1)$. From the latter two facts one can use the ordinary Szemerédi theorem (Theorem 11.1) to establish that

$$\Lambda_k(f_{U^\perp}, \ldots, f_{U^\perp}) = \Omega_{k,\delta}(1) - o_{N\to\infty;k,\delta}(1).$$

Since f_U is Gowers uniform, we can easily use (11.34) to then conclude

$$\Lambda_k(f_{U^\perp} + f_U, \ldots, f_{U^\perp} + f_U) = \Omega_{k,\delta}(1) - o_{N\to\infty;k,\delta}(1)$$

and Theorem 11.35 then follows since $0 \leq f_{U^\perp} + f_U \leq f$.

It thus only remains to prove Proposition 11.36. Here we follow the energy increment strategy already used to prove Propositions 10.36, 11.18, and 11.29. The first step is the following generalization of Lemma 11.14:

Lemma 11.37 (Soft inverse theorem) *[158] Let $f : \mathbf{Z}_N \to \mathbf{C}$ be a function bounded in magnitude by $\nu + 1$, and let $F = \mathcal{D}_{k-1}(f)$ be the dual function. Then $\|F\|_{L^\infty(\mathbf{Z}_N)} \leq 2^{2^{k-1}-1} + o_{N\to\infty;k}(1)$. Furthermore, if $\|f\|_{U^{k-1}(Z)} \geq \eta$, then $|\langle f, F\rangle| \geq \eta^{2^d}$.*

The key feature here is that even though f may be unbounded (or at least very large), the dual function F is bounded quite concretely. This is a consequence of the linear forms condition, which among other things provides a uniform bound for $\mathcal{D}_{k-1}(\nu+1)$ and hence for $\mathcal{D}_{k-1}(f)$.

One can then run the same energy increment algorithm used in Propositions 10.36, 11.18, 11.29, to convert any lack of uniformity in the f_U term into a dual function which is then added to a σ-algebra in order to increase the energy of the $f_{U^\perp}$ term. The only difficulty with executing this strategy is to ensure that $f_{U^\perp}$ stays bounded. This is accomplished by the following somewhat technical result.

Proposition 11.38 *[158] Let ν be a k-pseudo-random measure. Let $0 < \varepsilon < 1$ and $0 < \eta < 1/2$ be parameters. Then to every function $F : \mathbf{Z}_N \to \mathbf{R}$ bounded in magnitude by $\nu + 1$, one can construct a σ-algebra $\mathcal{B}_{\varepsilon,\eta}(\mathcal{D}_{k-1}F)$ with the following property: for any $K \geq 1$ and any $F_1, \ldots, F_K : \mathbf{Z}_N \to \mathbf{R}$ functions bounded in magnitude by $\nu + 1$, if we set $\mathcal{B} := \mathcal{B}_{\varepsilon,\eta}(\mathcal{D}_{k-1}F_1) \vee \cdots \vee \mathcal{B}_{\varepsilon,\eta}(\mathcal{D}_{k-1}F_K)$, then if $\eta < \eta_0(\epsilon, K)$ is sufficiently small and $N > N_0(\epsilon, K, \eta)$ is sufficiently large we have*

$$\|\mathcal{D}_{k-1}F_j - \mathbf{E}(\mathcal{D}_{k-1}F_j|\mathcal{B})\|_{L^\infty(\mathbf{Z}_N)} \leq \varepsilon \text{ for all } 1 \leq j \leq K. \tag{11.38}$$

Furthermore there exists a set Ω which lies in $\mathcal{B}$ such that

$$\mathbf{E}_{\mathbf{Z}_N}((\nu+1)1_\Omega) = O_{K,\varepsilon}(\eta^{1/2}) \tag{11.39}$$

and such that

$$\|(1-1_\Omega)\mathbf{E}(\nu - 1|\mathcal{B})\|_{L^\infty(\mathbf{Z}_N)} = O_{K,\varepsilon}(\eta^{1/2}). \tag{11.40}$$

The σ-algebras $\mathcal{B}_{\varepsilon,\eta}(\mathcal{D}_{k-1}F)$ are constructed very similarly to those in Proposition 10.38, the only real difference being that certain small atoms cause some difficulty and need to be placed in the exceptional set Ω. However these problems can be dealt with by taking η suitably small depending on K, ε, and then N suitably large depending on K, ε, η. The trickiest task is to establish (11.40). This ultimately comes down (using the Weierstrass approximation theorem as in the proof of Proposition 10.38) to establishing estimates of the form

$$\mathbf{E}((\nu-1)\mathcal{D}_{k-1}F_1 \cdots \mathcal{D}_{k-1}F_K) = o_{N\to\infty;k,K}(1)$$

whenever $F_1, \ldots, F_K : \mathbf{Z}_N \to \mathbf{R}$ are functions bounded in magnitude by $\nu + 1$. This estimate turns out to be achievable by application of the Gowers–Cauchy–Schwarz inequality, Hölder's inequality, and both the linear forms and correlation conditions; see [158].

Finally, we apply the energy increment argument and combine Lemma 11.37 and Proposition 11.38 as in the proof of Proposition 10.36 to obtain Proposition 11.36. Actually the energy increment argument here is slightly simpler than that in Proposition 10.36 as there is no arbitrary growth function F to deal with. As such one can use just a single loop iterative procedure rather than a double loop, which simplifies things slightly. On the other hand, the presence of the exceptional sets, and the unboundedness of several of the functions being manipulated, requires some additional care, in particular to ensure that one really does get a substantial energy increment at each stage in order to make the algorithm terminate in finite time (and to keep the quantity K appearing in Proposition 11.38 bounded by $O_\varepsilon(1)$).

Exercises

11.7.1 Suppose that one knew that $r_k(\mathbf{Z}_N) = o_{N\to\infty;k}(N \log\log N/\log N)$ for all $k \geq 3$. Derive the Green–Tao theorem as a consequence of this. (Hint: divide the primes from 1 to N into residue classes mod $P = \prod_{p<c\log N} P$ for some small absolute constant c, and use the pigeonhole principle (and Proposition 1.51) to conclude that the primes in one of these classes has density roughly $\log\log N/\log N$.)

11.7.2 Use Theorem 11.35 to prove a version of Theorem 10.18 for large cyclic groups $\mathbf{Z}_N$ and arbitrary k. (Hint: if B is a random subset of $\mathbf{Z}_N$ with expected density $\tau \geq N^{-\varepsilon}$ for some small $\varepsilon = \varepsilon_k > 0$, show using Chernoff's inequality that $\frac{1}{\tau}1_B$ is very likely to be k-pseudo-random.)

11.7.3 [158] Let $\nu : \mathbf{Z}_N \to \mathbf{R}$ be k-pseudo-random. Show that $\|\nu - 1\|_{U^{k-1}(\mathbf{Z}_N)} = o_{N\to\infty;k}(1)$. Conclude in particular that if $k \geq 3$, then one has the uniform distribution property

$$\mathbf{E}_{x\in\mathbf{Z}_N} 1_P(x)\nu(x) = \mathbf{P}_{\mathbf{Z}_N}(P) + o_{N\to\infty;k}(1)$$

for any arithmetic progression P. Thus pseudo-random measures must be evenly distributed in arithmetic progressions.

11.7.4 [158] Prove Lemma 11.37.

12

Long arithmetic progressions in sum sets

12.1 Introduction

One general theme throughout this book is that sum sets $A + B$ are more structured than arbitrary sets A, B, and in particular that iterated sum sets such as $lA = \{a_1 + \cdots + a_l : a_i \in A_i\}$ should get increasingly structured as l gets larger. One example of this phenomenon is Lemma 4.13, which shows that if A has small Fourier bias then lA quickly fills out the entire ambient group. (See also Exercise 4.3.12 for related demonstration of special structure of sum sets.) For another example, let A be a subset of $[1, n]$ for some large n and consider (as a measure of structure) the longest progression contained inside A. If A has no structure other than density, e.g. $|A| \geq 0.99n$, then there is not much we can say. Even the powerful quantitative version (11.23) of Szemerédi's theorem due to Gowers can only obtain an arithmetic progression of length $\Omega(\log\log\log\log\log n)$. For cubes the situation is somewhat better (and simpler); Lemma 10.49 guarantees that A contains a proper cube of dimension $\Omega(\log\log n)$, though this is still far from the maximal dimension $\Theta(\log n)$ of the cubes inside $[1, n]$.

The situation improves markedly with taking sum sets, though. First, if $A \subset [1, n]$ has cardinality at least $0.99n$, then it is easy to see that $A + A$ or $A - A$ contains an arithmetic progression of length $0.98n$. This is of course a rather extreme case, but more generally if $A \subset [1, n]$ is such that $|A| \geq \delta n$, then Bourgain's theorem (Theorem 4.47) shows that $A + A$ and $A - A$ contain proper arithmetic progressions of length at least $\exp(\Omega_\delta(\log^{1/3} n))$. For $3A$ and $2A - A$, Exercise 4.7.1 shows that these sets in fact contain proper arithmetic progressions of length $\Omega_\delta(n^{\Omega_\delta(1)})$, while Theorem 4.43 shows that these sets contain proper generalized arithmetic progressions of rank $O_\delta(1)$ and volume $\Theta_\delta(n)$. For $2A - 2A$, Chang's theorem (Theorem 4.42) gives similar results but with better dependence on δ.

These results however require A to be rather dense inside the ambient interval $[1, n]$; even Chang's theorem requires A to have density $\Omega(\frac{\log\log n}{\log n})$ in order to be non-trivial. If A is sparser than this, one can still ask what happens to sum sets such as lA when l gets large. One can show fairly easily (see exercises) that for fixed $A \subset \mathbf{Z}$ and l very large, lA essentially coalesces into a single long arithmetic progression, plus some negligible terms at the boundary. For other groups, the situation is slightly different (again, see exercises); note that unlike the situation with small l, Freiman homomorphisms are much rarer when l gets very large and one cannot identify the asymptotic behavior of lA as $l \to \infty$ for non-isomorphic ambient groups. See [260], [261] for some more advanced results in this direction.

Now we ask a more quantitative question. Suppose we are given positive integers l, m, n with $2 \leq m \leq n$ (the case $m = 1$ will be too degenerate to consider). If A is an arbitrary subset of $[1, n]$ with cardinality $|A| = m$, what can we say about the structure of lA, and more precisely, what is the largest arithmetic progression (or generalized arithmetic progression) one can find inside A? From the above discussion we expect to find quite a large progression when l is large. For instance, from the work of Lev [226] one has the following result:

Theorem 12.1 *[226] Let $A \subset [1, n]$ be such that $|A| > 2$ and A is not contained in any progression of step greater than 1. Let l be such that $l \geq 2(n-1)/(|A|-2)$. Then lA contains an interval $[m+1, m+n]$ for some integer m.*

In fact more precise statements are available; see [227]. An earlier result of Sárközy [307] established the following weaker result:

Theorem 12.2 *[226] There exists an absolute constant $C > 0$ such that the following holds. Let $A \subset [1, n]$ and $l \geq 1$ be such that $|A| \geq 2$ and $l|A| \geq Cn$. Then lA contains a proper arithmetic progression of length $\Omega(l|A|)$.*

We shall prove this theorem as a special case of more general results below.

We can phrase the above theorems in a different way. For any l, m, n with $2 \leq m \leq n$, we define $f(m, l, n)$ to be the largest integer such that, for every subset $A \subset [1, n]$ of cardinality $|A| \geq m$, lA contains a proper arithmetic progression of length $f(m, l, n)$. The question is now to determine the size of $f(m, l, n)$ for various values of m, l, n. Theorem 12.2 asserts that $f(m, l, n) = \Omega(lm)$ whenever $lm \geq Cn$. In fact we have $f(m, l, n) = \Theta(lm)$ in this case, as can be seen by considering the set $A = [1, m]$.

This gives a satisfactory answer to the question when m is large compared to n/l. It is now natural to ask whether this threshold n/l is sharp, and what happens to $f(m, l, n)$ for m below this threshold. It turns out in this case that the upper bound for $f(m, l, n)$ drops dramatically:

Lemma 12.3 (Upper bound on f) *Let $d \geq 1$ be an integer, and let l, m, n be positive integers such that $l \geq 2$ and $n \geq 2^d l^{d-1} m^d$. Then $f(m^d, l, n) \leq lm - l + 1$.*

Proof Let A be the rank d progression

$$A := [1, m]^d \cdot (1, 2lm, \ldots, (2lm)^{d-1}),$$

thus

$$lA = [l, lm]^d \cdot (1, 2lm, \ldots, (2lm)^{d-1}),$$

Then by summing the geometric series we see that $A \subseteq [1, n]$. From the base $2lm$ representation of the integers we see that the map $\phi : [l, lm]^d \to lA$ defined by $\phi(x_0, \ldots, x_{d-1}) = \sum_{j=0}^{d-1} x_j (2lm)^j$ is a Freiman homomorphism of order 2. The same argument shows that A is proper, so that $|A| = m^d$. From Proposition 5.24 we thus see that the length of the longest arithmetic progression in lA is the same as the length of the longest arithmetic progression in $[l, lm]^d$, which is clearly $lm - l + 1$. The claim follows. □

From this lemma (and the trivial observation that $f(m, l, n)$ is monotone increasing in m) we see that there exist constants c_d for $d = 1, 2, 3, \ldots$ such that $f(m, l, n) = O(lm^{1/d})$ whenever $m \leq c_d \frac{n}{l^{d-1}}$. Thus the upper bounds for $f(m, l, n)$ exhibit a thresholding behavior in m near the points $n/l, n/l^2, n/l^3, \ldots$. Somewhat remarkably, these thresholds are sharp up to constants:

Theorem 12.4 *[350] Let $d \geq 1$. Then there exists a constant $C_d > 0$ such that for any $l \geq 1$ and $A \subset [1, n]$ with $|A| \geq C_d \frac{n}{l^d}$ and $|A| \geq 2$, lA contains a proper arithmetic progression of length $\Omega_d(l|A|^{1/d})$.*

Note that Theorem 12.2 already gives the $d = 1$ case of this theorem. Combining this with the preceding discussion, we see that $f(m, l, n) = \Theta_d(lm^{1/d})$ whenever $C_d \frac{n}{l^d} \leq m \leq c_d \frac{n}{l^{d-1}}$. This settles the question of determining the magnitude of $f(m, l, n)$ as long as m is well away from the thresholds $n/l, n/l^2$, etc. and is not too small. The precise behavior near these thresholds is still unclear, and may be difficult to resolve.

We will prove Theorem 12.4 (and hence Theorem 12.2) in the following sections. A key observation is that up to constants, one only needs to consider the case when l is a power of 2, in which case one can view lA as an iteration of the doubling operation $A \mapsto A + A$. This gives the problem a certain dynamic flavor, in which we analyze the evolution of a set under the doubling map. We then discuss extensions and variants, in particular to restricted sum sets

$$l^*A := \{a_1 + \cdots + a_l : a_1, \ldots, a_l \in A, \text{ distinct}\}$$

and finite sum sets

$$FS(A) := \bigcup_{l=0}^{\infty} l^*(A) = \left\{ \sum_{x \in B} x : B \subset A, 0 \leq |B| < \infty \right\},$$

where we now allow A to possibly be infinite. This in particular will be used to resolve some conjectures of Erdős and Folkman on complete sequences; we also present some other applications.

Exercises

12.1.1 Let A be an additive set in a cyclic group $\mathbf{Z}_p$ of prime order. Show that $lA = \mathbf{Z}_p$ whenever $l(|A| - 1) \geq p - 1$, and that this condition is best possible. (Hint: use the Cauchy–Davenport inequality, Theorem 5.4.) Thus when $|A| \geq 2$, the iterated sum sets stabilize to the entire group after at most $(p-1)/(|A|-1)$ summations.

12.1.2 Let A be an additive set in a finite additive group Z. Show that there exists a subgroup G of Z such that lA is a coset of G for all sufficiently large l. If A contains 0, show that we in fact have $lA = \langle A \rangle$ whenever $l|A| \geq 2|\langle A \rangle|$, where $\langle A \rangle$ is the group generated by A. (Hint: quotient out by the symmetry group $\mathrm{Sym}_1(lA)$ and then apply Kneser's theorem, Theorem 5.5.) Thus in this case the iterated sum sets stabilize to a group after at most $2|Z|/|A|$ summations.

12.1.3 Let A is an additive set of integers. Show that if l is sufficiently large depending on A, then lA is a proper arithmetic progression of length $\Theta_A(l)$ together with at most $O_A(1)$ additional elements. (Hint: this statement is only non-trivial for very large l. Use the Chinese remainder theorem. It may be useful to reduce to the case when A has smallest element zero, and has no common divisor.)

12.1.4 Prove Theorem 12.2 in the case when l is extremely large compared to A and n.

12.1.5 Let A be an additive set in $\mathbf{Z}^d$ that contains the origin 0. Let B be the convex hull of A in $\mathbf{R}^d$, and let Γ be the sub-lattice of $\mathbf{Z}^d$ spanned by A. Show that for all large l we have

$$((1 - o_{l\to\infty;A}(1)) \cdot B) \cap \Gamma \subseteq lA \subseteq ((1 + o_{l\to\infty;A}(1)) \cdot B) \cap \Gamma.$$

How is this statement modified when A does not contain the origin?

12.1.6 Show that $f(m, l, n) \leq lm - l + 1$ for all $l \geq 1$ and $2 \leq m \leq n$.

12.1.7 Show that $f(m, l, n) = l$ whenever $n \geq (2l)^{m-1}$. (Hint: for the upper bound, consider $A = 2l\wedge[0, m-1] = \{1, 2l, (2l)^2, \ldots, (2l)^{m-1}\}$.)

12.2 Proof of Theorem 12.4

To prove Theorem 12.4 it turns out to be convenient to prove a stronger result. Observe in the example given in Lemma 12.3 not only contains an arithmetic progression, it in fact contains a much larger *generalized* arithmetic progression. This phenomenon turns out to be quite general:

Theorem 12.5 *[350] For any fixed positive integer d there is a constants $C_d > 0$ such that the following holds. For any positive integers n and l and any set $A \subset [1, n]$ satisfying $l^d|A| \geq C_d n$, lA contains a proper progression of rank d' and volume at least $\Omega_d(l^{d'}|A|)$, for some integer $1 \leq d' \leq d$.*

It is easy to see that this implies Theorem 12.4, and we leave this as an exercise. The example in Lemma 12.3 shows that one can have $d' = d$; the simple example $A = [1, m]$ also shows that one can have $d' = 1$. Of course intermediate values of d' are also possible.

We now prove Theorem 12.5. We begin with a version of this theorem for progressions.

Lemma 12.6 (Coalescence of progressions) *Let P be a proper progression of integers of rank at most d, and let $l \geq 1$ be an integer. Then lP contains a proper progression of rank d' and volume at least $\Omega_d(l^{d'}|P|)$ for some $1 \leq d' \leq d$.*

Remark 12.7 It is instructive to experiment with the sum sets lP for a proper progression P of rank d as $l \to \infty$, e.g. the progression $P = [1, m]^4 \cdot (1, N_1, N_1N_2, N_1N_2N_3)$ where $m < N_1 < N_2 < N_3$. At first, the sum sets lP will remain proper of rank d (and grow polynomially in size, like l^d). But at some point there will be a "collision", causing the sum set to essentially "coalesce" into a progression of rank $d - 1$ or less (and thus grow somewhat more slowly in l). After a finite number of collisions, the sum set will coalesce into a single arithmetic progression (plus negligible terms), at which point it only grows linearly in l. The proof of Lemma 12.6 below can be used to formalize this intuitive picture but we will not do so here as the notation required is somewhat complicated. This result is also closely related to Minkowski's second theorem (Theorem 3.30), as well as the other machinery in Section 3.5.

Proof We shall induce on d. The case $d = 1$ is obvious (indeed, lP is now an arithmetic progression of length $l|P| - l + 1$), so suppose $d > 1$ and the claim has already been proven for $d - 1$. We may assume that l is large depending on d since the claim is trivial otherwise (since lP contains a translate of P).

Let $C_d > 1$ be a large constant to be chosen later. Now let $k \geq 1$ be the largest integer such that $2^k \leq l/C_d$. If $2^k P$ is proper, then $|2^k P| = \Omega_d(2^{kd}|P|) = \Omega_d(l^d|P|/C_d^d)$ and the claim follows with $d' = d$ since lP contains a translate of $2^k P$. Now suppose that $2^k P$ is not proper. Let $1 \leq k' \leq k$ be the first integer such that $2^{k'} P$ is improper, then by arguing as before we see that $|2^{k'} P| \geq |2^{k'-1} P| = \Omega_d(2^{k'd}|P|)$. Applying Theorem 3.40 we see that $O_d(1)2^{k'} P$ contains a proper progression Q of rank $d-1$ and volume $\Omega_d(|2^{k'}P|) = \Omega_d(2^{k'd}|P|)$. Applying the induction hypothesis we see that $2^{k-k'}Q$ contains a proper progression of rank d' and volume at least

$$\Omega_d\big(2^{(k-k')d'}|2^{k'}P|\big) = \Omega_d\big(2^{(k-k')d'}2^{k'd}|P|\big) = \Omega_d(2^{kd'}|P|) = \Omega_d\big(l^{d'}|P|/C_d^{d'}\big)$$

for some $1 \le d' \le d-1$. If C_d is large enough, then lP will contain a translate of $2^{k-k'}O_d(1)2^{k'}P$ and hence a translate of $2^{k-k'}Q$. The claim follows. □

The above lemma allows us to split the problem of finding a progression of small rank in lA into two parts. First, we find a progression P of large rank in $l'A$ for some $l' < l$, and then use the above lemma to find a progression of small rank in kP, where $kl' \le l$. More precisely, we have

Proof of Theorem 12.5 Note from the hypotheses $l^d|A| \ge C_d n$ and $A \subset [1, n]$ that we can ensure that l is large depending on d, simply by choosing C_d large depending on d.

Let k be the largest integer such that $2^k \le l$. Thus

$$2^{kd}|A| \ge l^d|A|/2^d \ge C_d n/2^d.$$

On the other hand we have $2^k A \subset [1, 2^k n]$ and hence

$$|2^k A| \le 2^k n \le \frac{2^d}{C_d}2^{k(d+1)}|A|.$$

Now let $k' \ge 1$ be the smallest positive integer such that

$$|2^{k'}A| < 2^{k'(d+3/2)}|A|$$

then for C_d large enough we have $k' \le k$, and in fact

$$k' \le k - \Omega_d(\log C_d).$$

Set $A' := 2^{k'-1}A$. By construction of k', we have

$$|2^{k'-1}A'| \ge 2^{(k'-1)(d+3/2)}|A|$$

and hence

$$|A' + A'| \le 2^{d+3/2}|A'|.$$

By Exercise 2.3.14, we can find a subset $F \subset A'$ which is symmetric around some point $x/2$ such that $|F| = \Theta_d(|A'|)$ with doubling constant $O_d(1)$. Applying the Ruzsa–Chang theorem (Theorem 5.30), we see that $2F - 2F$ contains a proper progression of rank $O_d(1)$ and volume $\Omega_d(|A'|)$. By the symmetry of F, we see that $2F - 2F$ is a translate of $4F$, which is contained in $4A'$. Thus $4A' = 2^{k'+1}A$ contains a proper progression Q of rank $O_d(1)$ and volume $\Omega_d(|A'|)$. Now lA

contains a translate of $2^k A$, which in turn contains $2^{k-k'-1}Q$. Applying Lemma 12.6 we conclude that lA contains a proper progression P of rank d' and volume

$$|P| = \Omega_d\big(2^{(k-k')d'}|Q|\big) = \Omega_d\big(2^{(k-k')d'}|A'|\big) = \Omega_d\big(2^{(k-k')d'}2^{k'(d+3/2)}|A|\big)$$

for some $d' = O_d(1)$. On the other hand, since $lA \subseteq [1, ln]$, we have

$$|lA| \leq ln \leq 2^{k+1}n \leq \frac{2^{d+1}}{C_d}2^{k(d+1)}|A|.$$

Since $|P| \leq |lA$, we conclude that

$$2^{(k-k')d'}2^{k'(d+3/2)}|A| \leq O_d\left(\frac{1}{C_d}2^{k(d+1)}|A|\right)$$

and thus

$$2^{(k-k')(d'-d-1)} \leq O_d\left(\frac{1}{C_d}2^{-k'/2}\right).$$

Thus implies (for C_d large enough) that $d' < d+1$, and thus $1 \leq d' \leq d$ since d' is an integer. Since

$$|P| = \Omega_d\big(2^{(k-k')d'}2^{k'(d+3/2)}|A|\big) = \Omega_d(2^{kd'}|A|) = \Omega_d(l^{d'}|A|)$$

and the claim follows. □

Remark 12.8 The key trick here was to split up the long sum lA by expressing it as a binary tree of binary sums $2^kA + 2^kA = 2^{k+1}A$. The bounds we had on $|lA|$ forced one of the binary sums to have small doubling constant, at which point we could use an inverse theorem, in this case the Freiman cube lemma and the Ruzsa–Chang theorem. A similar trick was employed in Theorem 2.35; see also [42]. This method is sometimes referred to as the *tree argument.*

Remark 12.9 The above proof made use of some rather powerful theorems, including Theorem 3.40 and the Ruzsa–Chang theorem. However, it is possible to prove the above results without using such deep facts from additive geometry and Fourier analysis, instead relying on more elementary inverse theorems such as the $3k-3$ theorem (Theorem 5.11) and the Freiman cube lemma (Theorem 5.20). See [351], and the exercises below. This latter approach turns out to be more robust, in particular being able to deal with restricted sum sets l^*A.

Exercises

12.2.1 Show that Theorem 12.5 implies Theorem 12.4. (Hint: use Exercise 3.2.5.)

12.2.2 Using only the $3k-3$ theorem and the tree argument, show that if P is an arithmetic progression of integers and $A \subset P$ is such that $|A| \geq \delta|P|$, then there exists an positive integer $l = O_\delta(1)$ such that lA contains an arithmetic progression of length $\Omega_\delta(|P|)$.

12.2.3 Using only the preceding exercise and an iteration argument, show that if P is a progression of integers of rank d and $A \subset P$ is such that $|A| \geq \delta|P|$, then there exists a positive integer $l = O_{\delta,d}(1)$ such that lA contains a progression of rank at most d and cardinality $\Omega_{\delta,d}(|P|)$.

12.2.4 Without using Theorem 3.40, show that if P is a progression of integers of rank d, then there exists a positive integer $l = O_d(1)$ such that lP contains a *proper* progression of rank at most d and cardinality $\Omega_d(|P|)$. (You may wish to use Freiman's cube lemma, Theorem 5.20 or the variant in Proposition 5.35.)

12.2.5 [351] (Filling lemma) Using only the preceding exercises, show that if P is a progression of integers of rank d and $A \subset P$ is such that $|A| \geq \delta|P|$, then there exists a positive integer $l = O_{\delta,d}(1)$ such that lA contains a proper progression of rank at most d and cardinality $\Omega_{\delta,d}(|P|)$.

12.2.6 [350] Let $P = a + [0, N] \cdot v$ be a progression of rank d. Show that if P is not proper, then $|2P| \leq (1 - \frac{1}{2^{d+1}})|[0, 2N]|$. More generally, prove that

$$|kP| \leq \frac{O(1)^d}{k}|[0, kN]|$$

for all $k \geq 1$. Thus an improper progression becomes "increasingly improper" as one dilates it.

12.2.7 Using the preceding exercise, show that if P is a proper progression of integers of rank d such that $2P$ is not proper, show that there is a positive integer $l = O_d(1)$ such that lP contains a proper progression of rank at most $d - 1$ with cardinality $\Omega_d(|P|)$.

12.2.8 Use the above exercises to give alternate proofs of Lemma 12.6 and Theorem 12.5.

12.3 Generalizations and variants

There are various extensions of Theorem 12.4 and Theorem 12.5. An easy modification of the above arguments allows one to handle distinct summands:

Theorem 12.10 *[350] For any fixed positive integer d there is a constant $C_d > 0$ such that the following holds. Let $A_1, \ldots, A_l$ be subsets of $[1, n]$ of size m where l and m satisfy $l^d m \geq C_d n$. Then $A_1 + \cdots + A_l$ contains a progression of rank d' and volume at least $\Omega_d(l^{d'} m)$, for some integer $1 \leq d' \leq d$. In particular, $A_1 + \cdots + A_l$ contains an arithmetic progression of length at least $\Omega_d(lm^{1/d})$.*

From Lemma 5.25 we can also replace the integers by any other torsion-free additive group without difficulty. A more difficult strengthening is to work with the restricted sum sets $l^* A$ instead of lA. It is possible to adapt the above methods to deal with this case too:

Theorem 12.11 *[350] For any fixed positive integer d there is a constant $C_d > 0$ such that the following holds. For any positive integers n and l and any set $A \subset [1, n]$ satisfying $l \leq |A|/2$ and $l^d|A| \geq C_d n$, $l^* A$ contains a proper progression of rank d' and volume at least $\Omega_d(l^{d'}|A|)$, for some integer $1 \leq d' \leq d$. In particular $l^* A$ contains a proper arithmetic progression of length $\Omega_d(l|A|^{1/d})$.*

If we define $f^*(m, l, n)$ to be the obvious analog of $f(m, l, n)$ with lA replaced by $l^* A$, then we conclude (using Lemma 12.3 for the upper bound) that $f^*(m, l, n) = \Theta_d(lm^{1/d})$ whenever $C_d \frac{n}{l^d} \leq m \geq c_d \frac{n}{l^{d-1}}$ and $m \geq 2l$. (Note that $f^*(m, l, n)$ becomes vacuous when $m < l$, so the condition $m \geq 2l$ is fairly natural.)

Theorem 12.11 is significantly harder than Theorem 12.4 and we will not present it here. Let us, however, mention an important lemma, which can be viewed as a variant of Freiman theorem for subset sums. This lemma asserts that if $l^* A$ does not yield a proper GAP as claimed by the theorem, then A must contain a big subset which has a very rigid structure.

Lemma 12.12 *For any positive constants ν and d there are positive constants δ, α and d_1 such that the following holds. Let A be a subset of $[n]$, l be a positive integer and $n \geq f(n) \geq 1$ be a function of n such that*

$$\max\left\{\log^{10} n, (40 f(n) \log_2 n)^{1/3d}\right\} \leq l \leq |A|/2$$

and $l^d|A| f(n) \geq n$. Then one of the following two statements must hold:

- *$l^* A$ contains a proper GAP of rank d' and volume $\Omega(l^{d'}|A|)$ for some $1 \leq d' \leq d$;*
- *there is a subset $\tilde{A}$ of A with cardinality at least $\delta|A|$ which is contained in a GAP P of rank d_1 and volume $O(|A| f(n)^{1+\nu} \log^\alpha n)$.*

The function $f(n)$ can be seen as a *rigidity* parameter. The closer $l^d|A|$ is to n, the more rigid is the structure of $\tilde{A}$.

The case $d = 1$ is of special importance, being a generalization of the Theorem 12.2, and we isolate it as a corollary.

Corollary 12.13 *There exists a constant $C > 0$ such that whenever $A \subset [1, n]$ and $1 \leq l \leq |A|/2$ is such that $l|A| \geq Cn$, then $l^* A$ contains an arithmetic progression of length $cl|A|$.*

This result has the following consequence for subset sums:

Corollary 12.14 *[350] If A is a subset of $[1, n]$ of cardinality at least $C\sqrt{n}$ for a sufficiently large absolute constant $C > 0$, then the subset sums $FS(A)$ contains an arithmetic progression of length n.*

The first part of this result, with $\sqrt{n}$ replaced by $\sqrt{n \log n}$, was originally proven by Freiman [115]; we leave the deduction of this corollary from Corollary 12.13 as an exercise.

Another variant is to work in a cyclic group $\mathbf{Z}_n$ of prime order instead of in an interval $[1, n]$. Here, one can modify the preceding arguments to obtain

Theorem 12.15 *[350] For any $d \geq 1$ there exists $C_d > 0$ such that the following holds. For any additive set A in a cyclic group $\mathbf{Z}_n$ of prime order and any $l \geq 1$ with $l^{d+1}|A| \geq C_d n$ and $|A| \geq 2$, the set lA contains a proper arithmetic progression of length* $\min(n, \Omega_d(l|A|^{1/d}))$.

There are two differences between this theorem and Theorem 12.4. First the progression has length $\min(n, \Omega_d(l|A|^{1/d}))$ instead of $\Omega_d(l|A|^{1/d})$, but this is natural since lA cannot exceed n in size. Second the condition on l has been relaxed from $l^d|A| \geq C_d n$ to $l^{d+1}|A| \geq C_d n$. This is ultimately because the trivial bound $|lA| \leq ln$ which was used in the $[1, n]$ case can now be improved to the trivial bound $|lA| \leq n$. Otherwise the argument is essentially the same, and is left as an exercise.

Exercises

12.3.1 Prove Theorem 12.10. (Hint: use the tree argument with different sets at the leaves. You will need to replace the Ruzsa–Chang theorem by a different result, such as Theorem 4.43, or use the elementary approach sketched in earlier exercises.)

12.3.2 Deduce Corollary 12.14 from Corollary 12.11.

12.3.3 Let $\mathbf{Z}_n$ be a cyclic group of prime order, and let $f(m, l, \mathbf{Z}_n)$ be defined just like $f(m, l, n)$ but now with the sets A lying in $\mathbf{Z}_n$ instead of $[1, n]$. Show that $f(m, l, \mathbf{Z}_n) = n$ if and only if $l(m - 1) \geq n - 1$. (Hint: use Exercise 12.1.1.) Show also that for every $d \geq 1$ there exist constants $c_d, C_d > 0$ such that $f(m, l, n) = \Theta_d(lm^{1/d})$ whenever $C_d \frac{n}{l^{d+1}} \leq m \geq c_d \frac{n}{l^d}$.

12.3.4 Prove Theorem 12.15. (Note that the hypothesis that n is prime will prevent any "torsion" issues from arising in the progressions until the progressions become of size comparable to n, at which point one can proceed using the Cauchy–Davenport inequality instead.)

12.4 Complete and subcomplete sequences

An infinite set $A \subset \mathbf{Z}^+$ of positive integers is *complete* if its subset sums $FS(A)$ contain every sufficiently large positive integer. This notion is similar to, but distinct from, the concept of a base as studied in Chapter 1, as we allow sums of arbitrary length but require the summands to be distinct. The notion of complete sequences was introduced by Erdős in the early sixties and has since then been studied extensively by various researchers (see [89, Section 6] or [274, Section 4.3] for surveys). The center of this study is to find necessary and sufficient conditions for a sequence to be complete.

Intuitively, the denser a set is, the more likely it is to be complete. However, density alone is not sufficient; the even integers have density $1/2$ but are not complete. More generally, any set contained in an infinite arithmetic progression containing zero will not be complete. To deal with these cases, let us say that a set A is *subcomplete* if $FS(A)$ contains an infinite arithmetic progression $a + \mathbf{N} \cdot r = \{a, a + r, a + 2r, \ldots\}$. It is easy to see that these two notions are related as follows:

Lemma 12.16 *Let $A \subset \mathbf{Z}^+$ be infinite. Then A is complete if and only if A is subcomplete and $FS(A)$ intersects every infinite arithmetic progression in $\mathbf{Z}^+$.*

We leave this lemma as an exercise. The condition that $FS(A)$ intersects every infinite arithmetic progression in $\mathbf{Z}^+$ is a local condition that only depends on the residue classes that A occupies, together with their multiplicity; see exercises. In particular, this condition is typically quite easy to verify for standard bases such as the Waring bases $\mathbf{N}^\wedge k$ or the primes P. Thus we shall focus on the subcomplete property.

A simple example of Cassels [46] shows that there exist sets A of density $|A \cap [1, n]| = \Omega(n^{1/2})$ which are not subcomplete; see exercises. Remarkably, this example is sharp up to constants:

Theorem 12.17 *[350] There exists an absolute constant $C > 0$ such that every infinite set $A \subset \mathbf{Z}^+$ with $|A \cap [1, n]| \geq Cn^{1/2}$ is subcomplete. In particular, if $FS(A)$ also intersects every infinite arithmetic progression in $\mathbf{Z}^+$, then A is complete.*

We prove this result in the next section; the main tool is Corollary 12.14. The second part of this result was conjectured by Erdős [85] in 1962, and the first part by Folkman [103] in 1966. In [85] the second part was proven under the stronger hypothesis $|A \cap [1, n]| \geq Cn^{(\sqrt{5}-1)/2}$, while in [103] the first part was proven under the hypothesis $|A \cap [1, n]| \geq n^{1/2+\varepsilon}$ for any $\varepsilon > 0$ and sufficiently large n. This was lowered to $Cn^{1/2} \log^{1/2} n$ by Hegyvári [181] and Luczak and Schoen [241], using Sárkőzy's theorem (see the exercises).

There is an analog of the above results for infinite multisets $A = \{a_1, a_2, \ldots\}$ in $\mathbf{Z}^+$, where $a_1 \leq a_2 \leq \cdots$ are allowed to have repetitions, and define the finite sum sets

$$FS(A) := \left\{ \sum_{i \in I} a_i : I \subset \mathbf{Z}^+, \text{ finite} \right\}$$

in analogy with before, and define the notion of completeness and subcompleteness as above. In this case it is possible to have a density as large as $|A \cap [1, n]| = \Omega_\varepsilon(n^{1-\varepsilon})$ for any given $\varepsilon > 0$ (where of course we count multiplicity) and still not have subcompleteness (see exercises). Again, this example is basically sharp.

Theorem 12.18 *[350] There exists an absolute constant $C > 0$ such that every infinite multiset $A \subset \mathbf{Z}^+$ with $|A \cap [1, n]| \geq Cn$ is subcomplete. In particular, if $FS(A)$ also intersects every infinite arithmetic progression in $\mathbf{Z}^+$, then A is complete.*

This result was conjectured by Folkman [103], and is proven very similarly to Theorem 12.17, we leave it as an exercise for the next section where Theorem 12.17 is proved.

To end this section, let us disuss the finite version of completeness. We say that a subset A of $\mathbf{Z}_p$ (for a large prime p) is complete if $FS(A) = \mathbf{Z}_p$. Olson [265], answering a question of Erdős and Heilbronn, proved that if $|A| > 2\sqrt{p}$, then A is complete. The bound $2\sqrt{p}$ is essentially sharp. To see this, take $A = \{-k, -(k-1), \ldots, -1, 0, 1, \ldots, (k-1), k\}$, where k is the largest integer such that $\sum_{i=1}^{k} i < p/2$. Deshouillers and Freiman [70] showed that this is actually the only example, given that $|A|$ is sufficiently large. We call a set A of integers between 0 and $p-1$ *small* if

$$\sum_{a \in A} \|a/p\| < 1$$

where $\|z\|$ (as usual) is the distance from z to the closest integer. It is easy to check that a small set is not complete.

Theorem 12.19 *Let A be a subset of $\mathbf{Z}_p$ with more than $\sqrt{2p}$ elements. If A is not complete, then there is a non-zero element x of $\mathbf{Z}_p$ such that the set $x \cdot A$ is small.*

Szemerédi and Vu [349, 352] showed that it is possible to weaken the condition $|A| > \sqrt{2p}$ considerably by dropping a small subset from A.

Theorem 12.20 *Let A be a non-complete subset of $\mathbf{Z}_p$. Then there is a subset A' of A with at most $p^{.49}$ elements and a non-zero element x of $\mathbf{Z}_p$ such that $x \cdot (A \backslash A')$ is small.*

Exercises

12.4.1 Show that any infinite set $A \subset \mathbf{Z}^+$ of lower density strictly greater than $1/2$ is complete.

12.4.2 Prove Lemma 12.16. Generalize this to multisets.

12.4.3 Let $A \subset \mathbf{Z}^+$ be an infinite set, and for each positive integer N let $A_N^\infty \subset \mathbf{Z}_N$ be the set of all residue classes a mod N whose intersection with A is infinite, and let A'_N be those elements of A which do not lie in one of the residue classes in A_N^∞ (this set is automatically finite). Show that $FS(A)$ intersects every infinite arithmetic progression in $\mathbf{Z}^+$ if and only if $FS(A'_N) \bmod N + \langle A_N^\infty \rangle = \mathbf{Z}_N$ for all N. Note that one has $FS(A'_N) \bmod N = FS(A'_N \bmod N)$ if we view $A'_N \bmod N$ as a multiset; thus this criterion uses only the multiplicities of A modulo N rather than the actual values of A.

12.4.4 Consider an infinite sequence $A = \{a_1, a_2, \ldots\}$. Prove that if

$$\limsup_{i \to \infty} \left(a_i - \sum_{j=1}^{i-1} a_j \right) \to \infty, \qquad (12.1)$$

then A is not subcomplete.

12.4.5 [46],[351] Let $m := 10^4$ and let $A := \bigcup_{i=1}^{\infty} [m^{2^i}/4, m^{2^i}/2]$. Show that $|A \cap [1, n]| = \Omega(n^{1/2})$ for all n, and A intersects every infinite arithmetic progression in $\mathbf{Z}^+$ but that A is not subcomplete nor complete.

12.4.6 Modifying the previous example, show that for any $\varepsilon > 0$ there exists a multiset A with $|A \cap [1, n]| = \Omega_\varepsilon(n^{1-\varepsilon})$ which intersects every infinite arithmetic progression in $\mathbf{Z}+$ but is not subcomplete.

12.4.7 [181], [241] Let $A \subset [1, n]$ be such that $|A| \geq C n^{1/2} \log^{1/2} n$ for some large constant $C > 0$ Show that there exists $1 \leq i \leq O(\log n)$ and a set $A_i \subset [1, 2n]$ such that $2^i |A_i| = \Omega(Cn)$ and each element in A_i can be written as the sum of two distinct elements of A between 2^i and 2^{i+1} times. Use this and Theorem 12.2 to prove the first part of Corollary 12.14 with $n^{1/2}$ replaced by $n^{1/2} \log^{1/2} n$. By using the arguments of the next section, this establishes Theorem 12.17, again with $n^{1/2}$ replaced by $n^{1/2} \log^{1/2} n$.

12.5 Proof of Theorem 12.17

To prove Theorem 12.17, it is convenient to reduce the (infinitary) condition of subcompleteness to a finitary version. Let us say that a partition $A = A' \cup A''$ of a multiset A of positive integers is *good* if the following two properties hold:

- there is a number d such that $FS(A')$ contains arbitrary long arithmetic progressions with difference d;
- let $A'' = \{b_1 \leq b_2 \leq b_3 \leq \cdots\}$, then

$$\lim_{i\to\infty}\left(\sum_{j=1}^{i-1} b_j\right) - b_i = +\infty. \tag{12.2}$$

Thus A' enjoys a finitary version of subcompleteness, whereas A'' grows slower than a lacunary sequence. These two conditions imply subcompleteness:

Lemma 12.21 *[351, 241, 181] Any sequence A of positive integers which admits a good partition is subcomplete.*

Proof We begin with some reductions. First observe that we can remove finitely many elements from A'' without affecting the condition (12.2). In particular, we can remove any residue class $a + d \cdot \mathbf{Z}$ mod d which contains only finitely many elements from A''.

Let $A_d \subset \mathbf{Z}_d$ be the set of residue classes mod d which intersect A'' (and thus contain infinitely many elements from A'', by the above reduction). The group $\langle A_d \rangle$ spanned by A'' is a subgroup of $\mathbf{Z}_d$ and thus has the form $\langle A_d \rangle = d' \cdot \mathbf{Z}_d$ for some factor d' of d. In particular we see that every element of A'' is a multiple of d'. Observe that $FS(A'')$ must intersect every residue class in $\langle A_d \rangle$. Thus there exists a finite set $B \subset A''$ such that $FS(B)$ intersects every residue class in $d' \cdot \mathbf{Z}_d$.

Let $A''' := A'' \backslash B$, thus A''' still obeys (12.2). A simple greedy algorithm argument then shows that the subset sums $FS(A''')$ are *syndetic* (has bounded gaps); more precisely, there exists $L \geq 0$ such that given any positive integer n, we can find an element $m \in FS(A''')$ such that $0 \leq n - m \leq L$. Since $FS(A''')$ consists entirely of multiples of d', and $FS(B)$ intersects every residue class in $d' \cdot \mathbf{Z}_d$, we conclude that the set $(FS(A''') + FS(B)) \cap (d \cdot \mathbf{Z})$ is also syndetic. Since $FS(A')$ contains arbitrarily long progressions of length d, we conclude that $FS(A') + [(FS(A''') + FS(B)) \cap (d \cdot \mathbf{Z})]$ contains an infinite progression of length d. But since this set is contained in $FS(A)$, we see that A is subcomplete as claimed. □

We can now prove Theorem 12.17.

Proof of Theorem 12.17 We write $A = \{a_1 < a_2 < \cdots\}$, and split $A = A' \cup A''$ where $A' := \{a_{2m} : m \in \mathbf{Z}^+\}$ and $A'' := \{a_{2m-1} : m \in \mathbf{Z}^+\}$. It is easy to see using the hypothesis $|A \cap [1, n]| \leq Cn^{1/2}$ that the set A'' will obey (12.2) and we leave it as an exercise. Thus we only need to show that $FS(A')$ contains arbitrarily long arithmetic progressions of a fixed step d.

For each non-negative integer j, let $A'_j := \{a_{2m} : 2^j \leq m < 2^{j+1}\}$. Thus the A'_j partition A'. Also, the hypothesis $|A \cap [1, n]| \leq Cn^{1/2}$ implies that for all sufficiently large j we have $A'_j \subset [1, n_j]$ for some $n_j = \Theta(2^{2j}/C^2)$. From Corollary 12.14 we conclude (if C is large enough) that $FS(A'_j)$ contains a proper arithmetic progression P_j of length n_j for all j larger than some initial j_0. Note that $FS(A'_j) \subset [1, 2^j n_j]$, hence the step d_j of the progression P_j cannot exceed 2^j.

This is almost what we need, except that the progressions P_j do not have the same step. This however can be dealt with using the following elementary lemma, which follows from Exercise 3.6.5.

Lemma 12.22 (Coalescence of arithmetic progressions) *[349, 350] Let P_1, P_2 be proper arithmetic progressions of integers of length N_1, N_2 and step $d_1, d_2 > 0$ respectively, where $N_2 \geq 5d_1$ and $N_1 \geq 5d_2$. Then $P_1 + P_2$ contains a proper arithmetic progression of length $N_1 + N_2 - 2$ whose difference is $\gcd(d_1, d_2)$.*

Using this lemma we can see inductively that for j_0 sufficiently large, the set $P_{j_0} + \cdots + P_j$ contains an proper arithmetic progression of length $n_{j_0} + \cdots + n_j - O(j)$ and step $\gcd(d_1, \ldots, d_j)$ for each $j \geq j_0$. The steps $\gcd(d_1, \ldots, d_j)$ are decreasing positive integers and thus must eventually stabilize at some fixed d. Since $P_{j_0} + \cdots + P_j$ is contained in $FS(A'_{j_0}) + \cdots + FS(A'_j)$, which in turn is contained in $FS(A')$, and $n_{j_0} + \cdots + n_j - O(j)$ goes to ∞ as $j \to \infty$, the claim follows. □

The proof of Theorem 12.18 is similar and is left as an exercise.

Exercises

12.5.1 Show that the set A'' used in the proof of Theorem 12.17 obeys (12.2).

12.5.2 [350] Show that there is a constant C such that the following holds. If A is a multiset of positive integers in $[1, n]$ with $|A| \geq Cn$, then $FS(A)$ contains an arithmetic progression of length n. (Hint: Use Theorem 12.10.)

12.5.3 [350] Prove Theorem 12.18. (Hint: use the previous exercise as a substitute for Corollary 12.14.)

12.6 Further applications

In this section, we present a few short applications of Corollary 12.13, taken from [380]. For several applications of Theorem 12.2 we refer to [308, 309] and the references therein.

The following simple lemma will be useful. Let $\mathbf{Z}_n^\times$ denote the residue classes in $\mathbf{Z}_n$ which are coprime to n.

Lemma 12.23 *Let n be a positive integer and A be a multiset of k elements of $\mathbf{Z}_n^\times$ for some $1 \leq k \leq n$. Then $|FS(A)| \geq k$. In particular, if $|A| = n$, then $FS(A) = \mathbf{Z}_n$.*

Proof Observe that if $a \in \mathbf{Z}_n^\times$, then $|FS(A \cup \{a\})| = |FS(A) + \{0, a\}| \geq |FS(A)| + 1$ by Kneser's theorem (Theorem 5.5) or direct computation. The claim then follows by induction. □

12.6.1 Olson's problem

We say that an additive set A in a finite ambient group Z is *complete* if $FS(A) = Z$. In this section, we investigate the case when Z is a cyclic group $Z = \mathbf{Z}_n$.

A well-known result of Olson [265], mentioned in Section 12.4, shows that if n is a prime and $|A| > 2n^{1/2}$, then A is complete.

Theorem 12.24 *[265] If n is a prime and $A \subset \mathbf{Z}_n$ has cardinality $|A| \geq 2n^{1/2}$, then A is complete.*

Olson later extended his result to an arbitrary finite group [268], with the constant 2 replaced by a larger constant. Here we give a short proof for the case when G is cyclic.

Theorem 12.25 *There is a constant C such that the following holds. If n is a sufficiently large positive integer and $A \subset \mathbf{Z}_n^\times$ has cardinality $|A| \geq Cn^{1/2}$, then A is complete.*

Remark 12.26 The assumption that the elements of A are coprime with n is necessary. For instance, if n is divisible by 3 then it is possible to have an incomplete set of size $n/3$. Without the coprime assumption, the problem of bounding $|A|$ is known as Diderrich's problem. It has been proved that the sharp bound for $|A|$ is $p + n/p - 2$, where p is the smallest prime divisor of n (see [235] for the case of cyclic groups and [127] for the general case of arbitrary additive groups).

Proof For convenience, we identify the elements of A as positive integers in $[1, n-1]$. Let us split A into two components $A' \cup A''$ each of cardinality at least $Cn^{1/2}/2 - 1$. By choosing C large enough, we see from Corollary 12.14 that $FS(A')$ (viewed as a subset of $\mathbf{Z}$) will contain a proper arithmetic progression P' of length n. If the step d' of this progression is coprime to n, then it will cover all the residue classes of $\mathbf{Z}_n$ and we will be done, so suppose that this is not the case; then the quantity $d := \gcd(d', n)$ is larger than 1. Then $FS(A')$ intersects all the residue classes in $d \cdot \mathbf{Z}_n$, so it will suffice to show that $FS(A'')$ intersects all the residue classes in $\mathbf{Z}_d$.

Note that the largest element in $FS(A')$ (and hence in P) is at most $O(n^{3/2})$, hence d' (and d) are $O(n^{1/2})$. In particular, we see by choosing C large enough that $|A''| \geq d'$. By Lemma 12.23 $FS(A'')$ intersects all the residue classes in $\mathbf{Z}_d$, and we are done. □

We conjecture that one can have $C = 2$ in Theorem 12.25. Hamidoune [173] made the following general conjecture for arbitrarily finite group.

Conjecture 12.27 *Let G be a cyclic group of any order or a group (possibly nonabelian) of odd order, and let A be a subset of G such that, for every subgroup H of G, we have $|H \cap A| > 2\sqrt{|H|}$. Then A is complete.*

12.6.2 Monochromatic sum sets

Let $f(n)$ be the smallest number such that one can color $[1, n-1]$ by $f(n)$ colors so that n cannot be represented as sum of distinct numbers with the same color. Alon and Erdős [9] established the correct order of growth of $f(n)$ up to logarithmic factors.

Theorem 12.28 *For all large n we have $f(n) = O(\frac{n^{1/3}}{\log^{1/3} n})$ and $f(n) = \Omega(\frac{n^{1/3}}{\log^{4/3} n})$.*

It is conjectured [9] that the exact order of magnitude of $f(n)$ is closer to the upper bound. Combining Corollary 12.13 with the arguments from [9], one can have the following improvement

Theorem 12.29 *[380] For all large n we have $f(n) = \Omega(\frac{n^{1/3}}{\log n})$.*

Proof Let $c > 0$ be a small number to be chosen later, and color $[1, n-1]$ by at most $c\frac{n^{1/3}}{\log n}$ colors. It will suffice to represent n as the sum of distinct numbers of the same color.

From the prime number theorem (1.44) (or Exercise 1.10.4) there are $\Theta(\frac{n^{2/3}}{\log n})$ primes of magnitude $\Theta(n^{2/3})$. By the pigeonhole principle, we can thus find a monochromatic set A of primes of magnitude $\Theta(n^{2/3})$ of cardinality $|A| = \Theta(c^{-1}n^{1/3})$. Let us partition $A = A_1 \cup A_2 \cup A_3$ where each A_j also has cardinality $\Theta(c^{-1}n^{1/3})$. It would thus suffice to show that $FS(A) = FS(A_1) + FS(A_2) + FS(A_3)$ contains n.

Applying Corollary 12.13 with $l = \Theta(c^{1/2}n^{1/3})$ we see (for c small enough) that l^*A_1, and hence $FS(A_1)$, contains an arithmetic progression P_1 of length $\Omega(c^{-1/2}n^{2/3})$. Since the elements of l^*A_1 have magnitude at most $O(c^{1/2}n)$, we thus see the elements of P_1 do also, and that the step d of P_1 is at most $O(cn^{1/3})$.

Now the elements of A_2 are primes and are larger than d. In particular we can find a subset A_2' of A_2 with $|A_2| = d$ and all elements of A_2 coprime to d. By Lemma 12.23 we see that $FS(A_2')$ intersects all the residue classes in $\mathbf{Z}_d$.

Also note that the elements in $FS(A'_2)$ are quite small, having magnitude at most $O(dn^{2/3})$.

Now we add the long progression P_1 of step d to the small set $FS(A'_2)$ which intersects all the residue classes in $\mathbf{Z}_d$, and observe that the set $P_1 + FS(A'_2)$ contains an interval I of length at least $\Omega(c^{-1/2}n^{2/3})$. Indeed, if $P_1 = \{a, a + d, \ldots, a + md\}$, then one easily verifies that $P_1 + FS(A'_2)$ contains the interval $I := [a + O(dn^{2/3}), a + md]$ which has length $\Omega(md) = \Omega(m) = \Omega(c^{-1/2}n^{2/3})$, taking c suitably small of course.

Finally, observe that all the elements of A_3 have magnitude $\Theta(n^{2/3})$ and their total sum is $\Theta(c^{-1}n)$, which is larger than n for c small enough. Thus by the greedy algorithm one can subtract distinct elements of A_3 from n until one enters the interval I from the preceding paragraph. This shows that $n \in P_1 + FS(A'_2) + FS(A_3) \subset FS(A)$, as claimed. □

Remark 12.30 The proof was rather wasteful. We lose a factor of $\log n$ by reducing ourselves to the set of primes. On the other hand, the only thing we need is that our set contains enough primes in order to apply Lemma 12.23 to A_2. Note though that the elements with small prime factors coprime to n can be grouped into large color classes for which no subset sum can equal n (see the exercise), so the primes are in fact a large subset of the "useful" elements of $[1, n]$.

Exercise

12.6.1 [9] Prove the upper bound in Theorem 12.28. (Hint: experiment with the color class consisting of all the multiples of p, where p is a prime not dividing n that is not too large, the color class $[\frac{n}{k+1}, \frac{n}{k})$ for integers k that are not too large, and color classes consisting of elements whose total sum is less than n.)

Bibliography

[1] M. Ajtai, V. Chvátal, M. Newborn, and E. Szemerédi, Crossing-free subgraphs, *Annals of Discrete Mathematics* **12** (1982), 9–12.

[2] M. Ajtai, J. Komlós, and E. Szemerédi, A dense infinite Sidon sequence, *European J. Combin.* **2** (1981), (1), 1–11.

[3] M. Ajtai and E. Szemerédi, Sets of lattice points that form no squares, *Studia Scientarium Mathematicarum Hungarica* **9** (1974), 9–11.

[4] N. Alon, Combinatorial Nullstellensatz, Recent trends in combinatorics (Mátraháza, 1995), *Combin. Probab. Comput.* **8** (1999), (1–2), 7–29.

[5] N. Alon, Additive Latin transversals, *Israel J. Math.* **117** (2000), 125–130.

[6] N. Alon and M. Dubiner, A lattice point problem and additive number theory, *Combinatorica* **15** (1995), (3), 301–309.

[7] N. Alon, A. Shapira, Every monotone graph property is testable. In *STOC'05: Proceedings of the 37th Annual ACM Symposium on Theory of Computing*, Association of Computing Machinery (2005), 128–137.

[8] N. Alon, R. Duke, H. Leffman, V. Rödl and R. Yuster, The algorithmic aspects of the regularity lemma, *J. Algorithms* **16** (1994), 80–109.

[9] N. Alon and P. Erdős, Sure monochromatic subset sums, *Acta Arith.* **74** (1996), (3), 269–272.

[10] N. Alon and D.J. Kleitman, Sum-free subsets. In *A Tribute to Paul Erdős*, A. Baker, B. Bollobás and A. Hajnal (eds), Cambridge University Press (1990), 13–26.

[11] N. Alon, M. Nathanson and I. Ruzsa, The polynomial method and restricted sums of congruence classes, *J. Number Theory* **56** (1996), (2), 404-417.

[12] N. Alon and J. Spencer, *The Probabilistic Method*, Second edition, Wiley, (2000).

[13] G.E. Andrews, A lower bound for the volume of strictly convex bodies with many boundary lattice points, *Trans. Amer. Math. Soc.* **106** (1963), 270–279.

[14] B. Aronov, J. Pach, M. Sharir and G. Tardos, Distinct distances in three and higher dimensions, *Combin. Probab. Comput.* **13** (2004), (3), 283–293.

[15] A. Balog, Linear equations in primes, *Mathematika* **39** (1992) 367–378.

[16] A. Balog and E. Szemerédi, A statistical theorem of set addition, *Combinatorica* **14** (1994), 263–268.

[17] A. Baltz, T. Schoen and A. Srivastav, Probabilistic construction of small strongly sum-free sets via large Sidon sets. In *Randomizarion, Approximation, and*

Combinatorial Optimization, Lecture Notes in Computer Science **1671**, Springer, (1999), 138–143.

[18] B. Barak, R. Impagliazzo and A. Wigderson, Extracting randomness using few independent sources. In *Proceedings FOCS 2004*, IEEE Computer Society, (2004), 384–393.

[19] J. Beck, On the lattice property of the plane and some problems of Dirac, Motzkin, and Erdős in combinatorial geometry, *Combinatorica* **3** (1983), 281–297.

[20] W. Beckner, Inequalities in Fourier analysis, *Annals of Math.* **102** (1975), 159–182.

[21] F.A. Behrend, On sets of integers which contain no three terms in arithmetic progression, *Proc. Nat. Acad. Sci.* **32** (1946), 331–332.

[22] V. Bergelson, B. Host and B. Kra, Multiple recurrence and nilsequences, with an appendix by Imre Ruzsa. *Invent. Math.* **160** (2005), (2), 261–303.

[23] V. Bergelson and A. Leibman, Polynomial extensions of van der Waerden's and Szemerédi's theorems, *J. Amer. Math. Soc.* **9** (1996), 725–753.

[24] V. Bergelson and A. Leibman, Set polynomials and polynomial extension of the Hales–Jewett theorem, *Ann. of Math.* (Series 2) **150** (1999), (1), 33–75.

[25] S. Bernstein, Sur une modification de l'inéqualité de Tchebichef, *Annal. Sci. Inst. Sav. Ukr. Sect. Math. I* (1924).

[26] Y. Bilu, The $(\alpha + 2\beta)$-inequality on a torus, *J. London Math. Soc.* (Series 2) **57** (1998), (3), 513–528.

[27] Y. Bilu, Addition of integer sequences and subsets of real tori. In *Number Theory in Progress: Proc. Int. Conf. in Number Theory in Honor of A. Schinzel, Zakopane, 1997*, K. Győry, H. Iwaniec and J. Urbanowicz (eds), W. de Gruyter, (1999), 639–649.

[28] Y. Bilu, Structure of sets with small sumset, Structure theory of set addition. *Asterisque* **258** (1999), xi, 77–108.

[29] Y. Bilu, V. Lev and I. Ruzsa, Rectification principles in additive number theory, *Discrete Comput. Geom.* **19** (1998), 343–353.

[30] N. N. Bogolyubov, Sur quelques propriétés arithmétiques des presque-périodes, *Ann. Chaire Math. Phys. Kiev* **4** (1939), 185–194.

[31] B. Bollobás, Sperner systems consisting of pairs of complementary subsets, *J. Comb. Theory, Ser. A* **15** (1973), 363–366.

[32] B. Bollobás, *Combinatorics: Set Systems, Hypergraphs, Families of Vectors, and Combinatorial Probability*, Cambridge University Press, (1986).

[33] A. Bonami, Etudes des coefficients Fourier des fonctiones de $L^p(G)$, *Ann. Inst. Fourier*(Grenoble) **20** (1970) (2), 335–402.

[34] J. Bourgain, A Szemerédi type theorem for sets of positive density in $\mathbf{R}^k$, *Israel J. Math.* **54** (1986) (3), 307–316.

[35] J. Bourgain, Bounded orthogonal systems and the $\Lambda(p)$-set problem, *Acta Math.* **162** (1989), (3–4), 227–245.

[36] J. Bourgain, On arithmetic progressions in sums of sets of integers. In *A Tribute to Paul Erdős*, Cambridge University Press, (1990), 105–110.

[37] J. Bourgain, Estimates related to sumfree subsets of sets of integers, *Israel J. Math.* **97** (1997), 71–92.

[38] J. Bourgain, On the dimension of Kakeya sets and related maximal inequalities, *GAFA* **9** (1999), 256–282.

[39] J. Bourgain, On triples in arithmetic progression, *GAFA* **9** (1999), 968–984.

[40] J. Bourgain, Estimates on exponential sums related to the Diffie–Hellman distributions, *GAFA* **15** (2005) (1), 1–34.

[41] J. Bourgain, Mordell's exponential sum estimate revisited, *J. Amer. Math. Soc.* **18** (2005) (2), 477–4993.

[42] J. Bourgain and M. Chang, On the size of k-fold sum and product sets of integers, *J. Amer. Math. Soc.* **17** (2004) (2), 473–497.

[43] J. Bourgain, N. Katz and T. Tao, A sum-product estimate in finite fields, and applications, *GAFA* **14** (2004), 27–57.

[44] J. Bourgain and S. Konyagin, Estimates for the number of sums and products and for exponential sums over subgroups in fields of prime order, *C. R. Acad. Sci. Paris, Ser. I* **337** (2003), 75–80.

[45] T. Brown and J. Buhler, A density version of a geometric Ramsey theorem, *J. Combin. Theory Ser. A* **32** (1982) (1), 20–34.

[46] J.W.S. Cassels, On the representation of integers as the sums of distinct summands taken from a fixed set, *Acta Sci. Math. Szeged* **21** (1960), 111–124.

[47] A.L. Cauchy, Recherches sur les nombres, *J. École Polytech.* **9** (1813), 99-116.

[48] M. Chang, A polynomial bound in Freiman's theorem, *Duke Math. J.* **113** (2002) (3), 399–419.

[49] M. Chang, Erdős–Szemeredi sum-product problem, *Annals of Math.* **157** (2003), 939–957.

[50] M. Chang, A sum-product estimate in algebraic division algebras over **R**, *Israel J. Math.* **150** (2005), 369.

[51] M. Chang, Additive and multiplicative structure in matrix spaces, *preprint*.

[52] B. Chazelle and J. Friedman, Point location among hyperplanes and unidirectional ray-shooting, *Comput. Geom.* **4** (1994) (2), 53–62.

[53] S. Chen, On the size of finite Sidon sequences, *Proc. Amer. Math. Soc.* **121** (1994) (2), 353–356.

[54] H. Chernoff, A measure of the asymptotic efficiency for tests of a hypothesis based on the sum of observations, *Ann Math Stat.* **23** (1952), 493–509.

[55] S.L.G. Choi, The largest sum-free subsequence from a sequence of n numbers, *Proc. Amer. Math. Soc.* **39** (1973), 42–44.

[56] S.L.G. Choi, P. Erdős and M. Nathanson, Lagrange's theorem with $N^{1/3}$ squares, *Proc. Am. Math. Soc.*, **79** (1980), 203-205.

[57] F. Chung, The number of distinct distances determined by n points in the plane, *J. Combin. Theory Ser. A* **36** (1984), 342–354.

[58] F. Chung and R. Graham, Quasi-random subsets of $\mathbf{Z}_n$, *J. Comb. Th. A* **61** (1992), 64–86.

[59] F. Chung, R. Graham and R.M. Wilson, Quasi-random graphs, *Combinatorica* **9** (1989), 345–362.

[60] F. Chung, E. Szemerédi and W. Trotter, The number of different distances determined by a set of n points in the Euclidean plane, *Discrete Computational Geom.* **7** (1992), 1–11.

[61] J. Cilleruelo, New upper bounds for finite B_h sequences, *Adv. Math.* 159 (2001) (1), 1–17.

[62] J. Cilleruelo, I. Ruzsa and C. Trujillo, Upper and lower bounds for finite $B_h[g]$ sequences, *J. Number Theory* **97** (2002) (1), 26–34.

[63] K. Clarkson, H. Edelsbrunner, L. Gubias, M. Sharir and E. Welzl, Combinatorial complexity bounds for arrangements of curves and spheres, *Discrete Comput. Geom.* **5** (1990), 99–160.
[64] K. Costello, T. Tao and V. Vu, Random symmetric matrices are almost surely non-singular, *Duke Math. J.*, to appear.
[65] E. Croot, Long arithmetic progressions in critical sets, *J. Combin. Theory Ser. A* **113** (2006) (1), 53–66.
[66] D. da Silva and Y. Hamidoune, Cyclic spaces for Grassmann derivatives and additive theory, *Bull. London Math. Soc.* **26** (1994) (2), 140–146.
[67] S. Dasgupta, G. Károlyi, Gyula, O. Serra and B. Szegedy, Transversals of additive Latin squares, *Israel J. Math* **126** (2001), 17–28.
[68] H. Davenport, On the addition of residue classes, *J. London Math. Soc.* **10** (1935), 30–32.
[69] J.-M. Deshouillers, F. Hennecart and A. Plagne, On small sumsets in $(\mathbf{Z}/2\mathbf{Z})^n$, *Combinatorica* **24** (1) (2004), 53–68.
[70] J.-M. Deshouillers and G. Freiman, When subset-sums do not cover all the residues modulo p, *J. Number Theory* **104** (2004) (2), 255–262.
[71] J. Dieudonné, Une propriété des racines de l'unité, Collection of articles dedicated to Alberto Gonzalez Domiguez on his sixty-fifth birthday. *Rev. Un. Mat. Argentina* **25** (1970/71), 1–3.
[72] D.L. Donoho and P.B. Stark, Uncertainty principles and signal recovery, *SIAM J. Appl. Math.* **49** (1989), 906–931.
[73] F.J. Dyson, A theorem on the densities of sets of integers, *J. Lond. Math. Soc.* **20** (1945), 8–14.
[74] F.J. Dyson, Statistical theory of the energy levels of complex systems. I, *J. Math. Phys.* **3** (1962), 140–156.
[75] Y. Edel, Extensions of generalized product caps, *Designs, Codes, and Cryptography* **31** (2004), 5–14.
[76] G. Elekes, On the number of sums and products, *Acta Arith.* **81** (1997), 365–367.
[77] G. Elekes, Sums versus products in number theory, algebra and Erdős geometry, *Paul Erdős and his Mathematics, II* (Budapest, 1999), Bolyai Soc. Math. Stud., **11** János Bolyai Math. Soc., (2002), 241–290.
[78] G. Elekes and Z. Király, On the combinatorics of projective mappings, *J. Algebraic Combin.* **14** (2001) (3), 183–197.
[79] G. Elekes, M. Nathanson and I. Rusza, Convexity and sumsets, *J. Number Theory* **83** (2000) (2), 194–201.
[80] G. Elekes and I. Ruzsa, Few sums, many products, *Studia Sci. Math. Hungar.* **40** (2003) (3), 301–308.
[81] C. Elsholtz, Lower bounds for multidimensional sums, *Combinatorica* **24** (2004), 351–358.
[82] P. Erdős, On a lemma of Littlewood and Offord, *Bull. Amer. Math. Soc.* **51** (1945), 898–902.
[83] P. Erdős, On sets of distances of n points, *Amer. Math. Monthly* **53** (1946), 248–250.
[84] P. Erdős, Some remarks on the theory of graphs, *Bull. Am. Math. Soc.* **53** (1947), 292–294.
[85] P. Erdős, On the representation of large interges as sums of distinct summands taken from a fixed set, *Acta. Arith.* **7** (1962), 345–354.

[86] P. Erdős, Extremal problems in number theory. In *Proceedings of the Symp. Pure Math. VIII*, American Mathematical Society (1965), 181–189.

[87] P. Erdős, P. Frankl and V. Rödl, The asymptotic number of graphs not containing a fixed subgraph and a problem for hypergraphs having no exponent, *Graphs Combin.* **2** (1986) (2), 113–121.

[88] P. Erdős, A. Ginzburg and A. Ziv, Theorem in the additive number theory, *Bull. Res. Council Israel* **10F** (1961) 41–43.

[89] P. Erdős and R. Graham, Old and new problems and results in combinatorial number theory, *Monographies de L'Enseignement Mathématique* **28**, Université de Genéve, L'Enseignement Mathematique, Geneva, 1980.

[90] P. Erdős and M. Nathanson, Lagrange's theorem and thin subsequences of squares. In *Contribution to Probability*, J. Gani and V.K. Rohatgi (eds), Academic Press, (1981), 3–9.

[91] P. Erdős and E. Szeremédi, On sums and products of integers. In *Studies in Pure Mathematics; To the memory of Paul Turán*, P. Erdős, L. Alpar, and G. Halasz (eds), Akademiai Kiado–Birkhauser Verlag, Budapest (1983), 213–218.

[92] P. Erdős, Problems and results in additive number theory. In *Colloque sur la Théorie des Nombres, Bruxelles, 1955*,George Thone & Masson and Cie, (1956), 127–137.

[93] P. Erdős and L. Lovász, Problems and results on 3-chromatic hypergraphs and some related questions. In *Infinite and Finite Sets (Colloq., Keszthely, 1973; dedicated to P. Erdős on his 60th birthday)*, Vol. II, Colloq. Math. Soc. Janos Bolyai, **10**, North-Holland, (1975), 609–627.

[94] P. Erdős, C. Ko and R. Rado, Intersection theorems for systems of finite sets, *Quart. J. Math. Oxford Ser. 2* **12** (1961), 313–318.

[95] P. Erdős and R. Rado, Intersection theorems for systems of sets, *J. London Math. Soc.* **35** (1960), 85–90.

[96] J. Esary, F. Proschan and D. Walkup, Association of random variables with applications, *Ann. Math. Statist.* **38** (1967), 1466–1476.

[97] P. Erdős and P. Tetali, Representations of integers as the sum of k terms, *Random Structures Algorithms* **1** (1990) (3), 245–261.

[98] P. Erdős and P. Turán, On a problem of Sidon in additive number theory and some related problems, *J. London Math. Soc.* **16** (1941), 212–215.

[99] P. Erdős and P. Turán, On some sequences of integers, *J. London Math. Soc.* **11** (1936), 261–264.

[100] P. Erdős and P. Turán, On a problem of Sidon in additive number theory, and on some related problems, *J. London Math. Soc.* **16** (1941), 212–215.

[101] C.G. Esséen, On the Kolmogorov–Rogozin inequality for the concentration function, *Z. Wahrsch. Verw. Gebiete* **5** (1966), 210–216.

[102] R.J. Evans and I.M. Stark, Generalized Vandermonde determinants and roots of unity of prime order, *Proc. Amer. Math. Soc.* **58** (1977), 51–54.

[103] J. Folkman, On the representation of integers as sums of distinct terms from a fixed sequence, *Canad. J. Math.* **18** (1966), 643–655.

[104] C.M. Fortuin, P.W. Kasteleyn and J. Ginibre, Correlation inequalities on some partially ordered sets, *Comm. Math. Phys.* **22** (1971), 89–103.

[105] K. Ford, Sums and products from a finite set of real numbers, *Ramanujan Journal* **2** (1998) (1–2), 59–66.

[106] K. Ford, The distribution of integers with a divisor in a given interval, preprint.

[107] P. Frankl, R. Graham and V. Rödl, On subsets of abelian groups with no 3-term arithmetic progression, *J. Combin. Theory Ser. A* **45** (1987) (1), 157–161.

[108] P. Frankl and Z. Füredi, Solution of the Littlewood–Offord problem in high dimensions. *Ann. of Math. (series 2)* **128** (1988) (2), 259–270.

[109] P. Frankl and V. Rödl, The uniformity lemma for hypergraphs, *Graphs and Combinat.* **8**(1992) (4), 309–312.

[110] P. Frankl and V. Rödl, Extremal problems on set systems, *Random Struct. Algorithms* **20** (2002) (2), 131–164.

[111] G. Freiman, Inverse problems in additive number theory VI., On the addition of finite sets III, Izv. *Vysš. Učebn. Zaved. Matem.*, **28** (1962) (3), 151–157.

[112] G. Freiman, Inverse problems of additive number theory, VII. On addition of finite sets, IV. The method of trigonometric sums, *Izv. Vysš. Učebn. Zaved. Matem.*, **28** (1962) (3), 131–144.

[113] G. Freiman, *Foundations of a Structural Theory of Set Addition*, translated from the Russian. Translations of Mathematical Monographs, **37**, American Mathematical Society (1973).

[114] G. Freiman, What is the structure of K if $K + K$ is small? In *Number Theory (New York, 1984–1985)*, Lecture Notes in Math. **1240**, Springer-Verlag, (1987), 109–134.

[115] G. Freiman, New analytical results in subset-sum problem. In *Combinatorics and Algorithms* (Jerusalem, 1988). Discrete Math. **114** (1993) (1–3, 205–217.

[116] G. Freiman, Structure theory of set addition, *Asterisque* **258** (1999), 1–33.

[117] G. Freiman, H. Halberstam and I. Ruzsa, Integer sum sets containing long arithmetic progressions, *J. London Math. Soc.* **46** (1992), 193–201.

[118] G. Freiman, A. Heppes and B. Uhrin, A lower estimation for the cardinality of finite difference sets in R^n. In *Number Theory, Vol. I (Budapest, 1987)*, Colloq. Math. Soc. János Bolyai, **51** North-Holland (1990), 125–139.

[119] G. Freiman and W. Pigarev, The relation between the invariants R and T (in Russian), *Kalinin. Gos. Univ.* Moscow, 1973, 172–174.

[120] P.E. Frenkel, Simple proof of Chebotarëv's theorem, preprint. math.AC/0312398

[121] H. Furstenberg, Ergodic behavior of diagonal measures and a theorem of Szemerédi on arithmetic progressions, *J. Analyse Math.* **31** (1977), 204–256.

[122] H. Furstenberg, *Recurrence in Ergodic theory and Combinatorial Number Theory*, Princeton University Press (1981).

[123] H. Furstenberg and Y. Katznelson, An ergodic Szemerédi theorem for commuting transformations, *J. Analyse Math.* **34** (1979), 275–291.

[124] H. Furstenberg and Y. Katznelson, A density version of the Hales–Jewett theorem, *J. d'Analyse Math.* **57** (1991), 64–119.

[125] H. Furstenberg, Y. Katznelson and D. Ornstein, The ergodic theoretical proof of Szemerédi's theorem, *Bull. Amer. Math. Soc.* **7** (1982), 527–552.

[126] H. Furstenberg and B. Weiss, A mean ergodic theorem for $1/N \sum_{n=1}^{N} f(T^n x) g(T^{n^2} x)$. In *Convergence in Ergodic Theory and Probability (Columbus OH 1993)*, Ohio State Univ. Math. Res. Inst. Publ., **5** de Gruyter (1996), 193–227.

[127] W. Gao and Y.O. Hamidoune, On additive bases, *Acta Arith.* **88** (1999) (3), 233–237.

[128] R.J. Gardner, The Brunn–Minkowski inequality, *Bull. Amer. Math. Soc.* (N.S.) **39** (2002) (3), 355–405.

[129] R.J. Gardner and P. Gronchi, A Brunn–Minkowski inequality for the integer lattice, *Trans. Amer. Math. Soc.* **353** (2001) (10), 3995–4024.

[130] J. Garibaldi, *Erdős Distance Problem for Convex Metrics*, UCLA Ph.D. Thesis.

[131] D. Goldstein, R. Guralnick and I. Isaacs, Inequalities for finite group permutation modules, *Trans. Amer. Math. Soc.* **357** (2005), 4017–4042.

[132] D. Goldston and C.Y. Yıldırım, Higher correlations of divisor sums related to primes, I: Triple correlations, *Integers* **3** (2003) A5, 66pp.

[133] D. Goldston and C.Y. Yıldırım, Higher correlations of divisor sums related to primes, III: k-correlations, *preprint* (available at AIM preprints)

[134] D. Goldston and C.Y. Yıldırım, Small gaps between primes, I, *preprint.*

[135] I. Good, Short proof of a conjecture by Dyson, *J. Mathematical Phys.* **11** (1970), 1884.

[136] T. Gowers, Lower bounds of tower type for Szemerédi's uniformity lemma, *GAFA* **7** (1997), 322–337.

[137] T. Gowers, A new proof of Szemerédi's theorem for arithmetic progressions of length four, *GAFA* **8** (1998), 529–551.

[138] T. Gowers, A new proof of Szemerédi's theorem, *GAFA* **11** (2001), 465–588.

[139] T. Gowers, Quasirandomness, counting, and regularity for 3-uniform hypergraphs, *Comb. Probab. Comput.* **15** (1–2). (2006), pp. 143–184.

[140] T. Gowers, Hypergraph regularity and the multidimensional Szemerédi theorem, *preprint*.

[141] R. Graham, Complete sequences of polynomial values, *Duke Math. J.* **31** (1964), 275–286.

[142] R. Graham, V. Rödl and A. Rucinski, On Schur properties of random subsets of integers, *J. Numb. Theory* **61** (1996), 388–408.

[143] R. Graham, B. Rothschild and J.H. Spencer, *Ramsey Theory*, Wiley, (1980).

[144] B. Green, Edinburgh lecture notes on Freiman's theorem, *unpublished.*

[145] B. Green, The number of squares and $B_h[g]$ sets, *Acta Arith.* 100 (2001) (4), 365–390.

[146] B. Green, Some constructions in the inverse spectral theory of cyclic groups, *Comb. Prob. Comp.* **12** (2003) (2), 127–138.

[147] B. Green, Roth's theorem in the primes, *Annals of Math* **161** (2005) (3), 1609–1636.

[148] B. Green, On arithmetic structures in dense sets of integers, *Duke Math. Jour.* **114** (2002) (2), 215–238.

[149] B. Green, Arithmetic progressions in sumsets, *GAFA* **12** (2002), 584–597.

[150] B. Green, A Szemerédi-type regularity lemma in abelian groups, *GAFA* **15** (2005) (2), 340–376.

[151] B. Green, Finite field models in arithmetic combinatorics. In *Surveys in Combinatorics 2005*, B.S. Webb (ed), Cambridge University Press, (1995), 1–27.

[152] B. Green, The polynomial Freiman–Ruzsa conjecture, *unpublished.*

[153] B. Green, Long arithmetic progressions of primes, *preprint.*

[154] B. Green and I. Ruzsa, Sets with small sumsets and rectification, *Bull. London Math. Soc.* **38** (2006) (1), 43–52.

[155] B. Green and I. Ruzsa, Counting sumsets and sum-free sets modulo a prime, *Studia Sci. Math. Hungar.* **41** (2004) (3), 285–293.

[156] B. Green and I. Ruzsa, Counting sum-free sets in abelian groups, *Israel J. Math* **147** (2005), 157–189.
[157] B. Green and I. Ruzsa, Freiman's theorem in an arbitrary abelian group, *preprint.*
[158] B. Green and T. Tao, The primes contain arbitrarily long arithmetic progressions, *Annals of Math.*, to appear.
[159] B. Green and T. Tao, Restriction theory of the Selberg sieve, and applications, *preprint.*
[160] B. Green and T. Tao, An inverse theorem for the Gowers $U^3(G)$ norm, *preprint.*
[161] B. Green and T. Tao, Finite field analogues of Szemerédi's theorem, *preprint.*
[162] B. Green and T. Tao, Compressions, convex geometry, and the Freiman–Bilu theorem, *preprint.*
[163] J.R. Griggs, On the distribution of sums of residues, *Bull. Amer. Math. Soc.* **28** (1993), 329–333.
[164] J.R. Griggs, Database security and the distribution of subset sums in $\mathbf{R}^m$. In *Graph theory and Combinatorial Biology (Balatonlelle, 1996)*, Bolyai Soc. Math. Stud., **7**, János Bolyai Math. Soc., (1999), 223–252.
[165] J. Gunson, Proof of a conjecture of Dyson in the statistical theory of energy levels, *J. Math. Phys.*, **3** (1962), 752–753.
[166] R.K. Guy, *Unsolved Problems in Number Theory*, Springer-Verlag, (1994).
[167] G. Halász, Estimates for the concentration function of combinatorial number theory and probability, *Period. Math. Hungar.* **8** (1977) (3–4), 197–211.
[168] H. Halberstam and K. Roth, *Sequences*, Springer-Verlag, (1966).
[169] A.W. Hales and R.I. Jewett, Regularity and positional games, *Trans. Amer. Math. Soc.* **106** (1963), 222–229.
[170] H. Halberstam and H.E. Richert, *Sieve Methods*, Academic Press, (1974).
[171] P. Hall, On representatives of subsets, *J. London Math. Soc.* **10** (1935), 26–30.
[172] Y. Hamidoune, A.S. Lladó and O. Serra, On subsets with small product in torsion-free groups, *Combinatorica* **18** (1998), 529–540.
[173] Y. Hamidoune, Private communication.
[174] Y. Hamidoune and Ø. Rødseth, An inverse theorem mod p, *Acta Arith.* **92** (2000) (3), 251–262.
[175] Y. Hamidoune and G. Zemor, On zero-free subset sums, *Acta Arith.* **78** (1996), (2), 143–152.
[176] H. Harborth, Ein Extremalproblem für Gitterpunkte, *J. Reine Angew. Math.* **262/263** (1973), 356–360.
[177] D.R. Heath-Brown, Integer sets containing no arithmetic progressions, *J. London Math. Soc.* **35** (1987), 385–394.
[178] D.R. Heath-Brown, The number of primes in a short interval, *J. Reine Angew. Math.* **389** (1988), 22–63.
[179] D.R. Heath-Brown, Three primes and an almost prime in arithmetic progression, *J. London Math. Soc.* (2) **23** (1981), 396–414.
[180] D.R. Heath-Brown and S. Konyagin, New bounds for Gauss sums derived from kth powers, and for Heilbronn's exponential sum, *Quart. J. Math.* **51** (2000), 221–235.
[181] N. Hegyvári, On the representation of integers as sums of distinct terms from a fixed set, *Acta Arith.*, **92** (2000) (2), 99–104.
[182] H. Helfgott, Growth and generation in $SL_2(Z/pZ)$, *preprint.*

[183] G. Hoheisel, Primzahlprobleme in der Analysis, *Sitz. Preuss. Akad. Wiss.* **2** (1930), 1–13.
[184] B. Host, Progressions arithmétiques dans les nombres premiers (d'aprés B. Green et T. Tao), *Séminaire Bourbaki*, 57éme année, 2004–2005 (944).
[185] B. Host and B. Kra, Nonconventional ergodic averages and nilmanifolds, *Annals of Math.* **161** (2005), 397–488.
[186] Q. Hou and Z. Sun, Restricted sums in a field, *Acta Arith.* **102** (2002) (3), 239–249.
[187] M. Huxley, On the difference between consecutive primes, *Invent. Math.* **15** (1972), 164–170.
[188] A. Ingham, On the difference between consecutive primes, *Quart. J. Math. Oxford.* **8** (1937), 255–266.
[189] A. Iosevich, Curvature, combinatorics, and the Fourier transform, *Not. Amer. Math. Soc.* **48** (2001), 577–583.
[190] S. Janson, Poisson approximation for large deviations, *Random Structures Algorithms* **1** (1990) (2), 221–229.
[191] X. Jia, $B_h[g]$ sequences with large upper density, *J. Number Theory* **56** (1996), 298–308.
[192] X. Jia, On finite Sidon sequences, *J. Number Theory* **44** (1993) (1), 84–92.
[193] R. Jin, Freiman's conjecture and nonstandard methods, *preprint.*
[194] F. John, Extremum problems with inequalities as subsidiary conditions. In *Studies and Essays presented to R. Courant on his 60th birthday*, Interscience Publishers Inc., (1948), 187–204.
[195] J. Kahn, J. Komlós and E. Szemerédi, On the probability that a random ± 1 matrix is singular, *J. Amer. Math. Soc.* **8** (1995), 223–240.
[196] G. Katona, A simple proof of the Erdős–Chao Ko–Rado theorem, *J. Combin. Thy. Ser. B* **13** (1972), 183–184.
[197] N. Katz and G. Tardos, A new entropy inequality for the Erdős distance problem. In *Towards a Theory of Geometric Graphs*, J. Pach (ed), Contemporary Mathematics **342**, American Mathematical Society, (2004), 119–126.
[198] N. Katz and T. Tao, Bounds on arithmetic progressions, and applications to the Kakeya conjecture, *Math. Res. Letters* **6** (1999), 625–630.
[199] N. Katz and T. Tao, Some connections between the Falconer and Furstenburg conjectures, *New York J. Math.* **7** (2001), 148–187.
[200] A. Kemnitz, On a lattice point problem,*Ars. Combin.* **16** (1983), 151–160.
[201] J. Kemperman, On small sumsets in an abelian group, *Acta Math.* **103** (1960), 63–88.
[202] J. Kemperman, On complexes in a semigroup, *Indag. Math.* **18** (1956), 247–254.
[203] J.H. Kim and V.H. Vu, Concentration of multivariate polynomials and its applications, *Combinatorica* **20** (2000) (3), 417–434.
[204] J.H. Kim and V.H. Vu, Small complete arcs in projective planes, *Combinatorica* **23** (2003) (2), 311–363.
[205] B. Klartag and V. Milman, Isomorphic Steiner symmetrization, *Invent. Math.* **153** (2003), 463–485.
[206] D.J. Kleitman, On a lemma of Littlewood and Offord on the distributions of certain sums, *Math. Z.* **90** (1965), 251–259.
[207] D.J. Kleitman, On a lemma of Littlewood and Offord on the distributions of linear combinations of vectors, *Adv. in Math.* **5** (1970), 155–157.

[208] J. Komlós and M. Simonovits, Szemerédi's regularity lemma and its applications in graph theory. In *Combinatorics, Paul Erdős is Eighty, Vol. 2 (Keszthely, 1993)*, Bolyai Soc. Math. Stud., **2**, János Bolyai Math. Soc., (1996), 295–352.

[209] S. Konyagin and I. Laba, Distance sets of well-distributed planar sets for polygonal norms, *preprint*.

[210] A. Kostochka and B. Sudakov, On Ramsey numbers of sparse graphs, *Combinatorics, Probability and Computing* **12** (2003), 627–641.

[211] M. Kneser, Abschätzungen der asymptotischen Dichte von Summenmengen, *Math. Z* **58** (1953), 459–484.

[212] Y. Kohayakawa, T. Luczak and V. Rödl, Arithmetic progressions of length three in subsets of a random set, *Acta Arith.* **75** (1996) (2), 133–163.

[213] M.N. Kolountzakis, The density of $B_h[g]$ sequences and the minimum of dense cosine sums, *J. Number Theory* **56** (1996), 4–11.

[214] M.N. Kolountzakis, On the additive complements of the primes and sets of similar growth, *Acta Arith.* **77** (1996), 1–8.

[215] J. Komlós, On the determinant of (0, 1) matrices, *Studia Sci. Math. Hungar.* **2** (1967), 7–22.

[216] J. Komlós, On the determinant of random matrices, *Studia Sci. Math. Hungar.*, **3** (1968), 387–399.

[217] B. Kra, The Green–Tao Theorem on arithmetic progressions in the primes: an ergodic point of view, *Bull. Amer. Math. Soc.* **43** (2006), 3–23.

[218] M. Krivelevich, S. Litsyn and A. Vardy, A lower bound on the density of sphere packings via graph theory, *Int. Math. Res. Not.* **43** (2004), 2271–2279.

[219] M. Krivelevich and B. Sudakov, Pseudo-random graphs, *preprint*.

[220] I. Laba, Fuglede's conjecture for a union of two intervals, *Proc. Amer. Math. Soc.* **129** (2001), 2965–2972.

[221] I. Laba and M. Lacey, On sets of integers not containing long arithmetic progressions, *unpublished*.

[222] T. Lam, Graphs without cycles of even length, *Bull. Austral. Math. Soc.* **63** (2001) (3), 435–440.

[223] M. Laczkovich and I. Ruzsa, The number of homothetic subsets. In *The Mathematics of Paul Erdős*, R. Graham and J. Nešetřil (eds), Springer-Verlag, (1996), 294–302.

[224] T. Leighton, *Complexity Issues in VLSI*, Foundations of Computing Series, MIT Press, (1983).

[225] V. Lev, Structure theorem for multiple addition and the Frobenius problem, *J. Number Theory* **58** (1996) (1), 79–88.

[226] V. Lev, Optimal representations by sumsets and subset sums, *Journal of Number Theory* **62** (1997) (1) 127–143.

[227] V. Lev, On small sumsets in abelian groups. In *Structure Theory of Set Addition, Asterisque* **258** (1999), 317–321.

[228] V. Lev, Restricted set addition in groups. I. The classical setting, *J. London Math. Soc.* (Series 2) **62** (2000) (1), 27–40.

[229] V. Lev, Restricted set addition in groups. II. A generalization of the Erdős–Heilbronn conjecture., *Electron. J. Combin.* **7** (2000) (1), Research Paper 4, 10 pp.

[230] V. Lev, Restricted set addition in abelian groups: results and conjectures, *preprint*.

[231] V. Lev, Critical pairs in abelian groups and Kemperman's theorem, *preprint*.

[232] V. Lev and S. Konyagin, Combinatorics and linear algebra of Freiman's isomorphism, *Mathematika* **47** (2000), 39–51.

[233] V. Lev and P. Smeliansky, On addition of two distinct sets of integers, *Acta Arith.* **70** (1995) (1), 85–91.

[234] J. Liu and Z. Sun, Sums of subsets with polynomial restrictions,*J. Number Theory* **97** (2002) (2), 301-304.

[235] E. Lipkin, Subset sums of sets of residues, Structure theory of set addition, *Asterisque* **258** (1999), 187–193.

[236] A. Leibman, Host–Kra and Ziegler factors, and convergence of multiple averages. In *Handbook of Dynamical Systems*, vol. 1B, B. Hasselblatt and A. Katok (eds), Elsevier (2005), 745–841.

[237] J. Littlewood and C. Offord, On the number of real roots of a random algebraic equation III, *Mat. Sbornik* **12** (1943), 277–285.

[238] L. Lovász, *Combinatorial Problems and Exercises,* Second edition. North-Holland Publishing Co., (1993).

[239] L. Lovász and B. Szegedy, Szemerédi's regularity lemma for the analyst, *preprint.*

[240] D. Lubell, A short proof of Sperner's lemma, *J. Comb. Theory* (1966), 1–299.

[241] T. Luczak and T. Schoen, On the maximal density of sum-free sets, *Acta Arith.* **95** (2000) (3), 225–229.

[242] A. Macbeath, On the measure of sum sets II. The sum-theorem for the torus, *Proc. Cambridge Philos. Soc.* **49** (1953), 40–43.

[243] H.B. Mann, A proof of the fundamental theorem on the density of sums of sets of positive integers, *Annals of Math.* **43** (1942), 523–527.

[244] H.B. Mann, *Addition Theorems: The Addition Theorems of Group Theory and Number Theory*, Interscience, (1965).

[245] J. Matoušek, *Lectures on Discrete Geometry*, Graduate Texts in Mathematics, **212**, Springer-Verlag, (2002).

[246] L. Meshalkin. Generalization of Sperner's theorem on the number of subsets of a finite set, *Teor. Veroyatn. Primen* **8** (1963), 219–220.

[247] R. Meshulam, An uncertainty inequality and zero subsums, *Discrete Math.* **84** (1990), 197–200.

[248] R. Meshulam, On subsets of finite abelian groups with no 3-term arithmetic progressions, *J. Combin. Theory Ser. A.* **71** (1995), 168–172.

[249] R. Meshulam, An uncertainty principle for finite abelian groups, *preprint.* math.CO/0312407

[250] V. Milman, Entropy and asymptotic geometry of non-symmetric convex bodies, *Adv. in Math.* **152** (2000), 314–335.

[251] G. Mockenhaupt and T. Tao, Kakeya and restriction phenomena for finite fields, *Duke Math. J.* **121** (2004), 35–74

[252] L. Moser, On the different distances determined by *n* points, *Amer. Math. Monthly* **59** (1952), 85–91.

[253] L. Moser, Notes on number theory II. On a theorem of van der Waerden, *Canad. Math. Bull.* **3** (1960), 23–25.

[254] B. Nagle, V. Rödl and M. Schacht, The counting lemma for regular *k*-uniform hypergraphs, *Random Structures and Algorithms* **26** (2006) (2), 1–67.

[255] M. Nathanson, Sums of finite sets of integers, *Amer. Math. Monthly*, **79** (1972), 1010–1012.

[256] M. Nathanson, An inverse theorem for sums of sets of lattice points, *J. Number Theory*, **46** (1994), 29–59.

[257] M. Nathanson, *Additive Number Theory. Inverse Problems and the Geometry of Sumsets*, Graduate Texts in Mathematics **165**, Springer-Verlag, (1996).

[258] M. Nathanson, On sums and products of integers, *Proc. Am. Math. Soc.* **125** (1997), 9–16.

[259] M. Nathanson, Waring's problem for sets of density zero. In *Analytic Number Theory*, M. Knopp (ed), Lecture Notes in Mathematics **899**, Springer-Verlag, (1980), 302–310.

[260] M. Nathanson, Growth of sumsets in abelian semigroups, *Semigroup Forum* **61** (2000), 149–153.

[261] M. Nathanson and I. Ruzsa, Polynomial growth of sumsets in abelian semigroups, *J. Théor. Nombres Bordeaux* **14** (2002) (2), 553–560.

[262] M. Nathanson and G. Tenenbaum, Inverse theorems and the number of sums and products. In *Structure Theory of Set Addition, Asterisque* **258** (1999), 195–204.

[263] M. Newman, On a theorem of Cebotarev, *Linear and Multilinear Algebra* **3** (1975/76) (4), 259–262.

[264] K. O'Bryant, A complete annotated bibliography of work related to Sidon sequences, *Electronic Journal of Combinatorics* DS11, 39 pages, July 2004.

[265] J.E. Olson, An addition theorem modulo *p*, *J. Combinatorial Theory* **5** (1968), 45–52.

[266] J. Olson, A combinatorial problem on finite Abelian groups. I, *J. Number Theory* **1** (1969), 8–10.

[267] J. Olson, A combinatorial problem on finite Abelian groups. II, *J. Number Theory* **1** (1969), 195–199.

[268] J.E. Olson, Sums of sets of group elements, *Acta Arithmetica* **28** (1975), 147–156.

[269] J. Pach, Crossing numbers. In *Discrete and Computational Geometry (Tokyo, 1998)*, Lecture Notes in Comput. Sci. **1763**, Springer-Verlag, (2000), 267–273.

[270] J. Pach and G. Tardos, Isosceles triangles determined by a planar point set, *Graphs and Combinatorics* **18** (2002), 769–779.

[271] J. Pach and G. Tóth, Graphs drawn with few crossings per edge, *Combinatorica* **17** (1997), 427–439.

[272] A. Plagne, A new upper bound for $B_2[2]$ sets, *J. Combin. Theory Ser. A* **93** (2001) (2), 378–384.

[273] H. Plünnecke, *Eigenschaften und Abschätzungen von Wirkingsfunktionen*, BMwF-GMD-22 Gesellschaft für Mathematik und Datenverarbeitung, Bonn 1969.

[274] C. Pomerance and A. Sárkőzy, Combinatorial number theory. In *Handbook of Combinatorics Vol. 1*, Elsevier, (1995), 967–1018.

[275] T. Przebinda, Three uncertainty principles for a locally compact abelian group. In *Representations of Real and p-adic Groups*, Lect. Notes Ser. Inst. Math. Sci., Nat. Univ. Singap., **2**, Singapore University Press, (2004), 1–18.

[276] F.P. Ramsey, On a problem of formal logic, *Proc. London Math. Soc.* **30** (1930), 264–285.

[277] R.A. Rankin, Sets of integers containing not more than a given number of terms in arithmetic progression, *Proc. Roy. Soc. Edinburgh Sect. A* **65** (1960/1961), 332–344.

[278] Yu. G. Rešetnyak, New proof of a theorem of N.G. Cebotarev (in Russian), *Uspehi Mat. Nauk (N.S.)* **10** (1955) (3) (65), 155–157.

[279] C. Reiher, Kemnitz's conjecture concerning lattice points in the plane, *preprint.*

[280] A. Robertson and D. Zeilberger, A 2-coloring of $[1, N]$ can have $(1/22)N^2 + O(N)$ monochromatic Schur triples, but not less!, *Electronic Journal of Combinatorics,* **5** (1998), R19.

[281] V. Rödl, B. Nagle, J. Skokan, M. Schacht and Y. Kohayakawa, The hypergraph regularity method and its applications, *Proc. Nat. Acad. Sci.* **102** (2005), 8109–8113.

[282] V. Rödl and M. Schacht, Regular partitions of hypergraphs, *preprint.*

[283] V. Rödl and J. Skokan, Regularity lemma for k-uniform hypergraphs, *Random Structures Algorithms* **25** (2004) (1), 1–42.

[284] V. Rödl and J. Skokan, Applications of the regularity lemma for uniform hypergraphs, *Random Structures and Algorithms* **28** (2006), 180–194.

[285] C. Rogers and G. Shephard, The difference body of a convex body, *Arch. Math.* **8** (1957), 220–233.

[286] L. Rónyai, On a conjecture of Kemnitz, *Combinatorica* **20** (2000), 569–573.

[287] K.F. Roth, On certain sets of integers, *J. London Math. Soc.* **28** (1953), 245–252.

[288] K.F. Roth, Irregularities of sequences relative to arithemetic progressions, *IV. Period. Math. Hungar.* **2** (1972), 301–326.

[289] I. Ruzsa, On the cardinality of $A + A$ and $A - A$. In *Combinatorics (Keszthely, 1976), Coll. Math. Soc. Bolyai* **18**, Akadémiai Kaidó (1979), 933–938.

[290] I. Ruzsa, Arithmetic progressions in sumsets, *Acta Arith.* **60** (1991), 191–202.

[291] I. Ruzsa, On the number of sums and differences, *Acta Math. Hung.* **59** (1992), 439–447.

[292] I. Ruzsa, A concavity property for the measure of product sets in groups, *Fund. Math.* **140** (1992) (3), 247–254.

[293] I. Ruzsa, On the additive completion of primes, *Acta Arith.* **86** (1998) (3), 269–275.

[294] I. Ruzsa, Solving a linear equation in a set of integers, I, *Acta Arith.* **65** (1993), 259–282.

[295] I. Ruzsa, Generalized arithmetical progressions and sumsets, *Acta Math. Hungar.* **65** (1994) (4), 379–388.

[296] I. Ruzsa, Sum of sets in several dimensions, *Combinatorica* **14** (1994), 485–490.

[297] I. Ruzsa, Sums of finite sets. In *Number Theory: New York Seminar*, D.V. Chudnovsky, G.V. Chudnovsky and M.B. Nathanson (eds), Springer-Verlag, (1996), 281–293.

[298] I. Ruzsa, An infinite Sidon sequence, *J. Number Theory* **68** (1998) (1), 63–71.

[299] I. Ruzsa,A small maximal Sidon set, Paul Erdős (1913–1996), *Ramanujan J.* **2** (1998) (1–2), 55–58.

[300] I. Ruzsa, An analog of Freiman's theorem in groups. In *Structure Theory of Set Addition, Astérisque* **258** (1999), 323–326.

[301] I. Ruzsa, An almost polynomial Sidon sequence, *Studia Sci. Math. Hungar.* **38** (2001), 367–375.

[302] I. Ruzsa, A problem on restricted sumsets. In *Towards a Theory of Geometric Graphs*, J. Pach (ed), Contemp. Math., **342**, American Mathematical Society (2004), 245–246.

[303] I. Ruzsa, Sum-avoiding sumsets, *preprint*.
[304] I. Ruzsa and E. Szemerédi, Triple systems with no six points carrying three triangles, *Colloq. Math. Soc. J. Bolyai* **18** (1978), 939–945.
[305] R. Salem and D.C. Spencer, On sets of integers which contain no three terms in arithmetic progression, *Proc. Nat. Acad. Sci.* **32** (1942), 561–563.
[306] L. Santaló, Un invariante afin para los cuerpos convexos del espacio de n dimensiones, *Portugalie Math.* **8** (1949), 155–161.
[307] A. Sárkőzy, Finite addition theorems I, *J. Number Theory,* **32** (1989), 114–130.
[308] A. Sárkőzy, Finite addition theorems. II, *J. Number Theory* **48** (1994) (2), 197–218.
[309] A. Sárkőzy, On finite addition theorems. In *Structure Theory of Set Addition, Asterisque* **258** (1999), xi–xii, 109–127.
[310] A. Sárkőzy and E. Szemerédi, Über ein Problem von Erdős und Moser, *Acta Arith.* **11** (1965), 205–208.
[311] P. Scherk, An inequality for sets of integers, *Pacific J. Math.* **5** (1955), 585–587.
[312] L. Schnirelmann, Über additive Eigenschaften von Zahlen, *Annals Inst. Polyt. Novocherkassk* **14** (1930) 3–28; Math. Annalen **107** (1933) 694–690.
[313] T. Schoen, The number of monochromatic Schur triples, *European J. Combin.* **20** (1999), 855–866.
[314] I.D. Shkredov, On a problem of Gowers, *preprint*.
[315] I. Schur, Über die Kongruenz $x^m + y^m = z^m (modp)$, *Jber. Deutsch. Math.-Verein.* **25** (1916), 114–116.
[316] J. Shearer, A note on the independence number of triangle-free graphs, *Discrete Mathematics,* **46** (1983), 83–87.
[317] J. Shearer, A note on the independence number of triangle-free graphs II, *J. Combin. Theory Ser. B.* **53** (1991), 300–307.
[318] S. Shelah, Primitive recursive bounds for van der Waerden numbers, *J. Amer. Math. Soc.* **1** (1988), 683–697.
[319] S. Sidon, On B2-sequences, *Math. Annalen* **106**, (1932), 536–539.
[320] J. Singer, A theorem in finite projective geometry and some applications to number theory, *Trans. Amer. Math. Soc.* **43** (1938), 377–385.
[321] K.T. Smith, The uncertainty principle on groups, *SIAM J. Appl. Math.* **50** (1990), 876–882.
[322] H. Snevily, The Cayley Addition Table of $\mathbf{Z}_n$, *Amer. Math. Monthly* **106** (1999), 584–585.
[323] J. Solymosi, Note on a generalization of Roth's theorem. In *Discrete and Computational Geometry*, Algorithms Combin. **25**, Springer-Verlag, (2003), 825–827.
[324] J. Solymosi, On the number of sums and products, *Bull. London Math. Soc.* **37** (2005) (4), 491–494.
[325] J. Solymosi, On sumsets and product sets of complex numbers, *preprint*.
[326] J. Solymosi and V. Vu, Distinct distances in high dimensional homogeneous sets. In *Towards a Theory of Geometric Graphs*, Contemp. Math., **342**, American Mathematical Society (2004), 259–268.
[327] J. Solymosi and V. Vu, Near optimal bound for the distinct distances problem in high dimensions, *Combinatorica*, to appear.
[328] J. Solymosi and C.D. Tóth, Distinct distances in the plane, *Discrete Comput. Geom.* **25** (2001) (4), 629–634.

[329] J. Solymosi, G. Tardos and C.D. Tóth, The k most frequent distances in the plane, *Discrete Comput. Geom.* **28** (2002) (4), 639–648.

[330] R. Stanley, Weyl groups, the hard Lefschetz problem, and the Sperner property, *SIAM J. Alg. Disc. Math.* **1** (1980), 168–184.

[331] J. Spencer, Four squares with few squares. In *Number Theory, New York Seminar 1991–1995*, D.V. Chudnovsky et al. (eds), Springer-Verlag, 295–297.

[332] E. Sperner, Ein Satz über Untermengen einer endlichen Menge, *Math. Z.* **27** (1928), 544–548.

[333] Y. Stanchescu, On addition of two distinct sets of integers, *Acta Arith.* **75** (1996) (2), 191–194.

[334] Y. Stanchescu, On the structure of sets with small doubling property on the plane, *Acta Arith.* **83**, 1998, 127–141.

[335] Y. Stanchescu, On finite difference sets, *Acta Math. Hungar.* **79** (1998), 123–138.

[336] J. Steinig, On Freiman's theorems concerning the sum of two finite sets of integers. In *Preprints of the Conference on the Structural Theory of Set Addition*, CIRM, Marseille (1993), 173–186.

[337] S.A. Stepanov, The number of points on a hyperelliptic curve over a prime field, *Izv. Akad. Nauk SSSR Ser. Mater.* **33** (1969), 1171–1181.

[338] P. Stevenhagen and H.W. Lenstra Jr., Chebotarëv and his density theorem, *Math. Intelligencer* **18** (1996) (2), 26–37.

[339] A. Stöhr, Gelöste und ungelöste Fragen über Basen der natürlichen Zahlenreihe. I, II, *J. Reine Angew. Math.* **194** (1955), 40–65; 111–140.

[340] B. Sudakov, E. Szemerédi and V. Vu, On a question of Erdős and Moser, *Duke Math. J.* **129** (2005) (1), 129–155.

[341] Z.W. Sun, Unification of zero-sum problems, subset sums and covers of Z, *Electron. Res. Announc. Amer. Math. Soc.* , **9** (2003), 51–60.

[342] L. Székely, Crossing numbers and hard Erdős problems in discrete geometry, *Combin. Probab. Comput.* **6** (1997), 353–358.

[343] E. Szemerédi, On sets of integers containing no four elements in arithmetic progression, *Acta Math. Acad. Sci. Hungar.* **20** (1969), 89–104.

[344] E. Szemerédi, Integer sets containing no arithmetic progressions. *Acta Math. Hungar.* **56** (1990) (1–2), 155–158.

[345] E. Szemerédi, On sets of integers containing no k elements in arithmetic progression, *Acta Arith.* **27** (1975), 299–345.

[346] E. Szemerédi, Regular partitions of graphs. In *Problemés Combinatoires et Théorie des Graphes, Proc. Colloque Inter. CNRS*, Bermond, Fournier, M. Las Vergnas and Sotteau (eds), CNRS Paris (1978), 399–401.

[347] E. Szemerédi, On a conjecture of Erdős and Heilbronn, *Acta Arith.* **17** (1970) 227–229.

[348] E. Szemerédi and W. T. Trotter Jr., Extremal problems in discrete geometry, *Combinatorica* **3** (1983), 381–392.

[349] E. Szemerédi and V. Vu, Long arithmetic progressions in sum sets and the number of x-sum-free sets, *Proc. London Math. Soc.* **90** (2005) (2), 273–296.

[350] E. Szemerédi and V. Vu, Long arithmetic progressions in sumsets: thresholds and bounds, *J. Amer. Math. Soc.* **19** (2006) (1), 119–169.

[351] E. Szemerédi and V. Vu, Finite and infinite arithmetic progressions in sumsets, *Annals of Math.* (2) **163** (2006) (1), 1–35.

[352] E. Szemerédi and V. Vu, Olson theorem revisited, *preprint. See also [380].*
[353] G. Tardos, On distinct sums and distinct distances, *Adv. in Math.* **180** (2003) (1), 275–289.
[354] T. Tao, Finite field analogues of the Erdős, Falconer, and Furstenburg problems, *unpublished.*
[355] T. Tao, An uncertainty principle for cyclic groups of prime order, *Math. Res. Lett.* **12** (2005) (1), 121–127.
[356] T. Tao, Recent progress on the Restriction conjecture, *Park City lecture notes.*
[357] T. Tao, A quantitative ergodic theory proof of Szemerédi's theorem, *preprint.*
[358] T. Tao, Arithmetic progressions and the primes, *El Escorial lecture notes.*
[359] T. Tao, Szemerédi's regularity lemma revisited, *preprint.*
[360] T. Tao, A variant of the hypergraph removal lemma, *preprint.*
[361] T. Tao, Obstructions to uniformity, and arithmetic patterns in the primes, *preprint.*
[362] T. Tao, Product set estimates for non-commutative groups, *preprint.*
[363] T. Tao, A remark on Goldston–Yıldırım correlation estimates, *unpublished.*
[364] T. Tao and V. Vu, On random ± 1 matrices: Singularity and Determinant, *Random Structures Algorithms* **28** (2006) (1), 1–23.
[365] T. Tao and V. Vu, On the singularity probability of random Bernoulli matrices, *J. Amer. Math. Soc.*, to appear.
[366] T. Tao and V. Vu, Inverse Littlewood–Offord theorems, and the least singular value of random Bernoulli matrices, *preprint.*
[367] T. Tao and V. Vu, Littlewood–Offord problem in high dimensions, *preprint.*
[368] C. Tóth, The Szemeredi–Trotter theorem in the complex plane, *preprint.*
[369] P. Turán, On a theorem of Hardy and Ramanujan, *J. London Math. Soc.* **9** (1934), 274–276.
[370] J.G. van der Corput, Über Summen von Primzahlen und Primzahlquadraten, *Math. Ann.* **116** (1939), 1–50.
[371] B.L. van der Waerden, Beweis einer Baudetschen Vermutung, *Nieuw. Arch. Wisk.* **15** (1927), 212–216.
[372] P. Varnavides, On certain sets of positive density, *J. London Math. Soc.* **39** (1959), 358–360.
[373] R.C. Vaughan, *The Hardy–Littlewood Method*, Second edition, Cambridge Tracts in Mathematics **125**, Cambridge University Press 1997.
[374] T. Voight and G. Ziegler, Singular 0/1 matrices and the hyperplanes spanned by random 0/1 vectors, *preprint.*
[375] A.G. Vosper, The critical pairs of subsets of a group of prime order, *J. London Math. Soc.* **31** (1956), 200–205.
[376] V. Vu, High order complementary bases of primes, *Integers* **2** (2002), paper no. A12.
[377] V. Vu, Concentration of non-Lipschitz functions and applications, *Random Structures Algorithms* **20** (2002) (3), 262–316.
[378] V. Vu, On the concentration of multivariate polynomials with small expectation, *Random Structures Algorithms* **16** (2000) (4), 344–363.
[379] V. Vu, On a refinement of Waring's problem, *Duke Math. J.* **105** (2000) (1), 107–134.
[380] V. Vu, Some results on subset sums, *preprint.*
[381] V. Vu, On a question of Gowers, *Ann. Comb.* **6** (2002) (2), 229–233.

[382] T. Wooley, On Vu's thin basis theorem in Waring's problem, *Duke Math. J.*, **120** (2003) (1), 1–34.

[383] K. Wilson, Proof of a conjecture by Dyson, *J. Math. Phys.*, **3** (1962), 1040–1043.

[384] E. Wirsing, Thin subbases, *Analysis* **6** (1986), 285–308.

[385] K. Yamamoto, Logarithmic order of free distributive lattice, *J. Math. Soc. Japan* **6** (1954), 343–353.

[386] T. Ziegler, Universal characteristic factors and Furstenberg averages, *preprint*.

[387] J. Zöllner, *Der Vier-Quadrate-Satz und ein Problem von Erdős und Nathanson*, Ph.D thesis, Johannes Gutenberg-Universität, Mainz (1984).

[388] J. Zöllner, Über eine Vermutung von Choi, Erdős und Nathanson, *Acta Arith.* **45** (1985), 211–213.

Index